Advanced Communication Protocol Technologies:

Solutions, Methods, and Applications

Katalin Tarnay
University of Pannonia, Hungary

Gusztáv Adamis
Budapest University of Technology and Economics, Hungary

Tibor Dulai
University of Pannonia, Hungary

Information Science
REFERENCE

Senior Editorial Director:	Kristin Klinger
Director of Book Publications:	Julia Mosemann
Editorial Director:	Lindsay Johnston
Acquisitions Editor:	Erika Carter
Development Editor:	Hannah Abelbeck
Production Editor:	Sean Woznicki
Typesetters:	Mike Brehm, Keith Glazewski, Natalie Pronio, Jennifer Romanchak, & Deanna Zombro
Print Coordinator:	Jamie Snavely
Cover Design:	Nick Newcomer

Published in the United States of America by
Information Science Reference (an imprint of IGI Global)
701 E. Chocolate Avenue
Hershey PA 17033
Tel: 717-533-8845
Fax: 717-533-8661
E-mail: cust@igi-global.com
Web site: http://www.igi-global.com

Library of Congress Cataloging-in-Publication Data
Advanced communication protocol technologies : solutions, methods, and applications / Katalin Tarnay, Gusztáv Adamis, Tibor Dulai, editors.
p. cm.
Includes bibliographical references and index.
Summary: "This book explores the complications and solutions created by communication required between ever-expanding technologies, covering the fundamentals of protocol functions and protocol operations, the controlling protocols of ISDN and mobile networks, the evolution of IP-based protocols, and advanced solutions for routing, mobility and multimedia transmission"-- Provided by publisher.
ISBN 978-1-60960-732-6 (hardcover) -- ISBN 978-1-60960-733-3 (ebook) -- ISBN 978-1-60960-734-0 (print & perpetual access) 1. Telecommunication. I. Tarnay, Katalin. II. Adamis, Gusztáv, 1965- III. Dulai, Tibor, 1979-
TK5101.A35 2011
621.382'12--dc23
2011013392

British Cataloguing in Publication Data
A Cataloguing in Publication record for this book is available from the British Library.

All work contributed to this book is new, previously-unpublished material. The views expressed in this book are those of the authors, but not necessarily of the publisher.

Sándor Szabó, *Budapest University of Technology and Economics, Hungary*
Katalin Tarnay, *University of Pannonia, Hungary; Budapest University of Technology and Economics, Hungary*
Dragos Tuscan, *Åbo Akademi University, Finland*
Lai Xu, *SAP Research, Switzerland*

Table of Contents

Detailed Table of Contents

The increasing importance of fast and reliable communication has resulted in the creation of certain communication rules, which nowadays we call the communication protocols. In this chapter, a historical overview of early protocol development and deployment is presented, as a look-back to the evolution of communication protocols. Then the authors introduce the most important terms and notions of the protocol theory. To make it easier to understand these concepts, the authors use the INRES protocol as demonstration tool. In the next topic, the authors summarize the steps of the protocol engineering. At the end of the chapter, the authors give some guidelines for protocol classification.

This chapter gives a fundamental overview of communication protocol operation. The concept of layering is introduced along with the definitions of communication primitives, protocol and service data units, data encapsulation, and service access points. An example for layering – the OSI Reference Model – is also presented. The service provided by a protocol can be characterized based on the various distributed communication models and operational environments. The model of an operation environment along with its characteristics and the model of a communication node are introduced as well. Various communication problems, trigger conditions, and the purpose of the timer trigger condition are discussed. Protocol interaction scenarios are presented with respect to the number of participants and the reliability of the transmission medium. The characteristics of these dialogues are also introduced and discussed.

Communications networks connect people of the world in spite of the big geographical distance, have big role in our entertainment during leisure time activity, and help to arrange our everyday activities. Communications network's ubiquitous nature is a big challenge for network protocol developers. They have to offer different services in heterogeneous hardware and software environment. There are basic protocol functions, which have to perform usually appearing tasks in communication networks. Among these tasks, we highlight PDU creation, error handling, medium access, handling of load-related problems (flow control and congestion control), and PDU delivery's sub-tasks like finding the destination point by addressing and routing.

The purpose of this chapter is to provide a simple, understandable approach to testing and abstract syntax notation. This chapter is composed of four parts: the first part introduces the reader to the theory of testing and reviews the development of TTCN and ASN.1 notations. The second part discusses the main components of TTCN and ASN.1 with examples. The third part summarizes the future trends. At the end, a conclusion gives an overview of the fundamental mechanisms of this field.

This chapter is devoted to the classical ISDN signaling protocols. The main reason for which we deal with these protocols in details is that they are – with modifications and extensions – also used in GSM and UMTS systems. In this chapter, we present the DSS1 (Digital Subscriber Signaling System No.1) protocol, which is used for accessing the ISDN. We do not deal with the physical layer, but we discuss the upper two layers: LAPD (Link Access Procedure on D channel) and DSS1 3rd layer.The second part of this chapter deals with the SS7 (Signaling System No.7), which is used in the core network for signaling. We introduce the structure and the functional elements of SS7, and we go through the four levels of the protocol stack.

Starting in the 1990s, mobile communication became one of the most explosion-like developing areas of the telecommunications. The first worldwide used digital cellular radio system was the GSM. Primarily, GSM was suitable for voice communication and for circuit-switched data transfer at a very low bit rate only. This data transfer type was not only slow, but was quite expensive to the users as well. Aiming to eliminate the drawbacks of the data transfer of GSM, the packet-switched service (GPRS) was introduced. The goal of this chapter is to give an overview about the architecture of GSM and GPRS networks, and to introduce to the reader the protocols used for mobility management and call control at the core network. To make it easier to understand the way in which the protocols work, we give many examples to help the readers in his studies. We show how the location update takes place, how the calls are originated and terminated, and how the short messages are handled.

The aim of this chapter is to provide a brief yet comprehensive overview of the 3rd generation UMTS (Universal Mobile Telecommunication System) mobile network, with emphasis on its specific protocols. Therefore, in this chapter the basic operations and protocol structure of UMTS network are described. The main architectural changes compared to GSM are shown, as well as the basic operations and principles of the physical radio interface. The details of other relevant UMTS specific interfaces in the access network and their protocols are provided as well, along with the description of transport network solutions. The most relevant part of UMTS specifications, the radio interface protocols are also presented.

Nowadays, the Internet Protocol is facing the version change, IPv4 (the old version of IP) will be replaced by IPv6 (the new version of IP) in the long run. Although they are based on the very same concept, the two protocols are not compatible with each other. This chapter deals with the coexistence issues, which might arise due to the simultaneous existence of IPv4 and IPv6. On the other hand, this chapter also covers transition concepts: how IPv6 only hosts can communicate either over IPv4 only networks, or with IPv4 only hosts, in the Internet of the future.

Network mobility (NEMO) management concerns with the mobility management of an entire wireless mobile network to provide uninterrupted network connectivity to many mobile devices moving together in the mobile network. This is particularly important for ubiquitous computing, which commonly means anytime, anywhere computing and communication. Most of the 3G and entire 4G and beyond wireless communication technology is all-IP. This growing use of IP devices in portable applications has created the demand for mobility support for entire networks of IP devices. NEMO solves this problem by extending Mobile IP. Devices on a mobile network are unaware of their network's mobility; however, they are provided with uninterrupted Internet access even when the network changes its attachment point to the Internet. The main objective of NEMO is to provide continuous, optimal, and secure Internet access to all nodes and even recursively nested mobile sub-nets inside a moving network. Internet Engineering Task Force (IETF) is engaged in standardizing NEMO Basic Support protocol that ensures uninterrupted connectivity to nodes within a mobile network via a mobile router. This protocol extends the mechanisms utilized in the host mobility management protocol Mobile IPv6. There are few open problems remain to be addressed in NEMO. In this chapter, we will discuss about NEMO basic support protocols, its features, and other related issues.

This chapter focuses on Next Generation Networks (NGN) and shows how protocols employed in different segments of the telecommunication infrastructure interwork to make the provisioning of the so-called triple-play services, which covers data, voice, and television services. After the overview of the evolution of telecommunication networks, the definition and concept of NGN is interpreted, and the general architecture of NGN is presented with regard to the heterogeneous demands of services. The protocols used in different sections of the end-to-end communication relation are introduced from the point of view of service and transport functions of NGN. The main focus of the chapter is on the fixed access solutions and Ethernet based multi-service access and regional networks, but configuration and streaming protocols are covered as well.

This chapter gives an overview of one of the recent trends in telecommunication convergence: fix-mobile convergence (FMC). The implications of the trend are analyzed, and challenges for a converged network that is able to provide quadruple-play services are identified. The integration of fixed and mobile access to next generation networks (NGN) is investigated from the point of view of AAA (Authentication, Authorization and Accounting) protocols, and network and application layer mobility management solutions, and policy control with special respect to the access and core networks where the integration is going to take place. At the end, a short outlook is given to future mobile trends.

This chapter is committed to give a comprehensive overview of the Host Identity Protocol (HIP), to introduce the basic ideas and the main paradigms behind it, and to show how HIP emerges from the list of potential alternatives with its wild range of possible usability in next generation mobile architectures. The broad scale of feasible advanced mobility management proposals and scenarios together with the promising mobility management capabilities of HIP and its cryptographic identifier/locator separation technique will be introduced based on an exhaustive survey of existing mobility solutions designed for the Host Identity Protocol. This broad and up-to-date outline of advanced HIP-based mobility supporting schemes will guide the readers from the basics of HIP through the protocol's main functions to its complex feature set and power to create a novel Internet architecture for future mobility-centric communications.

The IP Multimedia Subsystem (IMS) provides advanced facilitates and a standardized service architecture for Next Generation Networks (NGN). IMS enables both cellular and fixed network operators to provide wide and attractive service portfolio to the customers. In this chapter, we provide an overview of IMS architecture and most popular IMS services. We discuss the basic components and mechanisms defined for IMS, the related IMS services, and we give a survey on the future of the evolving telecommunication networks. We examine the most popular IMS services such as Presence Service and Instant Messaging (IM), and the background these IMS services. The IMS system enables fixed-mobile convergence, and it bridges the gap between the fixed and mobile communications. To understand and use the new concepts and IMS services, it is important to know what the new IMS services and concepts bring to us. Our goal is to give a brief overview, to help to understand today's and tomorrow's relevant communication technologies and protocols.

The subject of this chapter is to present the TFRC protocol in the area of efficient wireless video transmission and its possible usage in cross-layer power management mechanisms. The basic aspects of TFRC operation are presented, along with the suitability of TFRC usage for video transmission. The chapter examines related work and presents several mechanisms for efficient wireless video transmission using TFRC that have been proposed. These mechanisms utilize cross-layer approaches for adaptation of the power transmission level of the sender and TFRC feedback information regarding the wireless connection status from the receiver for improved transmission statistics, and therefore user experience, without unnecessary power consumption.

In the last few years, the Internet throughput, usage, and reliability have increased almost exponentially. The introduction of broadband wireless mobile ad hoc networks (MANETs) and cellular networks, together with increased computational power, have opened the door for a new breed of applications to be created, namely real-time multimedia applications. Delivering real-time multimedia traffic over a complex network like the Internet is a particularly challenging task since these applications have strict quality-of-service (QoS) requirements on bandwidth, delay, and delay jitter. Traditional Internet protocol (IP)-based best effort service is not able to meet these stringent requirements. The time-varying nature of wireless channels and resources constrained wireless devices make the problem even more difficult. To improve perceived media quality by end users over wireless Internet, QoS supports can be addressed in different layers, including application layer, transport layer, and link layer. Cross-layer design is a well-known approach to achieve this adaptation. In cross-layer design, the challenges from the physical wireless medium and the QoS-demands from the applications are taken into account so that the rate, power, and coding at the physical (PHY) layer can adapted to meet the requirements of the applications given the current channel and network conditions. A number of propositions for cross-layer designs exist in the literature. In this chapter, an extensive review has been made on these cross-layer architectures that combine the application-layer, transport layer, and the link layer controls. Particularly, the issues like channel estimation techniques, adaptive controls at the application and link layers for energy efficiency, priority based scheduling, transmission rate control at the transport layer, and adaptive automatic repeat request (ARQ) are discussed in detail.

In the last couple of years, there has been a rapid increase in the number of users consuming media content through WebTV and IPTV services. The chapter aims at introducing the protocols used in and the application questions related to such services. First, a brief overview is given on an IPTV system. This introductory section is followed by the description of application layer transport protocols (Real-time Transport Protocol) and session management protocols (RTP Control Protocol and Real-time Streaming Protocol) of IPTV and WebTV services. Apart from the protocols themselves, implementation and application issues related to an IPTV and WebTV environment are also discussed.

This chapter presents an introduction to the main concepts and mathematical background relating to communications networks, Quality of Services (QoS), and Internet Protocol (IP) traffic nature, which subsequently are used in the measurements, analysis, and modeling of Voice over Internet Protocol (VoIP) traffic. In practice, we analyze network performance measurements and simulation results, in order to characterize the VoIP traffic nature. As results of this analysis, it is shown that VoIP jitter can be modeling by alpha-stable distributions and self-similar processes, with Short or Long Range Dependence (i.e., SRD or LRD). Furthermore, we investigate the packet loss effects on the VoIP jitter, and present a methodology for simulating packet loss on VoIP jitter. In this work, we propose a new model that allows us to relate the Hurst parameter (H) with the Packet Loss Rate (PLR) by means of a power-law function with three fitted parameters.

Numerous protocols were already introduced in the transport layer, which can be very different depending on the provided services. Beside the traditional TCP and UDP, new transport protocols (SCTP, DCCP) appeared in the recent years to overcome the limitations of the conventional protocols. The unique features of SCTP, like multihoming and multistreaming, makes this protocol very attractive for reliable data delivery of streams even in mobile environment. It can be also used for applications where monitoring and detection of loss is required. SCTP is the only transport protocol that is able to manage mobility issues and handle handovers in the transport layer. The multihoming feature allows an endpoint of a SCTP association to be mapped to multiple IP addresses, and change the delivery path according to the link conditions. The handover process is hardly influenced by several protocol

Section 4
Protocol Applications and Technologies

Time synchronization services are often required to support coordinated operation of the nodes in sensor networking applications, potentially containing hundreds or thousands of elements. The synchronization service must provide application-specific performance (in terms of accuracy and overhead), and must be scalable and robust. For the design of suitable algorithms the error sources of both the applied synchronization models and the physical devices must be understood, and taken into account. In this chapter, we will introduce various time synchronization models and show their potential accuracy and complexity. The error sources of real devices and communication channels will also be analyzed. Based on the models several synchronization primitives will be reviewed, and complex synchronization algorithms will be introduced and analyzed.

Sensor networks are built from tiny, resource limited nodes, which are able to communicate with each other, and thus provide distributed services. The quality of communication, and especially of routing, is one of the major enablers of good performance, energy efficiency, and longevity in resource-deprived sensor networks. Since the advent of wireless sensor networks, hundreds of routing algorithms, tailored especially for this domain, have been proposed in the literature. This chapter highlights the main ideas and illustrates how these solutions help to reach various design goals. After a general overview, the taxonomy of routing algorithms in sensor networks will be presented, and then the mainstream algorithms, with the greatest impact on the field, will be introduced and analyzed. Through typical routing algorithms, we will show how and what kind of quality of services can be provided for various application needs.

Radio frequency identification (RFID) is a group of technologies used to identify or track electronic devices (RFID tags or transponders), typically applied to or incorporated into products, animals, or even persons, using radio waves or electromagnetic fields. RFID systems have various applications, with the most prevalent one being enterprise supply chain management to improve the efficiency of inventory tracking and management, as RFID transponders complement bar codes (allowing the reading from several meters, even beyond the line of sight), and may one day even take their place. Other uses include payment services (for example public transportation, toll roads, season parking tickets, ski tickets, proximity smart cards), asset and inventory tracking, patient identification in hospitals, animal identification, library systems, anti-theft technologies, or even race timing.

Foreword

In our rapidly developing world when new technologies and new services come to the market day-by-day, it is of great importance that the controlling protocols of the new systems could also be developed as quickly as possible. The methodology of protocol development is the subject of protocol engineering. The book, "Advanced Communication Protocol Technologies: Solution, Methods and Applications," gives a detailed state of art description of protocol theory, protocol engineering, and formal languages needed in specification and testing of large-scale protocols. It also introduces the most important and most frequently used protocols in the areas of fixed and mobile telecommunication, Internet, multimedia, and special fields such as wireless sensor networks. This book gives a detailed overview of the protocols of nowadays' networks and an outlook for the next generation networks and applications.

In Section 1, the rules by which the protocols work and interact with each other are introduced, and the basic and advanced protocol functions like addressing, routing, flow and congestion control, security, mobility, and quality of service management are discussed. In this section, the steps of protocol development: formal description, verification, validation, implementation, and testing are summarized. In the rest of the book, the application areas of the protocols are displayed.

Section 2 is devoted to telecommunications. Here, the most important protocols used to control the ISDN, GSM, GPRS, and UMTS networks are shown.

In Section 3, the IP based protocols are discussed, including the convergence of fixed and mobile networks. This section gives an overview of network evolution, multimedia applications including video transmission, different mobility management solutions, and advanced routing methods.

In Section 4, special applications like sensor networks and RFID are introduced, and the protocols applied in these fields are shown.

The book, "Advanced Communication Protocol Technologies: Solution, Methods and Applications," could be very useful for university students as a textbook, for engineers who need theoretical background, and for scientists, who want to overview the latest applications of communication protocols. I also recommend this book to anyone else who is interested in this field.

Gyula Csopaki
October 2010

Gyula Csopaki *is a scientific research worker at the Budapest University of Technology and Economics (Hungary), Department of Telecommunications and Media Informatics. He received his M.Sc. degree as electrical engineer specialization of telecommunications at the Technical University of Budapest in 1969. He received his Ph.D. degree in 1989 in the field of computer science and computer engineering. He managed the development of several computer aided design software tools in the fields of digital system design and telecommunications protocols. He has more than 40 years of teaching experience in the fields of digital system design, computer science and telecommunications protocols. He acted as president of National Institute of Research and Development, a governmental agency responsible for the management and of the funding of research and development activities in Hungary.*

Preface

Technological developments are results of collaborations between scientists, but the collaboration itself is based on nothing but communication; experts are sharing their ideas about new ways of theorizing, defining, developing, and finally realizing products – either hardware or software – that may have a huge impact on our everyday life. It is interesting to note that communication tools like voice transmission, e-mail, and certain Web services themselves can also be products assisting in developing a complete bunch of new communication solutions. This is exactly what happened in the Conseil Européenne pour la Recherche Nucléaire (CERN) institute back in 1992, where scientists working hard on processing data coming from various measurement systems aimed to create a sort of network, in which the data could be forwarded and evaluated faster. They also needed a standardized frame, in which information could be stored in a somewhat uniformly determined way, to synchronize various types and versions of the collected data. The name, which was given to the final product of this development, is the World Wide Web – a merger of networks and databases heavily used today.

There are different approaches to describe what communication is and the way it is built up. However, it is evident that either human or computer communication must follow certain rules, which must adequately define the ways information is sent, processed, and received. For example, if we are unaware of saying a sentence in the right way, principally to keep the proper grammatical order of the words spoken, we will find that understanding us (the sender) will be quite problematic for the receiver. What is more, if we do not speak the language of the receiver at all, the communication will most likely be impossible. The very same thing happens if computers in a network cannot "speak" a common "language".

Our book, titled *Advanced Communication Protocol Technologies: Solutions, Methods and Applications* is about the rules of computer communication: the protocols. Simply speaking, the protocols ensure that information can be understood and processed by the computers in a network.

This book is composed of four sections. The first section deals with the fundamentals of protocol functions and protocol operations. The second section introduces the controlling protocols of ISDN and mobile networks. The third and largest section discusses the evolution of IP-based protocols, and shows advanced solutions for routing, mobility, and multimedia transmission. Finally, in the fourth section, special application fields are presented.

Section 1: Basics of Communication Protocols consists of four chapters. Chapter 1, "*Communication Protocols: An Introductory Overview,*" by Katalin Tarnay, University of Pannonia (Hungary); and Gusztáv Adamis, Budapest University of Technology and Economics (Hungary), gives an overview of protocol history starting with the very first usage of an optical telegraph system installed during the Napoleonic wars in Europe. It continues with the creation of ARPANET and the most frequently used Internet protocols. Then it introduces the reader to the protocol theory with the help of the INRES protocol, which is used for demonstration purposes. The main quantitative and qualitative protocol features, as well as the steps of the protocol engineering are presented. At the end of the chapter, the authors give some important guidelines for protocol classification. Chapter 2, "*Protocol Operation*," by Gábor Árpád Németh, Budapest University of Technology and Economics (Hungary), gives a fundamental survey of communication protocol operations. The concept of layering is introduced along with the definitions of communication primitives, Protocol Data Units (PDU) and Service Data Units (SDU), data encapsulation, and service access points. Chapter 3, "*Protocol Functions*," by Tibor Dulai, University of Pannonia (Hungary), presents the basic and advanced protocol functions to the reader. Therefore, the author highlights the Protocol Data Unit (PDU) creation, error handling, medium access, flow control, and congestion control. PDU deliveries based on addressing and routing are discussed. Chapter 4, "*Notations for Test Specification: TTCN3 and ASN.1*," by Szilárd Jaskó and Dániel Muhi, University of Pannonia (Hungary), provides a simple, understandable approach to protocol testing. First, testing of protocols in the field of telecommunication is discussed. Then notations used in the test specification, namely the Test and Test Control Notation (TTCN3) and the Abstract Syntax Notation One (ASN.1) are introduced.

Section 2: Telecommunications Protocols consists of three chapters. This section gives an overview about the most frequently used protocols both of the ISDN and of the mobile networks. Chapter 5, "*Signaling Protocols of Integrated Services Digital Networks*," by Gusztáv Adamis, Budapest University of Technology and Economics (Hungary) introduces the most important protocols used to control ISDN networks. First, the DSS1, the ISDN access protocol is discussed. This part of the chapter describes the message structure in detail as well as it shows how the messages are used for call establishment and release. The second part of Chapter 5 discusses the SS7, the core network protocol of the ISDN. Apart from the introduction to the architectural concept, a detailed description about the MTP and ISUP protocols can be found here. Chapter 6, "*Mobile Network Protocols of GSM and GPRS*," by Gusztáv Adamis, Budapest University of Technology and Economics (Hungary) and Chapter 7, "*UMTS: 3rd Generation Cellular Mobile Radio System*," by Péter Fazekas, Budapest University of Technology and Economics (Hungary) deal with the protocols used in the mobile networks. Chapter 6 provides an overview about the architecture of the GSM and GPRS networks, and discusses the protocols used in the core network in detail. The chapter focuses on SCCP, TCAP, MAP, BSSAP, and GTP protocols. To help to understand how these protocols are used, several typical scenarios, like call establishment, SMS transmission, and data transfer are presented as examples. Chapter 7 discusses the protocols used in the UMTS networks. After an overview about the UMTS architecture and the basics of the radio transmission, the chapter provides a detailed summary of the protocols used at various interfaces of the network: between the core network and the UTRAN, between the radio network controllers and the Node Bs, and at the radio interface.

Section 3: IP-Based Protocols consists of thirteen chapters. Chapter 8, "*IPv4 / IPv6 Coexistence and Transition: Concepts, Mechanisms and Trends*," by László Bokor and Gábor Jeney, Budapest University of Technology and Economics (Hungary), presents the methods how IPv4 and IPv6 can exist and work together. After introducing IPv6, the chapter deals with the transition methods from IPv4 to IPv6 like Dual Stack, and the several methods of Tunneling or Encapsulation and Translation. Chapter 9, "*Network Mobility*," by Arijit Ukil, Tata Consultancy Services (India), shows Network Mobility (NEMO), which is a Mobile IP extension for supporting movement of a complete network. The chapters describes the evolution from Mobile IP and Mobile IPv6 to NEMO and presents NEMO basic support protocol next to the architecture, QoS, route optimization, multihoming possibility, mobility management, security, and application of NEMO. Chapter 10, "*Protocols in Next Generation Networks*," by Róbert Horváth, Gábor Kovács, Budapest University of Technology and Economics (Hungary) and Zoltán Pap, Ericsson (Hungary), defines Next Generation Networks (NGN) and presents its main protocols and services related to fixed access. The protocols covered in this chapter are NGN's core network protocols, regional network and aggregation network protocols, especially Ethernet, along with service access and provisioning protocols. Chapter 11, "*Convergence of Fixed and Mobile Networks*," by Gábor Kovács, Gábor Árpád Németh, Budapest University of Technology and Economics (Hungary) and Zoltán Pap, Ericsson (Hungary), examines a trend in telecommunication: the fixed and mobile convergence (FMC). The chapter deals with the impact of FMC on the Next Generation Network architecture, the network and application layer mobility management solution, policy control, routing, AAA, and other security issues of FMC. Finally, it introduces long-term evolution of mobile telecommunications. Chapter 12, "*Host Identity Protocol: The Enabler of Advanced Mobility Management Schemes*," by László Bokor, Szabolcs Nováczki, and Sándor Imre, Budapest University of Technology and Economics (Hungary), introduces Host Identity Protocol (HIP), which is a promising protocol for separating the identification and location role of IP addresses. The chapter compares HIP with other protocols of the same purpose, and then shows several mobility management, privacy, and application solutions of HIP. Chapter 13, "*Overview of IP Multimedia Subsystem Protocols and Communication Services*," by Sándor Szabó, László Gyöngyösi, Károly Lendvai, and Sándor Imre, Budapest University of Technology and Economics (Hungary), shows an advanced subsystem which is relevant in Next Generation Networks and suits well to the convergence of fixed and mobile convergence: the IP Multimedia Subsystem (IMS). The chapter deals with the architecture, the main protocols (like SIP, SDP, RTO, RTCP, DIAMETER and MEGACO), and the most important services of IMS (like Presence Service, Instant Messaging and Circuit-Switched Combination Service). Chapter 14, "*The TFRC Protocol and Its Usage for Wireless Video Transmission*," by Christos Bouras, Vassilis Papapanagiotou, Giannis Zaoudis, Research Academic Computer Technology Institute, University of Patras (Greece); Kostas Stamos, Research Academic Computer Technology Institute, University of Patras and Technical Educational Institute of Patras (Greece), introduces TCP-Friendly Rate Control (TFRC) protocol. The chapter analyzes the effect of cross-layer management on wireless video transmission via TFRC. Results presented are related to the power consumption and transmission quality. Chapter 15, "*Cross-Layer Protocols for Multimedia Communications over Wireless Networks*," by Jaydip Sen, Tata Consultancy Services (India), goes into more details of cross-layer design for wireless multimedia transmission. Beside the transport layer, the data-link and the application layers are considered, and issues like channel estimation techniques, adaptive controls for energy efficiency, adaptive ARQ, transmission rate control, and priority based scheduling are investigated. Chapter 16, "*Session Management and Transport Protocols for Multimedia Services over IP Networks*," by László Lois and Ákos Sebestyén, Budapest University of Technology and Economics (Hungary), introduces application layer level protocols in the TCP/IP model, which care with session management and transport tasks of multimedia transmission. Especially the Real-time Transport Protocol (RTP), the RTP Control Protocol (RTCP), and the Real-time Streaming Protocol (RTSP) are highlighted. Their functions are introduced through a complex real-life application: IPTV. Chapter 17, "*A Solution for Evaluating QoS of Voice over IP: Measurements, Analysis and Modeling*," by Homero Toral-Cruz, Deni Torres-Román, and Leopoldo Estrada-Vargas, Center of Research and Advanced Studies (Mexico), presents a new model for VoIP QoS evaluation. After introducing the mathematical background of QoS concepts and IP traffic nature, jitter and packet loss behavior of VoIP is analyzed. Finally, the chapter proposes a model, which could be used well by other researchers of the topic. Chapter 18, "*SCTP: Solution for Transport Layer Mobility and Multihoming*," by Árpád Huszák and Sándor Imre, Budapest University of Technology and Economics (Hungary), introduces the Stream Control Transmission Protocol (SCTP) as a transport layer protocol, which is able to handle mobility management issues. In the chapter, SCTP's multihoming and multistreaming performances are analyzed using a native IPv6 UMTS-WLAN test environment. Regarding to SCTP's role, results can be used in future LTE-EPS architectures, as well. Chapter 19, "*IPv6 Routing in a Special Context: Serving Efficient Data Aggregation*," by Zoltán Kanizsai and Gábor Jeney, Budapest

University of Technology and Economics (Hungary), deals with IPv6 from routing aspects. Readers of this chapter can get familiar with the main unicast, multicast, and anycast IPv6 protocols. Finally, the authors introduce their ideas for aggregating measurement and multimedia feedbacks using the anycast routing protocols of IPv6. Chapter 20, "*Multiprotocol Label Switching Virtual Private Networks: Problems, Protocols, Possibilities*" by Jan Schankin and Eduardo Correira, Christchurch Polytechnic Institute of Technology (New Zealand), presents the basics of Multiprotocol Label Switching (MPLS) and its application in creating layer3 VPN over a shared physical channel (MPLS L3 VPN). After showing the protocols of this solution, the chapter summarizes its problems and possibilities. Finally, it describes the possible role of MPLS in the IP network of the future.

Section 4: Protocol Applications and Technologies consists of three chapters. Chapter 21, "*Time Synchronization in Wireless Sensor Networks*," by Gyula Simon and Gergely Vakulya, University of Pannonia (Hungary), familiarizes the reader with time synchronization services often required to support coordinated operation of the nodes in sensor networking applications, potentially containing hundreds or thousands of elements. The authors introduce various time synchronization models and emphasize the potential accuracy of the synchronization mechanisms that use these models. Chapter 22, "*Application-Driven Routing in Wireless Sensor Networks*," by Gyula Simon, University of Pannonia (Hungary), presents the sensor networks built from tiny, resource limited nodes, which are able to communicate with each other, providing distributed services. The chapter describes routing algorithms, important middleware services in sensor networks and the performance of the application, which is greatly affected by the quality of the utilized routing protocol. Chapter 23, "*Radio Frequency Identification*," by Róbert Schulcz and Gábor Varga, Budapest University of Technology and Economics (Hungary), discusses the history of Radio Frequency Identification (RFID) systems. In this chapter, the necessary components are listed, and systems are classified based on several parameters, in order to support planning and deployment of RFID systems.

We recommend *Advanced Communication Protocol Technologies: Solution, Methods and Applications* for experts, researchers, professors, and students working, teaching, and studying in the fields of telecommunications and Internet. We believe our book can be useful for those who have practical experiences to learn the theoretical background of communication, for those who are already familiar with the protocol theory to get an overview about various application areas, and this book can also serve as a textbook at universities.

Because we think the members of our target audience have different educational background and different practical experiences, we wanted to present some terms and certain protocols from different points of view, which is why the reader will find that some notions and applications are discussed in several chapters but in different context and approach.

Katalin Tarnay
University of Pannonia, Hungary

Gusztáv Adamis
Budapest University of Technology and Economics, Hungary

Tibor Dulai
University of Pannonia, Hungary

Acknowledgment

We would like to acknowledge the help of the six Editorial Advisory Board members and the twenty-six expert reviewers involved in the double-blind review process for all the twenty-three chapters of this book. Without their work and support, this book could not have been satisfactorily completed. We sincerely thank In Lee, Professor of Information Systems at Western Illinois University, Editor-In-Chief of Advances in E-Business Research (AEBR) Book Series; Jan Travers, Erika Carter, Christine Bufton, and Hannah Abelbeck and others from IGI Global, whose help and contributions throughout the whole process have been essential. We also thank Sarolta Dibuz from Ericsson, and telecommunications software expert Zsuzsa Harangozó for their brilliant ideas. We are grateful for the constructive work of Tamás András Zettner, who was assisting in organizing the whole project from the beginnings. Finally, we would like to express our gratitude to the thirty-six authors for their contributions and collaboration.

Katalin Tarnay
University of Pannonia, Hungary

Gusztáv Adamis
Budapest University of Technology and Economics, Hungary

Tibor Dulai
University of Pannonia, Hungary

Section 1

Basics of Communication Protocols

Chapter 1
Communication Protocols:
An Introductory Overview

Katalin Tarnay
University of Pannonia, Hungary

Gusztáv Adamis
Budapest University of Technology and Economics, Hungary

ABSTRACT

The increasing importance of fast and reliable communication has resulted in the creation of certain communication rules, which nowadays, we call communication protocols. In this chapter, a historical overview of early protocol development and deployment is presented as a look-back to the evolution of communication protocols. Then we introduce the most important terms and notions of the protocol theory. To make it easier to understand these concepts, we use the INRES protocol as demonstration tool. In the next topic, we summarize the steps of the protocol engineering. At the end of the chapter, we give some guidelines for protocol classification.

INTRODUCTION

This chapter provides an introduction into the world of protocols: the communication rules which are responsible for controlling various communication services. Communication rules are presented by introducing syntax (what sort of notation is used, what are the names of the messages, how the messages are structured, what are the components of the messages), semantics (what kind of messages are sent and expected as response in certain situations), and temporal rules (a strict and limited timing aspect, the encoded lifetime of every message sent). The author also presents various protocol classification methods, based on different points of view.

If we want to exactly determine the behavior of a living or a lifeless system, the interactions have to be in the center of our examinations, because actions and reactions follow each other. The communication protocols specify the rules and laws describing the interactions of computer,

DOI: 10.4018/978-1-60960-732-6.ch001

telecommunication, industrial etc. networks. It is very important to note that the protocols in this chapter always refer to communication protocols.

The notion of protocol was simultaneously used in 1967 in two networks, in the Advanced Research Projects Agency Network (ARPANET) and in the National Physical Laboratory Network (NPL-NET). NPL-NET was designed to elaborate physical data transmission and evaluation, and it was operating in Daresbury, United Kingdom. ARPANET, which came into general use in the United States, was created to handle communications among computer users. Both systems implemented a revolutionary method for data communication called packet switching, the transmission of data streams, which were composed of messages with unified lengths. Using the packet switching method, the size of the data packages could be set to require only an average store capacity in the node. Then the packet could be forwarded so it would not occupy the whole dedicated route but only a path between two nodes following each other. After various modifications and fine-tuning of the network itself, the descendant of ARPANET first mentioned as Internet in 1974, became used worldwide. Communication via the Internet is realized by a set of protocols called the Internet Protocol Suite also known as the Transmission Control Protocol/Internet Protocol (TCP/IP) stack, specified by the Internet Engineering Task Force (IETF).

The NPL protocols referred to the reference model of the Open System Interconnection (OSI) determined by the International Organization for Standardization (ISO). Based on the experience of working with the NPL protocols, the European Organization Conseil Européenne pour la Recherche Nucléaire (CERN) developed the World Wide Web (WWW) in 1992. The basic protocol component of the Web is the Hypertext Transfer Protocol (HTTP) providing excellent searching and browsing possibilities (Buchanan, 1999; Davies et. al. 1982).

The main reason why the use of computer networks and communication protocols became widespread is the coexistence of the Internet and the Web. Nowadays the Internet and the Web are used together.

The importance of protocols is considerably growing because of the so-called information boom. This concept concerns both the personal communication and the industrial processes. The computers and computer networks play an important role in our everyday life and they are also important in several areas, such as online libraries and dictionaries, mass production, computer-aided planning, e-business and e-learning, web trading, the stock exchange, advertisements, sensor networks, the GPS, and even the CERN fusion reactor.

In this chapter, an introductory overview is given about the basic syntactic, semantic, and temporal rules of protocol operation.

This chapter consists of four subchapters. After the introduction, the first subchapter contains a historical overview of communication protocols. First the optical semaphore system developed in the Napoleonic times is presented, then it is illustrated how the evolution of Internet and NPL protocols took place starting from basic physical measurements and gradually developing into very complex communication systems.

The second subchapter discusses the syntactic, semantic, and temporal rules of operation and specifies the qualitative and quantitative protocol parameters and features. The INRES protocol is used to illustrate the protocol operation.

The next subchapter deals with protocol engineering, showing the methods, languages, and tools thereof. The fourth subchapter presents guidelines for protocol classification. At the end, we draw the conclusion. The references are divided into two groups: the main references pertaining to this chapter are given first, which is followed by the enumeration of the papers for further reading. Finally, the key terms are explained.

BACKGROUND

In the background, we summarize the protocol history. We start with the early communication protocol implemented during the Napoleonic wars in Europe, and then we continue with the ARPANET and the National Physical Laboratory Network (NPL-NET).

Since ancient times many experiments have been carried out and many methods were used to transmit long distance messages, especially in times of war when the messenger had to run a long distance to carry the message. In these cases, the message transmission was risky and unreliable. The first reliable and technically well-supported successful practical solution was realized during the French revolution. In 1792, Napoleon charged Claude Chappe to create long distance communication between Paris and Lille (Holtzmann, 2003). Table 1 shows some of the codes. These cities were the nodes of the optical telegraph communication. Within the 190 km distance separating the two cities, they built 15 stations with a distance of 6 miles (roughly 10 kilometers) between each station to ensure visibility. The stations called semaphore telegraphs were built from a large horizontal beam called the regulator, equipped with two smaller wings called the indicators, mounted in both ends of the beams. The angles of the indicators and the position of the large regulator beam could be varied independently from each other in increments of 45 degrees, which allowed eight positions for each indicator and four positions for the regulator. This was sufficient for encoding 8x8x4=256 possible combinations.

In order to separate variable-length code groups from each other within a longer transmission, the last sign in each group was made with the unused indicator turned slightly away from the regulator on a 22.5-degree angle to clearly distinguish this "termination" sign from other possible semaphore positions. Apart from the correspondence signals, there were fourteen service or control signals known by all operators. They were used to signal the following conditions:

Table 1. Code examples of the Napoleonic telegraph system

Sign	Name
2H8	Activity (Paris)
1H1	Activity (Lille)
2H6	Error
3V7	Suspend- Fog
4V8	Suspend – Failure
5H5	Close

- Start of transmission.
- Synchronization: to instruct the stations along the telegraph line to resynchronize their clocks.
- Acknowledgement: after every transmission, the receiver sent back the received message to check its correctness. If the message was correct, it was forwarded to the next station. By this method, it could be ensured that the message was correctly deciphered at the final station.
- Conflict: to indicate that two messages arrived simultaneously at a station from opposite directions.
- Suspension of transmission for one/two hours.
- Error: to cancel the last transmitted sign.
- Idle: to indicate that the station or line closes.
- Minor failure: to indicate a small problem with the telegraph, or to indicate the temporary absence of the operator.
- Major failure: to indicate a serious problem, e.g. the telegraph is disabled and it needs outside assistance.

The appearance of the electrical telegraph superseded the Chappe telegraph but the latter was the first system to use a protocol in practice, which guaranteed an error-free message transfer from the source to the destination, and which exactly defined the transmission rules as e.g. acknowledgement or suspension.

Nearly two hundred years later, when several computers came to be connected to form a network, it was necessary to develop a protocol to ensure the communication between them. It is interesting that the rules of this protocol are basically similar to the set of rules used by Chappe.

The main points of similarity are as follows:

- Every message contains an address.
- Upon the completion of each message and before sending a new one, there was an acknowledgement, which assured the correctness of the messages.
- Message collision is handled.

ARPANET (Advanced Research Projects Agency Network) was the next step towards communication protocols, but the connection this time was realized among computers. The starting point for host-to-host communication on the ARPANET was a protocol, which defined the way a host computer could transmit messages to an ARPANET Interface Message Processor (IMP). The message format was designed to work unambiguously with a broad range of computer architectures. A message essentially consisted of a message type, a numeric host address and a data field. To send a data message to another host, the transmitting host would format a data message containing the address of the destination host and the user data being sent, and then transmit the message through a hardware interface. The IMP then delivered the message to its destination address, either by delivering it to a locally connected host, or by delivering it to another IMP. When the message was ultimately delivered to the destination host, the receiving IMP transmitted a RFNM (Ready for Next Message) acknowledgement to the sending host IMP. The message transfer and the number of messages between any nodes were controlled by the Network Control Program (NCP), which provided a standard method to establish reliable, flow-controlled, bidirectional communications links among different processes running on different host computers. In 1983, TCP/IP protocols replaced NCP as the ARPANET's principal protocol, and the ARPANET then became one component of the early Internet. Figure 1 shows the main protocol components of the Internet.

PROTOCOLS AS DYNAMIC DIALOGUES

The easiest way to demonstrate the operation of any communication protocol is to simply compare human dialogues with computer communication. This way, message exchanges between persons would be very similar to message exchanges between network nodes performing some communication task. The rules of this message exchange are applicable to the message format, to the message exchange itself and to the time assumptions. Again, if we consider dialogues, we find that the message is similar to a human word; the message format corresponds to the correctly spelled word, while the counterpart of the exchange process is the phrase or sentence told to each other. Furthermore, it should be noted that time assumptions are extremely important. Let us say, we are operating a computer network, and we want to monitor a network resource to guarantee the correct operation of the system. When monitoring, one needs a request to query a parameter value, and also needs a response, which includes it. The response time is normally limited; therefore this situation is also very similar to a human dialogue, where a person, who is asking a question, waits for the answer only for a certain amount of time (Budkowski, et. al.1998; König 1990; Tarnay 1991).

The essential function of protocols is to provide communication services among the users. The services can be either simple or complex. For example, warning is the task of a simple

Figure 1. Protocol components of the Internet

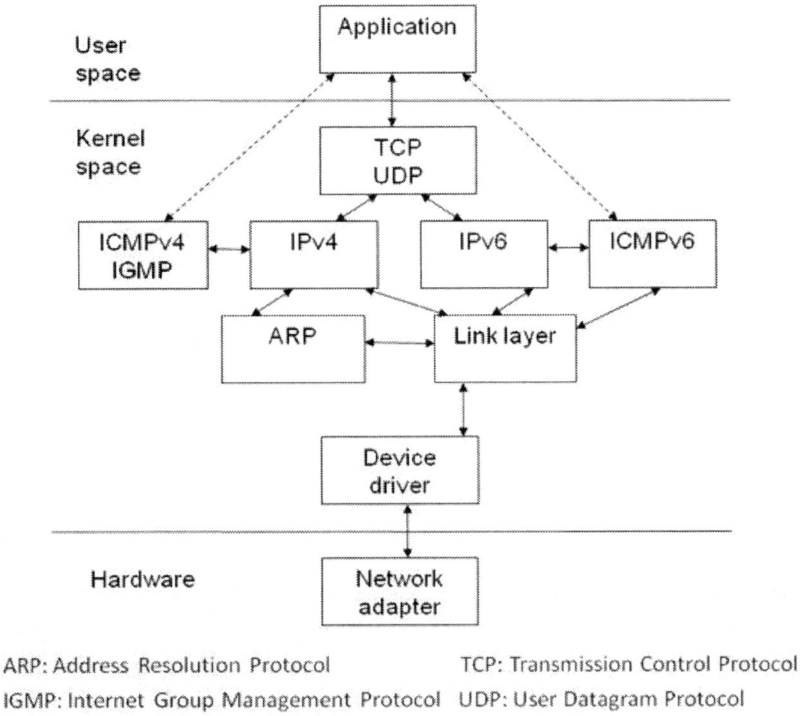

service, while complex services are composed of connection establishment, information exchange, and disconnection.

Protocols are the basic elements of network operation providing message exchanges among the network nodes. Every protocol is used for communication purposes. Certain semantic, syntactic, and temporal rules determine the message exchange itself. The syntactic rules define the format of the messages and their parameters, the semantic rules define the request-response pairs, while the temporal rules are responsible for the setting of the reaction time and the so called time to live interval.

Let us describe a simple example. According to conventional manners, if we meet someone, we would start the discussion by some kind of greeting: hello, good morning, good evening and so on. Our communication partner would say something very similar, and then we would start communicating with each other. The greeting "procedure" (also called a service) is a good example for describing the semantic rules: saying "hello" would normally result in answering with a "hello", which would fulfill the rule. Saying hello and then receive a "good-bye" would be rather strange and would normally end the communication, because it is a sort of breaking the semantic rule.

If the communication is established, the syntactic rules define the proper way of communication, namely the language itself. Saying, "How are you today?" is a sentence in English: it has a meaning and thus it follows a grammatical rule. Saying this sentence in an incorrect word order would result in breaking the syntactic rule and therefore would most likely result in a serious verbal communication problem. So, both partners must speak the same language.

However, the request-response time also plays an important role in any communication. Saying, "I feel fine, and how are you?" and then waiting, let us say, for an hour or so to get an answer would break the temporal rule, because normally we are "calibrated" to receive answers in seconds.

There are two types of messages, one is for control, and the other is for data transfer. The data transfer message also has a control field. (As an example, a data message can be seen in Figure 2.) The protocol messages generally have a maximal length, and if a message was longer than this maximal length, it should be cut into shorter parts. In this case, sequence numbers are used to be able to determine the order of the fragments of the message at the receiver side. This feature is often called segmentation and reassembling (SAR).

The main types of procedure and communication functions are:

- Connection establishment
- Connection release
- Data transfer
- Addressing
- Error handling
- Routing
- Flow control
- Assignment of resources

The time specification determines the time-out limits of waiting, i.e. the time interval, within which the system accepts an answer. Therefore, the protocol is defined by the formats, the procedures, and the time behavior.

During the transmission, messages may be damaged, or other problems may arise. The protocols often provide mechanisms to detect and correct the data transfer errors, like:

- A frame is lost
- A frame is doubled
- A frame arrives at a node with bad address
- The order of frames changes.

If the format of a received message is erroneous (invalid message), or if a syntactically correct message arrives at a time when it is not expected (improper message), an irregular answer may arrive. An example for the first case is when after saying: "How are you?" the answer is "Wefrto xid", and for the second case is when the answer is "The train departures at 5 o'clock". A timing error may cause a lost frame, or a message that is not replied to.

Figure 2. Example of a data message

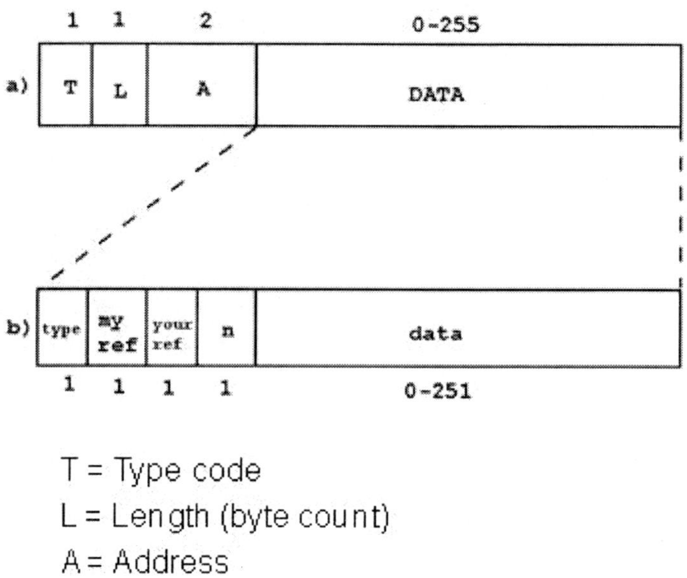

The protocol as a dynamic dialogue is well demonstrated with a simple example: the INRES protocol. By the help of this example, we will also introduce some definitions.

The Initiator-Responder (INRES) protocol is a simple protocol demonstrating the syntactic, semantic, and temporal rules. In our example, let us assume two nodes: an Initiator user and a Responder user (see Figure 3). The protocol itself is the dialogue between the two nodes.

The protocol operation of the INRES is asymmetric. (Of course, there are symmetric protocols as well.) Only the Initiator can send any data and the Responder may either receive or refuse it. First, the Initiator sends a Connection Request (CR) message. If the Responder accepts the request for connection, it answers with a Connection Confirmation (CC). If it denies the request, the response is a Disconnect Request (DR). In case of connection confirmation, the connection will be established, and data (DT) can be sent. After receiving the data, the Responder sends an Acknowledge message (ACK) back to the Initiator. This dialogue is presented as a time sequence diagram in the left side of Figure 4. Both the data and the acknowledgement have sequence numbers: 0, 1, 2, and so on. The Responder must include the same sequence number in the data acknowledgement as it received in the data message. The Initiator can send more data with new sequence number values. In this case, the Responder acknowledges each data unit by acknowledge messages (ACK) with the corresponding sequence number.

Only the Responder can terminate the connection by sending Disconnect Request message.

If after sending a Connection Request, no Connection Confirmation arrives within the time limit, the Connection Request is sent again. This procedure is repeated a given times depending on the predefined retransmission number (as seen the right side of Figure 4). These steps demonstrate the coexistence of the semantic and temporal rules.

Messages of a protocol are called Protocol Data Units (PDU). Every PDU between the Initiator and the Responder has a control field. In this field, a bit sequence characterizes the exact message type. In our example, there are five PDUs: CR, CC, DT, ACK, and DR, so three bits are enough to specify the message type. Apart from the mes-

Figure 3. INRES block diagram

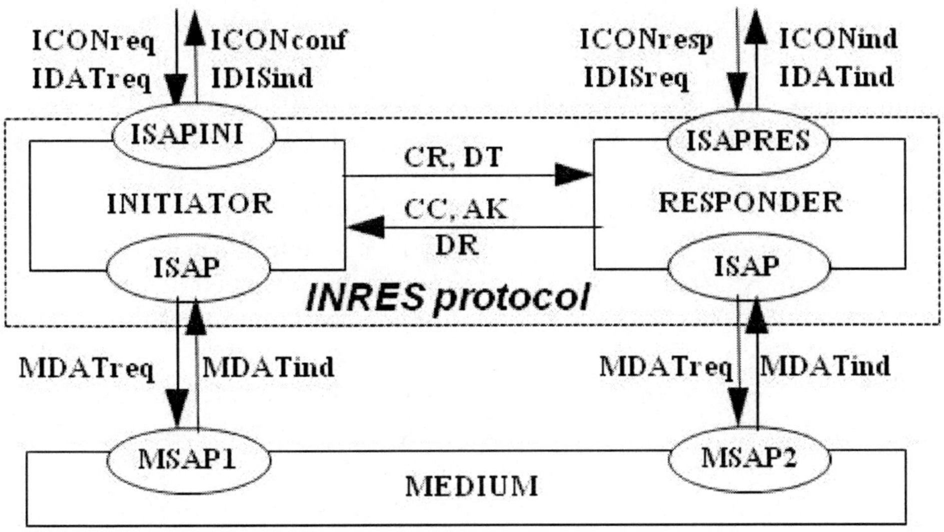

Figure 4. INRES time sequence diagram examples

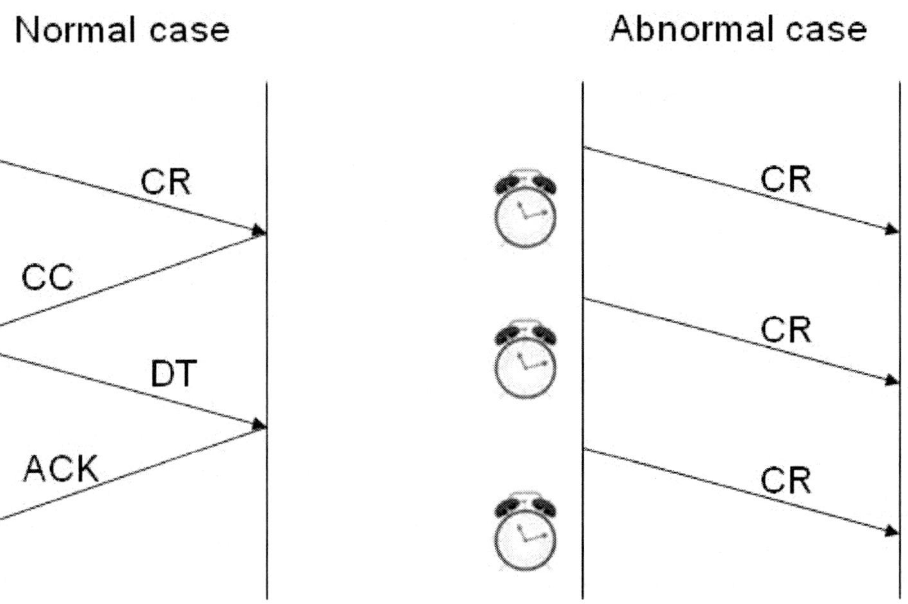

sage type, the control field of the DT and ACK PDUs contains another parameter, the Sequence Number. The only data transfer PDU of the INRES protocol is the DT message. Therefore, it contains a data field next to the control field.

The service offered by a protocol can be accessed through the so-called Service Access Points (SAPs). In INRES, there are two SAPs defined: ISAPINI and ISAPRES (INRES Service Access Point INItiator side and INRES Service Access Point RESponder side, respectively). The messages, by which the user of the protocol can communicate with a protocol (use a feature or be informed about an action) are called Abstract Service Primitives (ASPs).

There are four types of ASPs, in general:

- Request: used to request a service.
- Indication: used to inform an entity that an event has occurred.
- Response: used to reply for a request.
- Confirmation: used to inform an entity that a response arrives.

In INRES, the connection establishment is an acknowledged service, so all the four INRES connection (ICON) ASP types exist, while for data transfer (IDAT) and disconnection (IDIS) only the Request and Indication ASPs exist. (Note: the data transfer is acknowledged within the INRES, but the delivery is neither acknowledged by the Responder user, nor reported to the Initiator user.)

The PDUs are not exchanged directly between the pair entities of a protocol, but are transmitted through a communication medium. (More precisely: a protocol uses the transmission service of an underlying layer protocol – the layering concept will be introduced later.) Here a non-secure medium is assumed (e.g. the UDP protocol is of this kind in the Internet protocol family). The service of the medium can be accessed through MSAP1 and MSAP2 service access points (Medium SAP), with the help of the MDATreq (Medium DATA Request) primitive. The arrival of a message at the other SAP is indicated by the MDATind primitive.

Protocol features can be classified according to several viewpoints. One possibility is to classify them according to quantity and quality features. The quantity features can be given in bits, seconds, number of messages, percentages etc., whereas the quality features characterize the reliability, the logical correctness, as well as the identity of protocols.

The most frequent quantity features are:

- Length of data unit
- Window size
- Number of frames
- Segment size
- Number of data units per block
- Time duration of the connection establishment
- Time duration of the connection release
- Number of detected errors
- Ratio between control information and data field

There are many complicated relationships among the quantity features. The quantity features can be measured under operation by protocol analyzers and testers, as well as by the built-in network management center. The efficiency of the network can be increased by regulating it according to quantitative parameter settings.

The quality features of the protocols characterize the error-free operation (e.g. proper flow control in case of a node failure), the logical correctness (e.g. whether the protocol has a deadlock) as well as conformance and identity testing. Error-free operation is controlled by simulation in the course of designing and by measurements during implementation and operation. The logical correctness should be controlled only in designing; there are several verification methods for this.

One of the most important steps of implementation is the conformance test (testing the conformance of the realized protocol according to specifications). Conformance testing examines all three components of the protocols: formats, procedures, and timing.

The protocols have general and special quality features. The general features characterize every protocol, while the special features belong to some special protocol types. The most important general quality features are the following:

- Deadlock freeness
- Termination (guaranteed accessibility of end states)
- Completeness (the existence of a rule for the continuation in every possible situation)
- Stability (the protocol returns to the normal operation after an abnormal operation)

Good examples for special quality features are code conversion of presentation layer protocols, and setting of system requirements of real world terminals in a virtual terminal environment.

PROTOCOL ENGINEERING

The subject of the protocol technology is design, realization, and control. Steps of the realization process of a protocol can be followed in Figure 5.

The first step of protocol design is to prepare the textual ("informal") description of a protocol from the requirements. However, this description is often ambiguous and cannot be checked and processed automatically by computers. This is why protocols should be specified formally with a Formal Description Technique (FDT). An FDT is a special language or notation for protocol (or system) specification with a strong mathematical background. Nowadays, the most popular FDTs are:

- SDL (Specification and Description Language)
- UML (Unified Modeling Language)
- MSC (Message Sequence Charts) for system specification
- ASN.1 (Abstract Syntax Notation One) for data description

Figure 5. Steps of protocol engineering

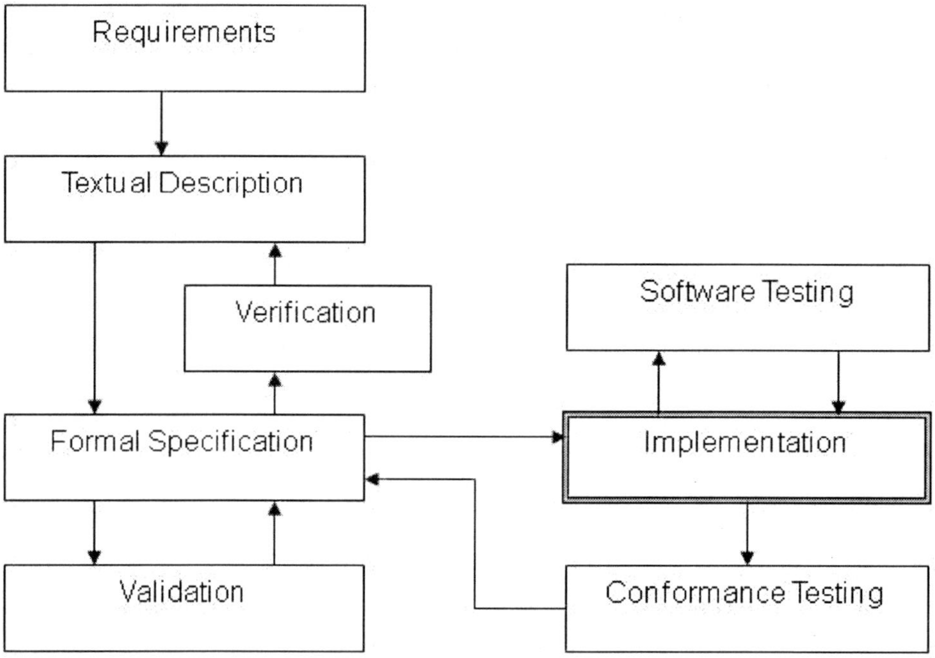

Of course, when the formal specification is ready, it must be compared with the requirements and/or with the textual description. This process is called verification.

The logical correctness of the protocol specification (e.g. deadlock freeness) should also be checked before the realization. There are different validation algorithms to check the logical correctness of a protocol. Depending on the result of the validation, the model should be modified or the requirements should be changed. The specification and the validation can be aided by computer simulation.

A validated specification can now be implemented. The implementation is the realization of a protocol specification in an executable form, taking into consideration the conditions and limitations of the target environment. The implementation can be carried out either in an ad-hoc or in a (semi-)automatic way. In an ad-hoc way, the protocol specification is implemented by a group of programmers in a similar way as other software products are developed, while the automatic implementation is actually just a translation between the FDT and the target language (e.g. C or C++), so this can be done by a cross-compiler. However, this process can never be fully automated, since the protocol specification hides many implementation-dependent questions. (This is why, the *same* protocol can be implemented on *different* systems.) A protocol specification just says, for example, that in a given situation a certain message is to be sent. However, it gives no hints, *how* this message can be sent in the target environment. Only skeletons of these environment-dependent parts can be prepared automatically, or they must be added manually. This is why, instead of automatic implementation, it is better to call the process semi-automatic implementation.

The protocol implementation must go through several tests. Of course, the testing process is easier in a case of a semi-automatic implementation, since in this case only the environment-dependent parts are written manually, the main part of the protocol implementation is generated automatically. *This is* the main advantage of the semi-automatic implementation.

First, the implemented protocol should be tested as software. Then, it should be checked, whether the protocol implementation fulfills the requirements, or in other words, whether it conforms to the specification. This check is called conformance testing (König, 2003; Csopaki 1999). The test sequences are written typically in TTCN (Testing and Test Control Notation) language. However, in cases of real protocols, we cannot state that a protocol implementation *completely* conforms to the specification, since the real protocols are too complex to test them during acceptable (or even finite) time. In practice, we can only say, that the protocol implementation conforms to the specification with a confidence level of, for example, 98%. (Most typically, the unexpected situations, like receiving invalid or improper messages, cannot be tested exhaustively, since there are too many of them.) Therefore, in networks, where protocol implementations of different vendors must interwork, interoperability tests should be carried out.

In some cases, it is necessary to perform performance tests, as well. During a performance test, we examine how the protocol implementation behaves under different traffic loads.

GUIDELINES FOR PROTOCOL CLASSIFICATION

This subchapter deals with points of protocol classification. There are many possibilities for classification; one of the simplest methods is selected here. The classification points are divided into five parts. The first part is the transmission medium, the second part is the transmitted content, the third part is the standardization forums, the fourth one is the communication functions, and finally the application-oriented protocols are introduced (Buchanan, 1999; Davies et. al., 1982). There are several protocols belonging to more classification groups.

First Part: Transmission Medium Used In Computer Networks

The transmission medium can be:

- wire, cable
- optical fiber
- radio, microwave
- satellite

Wired and Wireless Communication

Wired communication can be realized by an open wire line, by a coax cable, or by an optical cable.

Wireless communication forms are mainly radio and satellite communication.

Typical examples:

- Optical:
  - Example protocols:
    - OSRP (Optical Signaling & Routing Protocol) is an optical routing protocol similar to PNNI (Private Network-to-Network Interface) and MPLS (Multiprotocol Label Switching), which was developed for ATM networks.
    - FDDI (Fiber Distributed Data Interface) provides a 100 Mbps optical standard for data transmission in a local area network that can extend in range up to 200 kilometers (124 miles).
- Radio:
  - Example protocol: RLP (Radio Link Protocol) is an automatic repeat request (ARQ) fragmentation protocol used over a wireless (typically cellular) air interface
- Satellite:

   - Example protocol: GPS (Global Positioning System)

Second Part: Types of Transmitted Content

The transmitted content can be:

- Data:
   - Example protocol: X.25
- Voice:
   - Example protocol: VoIP (Voice over Internet Protocol) is a general term for a family of transmission technologies for providing voice communications over IP networks such as the Internet or other packet-switched networks.
- Image:
   - Example protocol: ITP (Image Transport Protocol)
- Video:
   - Example protocol: RTP (Real-time Transport Protocol)

Third Part: Basic Standardization Organizations

The protocols are standardized by different standardization forums. The three main groups are: ITU, IETF and special groups.

- ITU (International Telecommunication Union) is an agency of the United Nations, which regulates information and communication technology issues.
   - Example protocol: X.25
- IETF (Internet Engineering Task Force) develops and promotes Internet standards, dealing in particular with standards of the TCP/IP and the Internet protocol suite.
   - Example protocols: IPv4, IPv6, SNMP
- Special forums
   - WAP Forum
      - Example protocol: WTP (WAP Transport Protocol)
   - 3GPP (3rd Generation Partnership Project) is collaboration between groups of telecommunications associations, to make a globally applicable third generation (3G) mobile phone system specification
      - Example protocols: GSM, GPRS, UMTS protocols

Fourth Part: Communication Functions

The protocols should fulfill different communication functions and these functions can be divided into different communication layers. Some layers belong to the transmission medium, some other to the transmitted content. The number of layers can depend more or less on the application.

The functional layers were developed by the ISO and realized as the OSI (Open System Interconnection) Reference Model. The Reference Model (RM) defines seven layers. The IETF defined the TCP/IP protocol stack originally with four layers. The ITU-T also defined a layered protocol architecture for the Signaling System 7 used in the ISDN and mobile networks, but the SS7 recommendations use the term “level” instead of “layer”.

In this chapter, we just compare the ISO/OSI and the TCP/IP architectures. Here we describe the functionality of each layer briefly and give some example protocols. (An overview of the layering concept in general, the terminology, the interworking between layers, as well as a detailed description of the OSI layers can be found in chapter Protocol Operation. We deal with the SS7 protocols in chapters Signaling Protocols of

Integrated Services Digital Networks and Mobile Network Protocols of GSM and GPRS.)

The OSI layers are (See Table 2):

- Physical Layer
  - This layer supports the transparent transfer of any bit sequence through a physical communicating medium.
  - Example protocol: V.24
- Data Link Layer
  - It restricts the limitations of the physical connection. It permits detectable errors and supports the re-establishment of the connection. It eliminates the quality errors of the transmission or at least, restricts them.
  - Example protocol: LAPD (Link Access Procedure on D channel) used in ISDN
- Network Layer
  - This layer ensures transparent data transfer, selects the appropriate path, and routes the data through the selected path.
  - Example protocol: X.25 packet level protocol
- Transport Layer
  - It establishes connections among the end-users, optimizes the usage of the resources depending on the character of the communication and it allows the user to forget about matters of detail.
  - Example protocol: TP4 (Transport Protocol Class 4)
- Session Layer
  - It coordinates the cooperation between the communicating application processes.
  - Example protocol: SAP (Session Announcement Protocol)

Table 2. Seven layers of the OSI reference model

Application Layer
Presentation Layer
Session Layer
Transport Layer
Network Layer
Data Link Layer
Physical Layer

- Presentation Layer
  - This layer transforms the data syntax into a form, which the communicating application processes require.
  - Example protocols:
    - MIME (Multipurpose Internet Mail Extensions), which is an Internet standard that extends the format of e-mail
    - XDR (External Data Representation) is an IETF standard from 1995. It allows data to be wrapped in an architecture independent manner so data can be transferred between heterogeneous computer systems)
    - TLS (Transport Layer Security) and its predecessor, SSL (Secure Socket Layer) are cryptographic protocols that provide security for communications over networks such as the Internet
- Application Layer
  - This layer specifies the nature of the communication required by the user and ensures that the tools helping the application processes reach the OSI environment.
  - Example protocol: CMIP (Common Management Information Protocol)

is a protocol for network management, allowing communication between network management applications and management agents.

The layers of the TCP/IP protocol stack (Table 3) are:

- Link Layer
  - The Link Layer is the lowest layer in the Internet Protocol Suite. It defines the local network links and the interfaces and used to move packets between the Internet Layer interfaces of two different hosts on the same link.
  - Example protocols:
    - MAC (Medium Access Control)
    - FDDI (Fiber Distributed Data Interface)
    - Ethernet
- Internet Layer
  - The Internet Layer is used to transport packets between the originating and destination hosts across network boundaries (Internetworking). Provides addressing and routing.
  - Example protocols: IPv4, IPv6
- Transport Layer
  - In computer networking, the Transport Layer provides end-to-end communication between hosts either on the same or on different networks. Provides flow-control and error correction.

Table 3. Four layers of the TCP/IP protocol stack, as in RFC 1122

Application Layer
Transport Layer
Internet Layer
Link Layer

  - Example protocol: UDP (User Datagram Protocol) provides connectionless communications and does not guarantee packet delivery between transfer points.
- Application Layer
  - Application Layer contains all protocols and methods that fall into the realm of process-to-process communications.
  - Example protocol: IOTP (Internet Open Trading Protocol)

Fifth Part: Application-Oriented Protocols

Application-oriented protocols can be classified as management and client-server protocols.

- Management protocols
  - Example protocols:
    - CMIP (Common Management Information Protocol) is a protocol for network management. It provides an implementation for the services defined by CMIS (Content Management Interoperability Services), allowing communication between network management applications and management agents
    - IGMP (Internet Group Management Protocol) is a communication protocol used to manage the membership of Internet Protocol multicast groups.
    - SNMP (Simple Network Management Protocol) is a UDP-based network protocol. It is used to monitor network-attached devices for conditions that warrant administrative attention (Lee, 2009)

- Client – server protocols
    - Example protocols:
        - HTTP (Hypertext Transfer Protocol) is a networking protocol for distributed, collaborative, hypermedia information systems. HTTP is the foundation of data communication for the World Wide Web.
        - DNS (Domain Name System) is a distributed hierarchical naming system for computers, services, or resources connected to the Internet or to a private network. It associates various pieces of information with domain names assigned to each of the participants.
        - POP (Post Office Protocol) is an application-layer Internet standard protocol used by local e-mail clients to retrieve e-mail from a remote server over a TCP/IP connection.

CONCLUSION

This chapter gave a brief introduction into the advanced communication protocol technology. First, the historical overview of protocol evolution was discussed. Next, the protocol as a dynamic dialogue was presented. A simplified example, the INRES protocol was introduced to illustrate the protocol operation. Next, the protocol engineering steps and tools were described. Finally, the communication protocols were classified into groups from functional layers to different application areas.

REFERENCES

Buchanan, W. (1999). *Mastering networks. Houndmills, Basingstoke, Hampshire & London.* UK: MacMillan Press Ltd.

Budkowski, S., Cavalli, A., & Najm, E. (1998). *Formal description techniques and protocol specification, testing and verification.* Boston, MA/ Dordrecht, Germany/ London, UK: Kluwer Academic Publishers.

Csopaki, G., Dibuz, S., & Tarnay, K. (1999). *Testing of communicating systems: Methods and applications.* Dordrecht, The Netherlands: Kluwer Academic Publishers.

Davies, D. W., Barber, D. L. A., Price, W. L., & Solomonides, C. M. (1982). *Computer networks and their protocols.* John Wiley and Sons Ltd.

Holzmann, G. J., & Pehrson, B. (2003). *The early history of data networks.* Hoboken, NJ: John Wiley & Sons Inc.

König, H. (1990). *Kommunikationsprotokolle.* Berlin, Germany: Akademie-Verlag.

König, H. (2003). *Protocol Engineering Prinzip, Beschreibung und Entwicklung von Kommunikationsprotokollen.* Wiesbaden, Germany: B.G. Teubner Verlag.

Lee, I. (2009). *Handbook of research on telecommunications planning and management for business.* Hershey, PA & New York, NY: IGI Global.

Tarnay, K. (1991). *Protocol specification and testing.* New York, NY & London, UK: Plenum Press.

ADDITIONAL READING

Adamis, G., Horváth, R., Pap, Z., Tarnay, K. (2005). *Standardized languages for telecommunication systems*. Computer Standards & Interfaces.

Bochmann, G. (2000). *Testing of communicating systems: Tools and techniques. IFIP Advances in Information and Communication Technology*. Kluwer Academic Publishers.

Brzezinski, K. (2009). *Towards the methodological harmonization of passive testing across ICT communities*. InTech.

Jaskó, Sz., Dulai, T., Muhi, D., Tarnay, K. (2009). *Test aspect of requirement specification*. Computer Standards & Interfaces.

Kutvonen, L., König, H., & Tienari, M. (1999). *Distributed Applications and Interoperable Systems II*. Kluwer Academic Publishers.

Lee, I. (2009). *Open APIs and protocols for services and applications in telecoms*. IGI Global.

Lee, I. (2009). *Handbook of research on telecommunications planning and management for business*. IGI Global.

Murthy, C. S. R., & Manoj, B. S. (2004). *Ad-hoc wireless networks: Architectures and protocols*. Prentice Hall.

Robertson, P., Shrobe, H., & Laddaga, R. (2000). *Self-adaptive software*. First International Workshop. Oxford, UK.

Roe, P. R. W. (1995). *Telecommunication for all*. COST.

Schieferdecker, I., König, H., & Wolisz, A. (2002). *Testing of communicating systems XIV: Application to internet technologies and services*. Kluwer Academic Publishers.

Turner, K. J., Magill, E. H., & Marples, D. J. (2004). *Service Provision - Technologies for Next Generation Communications Systems*. John Wiley and Sons Inc.doi:10.1002/0470020504

Venkataram, P. (2004). *Communication Protocol Engineering*. Prentice-Hall of India.

KEY TERMS AND DEFINITIONS

Abstract Service Primitive (ASP): In a network layered architecture, when one layer requires another layer to carry out a service, the communication between the layers is carried out by service primitives. There are four categories of primitives: those that request a service, those that inform an entity that an event has occurred, those that represent responses to a service request, and those that confirm that the data associated with a request has arrived.

Formal Description: An unambiguous way of describing a (protocol) specification concisely. The abstraction level of the formal description extends from the network architecture to the dialogue realized by a protocol. It needs to be easy-to-handle and should have general validity.

Layer: The functional block of the Reference Model. There are seven layers in the OSI model, which identifies the functions of the layers and the relationships between them.

Open System Interconnection (OSI): This is a logical structure for network operations standardized by the ISO. Computers or devices that are OSI-compliant are able to communicate with each other. (See Reference Model)

Packet Switching: A method of transmitting messages through a communication network: long messages are divided into short packets, and they are then transmitted independently from each other. It is done by a packet store-and-forward system and no dedicated path is established, each packet contains a destination address and is passed from source to destination through intermediate nodes.

Protocol: The fundamental concept of the Reference Model, it defines the syntactic, semantic, and temporal rules of cooperation between two entities. A protocol contains the formal description of messages to be exchanged and the rules to be followed for two or more systems to exchange information.

Protocol Data Unit (PDU): Information that is delivered as a unit among peer entities of a network. A PDU contains control information, address information, and may contain (user) data.

Reference Model (RM): It is an abstract model, in which every real open system is described by an abstract open system as a whole, without preconditions for its internal organization. RM organizes the main functional communication tasks in seven layers.

Service Access Point (SAP): SAP is an identifying label for network endpoints used in Open Systems Interconnection (OSI) networking. When using the OSI Network Layer, the base for constructing an address for a network element is an NSAP address, similar in concept to an IP address.

Verification: Checking the logical correctness and proper operation of the protocol.

APPENDIX

Table of abbreviations

3G	3rd Generation
3GPP	3rd Generation Partnership Project
ACK	Acknowledgement
ARPANET	Advanced Research Projects Agency Network
ARQ	Automatic Repeat reQuest
ASN.1	Abstract Syntax Notation One
ASP	Abstract Service Primitive
CC	Connection Confirm
CERN	Conseil Européenne pour la Recherche Nucléaire
CMIP	Common Management Information Protocol
CMIS	Content Management Interoperability Services
CR	Connection Request
DNS	Domain Name System
DT	Disconnect Request
DT	Data
FDDI	Fiber Distributed Data Interface
FDDI	Fiber Distributed Data Interface
FDT	Formal Description Technique
GPRS	General Packet Radio Service
GSM	Global System for Mobile communication
GPS	Global Positioning System
HTTP	Hypertext Transfer Protocol
ICON	INRES CONnection
IDAT	INRES Data Transfer
IDIS	INRES DISconnection
IETF	Internet Engineering Task Force
IGMP	Internet Group Management Protocol
IMP	Interface Message Processor
INRES	Initiator-Responder protocol
IOTP	Internet Open Trading Protocol
ISAPINI	INRES Service Access Point INItiator side
ISAPRES	INRES Service Access Point RESponder side
ISDN	Integrated Services Digital Network
ISO	International Organization for Standardization
ITP	Image Transport Protocol
ITU	International Telecommunication Union
LAPD	Link Access Procedure on D channel
MAC	Medium Access Control

Communication Protocols

MDATind	Medium DATA Indication
MDATreq	Medium DATA Request
MIME	Multipurpose Internet Mail Extensions
MPLS	Multiprotocol Label Switching
MSAP	Medium SAP
MSC	Message Sequence Charts
NCP	Network Control Program
NPL	National Physical Laboratory
NPL-NET	National Physical Laboratory Network
OSI	Open System Interconnection
OSRP	Optical Signaling & Routing Protocol
PDU	Protocol Data Unit
PNNI	Private Network-to-Network Interface
POP	Post Office Protocol
PS	Packet Switch
RFNM	Ready for Next Message
RLP	Radio Link Protocol
RM	Reference Model
RTP	Real-time Transport Protocol
SAP	Service Access Point
SAP	Session Announcement Protocol
SAR	Segmentation And Reassembling
SDL	Specification and Description Language)
SNMP	Simple Network Management Protocol
SS7	Signaling System 7
SSL	Secure Socket Layer
TCP/IP	Transmission Control Protocol/Internet Protocol
TLS	Transport Layer Security
TP4	Transport Protocol Class 4
TTCN	Testing and Test Control Notation
UDP	User Datagram Protocol
UML	Unified Modeling Language
UMTS	Universal Mobile Telecommunications System
VoIP	Voice over Internet Protocol
WAP	Wireless Application Protocol
WTP	WAP Transport Protocol
WWW	World Wide Web
XDR	eXternal Data Representation

Chapter 2
Protocol Operation

Gábor Árpád Németh
Budapest University of Technology and Economics, Hungary

ABSTRACT

This chapter gives a fundamental overview of communication protocol operation. The concept of layering is introduced along with the definitions of communication primitives, protocol and service data units, data encapsulation, and service access points. An example for layering – the OSI Reference Model – is also presented. The service provided by a protocol can be characterized based on the various distributed communication models and operational environments. The model of an operation environment along with its characteristics and the model of a communication node are introduced as well. Various communication problems, trigger conditions, and the purpose of the timer trigger condition are discussed. Protocol interaction scenarios are presented with respect to the number of participants and the reliability of the transmission medium. The characteristics of these dialogues are also introduced and discussed.

INTRODUCTION

In the first chapter of the book, we have seen what the purpose of protocols is. A historical overview of the evolution, classification of protocols and the basics of models and formal methods were introduced. In this chapter, we will go further to get to know how these protocols work. An important aspect is to describe the environment where the communication protocols operate. The communication takes place in a distributed environment. This environment can be either symmetric or asymmetric, based on the associated roles of the participants. In a symmetrical environment, all participants of the communication have the same roles for communication, whereas in an asymmetrical environment there is at least one participant having different roles associated to it.

To realize the communication over the environment is a difficult task, therefore it is divided to different sub-problems. The introduction of these sub-problems allows us to see a given task at dif-

DOI: 10.4018/978-1-60960-732-6.ch002

ferent abstraction levels. We can either consider the communication between the participants as the transmission of the bits through given nodes or as a collection of data packets routed across selected nodes of the network. The set of more concrete functions can help to implement more abstract tasks, thus a task with a higher abstraction level uses subtasks from a lower abstraction level. The environment, where communication protocols operate can be described as the union of a transmission channel and the users of this channel. If we consider the functions of the lower levels with this transmission channel as a unit, we get a communication medium. This communication medium can be characterized based upon its performance and reliability, which affects the communication of higher-level tasks. The reliability – if required – can be ensured by the medium or it can be the task of its user. The different performance parameters of the participants and of the medium indicate that the entire information flow in the network must be controlled by a given task.

The communication scenario describes the interactions among the participants. The description must be unambiguous and coherent. The interactions between the participants can take place if the precondition of the given activity – which is described in the communication protocol – is fulfilled. To provide a consequent, conflict-free communication, a correct selection of these preconditions is necessary. The rules of communication can be analyzed through different protocol dialog scenarios. The dialogues can be discussed from different points of views, such as the number of participants of the communication, whether the receipt of a message has been acknowledged.

COMMUNICATION ARCHITECTURE

Distributed Environment

In a distributed environment, a set of nodes is interconnected with a certain medium, and form a network. The communication between these nodes is described with a communication protocol.

Two distributed architectures can be distinguished: client-server and peer-to-peer (Figure 1). In the client-server architecture, one or more nodes have a set of dedicated roles. The specialized nodes called servers provide a service to all the other nodes called clients or can control the communication. In peer-to-peer communications, all nodes implement an identical set of functions.

Layering

Describing communication in a distributed environment is a complex task; therefore, it is generally decomposed into sub-problems with the concept of layering. Layering organizes the protocol into

Figure 1. Client-server and peer-to-peer architecture

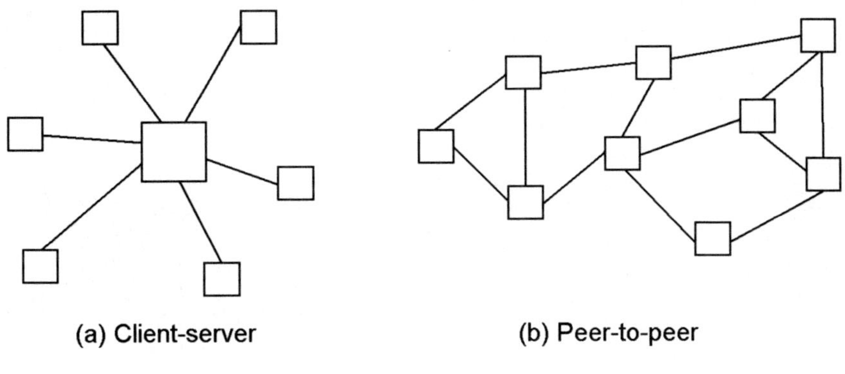

layers by forming communication interfaces from groups of closely related functions. These interfaces hide lower layer from higher layer. Higher layer tasks are the users of a lower layer interface; or in other words, a lower layer interface provides services for its users.

Consider the following example: a task manages a communication between two endpoints of the network. This communication uses a set of functions of a lower layer that is responsible for the communication between neighbor participants (Figure 2). Thus, the upper layer function uses the services of the lower layer functions, while it does not realize the details of the lower layer. The lower layer functions are transparent for higher layer functions.

When we define the different layers, many questions must be answered: How many layers should we use? What is the function of these layers? How should the communication interfaces be selected between these layers? The following aspects should be considered:

- Each layer should realize a well-defined sub-problem.
- Similar functions should be applied in the same layer.
- The different layers should represent different abstraction levels.
- The boundaries of the layers should be well defined. The number of communication interfaces and the data exchange through these interfaces should be minimized.
- There should not be too many layers defined otherwise the structure would be hard to handle.

The Layer Elements

Each layer consists of the same layer elements: entities, protocols, services and Service Access Points (SAP) – see Figure 3.

In each layer, one or more entities implement the functions of the layer. The protocol describes the set of rules – conventions or standards – within the layer. The layer-N protocol manages the cooperation of the layer-N entities. The interface is a boundary between two adjacent layers, through which these two layers can interact. The interface, where the set of functions provided by a logical layer-N entity can be accessed, is called N-SAP. The layer-(N+1) entities that access this set of functions are the N-service users. Each SAP has a distinctive identifier. The elements of communications between layer-N and the user in layer-(N+1) are the N-service primitives. The entity in layer-N does not only provide a service, but is a user of an (N-1)-service at the same time. It accesses the (N-1)-SAP with (N-1)-service primitives. From the point of view of layer-(N+1) the (N-1)-SAP is invisible, hidden by the N-SAP, hence the layer-N entity is an idealized medium.

In general, a layer-N entity has separate sets of functions for transmitting, receiving, and coding messages. In telecommunication networks, a layer-N entity has a management interface as well. This interface is used for operating the entity (startup,

Figure 2. Next hop and end-to-end communication

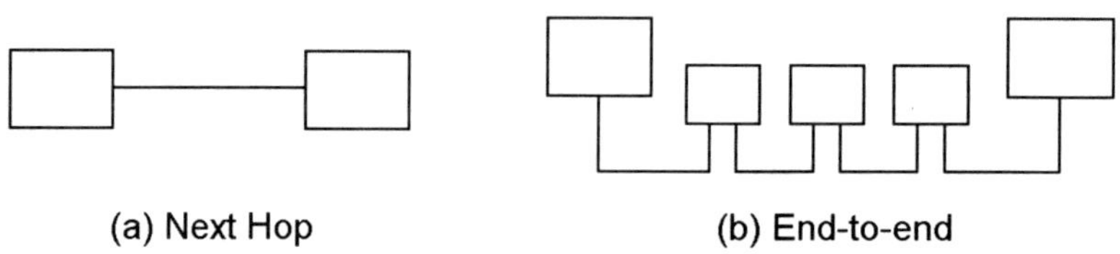

Figure 3. Layer elements

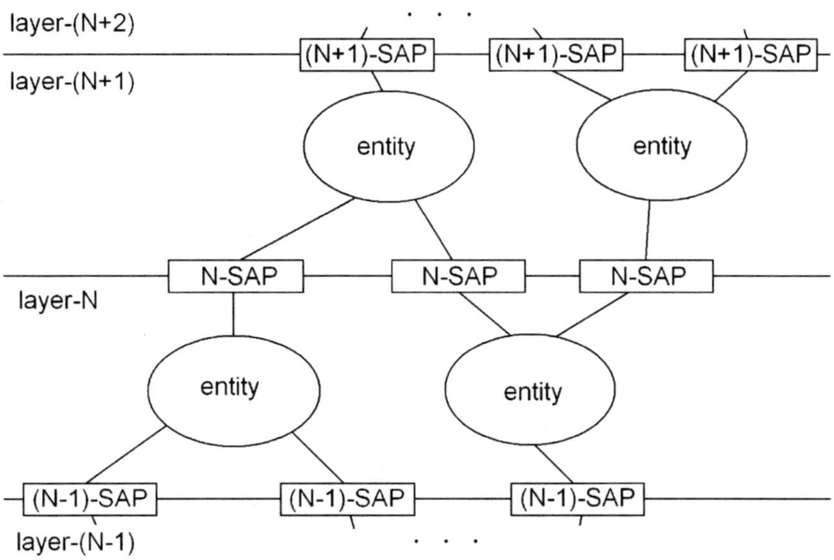

shutdown, recovery from emergency situations) or maintaining its configuration.

A layer-N entity of node *A* is in a logical relation with a layer-N entity of node *B*, if there is a communication relation between nodes *A* and *B*. In this case, these are called peer entities and the communication relation between them is specified by a protocol.

Protocol Messages and Service Messages

When investigating the communication between layers, there are two message formats: Protocol Data Units (PDU) and Service Data Units (SDU). The SDU of layer-N is a message, which is sent by the user of the services of the given layer-(N+1) and is transmitted semantically unchanged to layer-N of the communication peer. The entities in layer-N do not understand the SDU that is received from the higher, layer-(N+1) entities. The given layer-N keeps the received SDU unchanged and adds additional data to it – in the form of a header or a tail – that is required to implement its functions. Such an additional data might be an error checking information to provide error detection and error recovery or a network address to support routing functions. The original SDU that is received by layer-N with the additional data of layer-N is called the PDU of layer N. When a node of a network has received a message and this message is going upwards through layers, each layer takes off its additional data, reads it, and realizes its function using this additional data.

Therefore, the message format of layer-N encapsulates user (layer-(N+1)) messages (Figure 4). The layer-(N+1) messages may contain strings (bit string, byte string) that are valid tokens in the layer-N message structure. So before the encapsulation the higher layer message must be encoded. Then the layer adds a header and/or a tail to the encoded message and sends it. After the other participant receives this message, it removes its header and/or the tail and afterwards it transmits the message to the upper layer, which decodes it.

Based on the characteristics of the underlying medium, the optimal message size can be determined. In telecommunication, networks in case of data link layer protocols exact formula can be given to calculate the optimal message size based on physical layer parameters like the bit error ratio (Holzmann, 1991). Thus, the given data may be split into smaller pieces in order to achieve the optimal message size.

OSI Layering

The Open Systems Interconnection (OSI) Reference Model (ITU-T Recommendation, 1994) is an abstract description of communication and network architecture using layers. The OSI Reference Model is not a communication standard itself; it defines only a guideline how to develop such standards.

The International Organization for Standardization (ISO) suggests dividing the functions into seven layers. These layers – starting from the lowest – are the following:

1. Physical layer: This layer realizes the transfer of bit sequences using the physical communication medium. It defines the type of communication medium (such as a copper wire, a coaxial cable, an optical fiber, or a radio channel) and the physical and electrical specifications of this medium (such as voltages, cable specifications and pin layout.).
2. Data Link layer: The objective of this layer is to assure reliable data transfer between network entities by restricting the limits of the physical connection. It provides error detection and correction for the errors that may occur in the physical layer. This layer fragments the incoming data into data frames, transmits them in sequence, processes the acknowledgement frames sent by the receiver and realizes flow control.
3. Network layer: This layer performs typical network functions, such as routing and addressing. It selects the path between the source and the destination points and routes the data through this path. It provides transparent flow control for the transport layer.
4. Transport layer: This layer receives data from the session layer, splits them, and transmits this split data to the network layer. It ensures a transparent connection between the end users. It optimizes the usage of resources by the segmentation and de-segmentation of network connections while it hides the details from the users.
5. Session Layer: This layer controls the communication between the application processes. It supports their connections and data exchange.
6. Presentation Layer: This layer deals with the syntax of the transferred data, i.e., it transforms the different local syntaxes that are used in a heterogeneous system into a standard form that can be interpreted by any participant of the system.
7. Application Layer: This layer is the closest one to the end users: it provides an interface to them for reaching the OSI environment.

Figure 4. Message encapsulation

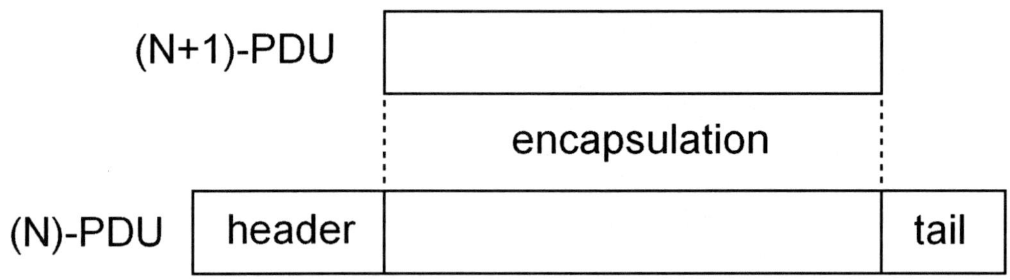

It describes the communication through this interface and the application processes. The typical functions of this layer are the following: communication partner identification, communication synchronization, and determination of the available resources.

The lower layers concentrate more on communication functions while the upper layers rather concentrate on data processing and information representation. The lower layers are the physical layer, the data link layer, the network layer and the transport layer. The upper layers are the session layer, the presentation layer, and the application layer.

The communication between the layers – or more precisely: the communication between entities that belong to a layer – can be discussed in two aspects. The logical communication is horizontal: a process on the nth layer of machine x sends a message to another process on the nth layer of machine y (Figure 5). The effective, physical communication is vertical, because each layer uses the service of a lower layer and provides a service to the upper layer. Actually, the nth layer sends a message to a lower, $(n-1)^{th}$ layer. Then this layer adds a header and/or a tail to the original message, and sends it to a lower layer and so on until the message is received by the first layer, which sends it to another machine. After the routing, the message is received by the first layer of machine y. Then, each layer removes its header or tail and sends the remaining message to the upper layer. This process is repeated until the process in the nth layer receives the original message.

By dividing the network communication architecture into smaller logical units the OSI model simplifies the way of designing network protocols. Moreover, different protocols can be used in the same layer if the interfaces of the layer remain unchanged. On the other hand, the layers have different sizes and importance: for example, the session layer is usually empty and the presentation layer is too small, while the data link and network layers are bigger, so they were split into sub-layers (Tanenbaum, 2003).

Model of the Operational Environment

Protocols operate in a distributed environment. This operational environment consists of the users of the communication service and a transmission channel responsible for message transfer. If we regard the functions of lower layers and this transmission channel as a single unit, we get a communication medium. The lower layers provide services for the upper layers that can be either reliable or unreliable. In case of unreliable service, the upper layers must assure reliability functions. For example, the Internet Protocol (IP) is unreliable and the Transmission Control Protocol (TCP) can provide reliability over the IP. If the lower layers provide reliable service, the implementation of the reliability functions is unnecessary in the upper layers. Although it is also possible, that a reliable layer is used above a reliable medium, for instance in the case of X.25. The reliability functions can be achieved using buffering: the sender buffers the outgoing messages, and if it realizes that a message is lost or damaged, it can resend the message from the buffer. Another interesting aspect of the communication medium is whether the messages arrive in a same order as they were sent. If not, then the upper layer must provide some ordering functions. The medium can be also investigated from the aspect of performance. The communication medium can be described numerically with the following characteristics:

- Delay: The delay is the time needed for the message to get from the sender to the receiver.
- Round-trip time: The round-trip time between two participants in a network is the time that elapsed from sending a signal from one participant to the other until the

Figure 5. Communication of the seven OSI layers

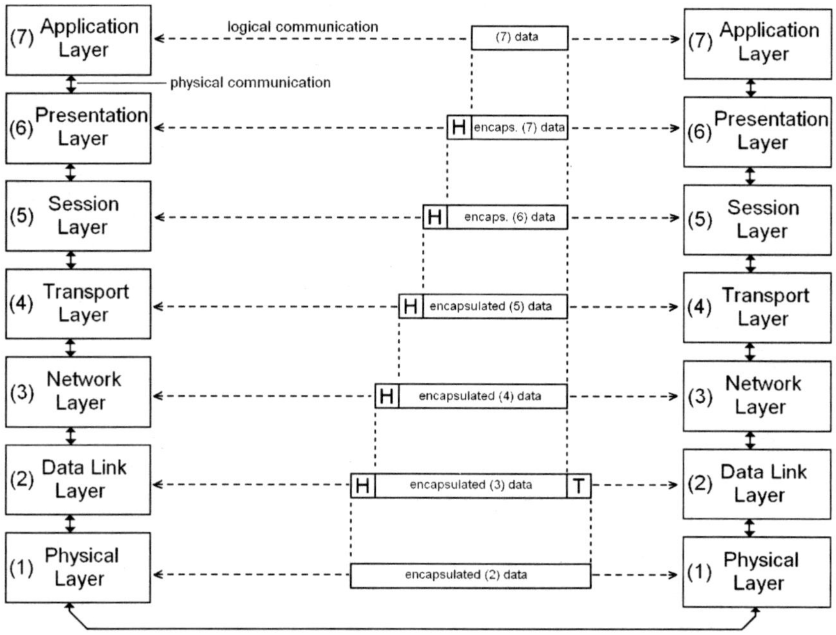

sender receives the returning signal. The signal is usually a data packet and the time duration between sending and receiving this packet is also referred to as ping time.

- Bandwidth: It is the amount of successful message transmissions of the communication medium during a time unit. It also called throughput. The unit of bandwidth is bits per second (abbreviated as b/s or bps).
- Insertion ratio: The insertion loss is the loss of signal power in the physical communication medium. It is defined as a ratio – called insertion ratio – of the original signal level and the signal level with a loss, and measured in dB.
- Loss ratio: It is the amount of messages lost during transmission compared to the amount of transmitted messages.
- Bit error ratio: The number of corrupted bits divided by the total number of transmitted bits. The packet error ratio describes the number of incorrectly transferred data packets correlated to the number of transferred packets.
- Reordering ratio: defines the number of messages that need to be retransmitted due to message loss or message delay compared to the number of all sent messages.

The performance of the network depends on its bandwidth, round-trip time and on different delays, while the loss ratio, the bit error ratio, and the reordering ratio inform us about the reliability of the medium.

Model of a Node

The node can be characterized based on the symmetrical or asymmetrical nature of communication. If the communication is symmetric, then the same set of rules can be used for every participant, i.e., both transmitter and receiver functions are present in every node. The peer-to-peer environment assumes symmetric communication, while the client-server does not. The block diagram of a node is presented in Figure 6. The receiver can either receive messages (SDUs) from the upper layer or PDUs from the medium. Furthermore, additional control messages may be exchanged through the management entity.

The layer management can realize basic message handling functions, such as the determination of the beginning and the end of the message. If required, error detection can also be implemented in the layer management in order to notify the sender about message error, and request the retransmission of the corrupted message. Error detection, for example, can be implemented by introducing checksum at the end of each message. Error detection can also be used to control management functions, for instance, a unit can observe the error ratio of a link and in case of high error ratio, it can decide to declare the failure of the link. Whenever retransmission is infeasible or inefficient, error correction might be applied. By encoding the data using an error-correcting code (ECC) the receiver can – to some extent – recover the original data.

A congestion control unit may be located between the transmitter and the receiver. It notifies the transmitter about congestion and restricts the number of outgoing messages. One task of the receiver block is to detect the congestion on the link and notify the congestion control unit periodically about this condition.

As discussed in the previous section, if reliability functions are required, then the transmitted messages need to be buffered. This buffering of the messages is also achieved by the congestion control unit. If the receiver detects with a status message – referred to as positive acknowledgement – that the transmitted messaged has arrived safely, it directs the congestion control unit to delete the buffered message. If the receiver gets a status message – called negative acknowledgement – that indicates that one of the transmitted messages is corrupted or lost, then it notifies the congestion control unit, which searches for the given buffered message and tells to the transmission unit to send it again. Another unit may also be applied for controlling the congestion unit: the timer unit. If the receiver does not get a message about the successful reception of a given message within a predefined duration after the transmitter sent it, then the timer may also direct the congestion control unit to search for the given message in the buffer and ask the transmitter to resend it. The presence of these units and status messages

Figure 6. The block diagram of a node

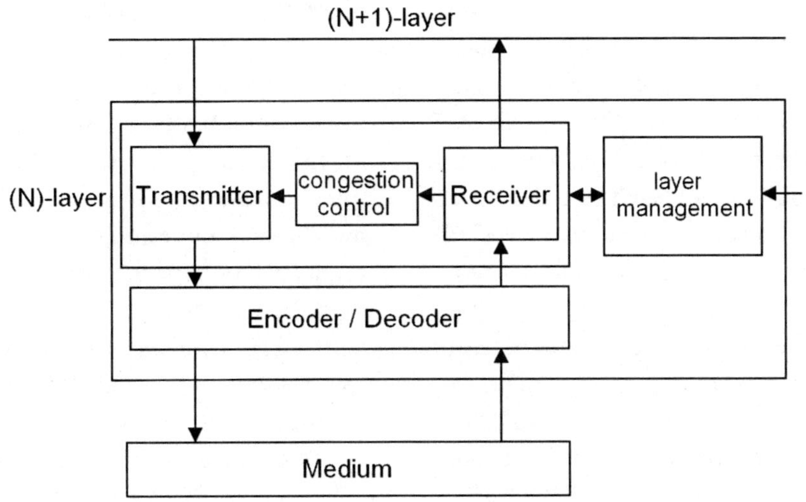

depend on the characteristics of the communication, which will be discussed later.

The encoder encodes the messages, which are sent to the medium, while the decoder decodes the messages, which are received from the medium.

The Purpose of Flow Control

Receiving a message often requires more time than sending a message. This is because the receiver participant has to interpret the given message and make a memory allocation for it (Holzman, 1991). Accordingly, in a distributed environment a receiver can be easily overloaded, if it is slower than the sender is or if the receiver has heavy traffic load from other participants in the network. Having an unreliable medium, messages may be lost and in case of buffering – when the upper layer tries to provide reliability functions – messages may be duplicated. To avoid these problems, a flow control is used for controlling the messages transmitted from the sender to the receiver. There are different types of flow control; an easy solution is when the number of messages sent by the sender is simply restricted by the receiver. A more sophisticated solution is to use a buffer before the receiver. The messages can be transmitted at an arbitrary speed by the sender until the number of messages step over a predefined upper bound, which tells the sender to adjust the speed of message transmitting to a lower rate. Thus, the main purpose of flow control is to make sure that messages are not sent faster than they can be processed by the receiver. Other functions may also be realized to enhance the efficiency of the communication system. Such functions can optimize the utilization of the resources or prevent links to become engorged. To continue with the given example – if the number of messages in the buffer before the receiver decreases underneath a predefined lower bound, it may notify the sender to increase the rate of messages in order to increase the utilization of resources. Of course, in case of an unreliable medium, these buffer status messages may be lost. This can cause the overflow of the receiver or a system halt caused by a deadlock situation, which is discussed later. Therefore, reliable message transfer must be provided. To avoid loss of these messages, the sender must keep track of the elapsed time. If the response does not arrive within predefined time duration, then the sender assumes that it is lost and asks the status of the buffer. The elapsed time from the sending of the message is measured by a timer referred to as retransmission timer. Timers will be discussed later. The transmitted messages may also be acknowledged by the receivers. The delay of acknowledgements may result in a race condition between the retransmission timer and the delayed message. This and the corruption of acknowledgements can also be resolved by flow control. Thus, the purpose of flow control is to manage the entire information flow in the network.

PROTOCOL DIALOGS AND CHARACTERISTICS

Protocol dialogs can be characterized by the number of participants, the number of messages exchanged, the reliability of the underlying medium, and whether a request message requires acknowledgement or not.

Number of Participants

Based on the number of participants in a protocol interaction, bilateral and multilateral communications can be distinguished. Bilateral communication is often referred to as unicast, while multilateral communication can be multicast, broadcast or anycast. In bilateral communication, the message is sent to a specific node of the network: this is a one-to-one connection between two participants. In multilateral communications, the message is sent to more possible destination nodes of the network. The purpose of multicast communication is to send messages to a specified destination group in a network. The multicast addressing creates a tree structure in order to reach the destination nodes and use each link exactly once. There are various algorithms to create such trees either in a centralized or in a distributed way (Diot, 1997). Then it duplicates the messages only in the nodes where the tree splits into different branches. The multicast communication is often used for example in streaming media applications. In case of the anycast communication, a message is also sent to multiple nodes of the network, but at a given time, only one of them is chosen to receive the message. These nodes are mainly the nearest ones. This scenario is used for example in load balancing, when the optimal route has been changed: in Internet Protocol Version 6 (IPv6). It selects a host in order to update the routing tables for a group of hosts. In case of broadcast communication, the message is delivered to all the participants in the network. Broadcasting is mainly used in local area networks (LANs). Multicast communication always assumes the acknowledgement from the receivers. Broadcasting sends the messages to every participant and no acknowledgements are required.

Communication Problems

In a fully automated system, the model of communication must be unambiguous and consistent. We must discuss some problems that may occur when these conditions are not satisfied: these are infinite loops, deadlocks, and livelocks.

An infinite loop is a situation, when a process is waiting for a condition that will never be fulfilled. Consider an example: we get the following workout plan from our coach: "STEP 1: Do 30 pushups, STEP 2: Do 50 sit-ups, STEP 3: Repeat this again!" Note that if we interpret the list of instructions word-for-word we will never finish since there is no instruction to stop.

The deadlock is an interoperability problem between the participants of the communication. To understand the deadlock, consider the following example: we have two stubborn uncles who do not speak to each other due to an old argument. Uncle A says: "I will apologize to B, after he apologizes to me". Uncle B says: "I will apologize to A, after he apologizes to me". Notice that they will never talk to each other, because uncle A is waiting for an event (B apologizes to A) to execute a process (apologize to B), but this event will never be fulfilled, since the precondition of this event is that A apologizes to B. Consequently, the deadlock is a situation, where two or more participants are waiting for each other. A common example in communication systems is when the receiver asks the sender to wait with the next transmission and sometime later, it notifies the sender to continue the transmission, but this notification is lost. If no reliability functions are implemented to recover this lost notification, then the receiver waits for the sender and the sender waits for the receiver forever.

The livelock is similar to the deadlock; however, instead of waiting for each other, the participants are continually processing regarding the other one without a sense: the specified task is not progressing. The following example describes a livelock: two people try to pass each other facing each other, but they always try to step aside in the same direction. The risk of livelock occurs when we would like to resolve a deadlock and force more than one participant to change. To prevent this we must force only one participant of a deadlock to take an action.

The correct behavior could not contain deadlocks, livelocks, or infinite loops.

Trigger Conditions, the Purpose of Timers

In a protocol dialog, we need trigger conditions to describe the preconditions of an activity. There are different trigger conditions, referred to as Automatic, User, External, or Timer Trigger.

- Automatic Trigger: An activity is starting automatically without any additional condition.
- User Trigger: Starting an activity requires human action. For example, a reservation program asks for data from the user.
- External Trigger: An external event is necessary to start the activity. For example, a query arrived to a process triggering it to answer.
- Timer Trigger: A timer is set with a given value. When it expires, it triggers the start of a given activity. For example, if the predefined time of a tender has elapsed, it triggers the jury to select the winner from the participants that applied to the tender.

Timers can help to avoid infinite loops and deadlocks. If none of the required activities has started within a predefined duration, the timer indicates the problem.

Timers can also provide reliability over an unreliable medium. Due to delay or message loss in the medium, retransmission of the message may be necessary. The timer provides this retransmission in the following way: The sender sends a message, sets a timer to a predefined duration, and starts it. The sender of the message is waiting for a confirmation whether the message has arrived to the receiver successfully. If there is no confirmation message until the timer expires – this event is called timeout – then the timer triggers the sender to transmit the message again. The sender of the message predicts the worst-case turn-around time for each acknowledgement message. This prediction is based on heuristics using the average round-trip delay to determine the worst-case turn-around time (Holzmann, 1991).

Timers can be used in flow control to delay the sending of a message. With proper timing, the delaying timer is able to prevent the receiver from being overloaded.

Timers are used to prove the time constraints according to the protocol specification are fulfilled (Hogrefe, 2001). For example, a timer is set to the upper bound of the required round-trip time. This timer starts, when a message is sent to another participant. The timer is reset immediately when the sender receives the response. If the required round-trip delay is over, then the timer expires. Another scenario is to set a timer to a given duration to measure the number of the successfully transmitted messages, in order to calculate the effective bandwidth.

Therefore, we can conclude that the main purposes of timers in a communication protocol are the following:

- To ensure reliable transmission over an unreliable medium
- To delay the sending of a message to prevent overload
- To check the performance of the protocol

Message Primitives

The specification of a protocol includes the definition of message identifiers and message structures. The message name can be mapped to one of four primitives that identify the phase of the interaction:

- A sends a message to B – this is a request message
- The request of A arrives at B – this is an indication message
- B acknowledges the receipt of the message – this is a response message
- The response of B arrives at A – this is a confirmation message

Protocol Dialogues

After we came to know some basic definitions and aspects, we are going to discuss the communication scenarios.

The scenario of the communication describes the interactions between the participants of the communication.

Figure 7. Time sequence diagram of bilateral communications

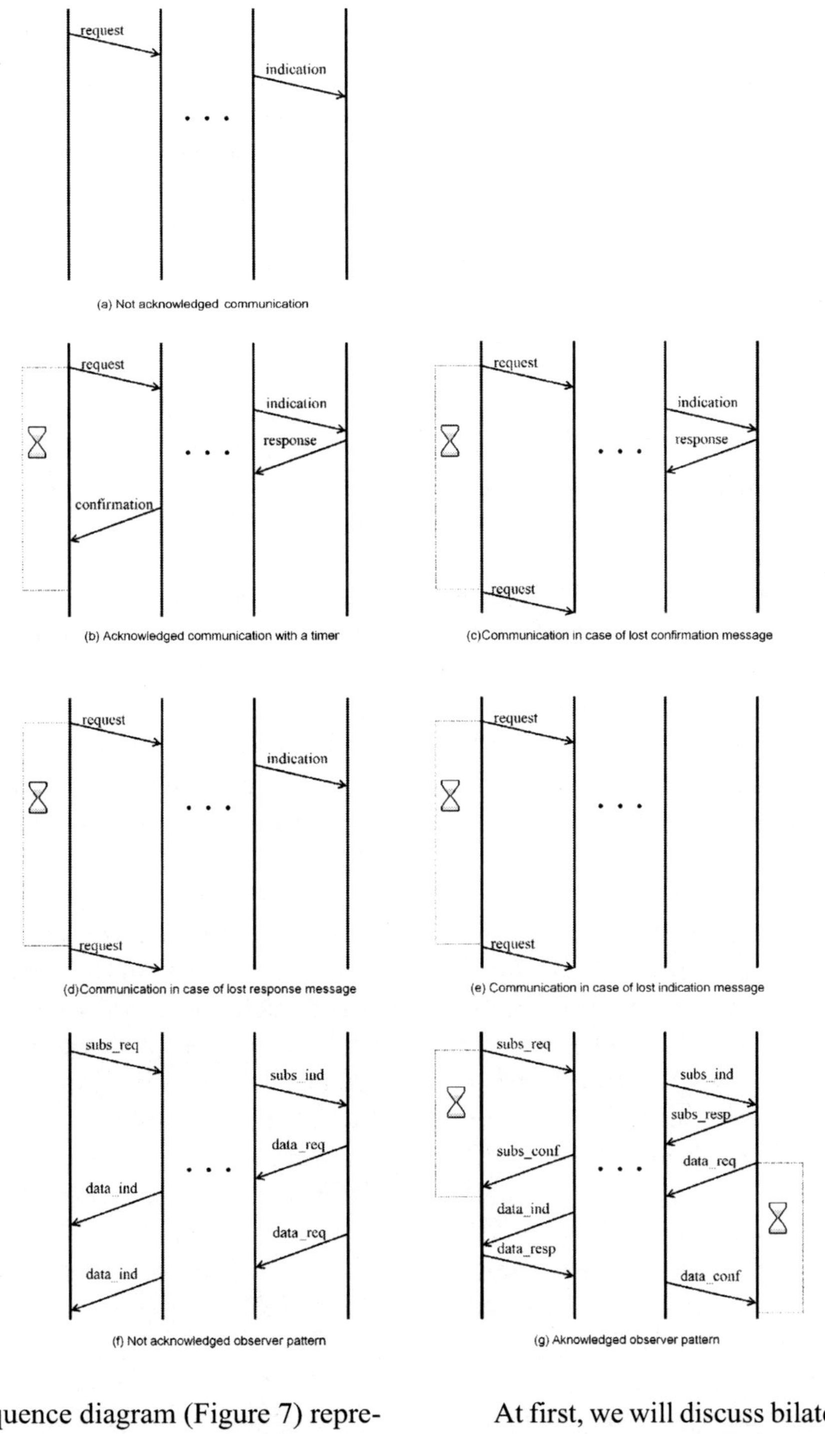

The time sequence diagram (Figure 7) represents the message sequence and the relationship between the primitives (Ellsberger, 1997). The three dots in the figure represent the communication medium.

At first, we will discuss bilateral communications. The first type of bilateral communication is a simple transmission between the two participants. The communication can be either confirmed or unconfirmed. The unconfirmed scenario is a one-way interaction between the two participants. In this case, the sender sends a request that arrives to the receiver as an indication message (Figure 7(a)), which triggers an activity. In the case of confirmed communication, the sender requires an acknowledgement. The dialog in this case is the following: A process sends a request to another process. After the message arrives as an indication message to the other process, it executes the required function and sends a response message. If the confirmation arrives to the sender, it learns that the information has been transmitted successfully (Figure 7(b)). However, several problems may occur: either indication (Figure 7(e)), the response (Figure 7(d)) or the confirmation (Figure 7(c)) message can be lost, therefore the sender sets a timer – called retransmission timer – to a given duration. If the given duration expires, the sender is warned that the message may have been lost and it should be retransmitted. When the sender buffers the outgoing messages, it is possible that different sending and receiving interactions overlap each other. Thus, the corresponding request, indication, response, and confirmation messages must be marked with an identifier, which unambiguously identifies the different interactions. Since these identifiers are made of fixed number of bits, and so they can only be selected from a given range, a solution is required, which determines and reuses those identifiers to ensure the correct behavior of the protocol. Fortunately, this assumption can be fulfilled easily: the sender sends the messages, and keeps track of the identifiers of messages, that have not yet been acknowledged. The identifiers of acknowledged messages can be used again. It is also possible that a participant sends a request to another participant who then sends back multiple messages. The sender of the request must know if there are any other messages remaining. One solution for this problem is that the latest message is a special end message. Another option is that the message contains a data field indicating, whether there are any more messages waiting for transmission. Of course, timers can also be used for indicating the end of the communication.

Another communication scenario is when a participant subscribes to another participant in order to get information interested in. This form is often referred to as the observer pattern. Examples of this case are the newsletters and event driven programming when the application is waiting for a key to be pressed or a mouse to be moved. This event can also be confirmed and unconfirmed. In case of the unconfirmed communication, the participant subscribes with a request, and then it is waiting for the data indication messages (Figure 7(f)). In the case of confirmed communication, both the subscribe-request and the data indication messages are acknowledged (Figure 7(g)). Timers are used to keep track of whether the confirmation messages have arrived in a given time. If no message arrives, it triggers the sender of the subscribe-request or data indication message to send the message again. Notice that in the case of an observer pattern, the subscribed participant may get multiple messages during the communication.

The observer pattern may be also used in multilateral communication. In the case of multicast-unicast communication, a participant is subscribed to get the specific data from multiple participants. In case of unicast-multicast communication, multiple participants are subscribed to a participant, while in case of multicast-multicast communication multiple participants are subscribed to multiple participants.

The multilateral communication can be divided into anycast, multicast, and broadcast scenarios as mentioned earlier.

In anycast communication, the anycast address is assigned to several nodes of the group, but only one participant of this group will communicate with the sender of the message at a time. The sender sends its request to one participant of this group. As the indication message arrives to the destination node, it responds. When the confirmation message arrives, the message can be sent to another participant of the group labeled with the given anycast address. The way, in which the specific destination participant is selected from the anycast group, depends on the anycast routing protocol (Ettikan, 2002). While the different messages can be lost as described in the case of bilateral communication, it is advisable to use a timer in the way discussed earlier. The communication described above is also referred to as unicast-anycast communication, because one participant of the network sends a request to a group of participants. Although the opposite – anycast-unicast – direction can also occur. For example, participant requests information from a group of participants, and later this group updates this piece of information.

To describe multicast communication we will discuss scenarios from the point of the sender as well as the receiver. In case of multiple senders and one receiver – also referred to as multicast-unicast communication – the following patterns could be observed:

- A participant is waiting for a message that can be sent by multiple participants. Thus, messages sent by different participants will race with each other. Only the first message will be served. The other messages may be discarded or processed later. An example for this dialogue is the fixed price auction: the seller has a book about communication protocols and is willing to sell it for $40. The first participant transmitting its request can buy the book.
- The receiver does not know the number of incoming messages – it stops waiting after a given time or after receiving a given number of messages. Then it selects one of the sender participants according to a given strategy (randomly, with a predefined hierarchy or based on the sent messages, etc.). An example for this kind of communication is the silent auction, when the bidders write their bid on a sheet of paper and then send them to the seller. After the predetermined time of auction elapses, the seller accepts the highest bid.

In case of a unique sender and multiple receivers – unicast-multicast communication – a participant sends messages to a group of participants in parallel. The receivers may already be known before the interaction or they may be selected during the conversation. Two different scenarios may occur depending on whether the receiver participants respond to the sent message or not. An example for unanswered unicast-multicast dialogue is when a company sends its advertisement to a group of people. In case of answered unicast-multicast communication, the sender may set a timer and wait for the responses until this timer elapses. An application area for such communication is the calling tender. Consider the following example: a town is looking for the best offer to build a bridge. It defines the required specifications and then calls a group of participants for the tender. The participants apply for the tender, and then the sender of the tender selects the winner offer. Another case of answered unicast-multicast communication is when the sender of the message is waiting for all of the participants – or a predefined subset of participants – to answer. For example, a company would like to produce clocks, so it sends a list of required components to other manufacturers. As soon as all of the required components have arrived, the company may start the production.

Multicast communication may also take place between multiple senders and receivers. Such multicast-multicast communication is the video conference, where a group of participants is talking to each other.

Another case of communication is when a sender sends a message to a receiver and the message contains a reference to a third participant. The third participant is a server, which provides services to the receiver participant. An example for this kind of communication is when a webpage redirects the web browser to another web page.

A special case of communication is when a participant forwards a message to another participant. This scenario is sometimes referred to as the relayed request pattern (Weske, 2007). This communication pattern often occurs in the peer communication relation, when it is broken by a proxy agent. The proxy is an intermediary participant, with the following main functions and purposes:

- Filtering and keeping track of communication. For example, a company filters its outbound communication in order to prevent data leak and monitors its employees to get to know, which websites are visited during working hours.
- Caching the commonly used functions in order to increase the efficiency of using resources. For example, a web proxy caches web pages from web servers.
- Hiding the participants of the network – mainly for security reasons.

Characteristics of Protocol Dialogues

When we specify a communication protocol, it has the following aspects that describe its characteristics:

- The protocol is reliable, if the correct delivery of a message is provided independently from the reliability of the operational environment, that is, the transport medium service. A protocol is unreliable if this statement is not true. Reliable protocols typically have greater overhead than unreliable protocols and therefore they are less scalable and less efficient. This is typically not a problem for bilateral protocols, but in case of multilateral protocols, it may be. Sometimes the fast speed of the communication is a requirement, while a little loss of data is permitted – for example in video streaming – so the participants use unreliable protocols, such as the User Datagram Protocol (UDP). The reliability of a protocol can be achieved by acknowledgement messages. There are two types of acknowledgement messages: positive and negative acknowledgements. The positive acknowledgement sent by the receiver notifying the sender that the message was received successfully, i.e. it was received without errors. For example, TCP uses positive acknowledgement messages. The negative acknowledgement informs the sender that some problems occurred during the communication. If the probability of transmission error is relatively high, then a protocol using only positive acknowledgements could be very inefficient: it forces the sender to wait until timeout after each lost or damaged message. If negative acknowledgements are used, then upon receiving an erroneous message, the receiver can immediately notify the sender to send this message again. The positive acknowledgement and timers are still needed in order to recover lost messages. However, many protocols – called negative acknowledge-based (NAK-based) protocols – only use negative acknowledgements: they pass all reliability functions to the receiver, so the sender never knows for sure, if the receiver gets the message. Thus these protocols trade reliability for efficiency: they offer lower level of reliability, i.e. they are not reliable in the sense defined above. Multicast protocols that are usually referred to as reliable multicast protocols are typically NAK-based protocols, thus they are not reliable in the traditional way.
- From the point of view of the communication model, a protocol can cover an end-to-end communication relation or just

communicate with a neighbor node. End-to-end protocols describe data transfer between network endpoints. The end-to-end communication function is provided by the transport layer in the case of the OSI model. For instance, UDP and TCP are end-to-end protocols, while the Point-to-Point Protocol (PPP) controls the communication of neighboring nodes.

- If different roles can be associated to nodes in the communication model, the protocol is asymmetric. On the other hand, if all nodes have the same set of roles, then the protocol of the communication must be symmetric.
- A protocol is said to be stateful, if it goes through different operational states during its operation. In this case, the history of preceding communicational events does affect the actual operation of the protocol. Otherwise, it is stateless. A stateful protocol can track the states of the communication session, while a stateless protocol supports less reliability features, but it has a lower overhead. For example, the File Transfer Protocol (FTP) is a stateful protocol, while HTTP and UDP are stateless protocols.
- A protocol is connection oriented, if a communication session needs to be established between the communicating peers before the message delivery can take place. Otherwise, the protocol is connectionless. Connection oriented protocols can be imagined as a tube, where the messages at the end of the tube arrive in the same order as they were sent, while a connectionless protocol is similar to mail delivery: the different messages can arrive in different order and through different routes. Connection oriented protocols have more overhead than the connectionless ones, due to the connection establishment and the termination steps, but in general they can handle real-time traffic more efficiently. For this reason, they are frequently used in backbone networks. In the case of connectionless communication, no prior agreement is required between the participants of the communication, so the sender of the message does not know whether the other participant is able to receive the data or not. Thus, more problems may occur during communication than in the case of connection-oriented protocols; and in the case of a reliable protocol, sometimes it may be necessary to resend the message. FTP and TCP are, for instance, connection oriented protocols, while HTTP and UDP are connectionless protocols. Usually connection oriented protocols are stateful and connectionless protocols are generally stateless. For example, when a protocol establishes a communication, it goes through different operational states while reserving resources along the communication path and this information is stored until the end of the communication.
- An important aspect of communication is the synchronicity. A communication is synchronous if a node has to wait for a reply message from a peer node and therefore it cannot progress with the operation until it arrives.

REFERENCES

Diot, C., Dabbous, W., & Crowcroft, J. (1997). Multipoint communication: A survey of protocols, functions, and mechanisms. *IEEE Journal on Selected Areas in Communications*, *15*(3), 277–290. doi:10.1109/49.564128

Ellsberger, J., Hogrefe, D., & Sarma, A. (1997). *SDL: Formal object-oriented language for communicating systems*. Englewood Cliffs, NJ: Prentice Hall.

Ettikan, K., & Abdullah, R. (2002). *An analysis of anycast architecture*. In Communications and Computer Networks Conference 2002. Cambridge, MA: ACTA Press.

Hogrefe, D., Koch, B., & Neukirchen, H. (2001). *Some implementation of MSC, SDL and TTCN time extensions for computer-aided test generation*. In SDL 2001: Meeting UML (pp. 168-181). Heidelberg, Germany: Springer.

Holzmann, G. J. (1991). *Design and validation of computer protocols*. Murray Hill, NJ: Prentice Hall.

International Telecommunication Union. (1994/07). ITU-T recommendation X.200: *I*nformation technology - open systems interconnection - basic reference model: The basic model.

Tanenbaum, A. S. (2003). *Computer networks*. Upper Sadle River, NJ: Prentice Hall.

Weske, M. (2007). *Business process management*. Heidelberg, Germany: Springer.

ADDITIONAL READING

Clark, D. D. (1988). The Design Philosophy of the DARPA Internet Protocols. Symposium proceedings on Communication Architectures and Protocols (pp. 106-114). New York, US: ACM.

International Telecommunication Union (1996/07). ITU-T Recommendation Q.703E.

Kurose, J. F., & Ross, K. W. (2008). *Computer Networking: A Top-Down Approach*. Reading, Massachusetts, US: Addison-Wesley.

Stevens, W. R. (1993). TCP/IP Illustrated: *Vol. 1. The Protocols*. Reading, Massachusetts, US: Addison-Wesley.

Tarnay, K. (1991). *Protocol Specification and Testing*. Budapest, Hungary: Akadémiai Kiadó.

KEY TERMS AND DEFINITIONS

Bilateral Communication: Communication between two participants also referred to as unicast.

Client-Server Architecture: One or more nodes of the network – referred to as servers – have a set of dedicated roles: they provide services to other nodes – called clients.

Entity: A well-defined unit, which implements one or more functions of a layer.

Layer: A functional block, which solves a given sub-problem.

Multilateral Communication: Communication between multiple participants of the network. The subclasses of multilateral communication are multicast, anycast, and broadcast.

Peer-to-Peer: All nodes of the network implement an identical set of functions.

Service: A function provided by a lower layer entity to an upper layer entity.

Service Access Point: A logical interface, through which a lower layer entity can provide services for an upper layer entity.

Timer: A type of a trigger condition. It is set to a given duration, and when it expires, it triggers the start of a given activity. It can be used, for example, to offer reliability over an unreliable medium.

APPENDIX

Table of Abbreviations

ECC	Error-Correcting Code
FTP	File Transfer Protocol
IP	Internet Protocol
ISO	International Organization for Standardization
LAN	Local Area Network
NAK	Negative Acknowledge
OSI	Open Systems Interconnection
PDU	Protocol Data Unit
PPP	Point-to-Point Protocol
SAP	Service Access Point
SDU	Service Data Unit
TCP	Transmission Control Protocol
UDP	User Datagram Protocol

Chapter 3
Protocol Functions

Tibor Dulai
University of Pannonia, Hungary

ABSTRACT

Communications networks connect people of the world in spite of large geographical distance, they play a big role in our entertainment during leisure time, and they help to arrange our everyday activities. Communications networks' ubiquitous nature is a big challenge for network protocol developers. They have to offer different services in heterogeneous hardware and software environment.

There are basic protocol functions, which have to do usually appearing tasks in communication networks. Among these tasks, we highlight PDU creation, error handling, medium access, handling of load-related problems (flow control and congestion control) and PDU delivery's sub-tasks like finding the destination point by addressing and routing.

After the introduction of basic protocol functions, this chapter deals with the advanced functions of communications protocols that are especially highlighted in network management and multimedia transmission. This chapter deals with the most important aspects of the network management: security management, performance management, accounting management, fault management, configuration management, and mobility management. During multimedia transmission over a well-managed network, another important question arises: How to ensure desired quality of service? This chapter presents methods for quality management, too. Finally, this chapter collects future trends of protocol functions based on the present directions of network technology development.

DOI: 10.4018/978-1-60960-732-6.ch003

INTRODUCTION

Computer and telecommunications networks started like supporting systems in the academic area, but in short time they became indispensable part of our life. We can send mails and arrange audio or video call with their help. They connect us in spite of the big geographical distance between us. We can listen to music, read news, watch movies, or television programs using the networks. They have big role in our entertainment during leisure time activity. We can do the shopping in e-shops, learn using e-learning materials, and do the banking functions from our living room. Networks help to arrange our everyday activities from the distance.

As both the kinds of offered services and the number of users of the network increases exponentially, the system has to deal also with the increasing heterogeneity. It requires sophisticated management methods. An important topic of this chapter is the advanced functions of network management. Solutions are collected based on the mostly used network management protocols (SNMP, CMIP), covering Fault, Configuration, Accounting, Performance, and Security issues.

Security issue is especially important topic of network protocols, because more and more people use communications networks, for example, to arrange their money transfer (e-shopping, e-banking, etc.). At this point, we have to mention the security gaps in some network protocols, which also influence the security of the communication (e.g. TCP-flood, DNS-spoofing, etc.).

Other proceeds of the popular use of networks are that people who change their location also want to access the information and services of the networks in increasing number. It requires that users should be connected continuously while they change their locations and the system should know their locations as well. Introduction of special methods of mobility management also takes part in our chapter of advanced protocol functions. Different types of handover and location update are the most important functions of mobility management. Especially Mobile IP and its complementary protocol, the Cellular IP care these functions in IP networks.

When we mention multimedia delivery over network, the quality of service (QoS) is an important measure of transmission success. Several methods have been introduced ensuring the desired quality of transmission. In real time, applications mainly jitter and congestion may debase the quality of service. Solutions for these problems are mostly implemented in network and transport layers. These functions are also covered in this chapter next to the basic protocol functions like addressing, PDU (Protocol Data Unit) creation, error handling, flow control, congestion control, medium access, and routing.

As we can see, different management methods play important role in protocol functions. After the presentation of basic protocol functions, this chapter intends to introduce the problems and advanced solutions of network, mobility, security, and service quality management of communications networks.

In other books, which deal with communication protocols, protocol functions are usually scattered between different network architecture (like OSI or TCP/IP) layers. Some functions re-appear in different layers. E.g., both data-link and transport layers deal with flow control, all the layers create their own PDU (usually with different names, like frame, packet, segment, block, etc.) and error handling is also a common protocol function of several layers. In this chapter, we do not follow this method. Our discussion is centered on tasks, which different protocol functions have to deal with. It is an unusual presentation type of protocol functions; however, our motivation for it is the presentation of the evolution of different solution methods for specific tasks of communication networks.

BACKGROUND

The topic of protocol functions is an important part of several books about communications networks (Tanenbaum, 2004; Schiller, 2000; Stallings, 2002; Anttalainen, 2003). However most books are organized around a special protocol stack (like TCP/IP), while others deal with a determined function (like network management or network security). In this chapter, the author intends to collect the most important protocol functions – both the basic and the advanced ones – of a usual communication networks and introduce them. We highlight PDU creation, error handling, addressing, routing, medium access, flow control and congestion control among basic protocol functions in the first part of the chapter, while the topic of advanced protocol functions includes the introduction of network management, network security, mobility management, and quality of service. The reason for collecting the latter ones as advanced protocol functions is their complexity. They are thought to have increasing importance because of the widening spread of heterogeneous mobile networks, which deal with multimedia content for helping user in nearly each part of life. In this chapter, different protocol functions are collected and briefly explained, which can be useful both for educational purposes and for protocol developers.

PROTOCOL FUNCTIONS: FROM BASICS TO ANVANCED FUNCTIONS

Communication protocols are developed for ruling a communication in semantic, syntactic, and time. They have role in establishing, maintaining and releasing connection or just simply sending and receiving a message. The role depends on the kind of the communication and its circumstances (e.g. channel). Network architectures separate network software entities – and also their communication rules, the protocols – into layers. As we move from the level of the communication channel to the user, the function of the protocols changes, too. Irrespectively of the network architecture, there are common functions, which have to be implemented for supporting the communication.

There are two main expectations:

- The receiver has to get the same information, which was sent her.
- The receiver should interpret the received information as the sender intended.

In this part, we review the main protocol functions, which help to fit to these expectations.

Protocol Data Unit Creation and Error Handling

As the transmission is never perfect – each channel is more or less noisy – there has to be a protocol function, which handles transmission failures. It can be done in two different ways: either by detecting errors or by correcting that. Both methods need some redundant bits added to a block of information bits. It means that protocols have to split bit stream into blocks of finite size. Depending on the protocol layer, these blocks have different names: frame, packet, etc. Commonly they are Protocol Data Units. One important function is the indication of the start and end of a PDU. Several methods have been developed for that: for example, timing, byte counting, byte stuffing, bit stuffing, and hurting the physical coding. When PDUs are prepared, error handling is easy: redundant bits have to be added to each PDU on the sender side. If it is an error detecting code, the receiver may ask for resending the PDU (like in case of the usually applied CRC), while when it is an error correcting code, the receiver may be able to correct the erroneously received PDU based on it.

Flow Control

There is another kind of problem with noisy channels: sometimes they lose PDUs (frame, packet, or segment). In this case, the receiver side does not face with an erroneous PDU, telling the truth it does not receive a PDU at all. This problem induced the usage of acknowledgement: when a receiver side gets a perfect PDU, it sends back an acknowledgement. In case of duplex communication, acknowledgement field is often included in a regular data PDU that is sent to the original sender – this technique is called piggybacking. As the waiting for the acknowledgement cannot last forever, the transmitter starts a timer when it sends a PDU. When the timer expires, the transmitter resends that PDU.

Suppose that an acknowledge message is lost on the channel. In this case, the timer at the sender side expires and he resends the message. It would be duplicated if the receiver side cannot differentiate between the two messages (the former message and the actual one). As a solution for this problem, sequence numbers were introduced. The sequence number field indicates which message has arrived from the sender (is it a duplication or not).

Since user requirements usually do not allow to send a message and wait for the acknowledgement before sending the next message (let's think about the delay of the channel – Round Trip Time shows it), it should be allowed to send more than one PDU onto the channel before getting acknowledgement on the first PDU. It is called pipelining. Sequence number and the acknowledgement fields help to answer the question: which PDUs are lost during transmission?

Sequence number is useful in another topic, too: it helps to prevent a quick sender from flooding a slower receiver. This function is called flow control and current protocols usually apply sliding windows for that. The changeable size sliding window on the sender side buffers those sent PDUs, which still are not acknowledged. The fixed size sliding window on the receiver side registers the PDUs, which are acceptable for the receiver. There are two different methods for sliding window flow control:

- Go back n: in this case, the size of the receiver's sliding window is 1. It means that receiver accepts PDUs only sequentially. This method increases the possibility of resends.
- Selective repeat: in this case the size of the receiver's sliding window is bigger than 1. It means that receivers may accept (and buffer) PDU arrived not in order. This method needs more memory on the receiver side.

Flow control prevents a fast sender from flooding a slow receiver by adjusting the size of the receiver's sliding window and manages the efficiency of the transmission by adjusting the size of the sender's sliding window.

Congestion Control

Sometimes the problem with the quantity of PDUs smites not the receiver side but the network. When there are too many PDUs arriving at an intermediate router, it has to throw some of them. It means that the sender's timer expires and the PDU will be resent. Two kinds of approaches are applied for this class of problems:

- First, the evolution of the congestion has to be obstructed.
- Second, if there is an existing congestion, it has to be eliminated as soon as possible.

The obstruction of congestion includes the applications of different politics. The possibility of congestion depends on:

- the timer settings (how long to wait for the acknowledgement)

- sliding window strategy (go back n or selective repeat)
- the applied routing algorithm
- error handling (error correction or error detection)

The goals are to minimize the PDUs in the network and equilibrate the network traffic. For reaching the latter goal, there are two popular algorithms: Leaky Bucket (Turner, 1986) and Token Bucket algorithms.

The elimination of existing congestion is the second topic of congestion control. It can be done by reducing the number of sent PDUs. Among several solutions choke packets are popular (choke packets are sent from the receiver to the sender), and when they are applied step-by-step at intermediate routers, the ease from problem is the quickest.

Addressing

An important question is to whom to send the PDU. The transmission of PDUs may happen in two different ways: either a connection is established first and then all the data of this connection is transmitted over the same link (virtual circuit), or for all of the PDUs the destination has to be determined separately (datagram-based service). In case of virtual circuits, the address of the destination is needed only in the connection establishment phase. After that, the ID of the virtual circuit is enough for identifying the destination. In case of datagram-based service all PDU has to include the address of the destination point. The destination point can be handled on different abstraction level. For example, data-link layer of computer networks handles addresses (e.g. in Ethernet networks it is the MAC address of network interfaces), telephone networks handle different level of addresses (MSISDN – the telephone number of a person, Global Title, etc.), transport layer of computer networks use TSAP (in case of IP networks it is called port), while perhaps the most popular address type is the IP address in the network layer of IP networks. Addresses are bit sequences, which identify unambiguously the destination point or point set. Addresses may refer to a single network interface or process, or to a set of them. Addresses referring to a single entity are called unicast addresses, referring to all the entities are called broadcast addresses, referring to a set of entities are called multicast addresses, while addresses, which point to a single entity from a set of similar entities are called anycast addresses. These groups increase the efficiency of addressing. For reaching the destination in a nearly optimal way destination address is used by routing protocol function.

In several functions, peer entities have to know each other. It means that not only the destination but source address is also has to be known by the processes. Both of them are often recognizable in the header of different protocols.

Routing

Another important function in network communication is routing: which path to choose from the sender to the receiver? Several algorithms were developed: some of them are static (e.g. shortest path algorithm based on Dijkstra's algorithm (1959) or flooding), while others take into account the changes of the network parameters and behave dynamically (e.g. distance vector routing (Bellman, 1957; Ford and Fulkerson, 1962) or link state routing). In case of dynamical routing, routers always update their routing table based on the information about the network they get. For reducing the size of routing tables, routers can be organized hierarchically. Hierarchical routing algorithms use this solution. There are some special routing algorithms: broadcast routing, multicast routing and mobile routing solutions for mobile hosts have their own specialty. Next to them, routing in ad-hoc networks needs also special algorithm, which is able to take into account the incidental connections.

Medium Access

In case of broadcast links, another question has to be answered: which host is entitled to use the common medium? Medium access algorithms give the answer. There are static methods: e.g. TDM (Time Division Multiplexing), FDM (Frequency Division Multiplexing), however when there is a station which has nothing to send, its time slot or frequency band remains unused. For increasing the efficiency, hosts have to be assigned dynamically to the channel. The simplest method was pure Aloha (Abramson, 1985), which let stations to send their data whenever they want (in the moment of their formation). Twice as efficient was Aloha with discrete time slots. The throughput of the channel was improved by introducing methods, which separated the time into the competition phase between hosts and the data transmission phase: when there is a transmission on the channel, everybody else has to remain silent. These algorithms let collisions happen only in the competition phase; however, they needed hosts, which are able to sense the carrier. Carrier Sense Multiple Access (CSMA) has many variants. The efficiency can be increased when in case of collision each concerned station stops the transmission. This method is called Collision Detection (CD). LANs usually apply Ethernet, which uses CSMA/CD.

Aloha and CSMA variants allow the existence of collisions. There are algorithms, which exclude collisions. Collision free algorithms do it by ruling the competition phase. In basic bit-map protocol, each station of the line has a timeslot in the competition phase and uses it for signing his intention for transmission. Naturally, this solution for competition may decrease the throughput of the channel in case of small traffic. Binary countdown protocol was developed for reducing the size of the competition phase (the bit-map). In the competition, each station having a frame to send transmits its own binary address. The station with the biggest valued address wins the right for transmitting his data.

In case of small traffic intensity, the efficiency of contention protocols is higher than that of collision free protocols however in case of high traffic intensity the state is the reverse. Based on this fact adaptive protocols were developed: adaptive tree walk protocol adjusts the size of concurrent stations' sets to the traffic intensity.

The above-mentioned medium access algorithms are based on stations that are able to sense the traffic on the line. It is not possible in case of mobile hosts when stations have different finite coverage. For mobile hosts (e.g. in case of IEEE 802.11) medium access protocols with collision avoidance were developed (e.g. MACA or MACAW). The goal was to determine by the sender whether the carrier is free in the receiver's coverage.

Other Basic Protocol Functions

In the previous parts, some usual problems of network communication were mentioned and different solutions were given for them. There are several more specialized problems to be solved by the network software:

- Fragmentation: In some cases at a border of a subnet, a block (packet) has to be fragmented.
- Versioning: its goal is to assure the cooperation between different versions.
- Connection establishment and release: in case of connection-oriented communication.
- Multiplexing and demultiplexing.
- Network security.
- Quality of Service: its importance is improved day-by-day.
- Application-specific problems: e.g. related to e-mail, file transfer, remote access, media transfer and coding, network management, domain name system.

ADVANCED PROTOCOL FUNCTIONS

In the followings, we introduce advanced protocol functions. Their existence was triggered by the increasing complexity and heterogeneity of networks and contents. The main advanced functions we deal with are network management, mobility management, network security, and quality of service (QoS) especially in case of multimedia transfer.

Network Management

Huge communication systems of our days are composed of different types of devices from different vendors. In this heterogeneity, the maintenance of reliable and continuous operation is not a simple task. Network management is responsible for doing that. For structuring management tasks in telecommunication networks, ITU created Telecommunications Management Network (TMN) (ITU, 2000), which contains 5 layers:

- the network's physical elements
- management of network elements
- network management
- service management
- business process management.

This hierarchical model could be valid for all kinds of communication networks. It shows that network management covers wide scale of responsibility with different complexity: from a single device to the whole business process.

Network management covers two main function sets: monitoring devices or processes and controlling them. Both of them is based on variables, which are stored in an information base and belong to a managed element of the network. In some cases, they are called managed object and the tree-organized information base has the name Management Information Base (MIB). The behavior of a managed device can be monitored by querying its variables from the information base. This request is issued by the network management system, while on the device's side there is a software entity called agent, who deals with the variables. Compatibility needs uniform handling of managed objects: they have standardized representation and transport rules (e.g. based on ASN.1 or described in Structure of Management Information – SMI). In case of controlling, management systems have to give new value to the appropriate managed object. Management operations help these procedures. There are operations to query and set the managed objects and for event report, too. The concrete Protocol Data Units depend on the protocol, which is used by the system for network management. Internet community prefers Simple Network Management Protocol (SNMP) (Stallings, 1998); while in telecommunication sector, Common Management Information Protocol (CMIP) (ITU, 1997) is still widely used.

Irrespectively of the used protocol, management functions are related to 5 main topics – they are often referred as FCAPS:

- Fault Management
- Configuration Management
- Account Management
- Performance Management
- Security Management

These aspects cover the fields of activity of the network management. Although they cannot be separated strictly, their essence can be characterized as follows:

The main purpose of the Fault Management is to detect, isolate, and – optionally – correct abnormal operation of a communication network (including its elements and its environment, too). Network errors include an invalid entry in the managed object or a network device out of order. In case of an error, an alarm process may be activated.

Configuration Management deals with the names and parameters of managed objects. Through the parameters the main goal of Configuration Management is to follow with attention every modification of the system (new device, installed software, removed element, etc.).

Account Management is for taking care of resource or service usage cost and billing. Based on the collection and evaluation of cost-related information, efficiency of business process can be improved.

Performance Management can trigger the previously mentioned improvement. Its task is to collect statistical information for increasing efficiency and improve overall performance by modification of the behavior of the system.

Security Management includes two main parts: security event detection and security services for communication. The former one deals with events, which may be harmful for the entire system security (e.g. physical attack or unauthorized usage). Security services for communication cover all the functions, which are related to safe communication (e.g. authentication, cryptography, data integrity).

As security of communication has emphasized importance, we devote a whole subchapter to this function.

Network Security

People apply communication networks in more and more fields of life. It caused that importance of communication's security increases day-by-day. Security is composed of two main factors: system security and data security. We deal with both of them in nutshell.

System security is mostly related to the physical elements of networks. Hardware-based questions can be:

- How to locate computers, routers, access points or other network devices?
- Who has access to the devices?
- What is the coverage of the wireless network device? (How to set the output power level? What kind of antenna do we use? How do we shield the signal? How to prevent interference?)
- Whom do we let connect to the network?
- Shall we use UPS? (Here we may mention all power management-related questions.)
- What can we do against fire, theft or bolt?

Considering system security, there are software-related aspects as well. What kind of traffic do we let in and out? Are there viruses, warms, or Trojans in our system? Are the used applications and protocols safe enough? In case of applications, we always should take care of the bug reports related to our installed software and install patches. Another important aspect is the security of the used protocols. Several intruders caused exploiting problems of well-known and widely used protocols. Two examples for that are SYN flood and DNS spoofing.

- SYN flood (Figure 1) is built on the well-known connection establishment method of Transmission Control Protocol (TCP). In tree-way handshake, first the client sends a SYN segment signing its intention for connection establishment. Accepting the request, the server sends a SYN+ACK message back, acknowledging the previous message. Finally, client answers with an ACK segment creating the other side of the duplex connection and finishing the connection establishment process. Denial of Service (DoS) attacks can be carried out as follows: first, the attacker sends a SYN segment (usually with a fake sender IP address) and does not answer with the sending of the third message. It results in reserved resources on the server side for a while. Starting several connections in parallel with the same server, its memory can be overloaded. A solution against this attack is the limitation of simultaneous connections from one address or the appropriate handling of half-opened connections. Recent network devices are prepared against this attack.

- The purpose of DNS spoofing is to misdirect a targeted computer with garbling a DNS record. The attacker places his own IP address with the desired domain name as a DNS record at a DNS server. It can be done simple by the creation of a fake reverse DNS record or by hacking the target's DNS server and placing there a fake DNS record. Either way is chosen, the targeted computer gets the attacker address and gets into connection with him instead of the originally desired computer. Naturally, this attack can be avoided by verifying the received IP address for example by asking the reverse DNS server (the use of tcp-wrapper is also a good verifier solution against DNS spoofing).

Data security involves the questions:

- Who has access to the data? Authentication deals with this aspect.
- Who can understand the received data? It is the topic of Cryptography.

The purpose of cryptography is to prevent an unwarranted person from being able to interpret the data. The two classes of cryptography algorithms are public-key algorithms (e.g. RSA) and symmetric key algorithms (DES, Twofish, RC4, AES, etc.). The main problem of symmetric-key algorithm is the key delivery, while public-key algorithms are much slower than symmetric-key solutions. Usually public-key algorithms are used for symmetric key delivery and after that; symmetric-key algorithms are used for encryption. (Stallings, 2006)

Several communication protocols build into themselves a cryptographic algorithm. When Hypertext Transfer Protocol (HTTP) is used above a security layer (SSL or TLS – see later in this chapter) it is called HTTPS. Telnet was improved to involve some security as SSH. Secure file transfer protocols are Secure FTP (SFTP) and Secure Copy (SCP). S/MIME is a secure version of Multipurpose Internet Mail Extension (MIME). S/MIME applies public key cryptography-based digital signature for ensuring message integrity, non-repudiation of origin and authentication. Moreover, S/MIME uses encryption for data security and privacy. Another method for encrypting

Figure 1. SYN flood

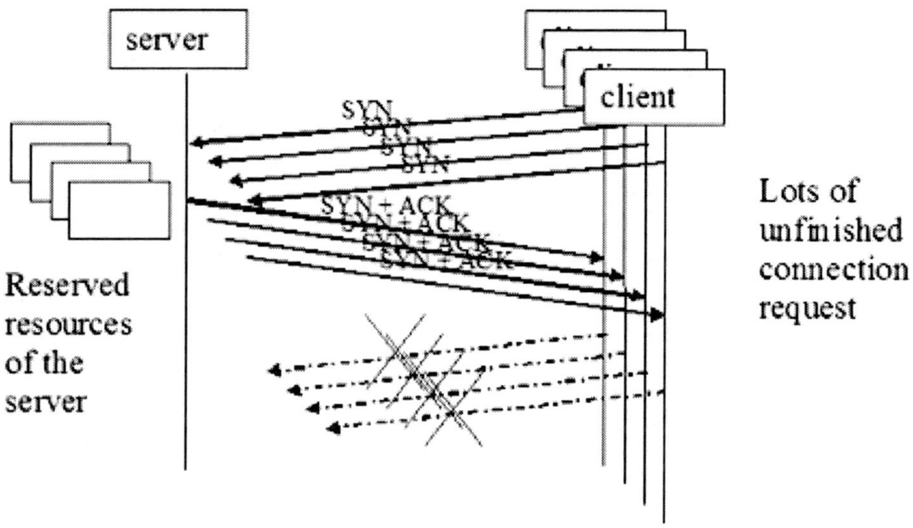

sent files, the content of the hard disk, mails, or instant messages is Pretty Good Privacy (PGP). This very popular method uses public-key cryptography for generating encrypted hash of the data (as message digest).

Instead of the simple encryption, secure frameworks were also developed for communication. Secure Sockets Layer (SSL) was created by Netscape Communications Corp. to offer secure connection establishment and after that to maintain the security of communication. Connection establishment of SSL is based on public-key methods, while for the communication on the secure channel SSL offers different symmetric-key algorithms (e.g. 3-DES) as encryption methods, message digest (e.g. SHA) as ensuring message integrity and different compressing solutions. The standardized variant of SSL is Transport Layer Security (TLS).

Other framework for communication security is IPsec. IPsec contains headers both for authentication and for encryption. Authentication Header (AH) and Encapsulating Security Payload (ESP) of IPsec offer message integrity, authentication, encryption, compression services. Moreover, IPsec can be efficiently used for defense against traffic analysis as well as for defense against message replay attacks. Using IPsec communicating parties may agree among each other about the used algorithms and parameters. IPsec is widely used for building Virtual Private Networks (VPN) between remote firewalls for ensuring a secure channel on a public network.

Authentication is also related to cryptography, an encryption method is used for certifying the party of the communication. There exist authentication solutions with symmetric-key cryptography (shared key algorithms) and public-key cryptography, too. For the latter one Public Key Infrastructure (PKI) is needed. One popular protocol for authentication using tickets (basically a shared-key algorithm) is Kerberos.

It can be seen that security solutions prepossesses nearly all segment of network communications and several different solutions were born to support it.

Mobility Management

In the last decade, the usage of handheld devices spread in a notable sort. When users communicate using a mobile device, it is not a special wish to change their own location. However, it may result in leaving the coverage of the original access point to the communication network, though usually access points (in mobile telecommunication they are called base stations or Node-Bs) are fixed devices and have limited coverage (cell). Mobile networks have to be able to pass a user from one network access point to another.

Suppose that a mobile device is in idle state and there is an incoming message (packet, call, etc.). How to find the targeted user? Unless the network knows the location of the called party, it should transmit the paging message in each cell of the network. It would result in tremendous traffic, whose main part is mostly unnecessary. This huge redundant traffic could be eliminated if the network knows the exact location of all users. That is why every cell change should be announced to the network. Telling the truth, in some cases it is not the best solution. When a handheld device is in idle state, the possibility of a cell change is much bigger than the possibility of a transmitted message. This recognition led to the creation of location areas in telecommunication sector. A location area is a small set of neighboring cells. When a mobile device is in idle state, it has to declare only location area changes to the network (instead of each cell change). This process is called Location Update. When a paging message arrives, it has to be transmitted by all the base stations in the location area of the user.

The second case is when the mobile device is in dedicated state: it receives or transmits packets or a call is in progress. This time the possibility of transmission is greater than that of a cell change. When all the messages are sent to more than one cell (to all cells in the location area), it would decrease the efficiency of the network. That is why when a handheld device gets into dedicated state (receives a paging message and accepts that) its location has to be refined to have cell accuracy. In case of a dedicated mobile station, all the cell changes have to be reported to the network. It is called Handover or Handoff (Schiller, 2000).

Both Location Update and Handover need the cooperation between the original base station and the actual base station of the mobile user. In case of telecommunication network, the responsible parts of these stations are called Home Location Register (HLR) and Visitor Location Register (VLR), respectively. When a mobile user leaves a location area, the VLR of the new location contacts the device. (Either the mobile device looks for VLR or VLR sends periodically a message searching for new users in the coverage area.) Based on this connection, the VLR gets information about the HLR's address of the user and contacts it. After authentication, VLR asks the HLR for user specific information (authority, user profile, additional services, etc.). As VLR is in possession of these data, the user is able to use the network in the foreign area, as he would be at home. Location Update and other aspects of the mobile communication can be read in more detailed in the chapter Mobile Network Protocols for GSM and GPRS in this book.

For supporting mobility in IP networks, Mobile IP protocol was developed. In Mobile IP, there are also dedicated elements of the network to help mobility management, like HLR and VLR in telecommunication networks. They are called Home Agent (HA) and Foreign Agent (FA). When an IP packet is sent to a mobile user who is not at his home network, the Home Agent catches the IP packet and encapsulates it in another IP packet, addressed to the Foreign Agent of the user. Because of the encapsulation, this method is called tunneling. When the Foreign Agent delivers the packet to the mobile user, the user unwrap the original IP packet and processes its content. After that – optionally – he may build a direct contact with the original sender.

Mobile IP has an extension: Cellular IP (CIP) offers efficient cell change in case of a set of neighboring cells: CIP introduced a tree-like hierarchical structure of CIP, routers (Figure 2). In the root, there is a special router, the gateway router, while the leaves of the tree represent access points with their cells. All the traffic in the cells of the whole tree is controlled by the gateway router. Its address is handled as the care-of-address of the tree: the gateway router is the Foreign Agent of the visitors in the tree's cells. When there is an inter-domain handover in a tree, only the gateway router has to be involved into the process and there is no need for additional communication with the original Home Agent. CIP also introduced a concept for set of cells driven by similar reasons as Location Area in telecommunication networks. It is called Paging Area. For an idle mobile device in a Paging Area, there is no need to start the handover process in case of intra-paging area cell change.

When a mobile device changes cell, it often has to leave the frequency of the previous cell and has to switch to a new frequency (neighboring cells usually differ in the set of frequencies used to minimize the chance of interference). Until the network device at the crossover point of the old and new path is not informed about the handover, the mobile uses the old frequency for transmission and messages may be lost.

There are three different handover types introduced: hard handover, semi-soft handover and soft handover.

Among the three methods, hard handover may lose the most of packets: in this case, the mobile device switches to the new frequency and all the packets sent onto the old frequency will be lost.

Semi-soft handover is for reducing packet-loss. When the mobile user switches to the new frequency, he sends a message to the crossover node asking it to register the new route to the mobile user. After sending this message, immediately switches back to the old frequency. During

Figure 2. Intra-domain handover in a CIP network

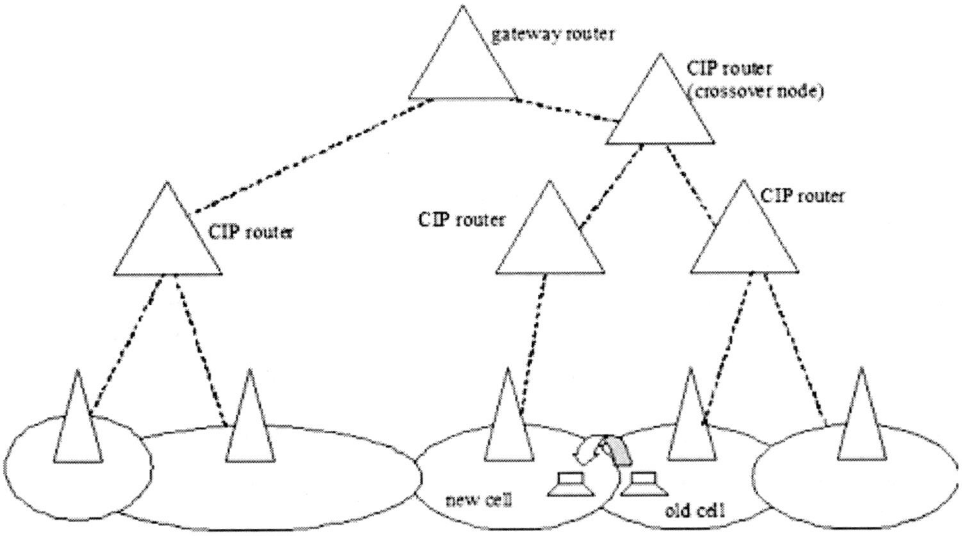

this process, some packets may be lost, however much less than in case of hard handover. When the crossover node gets the message of the mobile user, it starts to route the packets through both paths to the mobile user. After some time, the mobile station switches to the new frequency again, stays there, and sends another message to the crossover node for stopping the message duplication.

The purpose of soft handover is the prevention of packet-loss. The mobile user does not break the connection on the old frequency when switching to a new one. Instead of that, he retains the old connection while the new one establishes. This method requires the handling of – at least – two different frequencies in parallel; however, it does not cause any packet-loss at all.

A special case of handover is the switching between different service providers. In this case, apart from the user information, business information also has to be changed between the organizations. This type of mobility management is called roaming.

Quality of Service (QoS)

The role of QoS becomes more and more important as interactive or real-time transmissions happen on the network increasingly. At present, the most QoS-sensitive transmission type is perhaps the real-time multimedia transmission. This part of the chapter attempts to introduce methods for assuring the appropriate QoS for different types of services.

It is not possible to handle QoS independently of other protocol functions. There is a tight bundle between QoS and some basic or advanced protocol functions like routing or congestion control. Bad routing algorithms or false congestion control politics may cause huge unnecessary load on the network, which may influence QoS in the bad way. Several other functions may also be noticed related to QoS. The previously partly introduced protocol functions cared the quality problems of the network in an ad-hoc way. Real QoS methods need guarantee of the desired quality.

There are four main aspects, which are analyzed when quality of a service is determined: reliability, bandwidth, delay, and jitter (deviation of delay). Their importance and precedence are dependent on the actual transmission type, e.g. a file transfer requires high reliability and does not care with jitter or delay, while real-time video transmission (like a videoconference) is very sensitive of jitter and delay, but reliability is much less important for it. Generally stated, jitter is one of the biggest problems of real-time audio or video transmissions.

Transmissions may happen by virtual circuits or datagram. Ensuring appropriate QoS is much easier on virtual circuit basis than by the use of datagram, because resource reservation – which gives guarantee on the quality – is easier to be done on a predetermined static route. In the followings, we cover methods for both kinds of networks.

An obvious solution is the oversize of the network. However ensuring much more resources than the real average need can be a very expensive way of quality assurance.

There are three points where QoS can be influenced: at the sender host, at the receiver host and at the intermediate routers.

On the receiver side, buffering is a popular method for ensuring acceptable quality. This method does not influence bandwidth or reliability, but very efficient for fighting against jitter in case of real-time transmission. Its robustness depends on the receiver's buffer size. When we play media from a web page, media players usually buffer about 10 seconds of data compensating the possible jitter.

Maybe sender has the biggest responsibility in QoS: the whole transmission begins there. It should form its transmission stream to be the easiest to ensure the desired quality: the traffic should be smooth instead of sudden bursts. Of course, smoothened traffic causes congestion with less probability. Leaky bucket and token bucket algorithms – they were mentioned in case of congestion control – are for smoothing away the sudden bursts of traffic by their traffic shaping method.

The third place in the network for assuring QoS is the set of intermediate routers. The straightforward method for them is the resource reservation for the actual transmissions. Resource means the processor time, bandwidth and buffers. The lack of them results in thrown packets, cells, or undesired behavior of service. Resource reservation can be easily carried out in case of virtual circuits, where all the packets or cells – which belong to the same flow – follow the same path. When the expected traffic is determined and given as flow specification, the resource reservation is easy. Usually the source describes his offer as the flow parameters (e.g. maximal and minimal packet size, maximum date rate, capacity, and speed of the token bucket) in the flow specification. While sending it to the source, both the intermediate routers and the destination host may modify it in weakening direction. Finally, at the destination there is a flow specification, which is acceptable for all the involved parties. If remaining resources are not enough for the sender, the flow is not started.

Another function what has great influence on the quality of service is routing. In routers, two main questions may be arisen: how to schedule the input rows and which output interface to choose for the transmission. There are different solutions for scheduling of incoming packets:

- Fair queuing usually uses round robin algorithm: takes one packet from the following input queue and forwards it. Then goes to the next input queue.
- Not to favor hosts with longer packets, round robin can be modified: the basic of cycles is one byte instead of one packet.
- In weighted fair queuing different priorities can be assigned to different hosts: maybe a real-time transmission needs more bandwidth than e-mailing.

During routing, the desired service type also can be determined. Some packets need reliable transmission and do not care about delay, while others differ. Two main architectures were created by IETF for ensuring QoS during routing: integrated services (IS) and differentiated services (DS).

Integrated services are often called as flow-based services. IS deals with flows as whole entities and reserves resources on the whole route of a flow. Its disadvantage is that it needs preparations and maintains information about flows during transmission. The main protocol of IS is the Resource Reservation Protocol (RSVP), which was developed for ensuring optimal bandwidth reservation and management for multicast transmission between several senders and receivers even for dynamic changes in demands.

Class-based services of DS have wider practical use than flow-based services of IS. Its reason is that class-based services need mostly only local administration in routers. Class-based services use the Type of Service field of IPv4 packets, which describes how to care the packet. This field determines whether Reliability, Delay, or Throughput is important parameter for the packet and assigns a priority to it. Routers may handle each packet individually by analyzing the value of this field. IPv6 changed this field to describe Traffic Class of the packet. IETF predefined some traffic classes, like expedited forwarding or assured forwarding. In expedited forwarding, the rarely arriving expedited packets – which are handled in a separate output queue – have greater bandwidth than usual packets, while assured forwarding differentiates 12 traffic classes.

Another idea for differentiating traffic classes by routers is the use of MultiProtocol Label Switching (MPLS). In this solution, a standardized MPLS header is inserted in the packet, between the data-link and network layer protocol headers. This header – independently of all the other used protocols – describes the QoS class of the data unit (it can be IP packet, ATM cell, etc.) and indexes the routing table of routers. This field helps to fasten routing (in this context it is usually called as switching) and also reserving resources on the route. MPLS deals with traffic classes and not with individual flows; that is why data from different flows with the same traffic class – which belong to the same Forwarding Equivalence Class (FEC) – can be multiplexed on the level of MPLS.

Naturally, protocol developers realized that QoS is one of the most important topics in the future's network communication. It is marked with the fact that next to its Traffic Class field IPv6 header has also a Flow Label field. Its role is to create a pseudo virtual circuit in the datagram based subnet to facilitate ensuring the desired Quality of Service.

FUTURE RESEARCH DIRECTIONS

Protocol developers always intend to create solutions suiting for the new challenges of the age. As we follow user requirements with attention, the following properties of communication systems seem to be very important in the future:

- robustness
- efficiency
- self-adaptive, self-organizing and self-healing capabilities
- mobility
- flexibility
- ability for complex content delivery with appropriate quality
- security
- personalization

All of these properties should be available in a network, which is more and more heterogeneous. These requirements influence the future of network protocols and their functions. The main trend in the evolution of basic protocol functions is to reach more efficient operations. Companies may save money with more efficient operations; that is why this direction of development is strongly financed. Quicker routing algorithms, less congestions, smaller amount of thrown packets result in less resource requirements and spares money. Because in our world cost efficiency is one of the main goals, this direction of protocol evolution can be easily forecasted.

Related to the presented advanced protocol functions, trends seem to be as follows:

Network management moves to the direction of modularity and strengthened security. For example, the evolution of SNMP also shows it.

Another important trend in network management is the move from client-server architecture to peer-to-peer networks.

Central area of network security is cryptography. Its permanent purpose is to prevent attackers from decrypting data in a worthy way. The reliability and security of the algorithms (like RSA or AES) depends mainly on the size of the key.

Mobility management seems to move to confederate different mobile networks both for joining their advances and for ensuring wider coverage. Perhaps the main challenge of mobile networks will be security in the future.

Finally, trends for ensuring appropriate QoS seem to recall virtual-circuit based solutions even in datagram-based environments. Protocol developers realized that assured quality is not realizable in an ad-hoc, dynamically changing environment: some compromises have to be made. Its signs can be noticed in IPv6 header or in routing labels in MPLS.

CONCLUSION

In this chapter – after an introduction of basic protocol functions – advanced protocol functions are covered. These functions are generalized both for telecommunication networks and for computer networks. Advanced protocol functions were collected as the complex protocol functions, which play big role in our days and may have increasing importance in the future. Among advanced functions, this chapter presented network management, mobility management, network security, and Quality of Service. For each function its purpose is described, the related problems are mentioned and the solutions or the evolution of solutions are collected and introduced. The author hopes that this chapter helps to understand the necessity of different protocol functions and to realize the main points in their evolution.

REFERENCES

Abramson, N. (1985). Development of the ALOHANET. *IEEE Transactions on Information Theory*, *31*, 119–123. doi:10.1109/TIT.1985.1057021

Anttalainen, T. (2003). *Introduction to telecommunications network engineering*. Artech House.

Bellman, R. E. (1957). *Dynamic programming*. Princeton, NJ: Princeton University Press.

Dijkstra, E. W. (1959). A note on two problems in connexion with graphs. *Numerical Mathematic*, *1*, 269–271. doi:10.1007/BF01386390

Ford, L. R. Jr, & Fulkerson, D. R. (1962). *Flows in networks*. Princeton, NJ: Princeton University Press.

International Telecommunication Union. (1997). *Information Technology – open systems interconnection – common management information protocol: Specification*. Retrieved April 27, 2010, from http://www.itu.int/rec/T-REC-X.711-199710-I/en

International Telecommunication Union. (2000). *Series M: TMN and network maintenance: International transmission systems, telephone circuits, telegraphy, facsimile, and leased circuits*. Retrieved April 26, 2010, from http://www.itu.int/rec/t-rec-m.3400-200002-I/en

Schiller, J. (2000). *Mobile communications*. Addison Wesley.

Stallings, W. (1998). *SNMP, SNMPv2, SNMPv3 and RMON 1 and 2*. Indianapolis, IN: Addison-Wesley Professional.

Stallings, W. (2002). *Wireless communications and networks*. NJ: Prentice Hall.

Stallings, W. (2006). *Cryptography and network security: Principles and practice*. NJ: Prentice Hall.

Tanenbaum, A. S. (2003). *Computer networks*. NJ: Prentice Hall.

Turner, J. S. (1986). New directions in communications (or which way to the information age). *IEEE Communications Magazine*, *24*, 8–15. doi:10.1109/MCOM.1986.1092946

ADDITIONAL READING

Baker, F., Krawczyk, J., Sastry, A. (1997). RFC 2206 RSVP Management Information Base using SMIv2

Balakrishnan, R. (2008). *Advanced QoS for Multi-Service IP/MPLS Networks*. Wiley.

Berger, L., O'Malley, T. (1997). RFC 2207 RSVP Extension for IPSEC Data Flows

Blake, S., Black, D., Carlson, M., Davies, E., Wang, Z., Weiss, W. (1998). RFC 2475 An Architecture for Differentiated Services

Braden, R., Zhang, L. (1997). RFC 2209 Resource Reservation Protocol (RSVP) – Version 1 Message Processing Rules

Braden, R., Zhang, L., Berson, S., Herzog, S., Jamin, S. (1997). RFC 2205 Resource ReSerVation Protocol (RSVP) – Version 1 Functional Specification

Burke, J. R. (2003). *Network Management: Concepts and Practice, A Hands-On Approach*. Prentice Hall.

Cole, E. (2009). *Network Security Bible*. Wiley.

Davie, B., Charny, A., Bennett, J. C. R., Benson, K., Le Boudec, J. Y., & Courtney, W. (2002). *RFC 3246 An Expedited Forwarding PHB*. Per-Hop Behavior.

Dierks, T., Allen, C. (1999). RFC 2246 The TLS Protocol Version 1.0

Eastlake, D. (1999). RFC 2535 Domain Name System Security Extensions

Eberspaecher, J., Voegel, H. J., Bettstetter, C., & Hartmann, C. (2009). *GSM – Architecture, Protocols and Services*. Wiley.

Heinanen, J., Baker, F., Weiss, W., Wroclawski, J. (1999). RFC 2597 Assured Forwarding PHB Group

Jha, S., & Hassan, M. (2002). *Engineering Internet QoS*. Artech House.

Kent, S., Atkinson, R. (1998). RFC 2401 Security Architecture for the Internet Protocol

Kent, S., Atkinson, R. (1998). RFC 2402 IP Authentication Header

Kent, S., & Atkinson, R. (1998). *RFC 2406 IP Encapsulating Security Payload*. EPS.

Loshin, P. (2003). *IPv6: Theory, Protocol and Practice*. Morgan Kaufmann.

Mankin, A., Baker, F., Braden, B., Bradner, S., O'Dell, M., Romanow, A., Weinrib, A., Zhang, L. (1997). RFC 2208 Resource ReSerVation Protocol (RSVP) Version 1 Applicability Statement Some Guidelines on Deployment

Mao, W. (2003). *Modern Cryptography: Theory and Practice*. Prentice Hall.

Marchese, M. (2007). *QoS Over Heterogeneous Networks*. Wiley. doi:10.1002/9780470058763

Medhi, D., & Ramasamy, K. (2007). *Network routing: Algorithms, Protocols and Architectures*. Morgan Kaufmann.

Nichols, K., Blake, S., Baker, F., Black, D. (1998). RFC 2474 Definition of the Differentiated Services Field (DS Field) in the IPv4 and IPv6 Headers

Perkins, C. (2002). RFC 3220 IP Mobility Support for IPv4

Ramsdell, B. (1999). RFC 2633 S/MIME Version 3 Message Specification

Rosen, E., Viswanathan, A., Callon, R. (2001). RFC 3031 Multiprotocol Label Switching Architecture

Subramanian, M. (1999). *Network Management, Principles and Practice*. Pearson.

Wroclawski, J. (1997). RFC 2210 The Use of RSVP with IETF Integrated Services

KEY TERMS AND DEFINITIONS

Communication Protocol: Syntactic, semantic, and timing rules of communication between peer entities for offering a kind of service to the above layer.

Congestion Control: It is a protocol function for preventing or resolving congestion on the network.

Flow Control: It is a protocol function for preventing a quicker sender in flooding a slower receiver host.

Medium Access: In case of broadcast links it is a protocol function for determining who has right to transmit data.

Mobility Management: All the protocol functions belong to mobility management, which support users in changing their location maintaining the possibility for communication. Main topics of mobility management are location update, handover (handoff) and roaming.

Network Architecture: It is the set of communication protocols, layers, and interfaces.

Quality of Service: It describes a service considering user satisfaction. Reliability, delay, bandwidth, and jitter are its main elements. Different applications have requirements on different subset of the previously enumerated factors.

Routing: It is a protocol function for determining which output interface (which route) to use for sending data (packet or cell) to the destination.

APPENDIX

Table of abbreviations

ACK	Acknowledgement
AES	Advanced Encryption Standard
AH	Authentication Header
ASN.1	Abstract Syntax Notation One
ATM	Asynchronous Transfer Mode
CD	Collision Detection
CIP	Cellular Internet Protocol
CMIP	Common Management Information Protocol
CRC	Cyclic Redundancy Code
CSMA	Carrier Sense Multiple Access
CSMA/CD	Carrier Sense Multiple Access with Collision Detection
DES	Data Encryption Standard
DNS	Domain Name Service
DoS	Denial of Service
DS	Differentiated Services
ESP	Encapsulating Security Payload
FA	Foreign Agent
FCAPS	Fault, Configuration, Account, Performance and Security (management functions)
FDM	Frequency Division Multiplexing
FEC	Forwarding Equivalence Class
FTP	File Transfer Protocol
HA	Home Agent
HLR	Home Location Register
HTTP	HyperText Transfer Protocol
ID	Identification
IEEE	Institute of Electrical and Electronics Engineers
IETF	Internet Engineering Task Force
IP	Internet Protocol
IPsec	Internet Protocol security
IPv4	Internet Protocol version 4
IPv6	Internet Protocol version 6
IS	Integrated Services
ITU	International Telecommunication Union
LAN	Local Area Network
MACA	Multiple Access with Collision Avoidance
MACAW	Multiple Access with Collision Avoidance for Wireless
MIB	Management Information Base
MIME	Multipurpose Internet Mail Extensions

MPLS	MultiProtocol Label Switching
MSISDN	Mobile Subscriber ISDN Number
PDU	Protocol Data Unit
PGP	Pretty Good Privacy
PKI	Public Key Infrastructure
QoS	Quality of Service
RSA	Rivest, Shamir and Adleman (the developers of RSA public key cryptographic algorithm)
RSVP	Resource ReSerVation Protocol
SCP	Secure Copy
SFTP	Secure File Transfer Protocol
SMI	Structure of Management Information
SNMP	Simple Network Management Protocol
SSH	Secure Shell
SSL	Secure Sockets Layer
TCP	Transmission Control Protocol
TDM	Time Division Multiplexing
TLS	Transport Layer Security
TMN	Telecommunications Management Network
TSAP	Transport Service Access Point
UPS	Uninterruptible Power Supply
VLR	Visitor Location Register
VPN	Virtual Private Network

Chapter 4
Notations for Test Specification:
TTCN3 and ASN.1

Szilárd Jaskó
University of Pannonia, Hungary

Dániel Muhi
University of Pannonia, Hungary

ABSTRACT

The purpose of this chapter is to provide a simple, understandable approach to testing and abstract syntax notation. This chapter is composed of four parts: (a) The background of TTCN and ASN.1 (b) The main components of TTCN and the ASN.1 (c) Future research directions (d) The conclusion.

The first part reviews the development of testing TTCN and ASN.1. The main part discusses TTCN and ASN.1 with explanation examples. The future research directions subchapter provides possibilities most likely to appear in the future. The conclusion gives an overview of the fundamental mechanisms of this field.

INTRODUCTION

The number of telecommunication tools is growing day by day. Communication possibilities are almost endless. As a result, testing becomes an integral part of the developing process. In the beginning firms tried to solve this problem with the help of local solutions. It worked for a while, but soon the market became global, and the competition was tough. Incompatibility became a serious problem; however, the adherence to a set of standards could solve this problem. It is obvious that the testing process would also have to function according to a standardized way. In line with this thinking, the integration of standardized solutions gains strength. An important part of the integration process is the syntax notation. This notation has to be consistent during the developing and testing processes. For example, an error occurs with a simple integer data type, if 4 bytes

DOI: 10.4018/978-1-60960-732-6.ch004

are used during the development process, but only 2 bytes are used during the testing process.

The objective of this chapter is to introduce and explain testing, the Testing and Test Control Notation (TTCN) and the Abstract Syntax Notation One (ASN.1). Therefore, this chapter provides historical background on these notations and explains the fundamentals of testing in a simple and understandable way: starting with the definition of test, presenting the techniques of protocol testing, as well as the TTCN-3 and ASN.1. In this chapter, a simplified example – testing of the TCP connection process – demonstrates the operation of TTCN-3 and ASN.1, and the future trends part focuses on the expected new methods and solutions, while the conclusion summarizes the contents of this chapter.

BACKGROUND OF TTCN AND ASN.1

Testing and Protocol Testing

Testing is part of our everyday lives. This can be demonstrated with the help of a simple example - the common scenario of picking up a cup of tea from the table and drinking it. The action is so natural that we do not think about the parallel flows needed for it. Let us have a look at this situation from testing point of view. The human brain provides many parallel functions in the area of test and control. Two senses are used to pick up the cup: vision and touch. The hand is under control during the whole period. The brain controls the position of the hand and the strength of the grasp. If any control function of the brain is missing, medical science identifies it as an illness.

The example above helps us to understand the definition of testing: to test is to verify a process. This is a rather general description of testing. More exact definitions can also be found in this chapter. We can observe the following: The more information that exists about a function, we get more correct definition of "a process". Below are some questions worth considering:

- Where is testing needed?
- Why is testing needed?
- Who/What needs testing?
- Is testing always necessary?

The answer to the first question comes from the above stated definition of testing. In short, it means that tests have to be used in everywhere where something should be controlled. The main answer to the second question is to minimize the errors of a given system. The answer to the third question depends on who/what wants to control certain processes. The final response is clear: test flows can never be ignored.

Testing is a special process, because every test is unique. That is why a universal test process cannot be created. Difficulties arise from the complexity of the system being tested, the environment in which the test was performed, and the evaluation of the results – since we can test the simple process of picking up a cup, as well as a complex economical process. Basically, we have to know three things for building a test:

- Test goal
- Access point(s)
- Reliable method(s)

The test goal states exactly what needs to be controlled or checked. The tester can enter the system under test at the access point, and the test can be executed with the help of a reliable method.

Testing in Telecommunications

Telecommunication industry is very important nowadays. The competition is hard; therefore, technical innovations are needed daily in order to save or advance the position of a firm on the market. The tendency is that whoever creates good quality products faster and provides complex services can stay in the lead. For that reason, testing is very relevant in the area of telecommunication. The definition of "test" - in the context of machines/ products – is as follows: to use a machine, product, or substance to find out how well it works, or to find out more information about it. This definition is more correct than the previous, because we are given some information about the goal of the test. For example, we can test the following parts of a mobile phone:

- housing/keyboard deterioration
- display
- panel
- software
- protocol

There is a special method for testing each of the different parts. The exact definition of testing is the following: Running new or revised programs to determine if they process all data properly.

Protocol Testing

Before protocol testing we need to understand what a protocol is. The definition of a protocol is the following: Rules governing the transmission and reception of information. The term "protocol" is generic and is used for hundreds of different communication methods.

Protocols are hierarchal set-ups. This can be seen in Figure 1. There are two ways of communication. One is the physical, where real signals are used, and the other is the logical, where same level entities communicate. For example, if one protocol entity in the third layer sends a connection request message to another protocol entity in the same layer, it is a logical communication. This kind of message is called Protocol Data Unit (PDU). Of course, this layer cannot directly send the PDU. The protocol entity has to give the message to the lower layer. This means that the lower protocol entities provide service to the higher ones. The two layers can communicate with the help of SPs (Service Primitive).

Testing has to follow the same structure. (Figure 2)

The tester communicates with the IUT across interfaces. They are called Point of Control and Observation (PCOs). For full test, both the upper and lower tester functions will be used. Sometimes, only the upper or the lower tester function can be used. The testers also use Service Primitives (SP) to communicate with the IUT. The SPs are also given in the protocol definitions... Note that, if the IUT is embedded into a layered or distributed system, the PCOs that the tester is able to control

Figure 1. Protocol hierarchy

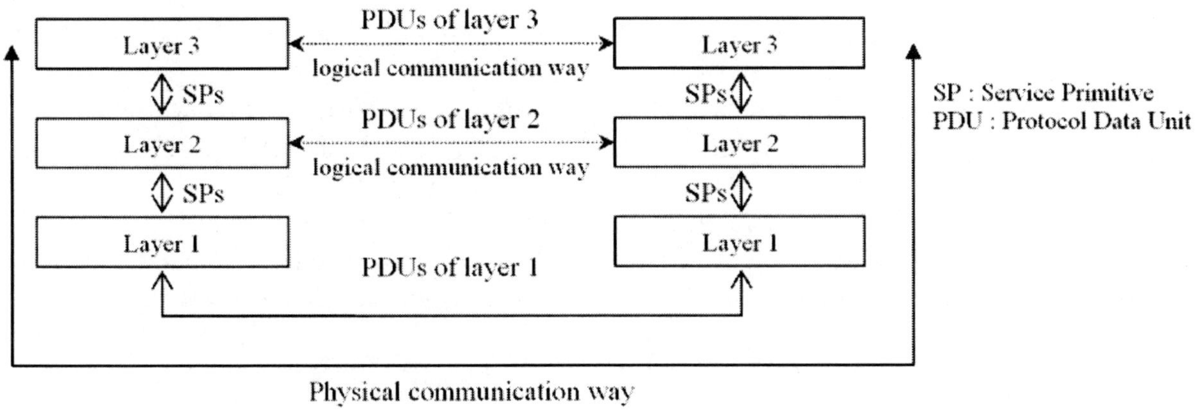

Figure 2. Protocol test structure

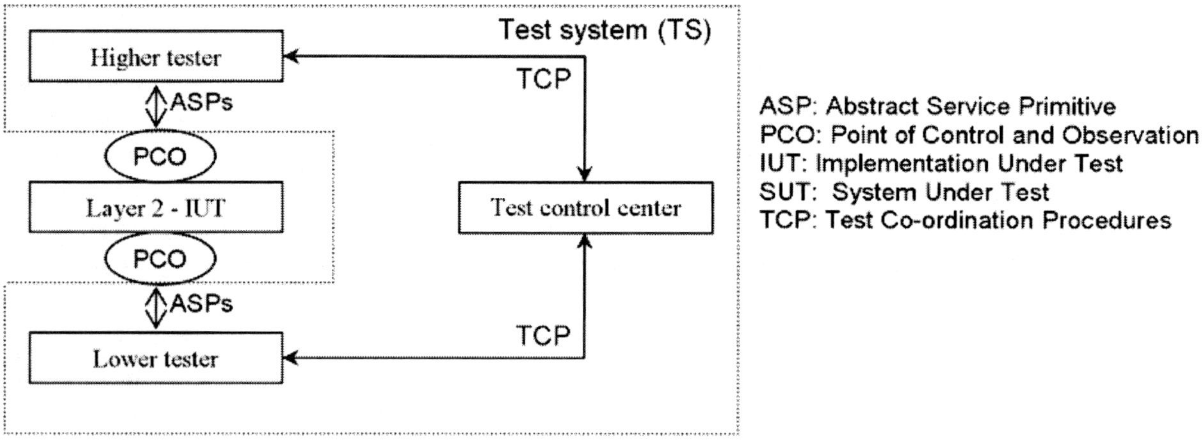

and to observe do not necessarily coincide with the implementation access point of the IUT.

TTCN (Testing and Test Control Notation) is the one of the standardized test languages in the field of telecommunication, and the latest version of TTCN is the third. (Baumgarten, 1994; Willcock 2005) An important test type is conformance test (Linn, 1994). It checks if the protocol confirms a given standard.

The History of TTCN

In the middle of the 1980s, some organizations for standardization (CCITT - International Telegraph and Telephone Consultative Committee and ISO - International Organizations for Standardization) and the telecommunication industry realized the need for a standardized formal testing language to ensure cooperation between different vendors' devices. It was necessary because heterogeneous telecommunication systems became widely used. To solve the problem of interoperability, the first version of Tree and Tabular Combined Notation (TTCN) was standardized in 1992 (ISO/IEC 9646-3 and CCITT Rec. X.292) and was recommended for all telecommunication companies to generate test cases. This language was especially created for testing protocols and not for usual software development purposes. TTCN became used widely in the telecommunication sector.

The first version of TTCN was extended as TTCN-2 in 1997. This testing language was more useful for testing concurrent systems than the previous version. Moreover, it was built up modularly, so it allowed reusing tests between different projects, and makes multi-user test suite development possible.

Giving up the tree and tabular format, the third generation of TTCN, Testing and Test Control Notation was standardized in 2001. TTCN-3 is better suited for testing 3rd generation protocols, where voice and (multimedia) data communications are dominant. It is also ideal for use in distributed applications. Test cases created in TTCN-3 core language can be presented in tabular, graphical or several other formats to the TTCN-3 user and can support different types of data (e.g. ASN.1) (Figure 3.).

Research has been done to connect the TTCN-3 to other description techniques. For example, one of them is UML (Zander, 2005). The last version of TTCN-3 is also specified in Z.160-Z.170 in ITU-T (International Telecommunication Union Telecommunication Standardization Sector) and ETSI ES 201 873-1 - ETSI ES 201 873-10 by ETSI (European Telecommunications Standards

Figure 3. Capabilities of TTCN-3

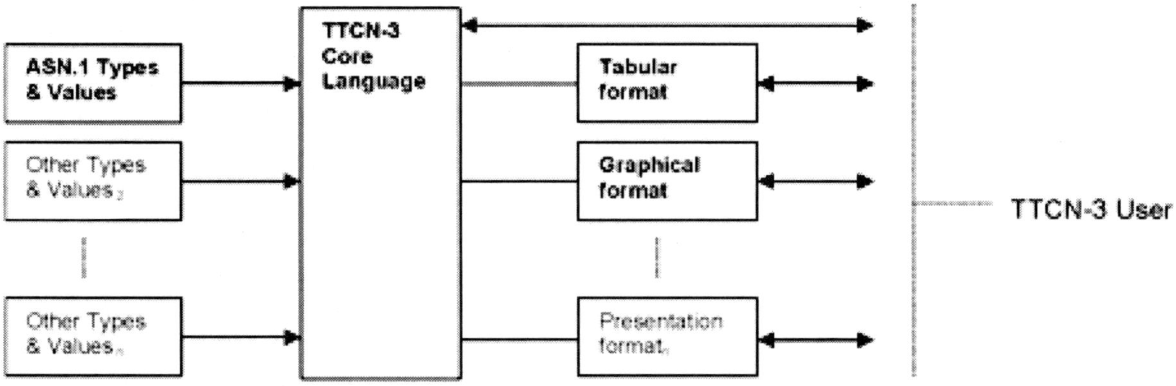

Institute). The last version of this standard family was published in 2007 by ITU-T and in 2009 by ETSI.

Abstract Data Structure

Parallel with the specification and description languages (Specification and Description Language (SDL), Message Sequence Charts (MSC), Unified Modeling Language (UML)) the Abstract Data and Syntax notation techniques were born as well. It is important, because faulty behavior can come from inconsistent data structure and usage. Consider the following: Imagine what a big problem can be caused if two communication partners use the same protocol, but one stores numbers into a 2-byte long variable and the other uses a 4-byte long variable. The problem is hidden until the 2-byte long variable does not roll over. Another advantage of abstract data structures is that they save time during protocol development, because they can be used in the whole development process from the requirement languages (e.g. Goal-oriented Requirements Language (GRL)) to testing (e.g. TTCN).

The History of ASN.1

In 1982, some engineers who developed applications for the application layer of computer networks faced the same problem. The data structures became complex, so reusability of these structures became important. James White and Douglas Steedman (Bell-Northern Research, Canada) developed the base of a new notation approach so that an encoding pattern could be given for the e-mail Message Handling System (MHS) protocols (White, 1989). X.400 was the official name of this solution by International Telegraph and Telephone Consultative Committee (CCITT). This organization made a more general standard than the X.400. It was called by X.409 ("Message Handling Systems: Presentation Transfer Syntax and Notation"). The X.409 was completely independent from the MHS systems, because it used more complex object treatment than was necessary. Therefore, it paved the way for a new trend. This idea was so brilliant that the other standardization organization started to examine it, and finally, John Larmouth (University of Salford, UK – retired) adopted it the world of Open Systems Interconnection Reference Model (OSI). Two documents were created that were the base of the ISO 8824 and the ISO 8825 documents. Those were published in 1987, and the common name is as follows: Abstract Syntax Notation One (ASN.1) and the Basic Encoding Rules (BER). In 1989, the CCITT published two recommendations, the X.208 (ASN.1) and the X.209 (BER), which replaced the X.409 standard. At that time, the main task of the ASN.1 notation was to help the developing process under the application layer.

In 1990, the ISO published a new version of the 8824 and 8825 documents. The difference was the CCITT (which later became the International Telecommunication Union - Telecommunication Standardization Sector (ITU-T)) and the ISO standard minimal at the time. A structural change supervened in 1994 into the new version of the standard. It was called ASN.1:1994, where 1994 is the year of final draft. The ISO adopted this version in 1995. This new concept had 4 main components, as follows:

- ITU-T Rec. X.680 (1994) | ISO/IEC 8824-1:1995 Specification of Basic Notation
- ITU-T Rec. X.681 (1994) | ISO/IEC 8824-2:1995 Information Object Specification
- ITU-T Rec. X.682 (1994) | ISO/IEC 8824-3:1995 Constraint Specification
- ITU-T Rec. X.683 (1994) | ISO/IEC 8824-4:1995 Parameterization of ASN.1 Specification

The structure and notation pattern of ASN.1 gave birth to the standard we know today. John Larmouth began work on the compatibility problem between the different types in 1995. His goal was to develop a solution that has a correct semantically description in ASN.1 for the software development. First, he named it "formal model", but at the end of the process, he changed it to "semantic model". There were numerous modifications in the 1997 version. For example, it introduced the UTF-8 (8-bit Unicode Transformation Format) character coding technique by ASN.1. This period was important for ASN.1 because its usage area expanded. It became used in the area of Open Document Management API (ODMI), Enhanced Language Of Temporal Ordering Specification (LOTOS), Transaction Processing, System management, X.400, and X.500 family. After that, ASN.1 jointed to the SC33 (Distributed Application Services), SC6 (Telecommunication and Information Exchange Between Systems) and the WG7 (3 and 4 layers of the OSI) documents and the developer solved the compatibility problem between the new and the version of 1990. Note that the ISO and the ITU-T have worked on this standard together since 1998.

The last version of the X.680, X.681, X.682, and the X.683 were published in 2002, but of course, the development of ASN.1 is continuous. Numerous amendments were attached to the standard. The last amendment was the PER (Packed Encoding Rules) in 2007 that belongs to X.680.

Currently the IETF works on three International Drafts related to ASN.1. Two of them add new modules to the ASN.1 and there are the Cryptographic Message Syntax (CMS) format, the PKIX certificate format, and many associated formats (Hoffman, 2010) (Schaad, 2010). The third one defines a translation method that is able to convert between different versions of ASN.1 (Wallace, 2010).

MAIN COMPONENTS OF TTCN AND ASN.1

Testing and Test Control Notation Version 3 (TTCN-3)

The Testing and Test Control Notation version 3 (TTCN-3) is a method for testing and specification. It is a part of the International Telecommunication Union Telecommunication Standardization Sector (ITU-T) recommendations Z-series. The last version of this standard family was published in 2007 by ITU-T, but two components (Z.169 and Z.167) came out one year later. The European Telecommunications Standards Institute (ETSI) has newer documents than the ITU-T. Many of those were published in 2009. TTCN-3 provides a solution to prepare a reusable, flexible, and effective test for the area of telecommunication and also for other areas. Its main usage areas are protocol testing, service testing, and module testing. For example, it can do the following kinds of tests: conformance testing, interoperability, robustness, regression, system, and integration testing. Its logic is based on the classical object oriented structure and it has two programming interfaces. These are the following: Tabular presentation Format (TFT), Graphical presentation Format (GFT). Of course, these can be expanded at will. The characteristics of TTCN-3 are the following:

- the ability to specify encoding information and other attributes (including user extensibility)
- well-defined syntax, interchange format and static semantics
- the ability to specify dynamic concurrent testing configurations
- operations for procedure-based and message-based communication
- the ability to specify data and signature templates with powerful matching mechanisms
- a precise execution algorithm (operational semantics)
- the assignment and handling of test decisions
- test suite parameterization and test case selection mechanisms
- value parameterization
- combined use of TTCN-3 with other languages
- different presentation formats (e.g. tabular and graphical presentation formats)

Core Language

The core language is a generalized text-based test language that provides a standardized interchange format of TTCN-3 test suites between TTCN-3 tools. Of course it has many links with other presentation formats (e.g. TFT, GFT), but neither can be used without the core language. Naturally, this property is not true for the core language, because it is usable by itself.

The core language follows the classical object oriented structure, but here the module is the top-level unit. This provides similar function like classes in Java. The module consists two parts: the definitions and the control part. The definitions part of a module specifies communication ports, constants, test components, functions, data templates, test cases and so on. The control part calls the test cases and controls their execution. A quick reference card can be founded here: http://www.blukaktus.com/TTCN3-QRC.html.

Modules

The module is the principle building block of the TTCN-3. It is able to define a complete executable test suite, but usually it is only a library. The module can handle different TTCN versions. An example can be seen here:

```
module TCP_connectionExample{
     [ModuleDefinitionPart]
 [ModuleControlPart]
}
```

A logical structure can be built into the module with the help of possibility of grouping and functions. These could be useful in complex modules, because it makes them clear and easy to develop. The control part of the module calls the test cases with the actual parameter and controls their execution. TTCN-3 will be shown via a part of the working flow of the Transmission Control Protocol (TCP). This part is the connection buildup process. TCP is a well-documented protocol, so its operation and structure will not be explained.

Test Specific Type and Values

This subchapter shows the test specific parts of TTCN-3 only, because the other part is similar to a common object oriented programming language.

Type and Template

First, the basic type has to be set. These basic elements are the parts of the TCP datagram.

```
module TCP_connectionExample{
...
type bitstring SequenceNumber length (32);
type bitstring AckNumber length (32);
type bitstring DataOffset length (4);
type bitstring Reserved length (6);
type bitstring URG length (1);
type bitstring ACK length (1);
type bitstring PSH length (1);
type bitstring RST length (1);
type bitstring SYN length (1);
type bitstring FIN length (1);
type bitstring Window length (16);
type bitstring Checksum length (16);
type bitstring UrgentPointer length (16);
type bitstring OptionalPart length (1..128);
type bitstring Padding length (0..31);
type bitstring Data;
...
```

The header of the TCP datagram has to be built up of these basic elements:

```
type record TCP_header{
      SequenceNumber SeqNum,
      AckNumber AckNum,
      DataOffset DataOffset,
      Reserved Res,
      URG urg,
      ACK ack,
      PSH psh,
      RST rst,
      SYN syn,
      FIN fin,
      Window win,
      Checksum chksum,
      UrgentPointer Urgpoint,
      OptionalPart Opt optional,
      Padding Pad optional,
      Data Data
}
```

Templates can be created from the types. This can be useful in the test configuration. For example in the case of connection establishment process of the TCP, *TCP_header* uses only the *ack* and the *syn* parts. Any more parts of the header are not relevant for this process. The notation "?" signs that a value of an element is not important, as we will see in the next example. The test process will use two templates: one is the initiation of connection (con_req_message) and the other is the establishment of the connection (con_resp_message). More information about the exact test flow can be read in the part of test configuration.

```
template TCP_header con_req_message{
        SeqNum:= ?,
        ...
        ack:= 0,
        psh:= ?,
        rst:= ?,
        syn:=1,
        ...
  }
  template TCP_header con_resp_message{
        ...
        ack:= 1,
        ...
        syn:=1,
        ...
  }
```

Communication Port

TTCN-3 needs some special additions for testing. One of them is the communication port. This is necessary for the communication between the test components or/and between the test system interface and the test components. This test tool provides two kinds of ports: One is message-based and the other is procedure-based. Every port has direction rule(s). TTCN-3 can handle three kinds of directions: *in*, *out* and *inout*. Of course, the directions have to be seen from the aspect of test component ports, where the test system interface controls the exceptions. For example, the *in* direction means that a message sending or a procedure call happens. The *out* direction is the inverse of it (message receive or exception catch). Lists can be created to control message or procedure signatures together to the wanted directions. Let us see an example that continues the example of TCP and includes two ports. One is the interface between the session layer and the application layer. The other is the interface between the session and the network layer. The name of the interface between the layers is Service Access Point (SAP).

```
type port SAPtoNetworkLayer message{
     inout  TCP_datagram;
}
type port SAPtoApplicationLayer message{
     inout  Data;
}
```

Both SAPs use *inout* direction. The *SAPtoNetworkLayer* ports TCP datagrams and the *SAPtoApplicationLayer* ports data for the protocols of the application layer.

The difference of notation between the message and the procedure is only one word as we can see:

```
type port Example procedure{
     in       Proc1;
     out      Proc2, Proc3;
     inout    Proc4, Proc5, Proc6;
}
```

This kind of notation is not used in the example of TCP.

Component Types

The component type can sign the port and other variable type utilization of a component. The name of each port and variable types has to be unique inside of the component type, but of course, every name becomes reusable in the different component types. The following example shows a declaration, wherein two ports are defined for testing the TCP connection.

```
type component TCP_connectionComponent{
     port     SAPtoNetworkLayer mtcPort;
     port     SAPtoApplicationLayer pco1;
}
```

Extensions can be defined in the TTCN-3. In this case, the extended type component derives from the parent type component. This extension is necessary for the timer function that uses the TCP via communication. The *Timer* variable is the timer and the *Counter* variable counts the missed number of attempts.

```
type component ExtendedWithTimer extends
TCP_connectionComponent{
     timer     Timer;
     var integer Counter;
}
```

Finally, a special port is defined that signs the gateway of the tester.

```
type component SystemType{
     port   SAPtoNetworkLayer systemPort;
}
```

Test Configuration

One test component has to be in the test configuration, but the TTCN-3 can handle the concurrent test configurations. The different test components can communicate with each other through correctly defined communication ports and a test system interface. The test system and its environment can only communicate via the test system interface. There is a special test component in every test configuration; its name is Master Test Component (MTC). The other test component(s) is called Parallel Test Component(s) (PTC(s)). The test system creates automatically the MTC at the start of every test case. Running of the MTC controls the execution time. It means that if the MTC terminates, the test cases execution will come to an end. Of course, this is not true for the PTC(s). An example can be seen in Figure 4.

Test components have to communicate via pre-defined ports. Each port is modeled as an infinite FIFO queue. It behaves like a general FIFO so it can store the incoming messages or procedure calls until the given component can process their ones. The test system interface(s) is a special "port(s)", because the MTC and the PTC(s) can connect to the observed implementation and/or system via those. If the test system tests an implementation that is called Implementation Under Test (IUT). If the focus is on the whole system, it is called System Under Test (SUT). In the simplest case, the SUT and the IUT are equivalent. The example TCP connection can be seen for a test case and an interface mapping:

```
testcase TCP_connectionTestCase() runs on
ExtendedWithTimer
                    system SystemType{
     map(mtc:mtcPort, system:systemPort);
     //Special behaviors
...
}
```

mtcPort and the *systemPort* maps into this test-case. It means that if a message goes via the *mtcPort*, it will go via the *systemPort* too. The special behaviors will be shown in the next sub-chapter.

Description of Special Behaviors

The common sequential programming method usually is not enough for test descriptions. In generally, it uses set of alternatives beside the general programming method. These are the followings: *alt* and the *interleave* statement.

The possible alternatives can be described with the help *alt statements*. It can be imagined like a tree graph where every leaf is a possible execution path. Initiation events can be defined in the *alt statement*. For example there should be a timer event and/or a receive/handling of communication event and/or the termination of parallel test component. Else alternative can be defined and it is always executed if no other alternatives could be. The alt statement uses a special mechanism to choose which alternative will be executed. It is called snapshot mechanism. A snapshot is taken when entering an alt statement. It considers every potential event that influences the decision and assigns one of them the execution. Tester sends a message based on *con_req_message* template and set the *Timer*. Then four alt statements can happen. First, the appropriate message is received based on *con_resp_message* template. In this case, the *Timer* is stopped, and the verdict is set pass. Second, any other message is received. *Timer* is stopped too, but the verdict is set fail in this case, because the connection is not established. Third, Timer expires and finally, all other cases. The verdicts are failed in both cases.

Figure 4. An example for the test configuration

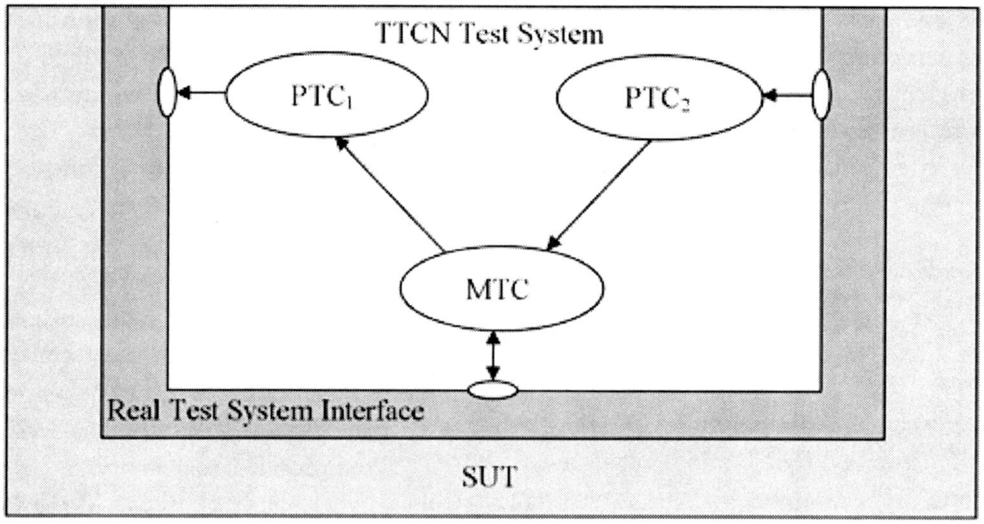

```
testcase TCP_connectionTestCase() run
on ExtendedWithTimer System systemType{
  ...
  mtcPort.send(con_req_message);
  Timer.start;
  alt{
     [] mtcPort receive(con_resp_message) {
              Timer.stop;
              setverdict(pass);
     }
     [] mtcPort receive {
              Timer.stop;
              setverdict(fail);
     }
     [] Timer.timeout {
              setverdict(fail);
     }
     [else] {
              setverdict(fail);
     }
  ...
}
```

The *interleave statement* is a simplification of the alt statement. In this case, there is no strict execution order of the different alternatives and its inner part. Therefore, more complex description has to be used if only alt statement is applied. The next example explains this. There are two interleaves events. One is when the session layer gets a data unit (*get_data*) form the network layer via the *mtcport*. In this case, it will send a message (*send_data*) and it will wait for the acknowledgement (*ack*). Another possibility is if it gets a data unit (*get_application_layer_data*) from the application data via *pco1* port. The session layer will send back a data unit (send_application_layer_data) to the application layer and it will wait for the *ok* message.

```
interleave {
    [] mtcport.receive(get_data){
             mtcport.send(send_data);
             mtcport.receive(ack);
    }
    [] pco1.receive(get_application_
    layer_data){
             pco1.send(send_application_
             layer_data);
pco1.receive(ok);
}
}
```

Both interleave events can run parallel. Therefore, the description of the same is the following if alt statement is used.

```
alt{
    [] mtcport.receive(get_data){
          mtcport.send(send_data);
          alt{
             [] mtcport.receive(ack){
                   pco1.receive(get_application_
                   layer_data);
                   pco1.send(send_application_
                   layer_data);
                   pco1.receive(ok);
              }
             [] pco1.receive(get_application_
             layer_data){
                   pco1.send(send_application_
                   layer_data);
                   alt{
                       [] mtcport.receive(ack){}
                       [] pco1.receive(ok) {}
}
}
}
      }
      []pco1.receive(get_application_layer_
      data){
          pco1.send(send_application_layer_
          data);
          alt{
             [] mtcport.receive(get_data){
                   mtcport.send(send_ data);
                   alt{
                       [] mtcport.receive(ack){}
                       [] pco1.receive(ok) {}
}
             }
             [] pco1.receive(ok){
                   mtcport.receive(get_data);
                   mtcport.send(send_data);
                   mtcport.receive(ack);
}
}
}
}
```

Control Part

Finally, the control part of the module should be set. Test cases can be run with the help of it. *TCP_connectionTestCase* will be set in the following example:

```
module TCP_connectionExample{
    ...
  control{
    ...
    execute(TCP_connectionTestCase());
    ...
}
```

Tabular Presentation Format (TFT)

The Tabular presentation Format of TTCN-3 (TFT) is defined by the ITU-T Recommendation Z.162 and the ETSI ES 201 873-2. The TFT is based on the core language of the TTCN-3. Its philosophy is in the nearest relationship with the programming interface of the previous version of TTCN. Of course, the syntax of TFT is different from that used in the previous version. In this case, an alternative of the core language is it preferably. The mapping between the core language and the TFT needs some transformation. These transformations can be classified to two groups. One is directly converted from the TFT to the core and the core language holds the same meaning. The transformation of an identical field is a typical example for it. The second group includes those transformation cases for which the core language does not have the same meaning. Some form elements (for example the comment or directive) is the typical example for this case.

Two tables demonstrate the description mode of TFT. These tables are the part of the TCP examples. Therefore, the meaning of *Testcase* table is the same as the previous *testcase* description in the core and the *Test Suite Control* table is the same as the control part of the module in the case of the core.

Graphical Presentation Format (GFT)

The Graphical presentation Format of TTCN-3 (GFT) is defined by the ITU-T Recommendation Z.163 and the ETSI ES 201 873-3. It is based on the ITU-T Recommendation Z.120 standard that is called Message Sequence Charts (MSC). Of course, GFT uses test specific extension over the MSC elements. The main part of these extensions is textual only that came from the core syntax. The other extensions only help us to read the GFT diagram easily. This solution provides us another programming interface where core can be programmed like in the case of TFT.

The same part of the example can be seen in GFT in Figure 5. Figure 5 has two parts. Part *a* demonstrates the description of the control part and part *b* shows the testcase.

Abstract Syntax Notation One (ASN.1)

The Abstract Syntax Notation One (ASN.1) is a standard notation for the specification of data values and types. It is a part of the International Telecommunication Union Telecommunication Standardization Sector (ITU-T) recommendations X-series (X.680-683). The last version of this standard family was published in 2008 by ITU-T. ISO standardized ASN.1 too; its name is ISO/IEC 8824 and 8825 series. ASN.1 tries to give us a general and unambiguous way to define types and values. It can represent almost any kind of data from the simple to the complex. A typical example for the simple data type is the integer, and for the value is 5. Each message in any protocol architecture is a typical example for the complex data. ASN.1 uses several basic types and their corresponding values and gives rules for combining them into more complex types and values. This capability is part of the original ASN.1 standard (Rec. X.680) and it works even as in the conventional programming languages.

Another important part of ASN.1 is the *encoding rules* that determine the exact bit-patterns to represent values of data structures when stored on a computer or transferred over a network. The encoding rules are the following in ASN.1: ITU-T Recommendation x.690| ISO/IEC 8825-1 specifies the *Basic Encoding Rules* (BER), *Canonical Encoding Rules* (CER), and *Distinguished*

Encoding Rules (DER). ITU-T Recommendation X.691 | ISO/IEC 8825-2 specifies the *Packet Encoding Rule* (PER). *XML Encoding Rules* (XER) is specified by ITU-T Recommendation X.693 | ISO/IEC 8825-4. Beside the encoding rules, the ASN.1 has an *Encoding Control Notation* (ECN) that signs the ITU-T Recommendation X.692 | ISO/IEC 8825-3 standard. The designer can create the formally specified syntax of a protocol with the help of ECN.

This subchapter does not have the chance to demonstrate the full syntax and encoding rules of ASN.1. Therefore, it contains a small part of it and concentrates on showing the possibilities of ASN.1.

Basic Notation

The basic syntax of ASN.1 is the following:

- Names
  Letters (case-sensitive), digits, and hyphen.
- Module names
  Start with upper case.
- Type reference names
  Start with upper case.
- Identifiers (for sequence components, choice alternatives, named bits, named numbers, and enumerations). Start with lower case.
- Value reference names

Start with lower case. It is common to use either or both of hyphens or upper case within names to separate parts of the name. (e.g. My-type or MyType) Single-line comments start with -- and end with -- or new line. Block comments start with /* and end with */. Block comments can contain other block comments and/or doublehyphen comments. For historical reasons, multiple single-line comments are often used instead of block comments.

The ASN.1 data type definition is mostly similar to a programming language. First, it disposes built-in types; on the other hand, it will provide the opportunity to create complex types from simple elements. This design capability is more flexible than that of most programming languages use. However, ASN.1 is not a programming language, so executable instructions (loop, conditional, etc.) do not exist. Note that ASN.1 makes distinction between the upper and the lower characters. There are some rules on how to write letters. Slightly over-simplification, all keywords have to contain only capitalized characters. The module and the type name should be capitalized to begin with, while all other identifiers (for example variables) should start with lowercase letters. The built-in ASN.1 types are the following: integer, boolean, real, bit string, octet string, enumerate, object identifier, any, and null. The ASN.1 does

Figure 5. (a) GFT description of control part. (b) GFT description of testcase

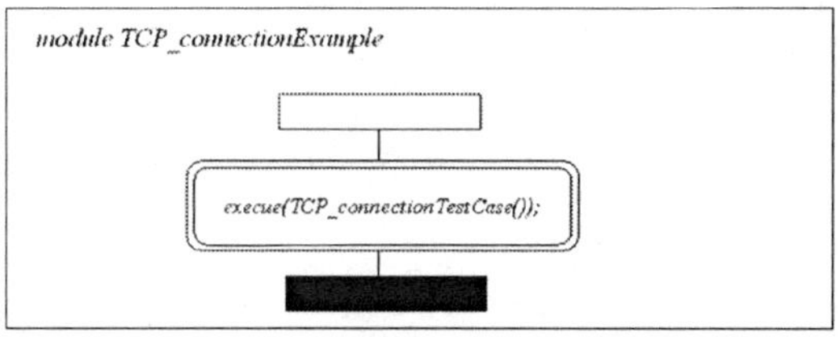

(a) description of module in GFT

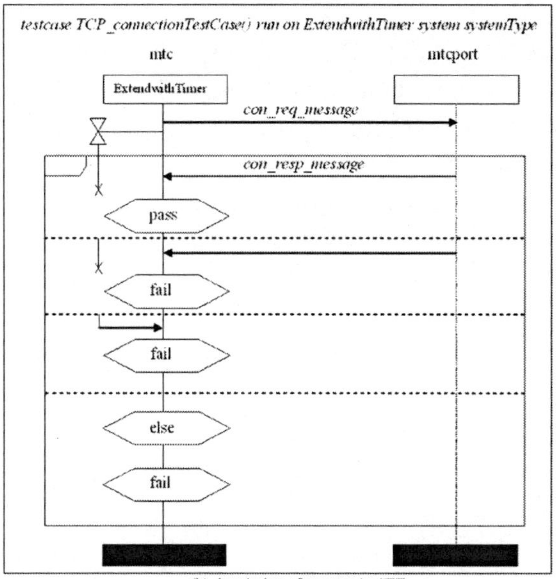

(b) description of testcase in GFT

Table 1. TFT description of a testcase

Testcase			
Name	TCP_connectionTestCase		
Group			
Purpose			
System Interface	SystemType		
MTC Type	ExtendedwithTimer		
Comments			
Local def. name	Type	Initial value	comments
Behavior			

```
...
mtcPort.send(con_req_message);
Timer.start;
alt{
        [] mtcPort receive(con_resp_message) {
                Timer.stop;
                setverdict(pass);
        }
        [] mtcPort receive {
                Timer.stop;
                setverdict(fail);
        }
        [] Timer.timeout {
                setverdict(fail);
        }
        [else] {
                setverdict(fail);
        }
}
...
```

not define the implementation of these types; so for example, in the case of integer, the value is not limited. These questions are decided in the coding rule.

A new (simple) type is created from simple types with the help of the renaming process. Symbol "::=" is used by the definition of the data-type. Note that the type name starts with capital letters. Let us see some example for it:

```
IsConnected::= Boolean
DestinationPort::= BIT STRING
NetworkError::= ENUMERATED
         {destinationHostUnreachable (0),
         destinationNetworkUnreachable (1),
         destinationPortUnreachable (2),
         noErrorMessage (3)}
```

The type of *IsConnected* can be used to store the status of the connection. The destination port can be stored in the type of *DestinationPort.* Finally, a new type of enumerate is seen that can store information about the network connection errors.

Structured types can be created with the help of the following structures: sequence, set, sequence of, set of, choice. The sequence consists of different types of records, while the SET defines different types of cluster elements. The main difference between them is that in the first case, the order of the elements is fixed, while in the second case it is free. The difference between the sequence and the sequence of and the set and the set of is, ~ of have to build up from the same type of queued

Table 2. TFT description of control part

Test Suite Control			
Module Name	TTCN_connectionExample		
Version	1.0		
Date	19 July 2010		
Base Standard Ref.	RFC-793		
Test Standard Ref.	RFC-793		
PICS Ref.	RFC-793		
PIXIT Ref.	RFC-793		
Test method (s)	Local		
Encoding			
Comments			
Local def. name	Type	Initial value	comments
Behavior			
… execute(TCP_connectionTestCase()); …			

records. In fact, sequence of is the typical array and the set of is the common set. Choice is similar to the union in C programming language. For example the header of the TCP is the following:

```
TCPheader::= SEQUENCE {
     seqNum BIT STRING (SIZE(32)),
     AckNum BIT STRING (SIZE(32)),
     dataOffset BIT STRING (SIZE(4)),
     res BIT STRING (SIZE(6)),
     urg BIT STRING (SIZE(1)),
     ack BIT STRING (SIZE(1)),
     psh BIT STRING (SIZE(1)),
     rst BIT STRING (SIZE(1)),
     syn BIT STRING (SIZE(1)),
     fin BIT STRING (SIZE(1)),
     win BIT STRING (SIZE(16)),
     chkSum BIT STRING (SIZE(16)),
     urgPoint BIT STRING (SIZE(32)),
     opt BIT STRING (SIZE(0..128)),
     padding BIT STRING (SIZE(0..31)),
     data BIT STRING
}
```

The sequence is not relevant in the case of set. Let us see an example for it. A typical – non-telecommunication – example is the recipe. The sequence of the ingredients is not important many times. For example, ham, egg and salt is the same as egg, salt and ham.

```
Recipe::= SET {
     ingredient1 ENUMERATED,
     ingredient2 ENUMERATED,
     ingredient3 ENUMERATED,
}
```

Basic Encoding Rules (BER)

ASN.1 can use more encoding rules. One of them (BER) will be shown in this subchapter. It may be used to derive the specification of transfer syntax for values of types defined using. The notation is specified in ITU-T X.680 – X.683 standards (ASN.1). First, we have a look at the encoding rule of BER.

The coding of each value must include the type of data (TAG), a length (Length) and finally, the value (Value). At complex recursive types, the value is more "trio". The possible values of class are the following:

- 00 = Universal (ASN.1-defined "integral" types)
- 01 = Application wide (a standard within the ITU-defined)
- 10 = Context-specific (e.g., an ITU standardized by a complex type within fields)
- 11 = Private (or Network. Provider-specific, do not use ISO or ITU)

The format should be two kind of the primitive (0) and the structured (1). A structured type means that the data field contains recursively additional values.

TAG has two kind of form: the short and the long. If the TAG number is less than 31, it will use the short type. In this case, the TAG consists of three elements: class (2 bits), format (1 bit) and the TAG number (5 bits). In this case, if universal class is used, the possible values of TAG numbers are the following:

- 1 = BOOLEAN
- 2 = INTEGER
- 3 = BIT STRING
- 4 = OCTET STRING
- 5 = NULL
- 10 = ENUMERATED
- 16 = SEQUENCE / SEQUENCE OF
- 17 = SET / SET OF

If the tag number is greater than or equal to 31, the tag follows the following rules:

- Bits 8 and 7 of the first octet specify the class of the tag as in low-tag-number tags (universal: '00', application-specific: '01', context-specific: '10', private: '11').
- Bit 6 of the first octet specifies whether the type is primitive or constructed as in low-tag-number tags (primitive: '0', constructed: '1').
- Bits 5–1 of the first octet all have the value '1'.
- The second and following octets give the tag numbering base 128, most significant digit first, with as few octets as possible, and with the bit 8 of each octet except the last set to '1'.

The length field can be coded in three ways: short form (length <= 127), long form and the indefinite form. The indefinite form is used in the case of complex types. In the short form, the value of the length field means how many octets follow the TAG and the length fields. Of course, it means that the maximum octet number is 127. The first octet of the long format gives the number of the length bytes (maximum 127 bytes and the bit 8 is one) and the other gives the length of the value. The indefinite is a special case and it gives a frame the value.

Two examples are seen in the followings. First, a simple value (79) is coded with the help of BER. A complex structure is coded in the second example that describes two numbers (79 and 102). In this case, the complex description builds up from primitive BER syntax.

Example 1: BER syntax (number 79) 02_H 01_H $4F_H$

From that TAG = 02_H it converts to binary => 00000010_2

- 00000010_2 it means the class is *universal.*
- 00000010_2 it means the format is *primitive.*
- 00000010_2 it means the value type is *integer*.

From that Length = 01_H it converts to binary => 00000001_2 It means that only one octet follows the *TAG* and the *Length* fields.

From that Value = $4F_H$ it converts to decimal => 79_{10}

Example 2: BER syntax (number 79 and 102) 30_H 06_H 02_H 01_H $4F_H$ 02_H 01_H 66_H

From that TAG = 30_H it converts to binary => 00110000_2

- 00000010_2 it means the class is *universal.*
- 00100010_2 it means the format is *structured.*
- 00010000_2 it means the value type is *sequence.*

From that Length = 06_H it converts to binary => 00000110_2 It means that 6 octets follows the *TAG* and the *Length* fields. Note that it has to be BER elements, so the next field is *TAG*.

From that TAG = 02_H it converts to binary => 00000010_2

- 00000010_2 it means the class is *universal.*
- 00000010_2 it means the format is *primitive.*
- 00000010_2 it means the value type is *integer.*

From that Length = 01_H it converts to binary => 00000001_2 It means that only one octet follows the *TAG* and the *Length* fields.

From that Value = $4F_H$ it converts to decimal => 79_{10}

From that TAG = 02_H it converts to binary => 00000010_2

- 00000010_2 it means the class is *universal.*
- 00000010_2 it means the format is *primitive.*
- 00000010_2 it means the value type is *integer.*

From that Length = 01_H it converts to binary => 00000001_2 It means that only one octet follows the *TAG* and the *Length* fields.

From that Value = 66_H it converts to decimal => 102_{10}

The *ASN.1 Complete*, the *ASN.1 Communication Between Heterogeneous Systems* books and the *Layman's Guide to a Subset of ASN.1, BER, and DER* technical note include more details about ASN.1. (Larmouth, 1999) (Dubuisson, 2000) (Burton, 1993)

Link between TTCN-3 and the ASN.1

The TTCN-3 and the ASN.1 are separated standards. The developer realized that these two standard relationships are important. These languages/notations were provided by several tools and platforms. Therefore, it should also appear in the standardization process. The ITU-T standardized it in the Recommendation X.167. The ETSI ES 201 873-7 standard is another approach of this problem. More standards deal with the fact that with the ASN.1 what other languages / notations can work together (such as Specification and Description Language (SDL) and ASN.1 or Message Sequence Charts (MSC) and ASN.1). It is important because the usage of ASN.1 can help the process of conversation from a language/ notation to another one. This technique avoids errors caused by type and variables.

FUTURE RESEARCH DIRECTIONS

Testing techniques of telecommunication will continue to evolve in the future. The relationship of the different standards (for example SDL, MSC, ASN.1, and the TTCN-3) will be stronger and various tools and software will be developed to convert from one to the other. This makes the developing and testing processes faster and more exact than now.

A further trend will be that the systems and the software will become more flexible and self-adaptive. It means that the test systems and processes will have to adapt to the challenges. A new generation of test tools will be born soon, these will be able to adapt to the changing environment and generate test cases automatically. All new solutions can connect to the core language of the TTCN-3 so this is a possible relation between the current and future techniques.

The self-adaptive test system needs mathematical background. Communicating Sequential Processes (CSP) can be a proper tool (Jaskó 2005)

(Hoare 1985). Therefore, the CSP and the telecommunication standards will be more linked than now. CSP seems to be a good choice, because it can describe the working flow of several processes in a compact mathematical way. These mathematical descriptions can describe the communication for example between the tested system and the tester. A possible integration process to the TTCN-3 is seen on Figure 6.

The innovation in this hierarchical structure is two new modules. They are the CSP intelligence module and the CSP module. CSP intelligence module has to solve the following functions: First of all this is connected to the implementation under test and it collects traces of the protocols that are used by the IUT. In the next step, the module recognizes the protocols and generates test cases from the traces. Finally, it builds a test description according to the control signal principle.

CSP module is a programming interface like the graphical format or tabular format. This module uses the control signal description method in CSP. Graphical description is a part of this module so users can see through the test description easily. Therefore, the CSP module is the programming interface between the test system and the TTCN-3 core language. One part of the test system is the CSP intelligence module. This CSP tool generates the test cases from the given traces. Of course, it has to include some kind of artificial intelligence, because the test generating process is very complex and this problem cannot be solved without multiple logical units. That is where the name of this module comes from. Human beings can control this automatic flow during "control flow" channels/interfaces.

CONCLUSION

This chapter gave overview of testing, Testing and Test Control Notation version 3 (TTCN-3) and the Abstract Syntax Notation One (ASN.1). First the fundamentals and the basic definitions were discussed to give an overview of testing and abstract syntax notations. Then the main components of TTCN and the ASN.1 were presented, starting with the TTCN-3, such as core language, Tabular presentation Format (TFT) and Graphical presentation Format (GFT). Two parts of ASN.1 were represented, the basic notation and the Basic Encoding Rules (BER). Examples helped the understanding of the notations. At the end of this chapter, we gave some possible future trends, such as the possible usage of self-adaptive test solutions.

Figure 6. The framework of a tester conception

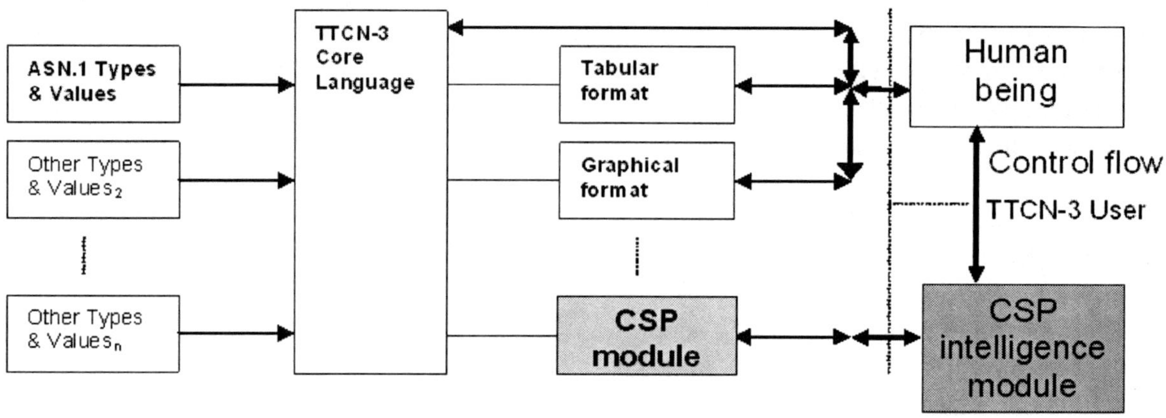

Standardized testing and the abstract syntax notation in the future can be combined with other standards. Better and more complex systems can be created if ASN.1 is used from the beginning of the development process and a standardized test tool is used to check the system's correctness.

REFERENCES

Baumgarten, B., & Giessler, A. (1994). *OSI conformance testing methodology and TTCN.* Elsevier Science Ltd.

Dubuisson, O. (2000). *ASN.1 communication between heterogeneous systems*. Morgen Kaufmann.

Hoare, C. A. R. (1985). *Communicating sequential processes*. Prentice Hall.

Hoffman, P., & Schaad, J. (2010). *New ASN.1 modules for PKIX.* IETF Internet-Draft.

Jaskó, S. (2005). A new mathematical formalism for the TTCN 3 core language. *Periodica Polytechnica Series . Electrical Engineering*, *49*(3), 1–12.

Kaliski, B. S., Jr. (1993). *A layman's guide to a subset of ASN.1, BER, and DER.* (NIST/OSI Implementors' Workshop document SEC-SIG-91-17).

Larmouth, J. (1999). *ASN.1 complete*. Morgen Kaufmann.

Linn, R. J., & Umit Uyar, M. (1994). *Conformance testing methodologies and architecture for OSI protocols* (pp. 93–152). IEEE Computer Society Press.

Schaad, J., & Turner, S. (2010). *Additional new ASN.1 modules*. IETF Internet-Draft.

Wallace, C., & Gardiner, C. (2010). *ASN.1 translation*. IETF Internet-Draft.

White, J. (1989). ASN.1 and ROS: The impact of X.400 on OSI. *IEEE Journal on Selected Areas in Communications*, *7*(7). doi:10.1109/49.44554

Willcock, C., Deiss, T., Tobias, S., Keil, S., Engler, F., & Schulz, S. (2005). *An introduction to TTCN-3*. Wiley. doi:10.1002/0470017317

Zander, J., Dai, Z. R., Schieferdecker, I., & Din, G. (2005). *From U2TP models to executable tests with TTCN-3 - an approach to model driven testing. Lecture Notes in Computer Science, 3502.* Berlin/Heidelberg, Germany: Springer.

ADDITIONAL READING

European Telecommunications Standards Institute. (1995). Methods for Testing and Specification; Protocol and Profile Conformance Testing Specification . *Standardization Methodology, ETS*, *300*, 406.

European Telecommunications Standards Institute (2007). Methods for Testing and Specification (MTS);The Testing and Test Control Notation version 3;Part 3: TTCN-3 Graphical presentation Format (GFT), ETSI ES 201 873-3 V3.2.1.

European Telecommunications Standards Institute (2007). Methods for Testing and Specification (MTS);The Testing and Test Control Notation version 3;Part 2: TTCN-3 Tabular presentation Format (TFT), ES 201 873-2 V3.2.1.

International Telecommunication Union (2004). Message Sequence Chart (MSC), Z.120-Z.129.

International Telecommunication Union (2004). Message Sequence Chart (MSC), Z.120.

International Telecommunication Union (2007). Specification and Description Language (SDL), Z.100-Z.109.

International Telecommunication Union (2007). Testing and Test Control Notation (TTCN), Z.160-Z.179.

International Telecommunication Union (2007). Testing and Test Control Notation version 3: TTCN-3 tabular presentation format (TFT), Z.162.

International Telecommunication Union (2007). Testing and Test Control Notation version 3: TTCN-3 graphical presentation format (GFT), Z.163.

International Telecommunication Union (2008). Abstract Syntax Notation One (ASN.1), X.680-X.699.

Internet Engineering Task Force (1981). Transmission Control Protocol (TCP), RFC-793

KEY TERMS AND DEFINITIONS

Abstract Syntax Notation: It is a flexible notation that describes data structures for representing, encoding, transmitting, and decoding data.

Conformance Testing: Conformance testing or type testing is testing to determine whether a product or system meets some requirements of a standard or definition that has been developed for interoperability.

Core Language of TTCN-3: It is an object oriented programming language that is created to control the test processes.

Graphical Presentation Format (GFT): It is a presentation format of TTCN-3 that is based on graphical elements. The core language of TTCN-3 can be programmed with its help.

Implementation Under Test (IUT): It refers to an implementation that is being tested for correct operation.

Protocol Data Unit (PDU): It is delivered as a unit between peer entities of a node and may contain control information, address information, or data.

Service Access Point (SAP): It provides the interface between layers. SP can cross the SAP.

Service Primitive (SP): It is delivered as a unit between neighboring layers and may contain control information, address information, or data.

System Under Test (SUT): It refers to a system that is being tested for correct operation.

Tabular Presentation Format (TFT): It is a presentation format of TTCN-3 that is based on tabular elements. The core language of the TTCN can be programmed with its help.

APPENDIX

Table of abbreviations

ASN.1	Abstract Syntax Notation One
BER	Basic Encoding Rules
CER	Canonical Encoding Rules
CMS	Cryptographic Message Syntax
CSP	Communicating Sequential Processes
DER	Distinguished Encoding Rules
ECN	Encoding Control Notation
ETSI	European Telecommunications Standards Institute
GFT	Graphical presentation Format
GRL	Goal-oriented Requirements Language
ITU-T	International Telecommunication Union Telecommunication Standardization Sector
IUT	Implementation Under Test
LOTOS	Language Of Temporal Ordering Specification
MHS	Message Handling System
MSC	Message Sequence Charts
MTC	Master Test Component
ODMI	Open Document Management API
OSI	Open Systems Interconnection
PCO	Point of Control and Observation
PDU	Protocol Data Unit
PER	Packed Encoding Rules
PTC	Parallel Test Component
SAP	Service Access Point
SDL	Specification and Description Language
SP	Service Primitive
SUT	System Under Test
TCP	Transmission Control Protocol
TFT	Tabular presentation Format
TTCN	Testing and Test Control Notation
TTCN	Tree and Tabular Combined Notation
UML	Unified Modeling Language
UTF-8	8-bit Unicode Transformation Format
XER	XML Encoding Rules

Section 2
Telecommunications Protocols

Chapter 5
Signaling Protocols of Integrated Services Digital Networks

Gusztáv Adamis
Budapest University of Technology and Economics, Hungary

ABSTRACT

This chapter is devoted to the classical ISDN signaling protocols. The main reason for which we deal with these protocols in detail is that they are – with modifications and extensions – also used in GSM and UMTS systems. In this chapter, we discuss the access (subscriber) and the core network signaling protocol stacks of ISDN: (a)The DSS1 (Digital Subscriber Signaling System No.1): We do not deal with the physical layer, but we discuss the upper two layers: LAPD (Link Access Procedure on D channel) and DSS1 3rd layer. (b) The SS7[1] (Signaling System No.7): Used in core network for signaling. We introduce the structure and the functional elements of SS7, and we go through the four levels of the protocol stack.

INTRODUCTION

In this chapter, we discuss the signaling systems which are used primarily in the ISDN: DSS1 is the signaling system used at the ISDN subscriber access points (at the user-network interfaces) while the SS7 in the ISDN core network.

DOI: 10.4018/978-1-60960-732-6.ch005

At the beginning we present the DSS1. We discuss in detail the LAPD, which is used to organize the exchange of signaling information between subscriber terminals and a local exchange as well as the DSS1 3rd layer which is used for call control between the network and the terminals.

After the discussion of the ISDN access protocols we turn our attention to the core network. We introduce to the reader the SS7 which is developed by the ITU-T as the signaling system of the international network, but the SS7 is used in general in the national networks as well. We show the functions of the Message Transfer Part and the ISDN User Part in detail. By the end of this chapter, the reader will be familiar with the ways in which the signals are routed in the networks and in which the network is managed, as well as with the typical signaling sequence used to establish a call in the core network.

But we have to clear some basic notions before we come to the details.

ISDN terminology uses the term "channel" to refer to the medium that carries information typically in a bidirectional way: A B (Bearer) channel carries the bidirectional voice signals of a call or bidirectional user data e.g. of an Internet connection. D (Data) channels are used mainly for bidirectional signaling purposes to control the connections of the B channels, but they can carry bidirectional low-speed user data as well. However actually a channel is a uni-directional structure. Circuits are defined as composed of two channels in opposing directions. So it is a misnomer to talk about "B channels" and "D channels", they should really be called as "B circuits" and "D circuits". But there is the "tradition" of talking of bidirectional channels, which is formally incorrect. Since this incorrect linguistic form does function within the telecommunications community, we will also use it later on.

In the telecommunications systems the signaling information can either be carried over the same circuits (in the same channels) as the voice or in (at least logically, but sometimes physically separated) dedicated signaling channels. In the latter case typically one signaling channel satisfies the signaling need of several voice/data channels, so this is why these systems are called common channel signaling systems (CCSS), while the others are the channel associated signaling systems (CAS). In ISDN only CCSS systems are used. In the core network the SS7 is obviously a common channel system (it is often referred to as Common Channel Signaling System 7), but – though it is not as obvious – the DSS1 at the user-network interface is also a CCSS system, since a D channel is common for several B channels. CAS systems are used e.g. in the analogue subscriber loops.

Digital Subscriber Signaling System No.1

The physical, logical, and procedural elements of communication on the user-network interface of the ISDN are jointly described as an access type. Two such types have been standardized: BRA (Basic Rate Access type) and PRA (Primary Rate Access type). Both operate on a number of bidirectional B channels transmitting data at the rate of 64 kbps in each direction, and a single, common bidirectional D channel. BRA operates on two B channels and a D-channel with the capacity of 16 kbps, while PRA operates on 30 (or in some countries 23) B-channels, with a D channel of 64 kbps. The representation of these channels at the level of the transmitted signal is the subject of the physical layer (L1) of DSS1, and, of course, is different for BRA and PRA accesses. This issue is, however, beyond the scope of this book (the reader is referred to (ITU-T, 1995) for BRA and (ITU-T, 1993/3) for PRA). The higher layers: L2 and L3 are applicable to both, BRA and PRA (with only minimal parameterization), and will be discussed jointly, with BRA as their application example.

Link Access Procedure on D Channel

The 2nd – commonly referred as data link – layer of the ISDN user-network connection is called a Link Access Procedure on D channel (LAPD) (ITU-T, 1997) and provides error free transmission of signaling messages between the exchange and the terminal(s) through the D channel of a Digital Subscriber Line (DSL). In this subsection we summarize the different signaling frame types of LAPD in details to show in a case of a relatively simple protocol, how standards define the messages, then we introduce to the reader the LAPD management procedure (TEI management).

General Frame Structure of LAPD

The signaling messages (frames) that are exchanged between a terminal and the network contain an integral number of octets and have the following format (Figure 1).

A frame starts and ends by a special bit pattern (01111110) called a flag. The opening flag is followed by a two-octet address field. After the address field a one or two octet long control field appears. In front of the closing flag, there is a two-octet Frame Check Sequence (FCS). In data transfer frames (as opposed to frames used purely for control), between the control field and the FCS an information field is placed. In general, the information field carries the message from the upper layer.

Flags and Bit-Stuffing

The flags are used to denote the beginning and the end of a frame. Therefore, the occurrence of the 0111110 bit sequence must be avoided in the body of the frame. Since we want to have a transparent transfer, it is not as easy as to simply exclude the transmission of this bit sequence. As the flag contains a sequence of six consecutive 1 bits, in the body of the frame the transmitter will insert a 0 bit after each sequence of *five* consecutive 1 bits. This technique is commonly referred as bit stuffing[2]. The receiver has to remove these extra 0 bits. Thus, the receiver counts the number of consecutively received 1 bits. After receiving five consecutive 1 bits, the receiver stops for a split of a second and checks the next bit. If it is a 0, it must be a stuffed bit, which is not a part of the message; therefore, it will be thrown away. If it is a 1, then a flag[3] has been found.

Figure 1. LAPD frame structure

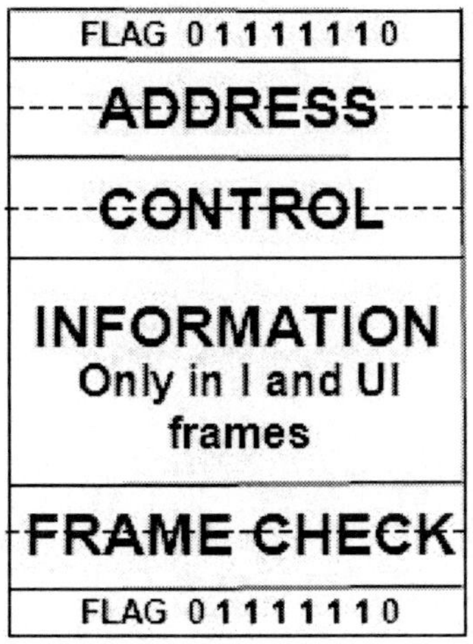

Address Field

The address field of a LAPD frame consists of two octets (Figure 2). The least significant bit is called address field extension bit (EA). It is encoded as 0 in the first octet and 1 in the second.[4]

The second bit of the first octet is the Command/Response (C/R) bit. The C/R bit shows whether a frame is a command or a response. The network side sends commands with C/R=1 and responses with C/R=0, while in the commands sent by a terminal the C/R=0 and in the responses the C/R=1.

The most significant six bits of the first octet of the address field is a Service Access Point Identifier (SAPI), which indicates the logical aim, or reason, for exchanging this frame. The SAPI identifies the point at which the services of the

Figure 2. LAPD Address field

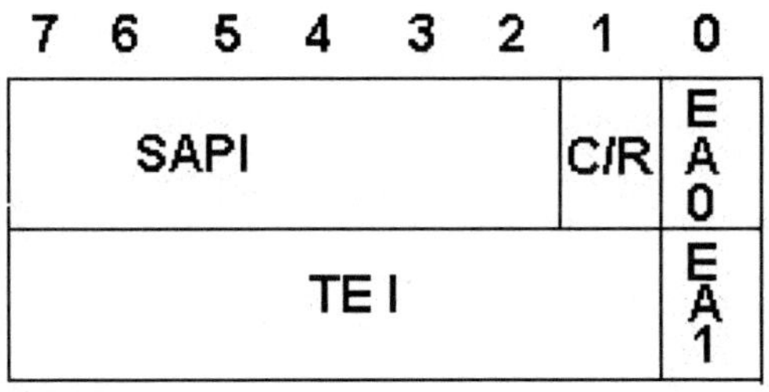

LAPD are provided and consequently it specifies the entity that should process the received LAPD frame. In LAPD, four SAPI values are defined: 0 for call control procedures (that is for the 3rd layer of DSS1), 12 for teleaction services, 16 for user packet communications, and 63 for management purposes[5].

It is interesting that already more than 20 years ago ISDN provided mechanisms for Internet access (SAPI=16), which later developed into the AO/DI (Always On/Dynamic ISDN) service. Since the D channel is always active, there can be a continuous connection between an ISDN customer and an information service provider. AO/DI uses 9.6 kbps of the 16 kbps D channel and switches to a circuit-switched B channel when wider bandwidth is required. It enables low-bandwidth traffic, such as e-mail, newsfeeds, credit card verifications, and automatic data collection, to be handled quickly and cost effectively. SAPI=12 can be used in a similar way, but for teleaction services, such as global home automation: security, climate control, etc.

The most significant seven bits of the second octet is the Terminal Endpoint Identifier (TEI). The TEI typically identifies a Terminal Equipment (TE), though a TE may have more than one TEIs, e.g. if it is a combined telephone and fax device. The TEI range of 0-63 is used for the non-automatic, and the range of 64-126 for the automatic TEI assignment user equipments. In the first case, the TEI is selected and set by the user, while in the second it is assigned by the network (exchange). The TEI=127 is dedicated for to indicate the broadcast messages that have to be received and processed by every TE.

Control Field Formats

The control field identifies the type of the frame. Three different frame types are specified:

- U (Unnumbered) frames that are mainly applied for to control the LAPD connection establishment or release and to carry unacknowledged information (e.g. TEI management) messages,
- I (Information) frames that are used to carry call control (3rd layer of DSS1) messages,
- S (Supervisory) frames, which serve to control the flow of I frame transmission.

The control field structure of the three frame types is different. Each frame type contains a Poll/Final (P/F) bit, but at a different position. In commands, P/F bit is referred to as P (poll) bit and in the case of responses as F (final) bit. The P bit is set to 1, if the command must be answered, and the F bit is set to 1 to indicate that the frame is transmitted as a reply for a poll command.

Control Field of a U Frame

The control field of a U frame consists of one octet (Figure 3). The least significant two bits are coded as 11 – this is the type code of the U frame. The 5th bit is the P/F bit, while the rest five bits are called as modifier (M) bits and dedicated to define the (sub)type of the U frame (Table 1).

Establishing of an acknowledged LAPD connection can be requested by sending an SABME[6] frame. If the connection can be established the answer is UA, otherwise it is DM. As soon as the connection is established, I frames can be sent.

Figure 3. Control field of a U frame

M	M	M	P/F	M	M	1	1

Table 1. LAPD U frame types

MMMMM	Acronym	Name	Meaning
01111	SABME	Set Asynchronous Balanced Mode Extended	Request to establish an acknowledged LAPD connection
01100	UA	Unnumbered Acknowledgement	Acknowledgement for SABME or DISC frames
01000	DISC	Disconnect	Request to disconnect a LAPD connection
00011	DM	Disconnected Mode	Negative acknowledgement for SABME: unable to establish an acknowledged LAPD connection
000000	UI	Unnumbered Information	Unacknowledged information transfer

To release the LAPD connection a DISC frame is to be sent, which must be acknowledged by an UA frame.

The only U frame possessing an Information field is the UI. It is applied to carry unacknowledged information. In LAPD, the discussed frame type is used for TEI management, but in a point-to-multipoint configuration the DSS1 3rd layer *Setup* message, if it is sent by the network to a terminal is also transmitted in a UI frame.

Control Field of an I Frame

The control field of an I frame consists of two octets (Figure 4). The least significant bit of the first octet is coded as 0 – this is the type code of the I frame, while in the second octet the P/F bit can be found at this position. The control field contains two seven-bit sequence numbers: in the first octet the sent number – N(S), and in the second one the received number – N(R).

N(S) is the sequence number modulo 128 of the transmitted frame, while N(R) is the sequence number modulo 128 of the next expected message. N(R) also serves as a positive acknowledgement for the frame N(R)-1 and all the previous.

Obviously the control field of an I frame is followed by an Information field carrying a DSS1 3rd layer (call control) message.

Control Field of an S Frame

The control field of an S frame consists of two octets (Figure 5). The least significant two bits of the first octet are coded as 01 – this is the type code of the S frame, while in the second octet the P/F bit stands for the least significant position.

Figure 4. Control field of an I frame

N (S)	0
N (R)	P/F

Figure 5. Control field of an S frame

0	0	0	0	S	S	0	1
N (R)							P/F

The most significant four bits of the first octet are not used (coded as 0000), the next two bits serve for to define the (sub)type of the S frame. (Table 2)

TEI Management Processes

In a BRA point-to-multipoint configuration (which is the typical configuration for the residential subscribers), a TEI must be assigned to a TE before it is used. In general, the TEI is assigned when a TE is connected to a DSL, although in some cases a terminal does not immediately request its TEI upon being plugged it. It may well remain "silent" and "invisible" until its user tries to initiate an outgoing call or an incoming call is to be handled. Then it quickly (and transparently to its user) requests its TEI value.

There are four processes belonging to TEI management:

- TEI assignment,
- TEI check,
- TEI removal
- TEI identity verification.

The messages used by the TEI management processes are UI frames, in which the SAPI is set to 63 (LAPD management) and TEI is set to 127 (broadcast) (Figure 6).

The structure of the Information part of the TEI management messages is the following:

- one-octet protocol identifier ($0F_{hex}$ – TEI management),
- two-octet reference number (RN). RN is used as transaction identifier in the messages,
- one-octet message type code (MT),
- one octet containing a seven-bit action indicator (A_i) plus a bit, which is encoded as 1. The A_i contains either the managed TEI or 127 that refers to all TEI values.

The TEI assignment process is always initiated by the TE by transmission of a Request message. The message contains a 16-bit random number (RAND) as the reference number of the request. In the case of a non-automatic TEI assignment device, the A_i contains the requested TEI value, while in the case of an automatic TE, this value is set to 127 indicating that any TEI value is accepted.

If the network side can satisfy the request, an Assigned message is sent as an answer. It contains the RAND copied from the request and the assigned TEI value. If the requested TEI is in use (in case of a non-automatic TEI assignment terminal) or there are no available TEI (in automatic cases) the answer will be a Denied message. In this message, the values of RN and A_i are the same as they were in the request.

Table 2. LAPD S frame types

SS	Acronym	Name	Meaning
00	RR	Receive Ready	Acknowledgement for an I frame OR Indication that the station is able to receive I frames again (when previously an RNR was transmitted)
01	RNR	Receive Not Ready	Indication that the station is temporarily unable to receive I frames (e.g. because the input queue is full)
10	REJ	Reject	Request for retransmission of all I frames from N(R)

Figure 6. TEI management message structure

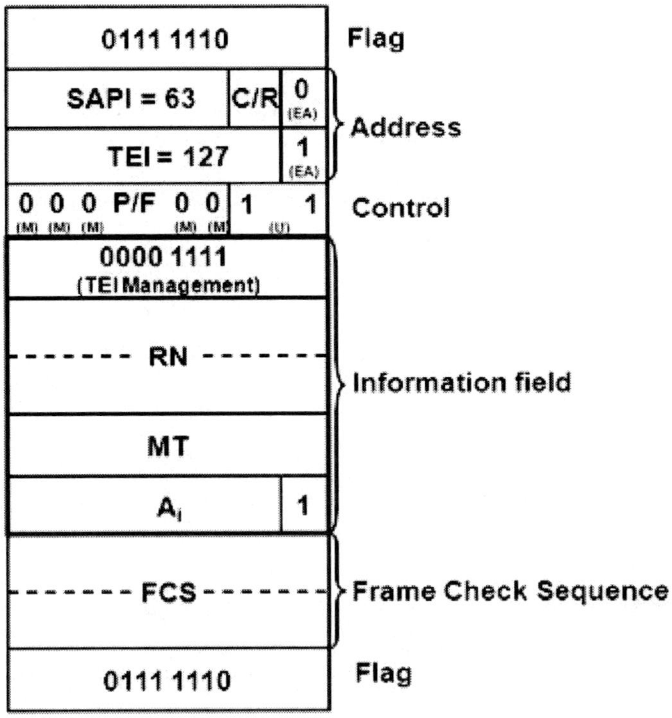

If an assignment request cannot be satisfied, the network side starts a TEI checking process by sending a Check Request message, which contains either the TEI value to be checked or the 127. In the latter case, the network aims to check all TEI values in use. If a terminal finds a TE suspicious since the TEI value assigned to that terminal is used by any other TE(s), it initiates a check process by sending a Verify message to the network. (The duplicate assignment of a TEI value may happen if two TEs are connected to the DSL roughly at the same time and they both generate the same RAND value in the Assignment request.)

If a TEI value is not used or used by more than one TEs, it will be removed by the network. In this case, the network sends a Remove message (twice to overcome the possible loss of the message). The Remove message contains the TEI value to be removed or 127 in the A_i field. In the latter case, all TEI values are to be withdrawn.

DSS1 3rd Layer: Call Control

The DSS1 3rd layer (ITU-T, 1998) is responsible for call control between an ISDN terminal and an ISDN exchange. In this subsection, we first explain the message structure used for call control then we introduce the typical process of a call set-up and release.

A DSS1 3rd layer message is carried in the Information field of a LAPD I (or, in certain well-defined circumstances, UI) frame. All messages start by a one-octet protocol identifier (08 for DSS1), which is followed by a Call Reference field. The Call Reference allows various signaling activities for multiple connections (calls) at the same user-network interface.

After the Call Reference, the message type code can be found. Depending on the type of the message and on the situation it is used in, a message must or may contain none, one or several additional Information Elements (IE).

In Figure 7, the typical sequence of messages used for a successful call establishment and release can be followed. The dashed lines indicate the layer 2 (LAPD) messages, while the continuous lines appear for the layer 3 messages.

When we want to make a call, a signaling connection must be established by a SABME/UA message exchange first. After that, call control messages can be sent in the Information field of I frames as discussed earlier in this chapter.

The first message of the call establishment sequence is a Setup message. This message contains a Bearer Capability (BC) information element that describes the desired bearer service characteristics of the connection. BC includes indicators, such as whether a connection is circuit or packet-switched, permanent or not, uni- or bidirectional furthermore the information transfer rate, modem type and so on. Setup also may present the Calling and Called Party's Addresses (telephone numbers).

In case the Setup message does not contain the Called Party's Address, the network sends a Setup Acknowledge message. As soon as the TE receives the message, it can start to transmit the dialed digits of the called number one-by-one in Information messages.

As the complete called number arrives (either in Information messages or in the Setup message), the network acknowledges it by a Call Proceeding message.

If the Setup message does not contain the called number, then the identifier of the voice channel is given in the Setup Acknowledge, otherwise this identifier is contained in the Call Proceeding message.

By analyzing the called number, the exchange determines who is to be the called party and starts to establish the voice path to it. In our example the called party belongs to the same exchange (local call) so the exchange starts to establish the connection at the called side as well.

The first message is a broadcast Setup message, which is transmitted in a UI frame. Since more than one TE can be connected to a DSL, first it has to be chosen the one, which will complete the call. That is why a broadcast message is sent, and the TE that recognizes its telephone number in the Setup message and is able to receive the call of the type indicated in the Bearer Capability parameter, will establish the layer 2 signaling connection by transmission of a SABME message. Network side acknowledges the SABME by a UA – and then the TE acknowledges the call attempt by a Call Proceeding message[7].

If the user of the TE is informed about the call (most typically when ringing starts), the TE sends an Alerting signal. Upon receiving it, the exchange connects the ringing tone to the B channel at the calling side and transmits the Alerting signal in the D channel.

When the call is answered, a Connect/Connect Acknowledgement message exchange takes place at both sides, and the B channel connection (the call) is established.

The release process can be initiated either by the calling or by the called party. (We have illustrated the first option.) The release process starts by a Disconnect message, which is forwarded to the called party. The Disconnect message contains a Cause information element that specifies the reason of the disconnection.

To make the release process faster, the continuation takes place in parallel at the two ends of the call, so the relative order of the rest of the messages can be arbitrary.

The Disconnect message is acknowledged by both ends, first by a Release message after that by a Release Complete message.

When the call is cleared and there is no further activity, the LAPD signaling connection is released by a DISC/UA message exchange.

In Figure 7, we have illustrated the steps of a successful call set-up. If the call cannot be estab-

Figure 7. DSS1 successful call set-up and release

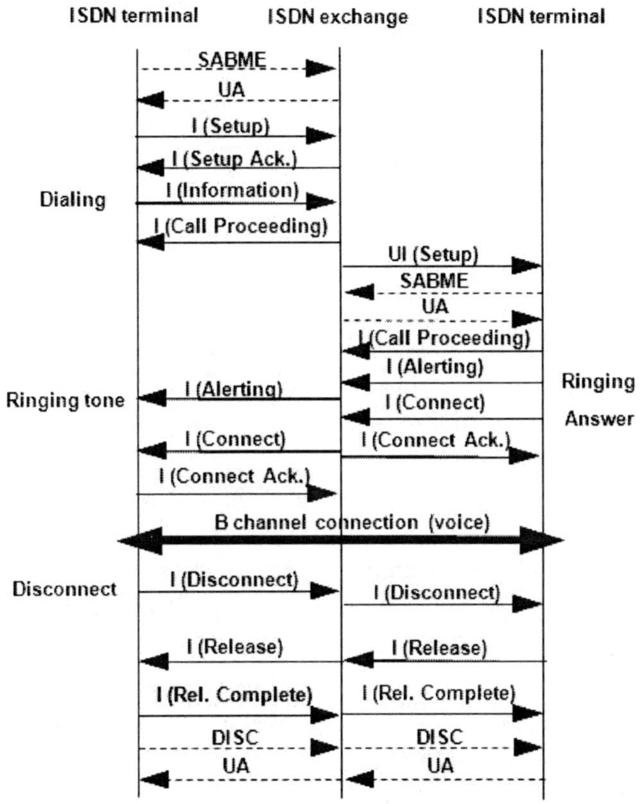

lished (e.g. the called party is busy, wrong number has been dialed, interworking problem arises, etc.) a disconnect procedure starts. In this case, the Cause information element in the Disconnect message carries the reason of the problem.

SIGNALING SYSTEM 7

Before the introduction of the ISDN, the core network of the telephone systems used channel associated signaling (CAS). In these systems the same lines were used both, for signaling and voice transmission. The former systems had several technical limitations (they were analogue, slow and hardly expandable) so a new signaling system had to be developed for the ISDN. The following requirements were formulated for the new signaling system: it should be digital, should support the signaling needs of voice and data calls and the ISDN supplementary services, should be open for controlling new, not yet existing services, should be easily expandable. However, apart from the fact that they have practical limitations, the CAS systems also have several theoretical limitations:

- the processing of a signal takes a relatively long time, while the voice line is idle,
- since in the CAS systems the signals and voice are coupled, it is not possible to transmit *non call-related* signals, in other words it is impossible to signal without a call.

Thus the idea arose to separate the voice and the signaling, which is the basics of the common channel signaling concept (Figure 8). Here all signaling is carried out over a separate network that is logically independent from the voice trunk network. This arrangement has several advantages. The most important ones are:

- the voice and signaling network may evolve independently (reliability, complexity, capacity),
- sophisticated control protocols may be used to control complex services and make it easier to introduce new services,
- not only exchanges, but also other functional elements can be connected to the signaling network, as controlling databases, network monitoring and maintenance systems, etc.
- it is possible to carry out non-call related signaling.

Since the processing time of a signal is typically larger by magnitudes than the transmission time, one channel in the signaling network can serve the signaling requirements of several voice circuits. The name of the concept came from this idea: a signaling channel is common for several voice circuits and several services.

Common channel signaling also has several disadvantages:

- a separate sub-network must be established and maintained,
- because of the signals travel in a different path than the user information (e.g. voice), sometimes it must explicitly be checked whether the user connection has been really established (call continuity control),
- since the signaling messages for different user connections are interleaved on a common signaling channel, each signal must contain an explicit identifier of its corresponding user connection.

The CCITT (now ITU-T) has specified the Signaling System 7 (SS7) as the signaling system of the international network, which is used also in many countries and by many service operators. (ITU-T, 1993)

Figure 8. Common channel signaling system concept

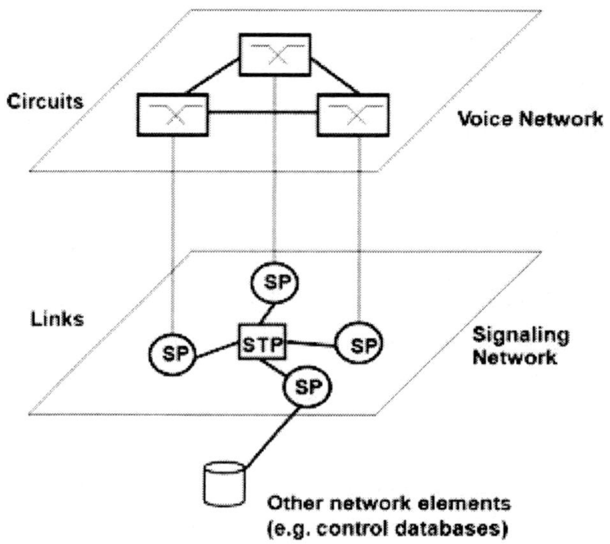

Structure of the SS7 Network

The nodes of the SS7 network are called Signaling Points (SPs). An SP can either have a Signaling End Point[8] function (SEP) or a Signaling Transfer Point (STP) function, or both. Through SEPs the telecommunication network elements (exchanges, control databases, etc.) can originate and receive signaling messages, while STPs are used to transfer signaling messages from one signaling point to another. Every exchange has a SEP by definition, but not every SEP locates in an exchange (there may be data nodes, such as Service Control Points of Intelligent Networks, etc.). A SEP is not automatically a STP (the message routing function may be missing) and there may be "stand-alone" STPs having no SEP function (Figure 8). The interconnections between the signaling points are referred to signaling links. SPs that are directly interconnected are called as adjacent signaling points. Typically, between two adjacent SPs several signaling links are used in parallel. These links are called a link set. A path that a message takes through the signaling network between the originating and destination points is called as a signaling route.

If between any two signaling end points, there exists the possibility of "direct" communication, it is said they have a signaling relation. In ISDN, if two exchanges are interconnected by voice trunks, signaling between them is needed, so they have a signaling relation. If there is a direct signaling link between these SEPs then we call it an associated mode of signaling, and if there is just an indirect signaling connection (containing one or several STPs) between the two SEPs of a signaling relation, it is called quasi-associated signaling mode. Even in the latter case the signaling network must guarantee that the signals of a given call or transaction must travel on the same path to keep the order of the messages. It is important to mention that, for the "normal" operation of the ISDN network, and for almost every supplementary service only the exchanges directly interconnected by voice circuits have a signaling relation and only this scenario is handled "by default"(without the need of the optional SCCP protocol – see later) by an SS7 network.

Figure 9. Recommended sub-network

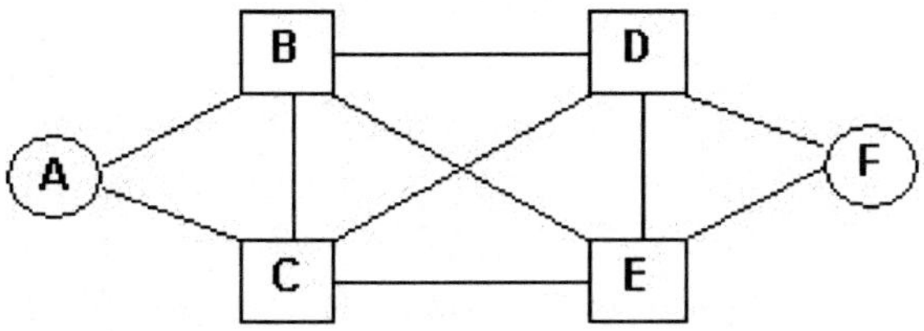

The signaling network must be reliable. Thus it is recommended to establish several (at least two) signaling paths between SEPs of a signaling relation. Where the reliability is critical (in the international network and in the national networks where the traffic is heavy), it is advisable to have a mesh network (Figure 9) between SEPs. As we can see, there are at least two outgoing signaling paths in every SP. The AB and AC, BD and BE etc. are called as alternative link sets, because under normal circumstances they work in a 50% load sharing. If one of them fails, the other will take its traffic over. Under normal circumstances, the vertical links (BC and DE) carry no traffic in A-F relation; they can only be used if there is no other alternative. (E.g. BC can only be used if both BD and BE fail.) It does not mean that the vertical lines are idle, since they can be parts of other relations, too.

To prevent deadlocks in the signaling network a message must not be sent backward. E.g. if a signal sent by A to F reaches STP D it may be sent to E (unless if the signal reached D via ED) or to F and must not be transmitted neither to B nor to C.

SS7 Protocol Architecture

The architectural overview of the SS7 protocols is shown in Figure 10. Similarly, to the well-known

Figure 10. SS7 protocol architecture

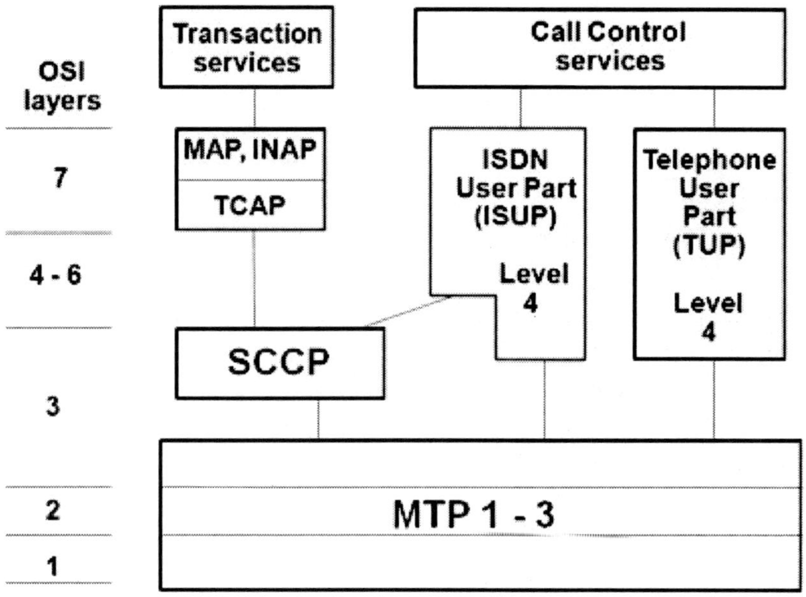

ISO OSI Reference Model, SS7 also has layered protocol architecture, but the number of the layers and the functionality of them are different from those of OSI Reference Model, so here we use the term "level" instead of "layer".

The lowest three levels of SS7, taken jointly, are called as Message Transfer Part (MTP). MTP is responsible for error-free transmission of signaling messages between any pair of originating and destination signaling points having a signaling relation of the same signaling network.

The first (the lowest) level describes the characteristics of a signaling data link. The second one provides reliable transfer of signals between two adjacent signaling points over a signaling data link. The third level deals with the message routing between user parts of any two signaling points of a signaling network, and provides the management of the signaling network. In cases of signaling network failures, it performs the reconfiguration of the signaling network to restore the signal transfer capability. These functions of MTP-3 are called, respectively, Signaling Message Handling (SMH) and Signaling Network Management (SNM).

The fourth level consists of the different users (User Parts – UPs) of the MTP.

The Telephone User Part (TUP) is responsible for controlling telephone calls, but it is currently of only historical interest. For controlling the ISDN voice and data calls and the supplementary services, the ISDN User Part (ISUP) is used. The ISUP replaces the TUP where it has been implemented.

The Signaling Connection Control Part (SCCP) extends the addressing capability of the MTP. By the help of the SCCP, it is possible to transmit signals between signaling points of *different* networks. There is no need for this kind of feature in the voice path establishment in ISDN networks, but it is often needed in mobile networks (e.g. when we want to send a short message or in the mobility management). SCCP provides both, connectionless and connection-oriented services. Connectionless services are used e.g. in controlling data bases queries in mobile networks, while the connection-oriented service is used e.g. to tunnel the subscriber protocol messages between the base stations and mobile switches in GSM networks.

The connectionless service of the SCCP is suitable for transmission of messages in a transparent way only. Therefore, another protocol must be used for the transaction management. The Transaction Capabilities Application Part (TCAP) controls the transactions and the operations within a transaction (e.g. to match the request and the reply of a data base query).

The TCAP users determine when and what kind of transactions are needed to invoke. Mobile Application Part (MAP), Intelligent Network Application Part (INAP), and CAMEL Application Part (CAP) are examples for TCAP user protocols.

In the rest of this chapter, we discuss the ISDN-related SS7 protocols (MTP and ISUP) in detail. The SCCP, TCAP, and MAP protocols will be explained in detail in the next chapter of this book, entitled Mobile Network Protocols of GSM and GPRS.

Message Transfer Part

The Message Transfer Part (ITU-T, 1993) is the tool for signal transmission between user parts. This subsection gives a detailed account of the three levels of the MTP.

MTP-1 Signaling Data Link Level

MTP-1 (ITU-T, 1988) provides the bearer for a signaling link, so it is responsible for the bit transfer. MTP-1 defines the functional, electrical, and physical characteristics of a signaling data link and the ways to access it. The SS7 is optimized for a full-duplex non-restricted transparent digital channel with the bit rate of 64 kbps. However, there are several other – even analogue – media types, which are allowed to be used as a signaling data link, although they are not used in practice.

MTP-2 Signaling Link Level

MTP-2 (ITU-T, 1996) is responsible for the error-free transmission of signaling messages between adjacent signaling points. Therefore, MTP-2 provides the framing, error detection and correction functions.

A structure of a signal unit can be seen in Figure 11.

A signal unit begins and ends with flags. Normally the opening flag of a signal unit is the closing flag of the preceding one. The bit pattern of a flag is 01111110. To avoid the appearance of this bit pattern inside the message, the same bit-stuffing method is used as we have discussed at LAPD (in the body of a signal a bit 0 is inserted after each sequence of five consecutive 1 bits).

In front of the closing flag, the signal unit contains 16 check bits (CK). If the receiver detects a signal with a wrong checksum (i.e., if the locally calculated checksum does not match the CK), it discards the signal.

Forward Sequence Number (FSN) is the sequence number modulo 128 of the transmitted Message Signal Unit (MSU).

Backward Sequence Number (BSN) is the sequence number modulo 128 of the last correctly received Message Signal Unit. The BSN serves as a positive acknowledgement for the corresponding and all the previous messages.

Figure 11. Structure of a signal unit

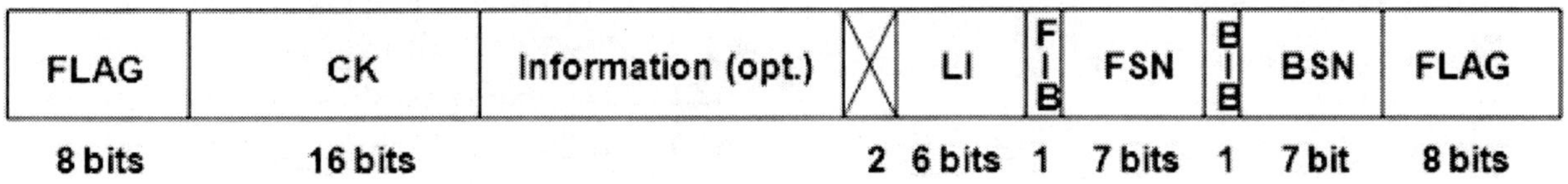

The Backward and Forward Indicator Bits (BIB and FIB, respectively) are used in the Basic Error Correction method to indicate and to correct an error. If the receiver detects a wrong MSU, it inverts the value of the BIB to request the repetition the transmitted messages, while the transmitter inverts the value of the FIB to indicate the beginning of the retransmission.

The six-bit Length Indicator (LI) shows the length of the Information field in octets if it is not longer than 62 octets. If the information field is longer, the LI is set to 63. However, the Length Indicator is actually used as a type code.

In the SS7, three different signal unit types are defined.

The Message Signal Unit (MSU) is the "useful" signal unit that carries a message from a user part or from the MTP management in the Information field. In this case, the length of the Information field is minimum six octets in practice[9].

The Link Status Signal Unit (LSSU) is used to indicate the status of the signaling link. The Information field of an LSSU is one octet long in practice[10]. The LSSUs are used to indicate when the state of the signaling link changes (e.g. when the link fails or if it is fixed after a failure) to make it possible for the Signaling Network Management function of the MTP-3 to reconfigure the signaling network if necessary.

The Fill-In Signal Unit (FISU) has no information field. In SS7, synchronous transmission is used: signals must be sent continuously. If no MSU or LSSU is to be sent, FISUs are transmitted.

As we could see above, the Length Indicator can make a difference between the three types of signal units as follows:

- FISU: Length Indicator = 0
- LSSU: Length Indicator = 1 or 2
- MSU: Length indicator > 2.

Error Correction Methods in MTP-2

There are two different error correction methods defined in MTP-2.

In the Basic Error Correction (BEC) method, the receiver can ask the retransmission of a faulty signal by a negative acknowledgement. To indicate the negative acknowledgement and the beginning of the retransmission the BIB and FIB are used, respectively. Initially, the FIB and the BIB have the same value of 0.

The transmitted MSUs contain forward sequence numbers. The transmitted but not yet positively acknowledged MSUs retain in the outgoing transmission buffer for a possible retransmission. The receiving side positively acknowledges the acceptance of one or more MSUs by setting the backward sequence number of the next signal unit transmitted in the opposite direction to be equal to the forward sequence number of the last correctly received MSU.

If a wrong (out-of-sequence) MSU is received, a negative acknowledgement is to be sent: the value of the backward indicator bit of the next signal unit transmitted in the opposite direction is inverted. (This new value of the backward indicator bit is maintained in subsequently sent signal units until a new negative acknowledgement is to be sent.)

When the value of the backward indicator bit in the received message is not the same as the value of the forward indicator bit in the last transmitted MSU, the transmitter starts to retransmit the MSUs from the MSU having a forward sequence number value of one more (modulo 128) than the backward sequence number in the negative acknowledgement message. When the retransmission starts, the forward indicator bit is inverted, thus it becomes equal again to the value of the backward indicator bit of the received signal units. (This new value of the backward indicator bit is maintained in subsequently sent signal units until a new retransmission starts.) The process is illustrated in Figure 12. Please note that a negative acknowledgement is not send immediately when the faulty message (marked by X on the Figure 12) arrives, because if a transmission error occurs, it would result a wrong checksum, and the receiver discards messages with wrong checksum without any further procession. Therefore, the problem can only be detected, when the first "good", but out-of-sequence message arrives.

This method is effective if the propagation delay of the link is not too large, that is if the link is not too long. On long (e.g. satellite) links the Preventive Cyclic Retransmission method is used. In this method, the sequence of MSUs that have been transmitted once but have not yet been positively acknowledged is cyclically retransmitted. The retransmission process continues until a positive acknowledgement for an MSU is received. At this time, that MSU and all the previous ones are dropped from the sequence. In this method the indicator bits are not used, because the propagation delay of the link is so much that the retransmission of messages can be started *before* an accidental negative acknowledgement arrives.

MTP-3 Signaling Network Level

The Signaling Network level (ITU-T, 1996/2) consists of two functions:

- Signaling Message Handling (SMH) is responsible for transmission of messages between any two signaling end points of the same signaling network
- Signaling Network Management (SNM) is responsible for reconfiguration of the signaling network if it is necessary (e.g. after a link failure).

Signaling Message Handling

Signaling Message Handling routes the signals between any two signaling points of the same signaling network[11]. The routing of the signals is based on the Signaling Point Codes (SPC). The SPCs are the identifiers, the "addresses" of the SPs. The length of a SPC is 14 bits. There are three different SPC formats used in the different types

Figure 12. Basic error correction method

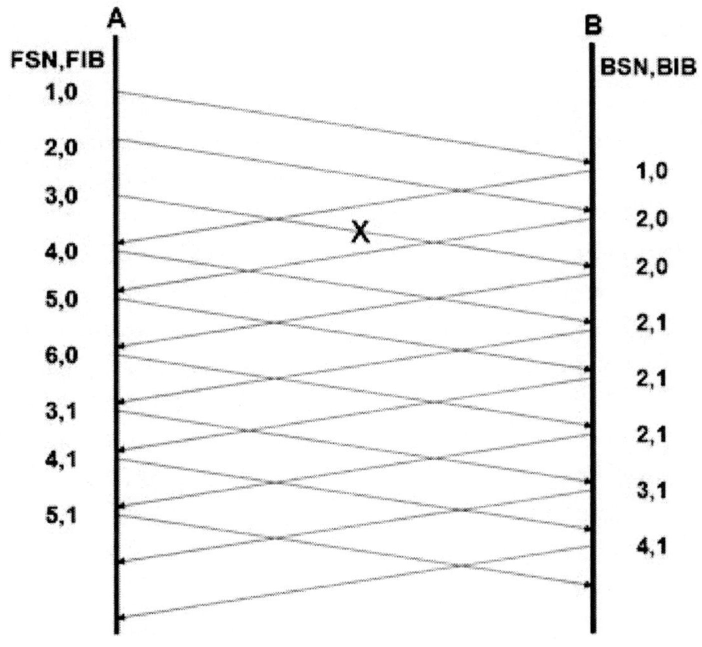

of signaling networks. The network of an operator is referred as to the national[12] signaling network. The network that interconnects the operators of a country is called the national interconnecting network, and the networks of the operators of different countries are interconnected by the international network. Because the signaling point codes have different structures in the different networks and the code, values are unique only within a given network and not globally, the gateway signaling points have as many SPCs as networks they interconnect, while the other SEPs have just one SPC.

The format of the SPCs used in the international network is specified by the ITU-T (ITU-T, 1999/5). The 14 bits of the International Signaling Point Codes (ISPC) are divided into three subfields. The upper 3+8 bits constitute the Signaling Area Network Code (SANC) while the lowest three bits identify SPs of a signaling area. The SANC codes are assigned by the ITU-T. Originally the upper 3 bits if an ISPC identified the zone (the continent), where the SP located, the next eight bits were the country (officially called as area) code while the rest three bits (the signaling point identifier) could differentiate among eight international gateways. This structure was suitable only when each country had just one operator, but as the telephony market opened for competition and new technologies – like GSM – were introduced, several new operators appeared in each country, therefore the eight signaling point identifiers became too few. Then the country could apply for a new area code. Sooner or later, there became continents where all area codes were assigned. Therefore, nowadays the SANCs have no internal structure; each country can have several different SANC values.

The format of the SPCs used in the national interconnecting network (National Interconnecting Signaling Point Code – NISPC) is specified by the NRA of the corresponding country. Most typically, an NISPC is divided into two parts. The first part identifies a network or service provider, while the second part indentifies the international gateway SPs of the given network or service provider.

There are no regulations for the format that can be used within the network of the different operators (National Signaling Point Code – NSPC), their internal structure and the assigned code values can be determined by the operators themselves.

Routing of Messages

The MTP-3 control field of an MSU contains two fields:

- a one-octet Service Information Octet (SIO), and
- a four-octet Routing Label (RL).

The routing label contains the Destination Point Code (DPC), the Originating Point Code (OPC) and a four-bit Signaling Link Selection field. DPC contains the address where the message is to be sent, while OPC contains the address of the sender. SLS is used to select one of the (possible) different paths between the originating and destination signaling points.

The SIO – apart from two not used bits – contains a Network Indicator (NI) and a Service Indicator (SI).

The Network Indicator determines the format of the point codes used in RL (00 – ISPC, 10 – NSPC, 11 – NISPC) and the Service Indicator identifies the user part, which the signal is to be delivered to (the most frequently used code values are: 0000 – SNM, 0011 – SCCP, 0101 – ISUP).

Let us examine the way in which the signaling points handle the signals (Figure 13). The DPC of an incoming signal is analyzed by the signaling message discrimination function in order to determine whether it is sent to the receiving signaling point. If the signal is destined to the receiving signaling point, the signaling message distribution analyzes the SI and delivers the signal to the corresponding user part or to the Signaling

Network Management function of the MTP-3. If the incoming signal is not destined to the receiving signaling point and the SP has the signaling transfer capability, the message is passed to the signaling routing function, which sends it on the appropriate outgoing link towards the destination point of the message. The signaling routing function chooses the appropriate outgoing link by analyzing the DPC and SLS fields. The signals, which are generated by the local user parts or by the local SNM, are treated by the signaling routing function in the same way.

Signaling Network Management

The Signaling Network Management functions provide procedures required to maintain signaling by a temporary reconfiguration of the signaling network in case of a disruption either in signaling links or at signaling points, and to restore the normal signaling conditions after the problem is fixed. The disruption may have the form of a complete loss of a signaling link or a signaling point, or of a reduced accessibility due to congestion. For example, in case of a link failure, the traffic of the faulty link should be diverted to an alternative link. The link failure may also result in unavailable signaling routes and this, in turn, may cause diversion of traffic at other signaling points in the signaling network (i.e. signaling points to which no faulty links are connected). (ITU-T, 1996/2)

A failure or a recovery from a failure results in a change of the status of the affected signaling link(s) and in certain cases the status of the route(s), too. A signaling link can be either *available* or *unavailable* to carry signaling traffic. An available signaling link becomes unavailable, if it is failed (a problem arose on the link itself, e.g. the error ratio has became too high, or even a complete loss of signaling capability has occurred), deactivated (when the operator takes it out of service), blocked (when not the link itself but its controlling processor has gone wrong) or inhibited (e.g. if a link should be tested, because it suffers too many failures in a short time period). An unavailable link becomes once again available if it is restored, activated, unblocked, or uninhibited, respectively. A signaling route and a signaling point may be considered available or unavailable.

The SNM contains three different signaling network management functions

- Signaling Traffic Management (STM),
- Signaling Link Management (SLM) and
- Signaling Route Management (SRM).

Figure 13. Signaling message handling

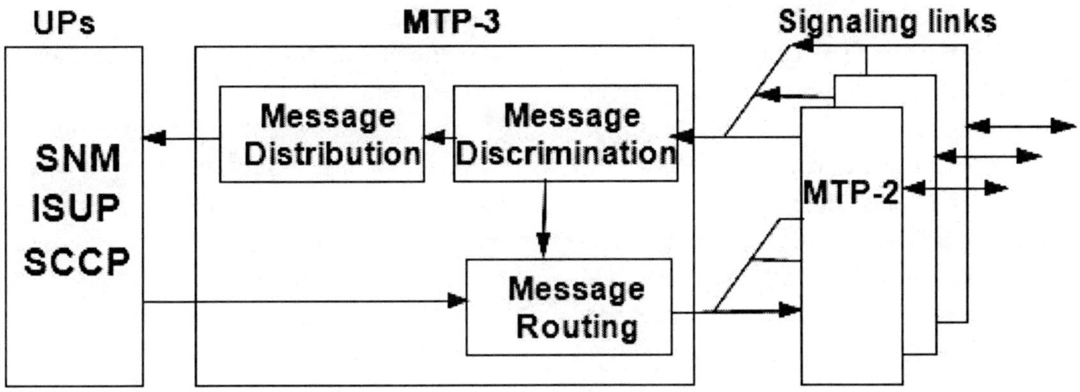

Signaling Traffic Management

Signaling Traffic Management is responsible for diverting the signaling traffic from an unavailable link or route, to reduce the quantity of the signaling traffic in case of congestion and to restore the original signaling conditions when a previously unavailable link or route becomes available again or when congestion ceases.

STM maintains tables describing the states of the links and the routes and collects the indicators of state changes on links and routes. A state change on a link may be indicated by either receiving an LSSU (e.g. link failure or processor outage at the remote end) or a command from an operator (e.g. to take the link out of service for a maintenance test) or the failure may be detected by the local link control (e.g. it measures too high error ratio). A state change on a route may be indicated either by receiving a signaling route management message from another SP (it has modified its routing that may involve also a route change in the concerned SP) or by a link state change (e.g. the last available link of the given route became unavailable).

When an availability state change arises on a link or on a route, the STM calls one of the following functions:

- Changeover: when a link becomes unavailable, changeover diverts its whole traffic to an alternative link (if any).
- Changeback: when a previously unavailable link becomes available again, Changeback restores its traffic that has been diverted by a Changeover.
- Forced Rerouting: when a route to a given destination becomes unavailable (e.g. due to a remote failure in the signaling network) Forced Rerouting diverts signaling traffic towards that destination to an alternative route outgoing from the concerned signaling point (if any). Forced Rerouting removes the traffic to a given destination from a link belonging to the route in failure and places it to other link, which belongs to the alternative route.
- Controlled Rerouting: in case a previously unavailable route becomes available again, Controlled Rerouting restores its traffic that has been diverted by a Forced Rerouting
- Signaling Traffic Flow Control: informs the local user parts when a SP becomes unavailable or congested to stop or reduce the traffic toward the concerned SP, respectively. When a previously unavailable SP becomes available again, the local user parts will be informed about that fact as well, therefore they can resume the traffic.
- Restart: when a signaling point is isolated from the network for some period, it cannot be told undoubtedly whether its routing data are still valid. Thus during the Restart process the restarting SP will exchange the routing data with the adjacent SPs.

Signaling Link Management

The signaling link management function is used to control the locally connected signaling links and link sets. It describes the actions to be taken to restore a failed link to be available for signaling again and the actions how to activate or deactivate a signaling link or link set.

Signaling Route Management

The signaling points can exchange information about the availability of the signaling routes with the help of the signaling route management procedures.

The Transfer Prohibited procedure is performed at a signaling transfer point when it has to notify one or more adjacent signaling points that they must no longer route messages toward a given destination via that STP, because it is unable to reach the destination in question. The reception of a Transfer Prohibited Message (TRP) may cause an initiation of a Forced Rerouting process.

The Transfer Controlled procedure is used at a STP to inform other SPs when congestion toward a given destination is detected.

The Transfer Allowed procedure is performed at a STP when it has to notify one or more adjacent signaling points that they may start again to route messages toward a given destination via that signaling transfer point, because it is able to reach the concerned destination. The reception of a Transfer Allowed Message (TRA) may cause an initiation of a Controlled Rerouting process.

The Signaling Route Set Test procedure is performed in a SP, which receives a Transfer Prohibited or a Transfer Controlled (TRC) message. With the help of a Signaling Route Set Test (SRST) message, the concerning SP periodically tests the sender of the TRP or TRC whether the traffic to a given destination is still prohibited or restricted, until a TRA message is received. The goal of this test procedure is to enforce the retransmission of a Transfer Allowed message that may have been lost due to a signaling network failure.

Case Study

In this subsection, we illustrate the behavior of the STM processes with the aid of a case study (ITU-T, 1993/3). Let us suppose that at first the AB then the BC link fails (Figure 14).

When A detects the failure of AB, it performs a Changeover process to divert the whole traffic of AB to AC. In parallel with it, B also detects the failure of BA and performs a Changeover to divert the whole traffic of BA to BC. During this changeover, B sends a Transfer Prohibited message with the parameter of A to C in order to inform C that (in case of a possible failure on CA) it should not send messages to A via B.

Figure 14. Case study: Failure of AB and BC links

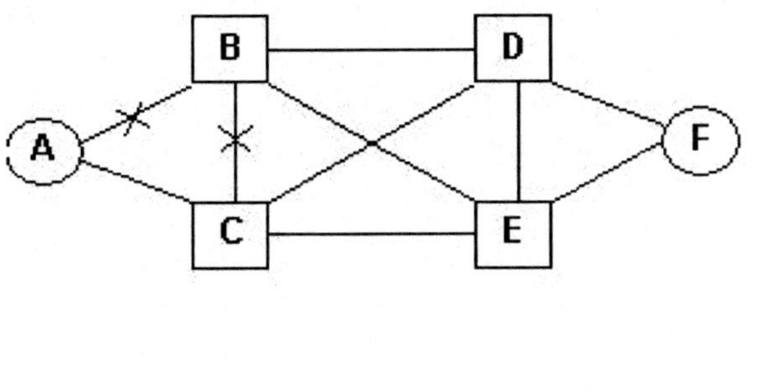

When C receives the TRP message, it starts to test B with periodical transmissions of Signaling Route Set Test messages.

When BC link fails, B recognizes that it cannot reach C any longer, and C also recognizes the unavailability of B. On the one hand B and C will inform their local User Parts about the unavailability of the opposite signaling transfer point by the help of the Signaling Traffic Flow Control process, and on the other, they will send TRP messages with the parameter of C and B, respectively to all their neighbors. When the neighbors receive the TRP messages, they start the periodical testing. When the second link fails, B detects the unavailability of A, too. It informs the local UPs about this fact and sends TRP messages with the parameter of A to all the available adjacent points (D and E). Apart from starting the periodical test, D and E will perform Forced Rerouting processes to divert the traffic destined to A from DB link to DC and from EB link to EC, respectively.

ISDN User Part

The ISDN User Part (ISUP) is responsible for the control of ISDN calls and for supervision of voice circuits (ITU-T, 1999/2). The ISUP can functionally be divided into four blocks (Figure 15).

Call Processing Control is responsible for to transmit and receive messages, which are related to establishment, supervision, and release of ISDN calls. However, we have to emphasize, that in general, not the ISUP, but the call control function of the exchange will decide which message is to be sent, and the call control function of an exchange will process the received messages and determine the action to be done.

Circuit Supervision Control treats the messages related to control the voice circuits themselves, e.g. to switch voice lines off (block) or on (unblock).

Figure 15. Functional block diagram of ISUP

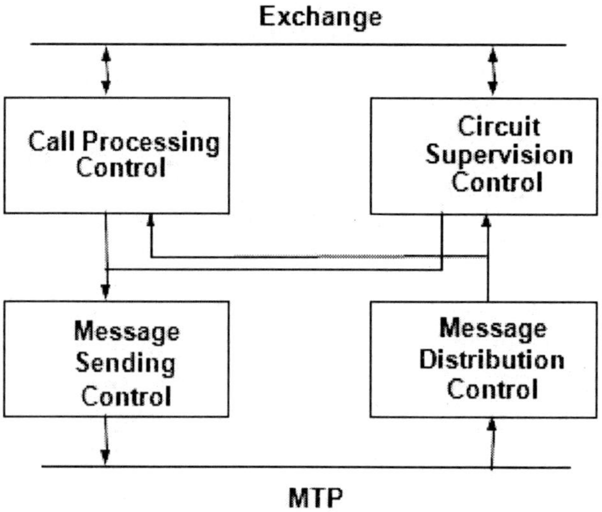

This functional unit is also under the control of the exchange.

Message Sending function collects the signals to be transmitted and passes them to the MTP, while the Message Distribution function receives the signals from the MTP and distributes them either to the Call Processing Control or to the Circuit Supervision Control based on the Message Type code of the received signal.

ISUP Message Structure

We can see the ordinary structure of an ISUP message (ITU-T, 1999/3) in Figure 16.

Each ISUP message starts with a two-octet Circuit Identification Code (CIC). CIC identifies the voice circuit, on which the controlled call takes place (in case of a call control message) or the one, which is controlled by the message (in case of a circuit supervision message).

The Message Type code follows the CIC. Each message must or may have parameters. The number and types of the parameters depend on the message type. Among the parameters, there are mandatory and optional ones. The mandatory parameters must be sent every time in the corresponding message while the optional ones just in specific situations. The mandatory parameters can be classified as fixed length or variable length parameters. To reduce the size of a message, after the message type code, the fixed length parameters are transmitted at first in a standardized order. Since the order of these parameters is standardized and their length is known, it is enough to transmit their value only. These parameters are followed by pointers to the variable length parameters in a standardized order and a pointer to the beginning of the optional part. Next to the pointers, the variable length parameters themselves are transmitted in a length and value notation.

It depends not only on the type of the message but also on the situation in which the message is sent, whether a particular optional parameter is sent or not. If no optional parameter is sent, then the pointer to the optional part is set to NULL, otherwise it points to the first optional parameter. To be able to recognize which optional parameters have been transmitted, their type code, length and value must also be sent. After the last optional parameter, an End of Optional Parameters (EOP) indicator[13] is to be sent.

Figure 16. Structure of an ISUP message

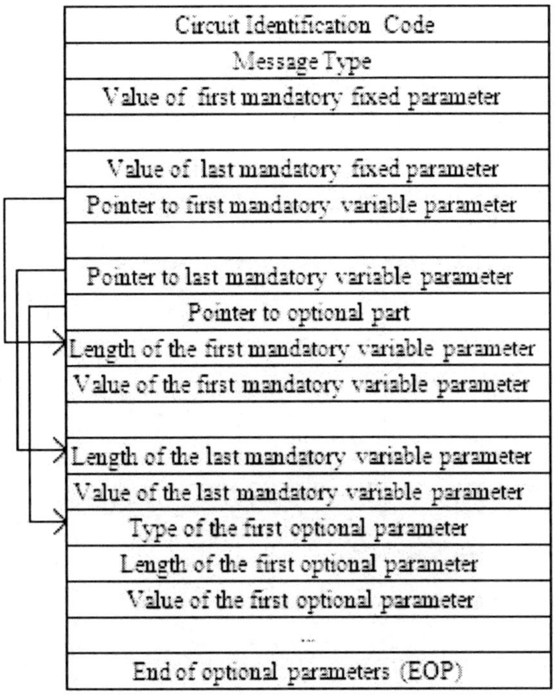

Call Control

We introduce the most important call control messages through an example (Figure 17). In the figure, we can follow the ISUP message sequence of a successful call set-up and release (ITU-T, 1999/4), and the interworking between the DSS1 and ISUP protocols.

When the calling exchange receives the last dialed digit, it analyzes the called number and decides where to direct the call. In our example, we are having an outgoing call towards another exchange. Therefore, the calling exchange reserves a voice circuit on a trunk toward the called exchange, and sends an Initial Address Message (IAM) on the signaling network. The IAM contains the identifier of the reserved voice circuit in the CIC field, and several other parameters describing the call to be established. The typical parameters of an IAM message are:

- Nature of Connection contains some general pieces of information in connection with the call: the number of satellite circuits used, echo control, call continuity check control.
- Forward Call Indicators describes the signaling requirements to be able to establish the call (e.g. ISUP is needed or not).
- Calling Party's Category determines the type of the call (ordinary calling subscriber, operator established call, priority call, data call, test call, payphone).
- Transmission Medium Requirement identifies the type of the requested voice circuit (speech, audio, unrestricted 64kbps digital, etc.).
- Called Party Number contains the dialed telephone number completely or partly. In case the dialed number is not complete, the rest of the digit(s) are transmitted in Subsequent Address Message(s) (SAM).

Figure 17. DSS1 and ISUP interworking

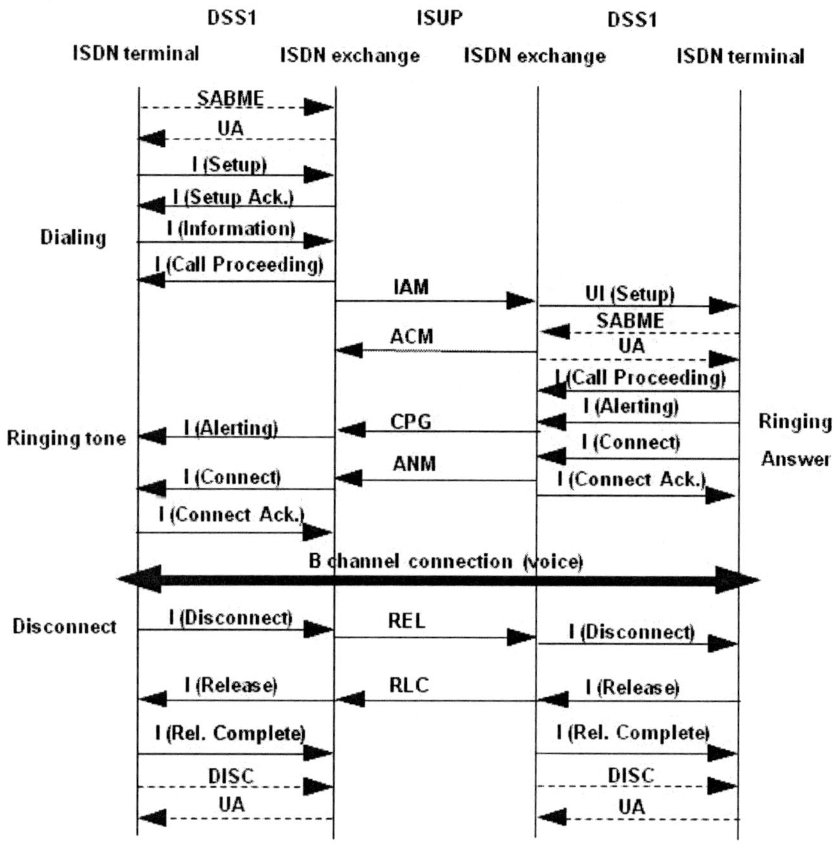

- Calling Party Number carries the telephone number of the caller and indicates whether the number can be presented on the display of the called party's terminal or not.

When the complete dialed number arrives (either in an IAM or in consecutive SAM messages), the answer is an Address Complete Message (ACM). There are two possible situations for the continuation. Most typically, the ACM is sent immediately after the called number has arrived to the destination local exchange (early ACM). Then the exchange analyzes the number, determines the state of the called party (free, busy, out of order, no such number). If the called party is free, the exchange starts to establish the call on the called DSL in the same way as we have discussed before.

When the called terminal alerts its user, the called exchange places the ringing tone onto the voice circuit and informs the calling exchange simultaneously by sending a Call Progress (CPG) message. CPG can be sent for several other reasons, thus it has an Event parameter describing the reason why the CPG is transmitted. The Event parameter has the value of Alerting in this case.

In some cases, the called exchange sends the ACM only when the called user has been alerted (late ACM). In this case, there is no need for the transmission of a CPG, but the dialed number is acknowledged later.

When the called party accepts the call, an Answer Message (ANM) is sent to the calling exchange and the voice line is connected through.

Figure 18. CFU signaling

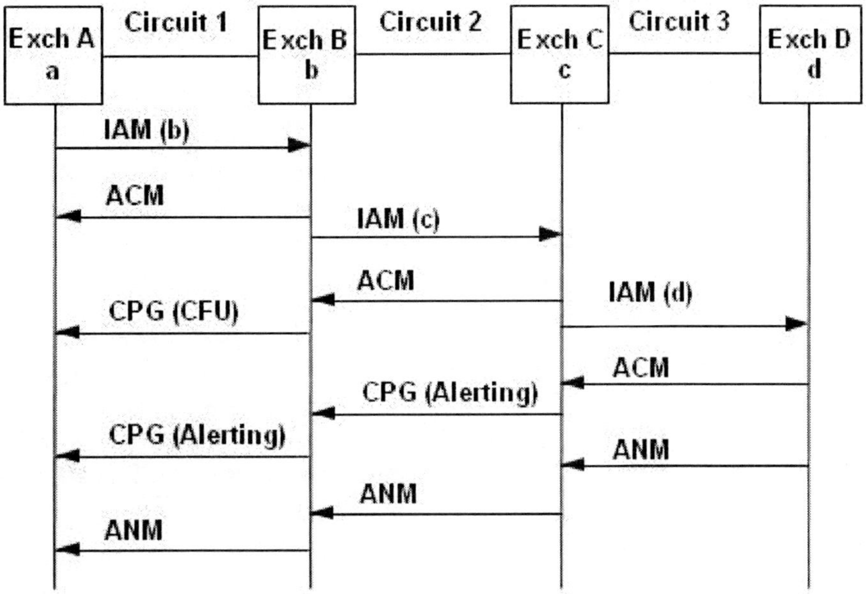

When ANM arrives in the calling exchange, the charging starts.

The release process can be initiated by both parties. In Figure 17, we illustrate when the caller wants to finish the call.

When a Disconnect message arrives on the calling DSL, the calling exchange sends a Release (REL) message to start the release process on the trunk line. The Release message contains a Cause Indicators parameter, which indicates the reason why the call is released. In our case, this parameter is set to the value of "Normal call clearing". The answer will be a Release Complete (RLC) message. We have to emphasize that in order to release the voice path as quickly as possible the release processes are always executed simultaneously on each circuit involved in the call.

In Figure 17, we have illustrated the steps of a successful call establishment. Should the call be not successful, a release process starts immediately when it becomes clear that the call cannot be completed. In this case the Cause Indicators parameter of the Release message informs the calling exchange about the reason of the failure (busy, not existing number, no idle voice circuit, requested service is not subscribed or not supported, interworking problem or protocol error detected, etc.).

As an illustration of how ISUP can control the ISDN supplementary services, let us consider the Call Forwarding Unconditional[14] (CFU) ISDN supplementary service (ITU-T, 1999).

CFU permits the served user to redirect the incoming calls to another number. In Figure 18, we can see that the caller from number *a* in exchange *ExchA* calls a subscriber number *b* in exchange *ExchB*. The called subscriber forwards the call to number *c* in exchange *ExchC*, where the call is forwarded again to number *d* in exchange *ExchD*[15]. At the beginning, the IAM message contains the called number (*b*) in the Called Party Number parameter. When *ExchB* detects the CFU is set, it includes in the ACM message two optional parameters:

- Call Diversion Information carries the fact that the call is forwarded unconditionally

- Redirection Number shows the number where the call has been forwarded to (*c*).

ExchB continues the call establishment by reserving a voice circuit to *ExchC*. However, the IAM message controlling this hop will differ from the one has been used before:

- Called Party Number contains the forwarded-to number (*c*), not the dialed number,
- Original Called Number contains the dialed number (*b*),
- Redirecting Number contains the number performing the redirection (*b*),
- Redirecting Information indicates that the call is forwarded unconditionally, and contains a counter with a value of 1, indicating that it is the first time when the call is forwarded.

When *ExchC* receives the call attempt and detects the CFU is set, it sends an ACM back to *ExchB* with setting Call Diversion Information to CFU and Redirection Number to *d*. These parameters are transmitted transparently to *ExchA* in a CPG message. The Event parameter of the CPG is now set to CFU.

The call establishment process continues at *ExchC*. It reserves a voice circuit to *ExchD*, sends an IAM, in which Called Party Number is set to *d*, Original Called Number is set to *b*, Redirecting Number is set to *c* and Redirecting Information is set to CFU with the counter set to 2.

When *ExchD* receives the IAM, it starts to establish the call on the called DSL by DSS1 messages. When an Alerting message arrives from TE *d*, *ExchD* sends a "normal" ACM back to *ExchC* indicating d is alerted. *ExchC* and e *ExchB* will relay this piece of information backward by CPG messages having the Event parameter set to Alerting. When the call is answered by *d*, ANM messages are sent.

Circuit Supervision

The circuit supervision messages are used to control the voice circuit themselves. Circuits and circuit groups can be blocked and unblocked. Blocking makes it possible for the exchange or the maintenance system to take a voice circuit or circuit groups out of service in order to detect a failure or to run tests, while unblocking brings a blocked circuit or circuit group back in service.

Since in SS7 the signals do not pass through the circuits, in systems having no inherent fault indication mechanism a call continuity check may be needed on a per call basis. If we want to check the continuity of a call, it can be required in the IAM message.

CONCLUSION

In this chapter, we have given an overview about the ISDN signaling protocols.

At first, we have discussed the DSS1 protocol, which is used on a DSL. We have presented a detailed description of the LAPD, which is responsible for transmitting signals between the network and the terminal(s) in an error-free way. The 3rd layer of DSS1 is responsible for the call control on a DSL. We have presented the process of a successful call set-up and release.

In the second part of this chapter, we have given an architectural overview about the SS7, which is defined by the ITU-T. We have discussed the Message Transfer Part and the ISDN User Part in details. MTP is used for signal transmission between any two signaling end points of the same signaling network, as well as it provides the network management functions. ISUP is responsible for the call control and the circuit supervision in the core network of ISDN.

REFERENCES

ITU-T. (1988). *Recommendation Q.702. Signalling data link*. Geneva, Switzerland.

ITU-T. (1993. 03). *Recommendation Q.700. Introduction to CCITT signaling system No. 7*. Geneva, Switzerland.

ITU-T. (1995. 11). *Recommendation I.430. Basic user-network interface - layer 1 specification*. Geneva, Switzerland.

ITU-T. (1996. 07). *Recommendation Q.703. Signalling link*. Geneva, Switzerland.

ITU-T. (1997. 09). *Recommendation Q.921. ISDN user-network interface - data link layer specification*. Geneva, Switzerland.

ITU-T. (1998. 05). *Recommendation Q.931. ISDN user-network interface layer 3 specification for basic call control*. Geneva, Switzerland.

ITU-T. (1999. 12). *Recommendation Q.732.2-5. Stage 3 description for call offering*. Geneva, Switzerland.

ITU-T. (1993/2. 03). *Recommendation Q.701. Functional description of the message transfer part (MTP) of signalling system No. 7*. Geneva, Switzerland.

ITU-T. (1993/3. 03). *Recommendation I.431. Primary rate user-network interface - layer 1 specification*. Geneva, Switzerland.

ITU-T. (1993/3. 03). *Recommendation Q.705. Signalling network structure*. Geneva, Switzerland.

ITU-T. (1996/2. 07). *Recommendation Q.704. Signalling network functions and messages*. Geneva, Switzerland.

ITU-T. (1999/5. 03). *Recommendation Q.708. Assignment procedures for international signalling point codes*. Geneva, Switzerland.

ITU-T. (1999/2. 12). *Recommendation Q.761. Signalling system no. 7 - ISDN user part functional description*. Geneva, Switzerland.

ITU-T. (1999/3. 12). *Recommendation Q.763. Signalling system No. 7 - ISDN user part formats and codes*. Geneva, Switzerland.

ITU-T. (1999/4. 12). *Recommendation Q.764. Signalling system no. 7 - ISDN user part signalling procedures*. Geneva, Switzerland.

KEY TERMS AND DEFINITIONS

B Channel (Bearer): B channel is a full duplex, digital channel with the bit rate of 64 kbps used for voice and user data transmission in ISDN systems.

Channel Associated Signaling: (CAS): CAS is a form of signaling, in which the signaling information is encoded and transmitted in the same channel as the voice or data.

Common Channel Signaling (CCS): CCS is a form of signaling, in which the signaling information is transmitted in a separate channel from the voice and data. In CCS systems (CCSS), a signaling channel is common for several voice or data circuits and for several services. Typical CCS systems in ISDN are SS7 and DSS1.

D Channel (Data): D channel is an ISDN channel, in which the control and signaling information is carried in general, but it can also be used for low-speed user data transmission. The bit rate of the D channel is 16 kbps on a basic rate interface, whereas it is 64 kbps on a primary rate interface.

Digital Subscriber Signaling System No. 1 (DSS1): DSS1 is a digital signaling protocol used in the D channel at the ISDN user-network interfaces for controlling the B channel (voice or data) connections.

Integrated Services Digital Network (ISDN): ISDN is a digital telecommunication network for simultaneous transmission of voice, video, data.

Signaling End Point (SEP): SEP is a kind of signaling point (SP) in SS7, through which the telecommunication network elements (exchanges, control databases, etc.) can originate and receive signaling messages

Signaling System No. 7 (SS7): SS7 is a common channel signaling system used in the core network of the telecommunication systems.

Signaling Transfer Point (STP): STP is a kind of signaling point (SP) in SS7 that transfers signaling messages from one signaling point to another.

ENDNOTES

1 SS7 is often referred also as Common Channel Signaling System 7 (CCSS7, CCS7, or simply C7).

2 This bit stuffing is not ISDN-specific, it is a general mechanism used in many telecommunications protocols, and its generic instance is to be found in the HDLC protocol.

3 To be more precise: if a sequence of more than six consecutive 1 is received, it indicates that the transfer of the message is aborted.

4 This technique is often used to indicate the end of a variable length parameter. The least significant bit of the last octet of the parameter is encoded as 1, while it has been 0 in all the previous octet(s). Though in LAPD the address field is always two octets long, LAPD is one of the variants of the HDLC protocol, and in the HDLC, protocol family the address field may have different length than two octets.

5 In HDLC protocol family SAPI=63 always indicates the management entity of the given protocol. In LAPD the SAPI=63 is used for TEI management.

6 Asynchronous mode: stations may send messages at any time, it is not necessary to transmit messages continuously. Balanced mode: both sides can send commands to the other (no master-slave connection). Extended: Seven bit long sequence numbers are used (in the original HDLC the sequence numbers were just three bit long).

7 A TE typically answers with Call Proceeding, but may skip this and answer with Alerting, or may skip even this and immediately answer the call automatically, by sending Connect.

8 SEP is not a standardized term, but we found it useful to introduce for a cleaner approach.

9 We have to mention that the standard considers a signal to be an MSU if its information field is longer than two octets. However, the length of the shortest possible message is six octets – as we will see later on.

10 The standard also allows a two-octet Information field in an LSSU, but it is not used in practice.

11 The MTP cannot send signals between signaling points of different networks. In the classical call control there is no need for this kind of signaling, because the two ends of a voice circuit always belong to the same network. However, in mobile networks, there may be a need for inter-network signaling (e.g., when we send an SMS from abroad), so this problem is solved by the SCCP protocol.

12 This name comes from the 1980s, when most commonly there was one telephone operator in each country.

13 EOP is also an optional parameter with a type code of 0, and it has neither a length nor a value.

14 There are conditional versions of the Call Forwarding service: Call Forwarding on Busy (CFB), Call Forwarding No Reply (CFNR). The signaling of the conditional versions is quite similar to that of the CFU.

15 In the examples, we illustrate the late ACM signaling option.

APPENDIX

Table of abbreviations

ACM	Address Complete Message
A_i	Action Indicator
ANM	Answer Message
AO/DI	Always On/Dynamic ISDN
BC	Bearer Capability
BIB	Backward Indicator Bit
BRA	Basic Rate Access
BSN	Backward Sequence Number
C/R	Command/Response bit
CAMEL	Customized Applications for Mobile networks Enhanced Logic
CAP	CAMEL Application Part
CAS	Channel Associated Signaling
CCS	Common Channel Signaling
CCSS	Common Channel Signaling System
CFU	Call Forwarding Unconditional
CIC	Circuit Identification Code
CK	Checksum
CPG	Call Progress (message)
DISC	Disconnect
DM	Disconnected Mode
DPC	Destination Point Code
DSL	Digital Subscriber Line
DSS1	Digital Subscriber Signaling System No.1
EA	Address field Extension bit
EOP	End of Optional Parameters
FCS	Frame Check Sequence
FIB	Forward Indicator Bit
FISU	Fill-In Signal Unit
FSN	Forward Sequence Number
GSM	Global System for Mobile communication
HDLC	High level Data Link Control
I	Information (frame)
IAM	Initial Address Message
IE	Information Element
INAP	Intelligent Network Application Part
ISDN	Integrated Services Digital Network
ISO	International Organization for Standardization

ISPC	International Signaling Point Code
ISUP	ISDN User Part
ITU-T	International Telecommunication Union Telecommunication Standardization Sector
LAPD	Link Access Procedure on D channel
LI	Length Indicator
LSSU	Link Status Signal Unit
MAP	Mobile Application Part
MSU	Message Signal Unit
MT	Message Type
MTP	Message Transfer Part
N(R)	Receive Number
N(S)	Sent number
NI	Network Indicator
NISPC	National Interconnecting Signaling Point Code
NRA	National Regulatory Authority
NSPC	National Signaling Point Code
NT	Network
OPC	Originating Point Code
OSI	Open Systems Interconnection
P/F	Poll/Final bit
PRA	Primary Rate Access
RAND	Random Number
REJ	Reject
REL	Release (message)
RL	Routing Label
RLC	Release Complete (message)
RN	Reference Number
RNR	Receive Not Ready
RR	Receive Ready
S	Supervisory (frame)
SABME	Set Asynchronous Balanced Mode Extended
SAM	Subsequent Address Message
SANC	Signaling Area Network Code
SAPI	Service Access Point Identifier
SCCP	Signaling Connection Control Part
SEP	Signaling End Point
SI	Service Indicator
SIO	Service Information Octet
SLM	Signaling Link Management
SLS	Signaling Link Selection code
SMH	Signaling Message Handling
SMS	Short Message Service

SNM	Signaling Network Management
SP	Signaling Point
SPC	Signaling Point Code
SRM	Signaling Route Management
SRST	Signaling Route Set Test
SS7	Signaling System No.7
STM	Signaling Traffic Management
STP	Signaling Transfer Point
TCAP	Transaction Capabilities Application Part
TE	Terminal Equipment
TEI	Terminal Endpoint Identifier
TRA	Transfer Allowed
TRC	Transfer Controlled
TRP	Transfer Prohibited
TUP	Telephone User Part
U	Unnumbered (frame)
UA	Unnumbered Acknowledgement
UI	Unnumbered Information
UMTS	Universal Mobile Telecommunications System
UP	User Part

Chapter 6
Mobile Network Protocols of GSM and GPRS

Gusztáv Adamis
Budapest University of Technology and Economics, Hungary

ABSTRACT

Starting in the 1990s, mobile communication became one of the most explosion-like developing areas of telecommunications. The first worldwide used digital cellular radio system was the GSM. Primarily, GSM was suitable for voice communication and for circuit-switched data transfer at a very slow bit rate only. This data transfer type was not only slow, but was quite expensive to the users as well. Aiming to eliminate the drawbacks of the data transfer of GSM, the packet-switched service (GPRS) was introduced.

The goal of this chapter is to give an overview about the architecture of GSM and GPRS networks, and introduce to the reader the protocols used for mobility management and call control at the core network. To make it easier to understand the way in which the protocols work, we give many examples to help the readers in his studies. We show how the location update takes place, how the calls are originated and terminated, how the short messages are handled, et cetera.

INTRODUCTION

In this chapter, we introduce to the reader the architecture of GSM. We explain the functions of the most important network elements. After the brief architectural overview, we discuss the protocols used in the Network and Switching Subsystem (NSS). At the end of this chapter, we show the new network elements as well as the core network protocol of the GPRS.

DOI: 10.4018/978-1-60960-732-6.ch006

Structure of the GSM Network

The user terminals of the mobile networks are the Mobile Stations (MS). An MS consists of the mobile phone itself (Mobile Equipment – ME) and a smart card identifying the subscriber (Subscriber Identity Module – SIM). Each SIM card has a worldwide unique identifier, the International Mobile Subscriber Identity (IMSI). The IMSI consists of a Mobile Country Code (MCC), a Mobile Network Code (MNC), and a Mobile Subscriber Identity Number (MSIN). Each MS has a telephone number referred to as Mobile Station ISDN Number (MSISDN).

A GSM network can be functionally divided into three subsystems:

- Base Station Subsystem (BSS),
- Switching and Management Subsystem (SMSS),
- Operation and Maintenance Subsystem (OMSS).

The BSS is the physical equipment used to give radio coverage to cells and contain the equipment needed to communicate with the MSs at the radio interface. The SMSS includes the functions required to handle signaling protocols by which calls and mobility of the subscribers are controlled. The OMSS performs functions related to system security (authentication and mobile equipment identity check) as well as it is in charge of remote operations and maintenance of the PLMN (ETSI, 1998). The description of the Operation and Maintenance Center (OMC) is out of scope of this chapter. The SMSS extended with the Authentication Center (AuC) and the Equipment Identity Register (EIR) from the OMSS is often referred to as Network and Switching Subsystem (NSS), thus we will use this term as well (Figure 1).

Base Station Subsystem

The Base Station Subsystem provides connections between the MSs and the GSM network through the radio interface. A BSS consists of the following elements:

- Base Transciever Stations (BTS),
- a Base Station Controller (BSC),
- a Transmission and Rate Adaptation Unit (TRAU).

Figure 1. GSM architecture and protocols

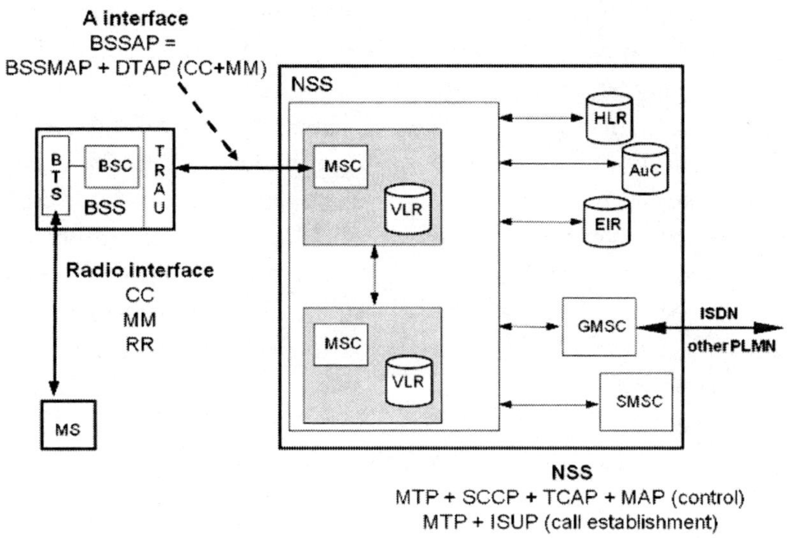

A BTS is the radio transmitter and receiver itself. A BSS can have up to 16 BTXs, controlled by a BSC. Apart from controlling the BTSs, the BSC performs a switching function between the traffic channels incoming from the mobile switch and the radio channels. Since the bit speed of the speech transmission is different at the radio interface and in the core network, we need a TRAU to perform the necessary conversion.

Network and Switching Subsystem

The NSS plays the central role in every mobile network. Various elements within the Network and Switching Subsystem perform the complete set of control and database functions, which are required to set up a call connection. The NSS contains the following main building elements:

- Mobile Switching Center (MSC) with Visitor Location Register (VLR),
- Gateway Mobile Switching Center (GMSC),
- Home Location Register (HLR) with Authentication Center (AuC),
- Equipment Identity Register (EIR),
- Short Message Service Center (SMSC).

The MSC – from a technical view – is quite similar to an ISDN exchange, but it has several additional extensions. The main task of an MSC is to provide the switching function. Since the subscribers of a mobile network may change their position, each MSC is built together with a database, the Visitor Location Register that contains the data of those subscribers who camp in the geographical area served by the MSC in question. The other important difference between an ISDN exchange and an MSC is that the latter is not in direct connection with the subscribers it serves, but through some base stations controls the radio resources. Between an MSC and a BSS locates the A interface. An MSC with an interface to other networks (to ISDN and/or to other GSM networks) is a Gateway MSC.

Each GSM network (Public Land Mobile Network – PLMN) has a database, the Home Location Register that contains the subscription information as well as the current location of every subscriber of the corresponding mobile operator. The Authentication Center is used to check those subscribers' identities who want to connect to the network and to provide information for ciphering on the radio interface as well. The AuC is integrated with the HLR in general.

Each ME has a unique identifier, the International Mobile Equipment Identity (IMEI), consisting of a type code and a serial number. The EIR maintains three lists about the MEs. The white list contains all approved types of MEs, the black list holds all IMEIs to be barred (either for technical reasons or because the corresponding ME is reported to be stolen), and the gray list keeps MEs to be traced.

The Short Message Service Center receives the mobile originated short messages and delivers them to their addressees.

Protocols of GSM

In the NSS of the GSM networks, the Signaling System 7 (SS7) protocols are used for signaling. We have introduced the SS7 architecture in details in Chapter V, "Signaling Protocols of Integrated Services Digital Networks". We have discussed the Message Transfer Part, which is used for signal transfer among various elements of the network and the ISDN User Part, which is used for call control. In this chapter, we focus on only those SS7 protocols that are used primarily in the mobile networks, such as SCCP, TCAP, and MAP. Later on we discuss the BSSAP, the special application part developed exclusively for signaling at the A interface (see Figure 1).

Signaling Connection Control Part

As we also discussed earlier in Chapter V, the MTP is suitable to transmit signals between any two signaling points of the same network only. To set-up a voice path of a call, there is no need for inter-network signaling, but in the mobile networks, many situations may arise, when we have to send a signal from a given network into another one. The most obvious example might be if we want to send an SMS from abroad. In this situation, a signal (the short message itself) must be sent from an MSC of the foreign network to the SMSC of our home network.

One of the most important features of the Signaling Connection Control Part (SCCP) is to provide the end-to-end signaling between signaling points (SP) of different networks (ITU-T, 2001). The routing of an inter-network signal cannot be based on the signaling point codes, since they are unique within the same network only. To solve this problem, the notion of the Global Title (GT) was introduced into the SCCP. A GT is a non-SS7 related address, most typically an E.164 telephone number. It will be the task of the SCCP to translate the GT into SPC.

Each functional unit in a network, which is supposed to be reached from a different network, must have a global title. In our previous example, the SMSC has a GT, which is a certain telephone number. (This is the number we have to set in our mobile at a menu point of "Message Center" in general.) This number will be inserted into the Called Party Address parameter of the signal, which contains the transmitted short message. The SCCP at MSC analyzes this GT, determines which will be the next SP on the way to the desired SMSC and writes the point code of that SP into the Destination Point Code parameter, which is used by the MTP to route this message. As soon as the message reaches that SP, the SCCP also analyzes the GT at that point, determines the next SP, and so on. The GT and SPC assignments are stored in general in a form of an assignment table at each SCCP. The situation is quite similar to the delivery process of an international letter. By analyzing the address on an envelope ("the GT"), the postal workers in each post office, at which the letter goes through, can determine the next post office ("the next SP"), to where the letter is to be sent.

The SCCP can differentiate among various applications (e.g. MSC, VLR, HLR, AuC, etc.) connected to the same SP. SCCP refers to these applications as subsystems. SCCP offers connectionless and connection-oriented services.

The basic connectionless class (Class 0) is suitable to transmit data units between any two SPs of the world in a transparent way. The in-sequence delivery connectionless class (Class 1) apart from the functionality of Class 0 guarantees that the messages transmitted between a subsystem pair preserve their sequence. (Messages of the so-called SCCP sequences use the same SLS; therefore, they are transmitted on the same path.) Class 1 also has segmentation and reassembling feature and it provides some management services as well.

The connectionless services use the Unitdata (UDT) message. UDT contains – apart from a Protocol Class – a Called Party Address (CdPA), a Calling Party Number (CgPA), and a Data parameter. CdPA contains in general the GT of the addressee and CgPA the GT of the sender. These addresses may contain subsystem numbers (SSN), too. Instead of GTs, we can use SPCs if the sender and addressee signaling points are in the same network. The Class 0 is used commonly to carry simple database queries and responses, while the Class 1 is used, if a transaction is more complex, e.g. contains several different queries, therefore we have to keep the order of the messages within that transaction.

With the help of the SCCP, we can control signaling connections (Figure 2) as well. We request the establishing of a signaling connection with a Connection Request (CR) message. The message contains a logical connection identifier (123456 in Figure 2) in the Source Local Reference (SLR) parameter. If a connection request is accepted, the receiving unit also assigns a connection identifier to it and answers by a Connection Confirm (CC) message. (Should the connection be not accepted, a Connection Refused (CREF) message is sent.) The CC contains the assigned connection reference value (789012 in Figure 2) in the Source Local Reference parameter, while the Destination Local Reference (DLR) parameter contains the connection reference value assigned by the other end (123456 in Figure 2). When a connection is established, both ends can send data in DataForm1 (DT1) messages. The connection can be released by a Released (RLSD) and Release Complete (RLC) message exchange. All the message types contain CdPA, CgPA and Data parameters. The basic connection-oriented class (Class 2) is suitable only for a connection control, while in the flow control connection-oriented class (Class 3) we can also control the maximal number of unacknowledged messages. Class 2 is used in the GSM at the A interface, but the Class 3 is not used.

Transaction Capabilities Application Part

As we could see above, the connectionless services of the SCCP are suitable simply to transmit a data unit from a subsystem of a SP to another one of another SP in a transparent way, but they do not offer any further control to manage the communication. The transaction management will be the function of the Transaction Capabilities Application Part (TCAP) (ITU-T, 1997).

With the help of the TCAP, we can call remote operations, most typically database queries. A transaction contains one operation in general, but there are more complex transactions, which contain several operations (e.g. when we start an operation at an ATM to inquire for the balance of our bank account, the bank executes an internal operation to get the PIN code before it reports the balance). Therefore, the TCAP has to identify not only the transaction itself, but the operations, too.

A TCAP message may have three parts:

- Transaction portion (mandatory),
- Dialogue portion (optional),
- Component portion (optional).

Transaction Portion

The transaction portion is used to identify the transaction with the help of transaction identifiers and it contains the message type code, as well.

A transaction starts with a Begin message. The originator of the transaction assigns a transaction identifier to the transaction that is transmitted in the Originating Transaction Identifier (OTID) parameter. The last message of a transaction is an End message containing the transaction identifier assigned by the invoker in the Destination Transaction Identifier (DTID) parameter.

In a more complex case in between the Begin and End messages, any number of Continue message pairs can be used. In this case, not only the

Figure 2. SCCP connection between two signaling points

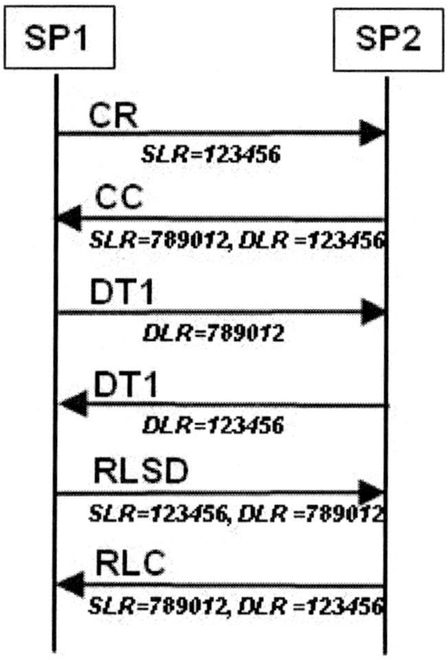

originator of the transaction generates a transaction identifier, but the remote side, too. In the Continue messages, the OTID contains the transaction identifier assigned by the message sender, while the DTID carries the transaction identifier assigned by the receiver of the message.

Should a fatal error occur during the transaction (e.g., not a valid message type code or a wrong transaction identifier has been sent, or a not implemented TCAP user protocol has been used) an Abort message is transmitted instead of an End. The Abort message identifies the reason of the problem.

Dialogue Portion

The most important function of the dialogue portion is to identify the "language" of the transaction (more precisely the TCAP version and the TCAP user protocol), if it is not known implicitly. In case we always use the same TCAP version and the same TCAP user protocol on a signaling link, there is no need for a dialogue portion. However, if different subsystems, which communicate on the same signaling link, use different TCAP versions and/or user protocols (e.g. MAP and INAP) we have to identify the used protocol and its version in the dialogue portion. The dialogue portion is included even in this case in the first message pair of the transaction only. The dialogue portion contains diagnostics information in Abort messages and may also contain user-to-user information.

Component Portion

The component portion is used to transmit the messages of the operations of the TCAP user protocol in a transparent way. These messages may be either invocations of remote operations or results of the operations. Though in general a component portion contains only one component, it may contain more components (i.e. more than one operation can be controlled by a single TCAP message).

Each component contains:

- the component type code (describing the phase of the operation execution),
- an operation identifier (Invoke ID),
- the type of the operation,
- the parameters (if any).

Since we can call more than one operations in a transaction, these operations are identified by a one octet long Invoke Identity. Each operation in a transaction must have a different Invoke ID.

The type of the component is one of the following:

- Invoke
- Return Result Not Last
- Return Result Last
- Return Error
- Reject

We can call a remote operation by an Invoke component. When the operation is executed successfully, the answer arrives in a Return Result component: if the answer fits in one message, then in a Return Result Last, otherwise the first segments of the answer travel in Return Result Not Last components, while the last part in a Return Result Last component.

If we do not recognize the request (e.g. the Invoke component is badly formatted or a parameter has an out-of-range value) then a Reject component is sent. Should the execution of the operation produce an error, a Return Error component – carrying the error code and optional diagnostics information – is transmitted instead of a Return Result Last.

Certainly, there is a relationship between the TCAP message types and the Component types. An Invoke component may be transmitted only in a Begin or a Continue messages, a Return Result Not last component can never be in an End message, and so on.

Mobile Application Part

The Mobile Application Part (MAP) is the protocol used for various controlling purposes in the NSS (3GPP, 2004). The MAP messages are sent in TCAP components. In this subsection, first, we discuss the most important controlling functions in the GSM network then we show the signaling sequences in the typical scenarios.

Security and Privacy

In the GSM networks, we have to prevent the interception of the calls and even the signaling. Therefore, at the radio interface the connections are encrypted. However, before switching to cipher mode the mobile station, which wants to connect to the network must be identified. Therefore, in the first message of every connection the identity of the subscriber must be provided. If this identity was the IMSI, though we would not be able to intercept the call itself, but by analyzing the first signal of the connection (which has not been encrypted yet) we would be able to determine the location of the subscribers. To reduce the possibility of this danger, at the radio interface the Temporary Mobile Subscriber Identity (TMSI) is used instead of the IMSI. The TMSI is a four octet long identifier, having no internal structure, no relation with the subscriber's identity, therefore it can be considered as a random number. When the mobile station connects to the network at the very first time, it uses the IMSI, since it has no other identity. However, in the encrypted phase of that connection the serving MSC assigns a TMSI to it. Next time, when the MS connects to the network it will use this TMSI instead of the IMSI. To increase the protection the TMSI value is changed from time to time.

To prevent an unauthorized access into the network (and to determine who will pay for the call!) the users must be authenticated before they can connect to the network (3GPP, 2001). By the authentication it can be checked, whether the subscriber identity provided by a mobile station corresponds to the inserted SIM. At the radio interface for authentication such a simple method as using a password or a PIN code is not secure enough, since intercepting once these personal codes breaks the protection. Therefore, we have to use a more sophisticated method (Figure 3); we have to check the identity of the subscriber *without* transmitting of its secret key through the radio interface.

When producing a SIM card, a 128 bit long secret key (Individual Subscriber Key – K_i) is burnt into it. When a mobile operator buys SIM cards, the vendor provides the IMSI-K_i assignments to the mobile operator in an off-line way. These assignments are stored in the AuC. When

Figure 3. Authentication: theoretical model

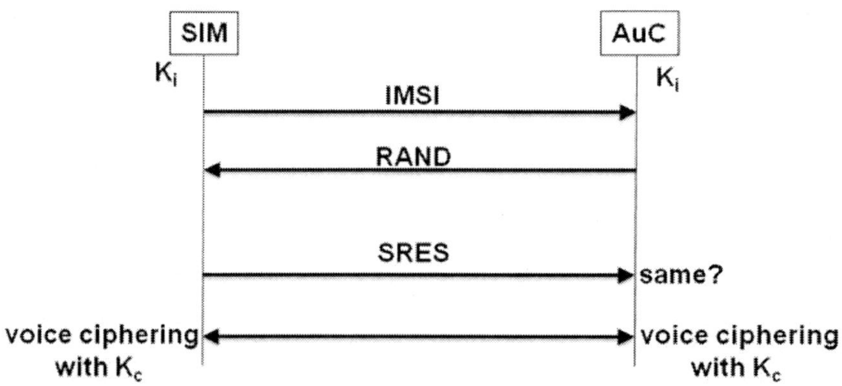

a mobile station wants to connect to the network, we have to check the K_i without transmitting it at the radio interface, i.e. the check has to be indirect. The Authentication Center sends a random number (RAND) to the mobile station that calculates a 32-bit signed response (SRES) by a secure algorithm called A3, and sends this SRES back to the Authentication Center. After this procedure, the AuC compares the received SRES with the correct answer. If they match, the mobile station can connect to the network, otherwise it cannot. In cryptography, A3 is called an asymmetric or a one-way function, since we can calculate SRES from K_i and RAND in an easy way, while the computation of K_i from RAND and SRES (that appear on the radio interface) is unworkable. As an example, let us suppose that we have a huge library. When somebody makes a mobile subscription, gets a book from this library (a book corresponds to a given K_i value). When the subscriber wants to connect to the network, the operator asks him a question like this one: "Tell me what is the content of #354 line of your book!" (The line number corresponds to the RAND and the content of the line to the SRES.) It is obvious that the correct answer can be determined if you really have the book in hand and the method to find the requested line is not too complicated. But should the line number and the content of the requested line be caught at the radio interface, it is impossible to find out, which book you have, since the library is really huge, it contains 2^{128} (3.4×10^{38}) different books!

From K_i and RAND another value, a 64 bit long ciphering key (K_c) is calculated by the algorithm of A8. K_c is used to encrypt the voice at the radio interface.

However, if the authentication and especially the encryption and decryption of the voice at the network side performed by the AuC, it would generate too much signaling. So in practice we draw the visited MSC/VLR into the authentication and the serving BTS into ciphering (Figure 4).

When a mobile station registers at a VLR, the VLR reports it to the AuC, which then generates a RAND and calculates the corresponding SRES and K_c values. The RAND, SRES and K_c together are referred to as an authentication triplet. AuC transmits five pieces of authentication triplets to the visited MSC/VLR in general. When the mobile station wants to connect to the network, the MSC/VLR sends a RAND to it and compares

Figure 4. Authentication: practical solution

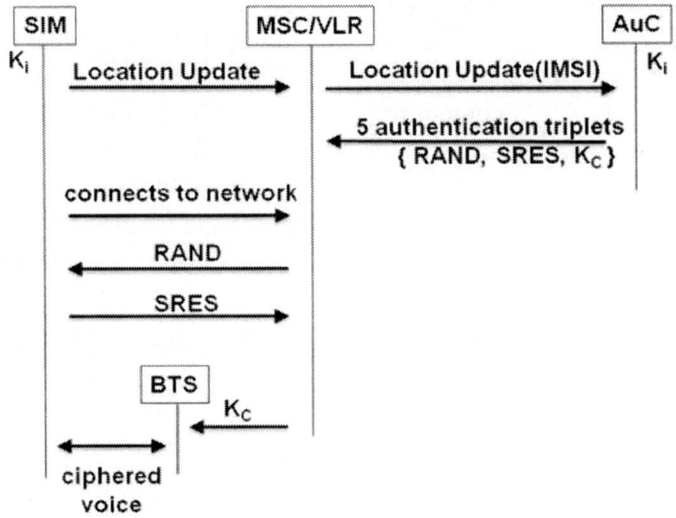

the received SRES bit-by-bit to the value received from the AuC. If the authentication is successful, the MSC sends the corresponding K_c value to the serving BTS where the ciphering takes place at the network side. When all the authentication triplets are used, the MSC/VLR requests another set from the AuC.

Mobility Management in NSS

If we did not deal with tracing the location of the subscribers, a called mobile station would have to be paged in *every* cell of the world, generating an unacceptably high utilization of the radio interface. To reduce the number of cells in which a called mobile station is to be paged, the network must know the location of the subscribers. Therefore, the MSs have to report their position to the network. This process is called as Location Update. On the other hand, if the mobile stations reported every cell change they make, the signaling need of the location updates would be unacceptably high. To make a balance between the signaling need of the paging and the location updates, the Location Areas (LA) are defined. An LA is a set of neighboring cells served by the same MSC/VLR. A mobile station reports to the network when it changes an LA, and when the MS is called, it will be paged only in the corresponding LA. At the network side the MSC/VLR knows in which LA the MS camps, and the HLR stores the identity of the serving MSC/VLR. Each LA has its own Location Area Identifier (LAI). A LAI consists of a MCC, a MNC and a location area code.

Location Update

If a mobile station wants to change a Location Area it performs a Location Update (LU) process (3GPP, 2001). If the new Location Area is served by the same MSC/VLR as the previous one, the MSC/VLR can manage the process by its own, so there is no signaling in the NSS, only at the radio and A interfaces. However, if the LA change is practically an MSC/VLR area change as well, then apart from the new MSC/VLR (where the mobile station wants to register to), the HLR, and the old MSC/VLR are also involved into the process. In the remaining part of this subsection, we only deal with the second case.

The Location Update is used to reach the following goals:

- the new MSC/VLR has to get the IMSI of the mobile station and an authentication triplet for the authentication, as well as the MSISDN and the service profile,
- the HLR must be informed about the new serving MSC/VLR area,
- the old MSC/VLR must be also informed about the change in order to clear the subscriber record from the VLR.

We have to differentiate the situation, in which the new MSC/VLR belongs to the same operator from the case when it does not (Figure 5). We suppose the MS uses the TMSI (and the LAI of that Location Area, in which the TMSI was assigned) at the radio interface. In the first case the new MSC/VLR can determine, which was the old MSC/VLR (the MSC/VLR – LA assignment may be known by each MSC of the same network) and the new MSC/VLR can ask the old one to send the authentication triplets as well as the IMSI of the MS. In the second case, when the new and the old MSCs belong to different networks, there is no possibility to the above-mentioned process, because the new MSC does not know the internal structure of the other network. Therefore, the new MSC must ask the MS to send its IMSI at the radio interface, and get authentication triplets directly from the AuC[1].

After a successful authentication, the continuation of the process is the same in both cases. The new MSC/VLR informs the HLR about the new location of the MS, by an updateLocation message, which contains the IMSI of the MS and the Global Title of the new MSC/VLR. The GT of

Figure 5. Location update in NSS

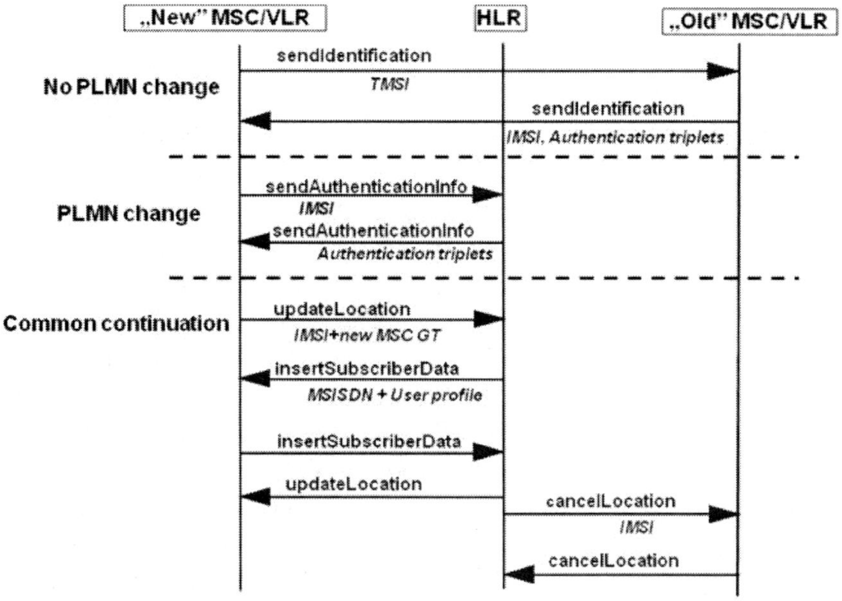

the serving MSC/VLR is the piece of information the HLR stores about the location of a MS, since this is the identifier which is needed to be able to send signals to the serving MSC/VLR (e.g. to acquire an MSRN during call set-up). Before the HLR acknowledges the updateLocation message, it sends the user profile of the MS in an insert-SubscriberData message. The user profile contains the MSISDN and several important pieces of information about the subscribed services and service restrictions. The user profile is stored in the serving VLR to reduce the need of signaling in the NSS, since this way if an MS wants to use a service (e.g. to receive a call during roaming) the MSC/VLR can accept or deny the requested service by its own without turning to the HLR every time.

Meanwhile the HLR informs the old MSC/VLR by a cancelLocation message which says that the corresponding subscriber has registered to a new MSC/VLR. Therefore, the old VLR can delete the user profile.

There are three other processes considered to belong to the mobility management in the NSS.

The Periodic Location Update is used to minimize the effect of a possible data base failure at a HLR. If a HLR loses the information about the location of a subscriber, the subscriber cannot be called until he wants to connect to the network by making a call, sending a short message, or performing a Location Update. To reduce this unavailability period, the following method was defined. At a data base failure, the HLR will be restored from a back-up copy. After the restoration, the HLR informs all the MSC/VLRs in which at least one of the subscribers of the corresponding HLR was known to be located at the time, when the back-up copy was created. The affected MSC/VLRs will inform the HLR about the location of these "lost" subscribers when they connect to the network. To enforce the connection at the radio interface, the MS performs a Periodic Location Update process, even if it does not move. The period can be set by the operator among 6 minutes and a bit more than one day. The typical setting time is about 3-4 hours.

To prevent the radio interface from the useless paging processes when an MS is not reachable, the MS logs off and on when it is switched off and on. These corresponding processes are called as IMSI detach and IMSI attach, respectively[2]. When an MS switches off, in the VLR a bit in the user data record of the affected subscriber is set, as a result of an IMSI detach procedure. This bit indicates that the subscriber shall not be paged when a call or a short message arrives. When the mobile is switched on, the IMSI attach procedure will reset this bit, and from that time the subscriber can be paged again.

Mobile Calls in NSS

In this subsection first we discuss in which way a mobile terminated (MT) call can be established in the NSS (ETSI, 1999), then we explain the management of the mobile originated (MO) calls, and finally we show the case of the PLMN internal MS-to-MS calls.

Let us start with the case when we call a mobile station from the ISDN. The dialed number (MSISDN) provides enough information (country code and national destination code) to determine the home PLMN of the called subscriber. Therefore, the ISDN can use the dialed MSISDN to establish the call up to the GMSC of the called network. The way in which the call is established up to this point is exactly the same as we discussed earlier, when we explained the ISUP. However, from the GMSC the dialed number cannot be used any longer, since it does not contain any information to find the MSC that serves the called party (visited MSC – VMSC). To be able to continue the call establishment from this point, another telephone number must be used, which identifies the VMSC instead of the subscriber. This temporary number is the Mobile Station Roaming Number (MSRN). There is a telephone number pool assigned to each MSC to act as MSRNs. These numbers cannot be called directly, but the network knows which number pool is assigned to a certain MSC, therefore these numbers can be used for routing the calls between the GMSC and the VMSC.

The GMSC has to get an MSRN from the VMSC. However, the GMSC does not know, which MSC acts as the VMSC, this piece of information is stored only in the HLR. Therefore, at first the HLR has to be interrogated. The HLR, in turn, instructs the VLR of the VMSC to provide an MSRN. The HLR forwards the MSRN to the GMSC, which can use this number later on instead of the dialed one to establish the voice path toward the VMSC.

We can follow the process in Figure 6 in details.

From the ISDN the call can be routed to the GMSC by using the dialed MSISDN. The gateway interrogates the HLR by a sendRoutingInfo message with the MSISDN as a parameter. The HLR sends a provideRoamingNumber message to the VLR of the VMSC. In this message, the parameter will be the IMSI. As a response, the VLR sends an MSRN to the HLR, and stores the IMSI-MSRN assignment in an internal table. The HLR passes the received MSRN to the GMSC. As soon as it happens, the gateway can start to establish the voice path with the help of ISUP messages. The only difference from the ISDN call set-up is that here the MSRN is used instead of the dialed MSISDN in the Called Party Number parameter of the IAM message. We have to draw the attention of the reader that the VMSC is not for sure in the same network as the GMSC, therefore if the called subscriber roams, an international call has to be established.

In the case of a mobile originating call, the situation is not as complicated as this. Let us assume that a mobile subscriber dials an ISDN number. The VMSC of the caller can route the call to the suitable GMSC, which can pass the call to the ISDN. During the call establishment, we use the well-known ISUP messages.

Figure 6. Mobile terminated call in NSS

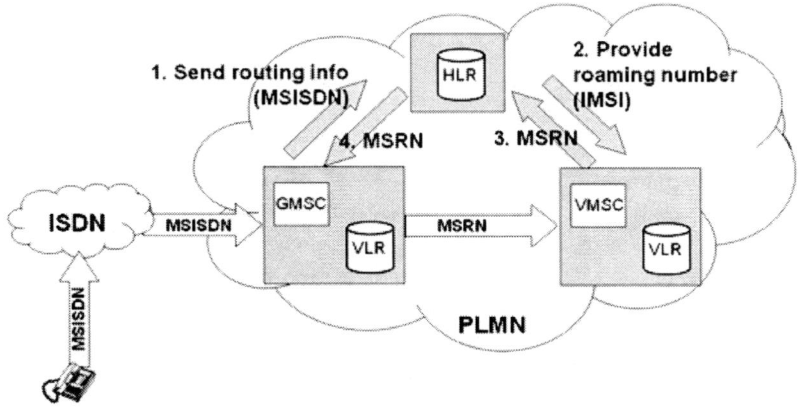

When a mobile subscriber calls another subscriber of a different PLMN, the situation is quite similar to the ISDN case. At the MO side, everything is the same if the "called network" is the ISDN or another PLMN, and similarly, at the MT side everything is the same if the call arrives from ISDN or from another PLMN. The "calling network" can route the call to the GMSC of the called PLMN by using the dialed number, and the MT part of the call will be exactly the same as in the situation when the call arrived from ISDN. The unique case is the PLMN internal mobile-to-mobile call. In this case, a mobile operator has two options. The calling MSC either can route the call to a GMSC and the GMSC handles the call in the same way as if it had arrived from an other network, or can act as a GMSC and interrogates the HLR directly. In the first case, the voice path will not be optimal, since it goes through the GMSC unnecessarily, while in the second one each MSC must have a gateway function.

Short Message Service

In this subsection we introduce the Short Message Service (SMS) and discuss the way, how a short message (SM[3]) is handled in the NSS.

The Short Message Service is a pure signaling service (it does not use voice circuits at all). SMS is a datagram service. In a datagram service, there is no direct connection between the sender and the receiver of a certain message. A datagram is sent to a center point and the center delivers it to the addressee. In our case, the center point is called as Short Message Service Center (SMSC). In reality, the Short Message Service is not only one service, but two. We differentiate mobile originated and mobile terminated services (MO SMS and MT SMS, respectively). MO SMS is responsible for sending a short message to the SMSC, while MT SMS is responsible for the delivery.

We can follow the steps of a successful SM transmission in Figure 7. In the figure the identifiers and network elements related to the sender are marked by "A" while those related to the addressee by "B".

The mobile originating short message transmission consists of only one message. The forwardSM message contains the telephone number of the addressee (MSISDN B) and the short message itself. On MAP level, there is no answer for it, but the transaction must be closed by an empty TCAP End message.

When the SMSC receives a SM, it tries to forward it to the addressee. At first, it has to find out, which is the VMSC of the addressee. Therefore, the SMSC interrogates the HLR by a sendRoutingInfoForSM message. The answer contains the Global Title of the VMSC/VLR B.

Figure 7. Mobile originated and mobile terminated SMS

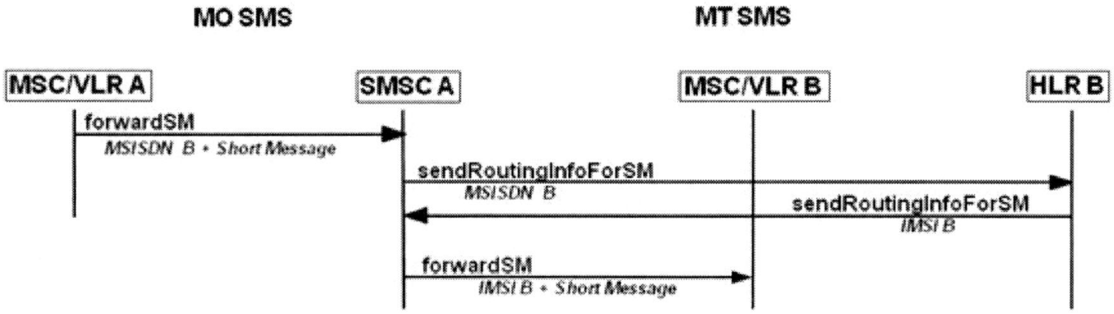

At this point, the SM can be delivered to the visited MSC of the addressee. If the SM is delivered successfully, the answer is also an empty TCAP End message.

In case when the SM delivery is not successful, for example because the MS is switched off (Figure 8) the TCAP End message contains a Return Error component describing the reason of the failure (in our case: Absent Subscriber). In the VLR B in the corresponding user data record, a bit is set to indicate a short message is waiting to be delivered. When the SMSC A receives the error report about the failed SM, it requests the HLR B to signal as soon as the addressee is available again by a reportSMDeliveryStatus message.

When MS B is switched on, the MSC/VLR B informs the HLR B that the mobile station is ready for SM. In turn, the HLR B sends an alertServiceCentreWithoutResult message to the SMSC A, indicating that the MS B is now able to receive short messages. Upon receiving this alerting message, the SMSC A gets routing information again from the HLR B, and delivers the SM to the VMSC B.

It may be asked why is there a request to get routing information for the SM again. It is superfluous, if the addressee has switched the mobile on in the same MSC/VLR area as it was switched off. However, if the addressee has moved into another MSC/VLR area, this second HLR interrogation is necessary to find out the new location of the subscriber. In this case, the HLR B will be informed about the availability of MS B by an updateLocation message instead of a readyForSM message.

Protocols of the A interface

The A interface is located between an MSC and a BSS. Being a network interface, the A interface uses the MTP with the SCCP as transport protocols. At the A interface the GSM-specific signaling is performed by a special SCCP user protocol, which is called as Base Station Subsystem Application Part (BSSAP) (3GPP, 2003). BSSAP itself can be separated into the Base Station Management Application Part (BSSMAP) and the Direct Transfer Application Part (DTAP).

BSSMAP is used either to control the BSS as a unit (e.g. to reset it) or to control its radio resources (e.g. to assign radio channels to calls). At the radio interface, the functional equivalent of the latter is the radio resource management (RR) protocol. (E.g. if the MSC requests from the base station to assign a traffic channel to a call by a BSSMAP message, the base station will, in turn, communicate the channel assignment with the mobile station by RR messages.)

The DTAP is used to carry the call control (CC) and mobility management (MM) messages exchanged between the NSS and an MS through the A interface in a transparent way. The base

Figure 8. Short message transmission: mobile station is switched off

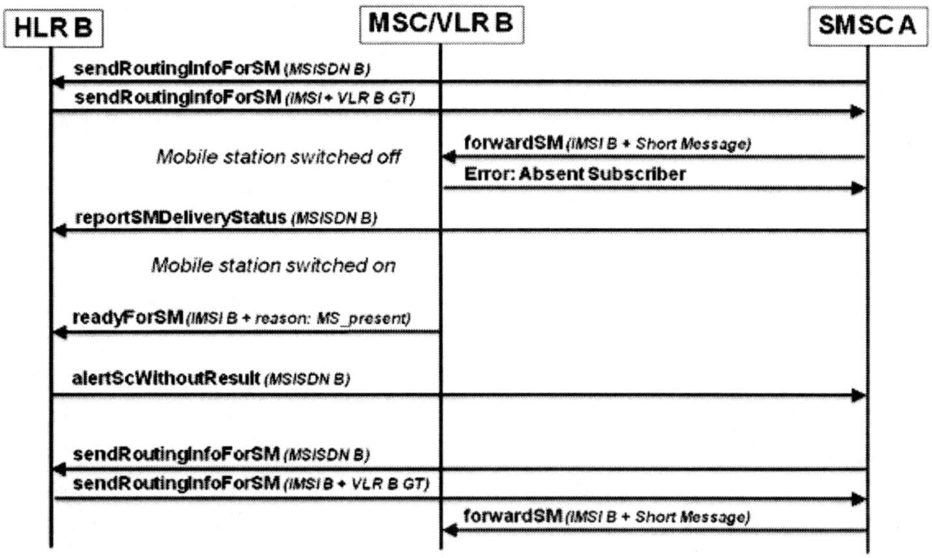

station does not process the DTAP messages it simply relays them through the radio interface to and from the mobile stations.

The MM messages are used – among others – for location update, IMSI attach and detach, authentication, TMSI reallocation, IMEI check and MO call request. The CC protocol is responsible for the call establishment and release, and can be considered as the variant of the DSS1.

The BSSAP messages are inserted into Data parameters of SCCP messages. The DTAP messages and the call-related BSSMAP messages are carried in connection-oriented (Class 2) SCCP messages; while the non-call related BSSMAP messages use the connectionless (Class 0) service of the SCCP. The only function of the BSSAP protocol is to differentiate between the BSSMAP and DTAP messages. BSSAP inserts an octet with the value of 0 in front of a BSSMAP message and an octet with the value of 1 in front of a DTAP message.

Signaling at the A Interface

In this subsection we deal with the basics of the signaling at the A interface through examples. First of all, let us show the steps of a signaling sequence at the A interface in general.

As we discussed earlier, the DTAP protocol carries the CC and MM messages in a transparent way. However, the very first message of a connection is an exception to this general rule. The initial message of a connection informs the MSC about the reason why a mobile station wants to connect to the network: Classmark Service (MO call) Request, Paging Response (MT call), Location Update or IMSI detach. At the A interface, the initial message is packed into a BSSMAP Complete Layer 3 Info message, that contains also the identifier of the cell in which the connection was initiated by the mobile station.

Regardless of the type of the connection, the next step is the authentication. If the authentication is successful, the MSC requests the base station to start the ciphering at the radio interface. In cipher mode, a new TMSI may be allocated and the IMEI may be checked.

The continuation will depend on the type of the connection.

In the following figures, we show the message sequences of a mobile originated and a mobile terminated call set-up as well as a call clearing. We display the most important parameters of the messages as well as the interworking between the A interface and the NSS protocols. The messages in parenthesis are optional.

Mobile Originating Call

When a mobile station wants to make a call, it sends a CM Service Request message (Figure 9). In this message, the mobile station uses the TMSI in general. (The mobile equipments differ in many characteristics, such as maximal transmission power, encryption algorithm, services supported. The classmark (CM) describes the features of a ME, therefore must be provided by the mobile station at the very beginning of each connection.) The CM Service Request is encapsulated in a Complete Layer 3 Info message that also contains the identity of the cell, in which the call attempt has been made.

The authentication is initiated by an Authentication Request message, which contains the RAND. From the RAND and the K_i stored on the SIM, the mobile station calculates the signed response that is sent back to the MSC in an Authentication Response message.

After a successful authentication, the MSC orders the BSS to start ciphering at the radio interface. Cipher Mode Command contains the ciphering key (K_c). The command is acknowledged by a Cipher Mode Complete message. The next two steps are optional, but in practice, they are typically executed. In cipher mode, we can allocate a new TMSI and check the IMEI. (The Identity Request message can be used to ask any identity of a mobile station, but it is most frequently used for IMEI check.) Due to the lack of space, we have not represented in the figure that after the receipt of the IMEI in the Identity Response message, the MSC sends a checkIMEI message to the EIR that responds whether the ME is black, gray, or white listed.

The continuation is quite similar to the DSS1 call establishment sequence discussed in Chapter V of this book.

The Setup message contains the requested bearer capability as well as the dialed number. At the A interface it is acknowledged by a Call Proceeding message, and then a radio channel is assigned to the call (obviously, this step is missing from the DSS1). Meanwhile the MSC starts to establish the voice path toward the GMSC by an ISUP Initial Address Message (here we suppose that the called party is a subscriber of another network). The Address Complete Message is sent after the user of the called side has been informed about the arrival of the call (most typically when ringing starts). The Alerting message at the A interface indicates that the ringing tone is switched on. When the call is answered, the voice path can be connected through after a Connect and Connect Acknowledge message exchange.

The signals described above travel at the A interface in Data parameters of connection oriented SCCP messages. Complete Layer 3 Info is transmitted in a Connection Request message. (The Connection Confirm, which acknowledges the connection request, does not contain a Data parameter, so it does not carry a BSSAP message.) All the following BSSAP messages in the call establishment sequence are transmitted in SCCP DT1 messages.

Mobile Terminating Call

In Figure 10, we assume that the mobile station is called from another network (actually from the ISDN).

As we could see before, when the call reaches the GMSC of the home PLMN of the called MS, the gateway has to get a roaming number (MSRN) to be able to continue establishing the call. Therefore, the GMSC requests the HLR to send routing information to the call, which, in turn, requests the visited MSC of the called party to provide a roaming number. As soon as the roaming number

Figure 9. Mobile originated call

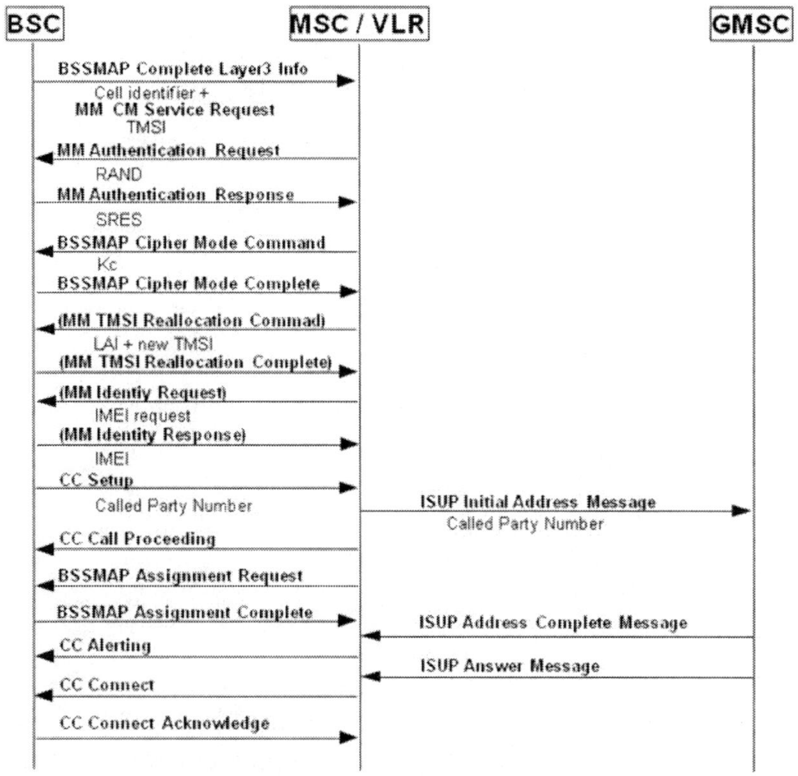

arrives to the GMSC, it can continue the establishment of the call by an IAM message using the MSRN instead of the originally called MSISDN in the Called Party Address parameter. Upon receiving the IAM message, the MSC instructs all the base stations of that location area, in which the called MS is supposed to be, to page the called party in each cell. At the A interface the Paging command is carried in a connectionless SCCP message (in a UDT). If the called MS is reachable, it connects to the GSM network by a Paging Response. At the A interface this message is packed into a Complete Layer 3 Info message with the cell identifier. This message travels in the Data parameter of an SCCP CR message. The CR is also acknowledged by an "empty" SCCP CC message (not shown in the figure). All the following BSSAP messages are transmitted in SCCP DT1 messages.

The continuation is similar to that we have discussed in the scenario of a MO call. The steps are: authentication, switching to cipher mode, the optional TMSI reallocation and IMEI check. After these steps, the MSC can actually start to establish the call. The first message is a Setup, containing – apart from the Bearer Capability parameter – the caller's telephone number, if the number presentation is enabled. The acknowledgement for the Setup is a Call Confirmed message (it differs from DSS1 where this message is called as Call Proceeding). The call confirmation is followed by a radio channel assignment. When the called mobile starts to ring, an Alerting message is transmitted and as soon as the call has been answered, the Connect and Connect Acknowledgement message exchange takes place.

Figure 10. Mobile terminated call

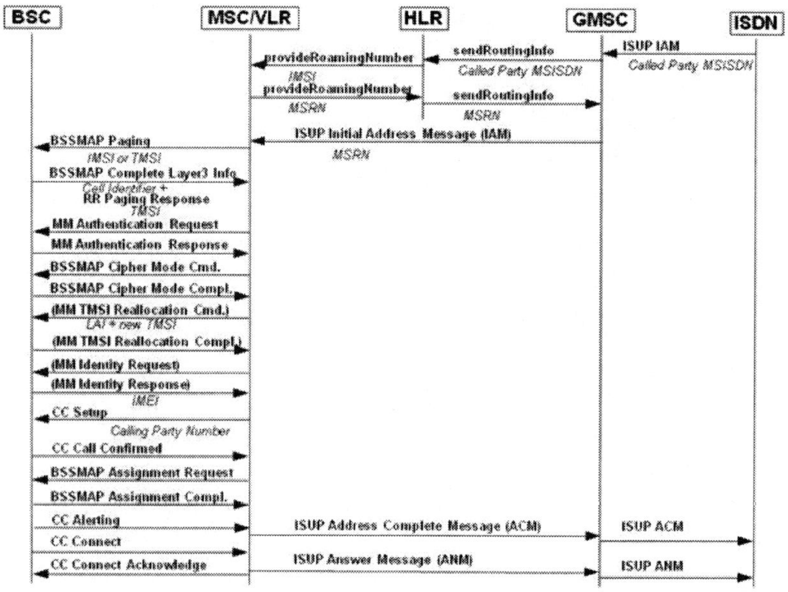

Disconnection of a Call

When either of the parties wants to make an end to a call a Disconnect message starts the disconnection process at the A interface (Figure 11). The MSC acknowledges this message by a Release message, which is acknowledged by a Release Complete message as the further step. As soon as the call has been released, the MSC instructs the base station by a Clear Command message to release the occupied traffic channel at the radio interface. As soon as the traffic channel has been released the base station answers by a Clear Complete message. This is the last message of the SCCP connection that also has to be released by an SCCP RLSD and RLC message pair. Since these messages do not contain BSSAP messages, they are not indicated in the figure.

GPRS

In this subsection, we give an overview about the General Packet Radio Service (GPRS) network elements, and we discuss the GPRS network protocol (GPRS Tunneling Protocol – GTP) in details, by examining several typical signaling scenarios.

The introduction of the GPRS made it possible to offer packet-switched data transmission in GSM networks. To provide this service on the one hand several additional new functional units must have been installed in the existing GSM networks, and on the other, the functionality of

Figure 11. Disconnection of a call

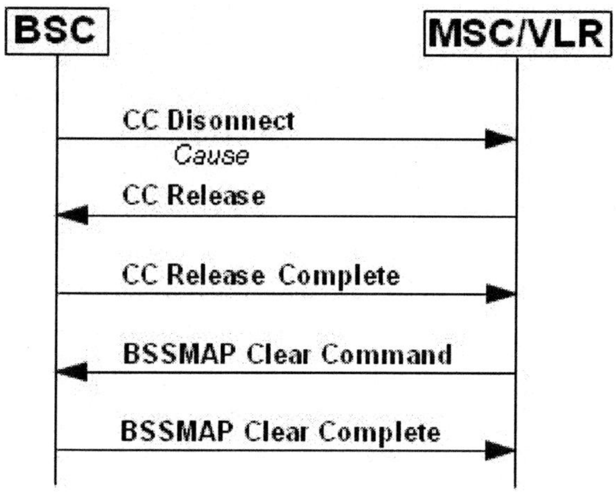

some other network elements must have been extended (Figure 12).

The GPRS networks are connected to the Packet Data Networks (PDN) by Gateway GPRS Support Nodes (GGSN). A PDN may be the Internet or a corporate network, and it is IP-based in general. The Serving GPRS Support Nodes (SGSN) route the data packets between the GGSN and the base stations. An SGSN is actually a powerful packet switch with an integrated VLR-like database, which temporarily stores the GPRS user profiles of the subscribers who are served by that particular SGSN. To retrieve these pieces of information from the HLR, the SGSN must have an interface (the G_r interface) toward the HLR. Therefore, HLR must store some new information elements in the user data records, especially the Global Title and the IP address of the serving SGSN and GPRS-related subscription information. The Border Gateway takes care of all inter-PLMN GPRS traffic, which is particularly important in case of roaming. A BG can be stand-alone equipment but in many cases, it is integrated in either an SGSN or a GGSN. The Charging Gateway (CG) collects charging information from the SGSNs (related to the use of the own network resources) and from the GGSNs (related to the use of the external PDNs) (Heine, 2002).

The BSS must have been extended, too. The new building block is the Packet Control Unit (PCU). The PCU serves as an interface between the BSC and the IP network, therefore it performs some code conversion functions, and it is responsible for the radio control functions, as well.

There are several new optional interfaces defined between the new GPRS units and the existing GSM databases. The G_s interface between an SGSN and a VLR makes it possible – among other functions – to perform common attach and detach processes to and from both the GPRS and GSM networks, and the common Location Update. The G_c interface between a HLR and a GGSN makes it possible for the latter to get routing information for the mobile terminated packet-switched connections directly from the HLR, not only through a bypass, involving an SGSN and its G_r interface. At the G_c, G_r, and G_s interfaces, SS7 is used, while the communication between the GPRS network elements is IP-based.

In the GPRS, networks there are three kinds of mobile stations. A Class A MS supports completely simultaneous operation of both packed-switched (PS) and circuit-switched (CS) services. Even if the MS is involved in a CS or in a PS connection it is able to monitor the paging channel of the other kind of service and react upon it. A Class B

Figure 12. GPRS architecture

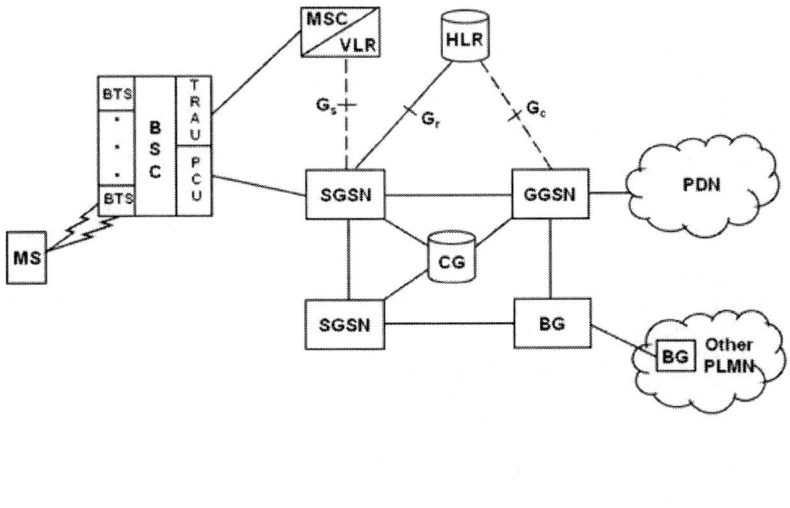

MS is able to monitor the paging channel of both services simultaneously, but if it is involved in a connection, it is no longer required to listen to the paging channel of the other service. A Class C mobile can be attached either to the CS or to the PS network. Therefore, all old-fashioned non-GPRS capable MSs are Class C devices, but on the other hand, all devices providing only mobile Internet access (not only the Internet terminals, but telemetric equipments, intelligent sensors, etc.) are also Class C MSs.

The nature of the packet-switched data transmission is quite different from that of the voice calls. While the voice calls require the continuous transmission of voice samples with the short period of 125s, the data connections are discontinuous. Between two data bursts, there is a long gap, in general. It is quite likely that in these gaps the MS changes its location, so it must be paged several times during a data call. To reduce the paging load at the radio interface, smaller areas are defined in which an MS is to be paged. These are the Routing Areas (RA). A RA is always a subset of a Location Area. The price we have to pay for the reduction of the paging load is the increased number of the location updates. A MS attached to the GPRS network must perform a GPRS Routing Area Update process at every time it changes a RA.

Another way to reduce the paging traffic is the introduction of GPRS mobility management (GMM) states (Figure 13). In Idle state, the MS is not connected to the GPRS network. If the MS performs a GPRS Attach process (connects to the network), it enters into the Ready state. In Ready state, the MS reports every cell change, so the SGSN knows exactly in which cell the MS is, therefore there is no need to page the MS if a user data is requested to be delivered to it. While the MS continuously transmits frames, it remains in Ready state. If the MS does not send a frame for a certain period determined by the Ready timer (typical duration is 44 ms), it switches to Standby state. In Standby state, the GPRS network knows the location of the MS at RA level. Therefore, if a downlink burst arrives, the MS must be paged in every cell of the corresponding RA, but after the reception of the first frame of the burst, the MS enters the Ready state again. An MS becomes Idle when it performs a GPRS Detach process (disconnects from the network) or if leaves the GPRS coverage area (implicit detach).

If the optional G_s interface between the SGSN and a VLR has been implemented, the circuit-switched paging load can also be reduced. If the MS is in Ready state it has to be paged only in one cell and if it is in Standby state, then only in a RA instead of a LA.

Figure 13. GPRS mobility management states

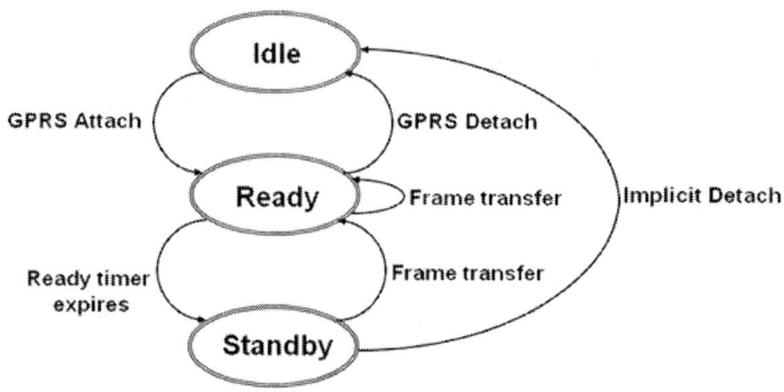

GPRS Tunneling Protocol

When an application on a MS sends an IP packet, the IP header contains the IP address of the MS and the IP address of the destination, e.g. an Internet address. In the core network the packet must be sent from the serving SGSN to a GGSN, both having its own IP address. Therefore, the IP packet received from the mobile station must be encapsulated into an additional packet. However, an additional problem arises: since the same path is used by many MSs, therefore a new protocol is needed to differentiate between packets from different MSs and between different applications running on the same mobile. The protocol used in GPRS to solve this problem is the GPRS Tunneling Protocol (GTP) over TCP or UDP. (3GPP, 2002)

GTP establishes a path between an SGSN and a GGSN or between two SGSNs if there is at least one active data connection (PDP Context) between these two nodes. A path is identified by the source and destination IP addresses and port numbers. For each PDP Context there is one GTP tunnel established in the corresponding path. A GTP tunnel provides the means to transmit packets in a transparent way between the nodes that may be in the same or in different GPRS networks. Each tunnel is identified by a Tunnel Identifier (TID) that consists of the IMSI of the MS and the Network Service Point Identifier (NSAPI). The NSAPI is allocated to the PDP Context by the MS to differentiate between the possible multiple data connections on the corresponding mobile station. Each tunnel contains two user data flows (one-one for each direction) and two signaling flows.

To illustrate that we have discussed so far, let us look at how we establish a mobile originated and a mobile terminated GPRS connection. In the following examples, we suppose connections between a MS and the Internet.

Mobile Originated GPRS Connection

Once the MS has attached to the GPRS network, it can establish a packet-switched connection by sending an Activate PDP Context Request (ACT_PDP_CT_REQ) message. This message contains – further to other parameters – the NSAPI, the requested Quality of Service (QoS), the requested PDN (in our case an Internet) address (Figure 14). The SGSN determines the suitable GGSN toward the requested address and establishes a GTP tunnel by sending a Create PDP Context Request (CT_PDP_CT_REQ) message. This message contains the MSISDN of the originating MS and the IP address of the SGSN. Do not forget that the tunnel identifier contains the IMSI of the originating MS and the NSAPI. The GGSN will authenticate the MS through a RADIUS server. As soon as the authentication is successful, the GGSN gets a dynamic IP address from a DHCP server, and transmits this IP address to the SGSN in a Create PDP Context Response (CT_PDP_CT_RSP) message. The IP address then will be assigned to the MS by an Activate PDP Context Accept (ACT_PDP_CT_ACC) message. Since from this point the MS has an IP address, the data transfer can start.

Mobile Terminated GPRS Connection

We can terminate a data call on a MS (Figure 15) only if it has a fixed IP address. The PDN (in our example the Internet) can route the IP messages to the suitable GGSN with the help of the fixed IP address of the MS. The GGSN – if the optional G_c interface to the HLR is installed – can interrogate the HLR for the IP address of the serving SGSN by a MAP sendRoutingInfoForGPRS message. In the reply, the HLR includes the requested IP address. (If the G_c interface is not established, the GGSN can perform the HLR interrogation indirectly by the help of an SGSN having a G_r interface to the HLR. In this case, the GGSN sends a GTP Send Routing Info for GPRS Request message to the

Figure 14. Mobile originated GPRS connection

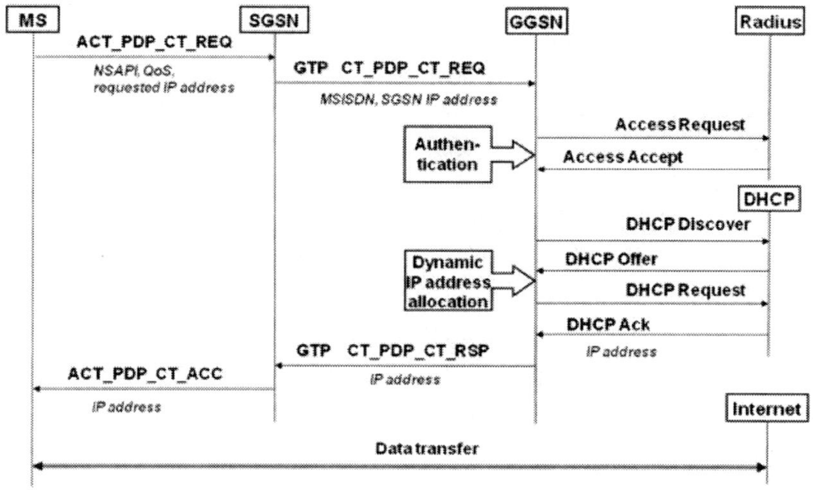

corresponding SGSN, which interrogates the HLR, and sends the result to the GGSN in a GTP Send Routing Info for GPRS Response message.)

As soon as the GGSN learns the IP address of the serving SGSN, it sends a GTP PDU Notification Request (PDU_NOT_REQ) message. Upon receiving this message, the SGSN pages the called MS. (We emphasize again, that the IMSI of the called MS – which is the parameter of the Paging message – is a part of the TID of the GTP message.) When the MS recognizes the Paging message, transmits an empty frame as a paging ac-

Figure 15. Mobile terminated GPRS connection

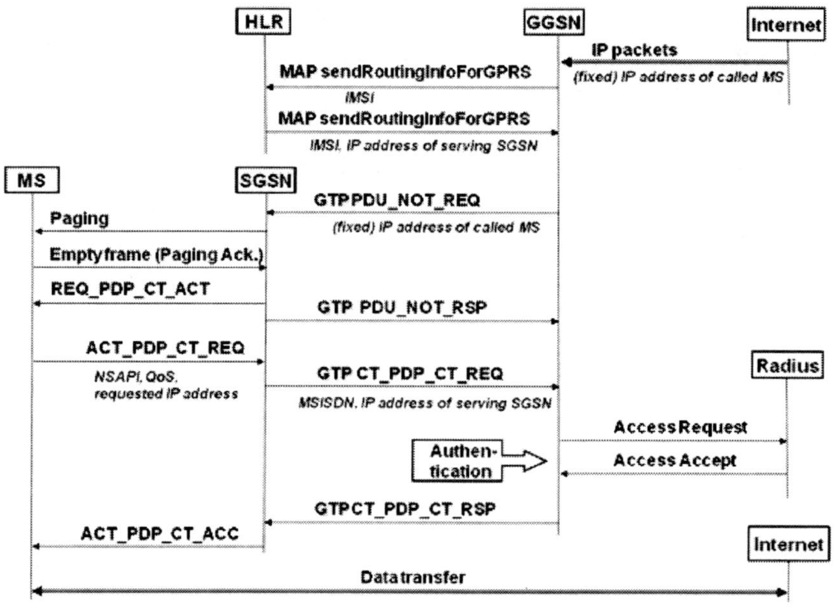

knowledgement and enters the Ready state. As the next step, the SGSN calls upon the MS to activate a PDP Context (to receive the GPRS call) by a Request PDP Context Activation (REQ_PDP_CT_ACT) message. In parallel with it, the SGSN informs the GGSN that the called MS has been paged by sending a GTP PDU Notification Response (PDU_NOT_RSP) message. The continuation is quite similar to the MO GPRS connection. The only difference is the lack of the dynamic IP address allocation, since the called MS has a fixed IP address.

CONCLUSION

In this chapter, we have given an overview about the GSM and GPRS networks and their protocols used in the Network and Switching Subsystem.

We have discussed in details the SS7 protocols, which are used primarily in the mobile networks, such as SCCP and TCAP. We have introduced the most important controlling functions of the GSM (Authentication, Location Update, Mobile Originated and Terminated call establishment, and SMS) and we have showed the ways in which the MAP protocol handles these scenarios. We have presented the protocols which are used at the A interface, between the NSS and the BSS. We have showed the call control scenarios as illustrations.

At the end of this chapter, we have presented the GPRS extensions to the existing GSM network and we have showed the way in which data calls can be originated and terminated in the GPRS as well.

REFERENCES

ETSI. (1998). *11). GSM 01.02. General description of a GSM public land mobile network*. France: Sophia Antipolis.

ETSI. (1999). *04). TS 100.524 (GSM 03.04). Signalling requirements relating to routing of calls*. France: Sophia Antipolis.

3GPP. (2001). *01). TS 0320. Security related network functions*. France: Sophia Antipolis.

3GPP. (2002). *12). TS 09.60. GPRS Tunnelling Protocol (GTP) across the Gn and Gp interface*. France: Sophia Antipolis.

3GPP. (2003). *09). TS 08.08. Mobile-services switching centre - base station system (MSC - BSS) interfaceL Layer 3 specification*. France: Sophia Antipolis.

3GPP. (2004). *03). TS 09.02. Mobile application part (MAP) specification*. France: Sophia Antipolis.

Heine, G. (2002). *GPRS signaling & protocol analysis*. Karlsruhe, Germay: INACON.

ITU-T. (1997. 06). *Recommendations Q.771-q.775. Specifications of signalling system no. 7 – transaction capabilities application part*. Geneva, Switzerland.

ITU-T. (2001). *Recommendations Q.711-Q.714. Specifications of signalling system no. 7 – signalling connection control part*. Geneva, Switzerland.

KEY TERMS AND DEFINITIONS

Base Station Subsystem Application Part (BSSAP): BSSAP is the protocol employed across the A interface of the GSM systems. It can be divided into two protocols: Base Station Subsystem Management Application Part (BSSMAP) is used to control the Base Station Subsystems (BSS), while Direct Transfer Application Part (DTAP) is used to transport Mobility Management and Call Control information between a BSS and an MSC.

General Packet Radio Service (GPRS): GPRS is a packet-switched mobile data service in the GSM and UMTS systems.

GPRS Tunneling Protocol (GTP): GTP is an IP-based communications protocol used to tunnel GPRS user packets between GPRS Support Nodes (SGSN and GGSN).

International Mobile Subscriber Identity (IMSI): IMSI is a unique number associated with all GSM and UMTS network subscribers. It is stored in the SIM inside the phone.

Mobile Application Part (MAP): MAP is an SS7 protocol, which provides an application layer for the various nodes in GSM, GPRS and UMTS core networks. MAP is used to access Authentication Centre (AuC), Equipment Identity Register (EIR), Home Location Register (HLR), Mobile Switching Center (MSC), Serving GPRS Support Node (SGSN), Short Message Service Center (SMSC) and Visitor Location Register (VLR).

Network and Switching Subsystem (NSS): NSS is the component of a GSM system that carries out call switching and mobility management functions. It is often referred to as the GSM core network.

Public Land Mobile Network (PLMN): PLMN is a land mobile network that is operated by a service provider.

Signaling Connection Control Part (SCCP): SCCP is an SS7 protocol that carries out – among many other functions – inter-network and connectionless signaling.

Transaction Capabilities Application Part (TCAP): TCAP is an SS7 protocol that controls multiple concurrent dialogs in NSS.

ENDNOTES

1 We have to mention that on one hand to reduce the signaling at the radio interface, the typical setting is to use the IMSI at Location Update and TMSI at all the other cases, and on the other hand a lot of operators use the second case (to get authentication triplets directly from the AuC) even when the first case (to get authentication triplets from the old MSC/VLR) could have been used.

2 The strange names of these procedures are due to an accident of the standardization; the procedures do not detach or attach the IMSI. IMSI detach is simply a log off, and IMSI attach is the log on procedure.

3 In fact, SMS is the acronym of the service itself, but in everyday life, we often use this acronym as a synonym of the short message. To differentiate between the two we use SM as the acronym of short message.

APPENDIX

Table of abbreviations

ACT_PDP_CT_REQ	Activate PDP Context Request (message)
ACT_PDP_CT_RSP	Activate PDP Context Response (message)
ATM	Automated Teller Machine
AuC	Authentication Center
BG	Border Gateway
BSC	Base Station Controller
BSS	Base Station Subsystem
BSSAP	Base Station Subsystem Application Part
BSSMAP	Base Station Subsystem Management Application Part
BTS	Base Transciever Station
CC	Connection Confirm (message)
CC	Call Control
CdPA	Called Party Address
CG	Charging Gateway
CgPA	Calling Party Address
CM	Classmark
CR	Connection Request (message)
CREF	Connection Refused (message)
CS	Circuit-switched
CT_PDP_CT_REQ	Create PDP Context Request (message)
CT_PDP_CT_RSP	Create PDP Context Response (message)
DHCP	Dynamic Host Configuration Protocol
DLR	Destination Local Reference
DNS	Domain Name System
DPC	Destination Point Code
DSS1	Digital Subscriber Signaling System No. 1
DT1	DataForm1 (message)
DTAP	Direct Transfer Application Part
DTID	Destination Transaction Identifier
EIR	Equipment Identity Register
GGSN	Gateway GPRS Support Node
GMM	GPRS Mobility Management
GMSC	Gateway MSC
GPRS	General Packet Radio Service
GSM	Global System for Mobile Communication
GT	Global Title
GTP	GPRS Tunneling Protocol

HLR	Home Location Register
IAM	Initial Address Message
ID	Identifier
IF	Interface
IMEI	International Mobile Equipment Identity
IMSI	International Mobile Subscriber Identity
INAP	Intelligent Network Application Part
IP	Internet Protocol
ISDN	Integrated Services Digital Network
ISUP	ISDN User Part
ITU-T	International Telecommunication Union Telecommunication Standardization Sector
K_c	Ciphering Key
K_i	Individual Subscriber Key
LA	Location Area
LAI	Location Area Identifier
LAPD	Link Access Procedure on D channel
LU	Location Update
MAP	Mobile Application Part
MCC	Mobile Country Code
ME	Mobile Equipment
MM	Mobility Management
MNC	Mobile Network Code
MO	Mobile Originated
MS	Mobile Station
MSC	Mobile (services) Switching Center
MSIN	Mobile Station Identity Number
MSISDN	Mobile Station ISDN Number
MSRN	Mobile Station Roaming Number
MT	Mobile Terminated
MTP	Message Transfer Part
NSAPI	Network Service Access Point Identifier
NSS	Network and Switching Subsystem
OMC	Operation and Maintenance Center
OMSS	Operation and Maintenance Subsystem
OPC	Originating Point Code
OTID	Originating Transaction Identifier
PCU	Packet Control Unit
PDN	Packet Data Network
PDP	Packet Data Protocol
PDU	Protocol Data Unit
PDU_NOT_REQ	PDU Notification Request (message)
PDU_NOT_RSP	PDU Notification Response (message)

PIN	Personal Identification Number
PLMN	Public Land Mobile Network
PS	Packet-switched
QoS	Quality of Service
RA	Routing Area
RADIUS	Remote Authentication Dial In User Service
RAI	Routing Area Identifier
RAND	Random Number
REQ_PDP_CT_ACT	Request PDP Context Activation (message)
RLC	Release Complete (message)
RLSD	Released (message)
RR	Radio Resource Management
SCCP	Signaling Connection Control Part
SGSN	Serving GPRS Support Node
SLR	Source Local Reference
SM	Short Message
SMS	Short Message Service
SMSC	Short Message Service Center
SMSS	Switching and Management Subsystem
SP	Signaling Point
SPC	Signaling Point Code
SRES	Signed Response
SS7	Signaling System 7
SSN	Subsystem Number
TCAP	Transaction Capabilities Application Part
TCP	Transmission Control Protocol
TE	Terminal Equipment
TID	Tunnel Identifier
TMSI	Temporary Mobile Subscriber Identity
TRAU	Transmission Rate and Adaptation Unit
UDP	User Datagram Protocol
UDT	Unitdata (message)
VLR	Visitor Location Register
VMSC	Visited Mobile Switching Center

Chapter 7
UMTS:
3rd Generation Cellular Mobile Radio System

Péter Fazekas
Budapest University of Technology and Economics, Hungary

ABSTRACT

The aim of this chapter is to provide a brief yet comprehensive overview of the 3rd generation UMTS (Universal Mobile Telecommunication System) mobile network, with emphasis on its specific protocols. Therefore, in this chapter, the basic operation and protocol structure of UMTS network is described. The main architectural changes compared to GSM are shown, as well as the principles of the physical radio interface. The details of other relevant UMTS specific interfaces in the access network and their protocols are provided as well, along with the description of transport network solutions. The most relevant part of UMTS specifications, the radio interface protocols, are also presented.

INTRODUCTION

The worldwide success of the GSM network during the 90s set the base for further development of cellular mobile telephony systems. The need for supporting packet switched transmission and higher bitrate was recognized early, thus GPRS (General Packet Radio Service) and later EDGE (Enhanced Data-rates for GSM Evolution) services based on GSM networks were standardized and deployed by the end of 90's. These allow higher bitrates than GSM, by means of allocating multiple timeslots to a single user's transmission and applying higher order physical modulation scheme in case of EDGE. Although Internet access with GPRS/EDGE mobile devices is possible, the supported bitrates of these systems are far less than today's requirements. The highest theoretical bitrate with

DOI: 10.4018/978-1-60960-732-6.ch007

GPRS is 160 kbps and 480 kbps with EDGE; however practically achievable useful rates are just around the third of these values. Moreover, although QoS (Quality of Service) classes and guarantees were introduced in GPRS, the allowed delay and delay jitter values are so high (and so are the practical delay values in existing systems) that real-time packet switched services cannot be used over this network.

Beside the mentioned shortcomings of GSM/GPRS/EDGE networks, the ever-increasing signal processing capability that can be integrated into a handheld terminal allowed for the definition of new radio interface that supports higher bitrates (note that the bottleneck in mobile communications is usually the radio interface, because of hardware limitations and available bandwidth). As it is known from Shannon's law, the achievable bitrate is directly proportional to the used physical bandwidth; thus, it was inevitable to define a system that uses much wider frequency band (compared to the 200 kHz in GSM based networks) for transmission.

Although early discussions about the 3rd generation of mobile systems were already initiated in ITU (International Telecommunication Union) at the beginning of 90's, 3G standardization speeded up in 1998, when ETSI (European Telecommunication Standardization Institute) decided that Wideband Code Division Multiple Access (WCDMA) would be the base of 3G radio interface and a new standardization body, 3GPP (3rd Generation Partnership Project) was formed with the aim of defining the global 3G standard. Currently 3GPP has 367 individual members (companies) form Europe, USA, Canada, Japan, China and the Republic of Korea.

The new 3G system was developed to fulfill initial requirements, originating from the existing deployment of cellular networks and the timely needs of mobile telecommunication services. The non-exhaustive list of requirements includes:

- interoperability with legacy GSM network, including the possibility of inter-system handover
- 3G system should inherit as much as possible from 2G systems (including existing infrastructure, equipment, protocols, etc.)
- up to 2 Mbps bitrate (indoor) for stationary customer, 144 kbps for outdoor user of pedestrian speed and 64 kbps for vehicular terminals
- definition and differentiation of different QoS classes according to typical communication services, with different delay and loss guarantees
- support of different and time-variable bitrates to different users, support of asymmetric upload and download traffic
- possibility of multiplexing different services (e.g. real-time video, voice and file transfer) over the same connection
- support of packet switched and circuit switched services, maximizing common networking functions serving both types of traffic.

According to the requirements above, the UMTS system features the following main properties, in comparison with GSM/GPRS/EDGE:

- 5 MHz wide channels instead of 200 kHz
- instead of Time Division Multiple Access (TDMA) Direct Sequence Code Division Multiple Access (DS-CDMA) based radio interface
- new radio access network to maintain the new radio interface
- shares core part of the network with GSM/GPRS
- support of four QoS classes
- DS-CDMA allows for a frequency reuse factor of 1 (the same frequency band can be used in neighboring cells)
- because of a reuse factor of 1, UMTS allows the so called "soft handover": a ter-

minal maintains its connection with two or more base stations simultaneously

- support of Frequency Division Duplex (FDD) and Time Division Duplex (TDD) operation modes, namely uplink and downlink communications might be separated in the frequency domain or in the time domain.

The first stable version of the 3GPP standards was published in early 2000 and is referred to as Release'99. The 3G system standardized by 3GPP is called Universal Mobile Telecommunication System (UMTS). However, standardization of UMTS did not stop then, later releases of 3GPP standards contained enhancements for increased bitrate services and all-IP network. The main advance is the definition of all-IP core network in Release 4. Standardization of High Speed Downlink Packet Access (HSDPA) radio and IP Multimedia Subsystem (IMS) core network solutions was done in Release 5, for delivering up to 14.4 Mbps physical bitrate on the radio interface and for supporting IP based multimedia applications by the network, respectively. Release 6 contains the High Speed Uplink Packet Access (HSUPA) extension of the radio interface and the advanced Multimedia Broadcast Multicast Service (MBMS). Release 7 defines further improvements of HSDPA and HSUPA (together: High Speed Packet Access, HSPA, the advanced versions are often mentioned as HSPA+), increasing the bitrate by means of introducing higher order modulation schemes and multiple antenna solutions. The main novelty of Release 8 and Release 9 (the last frozen release up until now) is the radio interface and core network, called Long Term Evolution (LTE) and Evolved Packet Core (EPC) respectively, although these also contain advances of HSPA by introducing dual carrier and parallel dual stream transmission operation.

The rest of this Chapter consists of the following. First, the system architecture of the UMTS system is described, and then the basic operation and parameters of the radio interface are introduced. This is followed by the description of radio protocols.

It has to be emphasized that throughout the subsequent descriptions FDD is assumed, as this mode is the most widely used in practice. Therefore, issues (functions, channels, sublayers, etc.) that are defined for TDD operation are not covered in this chapter. The information herein is based on the latest unchanged release of 3GPP standards, as given in the References section. This means that standard earlier than Release '9 is used if that is the latest unchanged version (and later releases are automatic updates of these).

SYSTEM ARCHITECTURE AND RADIO OPERATION

The System

The UMTS system is depicted in Figure 1. As it is apparent, the core network part of the system is inherited from GSM/GPRS, that is the main switches (Mobile Switching Center – MSC, Gateway MSC – GMSC) for Circuit Switched (CS) traffic and the routers (Serving GPRS Support Node – SGSN, Gateway GSN – GGSN) for Packet Switched (PS) traffic, along with the databases known from the description of the GSM system are used to serve UMTS subscribers as well. These networking elements together form the Core Network (CN) of the system. According to the above distinction of CS and PS traffic, the corresponding parts of the CN are called CS and PS domains. The standards define a third domain, namely broadcast (BC) domain of the CN; however, this is out of the scope of this chapter.

The roles of CN elements are the same as it was shown for GSM/GPRS network. It is worth noting that Release 4 of the standards define the handling of CS traffic on a per packet basis. For this purpose, instead of the MSC and GMSC devices two other elements, namely the MSC

Figure 1. UMTS system architecture

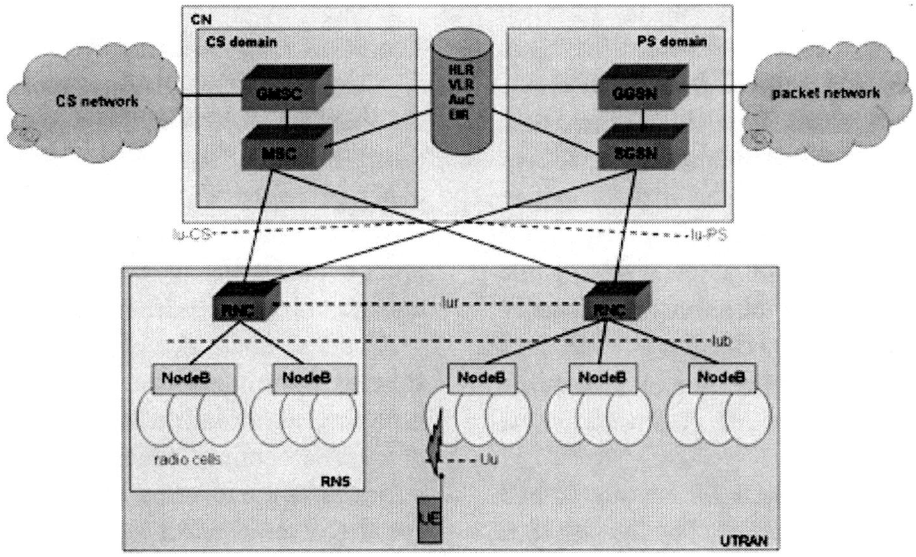

Server and Media Gateway are required. The MSC Server handles the control plane of the protocol stack, while Media Gateways "translate" the traffic of CS circuits into flow of IP packets.

The other main part of the network is the radio access system, which is called UMTS Terrestrial Radio Access Network (UTRAN, in other sources Universal Terrestrial Radio Access Network). The general technology covering the radio interface and access network is usually referred to as UMTS Terrestrial Radio Access (UTRA).

The UTRAN (3GPP 25.401) consists of several Radio Network Subsystems (RNS). Each RNS is controlled by a Radio Network Controller (RNC) device, which manages several base stations, referred to as Node B in UMTS terminology. The terminals attached to the system are called User Equipments (UE); these are further subdivided into the UMTS Subscriber Identity Module (USIM) cards and the Mobile Equipments (ME).

It is worth noting that in terms of physical range and cost, the UTRAN takes up the majority of the network. Actually, the CN part may reside in a server room, while the UTRAN must reach all areas to be covered by the system. It is apparent from the system architecture, that during the deployment of an UMTS network the whole UTRAN should be built, however existing base station sites and the high capacity transport network inherited from the GSM system might be utilized.

The role of RNC is of utmost importance to the system. This equipment is responsible for controlling the radio interface and the radio communication of customers. Although physically it may be placed far away from the actual base stations, yet radio protocols are operating between the RNC and the mobile: namely, RNC does the Layer 2 data processing for a connection. Data processing includes appropriate framing and retransmissions, among others. RNC is also responsible for distributing and scheduling radio resources (radio capacity), and maintains the transmission between the UTRAN and the CN. Load control and admission control is also performed in the RNCs. Considering that RNC usually resides physically far away from the base stations and mobiles (and actually accessed through a large scale transport network), this indicates that this kind of centralized intelligence of the UTRAN is not optimal, because significant delays could deteriorate performance.

Compared to the RNC, Node B operation is simple and restricted to maintain the physical connection, as well as to create the radio signal from the digital data received from the network. This latter mainly means digital signal processing, including channel coding, interleaving, spectrum spreading, and modulation. Fast transmit power control of terminals is also done here.

Regarding the terminals, USIM is responsible for the authentication of the subscriber (actually USIM represents the subscriber), while ME does all processing related to any CS or PS transmission/reception. USIM and ME communicate via the standard *Cu* interface.

The radio interface between the mobile and the Node B is the *Uu* interface. The *Iub* interface specifies the communication between the Node Bs and the RNCs. *Iur* is specified in order to allow inter-RNC communication that is needed when a customer is performing transmission in a soft handover state through two different base stations controlled by two different RNCs. The connection point between the UTRAN and the CN is the *Iu* interface. According to the actual endpoints in the CN and the type of traffic, it is further divided into *Iu-CS* and *Iu-PS* parts. For the sake of precision the *Iu-BC* interface should also be mentioned, that is defined for broadcast transmission. This does not significantly differ from the other two, and is out of scope of this chapter.

All these interfaces of the UMTS system are standardized open interfaces. This makes it possible to install and use equipment from different vendors within an operator's network. This is something considered to be natural regarding handhelds of different manufacturers, however, it should be valid for other networking equipments described above. This open nature and hence the competitiveness brought into mobile industry has already started with GSM, and is considered to be one of the main factors in the success of GSM.

Radio Interface Basics

The novel concept, compared to GSM was the application of DS-CDMA technology on the radio interface. However, this was not completely novel, as one of the 2G systems, IS95 (widely deployed in the US) already used DS-CDMA. In fact, the term "wideband" was introduced in order to distinguish the new radio interface from that of IS95 (that is indeed a narrowband system).

The very basic idea of DS-CDMA is that instead of transmitting a bit, a faster code sequence is transmitted during the bit period (see Figure 2). In wireless communications, the bits to be transmitted have the value of ± 1, the code sequences are also a series of ± 1 values. The sequence to be transmitted is the product of the bit and the codeword, hence depending on the value of the bit, the codeword or its negation is transmitted (in this context, the negation of 1 is -1 and vice versa). In CDMA, terminology one element of the codeword is called a chip; the rate at which the chips are transmitted is consequently referred to as chiprate.

The applied codes are the so-called Walsh-Hadamard, or Orthogonal Variable Spreading Factor (OVSF) codes. Different codewords of this family are orthogonal to each other, meaning that the scalar product (chipwise product and sum) of two codewords (± 1 sequences) is zero, while the scalar product of a codeword with itself is naturally equal to the length of the codeword. This property can be used to create channels by applying different codewords (hence, the OVSF codes are called *channelization* codes in UMTS): in the base station, the data for each physical channel is multiplied by different codes and then summed up; this sum is transmitted over the air. At the terminal side, knowing the connection's own channelization code, the whole summed up stream should be scalar multiplied by the own code. Due to orthogonality, all data with other channelization codes result in zeros after the scalar multiplication, only the own connection's data will peak out.

Figure 2. Basic CDMA mechanisms

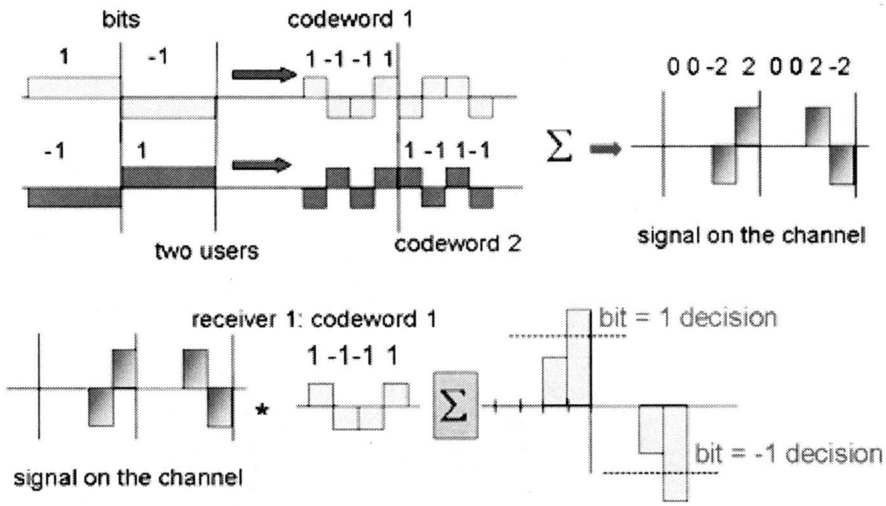

The set of codes can be represented as the nodes in a binary code tree, as shown in Figure 3. We denote node j (e.g. numbered from left hand side) at level i of the tree (where the root is considered to be level 0) by C_{ij}, representing a code of length 2^i The tree can be generated recursively: placing the length 1 code (a single 1) in the root, the two descendant codes of code j at level i are $C_{i+1,2j-1}=\{C_{i,j};C_{i,j}\}$ at the left branch and $C_{i+1,2j}=\{C_{i,j};-C_{i,j}\}$ at the right branch. Literally, this means that the left descendant of a code is created by doubling the codeword, while the right side descendant is created by means of putting the negation of the codeword behind the codeword.

This tree view and the method of constructing the tree have the advantage of showing the so-called code-blocking phenomenon. Namely, not all Walsh-Hadamard codes are orthogonal, but those that are not ancestors or descendants of each other. This means that in any moment, the codes allocated to users may not be arbitrary; they can only be orthogonal codes, to assure separation of different users' signals at the receiver. Although Figure 3 shows an example of a code tree with maximum code length of 16, in UMTS the applied lengths can vary from 4 to 512 in downlink and between 4 and 256 in uplink direction. A given channel is created by using a given channelization code (or the use of multiple codes simultaneously is also possible).

In UMTS, the chiprate on the radio interface is fixed 3.84 Mcps (megachips per second). Data is transmitted in radio frames of 10 ms (carrying 38400 chips each); these are further divided into 15 timeslots (each containing 2560 chips). As described above, to convey one bit of information, a codeword of length 2^n or its negation is transmitted. By considering these, it naturally turns out that user bitrate depends on the actual length of the channelization code used. Therefore, the variable and different bitrates to different connections are provided by assigning different channelization codes with different lengths and these may vary in time for a given connection. As an example, using a length 4 code results in 38400/4 = 9600 symbols to be transmitted, while using a length 64 one results in 38400/64 = 600 symbols in one radio frame. The term symbol is used, because UMTS actually uses Quadrature Phase-Shift Keying (QPSK) modulation, meaning that the transmitted channel symbol may have four possible values, hence carries two bits of information at the same time. In downlink direction, consequently the resulting number of transferred bits in a frame

Figure 3. OVSF codetree

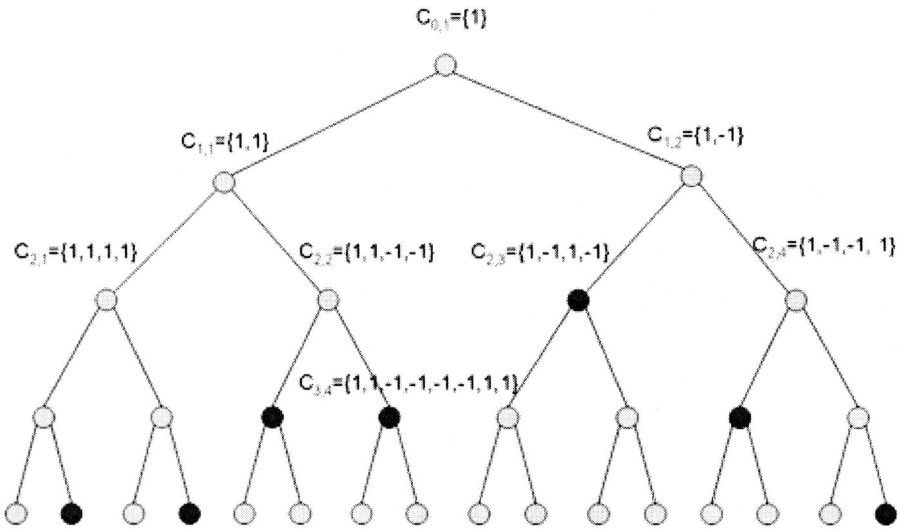

is calculated as $38400/2^{n-1}$ with a code of length 2^n. In uplink direction this factor of two is not present when talking about user data bits, that is because the four state of the modulation symbol jointly conveys physical layer signaling (taking the "half" of the symbol) information and user data. Physical layer signaling also reduces the number of effective user bits in downlink direction; however in this case signaling is multiplexed with user data in time.

From these figures above, it is apparent that the maximal physical layer download rate is 1920 kbps (including signaling), using a single length 4 OVSF code; this rate is halved as the length of the used code is doubled. In uplink, this value is 960 kbps. Apparently, the brave requirement of 2 Mbps transmission mentioned above can hardly be fulfilled; simultaneous use of more codes is needed for this. However, this was never implemented in practical UMTS systems, rather 384 kbps is the typical maximum bitrate (understood as useful bitrate, i.e. bitrate of user IP packet transmission) available. Later HSDPA enhancement of the network (see below) has increased the available rate significantly.

The concept of scrambling codes is also important in the UMTS system. The OSVF codes described above are only orthogonal (hence applicable to create different channels of different users) if they are transmitted chip-synchronously. This is the case in downlink by the very nature of the transmission of the Node B (data of all users is separately but synchronously multiplied by different channelization codes and then summed up). However, this precision in uplink synchronization would be very tough and expensive to implement, therefore in uplink, the OSVF codewords of different users arrive at the Node B with arbitrary shifts. This results in the codes being not orthogonal any more, meaning that the reception based on scalar multiplication would incorporate severe interference from other codewords. Consequently, OSVF codes cannot be used to create channels of different subscribers in uplink. Similar problem arises in downlink, when considering the transmission (and influence) of a neighboring cell on a given connection. It may happen that the neighboring cell uses the very same OSVF codeword for transmission; this would result in a mobile not being able to correctly receive its own data (this problem arises because frequency reuse 1 is used).

To overcome these problems, the use of scrambling codes is introduced. Scrambling codes are pseudo-random sequences, which feature very good cross-correlation properties: the scalar product of any fractions of two such codes remains close to zero (the term quasi-orthogonality is also used for this property).

Thus, in downlink direction, before transmission to the radio, the summed data is chipwise multiplied by a scrambling code, that is unique for the given cell and actually identifies that cell. As neighboring cells use different scrambling codes, the quasi-orthogonality allows the users to receive their data in downlink (but the traffic of neighboring cells will appear as interference). Similarly, in uplink direction, the scrambling code sequence is unique for a given terminal; hence, the traffic of different mobiles is distinguished by the scrambling sequences. The OVSF codes are used to create several channels (more traffic channels and signaling) of a given terminal. Naturally, this means that in uplink communications the traffic of other subscribers generates interference to a given customer.

Developments of the Architecture and Radio Interface

The basic UMTS architecture and radio interface of Release'99 was far from being optimal as compared to what a 5 MHz wide radio channel and the system could offer. Thus, as mentioned in the Introduction, further releases of the standards were published, defining advances for the network.

The all-IP core of Release 4 was already mentioned. The next release also contained changes in the UTRAN. Originally, UTRAN's transport technology is Asynchronous Transfer Mode (ATM), because ATM is originally designed to support both CS and PS traffic. Note that the physical network that carries the ATM payload (which actually carries the UMTS traffic of UTRAN) is not specified in the standards. Usually SDH (Synchronous Digital Hierarchy) or SONET (Synchronous Optical NETwork) over fiber is used for high-capacity aggregate backhaul links; also PDH (Plesiochronous Digital Hierarchy) links over fiber or point-to-point microwave connections is often used. Release 5 of the standards defined the UDP/IP based transport in the UTRAN, bringing UMTS network closer to the all-IP vision. With this solution, CS services are transferred over RTP/UDP/IP.

Release 5 also defined HSDPA. The goal was to increase physical bitrate on the radio interface. This is achieved by introducing 16-level modulation and clamping more (maximum 15) channelization codes of length 16 together, to form a single high-speed channel. With this solution, the achievable maximum bitrate at the bottom of the physical layer is 14.4 Mbps. HSDPA introduced other significant changes to the architecture. New sub-layers for the radio protocol stack were defined, with endpoints in the Node B and not in the RNC. Practically this means that HSDPA packet scheduling and fast retransmission is done by the base station. As these two cardinal operations had been brought close to the radio interface, response time and packet delays of the network were significantly improved. It can be stated that 3G networks reached success after the introduction of commercial services (often called mobile Internet). The vast majority of current 3G networks' traffic is Internet traffic over HSDPA access.

The architectural change of bringing intelligence to the Node B was further extended by some vendors and also in Release 7 standardization. That is, such Node Bs are available on the market that themselves include RNC functionalities (software RNC or emulated RNC concept).

The standardization of HSDPA continued towards achieving even higher physical bandwidths. Release 7 and 8 defined the 64-state modulation to be used and advanced multiantenna transmission that result in a physical rate of up to 42 Mbps.

Enhancements in uplink direction are also included in the standards. HSUPA of Release 6 enables up to 5.76 Mbps uncoded rate in the physical layer, that is achieved by means of creating high speed dedicated channel over multiple parallel channelization codes. Release 7 also incorporated the use of 16-state modulation in the uplink; hence, in theory the gross rate can be over 11 Mbps.

UTRAN PROTOCOLS

UMTS defines its transmission services as *bearers*. Bearers are defined in different layers and between different points in the network. Generally, an UMTS bearer service connects the CN gateway (either GGSN or GMSC) with the mobile terminal. This bearer service relies on the Radio Access Bearer (RAB), connecting the CN edge point (namely an SGSN or MSC) with the mobile terminal and the CN bearer, which is the connection within the CN (between the gateway element and the appropriate serving device). The RAB is usually further divided into Radio Bearer (RB, between the mobile and the RNC) and Iu bearer (between the RNC and the serving CN node). The role of UTRAN protocols is mainly the management of RABs (either signaling or data bearers), with all the necessary elementary procedures needed for this operation.

General Protocol Structure and Interfaces

The protocol structure of the UTRAN is divided into two main layers, the Radio Network Layer (RNL) and the underlying Transport Network Layer (TNL). The Radio Network Layer contains the actual UMTS specific protocols defined for the UTRAN, while the Transport Network Layer defines the protocols that carry UMTS specific messages and data. It is worth noting that Transport Network Layer is independent of and not specific to UMTS.

As usual, the protocol stack is also divided into horizontal planes: a control plane and a user plane (UP), both in the RNL and TNL. Note that control plane in TNL carries control signaling needed in the transport network, while control plane of RNL defines signaling needed for the UTRAN specific communication. However, signaling traffic of the RNL appears as any other user traffic in the TNL, thus it goes through the user plane of TNL.

These concepts and further details of the protocol structure on the Iu-CS interface is shown in Figure 4. The RNL part contains the control protocol, Radio Access Network Application Part (RANAP), that is the protocol actually responsible for management and signaling related to all user transmission over the Iu interface.

Iu Protocols

Iu interface connects UTRAN with the core network. Originally in Release'99 ATM was chosen to be the transport network technology of this interface. This was motivated by the fact that at that time IP technology was not well prepared for carrying CS services, while ATM was planned to support both PS and CS transmission from the beginning. However, ATM (with the required ATM Adaptation Layer) brings a serious overhead into the system (for example 5 byte header for each of the fixed 53 byte long ATM packets) in terms of data to be transmitted, system complexity (expensive ATM switches are needed) and signaling. Moreover, at the time of standardization the fail of ATM was already visible (in the contest between ATM and IP based transport technology). Hence, in Release 5 the IP based Iu transport layer was incorporated, even for CS services, allowing the IP or ATM transport option over Iu (3GPP 25.410). Figure 4 reflects this protocol architecture of the Iu-CS interface.

Note that the only UMTS specific layer in this structure is the RANAP protocol (Radio Access Network Application Part) in the control plane and the Iu user plane (UP) protocol layer. Furthermore, the sublayers of the TNL are standard IP and ATM

Figure 4. Iu-CS protocol stack (3GPP 25.410)

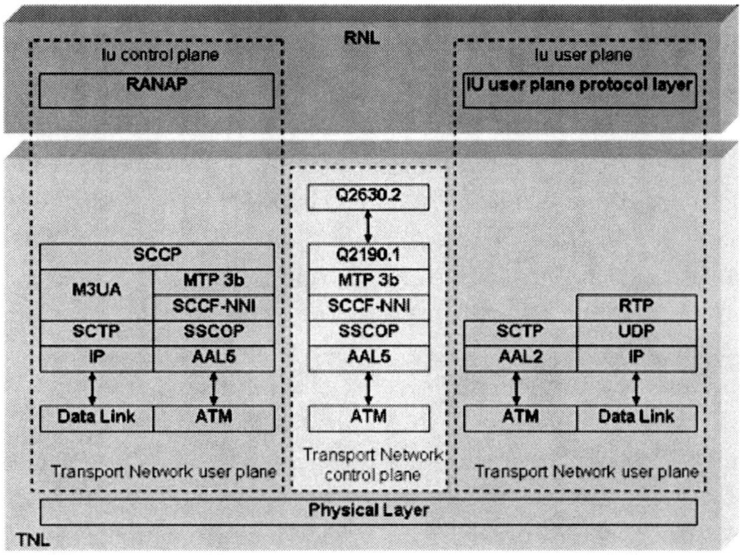

signaling protocols, which are further detailed in other chapters of this book.

The RANAP is responsible for the specific control functions that are required for UMTS communications. RANAP functions consist of one or more elementary procedures. Among several others, these functions include (3GPP 25.413):

- management of RABs (~ UMTS connections), including changing of RAB parameters, establishment/release of RABs
- relocation of the serving RNC of a RAB, without changing connection parameters
- overload control of the IU interface
- paging an UE, in case of mobile terminated connection arrival (i.e. finding the cell where the called UE dwells momentarily)
- control of the security mode in UTRAN, that is transmission of ciphering keys and setting operation mode for security procedures
- reporting the current location of an UE, from the RNC to the CN
- reporting unsuccessfully transmitted downlink data volume
- functions for management of MBMS transmission

Regarding the user plane protocol in RNL one objective was to remain independent of the core network domain and keep minimal dependency from the TNL. This enables flexible system evolution and deployment, as well as service migration between CN domains. Therefore, two modes of operation are defined and these can be configured on a RAB basis. These are the transparent mode and the support mode for predefined SDU (Service Data Unit) size. The mode of operation is set by the CN according to the RAB requirements and signaled to the UE as well.

Transparent mode simply means that user data is not affected in the user plane of RNL. Hence the CN and UTRAN communicate over the TNL as if RNL was not present (this refers to user plane data, naturally in the control plane RANAP is always operating). Generally, non real-time traffic uses transparent mode operation.

Support mode provides advanced services and therefore executes appropriate framing of user data into predefined sized Protocol Data Units

(PDU). These services are provided based on the following key functions of this layer

- frame handler function, which is responsible for creating the Iu user plane protocol data units (~ frames), creating and setting the correct values of the header parts of these frames, numbering the PDUs and checking for errors in the header
- procedure control functions, whose responsibilities include
    - rate control, that is setting the downlink data rate at the Iu interface
    - initialization of the support mode Iu data plane
    - handling of error events
    - time alignment, that controls the timing of data transfer towards the RNC over the Iu
- non access stratum data stream specific functions, that include detecting errors in the headers, classify frames according to their quality, create padding in the PDU if necessary

As it is depicted in Figure 4, several control and user plane sublayers are used in the TNL part; these are independent from the RNL protocols. As both ATM and IP based options are enabled, the number of used protocols is quite abundant.

In case of ATM option, the control plane information of the RNL is carried as packet switched user plane data in the TNL, as ATM Adaptation Layer (AAL) version 5 was originally developed for this reason. CS user data of the RNL is handled by AAL2 in the user plane of TNL, as AAL2 was originally developed for fixed rate circuit switched services. Control plane data of RNL is carried by broadband SS7 protocol stack over AAL5 (in the user plane of RNL), namely the Signaling Connection Control Part (SCCP), the Message Transfer Part (MTP3-b), Service Specific Co-ordination Function (SSCF) and Service Specific Connection Oriented Protocol (SSCOP). These latter two together is called Signaling ATM Adaptation Layer for Network-to-Network Interfaces (SAAL-NNI).

If the IP transport option is used, then CS user plane data is carried by the Real-time Transport Protocol (RTP), or optionally Real-time Transport Control Protocol (RTCP) is used, over UDP/IP. The RANAP control information is also carried by broadband SS7 protocols over the Stream Control Transport Protocol (SCTP), namely M3UA (Message Transfer Part, MTP3 User Adaptation) layer and SCCP.

Note that the control plane in the transport network is used only for ATM transport, for management of ATM connections, through explicit signaling with Q.2630.2 and Q2150.1, over the previously mentioned broadband SS7 protocols. Being a packet switched technology, IP transport does not need explicit signaling for connection management (as unlike in ATM, connections do not exist in IP), hence the control plane of TNL is not defined for IP transport.

Figure 5 shows the Iu interface protocol stack for the packet switched domain. Apparently, the structure is similar to the circuit switched interface, with the difference of using transport network layer protocols optimized for PS communications. Note that the RNL does not differ from the Iu-CS interface.

The choice of ATM or IP transport option is also present. In both cases, user plane data from the RNL is carried by GPRS Tunneling Protocol (GTP-U) over UDP/IP, as this is the solution inherited from the GSM/GPRS network. If ATM option is used, then the known IP over ATM solution is applied, namely IP packets are carried over AAL5. Regarding the control plane information of RNL, the same signaling protocols are used as in the Iu-CS case, but optionally the ATM transport choice might carry the IP based solution as well, with IP over ATM AAL5 solution.

Figure 5. Iu-PS protocol stack (3GPP 25.410)

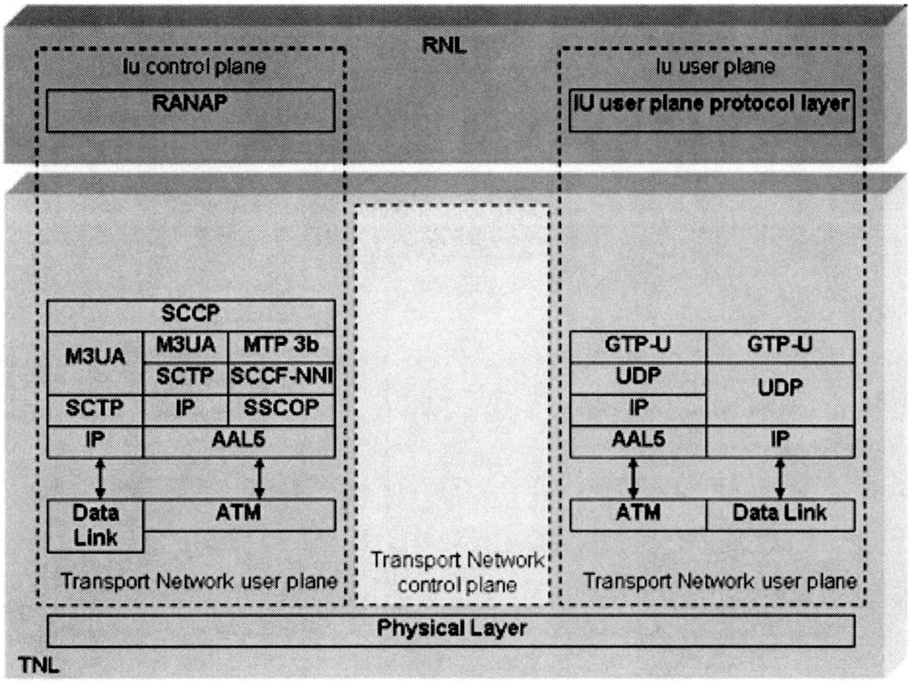

Iub Protocols

Iub interface connects Node Bs with RNCs in UTRAN and is described in general in (3GPP 25.430). Unlike Iu, this interface is common for PS and CS services. In UMTS, the radio access is defined between the RNC and the UE; hence, Iub interface should define the means of transmitting PDUs of the radio protocol stack, over the geographically most widespread network part connecting all the base stations to the system.

Figure 6 shows the Iub protocol stack. Its basic structure is the same as that of the Iu interface.

One main difference is that the control plane protocol of RNL is the Node B Application Part (NBAP). The other main difference is the user plane of the RNL. For this, the standards define the so-called *transport channels* on the radio protocol stack (see later). The frames of these transport channels are the PDUs of the RNL user plane on the Iub protocol stack. The transport channels are divided into two main groups, namely a common channel group (used by multiple customers, e.g. channel for paging, random access, shared data channel) and a dedicated channel group (used by a single UE). Hence, the roles of RNL protocols are also defined separately for common and dedicated channels. The framing protocols for all common and dedicated transport channels are the user plane protocols of the RNL. As it will be described in the subsequent Section, transport channels are understood as channels of the Medium Access Control (MAC) sublayer of the radio protocol stack. Therefore, in RNL user plane the frame protocol handles MAC frames as payload, and generates user plane PDUs that contain one or more MAC frames, extended by header.

The function of NBAP is to provide control for the Node Bs and to provide transport channels of the radio interface, as well. NBAP procedures include, but are not limited to the following functions (3GPP 25.433):

Figure 6. Iub protocol stack (3GPP 25.430)

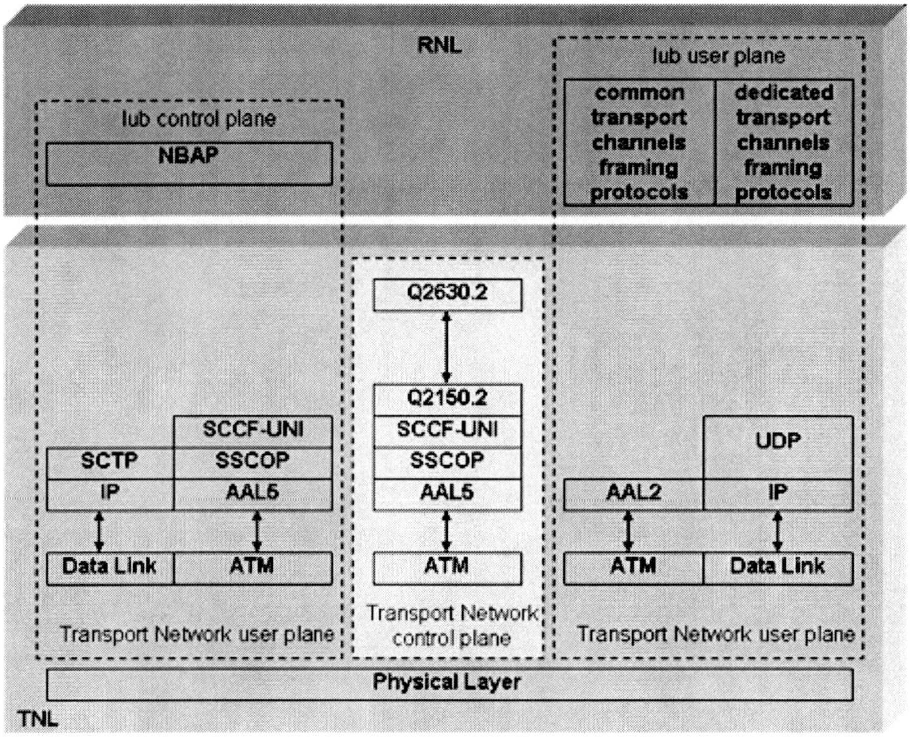

- common transport channel management, enabling the RNC to manage the configuration of common transport channels in a Node B
- system information management, i.e. the ability to manage the scheduling of system information to be broadcasted in a cell
- resource event management, that is the function to inform the RNC about the status of Node B resources
- measurements on common and on dedicated resources, which function initiates measurements on common or dedicated radio resources and enables the Node B to report the results of these
- radio link management and supervision, that allows the RNC to manage radio links and to report on radio link failures and its restorations
- physical shared channel management function, that allows the RNC to manage physical resources in the Node B belonging to the shared transport channels of HSDPA operation
- cell synchronization, allowing the synchronization of Node Bs or cells over the air interface

As far as the transport technology of the Iub interface is concerned, two possibilities, either IP or ATM are present as it is the case on Iu interface. Note that the link is between the RNC and the UE; hence, no tunneling is needed in case of IP transport from the RNC downstream in the network. It is also worth noting that since this is a common interface for CS and PS services, AAL2 is used for transmitting RNL user plane data. This indeed supports real-time CS traffic indeed, however imposes significant overhead, thus inefficiency for PS traffic. Another difference compared to Iu is the use of the User-Network Interface (UNI) version of SSCF signaling.

RADIO PROTOCOL STACK

The architecture of the radio protocols of UMTS is shown in Figure 7. The radio interface is layered into three protocol layers, the physical layer (L1), which performs the actual radio transmission; the data link layer (L2) and network layer (L3). Layer 2 is further divided into four sublayers: Medium Access Control (MAC), Radio Link Control (RLC), Packet Data Convergence Protocol (PDCP), and Broadcast/Multicast Control (BMC). BMC is not presented further in this Chapter. Layer 3 and RLC are split into control and user planes. PDCP and BMC are only defined in the user plane.

In the control plane layer 3 contains several sublayers, where the lowest sublayer (Radio Resource Control, RRC) has an interface with layer 2 and terminates in the UTRAN. The other sublayers for e.g. duplication avoidance and higher layer signaling such as mobility management or call control terminate in the CN or are not in the scope of UTRAN specifications, therefore not presented here.

UMTS Radio Channels

In UMTS, there are three types of channels defined corresponding to different layers of the stack. The so-called *logical channels* are between the RLC and MAC (services offered to RLC by the MAC) and can be viewed as *what* information is transferred (e.g. user data, paging information, etc.). The *transport channels* are between the MAC and PHY (services offered by PHY to MAC) and describe the format and characteristics (*how*) of the transmitted data. Within the PHY layer, *physical channels* are defined, too. These are the actual physical resources that actually physically carry the information of transport channels. Physical channels are not detailed in this chapter.

Logical channels are divided into two sets, data channels, and control channels, according to the type of information carried. The control

Figure 7. Radio protocol stack (3GPP 25.301)

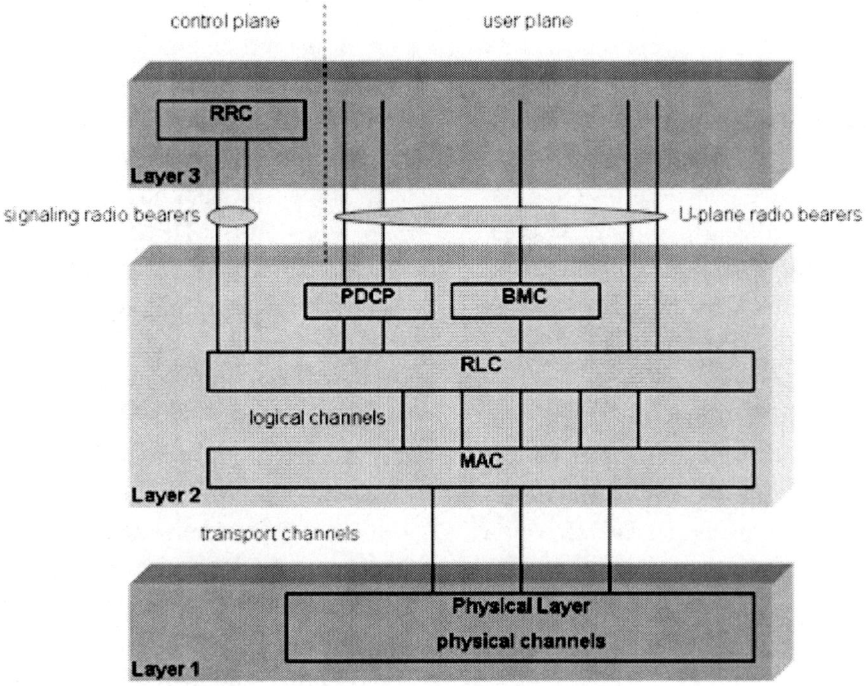

channels include Broadcast Control Channel (BCCH) for system broadcast information; Paging Control Channel (PCCH) for paging the UEs in case of incoming call; DCCH (Dedicated Control Channel) for conveying bidirectional dedicated control information between an UE and the network; Common Control Channel (CCCH) for transmitting control information valid for more or all UEs. Regarding logical traffic channels, there are Dedicated Traffic Channels (DTCH) for user traffic and Common Traffic Channels (CTCH) for multicast user traffic, whereas MBMS traffic is carried by MBMS point-to-multipoint Traffic Channel (MTCH). For MBMS two other control logical channels are defined. The point-to-multipoint MBMS Control Channel (MCCH) transmits control information from the network to a group of MBMS users. MBMS Scheduling Channel (MSCH) is point-to-multipoint, too, and is used for conveying scheduling information for one or several MTCHs.

As it was mentioned earlier, transport channels have two mayor types, namely common channels and dedicated channels. Common transport channel types are Random Access Channel (RACH) for initial access to the network; Forward Access Channel (FACH), which is a common downlink channel used for transmission of small amount of data or broadcast and multicast data; Broadcast Channel (BCH) for broadcasting system parameters into an entire cell; Paging Channel (PCH), used for transmitting paging messages, as well as changes of BCCH parameters; High Speed Downlink Shared Channel (HS-DSCH) is a downlink channel shared between UEs to transmit HSDPA traffic.

Dedicated transport channel types are the Dedicated Channel (DCH), which is used for serving one UE data in uplink or downlink, and the Enhanced Dedicated Channel (E-DCH), which is used for uplink HSUPA transmission for a UE.

UMTS Radio Protocols

The functions of the sublayers of the radio protocol stack are given in this section. First of all, MAC layer is reviewed, based on (3GPP 25.321). The main services that MAC offers to higher layers are data transfer (segmentation/reassembly is only done if HS-DSCH or E-DCH is used), (re)allocation of radio resources and MAC parameters and measurement reporting to RRC. These services are provided through the execution of some functions from the following non-exhaustive list:

- mapping logical channels onto appropriate transport channels
- selection of appropriate transport format for each transport channel from an assigned transport format set that depends on the instantaneous source rate
- handling priorities among different UEs (by dynamic scheduling of data transport) and between different data flows of a UE
- multiplexing/demultiplexing of higher layer PDUs into/from transport blocks delivered to/from the underlying (physical) layer both on common and dedicated transport channels
- traffic volume measurement on logical channels and reporting to RRC
- providing Hybrid ARQ (automatic Repeat reQuest) acknowledgements/retransmissions on HS-DSCH and E-DCH
- in-sequence delivery and segmentation/reassembly on high-speed channels

We should note that for different services several types of MAC entities are defined in the standards. For example, MAC-hs is used for HSDPA. Consequently, as for high-speed services some radio functions are placed into the Node B instead of the RNC, corresponding MAC entity types are running on the Node B.

The next layer of the stack is RLC (3GPP 25.322), that can provide services in transparent mode, meaning that upper layer PDUs are sent without adding any protocol information, but with the possibility of segmentation/reassembly. Unacknowledged data transfer mode is also defined, when upper layer PDUs are forwarded without guaranteeing delivery to the peer entity. This mode is capable of detecting erroneous data, and provides duplication avoidance and reordering of PDUs, but retransmission is not possible. The third operation mode of RLC is acknowledged transfer. In this case, errors of RLC PDUs are detected and notified towards the peer RLC entity, which is able to retransmit the erroneous PDU. Therefore, this mode is capable of providing error free transmission to the upper layer. In sequence or out of sequence delivery are both applicable in this mode (in this latter case reordering is done in upper layer), as well as duplication avoidance, which enables the maintenance of QoS The notification of unrecoverable errors is also supported. The RLC fulfils its roles by the following functionalities:

- segmentation and reassembly, to either fixed size (configurable) or flexible sized PDUs (in case of E-DCH or HS-DSCH channel)
- concatenation of payloads, padding of missing bits, transfer of user data
- error correction by means of retransmission in acknowledged mode
- flow control, that is how the receiver RLC entity controls the rate at which the sending peer RLC entity transmits PDUs
- sequence number check
- protocol error detection and recovery, ciphering for non-transparent modes, PDU discard

The following sublayer, PDCP of L2 is used mainly in order to provide header compression and decompression of IP data flows if PS data bearers are used (3GPP 25.323). The header compression methods are defined for IP communications and not specific to UMTS networks.

The most important sublayer of the radio stack in L3 is the RRC, defined by a more that 1700-page long standard (3GPP 25.331). RRC is in the control plane protocol in L3 and is responsible for signaling connections required for information exchange in UE-UTRAN communications. RRC messages are sent over so-called signaling radio bearers. Among others, the RRC performs the following functions:

- broadcasting system information to the UEs, provided by the Core Network, repeated on a regular basis
- establishment, re-establishment, maintenance and release of an RRC connection, as well as the same connection management functions for user plane radio bearers (connections)
- allocation, reconfiguration and release of radio resources (codes, power, timeslots) for the RRC connection, both for user plane and control plane
- connection mobility functions, such as handover or cell-reselection; initial cell selection; preparation of handover to other system (e.g. GSM) and cell or paging area update procedures
- paging and notification of incoming data transfer towards UEs
- QoS control of a given radio bearer, ensuring requested quality parameters
- reporting of UE measurement results, control of the reporting process
- outer loop power control, that is the setting of power level of a given radio bearer
- ciphering control, i.e. switching ciphering on/off between the UE and the UTRAN

CONCLUSION

This chapter covers the architecture and interfaces of the 3G UMTS system. First the architecture and the basic principles of the radio operation are given, summarizing the basics of CDMA, as well as the frame structure and physical resources on UMTS radio interface. As the system was target of constant further improvements after the publishing of the first set of UMTS standards, the main novelties of the later releases are also summarized. The major part of the system is UTRAN; hence, the main interfaces of UTRAN are also shown. The interfaces are presented through the protocol stacks implementing those interfaces, with emphasis on the UMTS specific protocol parts, but still describing the transport network options and solutions. Finally, the L2 and L3 solutions of the radio interface protocols are shown. Each sublayer is presented through its main purpose and functions.

REFERENCES

3GPP TS 25.301 V8.7.0. (2010-06). *3rd generation partnership project: Technical specification group radio access network - radio interface protocol architecture* (release 8).

3GPP TS 25.321 V9.3.0. (2010-06). *3rd generation partnership project: Technical specification group radio access network - medium access control (MAC) protocol specification* (release 9).

3GPP TS 25.322 V9.2.0. (2010-06). *3rd generation partnership project: Technical specification group radio access network- radio link control (RLC) protocol specification* (release 9).

3GPP TS 25.323 V8.5.0. (2009-12). *3rd generation partnership project: Technical specification group radio access network - packet data convergence protocol (PDCP) specification* (release 8).

3GPP TS 25.331 V10.0.0. (2010-06). *3rd generation partnership project: Technical specification group radio access network - radio resource control (RRC) - protocol specification* (release 10).

3GPP TS 25.401 V8.2.0. (2008-12). *Technical specification group radio access network: UTRAN overall description* (release 8).

3GPP TS 25.410 V7.0.0. (2006-03). *3rd generation partnership project: Technical specification group radio access network - UTRAN Iu interface: General aspects and principles* (release 7).

3GPP TS 25.413 V9.3.0. (2010-06). *3rd generation partnership project: Technical specification group radio access network - UTRAN Iu interface radio access network application part (RANAP) signalling* (release 9)

3GPP TS 25.430 V7.5.0. (2007-09). *3rd generation partnership project: Technical specification group radio access network- UTRAN Iub interface: General aspects and principles* (release 7).

3GPP TS 25.433 V9.3.0. (2010-06). *3rd generation partnership project: Technical specification group radio access network- UTRAN Iub interface node B application part (NBAP) signalling* (release 9)

ADDITIONAL READING

Dahlman, E., Parkvall, S., Skold, J., & Beming, P. (2008). *3G Evolution: HSPA and LTE for Mobile Broadband*. Burlington, MA: Elsevier.

Fares, A. (2009). UTRAN Signaling and Protocols BookSurge Publishing.

Holma, H., & Toskala, A. (Eds.). (2007). *WCDMA for UMTS – HSPA Evolution and LTE*. Chichester, England: John Wiley & Sons.

Johnson, C. (2008). *Radio Access Networks for UMTS: Principles and Practice*. Chichester, England: John Wiley & Sons. doi:10.1002/9780470727225

Kaaranen, H., Ahtiainen, A., Laitinen, L., Naghian, S., & Niemi, V. (2001). *UMTS Networks: Architecture, Mobility and Services*. Chichester, England: John Wiley & Sons.

Kreher, R., & Ruedebusch, T. (2005). *UMTS Signaling: UMTS Interfaces, Protocols, Message Flows and Procedures Analyzed and Explained.* Chichester, England: John Wiley & Sons.

KEY TERMS AND DEFINITIONS

Bearer: A connection, or service between two points of the UMTS system, e.g. radio bearer is a service between the user terminal and the Radio Network Controller, an Iu bearer is a connection over the Iu interface, etc.

High Speed Packet Access (HSPA): A set of solutions based on UMTS radio interface, that enables transmission bitrates over ten Mbps for a customer.

Iu Interface: The surface between the Core Network and UTRAN parts of the UMTS system that defines the format and protocol layers used to transmit circuit switched, packet switched, and broadcast information.

Iub Interface: The surface between the base stations (Node B) and the Radio Network Controller (RNC) equipment that defines the common format and protocol layers used to transmit information.

UMTS Channels: The UMTS standards define three types of channels, each within the radio protocol stack, namely *physical channels* meaning the actual physical resources used for a transmission; *transport channels* defining the means and format of the transmission; *logical channels* identifying the type of information to be transmitted.

UMTS Terrestrial Radio Access Network (UTRAN): The network consisting of the mobile terminals, radio base stations and Base Station Controller centers, this is the unique part of the UMTS system.

Wideband Code Division Multiple Access (WCDMA): The multiple access scheme applied in UMTS system, with the basic idea of distinguishing signals of different physical channels by means of sending information multiplied by unique codes.

APPENDIX

Table of abbreviations

3GPP	3rd Generation Partnership Project
AAL	ATM Adaptation Layer
ATM	Asynchronous Transfer Mode
BC	Broadcast
BCCH	Broadcast Control Channel
BCH	Broadcast Channel
BMC	Broadcast/Multicast Control
CCCH	Common Control Channel
CN	Core Network
CTCH	Common Traffic Channel
CS	Circuit Switched
DCCH	Dedicated Control Channel
DCH	Dedicated Channel
DS-CDMA	Direct Sequence Code Division Multiple Access
DTCH	Dedicated Traffic Channel
E-DCH	Enhanced Dedicated Channel
EDGE	Enhanced Data-rates for GSM Evolution
EPC	Evolved Packet Core
ETSI	European Telecommunication Standardization Institute
FACH	Forward Access Channel
FDD	Frequency Division Duplex
GGSN	Gateway GPRS Support Node
GMSC	Gateway MSC
GPRS	General Packet Radio Service
GTP	GPRS Tunneling Protocol
HSDPA	High Speed Downlink Packet Access
HS-DSCH	High Speed Downlink Shared Channel
HSPA	High Speed Packet Access
HSUPA	High Speed Uplink Packet Access
IMS	IP Multimedia Subsystem
ITU	International Telecommunication Union
LTE	Long Term Evolution
M3UA	Message Transfer Part, MTP3 User Adaptation
MAC	Medium Access Control
MBMS	Multimedia Broadcast Multicast Service
MCCH	MBMS Control Channel

ME	Mobile Equipments
MSC	Mobile Switching Center
MSCH	MBMS Scheduling Channel
MTCH	MBMS point-to-multipoint Traffic Channel
MTP	Message Transfer Part
NBAP	Node B Application Part
OVSF	Orthogonal Variable Spreading Factor
PCCH	Paging Control Channel
PCH	Paging Channel
PDCP	Packet Data Convergence Protocol
PDH	Plesiochronous Digital Hierarchy
PDU	Protocol Data Unit
PS	Packet Switched
QoS	Quality of Service
QPSK	Quadrature Phase-Shift Keying
RAB	Radio Access Bearer
RACH	Random Access Channel)
RANAP	Radio Access Network Application Part
RB	Radio Bearer
RLC	Radio Link Control
RNC	Radio Network Controller
RNL	Radio Network Layer
RNS	Radio Network Subsystem
RRC	Radio Resource Control
RTCP	Real-time Transport Control Protocol
RTP	Real-time Transport Protocol
SAAL-NNI	Signaling ATM Adaptation Layer for Network to Network Interfaces
SCCP	Signaling Connection Control Part
SCTP	Stream Control Transport Protocol
SDH	Synchronous Digital Hierarchy
SDU	Service Data Unit
SGSN	Serving GPRS Support Node
SONET	Synchronous Optical NETwork
SSCF	Service Specific Co-ordination Function
SSCOP	Service Specific Connection Oriented Protocol
TDD	Time Division Duplex
TDMA	Time Division Multiple Access
TNL	Transport Network Layer
UE	User Equipment
UMTS	Universal Mobile Telecommunication System

UNI	User-Network Interface
UP	User Plane
USIM	UMTS Subscriber Identity Module
UTRA	UMTS Terrestrial Radio Access
UTRAN	UMTS (or Universal) Terrestrial Radio Access Network
WCDMA	Wideband Code Division Multiple Access

Section 3
IP-Based Protocols

Chapter 8
IPv4 / IPv6 Coexistence and Transition:
Concepts, Mechanisms and Trends

László Bokor
Budapest University of Technology and Economics, Hungary

Gábor Jeney
Budapest University of Technology and Economics, Hungary

ABSTRACT

Nowadays, the Internet Protocol is facing the version change; IPv4 (the old version of IP) will be replaced by IPv6 (the new version of IP) in the long run. Although they are based on the very same concept, the two protocols are not compatible with each other. This chapter deals with the coexistence issues, which might arise due to the simultaneous existence of IPv4 and IPv6. On the other hand, this chapter also covers transition concepts: how IPv6 only hosts can communicate either over IPv4 only networks, or with IPv4 only hosts in the Internet of the future.

INTRODUCTION

IPv6 (Internet Protocol version 6) is the new version of Internet Protocol, which is expected to replace the old version, known as IPv4 (IP version 4). IPv6 is not too different from IPv4: it follows the same conceptual design. The basic operations of IPv4 and IPv6 do not differ from each other. However, IPv6 can be regarded as a descendant of IPv4; they are not compatible with each other. IPv4-only hosts or networks cannot directly communicate with IPv6 counterparts, due to the incompatible structure of the packets. Special transition mechanisms are needed to provide interoperability among them: they are described in this chapter.

DOI: 10.4018/978-1-60960-732-6.ch008

BACKGROUND

IPv4 is a standard (RFC791) from 1980. It was developed in an era when computers occupied huge rooms and networking was available at speeds of 300 bits/s. Since that time, many improvements have appeared, accepted, and employed, according to the needs of computer networks and communication elements (computers, routers, etc.) of the last few decades. These improvements are basically extensions (more RFCs) to the existing IPv4 standard. These RFCs describe the additional functionality (like multicasting) and the related protocols, which support the functionality. At last, in the beginning of 1990's the emerging IPv4 protocol suite resulted in an advanced communication protocol, which supports all the needs of Internet networks and nodes of that time.

However, it was visible that there is one property of the IPv4 protocol, which needs revision, but cannot be extended easily, since it is fixed deep in the roots of the protocol. It was addressing. An IPv4 address comprises four bytes (32 bits). Assuming uniqueness (that one IP address must be assigned to exactly one host in the whole world), this address space makes it possible to address a bit more than four billion hosts around the world. Four billion does not sound too much, as the number of people living on Earth exceeds 6.82 billion as of 10th March 2010. In the long run we can be sure that everybody will need Internet connection, and it is expected that more than one device will be connected to the Internet. Thus, in the long run, IPv4 will not be able to serve the whole humankind for sure.

Changing the addressing methodology in IPv4 would require to change not only all the RFCs which describe, or refer to addressing, but also all the hosts and network entities which are currently connected to the Internet. Clearly, this is not possible. Thus, Internet developers decided to release a new version of IP, namely IPv6.

IPv6 provides a huge address space, with addresses consisting of 128 bits. This address space could cover the surface of the Earth with an average density of 665 billion addresses/μm^2. That is, in every square micrometer, there could fit 150 pieces of the whole IPv4 Internet. This amount of addresses is expected to be enough for the long run, even if humankind starts to populate the space. Addressing is the main motivator for changing the version, and also the most important property of the new protocol.

Now let us see what other novelties can be found in IPv6. Basically, IPv6 is a conceptual copy of the IPv4 protocol, with all the functionalities existing in IPv4. That is, if a professional is familiar with IPv4, she really does not need big efforts to learn IPv6. The real difference is basically the integrity of the standard. At the creation of IPv6, the authors were aware of the existing functionalities of IPv4: they tried to integrate all functionalities of IPv4 extensions into the basic IPv6 standards. Thus, the IPv6 standards are more complete, and thus IPv6 implementations have all the functionalities that are available in advanced IPv4 implementations.

Of course, there are new concepts, which can be found in IPv6 only. However, it must be noted that they are the minority. As long as IPv4 is the main protocol of the Internet (there are more hosts use IPv4 today than the ones that use IPv6), all new functions of IPv6 will be backported to IPv4 for sure.

Overview of IPv6

As in IPv4, IPv6 packets are usually forwarded based on their destination address. Although IPv6 packet header contains a field called flow label (which makes it possible to route the packets based on the label, yielding label based routing), current implementations do not make use of this field. The forwarding entities are called routers.

The most important difference between the IPv4 and IPv6 packets is the structure of the header. The IPv6 header (Deering & Hinden, 1998) has been simplified to unburden routers in the network. Header checksum, which was an important component of IPv4 header, has been removed for two reasons. On one hand, nowadays networks do not produce often bit errors, so the lack of checksum is imperceptible. On the other hand, the decrease of the hop limit (or time-to-live, TTL) field would require the recalculation of this checksum, if it was there – yielding indeed a high resource save on the routers.

Fragmentation has been banned on routers, only source nodes can fragment IPv6 packets. This is also for the sake of unloaded routers. Fragmentation was problematic between links with different MTUs (Maximum Transfer Unit: the maximum number of bytes that can be put into one packet). In routers, which connected different MTU links, fragmentation was a continuous task: every packet arriving at the larger MTU link must have been fragmented to fit into the lower MTU link. Routers between different MTU links were overloaded just because IPv4 supported router level fragmentation. In IPv6, packets cannot be fragmented in the network. If the IPv6 packet arrives at a link, where it cannot be transmitted (it is larger than the MTU), then the IPv6 packet is simply dropped and the source of the packet is informed about this event. The source can then restart the transmission of packets with a lower size. It is up to the source to decide whether fragment these packets, or ask the application to produce smaller packets.

The header length field has disappeared, thanks to the fixing of header length, and the definition of extension headers. Specific flags have also been moved to the extension headers yielding a less complex IP header, which is easy to process and transmit by the network routers.

Considering addressing, the IPv6 address space is divided into three—four groups in two categories. First, based on the addresses, the IP address can be unicast, multicast, and anycast. The first two must be clear for everyone, unicast is to address a unique (one sole) station, multicast is to provide addressing for a set of nodes, where all the nodes in the set should receive the packets. The third class is anycast, which is an addressing method, where also a set of nodes is addressed, but here, only one of the nodes of this set should receive the packet. In other words, unicast is a one-to-one, multicast is a one-to-many communication scenario, anycast is a one-to-one-of-many type of communication.

The other classification method is based on routability of packets. Here, four possibilities exist; however, they can be distinguished in a structured manner. First of all, the addresses can be global, or local. Global addresses are used for global connectivity. Local addresses cannot be used in the public Internet: they must be used locally. However the boundary of locality arise another classification possibility: local addresses can be further divided into host-local addresses (which can be used only inside a host), link-local addresses (which are addresses possible to use only on the local link) and site-wide local unique addresses (which make possible to connect more than one links together, supporting communication between entities from different links).

Since the address space is huge enough, usually every node has more than one IPv6 addresses. It is quite usual that every interface of an IPv6 node has a link-local and a global IPv6 address. As in IPv4, it is not forbidden to have more than one global IP address: there are examples when the number of IPv6 addresses bound to an interface goes up even more.

Coexistence, Migration, Strategies

Since IPv6 header is completely different from the IPv4 header, and IPv6 addressing provides much wider space than IPv4, the two protocols are absolutely incompatible with each other. Without external support, IPv4-only hosts can communicate with IPv4 hosts and networks only, and vice versa: IPv6-only terminals can see only the IPv6 Internet. To migrate the two technologies, some kind of support is needed.

The most trivial (and most frequently used) migration strategy is to use both versions in the host, or the network. That is, having both IPv4 and IPv6 available the clients can see both the IPv4 and IPv6 Internet. Since this approach is based on a dual-stack implementation, it is also called as dual-stack approach. As long as one has connectivity to both the IPv4 and IPv6 Internet, the dual-stack approach is the best option to provide overall accessibility to users. Most dual-stack devices and operating systems try to use IPv6 first, and they only switch to IPv4, if IPv6 connectivity is not possible.

However, not all ISP's provide access to the IPv6 Internet nowadays. Although one can set up IPv6 addresses on their client machines, but they might not be able to see the IPv6 Internet, if the access network is not able to transmit them to the right place, or there is a network segment between the two IPv6 islands, which do not understand IPv6 packets. In this case, the IPv6 over IPv4 packet tunneling method could be useful. Tunnels can be built between IPv6 endpoints using IPv4 packets, and the IPv4 packets can carry the IPv6 traffic, embedded into IPv4 packets. Of course, IPv4 islands can also be interconnected through IPv6 by using inverse (i.e., IPv4 over IPv6) tunnels: this case is presumable when IPv6 will be the prevailing protocol but at some places, still IPv4 remains the IP version in use. Based on the nature of the endpoints, one can make distinction between router-to-router, node-to-router, router-to-node, and node-to-node tunnels. Tunnels have plenty of standardized forms; the next section provides a substantially detailed summary about the tunnels.

Next, there might be situations when the client has only an IPv6 address and no IPv4 support (because the network it connects to do not have IPv4 support, or the client itself lacks IPv4 implementation). Meanwhile, the user of this client still wants to be able to reach the IPv4 Internet, since there are many servers, which have IPv4 only. In this case, some trick is needed to provide connectivity for the user.

The question is, where and when IPv6 packets turn into IPv4 packets. Based on the layer of translation, one can list three basic alternatives: The network level (e.g. NAT-PT – Network Address Translation-Protocol Translation, or Stateless IP/ICMP Translation – SIIT), which translates IPv6 into IPv4 at purely IP level. The transport level (UDP, TCP) relays, which make the translation at the transport level. Finally, application gateway based solutions, which make conversion at the application level. In all cases, there must be a special entity in the network (the translator), which is dual-stack and can communicate with the host of interest and the IPv4 Internet as well. The drawback of such solutions is that end-to-end connectivity cannot be maintained any more.

The network level (e.g. NAT-PT, SIIT) translator has an implementation, which turns incoming IPv6 packets into IPv4 packets, and also marks them somehow to make the reconversion possible when the answer packets arrive. Transport level (e.g. TCP and UDP) relay translators set up separate dedicated connection to the IPv6 only host and the IPv4 addresses. Then these connections are bridged: every packet arriving on the one connection is forwarded to the other one and vice versa. We mentioned here two common transport protocols: but note that for each other transport protocols (e.g., DCCP, SCTP, etc.) a separate solution is needed (which means that transport level relaying is not as general as network level relaying). Finally, the application gateway (proxy) based solution is an application specific translator. For instance, installing a dual-stack web proxy provides automatic IPv4 connection to the IPv6 only hosts, if their web requests go through the web proxy. This is a very easy and simple solution, as long as all the applications are supported by a proxy: if the application proxy is missing, there will be no connectivity for that particular application.

There are also solutions for e.g., providing IPv6 connectivity to IPv4 only applications (using Bump-in-the-Stack – BIS, or Bump-in-the-API – BIA) which adds some extra layer in the hosts' networking stack in order to provide the possibility for IPv4 only applications to translate their outgoing packets into IPv6 packets. Here, there is no need for a new entity in the network; everything is done by the client itself. However, note that the change of the protocol stack is required for this kind of functionality. All the above-mentioned solutions are detailed in the next section.

TRANSITION CONCEPTS AND MECHANISMS

Techniques aiming to make IPv6 and IPv4 coexistence possible and provide an easy v4-v6 transition can be separated into the above drafted three main categories. Dual-stack techniques grant IPv4 and IPv6 coexistence in the same terminals and networks. Tunneling techniques provide transport schemes for IPv6 traffic over the existing IPv4 infrastructure and vice versa. Translation techniques make able IPv6-only nodes to communicate with IPv4-only nodes. Sections below describe the solutions available nowadays for each of these categories.

Dual Stack

Dual-stack terminals have complete support for both IPv4 and IPv6 protocol versions. The operation of a dual-stack terminal is twofold. On one hand if it communicates with an IPv6 host, the dual-stack terminal will behave like an IPv6 node. On the other hand, if it is in communication with an IPv4 node, the dual-stack terminal will behave like an IPv4 node. Many implementations have a certain configuration parameter in order to turn on or off one of the stacks, so such a terminal introduces three different modes of operation:

- If the IPv4 stack is turned on and the IPv6 stack is off, the terminal acts as an IPv4-only node.
- If the IPv6 stack is turned on and the IPv4 stack is off, the terminal acts as an IPv6-only node.
- If both the IPv4 and IPv6 stacks are on, the terminal acts as a dual-stack node and can use both protocols simultaneously.

An operating dual-stack terminal must have at least one address for each protocol version: IPv4 mechanisms (e.g., static configuration or DHCPv4) are applied to configure IPv4 address(es), and IPv6 mechanisms (e.g., stateless or stateful autoconfiguration) are used to configure IPv6 address(es).

For dual-stack v4/v6 and native IPv6 operation, the DNS must also be advanced as it can be requested to resolve names and addresses for both IP protocol versions. Therefore, modern IP systems require DNS resolvers, which are prepared to handle both IPv4 and IPv6 addresses. To achieve this functionality, the original DNS A record (which resolves IPv4 addresses) was extended with a DNS AAAA (quad-A) record designed to resolve IPv6 addresses.

This technique is very easy to use and provides flexible operation, not to mention that dual-stack protocol implementations are the basis of several other transition mechanisms. The most important disadvantage of the dual-stack technique is that two separate protocol stacks must be operated simultaneously and continuously. Network maintenance and configuration (e.g., address assignment, routing, etc.) must be managed for both protocols, often with separate commands depending on the protocol. It usually takes more time, and reserves more memory and CPU from the resources.

Tunneling or Encapsulation

Tunneling (or often called encapsulation) mechanisms are used to deploy a special IPv6 forwarding scheme on the top of an operational IPv4 infrastructure which either should not or cannot be improved to natively support IPv6. In case of tunneling techniques one specific protocol (IPv6) is encapsulated in the payload of another protocol (IPv4) and forwarded over the infrastructure of the protocol used as encapsulator (IPv4). The procedure of encapsulation has three main components:

- Encapsulation at the tunnel entrance
- Decapsulation at the tunnel exit
- Management of the tunnel

Based on these three main components, techniques in the below sections are able to transmit IPv6 traffic by encapsulating whole IPv6 packets in IPv4 datagram and tunneling them over an existing IPv4 infrastructure. Note, that not only IPv4 packets can be used as encapsulators but IPv6 packets as well: IPv4 over IPv6 or IPv6 over IPv6 tunnels are also possible to be built and managed for particular purposes. (Conta & Deering, 1998) defines the model and generic mechanisms for IPv6 encapsulation of IPv4, IPv6 and even AppleTalk, IPX, or other type of packets.

The main advantage of tunneling is that it gives totally free hand in the migration process: single hosts or even whole subnets can be connected to the IPv6 Internet through tunnels without the need of backbone or ISP support. Among disadvantages we can list: additional load on tunnel endpoints in means of CPU and memory consumption (due to packet encapsulation and decapsulation), increased overhead in the network (due to additional headers), introduced single points of failure in the network (i.e., tunnel endpoints), and hardened troubleshooting and network management (e.g., MTU size issues).

Configured or Static Tunneling

Configured or Static Tunneling is an IPv6-over-IPv4 tunneling technique where addresses of IPv4 tunnel endpoints are determined by explicitly given configuration information at the tunnel endpoints. These techniques operate bidirectional tunnels in order to provide virtual point-to-point link to the IPv6 networking layer using the pre-configured IPv4 tunnel addresses as the lower layer endpoints. In RFC4213 (Nordmark & Gilligan, Basic Transition Mechanisms for IPv6 Hosts and Routers, 2005) four types of configured tunneling are defined: router-to-router, host-to-router, host-to-host, and router-to-host. Configured tunneling can be applied in all of the above four cases, but it is most likely to be used router-to-router due to the need to explicitly configure the tunnel endpoints. Besides the possible use-cases, RFC4213 also details the configuration and discusses important issues like the determination of valid tunnel endpoint addresses, handling ICMPv4 or ICMPv6 errors, tunnel MTU and fragmentation questions, Neighbor Discovery over tunnels, IPv4 header construction, and security considerations.

Automatic Tunneling using IPv4-Compatible Addresses

Automatic tunneling techniques allow IPv6 or IPv4 nodes to communicate over an existing IPv4 infrastructure, without the need for preconfigured data on tunnel destination. In RFC2893 (Gilligan & Nordmark, 2000) authors define a scheme where the tunnel endpoint is determined by an IPv4-compatible destination address. In this scheme the prefix::/96 is separated for IPv4-compatible addresses, where the rightmost 32 bits of the IPv6 address stand for the IPv4 address of the destination. IPv6 packets destined to addresses from this range are automatically encapsulated in IPv4 packets and tunneled using the inherent IPv4 address as destination. RFC4213 obsoletes RFC2893 and removes the introduction of automatic tunneling using IPv4-compatible addresses, and refers to 6to4 (RFC3056) as the solution to follow.

6to4

In RFC3056 (Carpenter & Moore, Connection of IPv6 Domains via IPv4 Clouds, 2001) authors specify a scheme for IPv6 sites to communicate with each other over an existing IPv4 network without explicitly given tunnel endpoint information. 6to4 does not use IPv4-compatible IPv6 addresses as it has a proper IPv6 address format that includes the IPv4 address of the tunnel endpoint in the prefix such allowing automatic tunnel setup. In 6to4, the transport IPv4 network behaves as a unicast point-to-point link, and the native IPv6 segments communicate via 6to4 routers (i.e., 6to4 gateways): IPv6 packets are encapsulated and decapsulated here requiring at least one globally unique IPv4 unicast address. Only the gateways need to be 6to4 compatible, therefore no other changes have to be made to the nodes inside the 6to4 network. The prefix for the 6to4 protocol assigned by the IANA organization is: 2002::/16. It is important to notice, that if a host in a 6to4 network wants to exchange packets with a host in another 6to4 network, no tunnel configuration is needed: the tunnel entry point can take the IPv4 address of the tunnel exit point from the IPv6 address of the destination. Besides the above, a 6to4 relay router is needed for a successful communication with an IPv6 node in a remote IPv6 network. The relay router is a router configured for 6to4 operation and IPv6 connection. The relay router connects 6to4 networks to the native IPv6 network as the 2002::/16 prefix is announced into the native IPv6 network by such relays. Figure 1 shows the main 6to4 components and illustrates the communication scheme.

RFC3068 (Huitema, An Anycast Prefix for 6to4 Relay Routers, 2001) specifies a 6to4 relay router anycast address in order to optimize the configuration of 6to4 gateways which require a default route towards a 6to4 relay router on the IPv6 Internet.

Figure 1. 6to4 components and communication

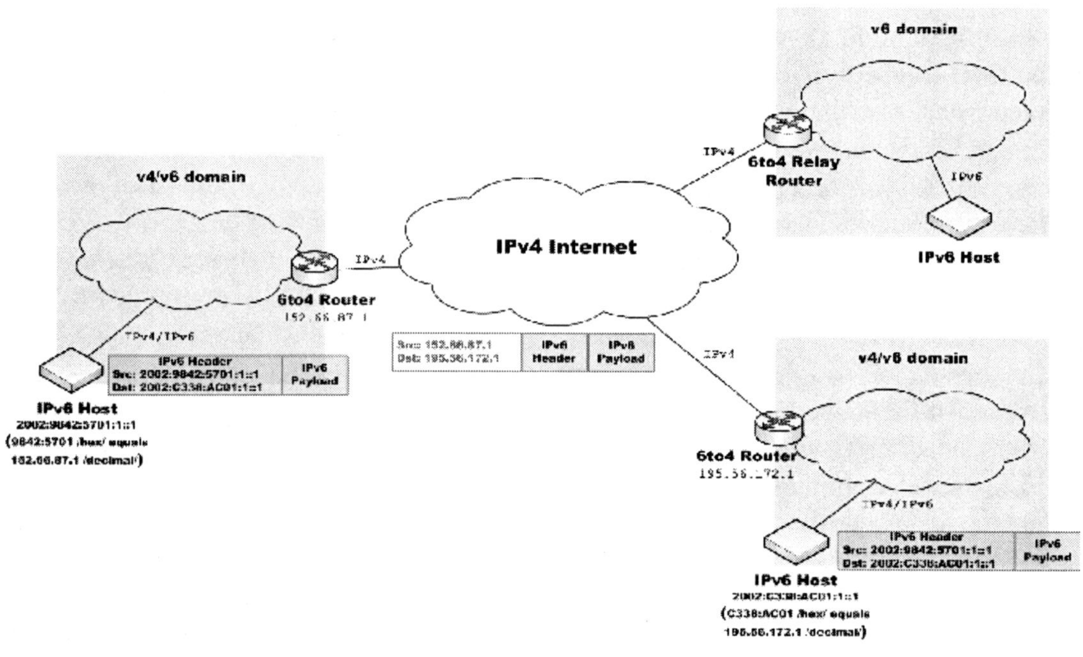

6over4

6over4 is defined in RFC2529 (Carpenter & Jung, Transmission of IPv6 over IPv4 Domains without Explicit Tunnels, 1999) and based on the multicast capability of the existing IPv4 network and uses the IPv4 infrastructure as a virtual subnet for IPv6 hosts and routers. This scheme is basically an IPv6-over-Ethernet subnetwork with the exception that the Ethernet's Layer 2 functionality is incorporated by IPv4 multicasting mechanisms. Therefore, the technique is also known as IPv4 multicast tunneling (see Figure 2). The main difference between 6over4 and 6to4 discussed in the above section is that the former allows full neighbor discovery with the IPv4 network acting as a virtual local area network (VLAN). 6over4 uses the underlying IPv4 multicast network as a Layer 2 media, thus this tunneling technique does not require special prefixes or address ranges as in case of the 6to4 method. Since 6over4 uses the IPv4 network as a single link with multicast capabilities, the IPv6 Neighbor Discovery (ND) procedures (e.g., address resolution and router discovery) will be performed as they would over any real physical link with multicast capabilities. However, there is no need for special prefixes, 6over4 hosts still need to form a number of addresses for ND operation (e.g., exchange of router solicitation, router advertisement, neighbor solicitation, neighbor advertisement messages). By forming these special addresses, a 6over4 host is able to use the IPv4 address of its interface in the same way as a native IPv6 host uses the 64-bit extended unique identifier (EUI-64) of its Ethernet interface. Since IPv4 multicasting is not a common feature on the Internet, it is very likely that 6over4 will not be widely employed.

4over6

RFC5747 (Wu J., Cui, Li, Xu, & Metz, 2010) introduces a mechanism for automatic discovery, creation and maintenance of IPv4-over-IPv6 tunnels via extensions to multiprotocol BGP. The aim of 4over6 is somehow the opposite as 6over4: as the continuously growing IPv6 islands introduce cases where connectivity with IPv4 networks crossing IPv6 transit backbones is desired, automatically connecting remaining segments of

Figure 2. 6over4 architecture

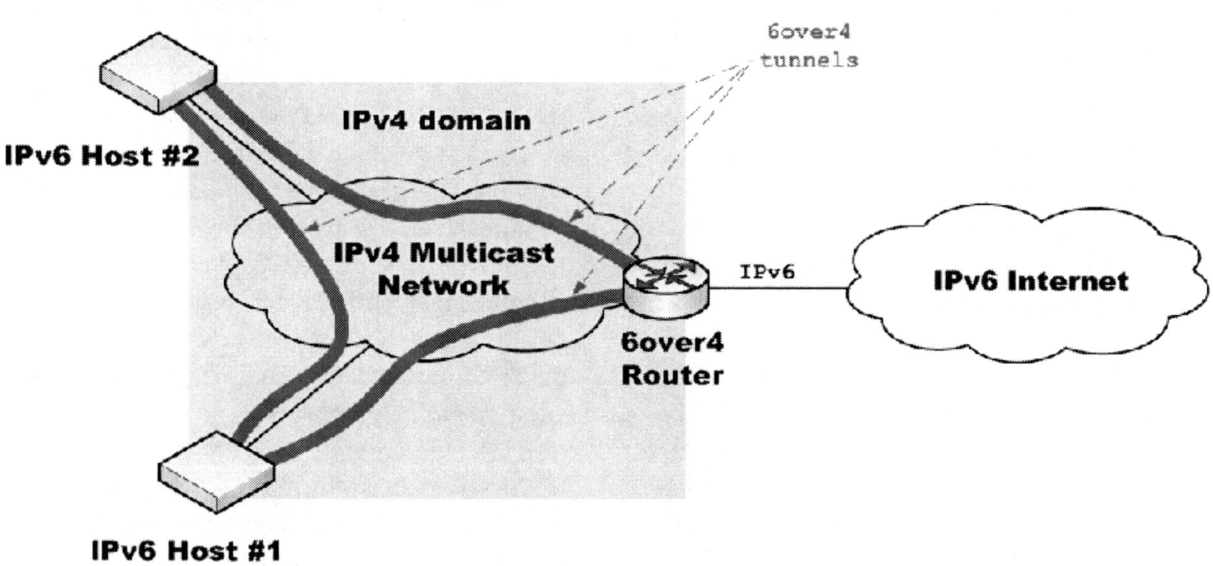

the IPv4 Internet across an IPv6-only backbone will be required. The mechanisms described in RFC5747 are dealing with this question based on experiments of real-life implementation in a large research IPv6 network in China (Wu J., Cui, Li, & Metz, 2006). Description of 4over6 concerns two main aspects: the control plane and the data plane. The control plane addresses the problem of how to set up the IPv4-over-IPv6 tunnel in an automatic and scalable fashion between a large number of edge routers, while the data plane focuses on the packet forwarding processes of encapsulation and decapsulation.

ISATAP

The Intra-Site Automatic Tunnel Addressing Protocol (ISATAP) is specified in RFC4214 (Templin, Gleeson, Talwar, & Thaler, 2005) aiming to provide IPv6 connectivity for dual-stack hosts over an IPv4-based networking infrastructure. This technique also uses the existing IPv4 network as one large link-layer architecture and allows the dual-stack hosts to automatically create tunnels and exchange data between themselves. ISATAP can be used regardless of whether the hosts have global or private IPv4 addresses. Addresses of this automatic tunneling mechanism embed an IPv4 address in the EUI-64 interface identifier in the following format: 64bitPrefix:16bitControl:5EFE:IPv4address. ISATAP interfaces form ISATAP interface identifiers using their IPv4 addresses and apply them to produce the ISATAP link-local addresses in order to make the technique able to perform standard IPv6 neighbor discovery mechanisms. Using this method, IPv6 nodes inside an IPv4 intranet can communicate with each other. If hosts want to communicate with IPv6 hosts outside the intranet (e.g., 6Bone hosts), a border router must be configured which can be an ISATAP router or even a 6to4 gateway. An important issue of this method is that all hosts in an ISATAP network need to support the ISATAP protocol.

ISATAP and 6to4 are considered as two really promising and already popular transition technologies evaluated and assessed within several real-life testbed experiments and projects like (Xiaodong, Yan, & Yumei, 2009), (Lee, Shin, & Kim, 2006) and (Hei & Yamazaki, 2004).

Teredo

Teredo is specified in RFC4380 (Huitema, Teredo: Tunneling IPv6 over UDP hrough Network Address Translations (NATs), 2006) as an IPv6 transition technology providing address assignment and automatic host-to-host tunneling for unicast IPv6 traffic in cases when IPv6/IPv4 nodes are placed behind IPv4 network address translators (NATs). Comparing Teredo with 6to4 and ISATAP we can summary that 6to4 makes IPv6 available over an IPv4 network using public IPv4 addresses, ISATAP helps deployment of IPv6 nodes within a site regardless of whether it applies private or public IPv4 addresses, and Teredo makes IPv6 available to nodes through any number of NAT layers using UDP-based tunneling. The Teredo architecture consists of a Teredo server (helping clients to access IPv6 networks), several Teredo clients (running on an IPv4/IPv6 dual-stack terminal in an IPv4 network behind a NAT), and Teredo relays (forwarding IPv6 traffic between a Teredo client and a host in the IPv6 network). The technique introduces a special prefix called Teredo Service Prefix (2001:0000::/32) which is announced by the Teredo relays to the outside world using conventional IPv6 routing mechanisms. A significant part of RFC4380 deals with how Teredo identifies the specific type of NAT deployed in the actual network and defines mechanisms for handling these various NAT types. Figure 3 introduces the Teredo architecture and the protocol's basic communication procedure.

However, Teredo solves numerous problems of IPv4-IPv6 transition, the current version of the standard does not work with symmetric NATs. In order to support Teredo for symmetric NAT tra-

Figure 3. Teredo architecture and communication

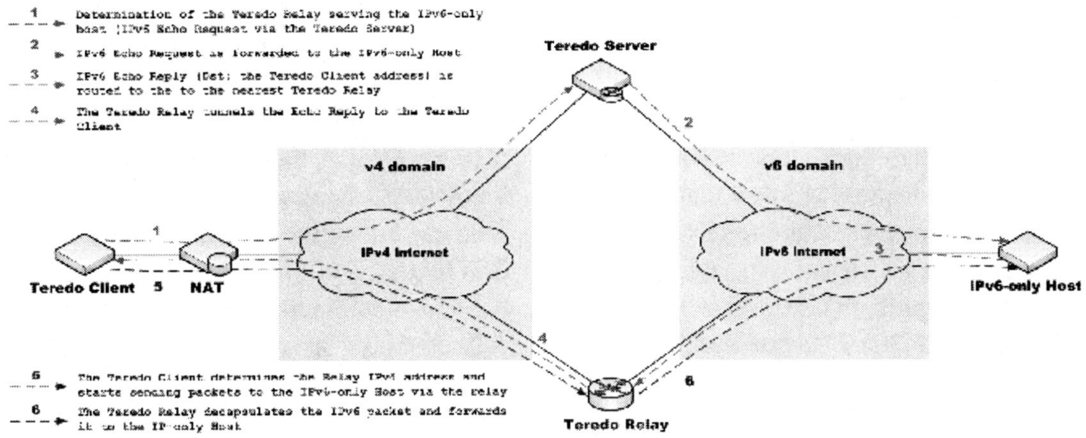

versal, (Huang, Wu, & Lin, 2006) proposed SymTeredo, which imposes minor modifications on the Teredo relay and the Teredo client components but also keeps compatibility with the standard protocol.

Forwarding Protocol 41

Most of the existing tunnel-based IPv4-to-IPv6 transition solutions assume that the client endpoint is an IPv6-compatible router. However, the number of installed and used IPv4-only NAT boxes/routers is still large, while client operating systems mostly support IPv6. This technique (Palet, Olvera, & Fernandez, 2003) relies on the fact that some NAT implementations allow the configuration of IPv6 tunnels from inside of the private LAN to routers or tunnel servers in the Internet, and uses this simple and rapid way to provide IPv6 hosts and networks located behind NATs with access towards the IPv6 Internet. A client (i.e., host or router) behind an IPv4-only NAT and with a private IPv4 address can use a Tunnel Broker (see later) or an IPv6 router to create an IPv6 tunnel. NAT implementations usually can be configured to forward packets based on the protocol value (41 in case of IPv6) in the IPv4 header, which makes possible to easily and quickly deploy IPv6. Forwarding Protocol 41 adds a special feature to IPv4-only NAT gateways, namely allowing seamless usage of IPv6 nodes behind them. The scheme focuses on home networks and allows of gradual migration from IPv4 to IPv6, while making the transition path simple and acceptable for the end users.

Silkroad

Silkroad is a novel mechanism proposed in (Liu, Wu, Jin, & Li, 2005). The technique manages IPv6 over IPv4 tunnels also for nodes sitting behind all existing NAT types, including symmetric NATs. Another difference from Teredo is that Silkroad does not need a special prefix for the operation. This scheme introduces two special entities: the Silkroad Access router helping Silkroad clients to gain IPv6 connectivity, and the Silkroad Navigator helping Silkroad Access routers to route between each other. The Silkroad IETF draft is in its early stage so here we do not go into more details of the specification.

Tunnel Broker

RFC3053 (Durand, Fasano, Guardini, & Lento, 2001) introduces the Tunnel Broker model in which IPv6 Tunnel Brokers are virtual IPv6 providers granting connectivity to the IPv6 Internet for users with IPv4 Internet connection. The motivation for the development of this model is to support early IPv6 adopters with a "shortcut" to an existing IPv6 network and with stable, permanent IPv6 addresses and DNS records. Users requesting IPv6 connection register with the Tunnel Broker that will further manage the establishment, maintenance, and removal of the tunnel on behalf of the registered users. Tunnel Brokers send configuration information to a Tunnel Server if there is a tunnel to establish, change, or remove, and in case of need, they also register the addresses of users in the DNS. Tunnel Brokers must be reachable through the IPv4 Internet, but IPv6 connection is not mandatory as the communication between Tunnel Brokers and Servers can go over both IP versions. However, Tunnel Servers are strictly dual-stack routers connected to the global IPv4/IPv6 Internet. When a Server receives configuration information from a Broker, it establishes, changes, or removes the server side of the tunnel. The Tunnel Broker can share the data load across several Tunnel Servers such achieving load sharing and load balancing if needed. Broker clients are a dual-stack hosts or routers connected to the IPv4 Internet. If such a client registers with the Tunnel Broker for an IPv6 connection, first it authenticates itself (e.g., using RADIUS procedures), then initiates the tunnel configuration in order to gain access to all IPv6 networks to which the Tunnel Server provides connection. This technique is particularly designed for small IPv6 networks and for single IPv6 hosts. As the scheme works only with public IPv4 addresses, Teredo or another mechanism must be applied if clients are behind NAT.

DSTM

Dual-Stack IPv6 Dominant Transition Mechanism or shortly called DSTM is a transition mechanism allowing the transport of IPv4 packets over an IPv6 dominant environment by using IPv4-over-IPv6 tunnels. DSTM specification (Bound, Toutain, & Richier, 2005) introduces a technique to assign a temporary IPv4 address to dual-stack nodes. The solution is designed to avoid the use of Network Address Translation (NAT) when allowing the communication between IPv6 nodes and legacy IPv4 nodes and their applications. DSTM is a transition technique that does not specify a new protocol but relies on existing ones. However, DSTM still introduces novel network entities (DSTM client, server, and border router) and also defines the behavior and properties of possible temporary addresses allocation mechanisms. A DSTM client is a dual-stack node, which acts as a tunnel endpoint and encapsulates IPv4 packets in IPv6 packets in order to send them to a DSTM Border Router. The Border Router acts as the other tunnel endpoint such decapsulates IPv6 packets and forwards IPv4 packets towards their (IPv4) destination. By caching the IPv6 path for the IPv4 address, the Border Router is able to forward incoming IPv4 packets and tunnel them to the appropriate client. The DSTM server manages the IPv4 address allocation mechanisms (e.g., by applying the Tunnel Setup Protocol defined in RFC5572) and configures the clients for the tunnel endpoint (e.g., IPv4 address assignment based on DHCPv6 and the DSTM Tunnel Endpoint option). By using DHCPv6 for configuration provision, DSTM is able to support IPv6-dominant infrastructures, which are much simpler and less costly than supporting dual-stack infrastructures.

VLANs for IPv4-IPv6 Coexistence

Motivated by traffic segregation demands, Virtual Local Area Networks (VLANs) are quite common nowadays, especially in enterprise networks. As the VLAN standard allows separate LANs to be spanned over a single bridged LAN based on e.g., insertion of virtual LAN tags into Ethernet frames, nodes of VLANs are able to communicate as if they were connected to the same broadcast domain, regardless of their actual physical location. Based on this principle RFC4554 (Chown, 2006) introduces a scheme where IPv6 can be routed in parallel with IPv4 in the network and IPv6 packets can be delivered at Layer 2 via VLAN technology. A parallel IPv6 routing infrastructure is such created beside the IPv4, and IPv6 links will be overlaid onto the IPv4 infrastructure without any changes to the legacy IPv4 environment.

Translation

Techniques based on address and protocol translation offer transition mechanisms in addition to dual-stack and tunneling schemes. The main advantage of this technique is that it provides transparent routing for IPv6 hosts in order to communicate directly with hosts in IPv4 networks and vice versa. The disadvantages are that translation-based schemes do not support the advanced features of IPv6 (such as end-to-end reachability and security) and that they present limitations on the design topology. In the following sections, we describe the most important translation methods designed for IPv4/IPv6 transition.

SIIT

The Stateless IP/ICMP Translation Algorithm (SIIT) is specified in RFC2765 (Nordmark, Stateless IP/ICMP Translation Algorithm (SIIT), 2000) as a set of algorithms which translate between v4 and v6 versions of IP/ICMP packet headers. The method is applicable for cases when an IPv4-only host wants to communicate with an IPv6-only host or vice versa. Separate translator boxes in the network deal with this task without requiring any per-connection state. SIIT also introduces a new address type known as the IPv4-translated address with the prefix format of 0::ffff:0:0:0/96. The host identifier part of such an address is an IPv4 address taken from a special pool and assigned to the IPv6 host that wants to communicate with IPv4 peers. Segments of upper-layer protocols like TCP and UDP can pass through the translator boxes relatively untouched as SIIT translation mechanisms are taking care of that TCP and UDP pseudoheader checksums would not get affected during translation. However, translation causes problems with applications embedding IP addresses in higher-layer protocol messages (e.g., FTP, SIP). Such scenarios can be solved by using application-specific application layer gateways (ALGs) in addition of the SIIT implementation. ALGs can extend other schemes (as SIIT or similar methods) and also can provide stand-alone solution in certain transition scenarios, like the proxy-based approach introduced in (Zhuang, Wu, & Xia, 2006).

BIS

RFC2767 (Tsuchiya, Higuchi, & Atarashi, 2000) defines the "Bump-In-the-Stack" (BIS) technique which is a variant of SIIT designed to support IPv4 applications in IPv6 dominant networks by performing IPv4/IPv6 translation in the end system. It means that the translator sits within the operating system of the end terminal such providing an interface between IPv4 applications and the underlying IPv6 network. The difference between the dual-stack and the BIS scheme is that in case of BIS hosts do not need IPv4 address(es): only IPv6 packets are transmitted and received over the network. This behavior is achieved by the BIS driver, which applies a pool of IPv4 addresses for internal communication with upper layer IPv4 applications and performs translation between IPv4 and IPv6 headers using the mechanisms defined in SIIT. The above functions are implemented in novel modules introduced in a BIS end system: name resolver (ensuring association of IPv4 addresses with the destination IPv6 application), address mapper (maintaining a table of the IPv6 addresses and the associated IPv4 addresses), and translator (dealing with the actual translation between IPv4 packets of BIS IPv4 applications and IPv6 packets received from the IPv6 network).

BIA

"Bump-in-the-API" (BIA) specified in RFC3338 (Lee S., Shin, Kim, Nordmark, & Durand, 2002) introduces almost the same mechanisms as BIS does, the only difference is that translation is performed internally between the IPv4 APIs and the IPv6 APIs. IPv4 application trying to communicate with an IPv6 host will initiate the API translator to intercept the socket API functions and to call the corresponding IPv6 socket APIs. A pool of IPv4 addresses is also used here for internal communication.

However BIS and BIA allows IPv4 applications to access IPv6 resources, neither of them is able to process the application-layer headers and such to translate e.g., IPv4 FTP and SIP applications to IPv6. This motivated authors of (Chen, Su, & Lin, NCTU SLT: a socket-layer translator for IPv4-IPv6 translation, 2005) to develop a BIA-based Socket-layer Translator scheme called NCTU-SLT for IPv6 transition of FTP and SIP applications. The method implements Application Layer Gateways (ALGs) and ALG Manager allowing dual-stack hosts to access IPv6 resources through IPv4 applications without any modification.

NAT-PT and NAPT-PT

RFC2766 (Tsirtsis & Srisuresh, 2000) introduces the Network Address Translation - Protocol Translation (NAT-PT) transition scheme which uses a pool of public IPv4 addresses for dynamic assignment to IPv6 hosts, and employs a stateful IPv4/IPv6 header translation on a special network device located at the boundary of the IPv4 and the IPv6 networks. This NAT device translates IPv6 packets into analogous IPv4 packets and vice versa, and such routes between an IPv6 network and an IPv4 network. NAT-PT reserves the pool of IPv4 addresses and translates the fields for IP Source addresses, IP, TCP, UDP, and ICMP header checksums.

An extension of NAT-PT is Network Address Port Translation - Protocol Translation (NAPT-PT), which further extends the original idea: in order to allow numerous IPv6 hosts to share one single IPv4 address for multiplexing multiple sessions on one address, transport identifiers (such as TCP and UDP port numbers and ICMP message types) are also translated in this technique.

The main benefit of NAT-PT and NAPT-PT is that no changes are required to end hosts because all the translation procedures are executed at the separate NAT device in the network. However, the mechanisms defined in RFC2766 seem to be convenient in several transition scenarios, serious issues exist with the standard. For example, NAT-based schemes cannot take full advantage of the enhancements offering by IPv6 offers, and it is really hard to maintain the big number of Application Level Gateways (ALG) needed in NAT devices to keep the widest scale of applications working correctly through the gateway. The raised problems are summarized in RFC4966 (Aoun & Davies, 2007) together with the conclusion that technical and operational difficulties resulting from these issues make it undesirable to recommend the usage of RFC2766 as a general purpose transition mechanism.

TRT

The Transport Relay Translator (TRT) is a translation mechanism defined in RFC3142 (Hagino & Yamamoto, 2001) in order to enable IPv6-only nodes to exchange TCP and UDP traffic with IPv4-only nodes. The operation of TRT is based on a transport layer relay that sits on a network border device and divides the transport layer communication path into two segments: the first segment utilizes IPv4 and second segment utilizes IPv6. In this technique, translation of IPv6 based TCP/UDP traffic to IPv4 (or vice versa) occurs at the transport layer, such IP layer intervention – and all the relating problems – is avoided. In case of TCP traffic, the relay terminates the session to the

Figure 4. Communication over NAT-PT

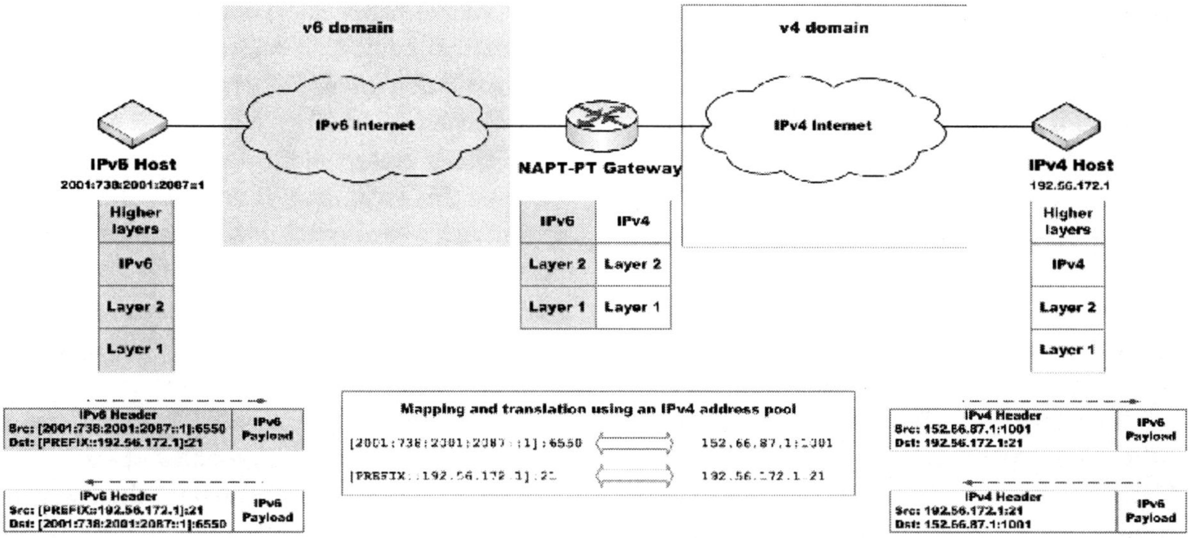

client (i.e., a TCP over IPv6 connection between the client and the relay is built up) and creates a new TCP connection on the other side towards the IPv4 application (i.e., a TCP over IPv4 connection between the relay and the destination is built up). After that, the translator translates between the two separate sessions. In case of UDP traffic, the translator purely translates and forwards the packet between the two segments.

SOCKS64

RFC3089 (Kitamura, 2001) outlines a SOCKS-based IPv6/IPv4 gateway mechanism that enables smooth communications between IPv6 and IPv4 application flows. SOCKS (Leech, Ganis, Lee, Kuris, Koblas, & Jones, 1996) is a protocol designed to provide a framework for TCP and UDP based client-server applications to transparently, securely and easily use the services of a network firewall. By applying the SOCKS mechanisms the SOCKS64 library intercepts the connection initiation packets (which is usually a DNS resolve request) on the client and responds with fabricated IP address for the session. After that, the library issues connection control calls to the SOCKS64 gateway, which will use the real IP address to transmit the packets towards the destination. The main advantage of this scheme is that it supports every possible IPv4/IPv6 combination and transition scenario. Since the operation is based on the existing SOCKS technique and the translation is accomplished without introducing more protocols, it is very convenient to use in environments that already employs SOCKS mechanisms. However, as SOCKS64 is an application-level translator, it only provides low throughput and requires stronger hardware. The mandatory client-side software upgrade is also considered as an important disadvantage of this technique.

FUTURE RESEARCH DIRECTIONS

Previous paragraphs of this chapter have summarized the currently existing standards for IPv4-IPv6 transition. However, these solutions are not perfect of course. Some publications like (Hirorai & Yoshifuji, 2006) and (Yan-ge, Shui-mu, & Jun-ying, 2009) describe the weaknesses of the proposed solutions and point out that the IPv4-IPv6 transition concepts and methods are still continuously evolving. There are plenty of proposals in the literature, which has not yet become a standard. It is almost impossible to show all the current research directions; however, in this section we try to overview the actual situation.

In addition to the IPv4-IPv6 transition mechanisms detailed in this chapter, there are other new methods currently proposed by scientists, which aim to provide seamless transition between the two versions. For instance, bi-directional mapping has been proposed by some researchers (AlJa'afreh, Mellor, Kamala, & Kasasbeh, 2008). The efficiency and performance of the proposal have been covered in (AlJa'afreh, Mellor, & Awan, 2009). Similar, proxy based solution is described in (Jankiewicz, Green, Fiuczynski, & Marc, 2006), and also an extension of NAT-PT with some more functionalities can be found in (Shi, Huang, Wang, Chen, Huang, & Cheng, 2007).

A possible alternative to the Bump-in-the-Stack and Bump-in-the-API methods is described in (Yuan-hang & Lijuan, 2009). Another migration proposal for IPv6 only hosts, with stateless mapping and scalable service for large networks can be found in (Zhu, Chen, Zhang, & Li, 2008).

A novel emerging area of Internet Protocol related research is the Host Identity Protocol (HIP), where hosts are identified by a cryptographic Host Identifier (and a hash-imprint of it called the Local Scope Identifier in IPv4 and the Host Identity Tag in IPv6) instead of the IP address, which is the case nowadays. Using HIP the IP address can be used only as a locator, and identity of nodes is described by the LSI or HIT. Using HIP helps also dual-stack workstations to have secure and quick handovers between IPv4 and IPv6 subnetworks (Varjonen, Komu, & Gurtov, 2009). A novel, HIP-based tunneling protocol called Escort was also proposed: Escort is an extension of Teredo that uses HIP in the background and provides connectivity also in the case of symmetric NATs (An, Luo, Li, Zhang, & Yan, 2009).

The next uncovered issue is how to handle IP level mobility in a dual-stack environment. That is a dual-stack node visits IPv4 only and IPv6 only subnetworks, the connectivity should be maintained, however it is not a trivial task. Mobile IP is a solution for mobility management in IPv4, Mobile IPv6 is for IPv6, but there is no solution for a mixed environment. Although we have not covered this area, there is an IETF draft, which describes how dual-stack nodes or networks should be served by Mobile IPv6/NEMO BS support (Soliman, 2007). Some performance tests have also been done for this mobile environment by (Xie & Narayanan, 2009).

Many articles about the mobility management, triangular routing and roaming between IPv6 and IPv4 networks have been published recently (Jayanthi & Rabara, 2010), (Govil, 2007), (Sanguankotchakorn & Jaiton, 2008), (Jeong, Shin, & Kim, 2007). Also extended mobility management solutions, like hierarchical mobility management is addressed in different articles, e.g., (Huang, Pan, Yu, & Chiao, 2005).

The central entity of future mobile networks will be the IP Multimedia Subsystem (IMS), which provides SIP connectivity for the clients. Among many other tasks, IMS will provide voice communication for the clients. Thus having IPv4 and IPv6 connectivity for IMS is an essential requirement of the system. There are many researchers working on the IMS and SIP IPv4-IPv6 issues nowadays, e.g., (Liu, Cao, & Li, 2008) and (Chen, Lin, & Pang, An IPv4-IPv6 translation mechanism for SIP overlay network in UMTS all-IP environment, 2005).

As usual, new technology is not as secure as technologies, which have been developed for longer periods. As long as one deals with new technologies, she should not forget about security issues. There are a number of articles dealing with security of IPv4-IPv6 transition, e.g., (Liu, Guan, Zheng, & Qu, 2009), (Sabir, Fahiem, & Mian, 2009), and (Taib & Budiarto, 2007).

In addition, the IPsec (IP security) extension of IP does not work in a NAT-PT environment, due to the fact that NAT-PT interrupts end-to-end communication with the translator device. A possible extension of IPsec to the NAT-PT case is described in (Peng, Zhou, Wang, & Yang, 2009).

The next group of research directions is about IPv4-IPv6 migration and transition in a multicast environment. Multicasting is a very promising technique in IPv4 and IPv6 networks to save network bandwidth by transmitting the same flow only once. However, in a mixed environment (where the flow goes through IPv4 and also IPv6 networks) the multicasting support is not self-explanatory. A good summary how multicast can be supported in such situations is given by (Xianfeng & Zheng, 2006).

There are many proposals to address multicast support in a mixed environment, e.g., using HIP as a supporting technology (Kovacshazi & Vida, 2007), or proposing a new network entity (gateway) for multicast operations, like in (Aoul, Negru, Nafaa, & Mehaoua, 2003).

At last, but not at least, applications can also help the migration process. Peer-to-peer (P2P) is the most bandwidth-consuming application, which drives network operators to improve their networks without limit. However, from the transition point of view, P2P gives a natural migration point, since every client can serve as an IPv4-IPv6 proxy. The details of P2P applied for IPv4-IPv6 transition is detailed in (Bi, Leng, & Wu1, 2007).

CONCLUSION

This chapter provides a summary of IPv4—IPv6 coexistence and transition concepts, mechanisms and trends. In this chapter, the reader could become familiar with the basics of IPv6, and the gap between the two versions.

To sum up, one can say that most of the existing transition mechanisms are devoted to the end-to-end transmission of IPv6 packets in an IPv4 only environment. It is not surprising, as long as IPv6 is deployed less widely compared to IPv4. One must not forget that IPv4 packet transmission over an IPv6 only environment is also possible.

It should be noted that although dual-stack was identified as one of the many approaches for transition, dual-stack plays an essential role in all the transition mechanisms. Without dual-stack, it would be impossible to connect the IPv4 and the IPv6 world. All transition mechanisms require at least one dual-stack device in the communication path (in the terminals, or somewhere in the middle), otherwise the unsupported version of IP is not seen by the end terminal. Dual-stack is the most trivial and the most important solution at the same time.

It is also important to emphasize that although IPv6 supports all the features of IPv4 (moreover, in most cases it provides improved support), the transition mechanisms cannot fully support the advanced network services. Most of the applications work well through transition entities; advanced network services, like mobility support, QoS support and communication security lacks a thorough transition scheme. That is, QoS of IPv6 packets will not work over an IPv4 only medium. Advanced mobility support (e.g. routing optimization) of IPv6 will not work over an IPv4 only medium. IPsec will not work if a migrating device/solution is present.

However, the question is (as always) how these transition mechanisms will work in practice. Basically, the effect of transition on different applications and services is the main issue here. As end-to-end connectivity is lost, unknown effects might exist. Although today we have all the tools to start to migrate to the IPv6 world into the IPv4 environment, for sure we will hear about problems when running real services over the migrating environment. Thus, we need patience and curiosity in the next few years to figure out how the applications and services behave in a real migrated environment.

REFERENCES

Al Ja'afreh, R., Mellor, J., & Awan, I. (2009). *A comparison between the tunneling process and mapping schemes for IPv4/IPv6 transition.* International Conference on Advanced Information Networking and Applications Workshops, 2009. WAINA '09 (pp. 601-606). Bradford.

AlJa'afreh, R., Mellor, J., Kamala, M., & Kasasbeh, B. (2008). *Bi-directional mapping system as a new IPv4/IPv6 translation mechanism.* Tenth International Conference on Computer Modeling and Simulation, 2008. UKSIM 2008 (pp. 40-45). Cambridge, UK.

An, H., Luo, W., Li, X., Zhang, X., & Yan, B. (2009). *A new IPv6 tunneling protocol: Escort.* International Symposium on Computer Network and Multimedia Technology, 2009. CNMT 2009 (pp. 1-6). Wuhan.

Aoul, Y. H., Negru, D., Nafaa, A., & Mehaoua, A. (2003). M3G: A mobile multicast multimedia gateway for seamless IPv4/IPv6 transition. *Lecture Notes in Computer Science*, 2839.

Aoun, C., & Davies, E. (2007, July). *Reasons to move the network address translator - protocol translator (NAT-PT) to historic status.* (IETF RFC 4966).

Bi, J., Leng, X., & Wu, J. (2007). Scalable IPv4/IPv6 transition: A peer-to-peer based approach. *Lecture Notes in Computer Science*, 406–416. doi:10.1007/978-3-540-71661-7_39

Bound, J., Toutain, L., & Richier, J. (2005, October). *Dual stack IPv6 dominant transition mechanism* (DSTM). IETF Internet-Draft. (draft-bound-dstm-exp-04).

Carpenter, B., & Jung, C. (1999, March). *Transmission of IPv6 over IPv4 domains without explicit tunnels.* (IETF RFC 2529).

Carpenter, B., & Moore, K. (2001, February). *Connection of IPv6 domains via IPv4 clouds.* (IETF RFC 3056).

Chen, W.-E., Lin, Y.-B., & Pang, A.-C. (2005). An IPv4-IPv6 translation mechanism for SIP overlay network in UMTS all-IP environment. *IEEE Journal on Selected Areas in Communications*, *23*(11), 2152–2160. doi:10.1109/JSAC.2005.856836

Chen, W.-E., Su, C.-Y., & Lin, Y.-B. (2005). NCTU SLT: A socket-layer translator for IPv4-IPv6 translation. *IEEE Communications Letters*, *9*(10), 865–867. doi:10.1109/LCOMM.2005.10001

Chown, T. (2006, June). *Use of VLANs for IPv4-IPv6 coexistence in enterprise networks.* (IETF RFC 4554).

Conta, A., & Deering, S. (1998, December). *Generic packet tunneling in IPv6 specification.* (IETF RFC 2473).

Deering, S., & Hinden, R. (1998, December). *Internet protocol, version 6 (IPv6) specification.* (IETF RFC 2460).

Durand, A., Fasano, P., Guardini, I., & Lento, D. (2001, January). *IPv6 tunnel broker.* (IETF RFC 3053).

Gilligan, R., & Nordmark, E. (2000, August). *Transition mechanisms for IPv6 hosts and routers.* (IETF RFC 2893).

Govil, J. (2007). *IPv6: Mobility management and roaming between IPv6 and IPv4.* International Conference on Convergence Information Technology, 2007 (pp. 1553-1558). Gyeongju.

Hagino, J., & Yamamoto, K. (2001, June). *An IPv6-to-IPv4 transport relay translator.* (IETF RFC 3142).

Hei, Y., & Yamazaki, K. (2004). *Traffic analysis and worldwide operation of open 6to4 relays for IPv6 deployment.* International Symposium on Applications and the Internet, (pp. 265-268).

Hirorai, R., & Yoshifuji, H. (2006). *Problems on IPv4-IPv6 network transition.* International Symposium on Applications and the Internet Workshops, 2006. SAINT Workshops 2006 (pp. 37-42). Phoenix, AZ.

Huang, L.-W., Pan, J.-Y., Yu, K.-S., & Chiao, C. (2005). *Hierarchical mobility management in IP4/IP6 co-existed networks.* International Conference on Wireless Networks, Communications and Mobile Computing, 2005, 2, (pp. 1023-1028).

Huang, S.-M., Wu, Q., & Lin, Y.-B. (2006). Enhancing Teredo IPv6 tunneling to traverse the symmetric NAT. *IEEE Communications Letters, 10*(5), 408–410. doi:10.1109/LCOMM.2006.1633339

Huitema, C. (2001, June). *An anycast prefix for 6to4 relay routers.* (IETF RFC 3068).

Huitema, C. (2006, February). *Teredo: Tunneling IPv6 over UDP through network address translations* (NATs). (IETF RFC 4380).

Jankiewicz, E., Green, D., & Fiuczynski, M. E. (2006). *IPv6 translation for IPv4 embedded systems.* 2006 IEEE Sarnoff Symposium, (pp. 1-4). Princeton, NJ.

Jayanthi, J., & Rabara, S. (2010). *IPv6 addressing architecture in IPv4 network.* Second International Conference on Communication Software and Networks, 2010. ICCSN '10 (pp. 461-465). Singapore.

Jeong, S., Shin, M.-K., & Kim, H.-J. (2007). *A handover scheme supporting IPv4-IPv6 traversal over network-based mobility management domains.* The 9th International Conference on Advanced Communication Technology, (pp. 1033-1036). Gangwon-Do.

Kitamura, H. (2001, April). *A SOCKS-based IPv6/IPv4 gateway mechanism.* (IETF RFC 3089).

Kovacshazi, Z., & Vida, R. (2007). *Host identity specific multicast.* Third International Conference on Networking and Services, 2007. Athens, Greece: ICNS.

Lee, S., Shin, M.-K., Kim, Y.-J., Nordmark, E., & Durand, A. (2002, October). *Dual stack hosts using "bump-in-the-API"* (BIA). (IETF RFC 3338).

Lee, S.-D., Shin, M.-K., & Kim, H.-J. (2006). *The implementation of ISATAP router.* 8th International Conference on Advanced Communication Technology (ICACT'06), (p. 1163). Phoenix Park, Republic of Korea.

Leech, M., Ganis, M., Lee, Y., Kuris, R., Koblas, D., & Jones, L. (1996, March). *SOCKS protocol version 5.* (IETF RFC 1928).

Liu, M., Wu, X., Jin, M., & Li, D. (2005, July). *Tunneling IPv6 with private IPv4 addresses through NAT devices.* IETF Internet-Draft. (draft-liumin-v6ops-silkroad-03).

Liu, T., Guan, X., Zheng, Q., & Qu, Y. (2009). A new worm exploiting IPv6 and IPv4-IPv6 dual-stack networks: Experiment, modeling, simulation, and defense. *IEEE Network, 23*(5), 22–29. doi:10.1109/MNET.2009.5274918

Liu, X., Cao, X., & Li, H. (2008). *An IPv4-IPv6 translation mechanism for IMS network.* 4th International Conference on Wireless Communications, Networking and Mobile Computing, 2008. WiCOM '08 (pp. 1-4). Dalian.

Nordmark, E. (2000, February). *Stateless IP/ICMP translation algorithm* (SIIT). (IETF RFC 2765).

Nordmark, E., & Gilligan, R. (2005, October). *Basic transition mechanisms for IPv6 hosts and routers.* (IETF RFC 4213).

Palet, J., Olvera, C., & Fernandez, D. (2003, October). *Forwarding protocol 41 in NAT boxes.* IETF Internet Draft. (draft-palet-v6ops-proto41-nat-03).

Peng, W., Zhou, Y., Wang, C., & Yang, Y. (2009). *Research on IPSec-based NAT-PT transition mechanism*. IEEE International Conference on Network Infrastructure and Digital Content, 2009. IC-NIDC 2009 (pp. 222-226). Beijing.

Sabir, M., Fahiem, M., & Mian, M. (2009). *An overview of IPv4 to IPv6 transition and security issues*. WRI International Conference on Communications and Mobile Computing, 2009. CMC '09, 3, (pp. 636-639). Yunnan.

Sanguankotchakorn, T., & Jaiton, P. (2008). *Effect of triangular routing in mixed IPv4/IPv6 networks*. Seventh International Conference on Networking, 2008. ICN 2008 (pp. 357-362). Cancun.

Shi, W., Huang, C., Wang, Q., Chen, Y., Huang, Y., & Cheng, Y. (2007). *A novel IPv4/IPv6 translation mechanism based on NAT-PT*. The 9th International Conference on Advanced Communication Technology, (pp. 1037-1041). Gangwon-Do.

Soliman, H. (2007, November). *Mobile IPv6 support for dual stack hosts and routers* (DSMIPv6). (draft-ietf-mip6-nemo-v4traversal-06).

Taib, A., & Budiarto, R. (2007). *Security mechanisms for the IPv4 to IPv6 transition*. 5th Student Conference on Research and Development, 2007. SCOReD 2007 (pp. 1-6). Selangor, Malaysia.

Templin, F., Gleeson, T., Talwar, M., & Thaler, D. (2005, October). *Intra-site automatic tunnel addressing protocol* (ISATAP). (IETF RFC 4214).

Tsirtsis, G., & Srisuresh, P. (2000, February). *Network address translation - protocol translation* (NAT-PT). (IETF RFC 2766).

Tsuchiya, K., Higuchi, H., & Atarashi, Y. (2000, February). *Dual stack hosts using the "bump-in-the-stack" technique* (BIS). (IETF RFC 2767).

Varjonen, S., Komu, M., & Gurtov, A. (2009). *Secure and efficient IPv4/IPv6 handovers using host-based identifier-locator split*. 17th International Conference on Software, Telecommunications and Computer Networks, (pp. 111-115). Hvar, Croatia.

Wu, J., Cui, Y., Li, X., & Metz, C. (2006). The transition to IPv6 part 1 - 4over6 for the China Education and Research Network. *IEEE Internet Computing*, *10*(3), 80–85. doi:10.1109/MIC.2006.67

Wu, J., Cui, Y., Li, X., Xu, M., & Metz, C. (2010, March). *4over6 transit solution using IP encapsulation and MP-BGP extensions*. (IETF RFC 5747).

Xianfeng, Y., & Zheng, C. (2006). *IPv4-IPv6 multicast transition technologies*. Retrieved April 15, 2010, from http://wwwen.zte.com.cn/endata/magazine/ztecommunications/2006year/no3/articles/200610/t20061010_162407.html

Xiaodong, Z., Yan, M., & Yumei, Z. (2009). *Research on the next-generation Internet transition technology*. Second International Symposium on Computational Intelligence and Design (ISCID '09), (pp. 380-382). Changsha, China.

Xie, J., & Narayanan, U. (2009). Performance analysis of mobility support in IPv4/IPv6 mixed wireless networks. *IEEE Transactions on Vehicular Technology*, *59*(2).

Yan-ge, C., Shui-mu, T., & Jun-ying, G. (2009). *IPv4/IPv6 intercommunication technology research*. International Conference on Environmental Science and Information Application Technology, 2009. ESIAT 2009 (pp. 626-630). Wuhan.

Yuan-hang, C., & Lijuan, H. (2009). *Transition from IPv4 to IPv6 based on socket applications*. International Conference on Networking and Digital Society, 2009. ICNDS '09, 2, (pp. 40-43). Guiyang, Guizhou.

Zhu, Y., Chen, M., Zhang, H., & Li, X. (2008). *Stateless mapping and multiplexing of IPv4 addresses in migration to IPv6 Internet*. IEEE Global Telecommunications Conference, 2008. IEEE GLOBECOM 2008 (pp. 1-5). New Orleans, LO.

Zhuang, Z., Wu, J., & Xia, Q. (2006). *Study & implementation of proxy-based IPv4/IPv6 transition mechanism*. Asia-Pacific Conference on Communications (APCC '06), (pp. 1-5). Busan, South Korea.

ADDITIONAL READING

Iljitsch van Beijnum. (2006). *Running IPv6*. Apress.

Joseph Davies, J. (2003). *Understanding IPv6*. Microsoft Press.

Popoviciu, C., Levy-Abegnoli, E., & Grossetete, P. (2006). *Deploying IPv6 Networks*. Indianapolis, USA: Cisco Press.

KEY TERMS AND DEFINITIONS

Dual-Stack: The dual-stack approach is the principal building block of IPv4/IPv6 transitioning. Dual-stack nodes maintain two separated protocol stacks operating in parallel in order to allow the end host or router to operate via both versions of the Internet Protocol.

Internet Protocol version 4 (IPv4): Is the core protocol of the Internet, and the most widespread version of the IP standard nowadays. IPv4 is described in IETF RFC 791 (September 1981), which replaced RFC 760, the earlier definition (January 1980). IPv4 is a connectionless protocol designed to use over packet-switched Link Layer networks (e.g., Ethernet). IPv4 operates on the best effort delivery model, meaning that there is no guarantee for delivery and proper sequencing.

Internet Protocol version 6 (IPv6): Is the next-generation Internet Protocol version designed to overcome the imperfections of IPv4. The main motivation for the redesign of IPv4 was the presumptive IPv4 address exhaustion. IPv6 was firstly introduced in December 1998 in RFC 2460. IPv6 is a conceptual copy of the IPv4 protocol with several modifications and extensions to the basic standard.

Translation: Translation is an IPv4/IPv6 transition technique designed to provide direct conversion between IPv4 and IPv6. Translation techniques may include transformation of both the protocol header and the protocol payload. The translation itself can be performed at several layers in the TCP/IP protocol stack, including IP, transport, and also application layers.

Tunneling (Encapsulation): Tunneling is a common and widespread scheme to interconnect compatible hosts or networks across incompatible infrastructures. According to its basic behavior, tunneling transfers payload protocol data units using encapsulation into a carrier protocol. In scenarios of IPv4/IPv6 transition, IPv6 protocol data unit is usually carried in an IPv4 packet as the payload. Encapsulation of the payload is executed at the tunnel entry point, while decapsulation is carried out at the tunnel exit point.

v4/v6 Transition Mechanisms: Transition mechanisms are techniques used to solve problems of IPv4 and IPv6 coexistence in devices and networks. For example, an IPv6-over-IPv4 tunnel is a transition mechanism allowing IPv6 devices to communicate over an existing IPv4 infrastructure. Such a mechanism deals with encapsulation of IPv6 packets into IPv4 packets and decapsulation of the IPv6 packets during the reverse procedure.

APPENDIX

Table of abbreviations

ALG	Application Layer Gateway
API	Application Programming Interface
BGP	Border Gateway Protocol
BIA	Bump-in-the-API
BIS	Bump-in-the-Stack
CPU	Central Processing Unit
DCCP	Datagram Congestion Control Protocol
DHCP	Dynamic Host Configuration Protocol
DNS	Domain Name Service
DSTM	Dual-Stack IPv6 Dominant Transition Mechanism
EUI	Extended Unique Identifier
FTP	File Transfer Protocol
HIP	Host Identity Protocol
HIT	Host Identity Tag
IANA	Internet Assigned Numbers Authority
ICMP	Internet Control Message Protocol
IETF	Internet Engineering Task Force
IMS	IP Multimedia Subsystem
IP	Internet Protocol
IPsec	Internet Protocol security
IPv4	Internet Protocol version 4
IPv6	Internet Protocol version 6
IPX	Internetwork Packet Exchange
ISATAP	Intra-Site Automatic Tunnel Addressing Protocol
ISP	Internet Service Provider
LAN	Local Area Network
LSI	Local Scope Identity
MTU	Maximum Transfer Unit
NAPT-PT	Network Address Port Translation – Protocol Translation
NAT-PT	Network Address Translation – Protocol Translation
NCTU	National Chiao Tung University
ND	Neighbor Discovery
NEMO BS	Network Mobility Basic Support
P2P	Peer-to-peer

QoS	Quality of Service
RADIUS	Remote Authentication Dial In User Service
RFC	Request for Comments
SCTP	Stream Control Transmission Protocol
SIIT	Stateless IP/ICMP Translation
SIP	Session Initiation Protocol
SLT	Socket Layer Translator
TCP	Transmission Control Protocol
TTL	Time-to-Live
TRT	Transport Relay Translator
UDP	User Datagram Protocol
UMTS	Universal Mobile Telecommunications System
VLAN	Virtual Local Area Network

Chapter 9
Network Mobility

Arijit Ukil
Tata Consultancy Services, India

ABSTRACT

Network mobility (NEMO) management is concerned about the mobility management of an entire wireless mobile network to provide uninterrupted network connectivity to many mobile devices moving together in the mobile network. This is particularly important for ubiquitous computing, which commonly means anytime, anywhere computing and communication. Most of the 3G and entire 4G and beyond wireless communication technology is all-IP. This growing use of IP devices in portable applications has created the demand for mobility support for entire networks of IP devices. NEMO solves this problem by extending Mobile IP. Devices on a mobile network are unaware of their network's mobility; however, they are provided with uninterrupted Internet access even when the network changes its attachment point to the Internet. The main objective of NEMO is to provide continuous, optimal, and secure Internet access to all nodes and even recursively nested mobile sub-nets inside a moving network. Internet Engineering Task Force (IETF) is engaged in standardizing NEMO Basic Support protocol that ensures uninterrupted connectivity to nodes within a mobile network via a mobile router. This protocol extends the mechanisms utilized in the host mobility management protocol Mobile IPv6. There are few open problems remain to be addressed in NEMO. In this chapter, we discuss about NEMO basic support protocols, its features, and other related issues.

INTRODUCTION

IP appears to be the base technology of future networks to provide all kind of services through different access technologies, both fixed and mobile. Nevertheless, IP was not designed for taking into account mobility of users and terminals, and in fact, IP does not support it, neither in IPv4 nor in IPv6. The IETF has defined some IP layer protocols that enable terminal mobility in IPv4

DOI: 10.4018/978-1-60960-732-6.ch009

(Perkins, 2002) and IPv6 (Johnson et al., 2004) networks. Nevertheless, these protocols do not support the movement of a complete network that moves as a whole by changing its point of attachment to the fixed or ad hoc infrastructure, which results in NEMO. To allow NEMO in practice, several protocols and methods were designed and evaluated. NEMO Basic Support (Devarapalli et al., 2005) is the most widespread network mobility protocol located in the IP layer, which inherits the benefits of Mobile IPv6 while staying away from the problems of the main approach such as protocol overhead and inefficient routing. This protocol ensures uninterrupted connectivity to nodes within a mobile network via a mobile router. This protocol extends the mechanisms utilized in the host mobility management in order to allow multi-homing and nested mobile networking (Ng et al., 2007). There are two important requirements must be met in network mobility service provisioning. For instance, a mobile user inside a bus moving within the coverage of its wireless Internet service operator may subscribe a VoIP service. He or she would like to be able to access the service by initiating and receiving calls while the bus moves. This global Internet reachability and ubiquity turns out to be the basic requirement of network mobility. Similar to Mobile IP and Mobile IPv6, it requires that the mobile user can always be reached by some unique identifier anywhere the bus roams. Another important requirement is that the communication quality, especially for VoIP like sessions must not be affected by handovers when the vehicle moves from one coverage area of a base station (BS) to another. Hence, the speed of QoS re-negotiation on the new route and fast and reliable packet redirection from old to the new path during handover becomes critical to assure no disruption and performance degradation of the ongoing services. In other words, this requires resources necessary to sustain the service quality of all ongoing communication instances be reserved in time at the new route to ensure a successful QoS handover, not just network connectivity establishment. NEMO basic support protocol is still in developing phase by IETF. There are some shortcomings, which need to be addressed. It has the side effect of increasing packet delivery overheads due to pinball routing and multi-layer encapsulation of data packets. Moreover, the protocol does not address the large handover latency that causes a large number of packet losses and, consequently, communication service interruption of all the mobile nodes belonging to the mobile network. Therefore, it can be observed that, there exist several issues regarding this evolving protocol, which are mostly open research problems, though some attempts were made to address these issues.

It can also be observed that NEMO should work properly with the security protocols in order for the protocol to actually be accepted. Some of the information that must be addressed includes authentication of the user, authorization of information to be sent or received, and also confidentiality of the data that is being sent and received. The mobile network basically had the same requirements as a normal network. The hosts have to be confirmed and identified before any information will be sent. In addition, all NEMO support messages that are transmitted in a network will have to have ample amount of security that needs to be pre-determined.

There are numerous applications and use cases of NEMO. We are particularly interested in onboard communication networks, where a number of users inside a vehicle use internet or other cellular infrastructure through a common mobile router, where the vehicle is moving. The vehicle can be public transport like train, airplane, ship or private transport. This facilitates ubiquitous computing, which is pioneered by Mark Weiser (Weiser, 1991).

The chapter is organized as follows. Firstly, we introduce Mobile IP (MIP) and the MIP operation. MIPv6 is also highlighted. After that, we describe how NEMO is evolved from MIP, which are the features of NEMO that makes it work seamlessly even when the entire network is moving from one point of attachment to another. We present the concept of NEMO and the NEMO basic support protocol in next section along with the details on the NEMO architecture and the characteristics of the NEMO architecture. We also discuss about IPv4, IPv6 and the path to NEMO. We outline the concept of multicast support in NEMO. After that, we discuss about the mobility management of NEMO, which mostly deals with multihoming, seamless handover and QoS related issues. Next, we address the route optimization issue in NEMO, which provides a mechanism to eliminate the inefficiency in tunneling packets over the Internet. After that, we highlight the security issues in NEMO. In the next section, we present the applications of NEMO and the possible use scenarios. We also highlight the future research scope related to NEMO. After that, we conclude the chapter.

MOBILE IP (MIP)

Mobile IP (MIP) is the key protocol to enable mobile computing and networking, which brings together two of the world's most powerful technologies, the Internet and mobile communication. MIP is the underlying technology for support of various mobile data and wireless networking applications. MIP is part of the bigger management functionality called mobility management. Mobility management is an important issue in wireless and eventually ubiquitous communication. Mobility can take different forms, like:

- User mobility: This is the mobility of the user, who uses the network. The user is able to continue to access the network services under the same user identity when the user moves. This includes the ability for a user to access network services from different terminals under the same user identity. World Wide Web (WWW) supports this feature.
- Host mobility: This is the mobility of the host or the mobile node. Host mobility is the ability for a user to continue to access the network when the host or terminal moves. Though host mobility solutions span all layers of the Internet layer hierarchy (link layer, network layer, transport layer, and/or application layer), the most popular one is the network layer solution, namely MIP.
- Network or sub-network mobility: This is the mobility of the entire network or the sub-network. This helps the user to remain connected even in the case of the network or the sub-network is moving along with the network. This is under the broader definition of service mobility (Held & Ziegert, 1999). Service mobility is the ability for a user to access the same services regardless of where the user is. NEMO supports network as well as service mobility.

In this section, we discuss the host mobility issue and the concept of MIP. In fact, NEMO is a logical extension of MIP to support network and service mobility (Held & Ziegert, 1999).

The challenge of MIP is to deliver IP-based applications to mobile terminals or users, even though, traditionally, IP-protocols have been designed with the assumption that they are stationary. The present IPv4 protocol does not have the support for mobile node mobility. IPv4 assumes that a node's IP address uniquely identifies the node's point of attachment to the Internet so that the node can be located on the network indicated by its IP address to receive the datagram destined to it. When a node needs to change its point of attachment without losing its ability to communicate, IPv4 employs one of the two following mechanisms:

- Node changes its IP address whenever it changes its point of attachment.
- Host-specific routes are propagated throughout much of the Internet routing fabric.

Both of these alternatives are mostly unacceptable. The first makes it impossible for a node to maintain transport and higher-layer connections when the node changes location. The second has obvious and severe scalability problems. So, a new scalable, mechanism is required for accommodating node mobility within the Internet. IETF introduced MIP based on Ipv4 (Perkins, 2002) to mitigate this problem, which enables nodes to change their point of attachment to the Internet without changing their IP address. MIP has been developed over several years at the IETF, initially for IPv4 and now for IPv6 as well. Mobile IP is the nearest thing to an agreed standard in IP-mobility. In MIP, two IP addresses are provided for each computer or mobile node: home IP address, which is fixed, and care-of IP address which is changing as the computer moves. When the mobile moves to a new location, it must send its new address to an agent at home so that the agent can tunnel all communications to its new address timely. A mobile host is identified by its home address, regardless of its current point of attachment to the Internet. When the mobile node is away from its home location, it possesses 'Care-of Address' (CoA), which is associated with the mobile's current location. MIP attempts to solve the mobility problem by storing a dynamic mapping between the home IP address, which acts as its permanent identifier, and the temporary care-of address. The key functional entity in MIP is the Home Agent (HA), which is a customized router that works out the mapping between a mobile's home and CoAs. Each time the mobile moves on to a new subnet, it obtains a new CoA and registers it with the HA. The fundamental philosophy of MIP is that a correspondent host (CH) can always send packets to the mobile node (MN). CH addresses the IP packets to the host's home address. HA intercepts them and performs IP-in-IP encapsulation to tunnel them to the mobile's CoA. At the other end of the tunnel, the original packet can be extracted or does the decapsulation by removing the outer IP header. This asymmetric routing is called triangle routing. It is generally far from optimal, especially in cases when the correspondent node is very close to the mobile node. Triangular routing could introduce long end-to-end packet delays and lead to inefficient use of network resource. Imagine the scenario when an MN moves to the same subnet, the CN will send the datagram all the way to the MN's HA, and its HA will then forward the datagram to its CoA. This process is shown in Figure 1. We can observe that there is a scope of route optimization. Route optimization (Ng. et al., 2007 & Perkins, 1997) is a technique that enables a CN to address packets to a mobile's current CoA directly so that these packets do not have to be first routed to HA of the mobile. Route optimization reduces the number of packets that have to experience triangular routing. The basic idea underlying route optimization is that the routes to MNs from their CNs can be improved if the CN has an up-to-date mobility binding for the MN in its routing table. With an updated binding, the CN will be able to send encapsulated datagram directly to the MN's CoA instead of relying on a possibly distant HA to do so. Before providing the detailed protocol description, few definitions are required to understand. These are defined entirely for MIP and NEMO as per (Perkins, 2002), for other protocols, these terms may be defined slightly different way.

- Mobile node (MN): a host or router that changes its point of attachment from one network or sub-network, without changing its IP address. A mobile node can continue to communicate with other Internet nodes at any location using its IP address.
- Correspondent node (CN): A peer with which a mobile node is communicating. A

Figure 1. Mobile IP operation

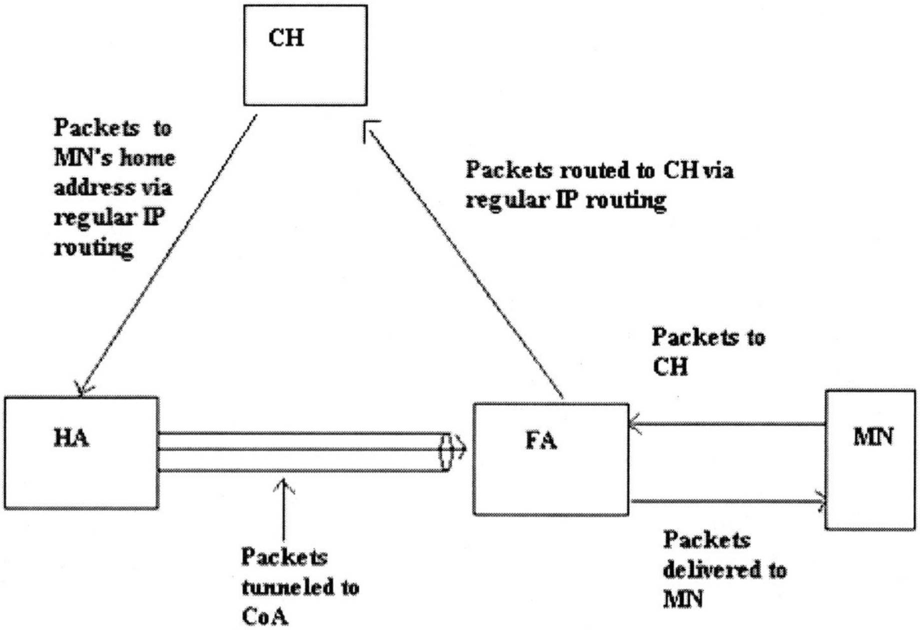

correspondent node may be either mobile or stationary.

- Home agent (HA): a router on a mobile node's home network which delivers datagrams to departed mobile nodes, and maintains current location information for each.
- Foreign agent (FA): a router on a mobile node's visited network, which cooperates with the home agent to complete the delivery of datagram to the mobile node while it is away from home.
- Home network (HN): a network, possibly virtual, having a network prefix matching that of a mobile node's home address. The standard IP routing mechanisms will deliver datagram destined to a mobile node's HA to the mobile node's Home Network.
- Home address (HA): an IP address that is assigned for an extended period of time to a mobile node. It remains unchanged regardless of where the node is attached to the Internet.
- Foreign network (FN): any network other than the mobile node's home network.
- Care of address (CoA): the termination point of a tunnel toward a mobile node, for datagram forwarded to the mobile node while it is away from home. The protocol can use two different types of care-of address: a "foreign agent care-of address" is an address of a foreign agent with which the mobile node is registered, and a "co-located care-of address" is an externally obtained local address, which the mobile node has associated with one of its own network interfaces.
- Co-located CoA: a co-located CoA is a CoA acquired by a mobile terminal through any method external to Mobile IP. For example, a mobile may use the Dynamic Host Configuration Protocol (DHCP) to obtain a temporary address dynamically. When a co-located CoA is used, the mobile terminal's HA tunnels the packets addressed to the mobile's home address directly to the

mobile itself; these packets do not have to go through any FA.

- Mobility binding: the association of a home address with a CoA, along with the remaining lifetime of that association.
- Tunnel:the path followed by a datagram while it is encapsulated. The model is that, while encapsulated, a datagram is routed to a knowledgeable agent, which decapsulates the datagram and then forwards it along to its ultimate destination.

MIP PROTOCOL OPERATION

There are two versions of Mobile IP: MIP for IPv4 and IPv6. We discuss MIPv4 operation only and then summarize the differences between MIPv4 and v6.

MIP performs five operations sequentially to achieve the seamless mobility of the roaming IP enabled. These are: agent discovery, movement detection, registration, service delivery and deregistration.

- Agent discovery: In order for a mobile terminal to communicate with a MIPv4 mobility agent (i.e. HA or FA), the mobile terminal will first need to know the agent's IP address. Typically, each mobile terminal is configured with the IP address of its home agent. This is because a mobile's home agent typically does not change. However, to allow a mobile's home agent to change while the mobile is away from home, a mobile may also discover a home agent dynamically. The process for a mobile terminal to discover the mobility agents and receive information from these agents is called MIPv4 Agent Discovery. There are two ways of finding agents, which use Internet Control Message Protocol (ICMP) discovery extension messages (Deering, 1991): (a) An agent advertisement message and an agent solicitation message. One way is by receiving agent advertisement messages periodically advertised from an FA. (b) Another is by sending out from an MN a periodic agent solicitation message until it receives a response from an FA. The MN thus gets the CoA, being dynamically assigned or associated with its FA.
- Movement detection: The MN needs to know when it enters a new IP subnet, i.e., when it may need to change its care-of address. The process for an MN to detect when it enters a new IP subnet is referred to as movement detection in MIPv4. MNs can use information in the received agent advertisement messages to detect movement.
- Registration: The MN registers the obtained CoA with its HA to get uninterrupted service. The registration process is performed directly from the MN or relayed by the FA to the HA, depending on whether the CoA is dynamically assigned or associated with its FA.
- Service delivery: During the period after the registration process and before the service time expiration, the service is continued as long as the MN is in the service area. Within the service period, the MN gets forwarded packets from its FA, which were originally sent from the MN's HA. Tunneling is the method used to forward packets from HA to FA and finally to MN.
- Deregistration: After the MN returns home, it deregisters its registered CoA in its HA. That is, the MN sets its CoA back to its HA. The MN achieves this by sending a registration request directly to its HA with the lifetime being set to zero. There is no need to deregister with the FA because the service expires automatically when the service time expires.

MIPv6

MIPv6 is designed to provide mobility support in an IPv6 network. It is very similar to Mobile IPv4 but takes advantage of various improved features of IPv6 to solve some of MIPv4's problems. Unlike MIPv4, MIPv6 does not use foreign agents. Recall that foreign agents in Mobile IPv4 provide two main functions: provide care-of addresses to visiting mobiles and help the mobiles detect whether they have moved into a new network and hence have to change its care-of address (i.e. movement detection). In an IPv6 network, mobiles use only co-located care-of addresses. Therefore, there is no need for a foreign agent to provide care-of addresses. Furthermore, standard Ipv6 facilities of IPv6 Neighbor Discovery (Narten et al., 1998) can be used to help IPv6 mobiles to detect movement. The IPv6 protocol and its attendant address configuration protocols like Neighbor Discovery and Stateless Address Auto-configuration form an almost perfect protocol basis for mobile networking. The basic idea, that a mobile node is reachable by sending packets to its home network, and that the HA sends packets from a home network to the mobile node's current CoA, remains the same. In addition, similar to the method used before (for IPv4), the HA encapsulates packets for delivery from the home network to the CoA. We summarize the difference of MIPv4 and MIPv6 in Table 1.

It is well known that traditional IP networks were not designed for mobile environments. Both in IPv4 and IPv6, IP addresses play two different roles. In fact, they are locators that specify, based on a routing system, how to reach the terminal using that address. On the other hand, IP addresses are also part of the end-point identifiers of a communication, and upper layers use the identifiers of the peers of a communication to identify it. Therefore, terminal mobility is the support that MIPv4 and MIPv6 need to provide. However, there exists also the need of supporting the movement of a complete network that changes its point of attachment to the fixed infrastructure, maintaining the sessions of every device of the network. In this case, the mobile network will have at least a router that connects to the fixed infrastructure, and the devices of the mobile network will obtain connectivity to the exterior through this mobile router. The IP terminal mobility solution does not support, as is now defined, the movement of networks. Because of that, the IETF NEMO WG was created to specify a solution, at the IP layer, to enable network mobility.

NEMO ARCHITECTURE AND BASIC SUPPORT PROTOCOL

The NEMO Working Group has established the NEMO Basic Support, which was set up to ensure

Table 1. MIPv4 vs. MIPv6

Key Features	MIPv4	MIPv6
Special router as FA	Yes	No
Support for route optimization	Part of the protocol	In extension
Ensure symmetric reachability between mobile nodes and its router at current location	No	Yes
Routing bandwidth overhead	More	Less
Decouple from Link Layer	No	Yes
Need to manage Tunnel soft state	Yes	No
Dynamic home agent address discovery	No	Yes

that the Internet would have continuous connectivity. The Basic Support group helps to solve the problems with bi-directional tunneling which are set up between MRs and the connection between the Mobile Network and the Internet along with the HAs. The NEMO Basic Support Protocol has been standardized for IPv6 and drafted for IPv4. Although, IPv6 and IPv4 are significantly different, the basic design is the same. Both versions are designed as extensions to Mobile IP. Mobile IP already provides HA and Mobile Node functionality. The NEMO Basic Support Protocol is a natural extension of the host mobility protocol, MIPv6. It specifies a mechanism that enables all nodes within a mobile network to be reachable via permanent IP addresses, as well as maintain ongoing sessions as the MR changes its point of attachment within the topology. This protocol, which runs on the MR and its HA, ensures uninterrupted connectivity to the mobile network nodes, without considering issues such as route optimization. There are few other requirements that lead to the development of NEMO basic support protocol. These are:

- The solution should not require modifications to any node other than MRs and HAs,
- The solution must be implemented at the IP layer level,
- All traffic exchanged between an MNN and a CN in the global Internet should transit through the bi-directional MR-HA tunnel
- MNNs must be reachable at a permanent IP address and name,
- The solution must maintain continuous sessions (both unicast and multicast) between MNNs and arbitrary CNs after IP handover of the MR,
- The solution must have multiple levels of nested mobile networks.

NEMO add MRs and the ability to bind a network prefix to a HA to MIP. NEMO concerns itself with the mobility management of a moving network or mobile network. A mobile network, which is composed of one or more IP subnets, moves as a single unit on the Internet topology. NEMO refers to a set of mobile hosts (MH) move collectively as a unit. It introduces a new communication paradigm of sets of mobile hosts with high mobility. In this paradigm, an entire mobile network changes its point of attachment to the Internet as one unit. All data packets sent to and from the mobile network are transmitted via one or more mobile routers (MR). A major advantage of network mobility is that it reduces the number of handovers of individual hosts and the power consumption of MH. NEMO is an extension of Mobile IP that enables an entire network to change its attachment point to the Internet. Whereas MIP solves the general issue of IP mobility, NEMO supports the mobility of the entire network. Mobile IP has the following characteristics:

- Enables Mobile Nodes to roam freely in IP Networks while remaining reachable at a single IP Address.
- Standardization of many important aspects completed.
- Creates significant overhead.

NEMO possesses the following characteristics:

- Enables networks to roam while remaining reachable and hiding the movement from their nodes.
- Decreases overhead by reducing amount of MIP-aware agents.
- Route Optimization though not standardized yet.
- Increases demands on agents.

Mobile IP enables a device to change its attachment point to the Internet without losing higher layer functionalities with tunneling between a Mobile Node (MN) and its Home Agent (HA). When an MN is on a foreign network, it obtains a local address, called a Care of Address (CoA). The MN then sends the CoA to its HA for binding. Once binding is complete, the HA intercepts and forwards packets that arrive for the MN to the MN via a tunnel to its CoA. Reverse traffic follows the same path through the tunnel to the HA for forwarding via standard routing on the Internet. As the MN moves to different foreign networks, it sends binding updates with its new CoA. Under Mobile IP, an MN may also utilize a Foreign Agent (FA). An MN can register with a FA at its current Internet attachment point. The FA will then assist the MN by performing tunneling functions on behalf of the MN. Under NEMO, an MR takes over the role of the MN in performing mobility functions. Nodes that are attached to an MR, Mobile Network Nodes (MNNs), are not aware of the network's mobility and do not perform any mobility functions. MR sends binding updates to its HA. However, binding updates from MR also contains the mobile network's network prefix. HAs will bind an entire network prefix to the MR's CoA and forward all packets for that network to the MR. The figure below depicts the basic architecture of NEMO and demonstrates the path of packets using NEMO. IP packets from a correspondent node (CN) that are destined for a node on a mobile network (MN) are delivered via standard routing on the Internet to the HA of that MN, where the HA tunnels the packets to the MR for delivery to the MNN. Reverse packets take the same path in the opposite direction; the MNN sends packets to the MR to be tunneled to the home agent and then sent out to the CN via standard routing on the Internet. Mobile Router and Home Agent are the main components of NEMO. This is shown in Figure 2.

The NEMO Basic Support Protocol defines two operating modes for NEMO: implicit and explicit. In implicit mode, BUs do not contain a Mobile Network Prefix option. Instead, the HA must determine the MR's network prefix by means outside of the NEMO Basic Support Protocol. When in explicit mode, all BUs contain one or more Mobile Network Prefix options. The HA is then able to use the Network Prefix option to set bindings for the MR. NEMO signaling is performed with extended MIP messages. Specifically, messages contain an additional router (R) flag to signal a Mobile Router instead of a Mobile Node. These messages are sent using the Mobility Extension Header in IPv6 and UDP control messaging in IPv4. The two major messages used by NEMO are Binding Updates (BUs) and Binding Acknowledgements. MRs use BUs to notify their HAs of a new CoA, thus new Internet attachment point. BUs contain the new CoA, the router flag, and an optional mobile network prefix, which is used to update the networks prefix information. Upon receipt of a BU, HAs send Binding Acknowledgement. Lastly, the NEMO Basic Support Protocol specifies the routing of

Figure 2. NEMO architecture

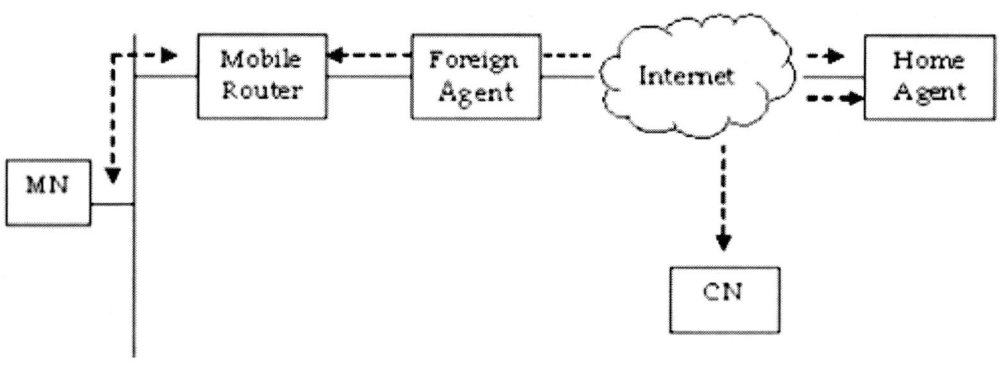

packets to and from the Mobile Network. Packets sent to an MNN from a CN are routed over the Internet using standard routing until they reach the HA. The HA intercepts packets and encapsulates them in a tunnel to the MR. The MR decapsulates packets and routes them to the MNN. Reverse traffic must be reverse tunneled to the HA before being routed to the CN. The Basic Support Protocol specifies bi-directional tunneling so that only MRs and HAs need to be aware of the network's mobility and to prevent firewalls from dropping packets. Some firewalls will drop packets to prevent spoofing - when the source address of packets (the Home Address) does not match the network from which they are being sent. The tunnel establishment concept is illustrated in Figure 3. With the NEMO basic support protocol, only the mobile router configures and performs prefix scope binding updates with its home agent when the mobile network moves. As the MNNs are not required to configure and register their CoAs as long as they are inside the mobile network, the signaling volumes in wireless as well as wired networks are significantly reduced.

IPv4, IPv6 and NEMO

IPv6 affords NEMO with many mechanisms that do not exist in IPv4. Thus, some of the tasks of NEMO are better suited for IPv6. In addition, the differences between NEMO on IPv4 and IPv6 tend to closely parallel the differences between MIPv4 and MIPv6 because NEMO is merely an extension of MIP. On IPv4, Mobile Routers must rely on Mobile IP Foreign Agents for tunneling packets to and from the HA. This means that any access router that a Mobile Router uses to attach to the Internet must be MIPv4 compliant. However, with NEMO Basic Support on IPv6, only the Mobile Router and Home Agent need to be aware of the MR's mobility. In this case, tunneling is performed directly between the HA and the MR without the need for a FA. Using IPv6 also affords NEMO with more efficient tunneling and more consistent security. Under IPv6, tunneling is performed by using the routing extension header. On the other hand, IPv4 tunnels IP packets through encapsulation, a process with a higher overhead. NEMO also utilizes IPSec for encryption and authentication of tunneled traffic. However, IPSec is an optional part of IPv4 implementation whereas IPv6 requires IPSec of all implementations.

Figure 3. NEMO tunnel establishment

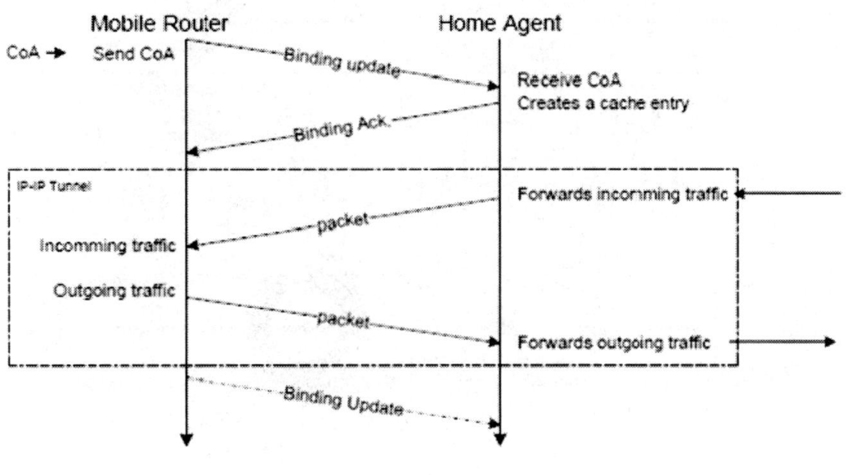

In Figure 4, we illustrate the difference between NEMO and MIPv6. In NEMO, network complexity brought back to the router. Here, only the mobile router manages mobility and maintains Internet access through one or several access technologies. Nodes behind the mobile router are standard IPv6 nodes with no mobility support mechanisms. Functionally, we can visualize the advantage or the performance gain achieved by NEMO when comparing Figure 5 and 6. It can also be observed that scalability is another factor that affects MIPv6 in moving vehicles totally infeasible. In MIPv6, no network aggregation effect is available and hence it is inefficient. Whenever the network attaches itself to a new point of attachment, each node has to perform Mobile IP operations. This would constitute a significant signaling overhead as each node has to first detect that it has moved to a new network and to configure a new IP address and send Binding Update to its HA. In addition, there is a provisioning overhead of security association establishment between the home agent and the mobile nodes. Furthermore, all hosts need to be aware of mobility. NEMO, on the other hand brings in the network aggregation effect, which improves efficiency. Now the mobility related signaling is solely handled by the special node on behalf of the mobile network and the rest of the nodes need not be concerned with mobility at all. This allows nodes to function as if they are "virtually home." In deployments such as transportation networks, provisioning individual Security Association between the HAs and the MNs can be avoided.

Multicast Support NEMO

IP multicast is an efficient way to send packets from one host to multiple hosts to support multicast traffic for mobile networks. In this case, the MR can use the bi-directional tunnel (BT) between the HA and the MR's CoA located in the visited network. Alternatively, a remote subscription (RS) to a multicast group within the visited network as described in MIPv6 is feasible. With respect to multicast traffic to and from mobile networks, the BT approach may prove inefficient in terms of non-optimal (triangular) routing, breech of the multicast nature of the flow, and limited scalability. The main disadvantage of applying RS for multicast services emerging or terminating within mobile networks is the required frequent re-construction of the multicast tree, especially if the traffic source is moving fast, resulting in

Figure 4. MIPv6 and NEMO

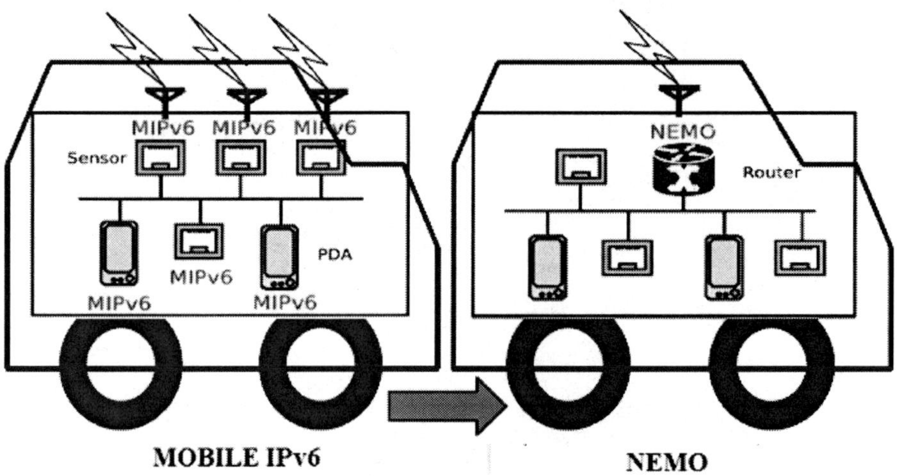

Figure 5. MIPv6 architecture

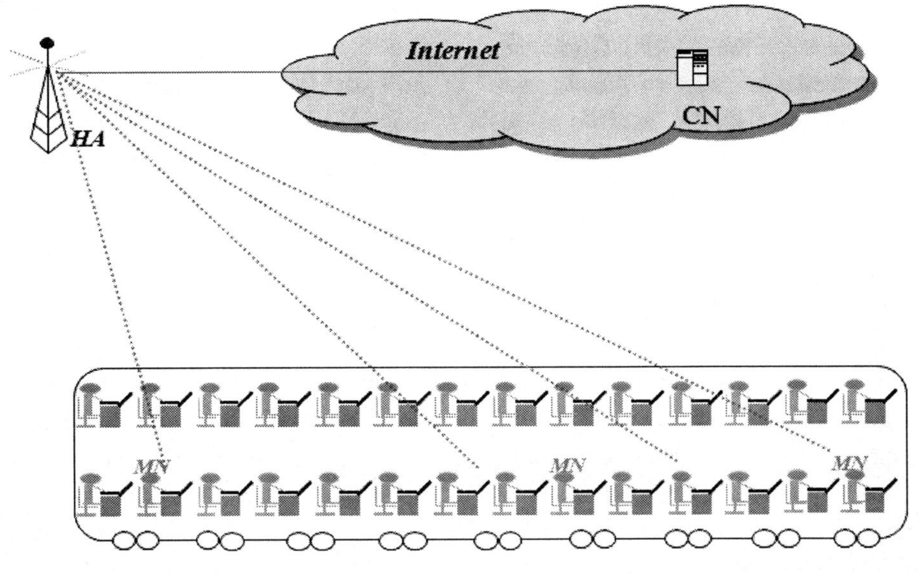

high latency and network traffic overhead. Upon subscription of a node within the mobile network to a multicast group or transmission of multicast traffic, the MR forwards the request or the traffic to the HA utilizing the MLD (Multicast Listener Discovery) (Vida & Costa, 2004) protocol. Subsequently the corresponding data traffic or group control messages are forwarded by the HA back to the MR. This proxy functionality of the HA is described in (Janneteau, et al., 2004). In case of reduced mobility of the sub-network detected by means of low handover (CoA change) rate, the MR initiates routing of multicast traffic via the remote access point.

Figure 6. NEMO architecture

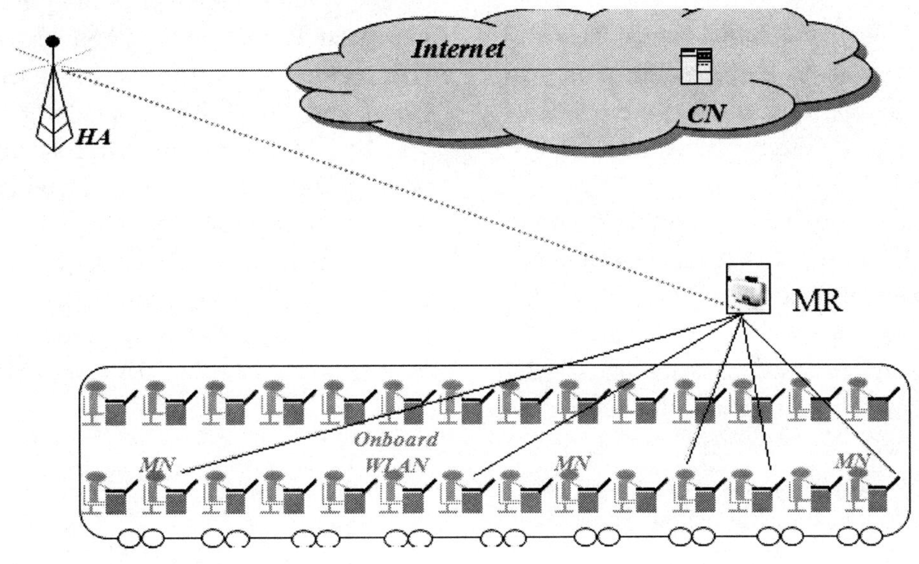

NEMO AND MULTIHOMING

By multihoming, we mean that a mobile network would have more than one point of attachment between the Internet and the MR. Multihoming technology improves the availability of host and network connectivity. It increases robustness of a mobile network system in terms of fault tolerance and load balancing. The goals and advantages of multihoming are:

- Ubiquitous Access: anytime, anywhere seamless access to the Internet.
- Redundancy/Fault-Recovery: when an address is not any more reachable, when a link goes down or a router has a failure, the reachability to the Internet can be provide by the others addresses, links or router. The continuity of the connectivity should be transparent for the applications.
- Load Sharing: the multi-homed host should be able to distribute upstream and downstream traffic between his interface routers.
- Load Balancing: distribute the traffic load to the multiple MRs or interfaces such that none of them are over or under loaded. This attempts to load the network in an optimal way.
- Traffic Policy: the multi-homed host/site should be able to define some policy to manage the network traffic for reasons of costs, traffic requirements, uses conditions, social policy.
- Preference Settings: some of the mobile nodes may prefer MRs or the interfaces depending on its capacity or availability. This distinguishes normal and premium users.
- Routing Policy: routing is defined as the ability for the user or the application to choose between interfaces for matter of cost, efficiency, real-time performance, etc. This could be established either statically or dynamically, and initiated either by the MR, the HA, or an MNN.

Currently, the term multihoming has no formal definition in the IETF community. It is used to indicate a situation where several routes are provided for network nodes to reach a particular correspondent. In the NEMO terminology, a mobile network is considered multihomed when either the mobile network is simultaneously connected to the Internet via more than one mobile router, or when a mobile router has more than one interface. This is shown in Figure 7. Users of small-scale networks need an easy method to improve network availability and to load balance several links. Multihoming technology is one of the solutions to improve availability. With the NEMO protocol, multihoming translates to simultaneous multiple bi-directional tunnels. In multi-homing architecture, an MR, which is a border router of the multihomed network, sets up several tunnels between the MR and the HA by multiple-CoA registration. An HA advertises the user's network prefix to Internet Service Providers (ISPs) via the routing protocol. If the HA has several multihomed networks, they can advertise an aggregated network prefix to ISPs. Therefore, the Internet routing entries do not increase one by one when the number of multihomed users is increased. Packets to multihomed users go to the HA, and the HA sends packets to the MR using CoA1 and CoA2. The HA selects the route, in which a CoA is used. The route selection algorithm is beyond the scope of this document. This can improve the availability of the user network and control traffic going from the ISP to the MR. In the basic architecture, HA1 is a single point of failure. In order to improve the availability of the user network, multiple HAs are needed.

NEMO AND SEAMLESS VERTICAL HANDOVER

NEMO introduces a new communication paradigm that provides sets of MH moved collectively as a unit with high-mobility. Efficient network mobility handover design is essential to meet the service requirements for real-time VoIP-like applications and seamless connectivity. There are two important requirements are to be met in NEMO service provisioning. For instance, an MH inside a bus moving within the coverage of its wireless Internet service operator may subscribe a VoIP or video streaming service. He or she would like to be able to access the service by placing and receiving calls or the requests while the vehicle moves. This global Internet reachability gives one of the basic requirements of network mobility. Therefore, seamless handover is becoming a major issue for all the NEMO projects taken up so far. Handover (or handoff) as a part of mobility management, is basically a process in which a mobile terminal changes its network attachment point. For example, a mobile may be handed over from one wireless base station (or access point) to another or from one router or switch to another. Roaming is the ability for a user to move into and use different operators' networks. Like other handover techniques in wireless mobile networks, NEMO handover can occur in two ways:

- Break-before-Make: In this case, MR first loses connectivity with its current access router, and becomes unreachable at its current CoA and then establishes connectivity with a new access router and acquires a new care-of-address. Here, the connectivity between MNNs and CNs is broken for the duration of the handover; therefore, the handover latency translates directly to packet loss. As long as an MR connects to only one network at a time and is limited to performing break-before-make handovers, handovers are likely to cause some degree of disruption to traffic. In case of MR handover, significant breakages in connectivity or Break-before-Make handover would potentially have an impact on a large number of nodes.
- Make-before-Break: In this case, the MR establishes connectivity with a new AR while still being connected to the old AR.

Figure 7. Multihoming

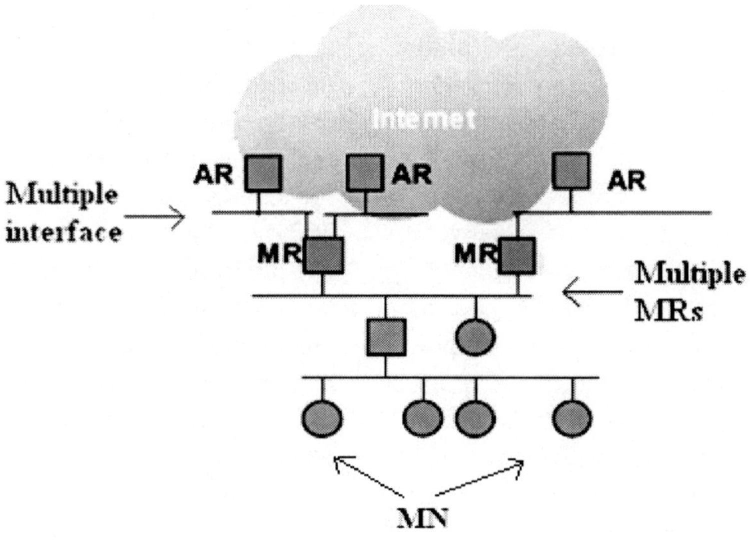

Here, the ongoing communication is not affected by the handover latency. There are many schemes have been proposed for make-before-break handovers. An easy way is to emulate make-before-break handovers by buffering and using a localized forwarding scheme that could be used in a network mobility setting in access-technology independent way. However, the potential performance benefits from the emulated make-before-break handovers are offset by the increased complexity required in the access network. In (Petander et al., 2005), using a dual interface MR, where one interface is utilized for data communications and the other interface for scanning, the possibility of achieving virtually loss-free handovers is shown. As the scanning and the handover are performed on a different interface than that of the active interface, this scheme enables make-before-break handovers and minimizes the packet losses incurred due to handover latencies.

In this section, we mainly concentrate on make-before-break handovers only, as seamless handover is becoming a necessary function for NEMO in the case of high mobility where handovers take place in high frequency. Seamless handover is one of the major challenges toward seamless and ubiquitous inter-networking and mobility access between heterogeneous access networks. NEMO handover is similar to MIPv6 handover. MIPv6 (Petander et al., 2005), was initially designed for single host mobility. It is preceded by a link layer handover and an IP layer network attachment procedure. With the development of NEMO, MIPv4 leads into MIPv4-NEMO. It aims to provide a transparent network mobility support to the attached MHs with almost no modification necessary to the existing protocol stacks on the MH. Its design however raises some problems. It requires traffic to and from the MH go through a bi-directional tunnel between the MR and its HA. This introduces extra overheads on packet header and tunnel processing, and additional routing legs, resulting in added transmission delays. Second, MIPv6-NEMO is a framework where many tasks should be performed to complete the handover process. Tasks such as the duplicate address detection (DAD) and re-registration operations as well as inflexible sequence between tasks (e.g., resulting in additional waiting time for route advertisement message) have negative effect on handover latency performance. Koodli (Koodli, 2005) proposed a method called fast handover for MIPv6 (FMIPv6) to improve the handover delay. This is an access-technology-independent way of emulating make-before-break handovers by buffering and using a localized forwarding scheme that could be used in a network mobility setting. In FMIPv6, the MR is required to complete a number of IP layer tasks, such as requesting the PAR for information about the NAR through a transaction of a Router Solicitation for Proxy (RtSolPr) and a Proxy Router Advertisement (PrRtAdv) message, configuring a CoA, sending a Fast Binding Update (FBU) message to the PAR, getting exchanged a Handover Indication (HI) and a Handover Acknowledgement (HACK) message between the Previous Access Router (PAR) and the New Access Router (NAR), and receiving a Fast Binding Acknowledgement (FBack) message before switching the link from the old network to the new one. However, this method may cause undesirable fluctuation of packet delays; it also has the problem of when should the Fast Binding Update (FBU) be sent. The use of optimized procedures such as Optimistic DAD (Moore, 2005) and Fast Router Advertisements (Kempf, et al., 2005) for network attachment can reduce the latency to half to two round-trip times between the mobile node and its access router. Link layer handovers from an old base station or access point to a new one and authentication procedures are other factors affecting the handover latency that can be improved using access-technology-specific optimizations (Ramani, et al., 2005). Hierarchical

Mobile IPv6 (HMIPv6) (Soliman, et al., 2005) is a well-referenced schemes for preventing the performance of a mobile node from being degraded during a handover. The approach used in this scheme is more concerned with reducing handover latency and packet losses by localizing signaling domain. In the NEMO basic support protocol, the root-MR sends a BU containing a new CoA to its home agent via the new AR during handover. After that, the home agent sends packets to the new CoA. To perform seamless vertical handover even under bad radio conditions, NEMO-Soft Handover (SHO) (Naoe et al., 2007) has been proposed as an IPv6 soft handover extension to NEMO-Basic Support. NEMO-SHO introduces packet bicasting and combining operations on an MR and a serving AR using two different wireless links during handover so that packet losses can be reduced if either link can successfully deliver a packet. It has a positive effect on UDP traffic and enhances UDP-supported NEMO. However, the performance gain for TCP traffic is limited and often detrimental because of spurious TCP retransmissions caused by the difference of bandwidth and transmission delay between the two links. The handover latency, namely, the sum of times taken by these operations, becomes sufficiently large, which leads to the loss of a large number of packets belonging to many MNNs. This may result in impairing the quality of multimedia communications. Many TCP extensions have been proposed to avoid spurious TCP retransmissions. In (Ludwig, et al., 2003), TCP-Eifel is proposed which detects whether a retransmission is spurious by using TCP timestamps option and then restores the size of Congestion Window to the previous state. The essential factor in causing the issue is that TCP congestion control is executed by a TCP sender that has neither an explicit handover notification nor information about the link capability on a target link. TCPHO (Wu, et al., 2007) thus introduces new TCP options that convey the information about the bandwidth on the target link to a sender so that it can quickly adapt the target link. NEMO on the other hand is aimed for providing seamless connectivity to large number of users in a moving (high-speed) vehicle. Therefore, it turns out that the schemes of TCP handover solutions are always suboptimal in the scenario NEMO operates. So, a near optimal solution is described in (Kaffle et al., 2006), which proposed a cooperative mobile router-based handover (CoMoRoHo) scheme that minimizes packet losses of a long vehicular mobile network during a handover. This scheme exploits the advantage of a multihoming environment of a long, vehicular mobile network that can have two or more mobile routers spatially separated by a certain distance. It makes the packet loss independent of the handover latency by establishing a local tunnel between an access router and a mobile router of the mobile network by using only one signaling message. The basic idea behind the CoMoRoHo scheme is to enable different MRs of a multihomed mobile network to access different subnets during a handover and cooperatively receive packets destined for each other. That is, when an MR is performing handover to a new subnet, it requests the access router of the old subnet to tunnel packets to some other MR that is still located in the old subnet. The other mobile router then forwards the packets to the mobile router that has requested the access router for tunneling. Similarly, when the mobile router completes its handover to the new subnet, it helps the other MR to perform a lossless handover to the new subnet.

NEMO AND QoS PROVISIONING

Along with the feature of seamless handover, QoS provisioning is an integral part of NEMO implementation. In fact, handover and QoS are inter-related. As it is pointed out that, handover brings about few QoS related issues like service connectivity, latency, packet drop, traffic loss. The important requirement is that the communication quality, especially for VoIP-like sessions should not be affected by handovers when the vehicle moves from one coverage area of a BS to another. Hence, the speed of QoS re-negotiation on the new route and fast and reliable packet redirection from old to the new path during handovers becomes critical to guarantee no disruption and performance degradation of the on-going services. This requires resources necessary to sustain the service quality of all on-going communication instances be reserved in time at the new route to ensure a successful QoS-handover, not just network connectivity establishment. It is known that many existing communication protocols fail to perform efficiently when groups of users move in unison and at high velocity. A classic example is the Mobile IP protocol, which, as it stands, leads to storms of binding updates at every IP handover. QoS is another aspect that could potentially produce signaling storms as a result of the combined effect of both high-speed vehicle motion and potentially hundreds of users simultaneously re-establishing QoS reservations at handover. When an MR attaches to a new BS, the aggregate QoS requirements of all existing communication instances through the MR must be submitted to the new BS to initiate resource allocation at the intermediate switching nodes along the new routing path from MR to each individual MH's CN. It is essential that the resources needed for the existing services must be reserved in time at the new cell to ensure a successful QoS-handover. Unlike mobility management, QoS has received very little attention from the NEMO research community, despite its importance in delivering consistently reliable services to users. Virtually all of the research carried out on QoS for mobile terminals has assumed the position of nodes within a wireless network to be randomly changing, such that the correlation of movement between any two users in the network is nil. Therefore, applying QoS solutions that have been designed based on this assumption may have a number of operational and scalability implications that degrades the performance of the protocol and the network as a whole. For example, use of the IETF differentiated services (DiffServ) protocol in a moving network with a large number of users could lead to severe service degradation of low-priority applications during periods of high network congestion for prolonged periods of time. The IETF integrated services (IntServ) protocol, on the other hand, requires that each node manages their own QoS, thus potentially resulting in a significant proportion of the wireless resources being lent to control signaling, and extra burden placed on routers within the access network. This is particularly the case during IP handovers, in which old QoS paths must be torn down and new ones established. One way to improve the scalability and robustness of QoS protocols within moving networks is to allow the MR to reserve resources on behalf of the nodes that it serves up to the MR's home agent. Using this approach, only a single state is required to be maintained in each access network router along the communication path, thus reducing excessive processing overheads, and facilitating seamless IP handovers. Currently there exist two NEMO-specific QoS protocols that use such a technique to reserve resources: On-Board RSVP (Malik et al., 2004) and NEMOR (Tais, et al., 2005). In On-Board RSVP, this is achieved by using extensions to the well-known RSVP protocol (Braden, et al., 1997), by the addition of a new object to the PATH and RESV messages. The NEMOR protocol, on the other hand, is based entirely on the NSIS signaling protocol, which accommodates the use of a per-class forwarding mechanism used within a stateful MR reservation ('virtual' RSVP tunnel) to control access to resources; on-Board RSVP allows only for best-effort treatment. NEMOR protocol works by allowing an MR located within a vehicle to maintain a single QoS reservation state on behalf of the mobile terminals that it serves. However, under bursty session requests arising from the influx of passengers embarking and disembarking a public transport vehicle, and when sessions are short-lived, the scalability of NEMOR is undermined, as the MR is still required to increase or decrease the QoS reservation accordingly. Shifting QoS management to the MR may well improve protocol efficiency during handovers. Under such conditions, control messages must still be sent to the access network to increase or decrease the resources allocated to the virtual RSVP tunnel, thus undermining the benefits provided by the protocol. A possible way to address this issue would be to over-provision resources to the virtual RSVP tunnel, eliminating the need for frequent re-allocation of resources. However, this method could prove to be resource-inefficient, particularly when sessions are frequently created and terminated, yet the overall resource requirement of the subnet, as a whole remains virtually constant, or during periods of high network congestion when resources are the most valuable. Traditional layered protocol reference model can no longer be efficient to support fast QoS-handover for network mobility service, where individual protocol layers are independently designed and functionally they do not cooperate. For example, during handover, the radio link attachment procedure to a new BS in a subnet may have been completed while the network layer handover procedure is still waiting for the reconfiguration of MHs' IP addresses. This inefficiency dictates a combined cross-layer mobility management and resource allocation design to reduce latency and packet loss during handovers. The QoS-handover architecture, HiMIPNEMO (Wei, et al.) combines the designs of routing and resource allocation with network mobility management and introduces the notion of foreign mobility agent to facilitate fast and reliable QoS handovers. It is a framework of routing and resource allocation with network mobility management. HiMIPNEMO introduces the notion of foreign mobility agent in a hierarchical backhaul packet forwarding architecture. Another effective QoS provisioning method in NEMO is QoS aggregation. While QoS aggregation improves protocol scalability by maintaining a single QoS state for potentially the entire moving network rather than for each individual node, the act of having to constantly maintain this state when sessions are frequently created and terminated undermines its benefit. QoS aggregation is the technique of introducing a time lag between a request for resources by a user and the actual reservation process. The architecture is shown in Figure 8. This allows multiple queued requests at the MR to be combined into a single request, thus reducing the frequency with which QoS requests are sent to the access network and presenting a saving to the network operator. The decision to aggregate queued resource requests is carried out by an aggregation policy at the MR, which may be either parameter driven or cost driven. The analysis of cost driven QoS aggregation in NEMO is studied in (Kamel, et al., 2007, 2008) under the case of bursty requests. It compares the expected cost efficiency and user waiting time with that of other previously proposed policies. It is shown that the cost driven policy reduces operator costs compared to other policies, whilst also reducing expected queuing times.

NEMO AND ROUTE OPTIMIZATION

It is discussed in earlier section that triangular routing is not optimal. All packets sent between an MNN and its CN must go through the bi-directional MR-HA tunnel. This sub-optimal routing results in various inefficiencies associated with packet delivery such as increased delay and bottleneck links leading to traffic congestion, which can ultimately disrupt all communications to and from the MNNs. NEMO basic support protocol does not yet include any mechanism for route optimization (RO). RO in NEMO means that one routing entity uses an IP tunnel to redirect the original packets to the other routing entity that is most closely located from the destination. RO is a mechanism that not only shortens the data delivery path between a mobile node and a CN, but also reduces the potential level of encapsula-

Figure 8. QoS aggregation

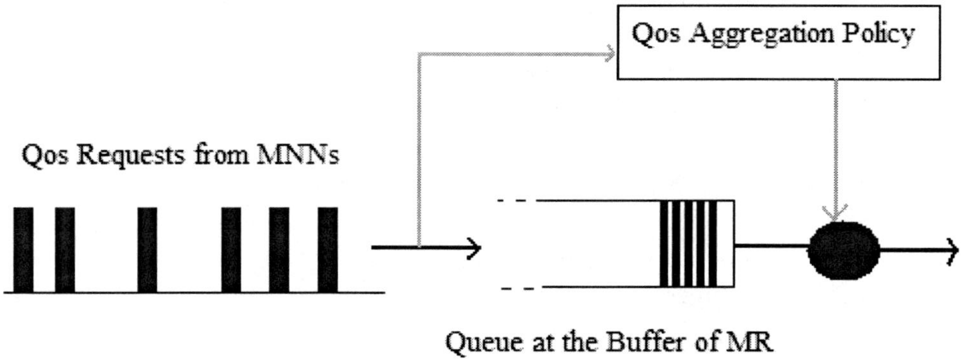

tion. Nevertheless, RO requires some route update signaling and/or additional information in the IP headers of data packets to enable packets to follow the optimal path and reach their destination intact. The generic consideration in designing an RO scheme is to use a minimum of signaling and/or additional information in the packet header. The use of MIPv6-type RO without any enhancement is inefficient because a different CoA process is used for mobile nodes in NEMO. In MIPv6, a mobile node's CoA gives the actual location of the mobile node and packets are forwarded directly to that CoA following RO. In contrast to MIPv6, the CoA of an MNN in NEMO does not provide the actual location of the MNN as the CoA belongs to the MNP of the mobile network, which is also moving. Although a standardized NEMO Route Optimization solution has yet to materialize, it is expected that NEMO RO (Ng. et al., 2007) will show some of the benefits mentioned below:

- Shorter Delay: RO involves the selection and utilization of a shorter and faster route to be taken up for traffic between an MNN and its CN. Therefore, in effect RO improves the latency of the traffic between the two end nodes. This may in turn lead to better overall QoS.
- Reduced Consumption of Overall Network Resources: Through the selection of a shorter route, the total link utilization for all links used by traffic between the two end nodes should be much lower than that used if RO is not carried out. This would result in a lighter network load with reduced congestion.
- Reduced Susceptibility to Link Failure: When a link along the bi-directional tunnel is disrupted, all traffic to and from the mobile network will be affected until IP routing recovers from the failure. An optimized route would conceivably utilize a smaller number of links between the two end nodes. Hence, the probability of a loss of connectivity due to a single point of failure at a link should be lower as compared to the longer non-optimized route.
- Greater Data Efficiency: Depending on the actual solution for NEMO RO, the data packets exchanged between two end nodes may not require as many levels of encapsulation as that in NEMO Basic Support. This means less packet overheads and higher data efficiency.
- Reduced Processing Delay: In a nested mobile network, the application of RO may eliminate the need for multiple encapsulations required by NEMO Basic Support, which may result in less process-

ing delay at the points of encapsulation and decapsulation.

- Avoiding a Bottleneck in the Home Network:- NEMO RO allows traffic to bypass the HAs. Apart from having a more direct route, this also avoids routing traffic via the home network, which may be a potential bottleneck otherwise.
- Avoid the Security Policy Issue: Security policy may forbid an MR from tunneling traffic of Visiting Mobile Nodes into the home network of the MR. RO can be used to avoid this issue by forwarding traffic from Visiting Mobile Nodes directly to their destinations without going through the home network of the MR.
- Avoid the Instability and Deadlock: A potential deadlock situation is when a HA is nested within a mobile network. RO may circumvent such deadlock situations by directly forwarding traffic upstream.

In Figure 9, we have illustrated the concept of NEMO RO. Mobile IP performs RO by utilizing a Binding Cache on the CN. Mobile IP nodes send Binding Updates with current CoAs to their CNs as they change attachment points to the Internet. Mobile nodes and CNs are then able to directly communicate using the CoA of the mobile node. There are several possible approaches to the NEMO RO problem; however, each has its own trade-offs. Such trade-offs include increased signaling overhead, longer handover delay, and the need to make additional devices such as CNs and MMNs aware of NEMO. A straightforward approach to RO in non-nested NEMO is for the MR to attempt RO with a Correspondent Entity. This can be viewed as a logical extension to NEMO Basic Support, where the MR would send Binding Updates containing one or more Mobile Network Prefix options to the Correspondent Entity. The Correspondent Entity, having received the Binding Update, can then set up a bi-directional tunnel with the MR at the current CoA of the MR and inject a route to its routing table so that packets destined for addresses in the Mobile Network Prefix will be routed through the bi-directional tunnel. Various other approaches to RO in NEMO have been proposed in the literature (Ng et al. 2007 & Na, et al., 2004). Most of them essentially require the introduction of new components or mechanisms in the existing networking system. They include generic overviews of various types of route optimization possibilities in NEMO (Ng, et al., 2007 & Ng., et al., 2005). In (Thubert et al. 2004), proposed the use of a new routing header, Routing Header (RH) type 4, also called a reverse routing header (RRH), for MNN-originated outbound packets, and a modified RH type 2 for inbound packets destined for MNNs. The RH type 4 collects the CoAs of all nested MRs, which are later included in the modified RH type 2 to reduce the number of nested encapsulations for inbound packets. This scheme, however, optimizes the path between the home agent and the MR serving the MNN, not between the CN and MR. Moreover; it requires MRs to modify packet headers, which would increase computational overheads. Na et al. (Na, et al., 2004) proposed that the MRs' HAs should assist RO between a correspondent router (CR), a router serving the CN, and the MR. However, this scheme has some disadvantages. It requires preservation of a soft state at the CR in addition to that at the MR, and it does not describe how to inform the CR promptly of changes in the MR's CoA. Route updating could be done faster if an MR instead of a home agent initiated the RO process when the network moved. Our scheme uses an MR-initiated RO approach. Ohnishi et al. (Ohnishi, et al., 2003) described RO problems relating to nested mobility. This scheme uses hierarchical configuration of MRs and enables the MNN's HA to send packets directly to the root-MR's CoA, skipping other MRs' HAs. By optimizing the route between the MNN's HA and the root-MR, but not between the CN and root-MR, this scheme again provides only a partial solution to RO problems.

However, there are certain issues related to the introduction NEMO RO. They are:

- Additional Signaling Overhead: the nodes involved in performing Route Optimization would be expected to exchange additional signaling messages in order to establish RO. The required amount of signaling depends on the solution, but is likely to exceed the amount required in the home Binding Update procedure defined in NEMO Basic Support. The amount of signaling is likely to increase with the increasing number of Mobile Network Nodes and/or Correspondent Nodes, and may be amplified with nesting of mobile networks.
- Increased Protocol Complexity and Processing Load: It is expected that NEMO Route Optimization will be more complicated than NEMO Basic Support. Thus, complexity of nodes that are required to incorporate new functionalities to support NEMO Route Optimization would be higher than those required to provide NEMO Basic Support. Coupled with the increased complexity, nodes that are involved in the establishment and maintenance of RO will have to bear the increased processing load. If such nodes are mobile, this may prove to be a significant cost due to the limited power and processing resources such devices usually have.
- Increased Delay during Handover: Due to the diversity of locations of different nodes that MNN may signal with and the complexity of NEMO RO procedure that may cause several rounds of signaling messages, a NEMO RO procedure may take a longer time to finish its handover than that in NEMO Basic Support.

SECURITY ISSUES IN NEMO

Security and privacy arc of thc most important issues, which need good attention. NEMO supports mobile networks travel on foreign, and possibly on untrusted networks when away from home. As MNNs are unaware of mobility, it is important that NEMO provides security while a network is away. In NEMO, the MR requires to allow the users from different domains to get Internet connectivity through it. In such settings, static trust relationships are lacking. Therefore, a variety of security threats arises. It must be admitted that until now not much research has gone to find a stable and widely accepted security solution for NEMO. In the NEMO Basic Support protocol (Devarapalli et al., 2005) the use of IPsec to protect signaling messages is recommended. Here, all signaling messages between the MR and the HA

Figure 9. NEMO route optimization

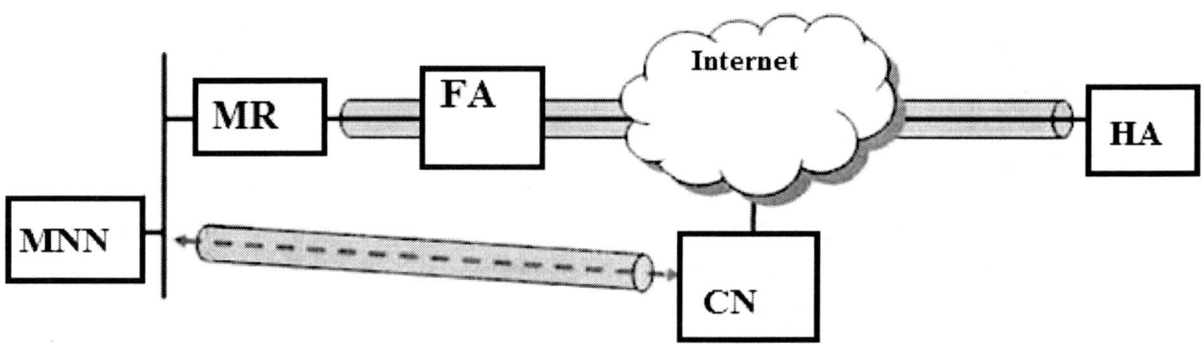

must be authenticated by IPsec (Kent, et al., 1998). IPsec provides data origin authentication, replay protection, data integrity, data confidentiality, and access control. It consists of security protocols, authentication and encryption algorithms, security associations, and key management. IPsec is optional for IPv4 but mandatory in IPv6. IPSec allows authentication of signaling messages as well as encryption of tunneled packets. NEMO on IPv4 does not make this strict requirement because IPSec is only an optional component of IPv4. NEMO Basic Support Protocol provides MMNs with location privacy. Due to the use of tunneling, the current location of MNNs is not revealed to CNs. First, both the MR and HA must check that all tunneled packets contain a source address that matches the originating network. This ensures that MNNs do not use the tunnel to perform IP spoofing attack. MR has to perform ingress filtering on packets received from the mobile network to ensure that nodes do not use the bi-directional tunnel to launch IP spoofing attacks. In particular, MR should check that the IP source addresses in the packets received belong to the mobile network prefix and are not the same as one of the addresses used by MR. If an MR receives an IP-in-IP tunneled packet from a node in the mobile network and it has to forward the decapsulated packet, it has to perform the checks on the source address of the inner packet, i.e. upon reception of a Binding Update, HAs must check that an MR is authorized to bind to a prefix before enabling forwarding to that prefix. HA should verify that packets received through the bi-directional tunnel belong to the mobile network. This check is necessary to prevent nodes from using the HA to launch attacks that would have otherwise been prevented by ingress filtering. The source address of the outer IPv6 header should be set to the MR's current CoA. The source address of the inner IPv6 header *must* be topologically correct with respect to the IPv6 prefixes used in the mobile network. Some research works have been done to further investigate the security issues in NEMO. In (Petrescu, et al), the security threats related to the NEMO protocol have been highlighted, where the authors identified signaling between the MR and the HA and nested mobility configurations as two main sensitive points of the protocol. Apart from these issues, Authentication, Authorization, and Accounting (AAA) are the other pillars for NEMO security system. Access control is vital in such wireless public access networks in order for any NEMO solution to be viable. A trusted identity may be an IP address or a certificate signed by a Certificate Authority (CA) which both the MR and MNN trust. Similarly, the MNN and CN are mutually trusted through the return routability test. The popularly used protocol for AAA over IP networks is the Remote Authentication Dial-In User Service (RADIUS) (Rigney, et al. 2000). The main purpose of RADIUS is to verify user name and password and to deliver configuration information detailing the type of service provided to the user.

NEMO APPLICATIONS AND TEST-BEDS

Global communication and reachability are one of the challenging requirements for next generation networking technologies. NEMO, with its capability of solving the problem of entire network movement faced by MIPv6, proves to be the ideal candidate to deliver the dream of ubiquitous computing and communication by Mark Weiser. It is known that when the entire network moves, MIPv6 cannot provide connectivity in an efficient manner. This in effect, results in poor connectivity and traffic loss. Therefore, NEMO finds its use cases as below:

- Ubiquitous Internet Access: "Always on-line" experience. Trains, buses, and airplanes may offer a mobile wireless network for their passengers. Mobile Devices of the passengers do not need any Mobility

Extensions. Only the network needs to move, no more handover overhead for each device.
- Personal Area Networks: Devices carried by people can connect to the Internet through the mobile phone. Photo Camera may upload pictures, needs only Bluetooth interface. Save energy consumption, size, and costs for devices.
- Smart Vehicles: Future cars may have many sensors and devices that benefit from connecting to the Internet. Traffic guiding systems may need sensor data. Integrated communication devices in the car. Entertainment devices in the car (digital TV/Radio, game consoles, etc.).

On this line, NEMO becomes one of the hot and happening areas to work in the field of mobile network, particularly in mobility management. Many projects, testbeds, and applications are built with NEMO. These projects are taken up both in industry and academia.

NEMO Platform for Linux (NEPL) is an implementation of NEMO on Linux by the GO-Core project of the Helsinki University of Technology (Helsinki University of Technology, 2006). NEPL adds NEMO extensions to the existing Mobile IPv6 for Linux (MIPL) implementation also from the GO-Core project. NEPL is working towards complete RFC 3963 compliance as well as additional functionality. Nautilus6 is currently working on Multiple Care-of Addresses and NEMO Prefix Delegation for NEPL.

Lancaster University has taken up the MANEMO (McCarthy et al., 2009) project. MANEMO stands for MANET + NEMO, or the integration of Mobile Ad-hoc Networking (MANET) and NEMO technologies. MANEMO is a broad term that recognizes the potential benefits of combining the localized, infrastructureless routing support of MANET with the global, persistent reachability properties of NEMO. To solve this problem, we consider few approaches: MANET-Centric MANEMO (MCM) and NEMO-Centric MANEMO (NCM).

Typically, MANET scenarios require support for a level of intercommunication between the mobile nodes that comprise the MANET. By this, we mean that when MANET nodes converge they often have a need to communicate directly with other nodes within the MANET. Examples of these kinds of scenarios include:

- Disaster Rescue
- Emergency Services
- Vehicle-to-Vehicle Communications
- Military Operations

New South Wales and Australian Research Council sponsored On-board Communication, Entertainment, And iNformation (OCEAN) (http://www.ocean.cse.unsw.edu.au/) project with collaborators such as the National ICT Australia (NICTA) and Boeing Airline Company, OCEAN recognizes the need for the extension of Internet services for public transportation systems and encompasses two research areas, namely networking and data management. The child project eMotion (Network in Motion) (Lan et al., 2005 & http://www.ocean.cse.unsw.edu.au/emotion/index.html) handles networking issues pertaining to providing global ubiquitous Internet access to passengers via MRs and wide area wireless access systems.

Nautilus6 working group (http://www.nautilus6.org &Tsukada, et al. 2008) is actively involved in developing NEMO solutions. It conducts researches into network mobility, as well as participated in reference implementations of the IETF Mobile IPv6 and NEMO Basic Support for BSD and Linux. It has successfully implemented multi-homing, seamless connectivity for NEMO.

The FleetNet (http://www.fleetnet.de/) project also aims to connect vehicles to the Internet. However, the MR-HA bi-directional tunneling architecture for NEMO is not the approach adopted when providing Internet connectivity to the devices in a vehicle in FleetNet. This project has introduced a novel network mobility architecture called MOCCA (MObile CommuniCation Architecture).

Another initiative worth mentioning is the ANEMONE testbed (http://www.ist-anemone.eu). The ANEMONE testbed is a pan-European MIPv6 and NEMO testbed for large-scale experiments. The ANEMONE testbed has the purpose of enabling the existing testbeds and extensions covering campus and metropolitan areas with a large spectrum of real end users for a pertinent evaluation of advanced services and applications. It offers a wide range of multimodal and open terminals such as laptops, PDAs, smartphones and an e-vehicle with a mobile router and local devices. Apart from acting as NEMO base Station, ANEMONE testbed also perform other mobility services like, MIPv6, Flow Bindings, and VoIP etc. The ANEMONE testbeds are successfully deployed in France, Finland, and Italy for carrying out the intended experiments and demonstrations.

We can find that with these kinds of projects and developments, mobility is becoming an insignificant issue. Therefore, anywhere, anytime computing or ubiquitous computing is not far from reality. Ubiquitous computing, or calm technology, is a paradigm shift where technology becomes virtually invisible in our lives. There are many constituents part of it, like context awareness, transparent interfaces, automated capture, and seamless communication. The last constituent, i.e. seamless communication is completely dependent on NEMO.

FUTURE RESEARCH SCOPE

It is very much clear from the above discussion NEMO is certainly a big challenge and going to replace MIPv6 in near future. Another vision, which will make NEMO the contender for a popular spot in the wireless network research community, is its strong tie with ubiquitous computing and communication. Ubiquitous computing enhances computer use by making many computers available throughout the physical environment, while making them effectively invisible to the user. It envisions an environment that seamlessly integrates digital and physical devices, where users can access digital data and applications from the environment as easily as accessing them through their personal computers. Since ubiquitous computing exists in the user's environment, the technology is sustainable if it is invisible to the user and does not intrude the user's consciousness. This requires functioning of multitude of devices in the environment to be oblivious to the users. Mark Weiser (Weiser, 1991) in his paper "The Computer for the Twenty-First Century" defines ubiquitous computing as a technology that "weaves itself into the fabric of everyday life until it is indistinguishable from it". Ubiquitous computing indeed is the future that the research community and visionaries are enthusiastically looking for. It is also the fact that NEMO is the solution for the ubiquitous access, which is one of the important components of ubiquitous computing and calm technology. NEMO already poised some advantages over MIPv6 like reduced complexity, reduced transmit power, increased manageability, superior scalability, reduced handover signaling. Still NEMO lacks in many fronts. For seamless connectivity, one device should have a continuous and transparent access to all the available networks, i.e. heterogeneity needs to be supported. In the traditional homogeneous wireless networks, mobility management operations are carried out by the link-layer specific mechanisms, which are usually not portable from one type of network to another. Therefore, mobility management in the newly emerging heterogeneous wireless networks is becoming a challenging issue for developing a ubiquitous computing system. An efficient mobility management scheme allows a mobile device to have always best-connected (ABC) to the Internet. It facilitates the mobile device to connect to different types of access networks that optimally satisfy the resource requirements of desired communication services. Therefore, in NEMO, which is basically IP layer specification, heterogeneity needs to be addressed in a very careful and optimized away. Until now, there is no such solution exists for NEMO. This is really a good scope for research.

Another research area is adaptivity in route optimization. NEMO decomposes the wireless connection between mobile nodes and access routers into two parts: one between the MN and MR and the other between the MR and access routers. This results in optimization of these two segments in some independent ways. MNs can find better wireless connectivity because of a shorter distance between them and MRs. Similarly, the MRs can have a high bandwidth dedicated wireless connection with the access routers. The MNs residing in a mobile network can communicate to one another. However, the current NEMO protocols do not provide an efficient way of such communications. This is an issue of local route optimization, which may enable mobile nodes to forward packets via an optimal path when communicating with other MNs of the same mobile network. Further extension of local route optimization may result in supporting a standalone mobile network, which will enable MNs to establish direct connections between them, even when the mobile network is not attached to the Internet infrastructure. This will, in fact, help in providing location-based services to MNs from the local servers located in the mobile network. Hence, local route optimization and supports for a standalone mobile network worth further study and research.

CONCLUSION

Network mobility (NEMO) management is concerned with managing mobility of an entire mobile network. The mobile networks are being designed to facilitate mobile devices to seamlessly access Internet by grouping the mobile devices and hiding their mobility on the Internet topology. The NEMO Basic Support protocol, developed by the Internet Engineering Task Force (IETF), outlines the basic operation of mobility management of mobile networks even when the entire network is moving. In this chapter, we have highlighted the issues of NEMO, also presented state of the art survey on those issues, and proposed solutions. We also emphasized the path from Mobile IP (MIPv4 and MIPv6) to NEMO and showed how NEMO inherited some features and solutions of MIP, particularly MIPv6, still hold its own philosophy of supporting the mobility of entire network. It is now clear that IP mobility is shifting towards NEMO. As new multimedia devices are coming and the technology moving so fast, people would like to get seamless connectivity even when the entire network is moving along. This in fact is a step forward to ubiquitous communication. Therefore, NEMO is truly becoming the de facto communication protocol for next generation communication. The IETF NEMO Working Group is chartered only to standardize solutions for basic network mobility; nevertheless, numerous members and many universities and industrial organizations have active participation in the research and test bed development on NEMO. We have also briefly discussed the scope of research in this field and the research communities' ambition of ubiquitous computing can be realized through NEMO.

REFERENCES

Anemone, I. S. T. (n.d.). Retrieved from http://www.ist-anemone.eu

Braden, R., et al. (1997). *Resource reservation protocol (rsvp) – version 1 functional specification*. (RFC 2205/2750).

Cheng-Wei, L., & Sun, Y. (2008). HiMIP-NEMO: Combining cross-layer network mobility management and resource allocation for fast QoS-handovers. *IEEE Proceedings,* (2282-2286).

Deering, S. (1991). *ICMP router discovery messages*. (RFC 1256).

Devarapalli, V., et al. (2005). *Network mobility (NEMO) basic support protocol*. (IETF, RFC 3963).

Fleetnet. (n.d.). Retrieved from http://www.fleetnet.de/

Held, A., & Ziegert, T. (1999). Service mobility - a new feature of mobile computing. *Integrated Computer-Aided Engineering*, *6*(2), 131–142.

Helsinki University of Technology. (2006). *Mobile IPv6 for Linux website*.

Janneteau, C. et al. (2004). *IPv6 multicast for mobile networks with MLDProxy*. (IETF Internet Draft, draft-janneteau-nemo-multicast-mldproxy-00.txt).

Johnson, D., Perkins, C., & Arkko, J. (2004). *Mobility support in IPv6*. (IETF RFC 3775).

Kaffle, V. P., Kamioka, E., & Yamada, S. (2006). CoMoRoHo: Cooperative mobile router based handover scheme for long-vehicular multihomed networks. *IEICE Transactions on Communications, 89*(10).

Kamel, G., Mihailovic, & Aghvami, A. H. (2008). Case analysis of a cost-optimal QoS aggregation policy for network mobility. *IEEE Communications Letters*, *12*(2), 130–132. doi:10.1109/LCOMM.2008.071662

Kamel, G., Mihailovic, A., Pangalos, P., & Aghvami, A. H. (2007). *Cost-optimal QoS aggregation for network mobility* (pp. 5006–5010). IEEE Globecom.

Kempf, J., Khalil, M., & Pentlan, B. (2005). *IPv6 fast router advertisement*. (draft-mkhalil-ipv6-fastra-05.txt), January 2005.

Kent, S., & Atkinson, R. (1998). *Security architecture for the Internet protocol*. (RFC 2401).

Koodli, R. (2005). *Fast handovers for mobile IPv6*. (IETF RFC 4068).

Lan, K., et al. (2005). *MOBNET: The design and implementation of a network mobility testbed for NEMO protocol*. 14th IEEE Workshop on Local and Metropolitan Area Networks.

Ludwig, R., & Meyer, M. (2003). *The Eifel detection on algorithm for TCP*. (IETF RFC3522).

Malik, M. A., Kanhere, S. S., Hassan, M., & Benatallah, B. (2004). On-board RSVP: An extension of RSVP to support real-time services in on-board IP networks. *Lecture Notes in Computer Science*, *3326*, 264–275. doi:10.1007/978-3-540-30536-1_31

McCarthy, B., Edwards, C., & Dunmore, M. (2009). *Using NEMO to support the global reachability of MANET nodes*. IEEE INFOCOM'09.

Moore, N. (2005). *Optimistic duplicate address detection for IPv6*. (draft-ietfipv6-optimistic-dad-05.txt), February 2005.

Na, J., et al. (2004). *Route optimization scheme based on path control header*. (IETF Internet Draft, draft-na-nemo-path-control-header-00.txt).

Naoe, H., Wetterwald,M. & Bonnet, C. (2007). *IPv6 soft handover applied to network mobility on heterogeneous access networks*. IEEE PIMRC, September 2007.

Narten, T., Nordmark, E., & Thomson, W. (1998). *Neighbor discovery for IP version 6* (IPv6). (IETF RFC 2461).

Nautilus6. (n.d.). Retrieved from http://www.nautilus6.org

Ng, C., et al. (2007). *Network mobility route optimization solution space analysis*. (RFC 4889).

Ng, C., Thubert, P., Ohnishi, H., & Paik, E. (2005). *Taxonomy of route optimization models in the NEMO context*. (IETF Internet Draft, draft-thubert-nemo-ro-taxonomy-04.txt).

Ocean, C. S. E. (n.d.). *Website*. Retrieved from http://www.ocean.cse.unsw.edu.au/

Ocean, C. S. E. (n.d.). *Emotion index*. Retrieved from http://www.ocean.cse.unsw.edu.au/emotion/index.html

Ohnishi, H., Sakitani, K., & Takagi, Y. (2003). *HMIP based route optimization method in a mobile network*. (IETF Internet Draft, draft-ohnishi-nemo-ro-hmip-00.txt).

Perkins, C. (1997). Mobile IP. *IEEE Communications Magazine*, *35*(5), 84–99. doi:10.1109/35.592101

Perkins, C. (Ed.). (2002). *IP mobility support for IPv4*. (IETF RFC 3344).

Petander, H., Perera, E., & Seneviratne, A. (2005). *Multiple interface handoffs: A practical method for access technology independent make-before-break handoffs*. NICTA Technical Report.

Petrescu, A., Olivereau, A., Janneteau, C., & Lach, H-Y. (2004). *Threats for basic network mobility support* (NEMO threats). (IETF Internet Draft, draft-petrescunemo-threats-01.txt), January 2004.

Ramani, I., & Savage, S. (2005). *Syncscan: Practical fast handoff for 802.11 infrastructure networks*. IEEE INFOCOM'05.

Rigney, C., Willens, S., Rubens, A., & Simpson, W. (2000). *Remote authentication dial in user service* (RADIUS). (IETF RFC 2865), June 2000.

Soliman, H., Catelluccia, C., Malki, K. E., & Bellier, L. (2005). *Hierarchical mobile IPv6 mobility management* (HMIPv6). (IETF RFC 4140).

Tais, M., & Labiod, H. (2005). *Resource reservation for NEMO networks*. International Conference Wireless Networks, Communications and Mobile Computing (pp. 232–237).

Thubert, P., & Molteni, M. (2004). *IPv6 reverse routing header and its application to mobile networks*. (IETF Internet Draft, draft-thubert-nemo-reverse-routing-header-05.txt).

Tsukada, M., Mehani, O., & Ernst, T. (2008). *Simultaneous usage of NEMO and MANET for vehicular communication*. 1st Workshop on Experimental Evaluation and Deployment Experiences on Vehicular networks.

Vida, R., & Costa, L. (Eds.). (2004). *Multicast listener discovery version 2 (MLDv2) for IPv6*. (IETF RFC 3810).

Weiser, M. (1991, September). The computer for the twenty first century. *Scientific American*, 94–104. doi:10.1038/scientificamerican0991-94

Wu, X., Chan, M. C., & Ananda, A. L. (2007). *TCP handoff: A practical TCP enhancement for heterogeneous mobile environments*. IEEE ICC 2007.

KEY TERMS AND DEFINITIONS

Foreign Agent (FA): It is a Mobile IP entity, which cares with mobile devices arrived to a foreign site of the network.

Handover: It is a method, which ensures undisturbed and continuous communication during cell (or channel) change.

Home Agent (HA): It is a Mobile IP entity, which handles data of the home network's own devices.

Mobile IP (MIP): It is a protocol, which makes mobility possible in IP networks.

Multihoming: It is a method for making devices useable in different networks without any re-configurations.

Network Mobility (NEMO): Capability of a network segment or subnet ensuring move and attach again to an arbitrary point of the routing infrastructure.

APPENDIX

Table of abbreviations

AAA	Authentication, Authorization and Accounting
ABC	Always Best Connected
BS	Base Station
BT	Bi-directional Tunnel
BU	Binding Update
CA	Certificate Authority
CH	Correspondent Host
CN	Correspondent Node
CoA	Care-of Address
CoMoRoHo	Cooperative Mobile Router-based Handover
DAD	Duplicate Address Detection
DHCP	Dynamic Host Configuration Protocol
DiffServ	Differentiated Services
FA	Foreign Agent
FBack	Fast Binding Acknowledgement
FBU	Fast Binding Update
FMIPv6	Fast handover for MIPv6
HA	Home Agent
HACK	Handover Acknowledgement
HI	Handover Indication
HMIPv6	Hierarchical Mobile IPv6
ICMP	Internet Control Message Protocol
ICT	Information and Communication Technology
IETF	Internet Engineering Task Force
IntServ	Integrated Services
ISP	Internet Service Provider
IP	Internet Protocol
MANET	Mobile Ad-hoc Networking
MCM	MANET-Centric MANEMO
MH	Mobile Host
MIP	Mobile Internet Protocol (Mobile IP)
MIPL	Mobile IPv6 for Linux
MIPv4	Mobile Internet Protocol (Mobile IP) version 4
MIPv6	Mobile Internet Protocol (Mobile IP) version 6
MLD	Multicast Listener Discovery
MN	Mobile Node

MN	Mobile Network
MNN	Mobile Network Node
MNP	Mobile Network Prefix
MOCCA	MObile CommuniCation Architecture
MR	Mobile Router
NAR	New Access Router
NCM	NEMO-Centric MANEMO
NEMO	Network Mobility
NEPL	NEMO Platform for Linux
NICTA	National ICT Australia
OCEAN	On-board Communication, Entertainment, And iNformation
PAR	Previous Access Router
PrRtAdv	Proxy Router Advertisement
RADIUS	Remote Authentication Dial-In User Service
RH	Routing Header
RO	Route Optimization
RRH	Reverse Routing Header
RS	Remote Subscription
RSVP	Resource Reservation Protocol
RtSolPr	Router Solicitation for Proxy
SHO	Soft Handover
WWW	World Wide Web

Chapter 10
Protocols in Next Generation Networks

Róbert Horváth
Budapest University of Technology and Economics, Hungary

Gábor Kovács
Budapest University of Technology and Economics, Hungary

Zoltán Pap
Ericsson, Hungary

ABSTRACT

This chapter focuses on Next Generation Networks (NGN), how protocols employed in different segments of the telecommunication infrastructure interwork to guarantee the quality for different service types. After the interpretation of the definition and concept of NGN, the general architecture of NGN is presented with regard to the heterogeneous demands of services. The protocols used in different sections of the end-to-end communication relation are introduced from the point of view of service and transport functions of NGN. The main focus of the chapter is on the fixed access solutions and Ethernet based multi-service access and regional networks, but configuration and streaming protocols are covered as well.

INTRODUCTION

Next Generation Networks is the response of the telecommunication industry including standardization bodies, telecommunication operators, and vendors to the increasing user demands of access to broadband services and mobility. NGN provides a single point of access for customers to data services, multimedia content, voice services, and virtual networking taking into account the heterogeneous set of intelligent user devices. From the point of view of telecommunication operators, the integration and upgrade of the infrastructure into an NGN leads to reduced costs and simplified maintenance.

Section 'Next Generation Networks' interprets the definition of NGN given by ITU-T (International Telecommunication Union), highlights the main transport functions and technologies of different segments of an NGN, and briefly describes the requirements of triple-play services on the NGN. In the section titled 'Core Network

DOI: 10.4018/978-1-60960-732-6.ch010

Protocols' the most important data plane and multicast protocols used in operators core networks are discussed. Section 'Regional network and aggregation network protocols and technologies' focuses on fixed Metropolitan Area Networks (MAN) and its dominant protocol: Ethernet; techniques for service separation, multicast and virtual networking are described. The subscriber's access to the service network in discussed in Section 'Service Access Protocols'. The next section titled 'Services in NGN' takes a look at NGN from the point of view of service functions, and protocols that take part in the service provisioning: IP (Internet Protocol) address configuration, user and service separation, service control, streaming protocols and authentication protocols. At the end of this chapter, a short outlook is given of near-future trends in telecommunication networking.

NEXT GENERATION NETWORKS: DEFINITION, ARCHITECTURE, AND SERVICES

The NGN Definition

The term NGN is defined by the ITU as follows (ITU-T, 2004a):

A Next Generation Network is a packet-based network able to provide telecommunication services and able to make use of multiple broadband, quality of service (QoS) enabled transport technologies and in which service-related functions are independent of the underlying transport-related technologies.

The interpretation of the definition is the following. In NGN the interconnection of two endpoints, addressing and data transfer is done by packet-based technology, which is practically IP. Service and transport functions are separated. The transport medium must support different types of services, all services work over IP, and the QoS can be managed. Network access provider is the operator who owns the network infrastructure; application service provider is the operator who provides services to end user over that network. There is no telephone or television network anymore; however, the NGN has interfaces to existing networks like PSTN or PLMN (Public Land Mobile Network).

The NGN Transport Architecture

The NGN architecture (ITU, 2004b) includes the five network segments with different sets of functions, and interface nodes and devices. Figure 1 shows a sample NGN architecture that is described in the rest of this section.

The end-user functions are implemented in the subscriber or home network, which is the property of the subscriber behind a residential gateway. There are public and private subscriber networks: any user can connect to public networks (especially public WiFi) and may or may not connect to private networks depending on the local policy. The interface to the transport infrastructure is the residential gateway. That device is a combination of the set-top-box and the ADSL (Asymmetric DSL) modem, and acts as the single access point for triple-play services.

First mile (in some sources referred as last mile) implements the access function of NGN. It connects residential gateways to the broadband access network. The dominant technologies employed on this subscriber line are from the physical and data link layers of the OSI (Open Systems Interconnection) reference model: the xDSL solutions, Passive Optical Networks (PON), Asynchronous Transfer Mode (ATM), Point-to-Point Protocol (PPP), and Ethernet in the case of fix access or WiFi or Global System for Mobile communications (GSM) or Universal Mobile Telecommunications System (UMTS) in the case of wireless access. The bottleneck in the end-to-end communication has been the first mile, however this problem loses its significance as that

Figure 1. A sample next generation network architecture

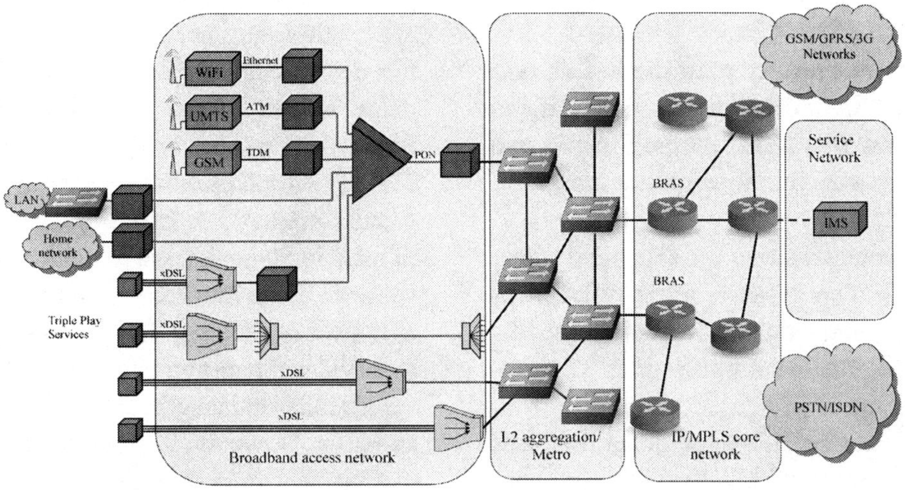

fixed operators deploy optical cable terminations closer and closer to the customers.

Metro access network (Metro Ethernet Forum, 2006; IEEE, 2007) aggregates the traffic of up to 100000 subscribers. It consists of two subnetworks: aggregation networks and regional networks that interconnect aggregation networks and the core network via IP Edge Nodes denoted with BRAS (Broadband Remote Access Server) in the figure. The main technologies are: Ethernet, VLAN (Virtual Local Area Network). In order to minimize traffic in the access network, operators deploy IPTV servers that broadcast the television services here to be as close as possible to subscribers.

The first IP-aware nodes in an end-to-end relation are the edge nodes that have an important role in NGN. Edge nodes reside on the border of the Metro access network and the core network. As all services operate over IP, this is the point, where service and transport control functions can be implemented. Transport control functions manage the resources of the access network, admit a new service connection according to a set of policy rules if there is still available capacity in the network, and dedicate resources to that connection. Service control functions validate the service connection request against the service profile of the subscriber, for instance if the subscriber has an SDTV (Standard-Definition Television) subscription, but tries to access an HDTV (High-Definition Television) service, then the request will be rejected.

Core network interconnects IP Edge Nodes and gateway nodes and Application Service Provider (ASP) nodes. A gateway node is an interface to mobile networks (GGSN – Gateway GRPS Support Node), PSTN, another NGN, or the backbone network. ASP provides IP-based services to end-users. The protocol stack of the core network has optical fiber and WDM (Wavelength Division Multiplexing) in the physical layer, gigabit Ethernet in the link layer, and IP over MPLS in the packet layer.

The service network implements application services and their management provided by operators to subscribers. The IP Multimedia Subsystem (IMS) responsible for service control is the main network element in this segment.

Quality Requirements on Service Functions

The heterogeneous service portfolio has diverse demands on the transport networks. Applications are usually classified based on their traffic and service requirements into four service classes:

- Applications sensitive for delay and transfer faults. This class is often referred as real time and contains applications like interactive video, compressed IPTV, or online gaming.
- Applications sensitive for delay, but less sensitive for transfer faults. Streaming media services like VoIP or IPTV are in this class where the delay of voice or raw video packets is more disturbing than the loss.
- Applications sensitive for data integrity, but less sensitive for delay, for example transfer of large files.
- Applications less sensitive for delay and data loss. This is the so-called best-effort class, the basic service of the IP. Services like Internet access, browsing, email are in this class.

These user requirements appear as end-to-end QoS parameters for the network like packet loss, delay, jitter that means delay variance, throughput that means the maximum bandwidth that can be assigned to the service.

Subscribers use several applications in their home network in parallel: download or upload data, peer-to-peer traffic, Internet browsing, and real-time transmission of voice or video. Therefore, the QoS requirements of applications cannot be handled separately, the capacity requirements of the compound user traffic must be investigated continuously. Table 1 shows the typical QoS parameter requirements of triple-play services. Voice services, like VoIP, are sensitive to end-to-end delay, which is a value calculated from the delay of all segments along the communication path. If this value is greater than 150 ms, some application may not work properly. The round-trip delay is calculated from the two end-to-end delay values of the communication relation. Subscribers notice the delay, if it is greater than 250 ms. Voice communication applications are usually prepared to handle jitter, a small buffer of usually 50 ms is used to compensate for delay variance. Data loss has smaller significance here as codecs can restore the stream up to 5% packet loss. Video streaming has similar QoS parameter value requirement with regard to end-to-end delay and jitter. However, it is sensitive for data loss if that video stream is compressed. Two classes of data services can be distinguished real-time and non real-time. These classes share the requirement of high integrity, the data loss ratio must be 0. Real-time data traffic is produced by for example interactive gaming, which has a very low bandwidth requirement (<1 kbps), however the end-to-end delay should be less than 200 ms like in the case of voice communication. On the other hand non real-time services have a high bandwidth demand like video services, however even a delay of 1s is tolerated.

The audio, video, and data services have different bandwidth requirements. The widest part of the bandwidth must be allocated to video services. Transferring the data of a single television channel requires in case of using MPEG-2 codec 20 Mbps in HDTV format and 6 Mbps in SDTV format along the whole path from the video server to the set-top-box. Voice and interactive data require much less bandwidth, however must still be separated from best-effort traffic, which is associated with the remaining bandwidth.

Until the worldwide deployment of optical fiber to the subscribers' home, the first mile remains the bottleneck of the end-to-end connection. Therefore, practically only a few number of television channels are delivered at a time to the set-top-box.

Table 1. QoS parameters of triple-play services

Information type	QoS parameter	Value range	Comment
Voice	end-to-end delay	<150 ms	Acceptable for most applications
		150-400 ms	Has impact on some applications
		>400 ms	Bad quality
	round-trip delay	<150 ms	Desired latency
		>250 ms	Subscribers notice the delay
		>500 ms	Impractical
	end-to-end jitter	<50 ms	Usually compensated with buffers
	data loss ratio	$<10^{-3}$	Phone quality
		>5%	Codecs cannot compensate
	bandwidth	5.3-8 kbps	Best codecs
		16 kbps	Phone quality
		64 kbps	PCM coding
		128 kbps	CD quality
Video	end-to-end delay	<250 ms	Acceptable
	end-to-end jitter	<200 ms	Adequate
	data loss ratio	$<10^{-2}$	For raw video
		$<10^{-11}$	For compressed video
	bandwidth	<2 Mbps	MPEG, H.261, H.120
		20 Mbps	compressed HDTV
		140 Mbps	TV
		>1 Gbps	raw HDTV
Data	end-to-end delay	1s	file transfer
	bandwidth	<10 Mbps	
	data loss ratio	$<10^{-11}$	
	end-to-end delay	<200 ms	interactive gaming
	bandwidth	<1 kbps	
	data loss ratio	0	

CORE NETWORK PROTOCOLS

MultiProtocol Label Switching

In NGN, IP over MPLS is used in the core networks for user and service separation, QoS support and traffic engineering. MPLS (Rosen & Viswanathan & Callon, 2001) defines virtual paths to support faster IP packet forwarding and QoS. On entering an MPLS domain (see Figure 2), the incoming IP packet is tagged with a with a four or eight byte long label in a Label Edge Router. In each Label Switch Router (LSR) along the Label Switch Path (LSP), the forwarding decision is made on this label without needing to examine the encapsulated IP packet itself. The LSR removes the old label, which determines the next hop, and assigns a new label to the packet for the next forwarding decision.

MPLS can be used to transport any protocol, however, in practice it is used exclusively together with different IP versions. The tagging takes place before the IP encapsulation, so ac-

Figure 2. A sample MPLS network

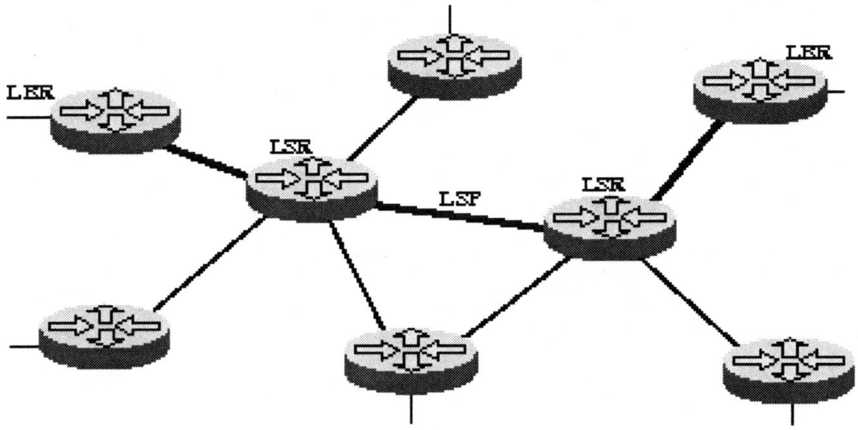

cording to the OSI reference model MPLS resides between the link layer and the packet layer. (see Table 2)

The 32 bits long MPLS header consists of four fields. The label value is the identifier of the virtual path; the forwarding decision is made on this value. The traffic class value is associated to a service class; it influences scheduling and discarding at a router. The bottom of stack bit is set to 1, if the label is the last value; otherwise, it is set to 0. The TTL field has the same functionality as in IP packets.

Packets arriving in an MPLS domain are associated with a Forwarding Equivalence Class (FEC). Packets in the same class are forwarded via the same path in the domain. In the LER the FEC of an incoming packet is determined based on fields the IP packet header (as source and destination addresses and the type of service field and possibly other QoS extension header fields) and the interface the packet arrived at, and that FEC is assigned to a label.

Table 2.

20 bits	3 bits	1 bit	8 bit
Label value	Traffic class	Bottom of stack	TTL

LSRs define, distribute and exchange label and FEC mapping information and establish LSPs by means of a Label Distribution Protocol (LDP). LDP (Andersson et al., 2007) defines LPSs by mapping network layer routing information to data link layer paths. After two LSRs have agreed on the LDP parameters, they reserve the required resources and establish an LSP, and assign a FEC to the path.

LDP has four message types: Discovery, Session, Advertisement, and Notification. LDP communication takes place over TCP (Transmission Control Protocol) except for Discovery messages that are sent over UDP (User Datagram Protocol). Discovery messages are sent periodically by LSRs to announce their presence. Session messages are used to create, maintain, or close a session between two LSRs. When two LSRs establish a session, they become LSR peers, and are able to exchange mapping information by advertisement messages. Notification messages are the means for signaling error information.

To support traffic engineering on an MPLS domain RSVP-TE can be used for selecting alternate paths for load balancing. RSVP-TE (Awduche et al., 2001) is an extension of RSVP (Resource Reservation Protocol) that supports the dynamic establishment of LSPs such that MPLS labels are associated with RSVP flows. The forwarding states of LSRs are set and resources along the LRP are reserved by RSVP-TE. Besides, RSVP-TE supports explicit routing that pre-determines the paths of label-switched RSVP-MPLS flows. Such an explicitly routed path can be configured by the operator, or computed based on QoS parameters and policy rules.

Multicast, IGMP, PIM

The high bandwidth demand of transferring streaming video services to multiple destinations simultaneously can be reduced by multicast. Multicast packets are sent only once, but replicated at network nodes of a multicast distribution tree. The only task of the source node is to set the multicast address as destination IP address, while receiver nodes can join or leave a multicast tree associated with that multicast address.

In IP networks, multicast Internet Group Management Protocol (IGMP) is used to manage the multicast group membership of a host (Cain et al., 2002). The host that wants to join a multicast group sends an IGMP Join message to neighboring multicast routers. To clear membership the host must send an IGMP Leave message.

IGMP messages are encapsulated in the data field of IP packets. The message structure begins with the type of message that can be join, leave or membership report. The maximum response time field, which is used only in join messages and ignored otherwise, sets a timer for the response. The checksum protects the first two fields. The group address field is set to the multicast IP address of the group the host wants to join or leave. (see Table 3)

In the core network, the Protocol Independent Multicast (PIM) (Estrin et al., 1998) family is used for routing multicast traffic. It operates over an existing unicast infrastructure, and although it is called a multicast routing protocol, it uses the unicast routing table to perform the reverse path forwarding (RPF) check function instead of building up a completely independent routing table. PIM does not send and receive multicast routing update messages between routers. PIM has three operation modes: sparse mode (SM), dense mode (DM) and source specific multicast (SSM).

PIM-SM uses a pull model to deliver multicast traffic; it assumes that no hosts want the multicast traffic unless they specifically ask for it. It uses a unidirectional shared tree to distribute the information about active sources. Depending on the configuration options, the traffic can remain on the shared tree or switch over to an optimized source distribution tree. The traffic will be forwarded to networks with active receivers that have explicitly requested the streaming multicast data. PIM-SM operates by defining Rendezvous Points (RP). The source of the multicast traffic sends the stream to an RP. The receiver registers itself at an RP. Once the data stream begins to flow from the sender to the RP and from the RP to the receiver, routers along the path will optimize the path automatically and remove unnecessary hops by building a shared tree.

PIM-DM is the brute force method to deliver multicast data to receivers. It uses a push model to flood multicast traffic to every corner of the network. This method is practical, if there are active receivers in every subnet in the network.

In PIM-SSM, multicast groups are replaced with multicast channels. Only the source is able

Table 3.

8 bits	8 bits	16 bits	32 bits	
Type	Max response time	Checksum	Group address	Data

to transmit data to that channel. This means that the forwarding mechanism uses a source bases forwarding tree. This method is practical for inter-domain use and allows source filtering.

REGIONAL NETWORK AND AGGREGATION NETWORK PROTOCOLS AND TECHNOLOGIES

The dominant technology in these segments of the NGN architecture is Ethernet (IEEE, 2008) because of its flexibility and cost effectiveness. More than 90% of the traffic of WANs (Wide Area Network) or MANs (Metropolitan Area Network) end at an Ethernet interface (IEEE, 2007).

Ethernet

The Ethernet technology covers two sublayers of the OSI data link layer, the physical layer access (Media Access Control – MAC) and framing (Logical Link Control – LLC). The MAC sublayer provides connectionless datagram services to the LLC sublayer.

The Ethernet frame is at most 1542 octets long. It begins with a 7-octet preamble of alternating 1 and 0 bits used for carrier sensing in Carrier Sense Multiple Access Collision Detection (CSMA/CD), and a Start-of-Frame Delimiter with the bit sequence of 10101011 that signals the beginning of the data. The destination and source MAC addresses identify the destination and source interfaces of this packet. The two-octet Type or Length field defined either the length of the payload or the type of the frame if it is used for link layer signaling, or VLAN encapsulation. The length of effective payload is between 46 and 1500 octets. The checksum of 4 octets covers the bits from the destination address to the end of the payload. After the frame, 12 idle octets are transmitted. (see Table 4)

Ethernet forwarding is done by Ethernet switches based on MAC address tables. When a frame arrives at a port of the switch, the source address of that frame is associated to that port in their forwarding table. When a switch receives a frame with a destination address with no entry in the MAC table, the frame is broadcast to all port except for the one it arrived at.

Traditionally LANs used a half-duplex shared medium and CSMA/CD. Today most Ethernet LANs are based on a point-to-point, full-duplex switched architecture, where, to be able to send and receive simultaneously, the full-duplex operation is achieved by two independent communication channels, one in each direction. Switched Ethernet has now evolved to include many advanced features, including separation of users into logical broadcast domains, priority between user traffic, security, and link aggregation.

Virtual LANs

Ethernet is a connectionless technology, so it is not possible to create a virtual channel identified by a pair of MAC addresses. While switches limit the collision domain into one switch port, a switched LAN is still one broadcast domain so there is still an issue of broadcast frames being flooded to all end user devices.

The Virtual LAN (VLAN) technique (IEEE, 2005) partitions a switched network into multiple broadcast domains. It introduces logical connectivity over the existing physical connectivity in the network by adding a VLAN header to the Ethernet

Table 4.

7 octets	1 octet	6 octets	6 octets	2 octet	46-1500	4 octets	12 octets
Preamble	SFD	Dst MAC	Src MAC	Type/LEN	Payload	CRC	gap

frame header. The TPID field that plays the role of the Ethernet Type field is set to 0x8100 and that identifies the frame as VLAN tagged frame, and two other octets are inserted between that field and the real Ethernet Type field. The PCP is a three bits long field holding a priority value for the frame that can be used to differentiate service classes. The CFI is 0 in Ethernet networks. The VLANID field identifies the virtual LAN of this frame. (see Table 5)

This logical layer limits the broadcast domain and influences the forwarding, if the port has been grouped into a VLAN, the switch forwards only to those ports that are members of that VLAN.

Beside the QoS support provided by the priority bits there are multiple uses of VLANS in NGN. VLANs improve network performance as it reduces the need for nodes that handle traffic not destined to them. Since VLAN traffic is restricted to the nodes of that VLAN and therefore only authorized nodes receive the data, it enhances security as well.

For a WAN or MAN serving hundred thousands of residences the 12 bits that means altogether 4096 VLAN identifiers may not be enough. The IEEE 802.1QinQ (Juniper, 2009) gets rid of this limitation by adding a second VLAN header to the frame. The VLAN tags are inserted between the Source Address and Ethernet Type fields of the packet. The last added VLAN tag is the one closest to the Source Address field and this is the one to be processed and removed first. (see Table 6)

Spanning Tree Protocols

The regional and access network segments are composed of several Ethernet broadcast domains. In the case of a set of interconnected Ethernet networks like that, it may possible that a packet gets stuck between two domains because of the badly configured forwarding tables. To ensure that there are no loops in the forwarding path from any node to any other the family of Spanning Tree Protocols (STP) has been introduced (IEEE, 2004a).

The STP creates a spanning tree over the whole data link layer domain, the links included in the spanning tree are enabled, and those links that are not included are disabled. STP requires initialization, the root node of the spanning tree protocol must be selected, and port costs in every switch in the domain must be set. Each port of every switch keeps track of its cost to the root node and its state. Each port periodically sends a Bridge Protocol Data Unit with its current status and root path cost to let neighboring switches know its status. The cost is the smallest sum of costs of traversed ports along the path towards the root node, the state determines if the port is blocked or can be used for forwarding.

STP had been found to respond slowly to topology changes, so Rapid STP was developed for faster convergence of link states.

Multiple STP extends the functionality of STP with regard to VLANs. It maintains a separate spanning tree for each VLAN to provide better load balancing.

Table 5.

16 bits	3 bits	1 bit	12 bits
TPID	PCP	CFI	VLANID

Table 6.

DA	SA	ET2	VLAN	ET1	VLAN	ET	data

Multicast, IGMP

Since the video servers usually reside in the Ethernet domain of the NGN rather than behind the IP edge node, the IGMP request messages must be processed in the Ethernet switches. This is done by the IGMP snooping technique that filters IGMP join and leave messages and keeps track of relations between listeners and multicast servers to minimize multicast traffic in the network.

Ethernet Services

Metro Ethernet services (Metro Ethernet Forum, 2006) refer to a technology being standardized at multiple forums: Metro Ethernet Forum, ITU, and IETF. The services are accessible at a standardized UNI (User Network Interface), which is the border between the operator and the subscriber networks. Virtual Ethernet connection is established between the Ethernet UNIs, which means point-to-point, point-to-multipoint or multipoint-to-multipoint associations between two or more UNIs. Services have two types: private and virtual private services. In the case of private services, the subscriber had dedicated resources; in the case of virtual private services more subscribers share dedicated resources.

The E-Line service by Metro Ethernet Forum's and Ethernet (Virtual) Private Line by ITU are similar to Frame Relay or private leased lines. The UNIs of the E-LAN service by MEF or Ethernet (Virtual) Private LAN by ITU appear to be on the same LAN from the point of view of customers. In the simplest form theses services provides symmetric best-effort connection, however in complex forms connection parameters like CIR (Committed Information Rate), CBS (Committed Burst Size), EIR (Excess Information Rate), delay, packet loss and jitter can be set.

SERVICE ACCESS PROTOCOLS

In NGN, the first mile connects the Ethernet based aggregation network and the residential gateway. The most characteristic protocol of this section is the Point-to-Point Protocol (PPP) that establishes a session over the subscriber line to the Internet Service Provider (Simpson et al, 1994). If the first mile is ATM-based, then PPP over ATM (PPPoA) is used, and if the first mile is Ethernet based, then the PPP over Ethernet (PPPoE) is used.

PPP established a full-duplex bi-directional channel to transfer packets in order between two peers. PPP has a HDLC-like frame structure. It is bounded by flags of the bit pattern 01111110. All bits of the one-octet address field are constantly set to 1. The value of the one-octet control field is set to hexa 3. The protocol field identifies if the data field of this frame carries management information or encapsulates higher layer data. The data field holds management or higher layer information. The FCS field is a two-octet checksum that covers the octets from the address field to the end of the data field. (see Table 7)

PPP has three operation modes. It can encapsulate higher-level data, or it can manage the data link layer connection (Link Control Protocol – LCP), or it can manage the network layer connection (Network Control Protocol – NCP). LCP is used for establishing, configuring, maintaining, and terminating a connection including the authentication with either the Password Authentication Protocol (PAP) or the Challenge Handshake Protocol (CHAP), and the Link Quality Monitoring (LQM) and compression. NCP does the same but for network layer protocols like IP.

PPP authentication has two formats. The simplest one is PPP, where on a PAP authentication request by the server, the client sends back the user name and the password unencrypted. CHAP is more complex, it uses three interactions to authenticate the user. Both the server and the client maintain a table with chap secrets. First, the server asks the client for chap challenge and

Table 7.

1 octet	1 octet	1 octet	2 octets		2 octets	1 octet
Flag	Address	Control	Protocol	Data	FCS	Flag

gives random number. The client looks up the chap secret and computes the response value based on that and the random number, then send that value back to the server. The server validates the received value based on the chap secret and the random number sent previously and either accepts or rejects the connection based on the result of comparison. PPP callback is initiated by the client, and after a successful authentication, the server terminates the connection, calls back, and re-authenticates the client.

SERVICES IN NGN

Device Configuration, Service Profile

The service related functions of NGN are deployed in the first IP node towards the subscriber that is in the Edge Node, however, when a user device has just been turned on, it is not yet able to access a service. For the efficient management of the IP address domain with regard to multiple devices in home networks, operators prefer the automatic configuration of user devices that includes configuration of the IP addresses of the device and service nodes. In case of automatic configuration, the user device obtains an IP address and learns the address of the Edge Node where its service profile is available. First, the user device must catch a router advertisement ICMP (Internet Control Message Protocol) message or challenge it with a router solicitation ICMP message. Since router advertisement messages are mapped to broadcast messages on the data link layer, to avoid broadcast storms in the WAN/MAN, these are generally disabled, so the user device must use the latter option. The router advertisement message that comes in response comes from one of the edge routers and carries the address of the DHCP (Dynamic Host Configuration Protocol) server and possibly other configuration options. Then DHCP interaction is used to set up the configuration of the device.

DHCP can set configuration parameters of client devices as controlled by system administrators in the network policies (Droms, 1997). The DHCP architecture consists of two peers: a DHCP client and a DHCP server. The DHCP server is a persistent storage of network information; it maintains a table on the clients and their configuration parameters. Clients are usually identified by their hardware (MAC in the case of Ethernet) addresses. During automatic configuration, the client device sends a DHCP discover message to a DHCP server. On each discover message, the server first retrieves the appropriate configuration parameters of the client from the network policy NGN service. Then it assigns an IP address to the physical address, stores these data in its table. Finally, it returns the configuration data to the client in a DHCP offer message. The address assigned to the client is periodically renewed by the server until the client sends a DHCP Release message to initiate the unbinding of its configuration parameters.

The message structure of a DHCP message consists of fourteen fields. The first field is the operation code. The next two fields carry information about the type and the length of the hardware address. The hop field is set when the DHCP configuration does not arrive directly from the server, but comes through a DHCP relay. The xid field is a random number set by the client to identify the transaction. The next four fields, which are ciaddr, yiaddr, giaddr and chaddr, carry the IP addresses of the client when the address is already bound, the client when the address is being assigned, the next server to use and the relay agent respectively. The sname is the host name of the server, and file is the file name of the boot file. The last field is the options field that may contain various DHCP options. (see Table 8)

After the DHCP interaction, the user device is able to contact the edge node, access its service profile, and access the service.

User and Service Separation

In a multi-service multi-provider (MSMP) environment, traffic separation must be implemented not only among customers, but also among different services and even among service providers. In the MSMP environment one customer can access more services from different providers at the same time, so service and provider identification are essential, while traffic belonging to different services and providers must be separated (Juniper Networks, 2009).

User separation can be easily solved with either using PPP over Ethernet or assigning a dedicated VLAN to each customer's residential gateway. However, these two solutions create a pipe between the user device and the edge node making the efficient provisioning of downstream multicast services like IPTV impossible.

The VLAN based traffic separation can have two bases: in the customer VLAN (C-VLAN) model each subscriber can have a dedicated VLAN, in the service VLAN (S-VLAN) model each service can have a dedicated VLAN so multiple users share the same VLAN to access a service. In the first case all traffic must go through the edge router, so all traffic engineering intelligence must be deployed there. Its major weakness is the network utilization as downstream multicast is transferred to each subscriber separately; and the number of available VLAN IDs is a limited resource. The S-VLAN model groups the traffic of different services together, which is very effective for the delivery of IPTV traffic. Its weaknesses are the lack of support for pricing after usage and that services are separated in end-to-end relation, that is, the residential gateway has dedicated a port for each service.

Figure 3 shows how the QinQ tunneling technique can be used in Ethernet access networks to separate both users' and services' traffic by using both the C-VLAN and the S-VLAN tunneling. In the relation between the subscriber on the left hand side of the picture and the edge router on the right hand side, Ethernet packets are tagged with the Customer-VLAN (C-VLAN) header. The access multiplexer aggregates the traffic of the same class from several subscribers and in the relation between the access multiplexer and the edge router the Ethernet packets are tagged with the Service-VLAN (S-VLAN) header.

Alternatively, if neither the PPP nor the VLAN tagging technique is used, security and traffic engineering problems arise. As the access and aggregation networks are Ethernet based, clients on the same subnet, behind the same access node or access multiplexer, may be able to listen to each other's traffic. Another problem is that simple address resolution (ARP) requests are transformed into broadcast messages in the Ethernet domain that floods the network causing so called broadcast storms.

Therefore, an important challenge of access provisioning is the separation of different customer's traffic from each other. This can

Table 8.

8	8	8	8	32	16	16	32	32	32	32	512	1024	*
op	type	hlen	hop	xid	secs	flags	ciaddr	yiaddr	giaddr	chaddr	sname	file	ops

Figure 3. QinQ VLAN tunneling in Ethernet access network for simultaneous user and service separation

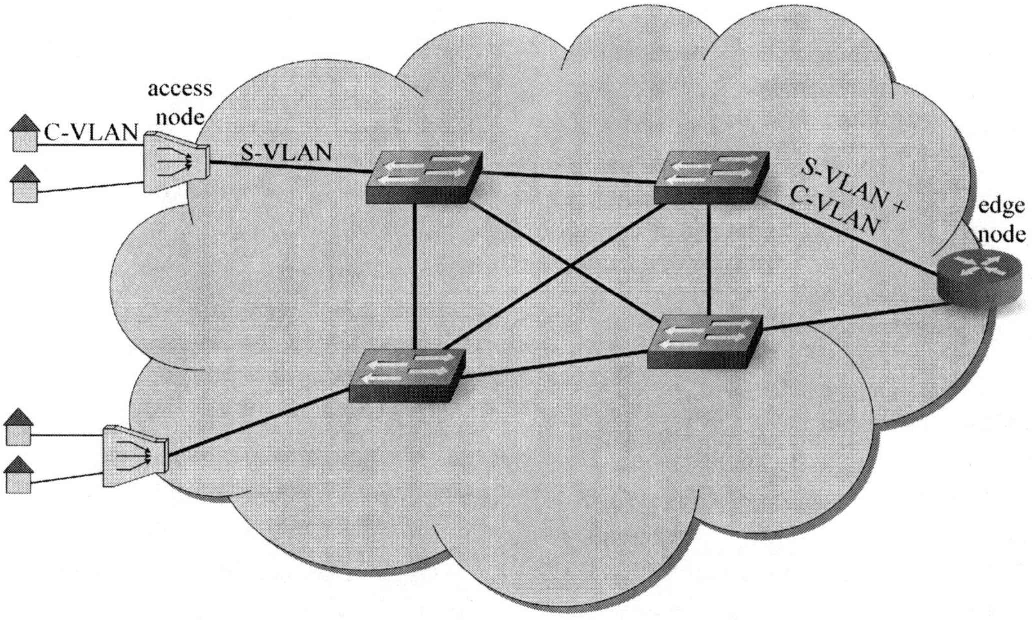

be achieved by forced forwarding customers' upstream broadcast and in-bound management messages towards the access edge node (Melsen & Blake, 2006), so each user is in a point-to-point relation with the edge node. In this technique, only the edge routers' MAC addresses are visible to the subscriber, and all other MAC addresses are disabled as upstream destination address by the first Ethernet bridge after the access line. The upstream broadcast messages are caught, filtered, and handled by replacing them with a unicast request to a server that can perform the address resolution.

Service Session Control in NGN

The Session Initiation Protocol (SIP) is used for initiating, modifying, and terminating interactive multimedia sessions between two or more endpoints (Rosenberg et al., 2002) referred as User Agents. It primarily targeted packet based voice services as VoIP, but its area of application is no longer limited to that field. SIP is textual, human readable client-server protocol, and the addressing employed is a URL scheme. The structure of SIP messages is analogous to HyperText Transfer Protocol (HTTP) messages. The first row of a SIP request message begins with the method that identifies the operation. The most frequently used methods are REGISTER, used for registering the location of a given endpoint, INVITE, used for starting a SIP session, ACK, used for acknowledging a SIP session establishment, CANCEL, used for interrupting the establishment of a SIP session, BYE, used for ending the SIP session and OPTION, used for requesting the available capabilities of a server. The second field in the row is a Uniform Resource Identifier (URI) that identifies the resource to access. The last field of the head row is the SIP version. Just like in the case of HTTP a set of options follow this head line, one option is one line long. Options frequently used are: from, to, call-id, cseq, contact, via. The body is optional, and may carry Session Description Protocol datagrams with information about the current multimedia flow (name, purpose, IP address, ports, and codecs). (see Table 9)

The head line of the SIP response message carries information on the status of the request operation. The first element of this row is the SIP version, then a three-digit status code follows that can mean six different classes of responses INFORMATIONAL (1xx) for providing provisional information, SUCCESSFUL (2xx) for informing successful communications, REDIRECTION (3xx) for informing about redirection, REQUEST FAILURE (4xx) for informing about invalid message format, SERVER FAILURE (5xx) for informing about an intermediate server error, or GLOBAL FAILURE (6xx) for informing about a network error.

The SIP architecture can contain optionally registrars, location services, proxy servers, and redirect servers. Registrar is a user availability service for users for communicating that they are present in the network. The presence and location information is stored by the SIP location service. The SIP proxy server supports user authentication, authorization for services and the routing of requests to the current location of a participant. The proxy is the place where network policies can be implemented. The service proxy acts on behalf of user agents, update and redirect SIP requests to the recipient or to another proxy closer to the recipient.

NGN architectures offering voice and video session based services rely essentially on two network elements: a Session Border Controller (SBC), and a Softswitch. The SBC is usually positioned at the edge between a regional (or access) network and a core network, and supports policy management on the border of an operator's network domain and security, connectivity and QoS. Softswitches are software based network components with the role of a traditional telephone switch that can interconnect PSTN and packet based voice calls. These entities can hence support the migration to NGN. A third element, the Media Gateway (MGW), is required if connection with legacy networks (normally with the PSTN) is desired.

Protocols for Audio and Video Service Provisioning

For end-to-end transfer of real-time unicast or multicast audio (VoIP) and video (IPTV) content over the Internet the Real-time Transport Protocol (RTP) can be used in conjunction with RTP Control Protocol (RTCP). RTP is responsible for the data transport and RTCP provides monitoring and control functions (Schulzrinne et al., 1996). Both RTP and RSTP operate usually over UDP, and the port numbers used are negotiated during the call setup with SIP.

RTP payload is transferred in RTP packets. The RTP header consists of ten fields. It begins with a two-bit version (V) field. The padding (P) bit is set if the data contains padding octets at the end. The extension (X) bit is set if there is an extension header after this header. The CC field contains the number of CSRC identifier records at the end of the header. The marker (M) bit shows if the payload is marked at significant events. The payload type field identifies the format of the content. The sequence number (SN) supports flow control, the timestamp field (TS) provides the basis for synchronization. The SSRC field contains

Table 9.

Method	Request URI	Version	Options	Body

a random identifier that selects synchronization source. CSRC is a list of source identifiers of contributors. (see Table 10)

RTCP has four main services. The first one is to give feedback on the quality of data to support flow and congestion control protocols. The second one is to identify streams by a canonical name, so traffic from or to the same participant can be separated. The third one is the support of session bandwidth management taking the number of participants into consideration. The fourth one is to carry session control information.

The RTCP message header format consists of five fields. The version field identifies the RTCP version. The padding (P) bit is set if the data contains padding octets at the end. The reception report count (RRP) field has the value of the number of report blocks in the data field of this packet. The packet type can be Sender Report, Receiver Report, Source Description, Bye or App. Sender Report packets are sent by each active sender and contain transmission statistics. Receiver Reports are sent by non-active senders and contain reception statistics. The Source Description carries the canonical name value. The Bye message is used for terminating the participation. The App message carries application specific extensions to RTCP. The length field is set to the length of the packet in 32-bit units minus one. (see Table 11)

Authentication, Security

An important question in case of NGN is whether a subscriber is allowed to access a certain service or not. If the access network is Ethernet, the IEEE 802.1X standard (IEEE, 2004b) gives a simple solution: each subscriber is identified by the port of the access multiplexer, which the user's residential gateway is connected to. For this port based network access control, the standard provides a framework of a client-server authentication protocol that restricts unauthorized hosts from connecting to a LAN through public accessible ports. The host (supplicant) communicates with the switch (authenticator) and requests access to the network. The switch will relay the host credentials to an 802.1X authentication server, typically a RADIUS server, before granting or refusing access to the network. The use of switched Ethernet in combination with VLANs, to limit the extent of and to broadcast domains, in home networks will also enhance security.

CONCLUSION AND FUTURE RESEARCH DIRECTIONS

In the first decade of this century the PSTN to NGN migration continues. In the first phase, the two networks are interconnected, in the final phase

Table 10.

2	1	1	4	1	7	16	32	32	Nx32
V	P	X	CC	M	PT	SN	TS	SSRC	CSRC

Table 11.

2	1	5	8	16
Version	P	RRP	Packet Type	Length

NGN takes over both service session control and data delivery functions in the end-to-end relation.

NGN separates service and transport functions. New business roles can appear along this separation. A telecommunication operator can become transport provider or service provider or both. If an operator plays both roles, the same services can be accessed from different types of its access networks: fixed Ethernet or wireless access networks like UMTS, and can therefore provide mobility services beside the triple play services to its subscribers.

Fixed access is going towards the fiber to the home (FTTH) architecture, which gets rid of the bottleneck on the first mile by deploying optical fiber to residential gateways.

This chapter intended to give an overview of NGN architectures and the most important protocols and technologies employed to provide triple play services for broadband subscribers. The chapter focused on fixed access, mobility aspects are considered in the Fix Mobile Convergence chapter.

REFERENCES

Andersson, L., et al. (2007 October). *RFC 5036 LDP specification*. Retrieved from http://www.ietf.org/rfc/rfc5036.txt

Awduche, D., et al. (2001, December). *RFC 3209 RSVP-TE: Extensions to RSVP for LSP tunnels*. Retrieved from http://www.ietf.org/rfc/rfc3209.txt

Cain, B., et al. (2002, October). *RFC 3376 Internet group management protocol*. Retrieved from http://www.ietf.org/rfc/rfc3376.txt

Droms, R. (1997, March). *RFC 2131 dynamic host configuration protocol*. Retrieved from http://www.ietf.org/rfc/rfc2131.txt

Estrin, D., et al. (1998, June). *RFC 2362 protocol independent multicast-sparse mode (PIM-SM): Protocol specification*. Retrieved from http://www.ietf.org/rfc/rfc2362.txt

IEEE. (2004). *IEEE 802.1D local and metropolitan area networks: Media access control (MAC) bridges*. Retrieved from http://standards.ieee.org/getieee802/download/802.1D-2004.pdf

IEEE. (2004). *IEEE 802.1X local and metropolitan area networks: Port-based network access control*. Retrieved from http://standards.ieee.org/getieee802/download/802.1X-2004.pdf

IEEE. (2005). *IEEE 802.1Q local and metropolitan area networks: Virtual bridged local area networks*. Retrieved from http://standards.ieee.org/getieee802/download/802.1Q-2005.pdf

IEEE. (2007). *IEEE 802-2001 local and metropolitan area networks: Overview and architecture*. Retrieved from http://standards.ieee.org/getieee802/download/802-2001.pdf

IEEE. (2008). *IEEE 802.3-2008 local and metropolitan area networks: Carrier sense multiple access with collision detection (CSMA/CD) access method and physical layer specifications*. Retrieved from http://standards.ieee.org/getieee802/download/802.3-2008_section1.pdf

ITU-T. (2004). *Recommendation Y.2001 general overview of NGN*. Retrieved from http://www.itu.int/rec/T-REC-Y.2001/en

ITU-T. (2004). *Recommendation Y.2011 general principles and general reference model for next generation networks*. Retrieved from http://www.itu.int/rec/T-REC-Y.2011/en

Juniper Networks. (2009). *VLAN design for IPTV/multiplay networks*. Retrieved from www.juniper.net/us/en/local/pdf/whitepapers/2000186-en.pdf

Melsen, T., & Blake, S. (2006, June). *RFC 4562 MAC-forced forwarding: A method for subscriber separation on an Ethernet access network*. Retrieved from http://www.ietf.org/rfc/rfc4562.txt

Metro Ethernet Forum. (2006). *Metro Ethernet services – a technical overview*. Retrieved from http://metroethernetforum.org/PDF_Documents/metro-ethernet-services.pdf

Rosen, E., Viswanathan, A., & Callon, R. (2001). *RFC 3031 MPLS architecture*. Retrieved from http://www.ietf.org/rfc/rfc3031.txt

Rosenberg, J., et al. (2002, June). *RFC 3261 SIP: Session initiation protocol*. Retrieved from http://www.ietf.org/rfc/rfc3261.txt

Schulzrinne, H., et al. (1996, January). *RFC 1889 RTP: A transport protocol for real-time applications*. Retrieved from http://www.ietf.org/rfc/rfc1889.txt

Simpson, W. (1994, July). *RFC 1661 the point-to-point protocol*. Retrieved from http://www.ietf.org/rfc/rfc1661.txt

KEY TERMS AND DEFINITIONS

Access Network: The access network is an Ethernet network that interconnects access lines and IP Edge Nodes where most of the service functions reside.

Best-Effort Data Service: The transmission medium does its best to deliver data traffic from one node to another in a network; the data may be delayed or even lost. The Internet Protocol provides such best-effort connectivity.

First (Last) Mile: The access line between the aggregation network and the residential gateway.

IPTV: Internet Protocol based digital television service delivery over packet switched networks. The IPTV service can be live broadcast or Video-on-Demand (VoD). This service is sensitive for jitter (delay variance), but less sensitive for packet loss.

Next Generation Networks: A telecommunication network architecture that is able to deliver triple-play services to end users.

Triple-Play: Television, voice, and data transmission services provided for a user by a telecommunication operator.

Voice over Internet Protocol (VoIP): Voice traffic is encapsulated in Internet Protocol packets and delivered over packet switched networks. This service is sensitive for jitter (delay variance), but less sensitive for packet loss.

APPENDIX

Table of abbreviations

ADSL	Asymmetric Digital Subscriber Line
ARP	Address Resolution Protocol
ASP	Application Service Provide
ATM	Asynchronous Transfer Mode
BRAS	Broadband Remote Access Server
CBS	Committed Burst Size
CHAP	Challenge Handshake Protocol
CIR	Committed Information Rate
C-VLAN	Customer VLAN
CSMA/CD	Carrier Sense Multiple Access Collision Detection
DM	Dense Mode
EIR	Excess Information Rate
FEC	Forwarding Equivalence Class
FTTH	Fiber to the Home
GGSN	Gateway GRPS Support Node
GSM	Global System for Mobile communications
HDTV	High-Definition Television
HTTP	HyperText Transfer Protocol
IGMP	Internet Group Management Protocol
IMS	IP Multimedia Subsystem
IP	Internet Protocol
ITU-T	International Telecommunication Union Telecommunication Standardization Sector
LCP	Link Control Protocol
LDP	Label Distribution Protocol
LLC	Logical Link Control
LQM	Link Quality Monitoring
LSP	Label Switch Path
LSR	Label Switch Router
MAC	Media Access Control
MAN	Metropolitan Area Network
MGW	Media Gateway
MPLS	Multiprotocol Label Switching
MSMP	Multi-Service Multi-Provider
NCP	Network Control Protocol
NGN	Next Generation Network

OSI	Open Systems Interconnection
PAP	Password Authentication Protocol
PIM	Protocol Independent Multicast
PLMN	Public Land Mobile Network
PON	Passive Optical Networks
PPP	Point-to-Point Protocol
PPPoA	PPP over ATM
PPPoE	PPP over Ethernet
PSTN	Public Switched Telephone Network
QoS	Quality of Service
RP	Rendezvous Points
RPF	Reverse Path Forwarding
RRP	Reception Report Count
RSVP	Resource Reservation Protocol
RSVP-TE	Resource Reservation Protocol Traffic Engineering
RTCP	RTP Control Protocol
RTP	Real-time Transport Protocol
SBC	Session Border Controller
SDTV	Standard-Definition Television
SIP	Session Initiation Protocol
SM	Sparse Mode
SSM	Source Specific Multicast
STP	Spanning Tree Protocols
S-VLAN	Service VLAN
TCP	Transmission Control Protocol
UDP	User Datagram Protocol
UMTS	Universal Mobile Telecommunications System
UNI	User Network Interface
URI	Uniform Resource Identifier
VLAN	Virtual Local Area Network
VoD	Video-on-Demand
WAN	Wide Area Network
WDM	Wavelength Division Multiplexing

Chapter 11
Convergence of Fixed and Mobile Networks

Gábor Kovács
Budapest University of Technology and Economics, Hungary

Gábor Árpád Németh
Budapest University of Technology and Economics, Hungary

Zoltán Pap
Ericsson, Hungary

ABSTRACT

This chapter gives overview of one of a recent trend in telecommunication convergence: fixed-mobile convergence (FMC). Its types, implications, and main challenges are identified and analyzed with special respect to the protocols employed. The future integration of fixed and mobile access to next generation networks (NGN) is investigated from several aspects, such as AAA (Authentication, Authorization, and Accounting) protocols, network and application layer mobility management solutions, and policy control. The focus is on access and core networks where the convergence is going to take place. At the end, a short outlook is given to near future mobile trends.

INTRODUCTION

Trends of the first decade of the twenty-first century point in the direction of a world of portable and personalized services. The subscribers would like to access their subscribed services from any location at any time and expect similar interfaces, similar connection quality, similar reliability and high level of security regardless of access type (mobile or fixed) they use. When subscribers can be served via multiple access technologies, for instance while on the move, they expect the network to find the best possible access type. This leads to a single user device that can access the same set of services everywhere.

The fixed-mobile convergence (FMC) has benefits for both fixed and mobile operators. Providing triple-play services to residential users has become reality by the first decade of this century. The availability of services of a fixed operator is however constrained to the subscription point; services are tied to the household. To be able to

DOI: 10.4018/978-1-60960-732-6.ch011

stop losing subscribers to mobile operators fixed operators seek quadruple play, which integrates mobility into the triple play services. Mobile operators on the other hand target potential sources of growth; they seek the cost efficient provisioning of the service set familiar for subscribers from home.

When designing the architecture of the converged network several fundamental requirements must be taken into account. The key requirement is the consistent user experience that must be provided independently from the access type; a uniform service set with the same quality measures must be provided to a subscriber even while having an active service session. Therefore, the transfer of service sessions between access types and between operators must be investigated. When the subscriber is on the move, it must be checked if the subscriber has sufficient privileges to connect to an access network and it must be made sure that the network is able to meet the QoS (Quality of Service) requirements of the user's contract. Attaching to a new network connection point triggers authentication, authorization and accounting (AAA) processes where the sensitive personal data exchanged in form of the user profile must be handled with high level of security.

The chapter gives an overview of the main protocols that play role in FMC and is organized as follows. The 'Background' section summarizes FMC and its main challenges for NGN operators. In the next section, first the most important terms of FMC are defined; then the impact of FMC on the NGN architecture is discussed. The section titled 'Mobility support, nomadism, session continuity, handover' reviews network and application layer mobility management solutions that can be considered to provide nomadic services and continuous mobility. The section on policy control takes a look at the management plane functions of converged NGNs. The 'Roaming' section raises traffic routing problems and authentication issues to be solved for subscribers away from their home domain. In the section 'AAA and security', protocols and techniques that can be employed in a converged NGN are introduced. The '3GPP Long-Term Evolution' section gives an outlook on the evolution of mobile telecommunications for the next few years. Finally, future trends and open problems are summarized.

BACKGROUND

The FMC has four driving forces: the service based interconnection of networks, the transport network convergence, the service plane convergence, and the terminal convergence. The service-based interconnection has begun to become possible with the appearance of mobile telecommunication networks. The service provided is the same: it is possible make a phone call from fixed network to mobile network and vice versa. However, this integration has not yet been achieved for all services, for instance video calls. The first real step towards FMC is the network based (transport plane) convergence, where a common IP based network infrastructure is used from different access technologies. The user centric point of view, that is the service plane convergence, means that the user can access any subscribed service from different terminals and the service is adapted to the capabilities of that terminal. The terminal based convergence is supported by high-tech terminals that can use several access technologies. The ultimate convergence, when all these four conditions are met, is an IMS (IP Multimedia Subsystem – see Chapter XIII) based NGN with hybrid terminals.

From the telecommunication operators' point of view, simultaneous fixed and mobile access leads to a competitive advantage. The mobile operator may become a fixed virtual operator. The fixed operator may become a mobile virtual operator and extends its fixed network with different types of wireless access technologies. As result integrated operators emerge that own a fixed broadband NGN infrastructure and IP service platform of high utilization and offer a large scale of access technologies as xDSL, GSM (Global System for Mobile communications, UMTS (Universal Mobile Telecommunications System), WLAN (Wireless Local Area Network), WiMax. However, this convergence implies several challenges as well:

- Users would like to access subscribed services from any location. This leads to the personalization of network services, the contracts are no longer going to be made between operators and households, but between operators and individuals. Hence, the personalized subscription management must be located in a network function shared by fixed and mobile networks. The connection point and therefore the access type used may change from connection to connection or even during an active service session, so operators must ensure service delivery for any access method.
- Single subscription for home, enterprise, and mobile access requires user identities and profiles to be shared for all operators.
- Technological questions regarding mobility between different types of accesses must be resolved. The mobility can be nomadic, or if continuous, the issues of session continuity must be addressed as well.
- Heterogeneous network and service access results in additional requirements to the existing models of the AAA solutions. These requirements can be quite complicated depending on the scenario of how the service is being accessed.

Just like in the case of NGN, the development of standards is being done in multiple organizations: 3GPP (3rd Generation Partnership Program), ATIS (Alliance for Telecommunications Industry Solutions), Broadband Forum, IEEE (Institute of Electrical and Electronics Engineers), ETSI (European Telecommunications Standards Institute), IETF (Internet Engineering Task Force), ITU (International Telecommunications Union). Each of these standardization bodies have technical or regional focus area and have the main telecommunication operators and vendors among their members. The research efforts of these organizations sometimes point in diverse directions; the coordination is resolved by cross-contribution of individual members. 3GPP develops technical reports on mobile access networks and IMS based session control architecture, which allows IP based sessions to be established between heterogeneous wireless access networks. Broadband Forum and the TISPAN (Telecommunications and Internet converged Services and Protocols for Advanced Networking) project of ETSI approach FMC from the perspective of fixed operators, and develop architectures that extend IMS to fixed networks. The standardization of NGN is taking place in several work groups of ITU-T. Most of the Internet standards used come from IETF.

OVERVIEW AND ARCHITECTURE OF FIXED-MOBILE CONVERGENCE

Types of Fixed-Mobile Convergence

Four types of convergence can be distinguished when talking about FMC: device convergence, service convergence, network convergence and user data convergence.

Device convergence means that user terminals are able to connect to different types of access networks even simultaneously. Most mobile devices have not only GSM/UMTS interfaces, but are able to get an IP address via the Wi-Fi or Bluetooth interface as well. No mobile phones are sold nowadays without a built-in camera and a display with high enough resolution for SDTV (Standard Definition TeleVision), so these devices are able to utilize the extended audio, video and data service set. Notebooks and netbooks are equipped beside the traditional fixed Ethernet interface with a Wi-Fi interface for wireless access.

The convergence of services allows the subscriber to have the same experience when accessing a service with irrespectable of the current access type, and to see the same set of services are available on different types of terminals.

Network convergence merges the core networks of current fixed and mobile networks into a single IP based NGN core where session management is handled by IMS. This core network can be accessed from various wired and wireless technologies.

Users expect to have a single subscription; however, their personal data are stored and maintained in several operator domains and even in several entities of the same operator domain, which makes the coordination of services difficult. Therefore, converged user data are to be stored in a single repository.

Types of Mobility

Before going into the details of FMC, a small set of definitions is given to help the understanding of challenges and have a common language. The definitions include nomadism, different types of mobility, roaming, session continuity, and handover.

Nomadism is according to the definition of ETSI TISPAN (ETSI, 2006) the "ability of the user to change his network access point on moving; when changing the network access point, the user's service session is completely stopped and then started again, i.e. there is no session continuity or handover possible." This means that nomadism is a discrete mobility that allows users with fixed broadband subscription to establish connectivity and use their subscribed services in different access points even when away from home. The meaning discrete is that a service session does not survive the change of network connection points. Therefore, the first step of FMC from fixed operators' aspect is the support for nomadism.

Mobility can be regarded as the extension of nomadism with the maintenance of established sessions while the user relocates or is on the move. There are five mobility types that must be taken into account are:

- personal: the subscriber changes the terminal used for network access at different locations. The identified user can access the services in the user's service profile at any terminal and the network can support the provisioning.
- terminal: the terminal can access the telecommunication services at any location and the network is able to locate the terminal
- service: the user can access subscribed services at any network attachment point and any terminal
- session: the subscriber's active sessions are maintained while changing terminals in a Personal Area Network
- network: the point of attachment is on move itself, for instance the network in a train, ship or plane, or a mobile ad hoc network

It is possible for the user to leave the service area of his NGN service provider by means of terminal or user mobility. This situation is referred as roaming, which is defined by ETSI TISPAN (ETSI, 2006) as "the ability of the users to access services according their user profile while moving outside of their subscribed home network, i.e. by using an access point of a visited network. This requires the ability of the user to get access in the visited network, the existence of an interface between home network and visited network, as well as a roaming agreement between the respective network operators." Roaming has major implications on AAA and policies: The user's access rights and service profile including service parameters have to be pre-negotiated between the home and visited operators, and accounting information must be exchanged after the roaming session is terminated.

For streaming services as voice call, the maintenance of service sessions has special importance; the session must survive the change of access point or even type. Session continuity is defined in ETSI TISPAN (ETSI, 2006) as "the ability of a user or terminal to change the network access point while maintaining the ongoing session. This may include a session break and resume, or a certain degree of service interruption or loss of data while changing to the new access point."

FMC Evolution

A strategic goal of the telecommunication industry from the second half of the 1990's is to establish an all-IP network referred as NGN. On the way to achieving this goal, three FMC scenarios can be distinguished: convergence on enterprise level, Unlicensed Mobile Access and IMS based solutions.

Enterprise level solutions represent the first scenario of the FMC. The convergence is limited to a private enterprise network; hence, the network infrastructure of operators remains unchanged. Mobility, which means handover between a mobile phone and the desktop phone in this case, is managed within the enterprise network by a Private Branch Exchange (PBX) extension.

The UMA (Unlicensed Mobile Access) technology makes use of dual-mode user devices, and is able to tunnel GSM or GRPS data to the mobile core over an IP based network accessed via the Wi-Fi or Bluetooth interface of the device. On the side of the mobile network, a UMA Network Controller (UMC) processes the encapsulated voice and data packets and forwards them to the mobile core network. UMA has been included in the Generic Access Network specification of 3GPP. The wireless technology used as transport on the air interface is transparent for the UMA architecture, which is able to provide roaming and seamless mobility between IP based cellular networks. From the perspective of operators, the weakness of UMA is the lack of presence of an operator controlled residential gateway.

The third and ultimate FMC solution is the IMS controlled all-IP network discussed in the next section. IMS has been developed by 3GPP primarily for the GSM to 3G transition; however, it has been adopted for fixed networks as well in ETSI TISPAN. The main advantages of employing IMS in FMC architecture are the efficient session control, service management, QoS management and IP based charging. Its weaknesses are the not yet clear business cases and the high cost of deployment, and whether its net value is worth that.

FMC Architecture

Figure 1 shows a high-level view on an FMC-aware NGN architecture (ITU-T, 2004) from the perspective of transport network convergence. Users can access their service profile from multiple drop technologies. Network functions as policy control, AAA and partly mobility support that are common for any access technology, which can be both fixed or wireless, can be organized into a layer. The common set of service functions are implemented in the service layer that can be IMS (IP Multimedia Subsystem) based.

Fixed and mobile networks differ in the access network; the core network is common in NGN. The network-based convergence imposes changes in particular the frameworks for AAA, IP address configuration, and policy control of the pure fixed NGN. FMC also introduces two completely new areas to the fixed network domain: mobility management and roaming. A subscriber has a service profile by a fixed or mobile service provider and wants to access services independently from the location and terminal used. An NGN that supports FMC must therefore meet the following requirements (Melander et al., 2007):

- Not only mobile-to-mobile handover must be supported, but also fixed-to-fixed, fixed-to-mobile, mobile-to-fixed handover. The

Figure 1. High level view of a converged NGN architecture

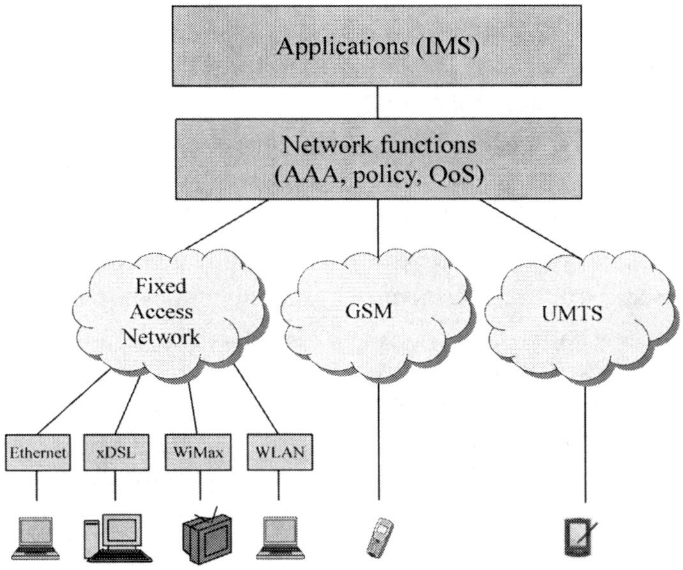

handover can take place within the same type of access network or it can be an inter access system handover.

- To provide session continuity the localization of the user must be up to date, and the authentication of users between fixed and wireless networks must be carried out with the same way. There must be user terminals that are able to manage several connectivity networks.
- Not only mobile, but also fixed access network must keep track the subscribers' location. Roaming makes it possible for the subscriber to access (a subset of) his own services when away from the home network. Roaming is defined by agreements between fixed and fixed, fixed and mobile, and mobile and mobile operators. The most important aspects of roaming are the AAA and addressing.
- The network must have its own policy management for users, services, and networks. For instance, when an HDTV (High Definition Television) stream is transferred from a television set to a mobile phone its quality has to be degraded as determined by the network policies.

MOBILITY SUPPORT, NOMADISM, SESSION CONTINUITY, HANDOVER

For streaming and interactive services, the change of network attachment point is required to be unnoticeable to the end user. However, the re-authentication process and the session transfer is time consuming.

The continuity of the session can be implemented in three ways. Network layer mechanisms such as the different versions of Mobile IP can be used to hide the network attachment point from the application layer. Alternatively, application layer signaling by means of SIP (Session Initiation Protocol) as used in VoIP and IMS can allow the redirection of media flows. The third alternative is to use an oracle, a so-called mobility manager, in the access networks to support network-triggered handover. This entity keeps track of terminals and resources in the network, and instructs the terminal when the handover shall be carried out.

SIP Based Session Continuity

In NGN SIP is the main control plane protocol for voice and video services. SIP offers two methods for session continuity: re-INVITE and REFER. SIP mobility has two major weaknesses. One is its performance, neither of these solutions have the performance to support real-time services. The second weakness is the lack of privacy, the party who is not on the move receives information about the movement of the moving party.

A possible solution to support SIP based session continuity is the deployment of Session Border Controllers (SBC) on the border of regional and core networks. The role of SBC is to support QoS, policy management and security. In the case of pure fixed NGNs this network element is already present, and its functionality has to be extended with the implementation of the SIP B2BUA (Business to Business User Agent) function and RTP proxy functions. This way the SBC can hide the movement of a terminal within the same operator domain from the rest of the network. If an authenticated terminal is on move, it sends a re-INVITE, which is intercepted by the SBC. From this point on, two SIP sessions exist between the SBC and the terminal until the authentication process for the new access is completed.

Mobile IP Based Session Continuity

Mobile IP is a family of network layer protocols that make it possible for a Mobile Node to migrate from its home network to a visited network without changing its visible IP address. The routing mechanism used is independent from the current foreign IP address of the Mobile Node. The visited network can be outside of the administrative domain of the home network, in the context of NGN this means a different WAN/MAN.

From the point of view of the converged NGN architecture, four mobility cases that require the change of attachment point (handover) can be considered. When the handover takes place between two access points within the same access network, it may be transparent to network or higher layers. Such handover is when the Mobile Node attaches to a new wireless access point in the same network domain (left movement in Figure 2). When the handover takes place between different access points in different operator's core networks, it affects the network layer as well, and the application layer might also demand re-authentication and the re-negotiation of QoS parameters. A third version of handover is when all network connections must be broken because of the change of access type (middle movement in Figure 2). This happens when the Mobile node migrates from a wireless access point to a fixed access point, or moves between fixed access points connected to the same operator's core network. If the user migrates between different operator's access networks (right movement in Figure 2), application layer support is necessary.

In Mobile IP, the handover latency that influences the quality of real-time services is composed of several smaller interaction times: router discovery in the new visited domain, address configuration, authentication and authorization and address binding. Local proxy agents can decrease this time requirement by keeping the old connection alive until the new connection is established. Generally, Mobile IP offers smaller disruption time, and hence better performance than SIP, nevertheless the delay is still in the same order of magnitude (Kwon et al, 2002).

The two MIP variants are considered to overcome the high latency of inter-provider handover are PMIPv6 (Proxy MIP) and DSMIP (Dual-Stack MIP). PMIP employed in 3GPP architectures is a network mobility based solution, the mobile node itself does not take part in the signaling of mobility management. Mobility is hidden from the mobile node by a proxy agent called Mobile Access Gateway (MAG). MAG is deployed in the access router of an access domain. An agent called Local Mobility Anchor (LMA) plays the role of the Home Agent and defines a PMIP domain over

Figure 2. Handover scenarios in FMC

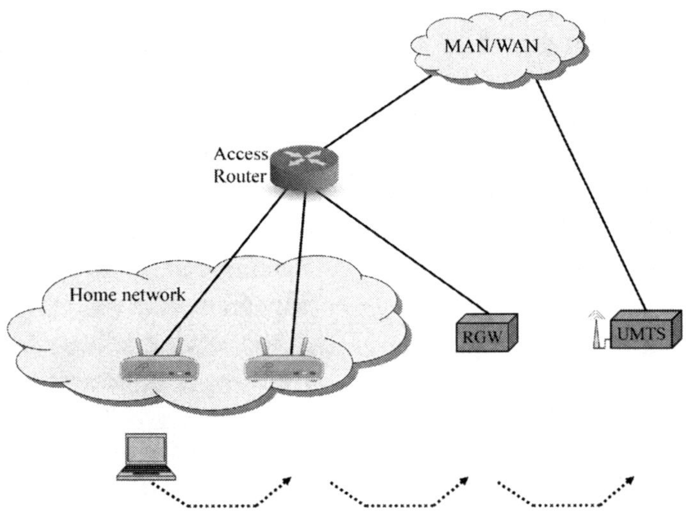

several access domains. An IP tunnel connects the current MAG to the LMA. As on a move of the user only the MAG associated with the Mobile Node changes, the network is able to ensure that the authenticated Mobile Node is always on its home network, and the home network prefix follows the Mobile Node.

As IPv4 is generally used in today's access networks, there must be a transition phase to IPv6, which is supported by Dual Stack Mobile IPv6 (DSMIP). DSMIP extends the binding management of MIPv6 variants such that it can create bindings for both MIPv4 and MIPv6 home addresses. When updating a binding, DSMIP establishes two tunnels to the Home Agent depending on the type of the visited network. For instance in an IPv6 domain an IPv4 over IPv6 tunnel and an IPv6 over IPv6 tunnel are created, and the communication takes place with the tunneled packet.

Mobility Manager

The driving force behind the mobility manager solution is that both Mobile IP and SIP mobility management solution have a disruption time too large for seamless inter-domain handover in the case of conversational services (3GPP, 2006a; 3GPP, 2006b). The mobility manager manages and stores information about user devices: identities, mobility state and security parameters. The handover decision is made within the mobility manager based on these data, and the user device is re-authenticated and triggered to perform the handover.

The idea is that the each type of access network is able to manage the mobility of users. Existing mobile network (GSM, GPRS, UMTS) do already have this support, in FMC this has to be extended to fixed networks. The terminal must be able to manage a double connectivity network one for passive and the other one for active connection: During the network authentication (of the passive connection), the active session must be able to receive data. Therefore, the user terminal must be able to manage several connectivity networks and the continuity of session, and it must also be able to manage several IP addresses.

The operation of this solution is the following. When a user device connects to an access network, it is authenticated and registered at the mobility manager. The terminal continuously informs the mobility manager about the network quality for instance the measured signal strengths. At the same time, the access network provides information to the mobility about the available resources. The mobility manager triggers a terminal to change access type if the connection is degraded. The decision is made on the service profile of the user, the measured signal qualities, and the resources available in other access networks. After connecting to the new access network, the user device registers itself at the mobility manager of the new access and transfers its active connections. In the meantime, the old active connection is closed.

ROAMING

In NGN as the service and transport functions are separated, it is more difficult to support roaming as in pure mobile networks. The most important task is to set up a data connection from the visited network to the service's source nodes. The traffic can be routed directly from the visited network to the service source or tunneled through the home operators service nodes as in the case of Mobile IP.

When roaming, there are multiple interactions between the user or terminal and the authentication subsystem, which are typically done in a single step in fixed networks in non-roaming scenarios. The most straightforward is the authentication, authorization and accounting for the use of resources in the visited network domain. This procedure includes the retrieval of the customer's service profile from the home network to the visited network. A second authentication interaction takes place before the user tries to access a network service while away and must be authenticated to receive an IP address. After the first two authentication processes the subscriber is authorized to access his services. The third interaction is required for the authorization of the subscriber to use a specific service. These first two types of authentication are already distinguished in 3GPP and WiMAX networks. Therefore, in the long term, this separation must be made in fixed networks as well.

After authentication has taken place, services like Internet connectivity can be used immediately, require no further actions to be taken. On the other hand, IMS services require SIP signaling to set up the service session. The standard scenario in IMS is that signaling goes through the P-CSCF (Proxy Call and Session Control Function) in the visited network to the S-CSCF (Serving Call and Session Control Function) in the home network. Thus, the service is controlled from the home network. A roaming relation is required between the service networks of the operators to document the details of the service delivery to roaming users and the signaling arrangements.

Location based services and emergency services are exceptions in roaming and must be handled separately from regular services like the voice service. These be controlled from the visited network, and therefore no roaming agreement is required. Emergency services are provided by local E-CSCFs (Emergency Call and Session Control Functions) that have the ability to interpret the location information obtained from a local CLF (Connectivity Session Location and Repository Function) and route the call to the correct Public Safety Answering Point. The home E-CSCF cannot interpret the location information from the visited CLF and vice versa.

AAA AND SECURITY

This section gives an overview of authentication, authorization and security protocols, and solutions used in a converged NGN.

AAA is an important part of network security architecture. It provides functions to make sure that network resources are used only by authorized end-users. In fixed NGN access networks, customer subscription is tied to the line connecting the user network to the access network and there is no explicit end-user authentication (Mickelsson et al., 2006). In FMC scenarios, the subscriber may use the network resources of the visited operator's domain, so before granting access the two operators must negotiate if the subscriber has the sufficient rights.

Radius

The RADIUS (Remote Authentication Dial In User Service) protocol operates in centralized client-server architecture; the messages are encapsulated in UDP packets (Rigney et al., 1997). The client is normally a Network Access Server responsible for passing user information to designated RADIUS servers, and then acting on the response returned. This usually takes place when the user requests the establishment of a Point-to-Point Protocol (PPP) session. RADIUS servers receive an Access Request, authenticate the user, and on successful authentication return configuration information necessary for the client to deliver service to the user in an Access Accept message. The configuration information can contain the IP address of the user, the validity period of the authenticated session, QoS, and VLAN parameters. On unsuccessful authentication Access Reject message is replied, and if further information like a password from the user is required, Access Challenge is returned. The header of RADIUS messages contains four fields. The code is the type of message, and the identifier is used for distinguishing request and reply messages. The length field gives the length of the message including the header. The authenticator field is used for authenticating the reply from the server, and hiding passwords. (see Table 1)

Diamater

Diameter (Calhoun et al., 2003) is another AAA protocol that extends RADIUS with support for IP mobility, security, and roaming. The Diameter protocol transfers attribute value pairs over SCTP or TCP in a negotiation between a Diameter Client and a Diameter Agent; the message can be forwarded by Diameter Relay Agents redirected in Diameter Redirect Agents and proxied in Proxy Agents. The diameter header format consists of seven fields. Diameter Translation Agent can convert between two protocols as Diameter and RADIUS. The version field identifies the Diameter protocol version used, and it is always set to 1. The length is the total length of the message including the header. The command flags field contains bits to signal if the message is request or response, whether it can be proxied, whether it is an error message, and whether it is a retransmitted message. The command code field is the type of the message. The application identifier field is used to identify the application the message is applicable for. The Hop-By-Hop identifier is a unique number from the point of view of the sender to aid the pairing of request and response messages. The End-To-End identifier is a unique number from the point of view of the sender to help the detection of duplicate messages. (see Table 2)

After this header, a set of attribute value pairs follow. The AVP code and the optional Vendor identifier identify the value uniquely. Three AVP flags are defined: the V flag defines whether the vendor identifier must be used; the M flag defines if the attribute is mandatory, and the P flag signals the need for end-to-end encryption. The length field is the length of this attribute value pair in octets. (see Table 3)

Radius and diameter are used not only for user authentication, but also in roaming between two operators to allow visiting users to be authenticated at the home authentication server.

Table 1.

8 bits	8 bits	16 bits	128 bits
Code	Identifier	Length	Authenticator

IPSec

IPSec (Kent & Seo, 2005) is a collection of protocols that secure IP communication by authenticating and/or encrypting every IP packet in a data stream exchanged by two network nodes. IPSec provides access control and uses the IP Authentication Header to authenticate the source, and the Encapsulating Security Payload to authenticate the source and provide confidentiality.

In a roaming scenario when the Mobile Station visits an untrusted access domain, it must establish an IPSec tunnel to its Home Agent and Foreign Agent to be able to carry out a successful authentication process.

Extensible Authentication Protocol (EAP)

In a converged NGN, several network elements take part in the authentication process. In fixed network, the first participating device is the access multiplexer, and in wireless LANs, the access points can play this role. These devices may not need to perform the authentication, but have to detect and forward the authentication messages.

The EAP is an authentication framework (Aboba et al., 2004) supporting multiple authentication methods. EAP typically runs directly over data link layers, e.g. PPP or IEEE 802.1X, without requiring IP. Typically, there are three entities involved in EAP signaling: the authenticator, which is the end of the link initiating EAP authentication, the peer, which is the end of the link that responds to the authenticator, and the backend authentication server, which provides authentication service to the authenticator.

EAP operates in a multiplexing model with four components. The lowest layer is responsible for the transport between the peer and the authenticator. The EAP layer that operated over the transport layer provides functions for detecting inserted messages and retransmission. The peer and authenticator layer provides authenticator or peer functions for the network element. The methods layer implements authentication algorithms.

The EAP packet format has the following structure. The packet begins with a code field that identifies the type of the EAP packet, which can be request, response, success, or failure. The identifier is a number to aid the pairing of request and response messages. The length field indicates the length of the packet in octets. The format of the data field depends on the code field. If the code is either request or response, then the first octet of the data field is the type of request or response specified by the EAP version used. Success and failure packets do not have a data field. (see Table 4)

For fixed access lines the IEEE 802.1x Standard defines a port-based network access control which makes use of the physical access characteristics

Table 2.

8 bits	24 bits	8 bits	24 bits	32 bits	32 bits	32 bits
Version	Length	Flags	Code	App-ID	HBH-ID	E2E-ID

Table 3.

32 bits	8 bits	24 bits	32 bits	
AVP code	AVP flags	AVP length	Vendor-Id	Data

of IEEE 802 Local Area Networks (LAN) infrastructures. A port of the access switch can be used to uniquely identify the devices attached to it because of the point-to-point first mile connection. (see Figure 3)

CAPWAP

The Control And Provisioning of Wireless Access Points (CAPWAP) protocol is limited for the IEEE 802.11 WiFi access case (Calhoun et al., 2009). This authentication method to be applicable assumes the WiFi access point, that is, the Wireless Termination Point (WTP), resides in the residential gateway, and the Access Controller (AC) function is implemented in the access node. The authentication takes place with an interaction between the AC and the WTP using the CAPWAP protocol.

The CAPWAP protocol establishes a wireless binding on the data plane between ACs and WTPs. The first phase of the operation is the discovery phase where the WTP searches for ACs. Then the WTP selects an AC and establishes a Datagram Transport Layer Security (DTLS) session in that relation. The establishment of the DTLS session begins with the setting of a cookie on the WTP, and then the WTP authorizes itself. The next phase is the exchange of configuration information that updates the configuration of the WTP. After this connection setup, all packets encapsulated within CAPWAP are transmitted in an authenticated and secured relation.

PANA

PANA (Protocol for carrying Authentication for Network Access) (RFC5191) is a link-layer agnostic network access authentication protocol that runs between a client that wants to gain access to the network and a server on the network side. It uses IP to carry different messages of EAP authentication methods between the client and the server.

PANA is low layer for EAP and uses IP to carry "EAP authentication methods" between a client node and a server in the access network, which is the main difference with IEEE 802.1x that relays on IEEE 802.3 link layer to carry the EAP authentication. Besides, PANA also differs from IEEE 802.1x in the access control mecha-

Table 4.

8 bits	8 bits	16 bits	
Code	Identifier	Length	Data

Figure 3. EAP over LAN authentication

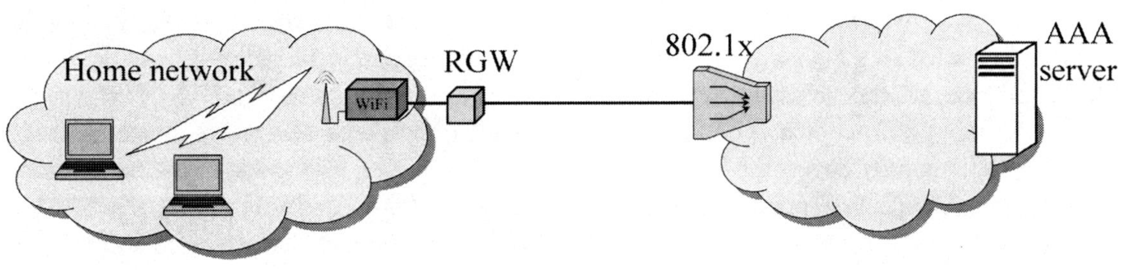

nism definition. In addition, while IEEE 802.1x provides network access control based on ports, PANA needs to interact with other protocols to provide the network access control.

In PANA, the access control takes place in the Enforcement Point (EP), which can be the access node in the case of a converged NGN. This entity is responsible to allow access to authorized clients while preventing access by others. An EP learns the attributes of the authorized clients from the PANA Authentication Agent, and uses both non-cryptographic and cryptographic filters to selectively allow and discard data packets. These filters may be applied either to the link layer or to the IP-layer.

Being IP based, the PANA client (PaC) needs an IP address to be configured prior to the authentication (done either statically or dynamically). After the successful PANA authentication, and depending on the deployment scenario, the PaC may need to reconfigure its IP address or configure additional IP addresses. The IP address configuration of visitors' Mobile Station is either performed as part of the authentication procedure (the PANA solution) or following directly after (the CAPWAP and 802.1AE/af solutions). The Mobile Station of the visitor requests an IP address using DHCP or as part of Mobile IP signaling. The problem of forwarding the DHCP messages to the correct DHCP server in the appropriate operator's domain can be solved by the RADIUS server in the operator's core network by including the IP address of the DHCP server in the RADIUS Access-Accept.

POLICY CONTROL AND CHARGING

Policy can be regarded as a management action done in a certain network state in the management plane, while the policy enforcement is done in the control plane. Policy control covers the definition of policy rules for network elements applied when some policy conditions are met, and the enforcement of these rules at the network.

Policy has important role in providing QoS, charging and roaming. Policies of a network domain can be into four classes. Two classes are implied by the NGN architecture, two other classes are implied by the contracts of the network operator. Network and service policies reflect the network and service functions of NGN. Network policies include QoS, call admission control, resource management, and internal signaling policies. Service policies define requirements on user devices to be able to access a service. User policies are derived from subscriber contracts describe and identify the user and personalized configuration parameters for services, and business policies are related to roaming and service level agreements (SLA).

Figure 4 shows a possible architecture for policy control in a converged NGN. The user device is in connection with the network entity called Access Node (AN). The Access Edge Site (AES) is responsible for retrieving the user information from the user repository of the Identity Provider, and enforce roaming agreements and service policies. Network policies are enforced in the resource manager, which reserves the resources necessary for serving the user in the access network. The four policy manager entities (Business Policy Manager (BPM), Service Policy Manager (SPM), Network Policy Manager (NPM), and User Policy Manager (UPM)) collaborate to provide policy control functions.

Business policies affect service policies, the service level agreement between two operators determine the available service quality available for a roaming user. Network policies map service polices on the network based on the current network status. In an FMC network, two functional entities take part in the policy execution. A service controller entity (SPDF in ETSI TISPAN architecture) compares service requests against the service policies, and sends a request to the other entity, a resource controller (RACF in ETSI TISPAN architecture). The resource controller performs the validation of the request against the user policies and network policies, and makes a decision based on the current network status about the provisioning of the service regarding the requested QoS parameters.

In a nomadic or roaming situation, inter-domain policies can be controlled two ways. The policies can be managed and executed in the visited network domain. In this case, when the roaming user is detected, policy control information is exchanged between the home and the visited network domains. Alternatively, each decision can be directed back to the home network, where the policy decision can be made. The other three policy types, i.e. user, network and service policies, are managed on the elements the visited access network.

The main protocols involved in policy control frameworks are the RADIUS-CoA method, DIAMETER, GTP' (GPRS Tunneling Protocol), SOAP (Simple Object Access Protocol) and IPSec. The CoA (Change of Authorization) extension of RADIUS allows the change of session characteristics without having to terminate the ongoing session. In GSM and UMTS networks the GTP' is used to carry charging data records.

USER DATA MANAGEMENT

3GPP identified that user information is scattered across several architectural entities in an FMC domain (Alcatel-Lucent, 2010). For the efficient service control, user information is to be stored in a single functional unit called User Data Repository (UDR), which has a uniform interface towards its users. In the converged network, fixed and mobile operators in trusted relationship can share a common UDR.

3GPP LONG-TERM EVOLUTION

Overview

Mobile broadband data connections are becoming increasingly popular. Some studies suggest that by

Figure 4. Policy control architecture

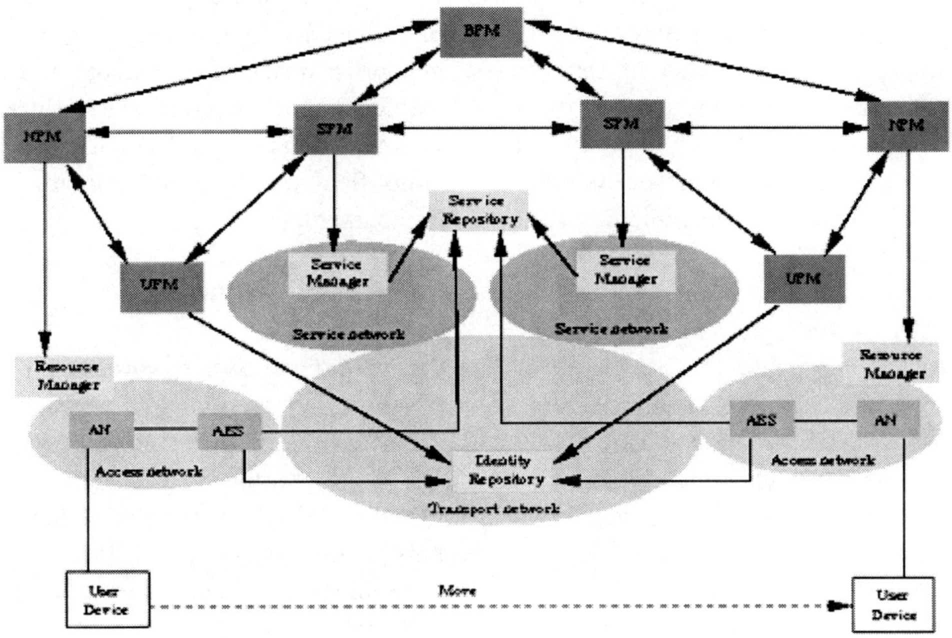

2015 around 80 percent of broadband subscriptions will be mobile with only 20 percent fixed access. The rapid increase in interest is driven by multiple factors, with arguably the most important being the advent of affordable mobile computing equipment. Most people already use mobile phones and many connect their notebooks over wireless LANs as well, and the new generation of smartphones and tablet PCs are designed to utilize constant broadband connection.

Consumers appreciate the benefits of mobile broadband. Users can spend less time setting up the WLAN connection, worrying about security or losing coverage and more time actually using the service. In many cases, mobile broadband can compete with fixed broadband on price, performance, security, and convenience.

The step towards full mobile broadband is straightforward, especially with LTE, which provides for even better user experience. It will enhance more demanding applications, such as mobile video and TV, advanced games, and professional services. With mobility, many of these applications even become more valuable to users.

To support increasing demand a new mobile standard is required beyond 3G. The most important objectives set at the start of the LTE standardization project was the improvement of spectral efficiency, improvement of quality of service, flexible use of carrier bandwidths and improved interoperation with other standard communication systems. The standardization was driven by 3GPP. The 3GPP is a collaboration agreement, established in December 1998, which brings together a number of telecommunications standards bodies, known as organizational partners. The current organizational partners include the Association of Radio Industries and Businesses (ARIB), China Communications Standards Association (CCSA), ETSI, ATIS, Telecommunications Technology Association (TTA) and Telecommunication Technology Committee (TTC).

In contrast to *HSPA* (High Speed Packet Access), which assumes strong backward compatibility requirements towards 3G, LTE's radio access technology significantly differs from the 3G solution (CDMA - Code division multiple access). LTE uses orthogonal frequency division multiplexing (OFDM) that ensures efficient transmission of high bandwidth data even with multipath propagation. It provides increased peak data rates: 100Mbps downlink and 50Mbps uplink bandwidths are required but the technology allows for speeds more than 300Mbps. Radio Access Network (RAN) round-trip times are reduced to less than 10 ms. Scalable carrier bandwidths of 20MHz, 15MHz, 10MHz, 5MHz, 3MHz and 1.4MHz are supported.

LTE supports wide range of terminals - In addition to mobile phones, many computer and consumer electronic devices, such as notebooks, ultra-portables, gaming devices and cameras, will incorporate embedded LTE modules. LTE supports handover and roaming to existing mobile networks, therefore LTE devices will have wide mobile broadband coverage from day one.

To support the new services enabled by LTE the 3GPP is also defining an IP-based, flat core network architecture. This architecture is defined as part of the System Architecture Evolution (SAE) effort (3GPP, 2006b). The LTE–SAE architecture and concepts have been designed for efficient support of mass-market usage of any IP-based service. The architecture is based on an evolution of the existing GSM/WCDMA core network, with simplified operations and smooth, cost-efficient deployment.

LTE Requirements

The requirements set towards LTE (3GPP, 2009) covers the following areas: radio interface throughput, latency, terminal state, spectrum flexibility, and interworking with UMTS.

The LTE radio interface is required to support an instantaneous downlink (from network to terminal) peak data rate of 100 Mb/s within a 20MHz downlink spectrum allocation and an instantaneous uplink (from terminal to network) peak data rate of 50 Mb/s within a 20 MHz uplink spectrum allocation. This corresponds to a spectrum efficiency of 5 bits/s/Hz for the downlink, and 2.5 bits/s/Hz for the uplink.

User data latency should be less than 5 ms in a non-loaded condition (assuming a single data stream) between the access gateway and the user terminal, including the transfer delay between nodes within the Access Network and the transmission over the radio interface. This is a very constraining requirement, but it guarantees that the LTE Access Network will be suitable for the most demanding existing and upcoming real-time services.

The requirement on terminal state is also referred to as control plane latency or efficient Always-On support. In classical telephony, it was generally assumed that the end-user terminal could only be into two different states:

- The IDLE mode: a state in which the mobile is reachable but no communication sessions are active.
- The ACTIVE mode: corresponding to the active phase of a session. In this mode, data are actively transmitted to and from the user terminal, as in a voice call or circuit-switched data session.

These two states are in line with the classical dial-in communication mode; however, with the emergence of new packet services characterized by bursts of activity separated by silent periods of time, the need for a new terminal mode rose. The Standby mode was an answer to this need, filling the gap between the IDLE and the ACTIVE modes. This state was initially defined in the scope of 2G GPRS, and re-used from the first version of 3G UMTS specifications. In such a state, the user terminal is silent but always connected, furthermore it allows for quick transition to the ACTIVE mode in case end-user session activity is resumed.

In LTE the transition from IDLE (or Standby) to ACTIVE state shall be less than 100 ms (this time does not include the time needed for paging procedures, in the case of mobile-terminated sessions). The transition between ACTIVE and STANDBY states shall be less than 50 ms.

One of the major operational constraints of the UMTS network is about spectrum flexibility. As an improvement to this situation, LTE shall be able to operate in spectrum allocations of different sizes, including 1.25, 2.5, 5, 10, 15, and 20MHz, for both uplink and downlink. The intention is to allow flexible deployment and operation of spectrum according to the expected traffic and service requirements (3GPP, 2006c).

In addition, operation in paired and unpaired spectrums shall be supported, so that the system supports asymmetric services (such as content streaming, or broadcast/multicast services) in the most efficient way.

From the beginning of UMTS standardization, inter-working between new and old technologies (with existing UMTS) has always been a subject of concern. Interruption time for handover between LTE and 2G/3G shall be less than 300 ms for real-time services and 500 ms for non-real-time services.

LTE-Advanced

LTE-Advanced (or IMT-Advanced) is the most recent - preliminary - mobile communication standard submitted to ITU-T in 2009 by 3GPP (3GPP, 2009b). It is planned to improve over (version 1 of) LTE on several fronts including:

- Utilization of ultra wide bandwidth up to 100 MHz of spectrum
- Increased peak data rates up to 1 Gbit/s
- Ability to leverage advanced topology networks
- Concepts for Relay Nodes

LTE-Advanced should be compatible with first release LTE equipment, and should share frequency bands with first release LTE.

FUTURE DIRECTIONS AND CONCLUSION

The high level NGN architecture has already the ability to provide generalized mobility for consistent and ubiquitous provisioning of services to users. It defines a common IP based network infrastructure accessible from a various set of fixed and mobile access types. IMS is able to provide a service control platform that is independent from the type of the access network making a uniform service platform for converged NGNs.

Roaming plays the central role in the fixed mobile convergence. It makes additional agreements necessary; those have to exist not only between mobile operators, but between fixed and fixed, and fixed and mobile operators as well. Service access while roaming influences the data plane routes, mobility management, policy control and authentication interactions. Roaming makes also the authentication of a subscriber more complex, the connectivity, the service access, and the application access have to be authorized separately in the visited domain.

The main challenges in AAA are the integration of the heterogeneous authentication methods employed by operators in their network domains, and the avoidance of multiple or multi-stage authentication, which can be based on the trusted relation between operators or the trusted status of access networks.

From the technological point of view, the biggest challenges are the efficient data plane solution for roaming scenarios and the efficient delivery of multicast streaming services to the visited network.

The standardization of FMC is taking place in parallel in several institutes that approach FMC from different perspectives. The coordination of research efforts and the synchronization of results remain a challenge.

REFERENCES

Aboba, B., et al. (2004, June). *RFC 3748 extensible authentication protocol*. Retrieved from http://www.ietf.org/rfc/rfc3758.txt

Alcatel-Lucent., et al. (2010, February 11). *Beyond FMC interworking – UDC to FMC*. 3GPP/BBF FMC Workshop, February 18-19, San Francisco, USA.

Calhoun, P., et al. (2003, September). *RFC 3588 diameter base protocol*. Retrieved from http://www.ietf.org/rfc/rfc3588.txt

Calhoun, P., et al. (2009, March). *RFC 5415 control and authentication of wireless access points (CAPWAP) protocol specification*. Retrieved from http://www.ietf.org/rfc/rfc5415.txt

Chiba, M., et al. (2003, July). *RFC 3576 dynamic authorization extensions to remote authentication dial in user service* (RADIUS). Retrieved from http://www.ietf.org/rfc/rfc3576.txt

ETSI. (2006, February). *Technical report 180 000 telecommunications and Internet converged services and protocols for advanced networking (TISPAN)- NGN terminology*. Retrieved from http://www.tech-invite.com/Ti-etsi-standards-TISPAN.html#tr-180-000

3GPP. (2006). *3GPP system architecture evolution: Report on technical options and conclusions*. Retrieved from http://www.3gpp.org/ftp/Specs/html-info/23882.htm

3GPP. (2006). *S2-060730 inter-system mobility alternate solutions*.

3GPP. (2006). *Physical layer aspect for evolved universal terrestrial radio access* (UTRA). (3GPP TR 25.814).

3GPP. (2009). *Requirements for evolved UTRA (E-UTRA) and evolved UTRAN* (E-UTRAN) (release 9). (3GPP TR 25.913).

3GPP. (2009). *Requirements for further advancements for evolved universal terrestrial radio access* (E-UTRA) (LTE-advanced) (release 9). (3GPP TR 36.913).

Gundavelli, S., et al. (2008, August). *Proxy mobile IPv6*. Retrieved from http://www.ietf.org/rfc/rfc5213.txt

ITU-T Y.2001. (2004). *General overview of NGN*. Retrieved from http://www.itu.int/rec/T-REC-Y.2001/en

Kent, S., & Seo, K. (2005, December). *RFC 4301 security architecture for the Internet protocol*. Retrieved from http://www.ietf.org/rfc/rfc4301

Kwon, T. T., et al. (2002, October). *Mobility management for VoIP service: Mobile IP vs. SIP. IEEE wireless communications.*

Melander, B., et al. (2007). *FMC support in fixed access architecture, technical report*. Retrieved from http://www.ist-muse.org/Deliverables/TF1/MUSE_DTF1.8_v1.0.pdf

Mickelsson, H., et al. (2006 December 11-14). *Nomadism/FMC use cases and AAA impact*. Geneva, Switzerland: BroadBand Europe.

Motorola. (2008). *LTE inter-technology mobility – enabling mobility between LTE and other access technologies*. Retrieved from http://www.motorola.com/staticfiles/Business/Solutions/Industry%20Solutions/Service%20Providers/Network%20Operators/LTE/_Document/Static%20Files/InterTech%20Mobility%20White%20Paper.pdf

Rigney, C., et al. (1997, April). *RFC 2138 remote authentication dial in user service*. Retrieved from http://www.ietf.org/rfc/rfc2138.txt

KEY TERMS AND DEFINITIONS

AAA: A set of network functions related to authentication, authorization and accounting.

Continuous Mobility: The ability of a mobile user, terminal, or network to change location while media streams are active.

LTE: A 3GPP project that targets the introduction of the 4th generation radio communication technologies.

Nomadism: The ability of the user to change his network access point on moving; when changing the network access point, the user's service session is completely stopped and then started again.

Policy Control: A set of network functions that allow the network and service providers to control the behavior of the network or a service at various granularities.

Roaming: The ability of the users to access services according their user profile, while moving outside of their subscribed home network, i.e. by using an access point of a visited network.

Session Continuity: The ability of a user or terminal to change the network access point while maintaining the ongoing session. This may include a session break and resume, or a certain degree of service interruption or loss of data while changing to the new access point.

APPENDIX

Table of abbreviations

2G	2nd Generation
3G	3rd Generation
3GPP	3rd Generation Partnership Project
AAA	Authentication, Authorization and Accounting
AC	Access Controller
AES	Access Edge Site
AN	Access Node
ARIB	Association of Radio Industries and Businesses
ATIS	Alliance for Telecommunications Industry Solutions
AVP	Attribute Value Pair
B2BUA	Business to Business User Agent
BPM	Business Policy Manager
CAPWAP	Control And Provisioning of Wireless Access Points
CCSA	China Communications Standards Association
CDMA	Code Division Multiple Access
CLF	Connectivity Session Location and Repository Function
CoA	Change of Authorization
DHCP	Dynamic Host Configuration Protocol
DSMIP	Dual-Stack Mobile IP
DTLS	Datagram Transport Layer Security
EAP	Extensible Authentication Protocol
E-CSCF	Emergency Call and Session Control Functions
EP	Enforcement Point
ETSI	European Telecommunications Standards Institute
FMC	Fixed-Mobile Convergence
GPRS	General Packet Radio Service
GSM	Global System for Mobile Communications
GTP	GPRS Tunneling Protocol
HDTV	High Definition Television
HSPA	High Speed Packet Access
IEEE	Institute of Electrical and Electronics Engineers
IETF	Internet Engineering Task Force
IMS	IP Multimedia Subsystem
IMT	International Mobile Telecommunications
IP	Internet Protocol
IPsec	Internet Protocol security

IPv4	Internet Protocol version 4
IPv6	Internet Protocol version 6
ITU	International Telecommunication Union
ITU-T	International Telecommunication Union - Telecommunications
LAN	Local Area Network
LMA	Local Mobility Anchor
LTE	Long Term Evolution
MAG	Mobile Access Gateway
MAN	Metropolitan Area Network
MIP	Mobile IP
NGN	Next Generation Networks
NPM	Network Policy Manager
OFDM	Orthogonal Frequency Division Multiplexing
PaC	PANA client
PANA	Protocol for carrying Authentication for Network Access
PBX	Private Branch Exchange
P-CSCF	Proxy Call and Session Control Function
PMIP	Proxy Mobile IP
PPP	Point-to-Point Protocol
QoS	Quality of Service
RADIUS	Remote Authentication Dial In User Service
RAN	Radio Access Network
RFC	Request for Comments
RTP	Real-time Transport Protocol
SAE	System Architecture Evolution
SBC	Session Border Controllers
S-CSCF	Serving Call and Session Control Function
SCTP	Stream Control Transmission Protocol
SDTV	Standard Definition TeleVision
SIP	Session Initiation Protocol
SLA	Service Level Agreements
SOAP	Simple Object Access Protocol
SPDF	Service Policy Decision Function
SPM	Service Policy Manager
TCP	Transmission Control Protocol
TISPAN	Telecommunications and Internet converged Services and Protocols for Advanced Networking
TTA	Telecommunications Technology Association
TTC	Telecommunications Technology Committee
TV	TeleVision
UDR	User Data Repository

UMA	Unlicensed Mobile Access
UMC	UMA Network Controller
UMTS	Universal Mobile Telecommunications System
UPM	User Policy Manager
VLAN	Virtual Local Area Network
VoIP	Voice over IP
WAN	Wide Area Network
WCDMA	Wideband Code Division Multiple Access
Wi-Fi	Wireless Fidelity
WiMax	Worldwide Interoperability for Microwave Access
WLAN	Wireless Local Area Network
WTP	Wireless Termination Point
xDSL	Digital Subscriber Line technologies

Chapter 12
Host Identity Protocol:
The Enabler of Advanced Mobility Management Schemes

László Bokor
Budapest University of Technology and Economics, Hungary

Szabolcs Nováczki
Budapest University of Technology and Economics, Hungary

Sándor Imre
Budapest University of Technology and Economics, Hungary

ABSTRACT

This chapter is committed to give a comprehensive overview of the Host Identity Protocol (HIP), to introduce the basic ideas and the main paradigms behind it, and to show how HIP emerges from the list of potential alternatives with its wild range of possible usability in next generation mobile architectures. The broad scale of feasible advanced mobility management proposals and scenarios, together with the promising mobility management capabilities of HIP and its cryptographic identifier/locator separation technique, will be introduced based on an exhaustive survey of existing mobility solutions designed for the Host Identity Protocol. This broad and up-to-date outline of advanced HIP-based mobility supporting schemes will guide the readers from the basics of HIP through the protocol's main functions to its complex feature set and power to create a novel Internet architecture for future mobility-centric communications.

INTRODUCTION

Actual trends in mobile telecommunication show rapid growth of Internet related applications, and ever-growing demand for them. The phenomenon of convergence in means of communication protocols, services, and terminals accelerates this process: mobile applications are able to become more and more popular; users are willing to access Internet resources from their portable devices seamlessly, anytime and anywhere. The continuously growing number of mobile users generates an increasing demand for more widespread and more sophisticated support of mobility, such

DOI: 10.4018/978-1-60960-732-6.ch012

creating serious challenges for today's Internet architecture, which is the TCP/IP stack.

The Internet Protocol (IP) itself was designed in the 1970's, when all hosts of the early Internet were connected using wires: they were fixed hosts, not able to change their network point of attachment. That is why the basics of TCP/IP systems were not designed with any kind of mobility in mind. However, nowadays users are much rarely interconnected by wires: a remarkable mass of modern Internet devices are mobile and thus require the support of frequent changes in their network point of attachments. The shortcomings, which make this support hard to provide, come from the early days of the Internet. The most important one is the double role of IP addresses. On one hand, an IP address identifies the host on the global network: all communication sessions initiated from or terminated at a given terminal is identified by its IP address. On the other hand, IP addresses have a topological locator role too: a special network identifier belongs to IP addresses telling the position of the node on the Internet. In other words IP addresses have dual significance (i.e., being identifier and locator at the same time), thus becoming semantically overloaded. These two roles make things complicated and inconvenient when the host starts to move: if the node changes its network point of attachment (and thus its IP address), active communication sessions (which are mostly connected to the TCP/IP numbers) are interrupted and even lost in many cases. Obviously, users want ubiquitous connection with seamless handovers and uninterrupted sessions, so engineers started to find an answer here.

One of the many solutions for the above-mentioned problem is a brand new protocol, which is called Host Identity Protocol (Moskowitz & Nikander, Host Identity Protocol (HIP) Architecture, 2006). HIP is a novel approach, which decouples IP addresses from applications by proposing a new, cryptographic namespace to identify hosts or other network entities while IP addresses will remain to act as pure locators. In this architecture, separation of identifier and locator roles of IP addresses (i.e., ID/Loc split) is provided: transport level connections are no more bound to IP addresses but to permanent identifiers, which remain the same for quite a long time. Therefore, HIP provides sophisticated and secure mobility/multihoming support, and creates a powerful toolset as the basis of several advanced mobility management schemes and extensions.

The rest of this chapter is organized as follows. The next section gives a general overview of the ID/Loc splitting paradigm together with the introduction of Host Identity Protocol, its fundamentals and main instruments. Some alternatives of HIP are also described here in order to present a wider view on the ID/Loc separation concept and its possible ways of implementation. This is followed by the discussion of HIP's built-in mobility/multihoming support and the introduction of several enhancements, improvements, and applications designed to extend the main standard with advanced capabilities in the area of mobility management. At the end of the chapter, we will discuss the expected future trends in HIP-based mobile telecommunication and present our concluding remarks.

BACKGROUND

Current IP networks are based on two basic kinds of namespaces. On one hand, there are human readable domain names, which can be resolved to IP addresses by Internet applications via Domain Name System (DNS) lookups. DNS provides fast queries but it is not designed for fast updates and quick retrieval of dynamic information. On the other hand, there are IP addresses used in the network layer as *locator* (Loc) for packet routing purposes and also used as *identifier* (ID) in upper layers to refer to the host or a particular communication session. The inseparable bond between the locator and identifier functions of IP address makes it inconvenient or even impossible to design efficient and scalable mobility, multihoming, traffic engineering, and routing and security solutions. Supporting heterogeneous network layer protocols or different locator families is also limited because of the same reason. The general concept of ID/Loc separation aims to eliminate the above problems and limitations by splitting the two roles of IP addresses and such allowing network layer to change locators without interfering with upper layer procedures. This separation makes the routing infrastructure more scalable, and by introducing a mapping function between identifiers and locators, a natural and effective support of mobility and multihoming can be provided.

The concept gains more and more popularity: several different approaches exist for ID/Loc separation (e.g., LISP, LIN6, SHIM6, GLI-Split, or HIP) and it also has recently been introduced in the standardization activities of the ITU Telecommunication Standardization Sector (ITU-T) for integration in future network architectures (Y.ipsplit, 2009), (Y.ipv6split, 2009). The common in all the above standards, drafts, and recommendations is the separation of identifiers from locators and applying a dynamic mapping mechanism between them, making the duplicate role of IP addresses disappear. They either use distinct namespaces for both (i.e., ID and Loc) functions or provide an architecture where the nature of the split is operational. Before introducing Host Identity Protocol –which takes the ID/Loc separation paradigm even further by using special self-certifying identifiers as host identities– we first briefly overview some of the most important alternatives and HIP-like technologies in the below sections.

Locator/ID Separation Protocol (LISP) (Farinacci, Fuller, Meyer, & Lewis, 2010) was one of the first instances of protocols providing practical implementation of the ID/Loc concept. LISP itself does not fully compliant with the locator identifier philosophy as Endpoint Identifiers are still represent some location information even if they are limited in scope.

LIN6 (Teraoka, Ishiyama, & Kunishi, 2003) is the application of the Location Independent Network Architecture (LINA) for IPv6. LINA describes a framework for separating the locators from identifiers and LIN6 is a specialized form of LINA as it uses the IPv6 environment to act according the LINA framework.

Level 3 Shim for IPv6 (SHIM6) (Nordmark & Bagnulo, 2009) is a protocol primary defined for providing multihoming support for the IPv6 internet based on the locator identifier separation idea. Later it has been extended to natively support host mobility.

Global Locator, Local Locator, and Identifier Split (GLI-Split) (Menth, Hartmann, & Klein, 2010) is one of the most recent locator identifier splitting architectures. Here, the dual role of IP addresses is shared between three representations: a global and a local locators and an identifier. It facilitates provider changes, network renumbering, multihoming and multi path routing as well as traffic engineering and advanced mobility support and incremental deployment.

As the above schemes, Host Identity Protocol also separates the different roles of IP addresses. In HIP, IP addresses continue to act as pure locators, while HIP will introduce a novel, globally unique namespace (the Host Identity namespace), which is a special pool of identity representations called Host Identifiers (HIs). The elements of the Host Identity namespace are public keys of asymmetric key pairs (i.e., self-certifying cryptographic names) used to identify nodes and to integrate strong security features such as authentication, confidentiality, integrity, and protection against certain kind of Denial-of-Service (DoS) and Man-in-the-Middle (MitM) attacks. Furthermore, based on the cryptographic HIs special certificates can be generated by the nodes for secure signaling (Nikander & Arkko, Delegation of Signaling Rights, 2004) or even identity delegation (Herborn, Huber, Boreli, & Seneviratne, 2007), offering enormous resource savings, effective session mobility and other promising application possibilities in wireless and mobile environments. However, HIs are rarely used in actual HIP protocol packets, instead hash representations called the Host Identity Tag (128 bit long for global, IPv6-based communication) and Local Scope Identifier (32 bit long for local usage and IPv4 compatibility) are applied. HIP related signaling information is passed by HIP headers having a form of a standard IPv6 extension header. Every HIP compatible node has at least one HI and implements the functions required to handle the new namespace and the relevant mechanisms. Therefore, the scope of the protocol includes the modifications and new methods designed to integrate the concepts of HIP into the existing Internet architecture. As Figure 1 shows, these functions, and TCP/IP extensions form a new protocol layer, which resides between the transport and network layer in the TCP/IP reference model (Moskowitz & Nikander, Host Identity Protocol (HIP) Architecture, 2006).

The basic function of HIP is to set up Host Identifier based connections between nodes and to map HIs to IP addresses and vice versa. A HIP association can be established between two nodes (i.e. an Initiator and a Responder) by a four way end-to-end security handshake called the Base Exchange (BEX) (see Figure 2).

The BEX performs mutual authentication based on the peers' asymmetric keys and implements a Diffie-Hellman key exchange to create symmetric keys for later payload encryption. Additionally, a special puzzle-solution mechanism is applied to protect the responder against certain DoS attacks. As a result of a successful HIP Base Exchange an IPsec Security Association pair is created between the peers where SAs are bound to HIs instead of IP addresses (Moskowitz, Nikander, Jokela, & Henderson, 2008). After the BEX, payload data is passed between the peers using the Encapsulating Security Payload (ESP) through a special ESP tunnel. A new transport mode of ESP was designed especially for HIP (Jokela, Moskowitz, & Nikander, Using the Encapsulating Security Payload (ESP) Transport Format with the Host Identity Protocol (HIP), 2008). This so-called Bound-End-to-End-Tunnel (BEET) mode integrates the ESP tunnel mode with the low overhead transport mode. Using BEET mode the outer IP header of the ESP packet holds the IP addresses of the peers but the inner header is missing. Instead, the Security Parameter Index (SPI) is used to identify the correspondent HIP association by reception at the

Figure 1. The host identity layer

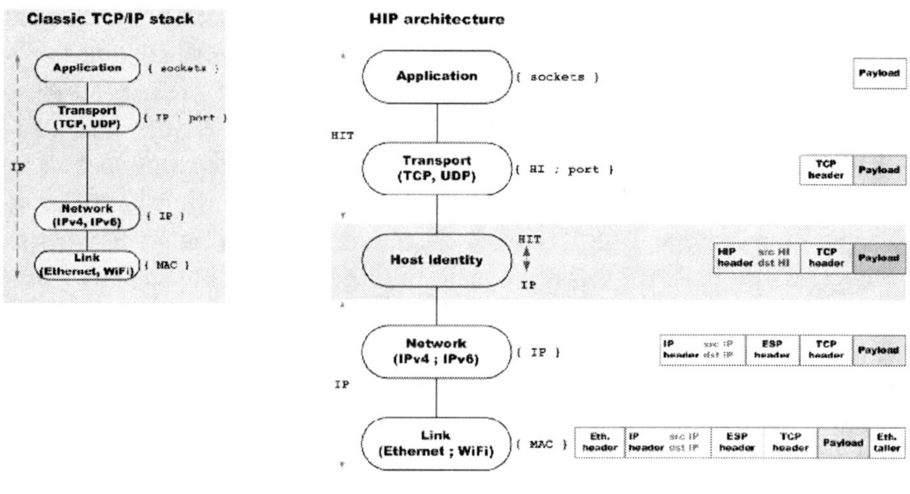

Figure 2. HIP base exchange

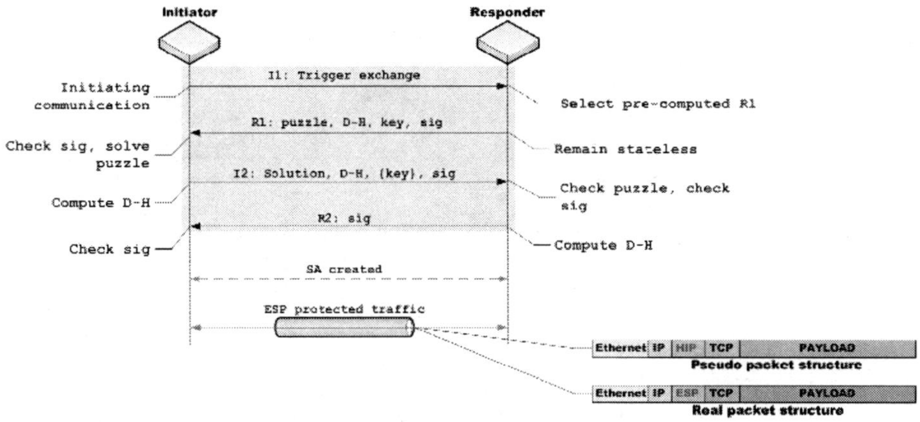

destination. Thanks to the BEET mechanisms HIP related signaling information (i.e., HIP header with source and destination HITs, and HIP parameters) must be applied only to HIP control packets but not in case of data transfer messages.

During HIP operation IP addresses (i.e., locators) are intended to be used mostly for "on-the-wire communication" between peer hosts, while upper layer protocols and applications use HIs (or HITs) and port numbers instead. This implies the need of some method to translate domain names to HIs. Using the existing infrastructure of Domain Name System (DNS) for this translation is quite straightforward. Therefore (Nikander & Laganier, Host Identity Protocol Domain Name System Extension, 2008) designed a new resource record for the DNS, and laid down how to use it with the Host Identity Protocol. This novel resource record allows a HIP node to store its Host Identity and other relevant information (e.g., HIT) in the DNS.

The above overview on the main concepts and mechanisms of HIP is devoted to help the reader to get better understanding on the base protocol. This enables one to easily comprehend the basic and also the advanced HIP-based mobility management concepts and schemes to be discussed in the upcoming paragraphs (see Table 1 for summary).

MOBILITY AND MULTIHOMING IN THE HOST IDENTITY LAYER

The Host Identity Protocol was created with basic host mobility and simple multihoming support in mind as important part of the main design principles. However, the flexibility of the protocol enables the design of advanced mobility management schemes built on the top of the base specification.

Basic HIP Mobility and Multihoming Support

The base specification of HIP describes a secure locator update procedure, which we describe here in detail. The procedure is used to maintain the HIT-IP mappings between the communicating peers (Nikander, Henderson, Vogt, & Arkko, 2008). The mobile endpoint informs partners that its location has changed. Inherited from the key idea of HIP the update procedure does not affect higher layer connections. The procedure is transparent for all established connections of the transport or application layers. This property makes HIP an exciting ground to develop sophisticated mobility schemes or use it to handle more complicated and advanced mobility scenarios like micromobility (Bokor, Nováczki, & Imre, A Complete HIP based

Table 1. Protocols and schemes supporting HIP-based mobility management

Protocol/Scheme	Main features	Reference
RFC 5206	-Provides a secure locator update procedure in the most simple mobility and multihoming scenarios	-Nikander, Henderson, Vogt, & Arkko, 2008
RFC 5204	-Introduces a permanent anchor point for mobile nodes	-Laganier & Eggert, Host Identity Protocol (HIP) Rendezvous Extension, 2008
HIP Signaling Delegation	-Defines the HIP mechanisms for secure delegation of (mobility) signaling rights	-Nikander & Arkko, Delegation of Signaling Rights, 2004
The original idea of integrating micromobility and HIP	-Focuses on the security issues of HIP-based micromobility but does not consider protocol details regarding the operation and the mobility support	-Ylitalo, Melén, Nikander, & Torvinen, Re-thinking Security in IP Based Micro-Mobility, 2004
μHIP	-Defines the first complete solution for HIP-based micromobility management by introducing the concept of local rendezvous servers (LRVS) -Introduces paging based on HIT specific multicasting	-Bokor, Nováczki, & Imre, A Complete HIP based Framework for Secure Micromobility, 2007
P-HIP	-Describes a HIP paging extension where LRVSs are controlling paging area,, and paging request messages are broadcasted towards all access routers in the paging area in order to find an idle MN	-Yang S., Qin, Luo, & Zhang, 2008
mHIP	-Introduces hierarchy in the HIP micromobility domains by defining two different types of nework agents: mHIP gateways and mHIP routers	-So & Wang, 2008
DH-HIP (Dynamic Hierarchical HIP)	-Introduces a three level architecture of rendezvous servers for HIP micromobility as Rendezvous Server (RVS), Gateway RVS (GRVS) and Local RVS (LRVS)	-Yang, Qin, & Yang, 2007
eHIP (Early Update for HIP)	-Improves handover latency by applying two basic concepts: the hierarchy and the anticipation of the location update process of HIP	-Aydin, Chaouchi, & Zaim, 2009
HIP-PMIP	-Merges the advantages of network-based micromobility management of Proxy Mobile IPv6 (PMIPv6) with HIP's macromobility management, security, vertical handover and multihoming capabilities	-Iapichino, Bonnet, Herrero, Baudoin, & Buret, 2009
The first work touching mobility of HIP networks	-Discusses the basics of a possible NEMO solution that handles HIP aware mobile networks	-Ylitalo J., Re-thinking Security in Network Mobility, 2005
HarMoNy	-Defines a hybrid HIP-MIPv6 based network mobility management scheme enabling HIP based MNNs to communicate with MIPv6 based MRs	-Herborn, Haslett, Boreli, & Seneviratne, 2006
HIP-NEMO	-Extends HIP with the concept of mobile Rendezvous Server (mRVS) in order to handle moving networks (can be considered as the first complete and pure HIP-based NEMO solution)	-Nováczki, Bokor, & Imre, A HIP based Network Mobility Protocol, 2007 -Nováczki, Bokor, Jeney, & Imre, 2008
HIPMR (Host Identity Protocol-based Mobile Router)	-Proposes a HIP-based NEMO solution based on signaling delegation between MNNs and the HIP enabled mobile router (an alternative to HIP-NEMO)	-Melen, Ylitalo, Salmela, & Henderson, 2009 -Ylitalo, Melé, Patrik, & Petander, 2008
Simultaneous Multi-Access extension to HIP	-Proposes a method where two HIP nodes, of which at least one is multihomed, can separate different upper layer data flows (e.g., TCP and UDP) between peers and allow flows to use different interfaces at the multihomed node	-Pierrel, Jokela, & Melen, Simultaneous Multi-Access extension to the Host Identity Protocol, 2006
HIP-based per-application mobility management	-Defines a complete HIP-based per-application mobility management platform	-Bokor, Zeke, Nováczki, & Jeney, 2009
HIP extension for simultaneous end-host mobility support	-Defines an extension to HIP which improves mobility management of the core protocol with simultaneous end-host mobility support.	-Hobaya, Gay, & Robert, 2009
IPv4/v6 handovers using Host-Based Identifier-Location Split	-Makes cross-family handovers possible as in this scheme peers can inform each other about all available addresses they are reachable at	-Varjonen, Komu, & Gurtov, Secure and Efficient IPv4/v6 Handovers Using Host-Based Identifier-Location Split, 2010
Location Privacy Framework with Mobility Using HIP	-Introduces a complete location privacy framework with mobility support using HIP	-Maekawa & Okabe, 2009
I-HSIP	-Presents an integrated HIP-SIP based mobility management architecture	-Li, Chen, Su, Jin, & Zeng, 2009
SHIP	-Describes a cross-layer mobility management scheme based on SIP and HIP	-Yang, Zhou, Qin, & Zhang, 2009
Hi3 (Host Identity Indirection Infrastructure)	-Upgrades basic HIP mapping system (RVS) to a map-request forwarding overlay	-Gurtov, Korzun, Lukyanenko, & Nikander, 2008

Framework for Secure Micromobility, 2007), network mobility (Nováczki, Bokor, Jeney, & Imre, 2008) or per-application mobility (Bokor, Zeke, Nováczki, & Jeney, 2009). The update sequence is illustrated on Figure 3.

This is the most simple mobility scenario specified for HIP. There are two HIP capable nodes, which have established communication sessions. Note that their higher layer connections are bound to HITs instead of IP addresses. In case the IP address of the mobile node is changed, it will trigger a HIP update procedure by sending an UPDATE message to its peer(s). This delivers the new location information (loc) and informs the peer if the mobile wants to update the security parameters (esp). If there is a need for refresh, the mobile also sends the updated parameters (D-H). The update procedure is proved to be protected against security attacks. On one hand, all the messages are digitally signed by the peers; this authenticates the senders. On the other hand, there is a built in protection against distributed Denial-of-Service attacks. The second and the third message of the update procedure implements this. The peer node receiving the first UPDATE packet verifies the signature and answers with another UPDATE packet. This includes information to update the security parameters (esp and D-H) and a data block that contains a nonce (e_req). This must be echoed back by the mobile node in the third UPDATE packet (e_res). This simple echo request-response sequence verifies the new address of the mobile node.

A related functionality of HIP is host multihoming. In case of multihoming, the HIP node has more than one physical interfaces and/or global addresses. However, the update procedure described above is used to update the primary loca-

Figure 3. The HIP UPDATE procedure

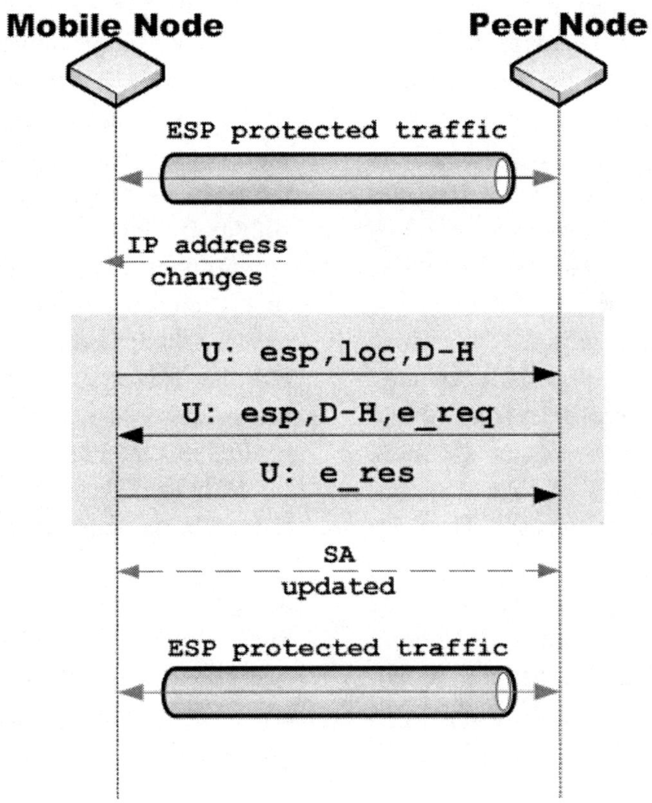

tor of HIP nodes; a multihomed node can inform its peers about secondary locators it is reachable at. It is recommended to use different SAs for different interfaces and/or addresses. To do this, a multihomed HIP host creates a new inbound SA and a corresponding SPI. This is also managed by the update procedure. The first UPDATE packet should hold an ESP_INFO parameter having the NEW SPI field set to the newly created SPI value and setting the OLD SPI field set to zero. The packet also contains a LOCATOR parameter that indicates the new address-SPI mapping and the old one as well. Peers will use the primary locator as long as it is available and can switch to one of the secondary locators upon loss of connection.

The above introduced update procedure for mobility and multihoming can handle locator changes in case there are ongoing HIP sessions between the endpoints. However, this is not a solution for initial reachability of mobile nodes and cannot cope with simultaneous mobility of endpoints. The problem to solve in these scenarios is how to provide a permanent anchor point for mobile nodes that makes it able to reach them no matter what their actual location is. There is an extension in HIP standards that introduces such an anchor point (Laganier & Eggert, Host Identity Protocol (HIP) Rendezvous Extension, 2008). This is called the Rendezvous Server (RVS). Figure 4 shows the service the RVS provides for mobile nodes to handle scenarios described above.

The basic idea is that there is an RVS, which is known to every potential peer nodes by e.g., DNS queries. The RVS stores the actual HIT-IP mapping for registered mobile nodes. The Base Exchange is assisted by the RVS to enable connection establishment for the peers. Here we describe the sequence in detail. First, the mobile node has to register itself at the RVS to use the offered service (Laganier, Koponen, & Eggert, Host Identity Protocol (HIP) Registration Extension, 2008). This creates an entry in the RVS database that holds the HIT-IP mapping for the mobile. The entry is updated time-to-time by the mobile if its IP address changes. After registration the mobile informs the DNS indicating its serving RVS. At this point, any potential peer can initiate a HIP connection with the mobile. The peer performs DNS queries to get to know the serving RVS of the mobile it wants to reach. This is a two-stage query. First, the peer asks the IP of the mobile node indicating its domain name. The DNS returns the domain name of the RVS. In the second stage, the peer asks the IP address of the RVS and the DNS returns its IP address. Now the peer can trigger the Base Exchange by sending the I1 message. This is delivered to the RVS. The anchor point knows the actual IP address of the mobile node and modifies the I1 message accordingly. The RVS also attaches an additional data field to the message that identifies the original sender of the message (FROM(IP_{PN})). The message is delivered than to the mobile, which continues the Base Exchange by sending the R1 message. This also contains an additional parameter, which verifies that the I1 message was forwarded by the RVS (VIA(IP_{RVS})). Finally, the connection setup finishes in the regular way without the inclusion of the RVS. Note that the RVS is used only in the connection setup. Any other communication (signaling or data) between the peers is transferred in the direct path. The similar process must be followed when the endpoints are changing IP addresses parallel. In this case, the HIP connection is broken and must be reestablished. There is however an extension for HIP that solves simultaneous end host mobility in a more sophisticated way. This method is described in a later section.

We have described the standard procedures of HIP mobility management. Thanks to their flexibility, it is easily extendable to handle complex mobility scenarios in an efficient and secure way. Such advanced mobility management solutions based on the already standardized location update and RVS mechanisms are introduced in the next sections.

Figure 4. Operation of HIP RVS mechanisms

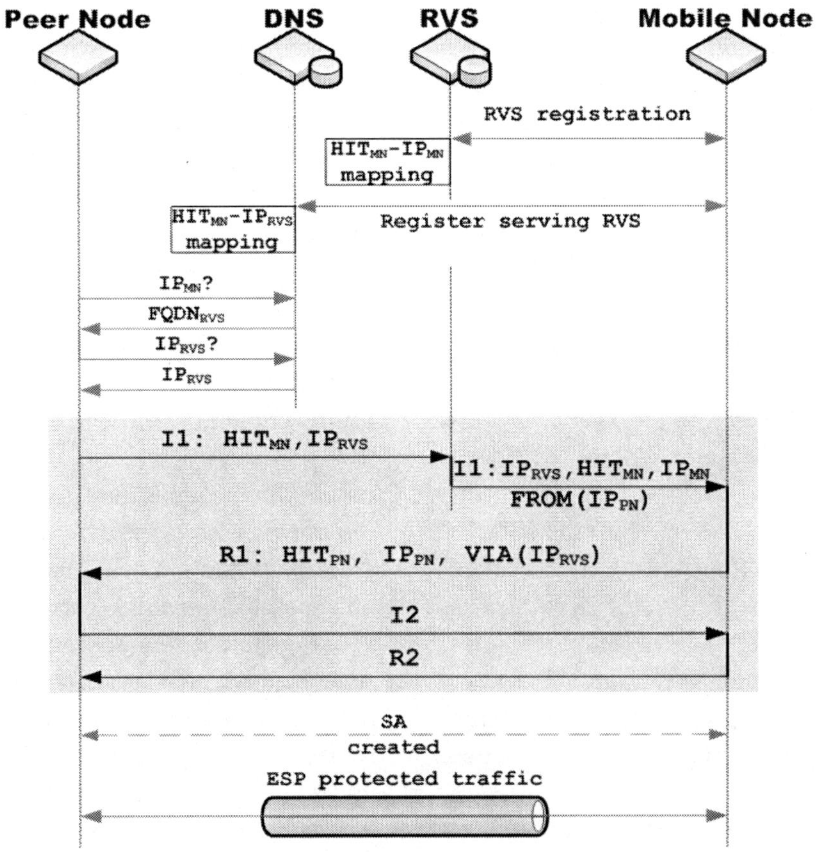

Micromobility

Access systems of modern mobile architectures commonly consist of one or more domains providing users with topologically valid IP addresses and connecting users to the IP core through the wireless interface. These domains of wireless access networks can be aggregated and special protocols can be assigned for local mobility management of the group of domains in order to offer fast and seamless handover control over a limited geographical area. In such cases, we speak about micromobility, and the aggregated group of domains is called micromobility domain. Mobility management solutions of such structures (i.e., micromobility protocols) are specially designed for environments where mobile terminals change their network point of attachment so often that the general mobility management protocols (i.e. macromobility solutions) originate excrescent overhead and ineffective operation.

Basic HIP mobility mechanisms are only for macromobility support and further extension of the base protocol is needed for micromobility. The original idea of integrating micromobility with HIP was presented by authors of (Ylitalo, Melén, Nikander, & al., Re-thinking Security in IP-based Micro-Mobility, 2004) and (Ylitalo, Melén, Nikander, & Torvinen, Re-thinking Security in IP Based Micro-Mobility, 2004) where a secure micromobility management scheme based on Lamport one-way hash chains and secret splitting techniques was introduced for IP networks. However, this scheme is not built on an effective and intact micromobility model as it only focuses on the security issues and it also does not consider protocol details regarding the operation and the mobility support. Moreover, in this method MNs still need to update their location information at the RVS during the handover, therefore the scheme cannot fulfill the requirements of micromobility architecture: it is only a partial answer for the complex problem.

The first complete solution for HIP-based micromobility management was presented in (Nováczki, Bokor, & Imre, Micromobility Support in HIPHIP: Survey and Extension of Host Identity Protocol, 2006), and (Bokor, Nováczki, & Imre, A Complete HIP based Framework for Secure Micromobility, 2007) where the authors introduced a novel HIP gateway entity called the Local Rendezvous Server (LRVS) which is responsible for managing HIP Mobile Nodes (MNs) in a given domain (see Figure 5 for details). LRVS gateways provide HIP registration service for users in the domain, and also introduce an IP address mapping function which is used to attach the MNs to the μHIP access network by registering the local locators (IP_L) of MNs. IP_L is valid only in the given domain and the LRVS is responsible for mapping every IP_L to a globally routable address (i.e., global locator, IP_G) chosen from a private address pool. IP_G is used to register the MNs at their RVSs and to deliver packets outside the micromobility domain during further communication sessions.

The basic operation of this architecture starts with an initialization mechanism right after the MN joined the domain. First the MN physically connects to one of the access routers (AR) of the domain, then gets the local locator based e.g. on IPv6 stateless autoconfiguration. After this, the MN either may actively initiate a HIP service discovery procedure (Jokela, Melen, & Ylitalo, HIP Service Discovery, 2006) or passively wait for a service announcement in order to detect the LRVS service provided in the visited micromobility area. In some way, the MN will be informed about the HIT and the IP address of the LRVS responsible for its actual domain, and will register itself to the LRVS with the Base Exchange sequence. During this service discovery and registration procedure, the LRVS not only registers the MN's HIT with the new IP_L, but maps IP_L with the globally routable address (IP_G) as well. After the MN successfully registered at the LRVS (with the HIT_{MN}-IP_L-IP_G triplet), it needs to perform the update and/or registration procedures at its RVS and current CNs (with the HIT_{MN}-IP_G pair) in order to be reachable for the current and possible future communication partners. Therefore, the MN – strongly relying on the self-certifying cryptographic identifiers provided by HIP and on the mechanisms introduced by (Nikander & Arkko, Delegation of Signaling Rights, 2004) – delegates its signaling rights to the LRVS at which it is registered. The appropriate certificates are sent during the service discovery phase, resulting that the LRVS will own the rights to signal on behalf of all MNs in the micromobility domain under its authority. In possession of these delegated rights, the LRVS is able to securely register or update to the RVSs and CNs on behalf of the MNs with the IP_G global locators assigned to them.

During μHIP connection establishment between already initialized MNs initiator starts the Base Exchange and sends the first packet (I1) which will be intercepted by the LRVS in the initiator's domain. This LRVS changes the source IP address of the I1 packet to the initiator's global locator and sends the packet to the RVS of the responder. The RVS forwards the packet towards the responder's registered global locator thus reaching its serving LRVS which knows the actual location of the responder, so the packet can be forwarded by changing the destination IP address of the I1 packet to the responder's actual local locator. The Base Exchange continues in the regular way, without the inclusion of the RVS, but with the address mapping function of the two LRVSs. This message flow builds up an active HIP

Figure 5. The μHIP architecture

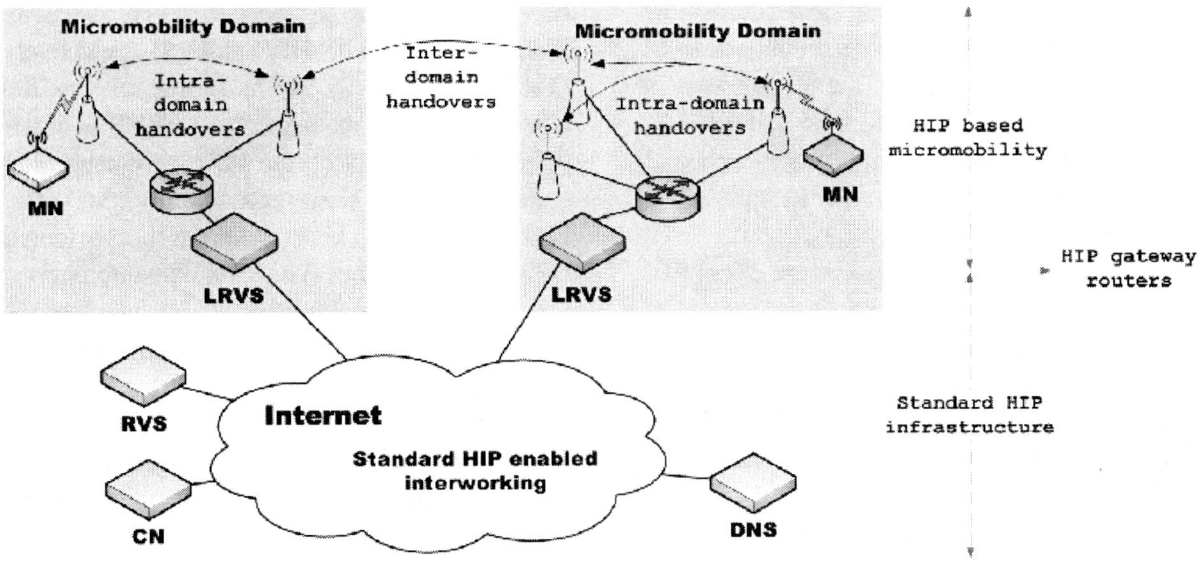

association between the initiator and responder, so they can begin sending data packets to each other.

In case of intra-domain handovers the MN will receive a new IP_L from the new Access Router belonging to the serving LRVS. In this case the MN – realizing the change of its IP address – uses (Laganier, Koponen, & Eggert, Host Identity Protocol (HIP) Registration Extension, 2008) to update its registration (and if needed its delegation certificate as well) with its new IP_L at serving LRVS. It is important to note that neither the CNs of the mobile node nor the RVS has to be informed about the movement as the address changes are locally handled by the proposed micromobility extension. The movements of nodes are completely hidden from the outside world resulting in less signaling overhead, packet loss and handover latency.

In case of inter-domain handovers, the mobile node moves between local administrative domains thus invoking the macromobility management procedures. Arriving at the new domain, the node will receive a new local locator, and will discover the service parameters of the new LRVS. After the MN realized that it leaved the previously used micromobility domain by entering a new one and learned the HIT and IP address of the new LRVS, it performs the initialization mechanism. Since MN changes its old LRVS, it also has to update its RVS and all the correspondent nodes with ongoing communication.

(Bokor, Nováczki, & Imre, A Complete HIP based Framework for Secure Micromobility, 2007) proposes a paging extension to μHIP in order to effectively locate a particular mobile node whose current position is not accurately known in the network, to reduce signaling costs and to save power consumption of MNs. The solution is based on the HIT specific multicasting (HISM) proposal (Kovácsházi & Vida, 2007) and implements an efficient location tracking that distinguishes between active and idle MNs, applies multicast group IP addresses for identification of paging areas (PA) in the system, and uses HISM mechanisms for passing paging messages to MNs through the multicast tree (built and maintained based on HIT information) of the MNs actual paging area.

(Yang S., Qin, Luo, & Zhang, 2008) also describes a HIP paging extension called P-HIP where LRVSs are controlling paging areas identified with the LRVSs' HIT, and – as the network always knows the current paging area of the mobile node – paging request messages are broadcasted towards all access routers in the paging area in order to find an idle MN. The main feature of P-HIP is similar to the HISM based paging scheme: only active MNs must update their locators when they move within the same paging area.

Several extensions of μHIP were published using the original proposal as a basis for further optimization of HIP mobility mechanisms. Authors of (So & Wang, 2008) introduced mHIP, which introduces hierarchy in the micromobility domains by defining two different types of network agents: mHIP gateways and mHIP routers. The main roles of these agents are to handle HIP-based signaling messages of the intra-domain handovers, and to re-direct the connections to the correct location. All mHIP agents in this scheme can sign the message on behalf of the whole group. The mHIP gateway is basically an LRVS: it is the root of all mHIP routers in a domain keeping the location information of the MN inside the domain. The mHIP router is a HIP layer router that redirects HIP sessions to the MN's current location. mHIP routers are also able to handle the intra-domain handover signaling inside the domain so that the handover latency and load of the mHIP gateway can be further reduced. Description of operation in multihoming scenario is also included in mHIP.

(Muslam, Chan, & Ventura, 2009) designed an extension aiming to improve intra-domain communication of μHIP. The scheme also presents a new entity called Co-Agent (Co-A) for each domain. Co-A is responsible for managing MNs in the given domain and for acting as virtual Mobile Nodes and Corresponding Nodes during both intra- and inter-domain handovers in order to reduce the number of control messages required in mobility management. In this framework, the LRVS still manages the connectivity between domains' access networks and the Internet.

DH-HIP (Dynamic Hierarchical HIP) is also a HIP-based micromobility management scheme introducing a three level architecture of rendezvous servers as Rendezvous Server (RVS), Gateway RVS (GRVS) and Local RVS (LRVS), respectively (Yang, Qin, & Yang, 2007). In this scheme, the size of a domain managed by a LRVS is determined dynamically by the MN itself, according to the packet arrival rate and mobility status after LRVS selection. The network in the presented DH-HIP architecture is divided into autonomous and administrative domains, where autonomous domains may consist of several administrative domains. LRVSs are managing the administrative domains, while GRVSs are responsible for the autonomous ones. GRVS nodes are dealing with the control of the communication between the LRVS and MNs.

eHIP (Early Update for HIP) improves handover latency by applying two basic concepts: the hierarchy and the anticipation of the location update process of HIP (Aydin, Chaouchi, & Zaim, 2009). In addition to the introduction of a hierarchical domain approach, this scheme also extends μHIP with an early update mechanism in which MNs obtain their new IP addresses from the network they are about to enter and make their registration before the actual handover process. Such an early update handover can be triggered based on different parameters applied to the handover decision engine of the MN.

Authors of (Iapichino, Bonnet, Herrero, Baudoin, & Buret, 2009) and (Iapichino & Bonnet, Host Identity Protocol and Proxy Mobile IPv6: A Secure Global and Localized Mobility Management Scheme for Multihomed Mobile Nodes, 2009) propose to merge the advantages of network-based micromobility management of Proxy Mobile IPv6 (PMIPv6) with HIP's macromobility management, security, vertical handover and multihoming capabilities. This integrated HIP-PMIPv6 approach results system architecture with double benefits. On one hand, it incorporates an efficient micromobility extension of HIP. On the other hand, it supports PMIP with macromobility management skills in a framework where users can securely use the different access technologies for connecting their multihomed devices and also can move sessions from one interface to another one without breaking the already established secure associations, such being connected to the always best network available.

Network Mobility

Network mobility (NEMO) is one of the latest scenarios discovered in the mobility management research area. Protocols addressing networks in motion has to deal with mobility of whole networks, not only single endpoints. Moving networks are usually consisting of one or more mobile router (MR) and several different mobile network nodes (MNN) connected to the MR. The MR is the gateway to reach the rest of the network for MNNs. All data and control signaling sent by the MNNs are traveling through the MR. The task of the MR is to provide uninterrupted connection for MNNs and handle mobility situations so that it is transparent to the MNNs. MNNs are not limited to single end terminals, they can also be whole mobile networks such creating nested NEMO structures. Such a nested network may contain multiple levels of moving networks attaching and detaching dynamically each other. To handle such complicated mobility scenarios there is a need for a very effective, scalable, flexible, and secure solution. Host Identity Protocol provides a strong basis for such an extension.

The first proposal addressing HIP-based NEMO is described in (Ylitalo J., Re-thinking Security in Network Mobility, 2005). The authors discuss the basics of a possible NEMO solution that handles HIP aware mobile networks. While this chapter is the first work touching mobility of HIP networks it does not get into describing the details of the solution. Later the authors have further developed their ideas, which also will be discussed in this chapter.

A hybrid HIP-MIPv6 based network mobility management scheme has been proposed by (Herborn, Haslett, Boreli, & Seneviratne, 2006). It enables HIP based MNNs to communicate with MIPv6 based MRs. NEMO signaling is performed by the MR and is based on MIPv6 bindings. Inside the moving network, HIP bindings are used between the MR and the MNNs.

HIP-NEMO (Nováczki, Bokor, & Imre, A HIP based Network Mobility Protocol, 2007), (Nováczki, Bokor, Jeney, & Imre, 2008) extends μHIP to be able to handle moving networks. This scheme can be considered as the first complete and pure HIP-based NEMO solution. A new HIP capable network entity is introduced called the mobile Rendezvous Server (mRVS). The mRVS provides continuous connectivity for the served MNNs and also for other moving networks connected to it. Moreover, the mRVS is in charge to be the signaling proxy for MNNs. The MNNs delegate their signaling rights to the mRVS thus it can control any mobility scenarios seamlessly to MNNs. Figure 6 shows the initial steps of a HIP-NEMO based moving network. First of all the mRVS discovers all LFNs, which are registering themselves at the mRVS and delegating their signaling rights to it. The mRVS maintains a mapping between the HIT of the LFNs and their IP address valid inside the moving network. This mapping entry is than associated with an IP address assigned by the mRVS. The purpose of this address is exactly the same as in case of μHIP. It is valid outside the moving network and is exchanged with the LFNs actual IP address by the mRVS. After the initial registration, the mRVS does the necessary signaling for the LFNs. It informs the RVS and the DNS as shown in Figure 6 as step 2 and 3. The RVS creates a special mapping that contains the HIT and global IP address of the LFN and is associated with the HIT of the mRVS. This latter association is useful when the moving network changes its network point of attachment. The mRVS has to perform only one update sequence indicating its new IP address and HIT. The RVS automatically refreshes all other entries that are associated with the HIT of the mRVS. Finally the DNS is informed about the serving RVS.

As these initial steps are completed, it is possible to trigger a communication session with one of the LFNs in the mobile network. Figure 7 can be used to track the process. The peer node queries the DNS and gets the IP address of the RVS serving the LFN. The first packet of the Base Exchange is sent to the RVS, which forwards the packet towards the mRVS according to its mapping. The destination IP address is the one assigned by the mRVS to the LFN at the registration phase. The I1 packet is intercepted by the mRVS and the destination address is changed to the one of the LFN valid inside the moving network. Finally, the LFN receives the message and answers with the second Base Exchange message (R1). The source address is changed at the mRVS to the globally reachable one and sent directly to the peer. The connection setup finishes in the regular way. Note that all the packets (signaling or data) send by or addressed to the LFN are intercepted by the mRVS. This entity changes the source or the destination addresses depending on the direction of the communication. Moreover, the mRVS records this connection to be able to update it upon movement of the network on behalf of the LFN.

The key issue in field of mobile networks is how nested mobile subnetworks are handled. Nested NEMOs are challenging problems as they usually raise serious scalability problems. In HIP-NEMO, nested subnetworks are managed by special inter-mRVS signaling. A bi-directional IP tunnel is built up between mRVSs. All signaling and data addressed to nested networks is injected through these tunnels. Moreover, signaling between mRVSs manages the introduction of new nested subnets as well as nested subnet handover situations. The detailed description is out of scope of this summary.

An alternative solution that uses HIP to handle moving networks is described in (Melen, Ylitalo, Salmela, & Henderson, 2009), (Ylitalo, Melé, Patrik, & Petander, 2008). The proposal is based on signaling delegation between MNNs and the HIP enabled mobile router. After the initial registration, all the signaling needs of MNNs are performed by the mobile router. The main difference compared to the above solution is that there is no need to change the source and destination addresses of packets passing the boundary of the moving network. Instead, a special NAT functionality is implemented in the MR. The MR intercepts packets passing through it and checks the ESP header. This contains a special parameter; the

Figure 6. HIP-NEMO network initialization

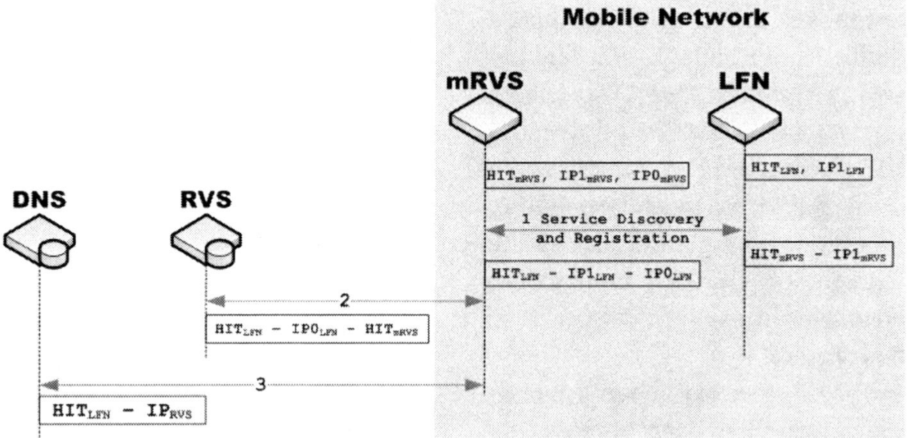

Figure 7. HIP-NEMO connection establishment

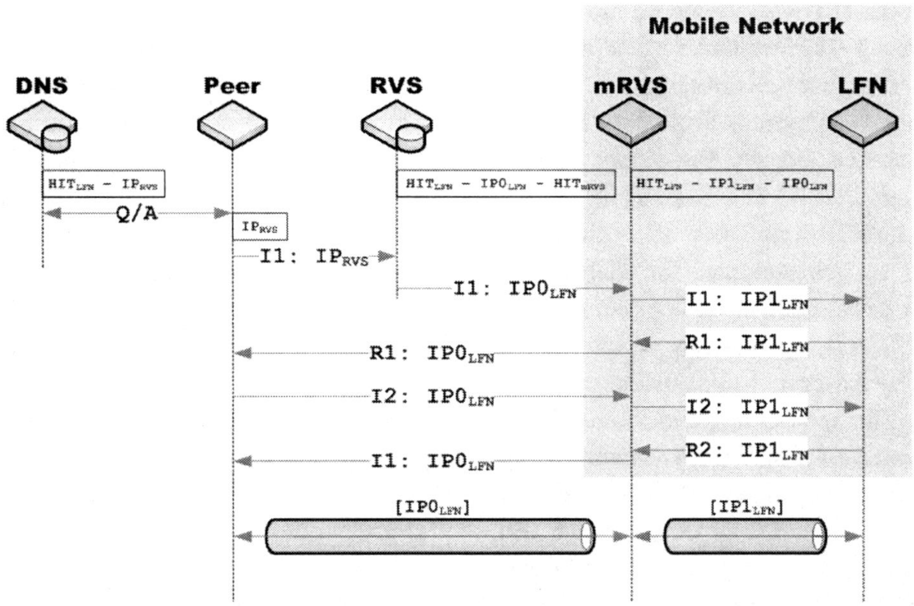

Security Parameter Index (SPI) that can be used to identify nodes behind the MR. This maintains an association between the SPI values and the IP address of the MNNs. Based on this information the packets can be routed to the proper endpoint. The proposal can also handle nested mobile networks in a scalable manner.

Flow Management and Per-Application Mobility

With the advance of heterogeneous access structures – and considering that users are running multiple applications simultaneously –, the traditional per-host or network mobility management technique cannot be the optimal solution for handling connection changes. Instead, the concept of per-application mobility management is to be introduced, where a dedicated interface (i.e., access network) is selected for each application according to its QoS requirements and the actual networking conditions.

Aiming to benefit from this novel concept in practice, authors of (Pierrel, Jokela, & Melen, Simultaneous Multi-Access extension to the Host Identity Protocol, 2006) and (Pierrel, Jokela, Melen, & Slavov, A Policy System for Simultaneous Multiaccess with Host Identity Protocol, 2007) proposed firstly a method where two HIP nodes, of which at least one is multihomed, can separate different upper layer data flows (e.g., TCP and UDP) between peers and allow flows to use different interfaces at the multihomed node. This HIP extension basically describes mechanisms for flow identification and carrying filter rules (i.e., policy transfer) in HIP signaling.

In (Bokor, Zeke, Nováczki, & Jeney, 2009) authors go even further by defining and evaluating a complete HIP-based per-application mobility management platform. This platform consists of a Monitoring/Mediator Agent (collecting protocol specific information from different layers for cross-layer per-application mobility decisions), an Application Profile DataBase (storing and maintaining profile attributes of ongoing application sessions for decision support), a Decision Engine (processing inputs from the above entities and making mobility decisions), and a HIP Agent.

The HIP Agent is responsible to initialize mobility procedures in the HIP layer based on the control information sent by the Decision Engine, and also to maintain per-application bindings between the host machine and its partners. From the HIP point of view, multihomed Security Associations are the protocol elements, which can make a HIP system to be able to handle mobility in an application-wise manner. In consequence, the HIP Agent implements a certain SA grouping scheme and a modified UPDATE mechanism, which are the keys to the HIP-based per-application mobility management. The approach is end-to-end based and uses spanned SAs between communicating peers as basis for operation, therefore standards of HIP DNS and RVS extensions are left intact, and only SA handling, packet processing and UPDATE procedure are slightly modified in the scheme.

In (Fekete, 2009) the author introduces a policy based flow management system for multihomed HIP hosts focusing on the definition of a simple language to specify policies that influence the source and destination IP addresses of outgoing packets. The solution makes possible to express policies for selecting IP addresses based on different QoS parameters (e.g., bandwidth, RTT) or other kind of access characteristics like the cost of usage, and requires only minimal changes to the Host Identity Protocol.

Simultaneous End-Host Mobility

In (Hobaya, Gay, & Robert, 2009) the authors define an extension to HIP, which improves mobility management of the core protocol. Simultaneous end-host mobility is the problem to be solved in this work, as such scenarios result broken HIP sessions that have to be reestablished: as both endpoints are moving away from their location more or less parallel, the UPDATE messages cannot reach their destination. The messages are delivered to the old locations of the peers, and the partners will lose each other and they have to restart their common session(s). This extension of HIP is designed to avoid situations described above. The proposal defines that mobile nodes are configured to send all UPDATE messages to the serving RVS of the correspondent nodes. As mobiles nodes always update their RVS upon mobility the latter will always have an up-to-date mapping between the HIT and IP address of mobile nodes. UPDATE messages sent to the RVSs are never lost and are delivered to the correct location. Note that opposed to the standard HIP mobility framework, this extension suggest that RVSs should be always included in the update process. Nodes do not have to wait the regular update process to fail first but they can immediately involve RSV entities. This of course ensures the delivery of UPDATE packets but introduces unnecessary delay in non-simultaneous mobility scenarios.

Dual-Stack Mobility

Host Identity Protocol has nice properties when considering IPv4 and IPv6 interworking but lacks support of cross-family handovers. The problem within the base specification is that when locators are included in the R1 and I2 packets of the Base Exchange with the preferred bit (i.e., this indicates address the corresponding peer prefers to use) set. This forces the peer to immediately switch to the preferred locator and makes it impossible to handle alternative locators. One modification (Varjonen, Komu, & Gurtov, Secure and Efficient IPv4/v6 Handovers Using Host-Based Identifier-Location Split, 2010), (Varjonen, Komu, & Gurtov, Secure and Efficient IPv4/IPv6 Handovers Using Host-Based Identifier-Locator Split, 2009) proposed for HIP suggests that the locators should be sent in the Base Exchange with preferred bits unset. Locators communicated this way should be considered as alternative locators, which do not have to be used immediately. This makes cross-family handovers possible because peers can inform each other about all available addresses they are reachable at. Cross-family handovers are having high importance as they can aid transition from IPv4 towards IPv6. This makes coexistence and transition less painful.

Location Privacy

With traditional mobility management protocols, mobile nodes are informing their peers if they are available at a new location. However, this communication is rarely protected in a sufficient way and does not provide location privacy for the users. If a third entity captures location update messages, it can locate the sender and trace its movements, which most users want to avoid. A recent extension (Maekawa & Okabe, 2009) to HIP enables to empower the base protocol with location privacy features. The proposal combines two solutions to enable complete location privacy for mobile HIP users. An earlier solution (Matos, Santos, Girao, Liebsch, & Aguiar, 2006), (Matos, Santos, Sargento, Aguiar, Girao, & Liebsch, 2006) defined a limited location privacy framework for HIP while BLIND (Ylitalo & Nikander, BLINDYlitalo & Nikander, BLIND: A Complete Identity Protection Framework for End-Points, 2006) provided a much enhanced solution but without mobility support. This novel scheme integrates the strong points of the above two methods and introduces complete location privacy with mobility. The key idea is to use Temporary Host Identifiers (THI) for location update purposes. THIs are constructed so that capturing them provides no information about the location of the user.

HIP in 3G and Beyond Mobile Architectures

The use of Host Identity Protocol in telecommunication architectures such as 3G UMTS, LTE/EPS, etc., can provide number of benefits and possible achievements but also can cause a number of issues for the network operator. Here we summarize aspects of HIP in 3G and beyond systems by means of mobility management, focusing on HIP-SIP integration and network interface selection. A wider survey on issues arising when a network operator wants to deploy HIP for its own customers is presented by authors of (Dietz, Brunner, Papadoglou, Raptis, & Kypris, 2005).

Today and future telecommunication networks likely rely on the wide-scale capabilities of IP Multimedia Subsystem (IMS) and Sessions Initiation Protocol (SIP) protocol. Authors of (Rothenberg, Wong, Verdi, & Magalhae, 2008) state that IMS and SIP can be seamlessly supported and can clearly benefit from any identifier/locator split architecture in terms of security, mobility and multihoming based on transparently using permanent cryptographic node identifiers decoupled from the actual network locators. Recent standardization activities of the ITU Telecommunication Standardization Sector (ITU-T) also encourage operators for integration of the ID/Loc separation concept into future network architectures (Y.ipsplit, 2009), (Y.ipv6split, 2009). Based on the above and according to (Henderson, 2004) it can be concluded that the main benefit a HIP-enabled IMS/SIP networking infrastructure would offer is advanced and more efficient mobility management (such as micromobility or network mobility management), possible integration of rendezvous servers with SIP proxy and redirection servers, and NAT traversal.

Author of (Heikkinen, 2008) tries to provide a HIP-SIP integrated IMS architecture in order to allow enhanced interaction that takes into account the liability and reliable identification needs of the different entities partaking in communication, i.e. the user, the access operator, and also the home network. This scheme allows non-repudiative accounting records to be made, meaning that if the user has used the service, this cannot be denied afterwards. Similarly, if the user does not seem to be honest in its service usage in terms of compensation, the service will no longer be provided.

In (Camarillo, Mas, & Nikander, 2008) authors present a complete HIP-SIP interworking framework where SIP-based services between HIP-enabled hosts take advantage of the mobility, security, and multihoming capabilities of HIP. Additionally, some SIP-based services can also benefit from the IPSec-based data encryption provided by HIP.

A HIP based mobility management extension of 3G UMTS architecture is presented in (So & Wang, 2006), where the focus is on heterogeneous UMTS/WLAN access environments. Implementation work is also in progress regarding HIP-SIP integration: (Wagner, 2006) presents the first measurement results of a real-life HIP-SIP interworking system.

The I-HSIP scheme (Li, Chen, Su, Jin, & Zeng, 2009) also presents an integrated HIP-SIP based mobility management architecture for next generation mobile wireless networks in order to support secure and efficient mobility services for both real-time (running over RTP/UDP and SIP) and non-real-time (applying TCP and HIP) applications. This architecture merges the similar operations and entities of HIP and SIP (such as SIP server and HIP RVS), aiming to minimize the system signaling cost and to improve handover performance.

SHIP (Yang, Zhou, Qin, & Zhang, 2009) is a cross-layer mobility management scheme based on SIP and HIP. SHIP focuses on handover optimization: by the way, of integration a one-suite protocol stack comes into existence that provides seamless mobility and fast handoff. The basic idea behind SHIP is that the MN and the CN use their HITs to establish the SIP sessions, and the MN uses the HIP update procedures instead of the SIP location update scheme in case of handovers.

Also a very active research area of HIP-SIP interworking in future communication architectures is the use of HIP for the design of a peer-to-peer version of SIP (P2PSIP) (Bryan, Matthews, Shim, Willis, & Dawkins, 2008). Several proposals suggest either to use HIP directly e.g., (Hautakorpi, Camarillo, & Koskela, 2007), or just borrow ideas from HIP like (Cooper, Johnston, & Matthews, 2007).

FUTURE RESEARCH DIRECTIONS

As described above in a locator/ID separating system there are at least two namespaces. One includes routing locators (RLOCs) while the other collects endpoint identifiers (EIDs). Some representations of the core separation idea are using intermediate mapping systems to connect the two namespaces. In these systems, the EIDs are used to locate endpoints in the edge network while RLOCs are used to route packets between edge systems. The actual association between EIDs and RLOCs are stored by the mapping system. The association is updated when a host moves from one edge network to another. In this case, the permanent EID is mapped to a new RLOC. The routing architecture determines whether the endpoints (e.g., in HIP) or intermediate entities (e.g., in LISP) at the border of edge networks are in charge to query the mapping system to get the actual RLOC of a packet's destination. The latter case imposes a complex challenge towards the mapping systems, which has to be fast in providing the actual mapping. Packets are congested or dropped at these middleboxes until the mapping is not resolved by the mapping system. There are many recent researches focusing on such mapping systems, designing them and identifying the requirements they must meet (e.g., fast resolution, resilient to outage and security).

Mapping systems can be classified into four major categories (Menth, Hartmann, & Hoefling, Mapping Systems for Loc/ID Split Internet Routing, 2010). (1) Direct Map-Bases (DMBs) are the straightforward solutions to the problem. They store the entire mapping database on a single server containing the same database. In a bit sophisticated solution, the database is divided among the servers and a central entity ensures that they receive requests falling into the address space they are responsible of. The RVS system of HIP combined with DNS support is an example. (2) In Map-Request Forwarding Overlays (MRFO), the mapping database is distributed among mapping servers and an overlay networks is responsible to route mapping requests to the correct destination. The sender of the request must be aware of an overlay entry point but not the exact server storing the requested mapping. A recent proposal (Luo, Qin, & Zhang, 2009) implements such an architecture utilizing distributed hash tables. (3) Two Level Mapping Systems (TLMS) is a more general approach. It defines button and top level mapping system. Button level systems are storing direct EID-RLOC mapping. The top-level system has information about the distribution of the address space among button level systems. Both top level and button level systems can be implemented as DMBs or MRFOs. The requesting entity first uses the top-level system to learn which button level system can resolve the requested mapping. Then the button level system provides the EID-RLOC information. A TLMS proposal is FIRMS (4). Finally, DNS-Based Mapping Systems (DBSMS) can work with hierarchical EID and functioning quite similar to DNSs today (Yang, Luo, Qin, & Zhang, 2009).

As mentioned above the mapping architecture of HIP is a simple DMB. The Host Identity Indirection Infrastructure (Hi3) is a proposal that upgrades HIP mapping system to a MRFO (Gurtov, Korzun, Lukyanenko, & Nikander, 2008). Such upgrade alone has many advantages but Hi3 provides much more. This new architecture is derived from HIP and the Internet Indirection Infrastructure (Stoica, Adkins, Zhuang, Shenker, & Surana, 2002). The latter is an overlay network abstraction that provides rendezvous based communication for endpoints. Hi3, which compared to its building blocks, inherits the best from both approaches. While improving the performance of i3 it enhances flexibility and security of HIP. As i3 is an overlay infrastructure, it can provide a secure MRFO for HIP control messages. As data packets in Hi3 are routed in an end-to-end manner, the proposed architecture can be considered as a secure control plane for HIP based communication sessions. The separation of data and control plane makes network architectural design easier while inherited from HIP mobility and multihoming are available as a generic feature. Moreover, thanks to the decentralized overlay the system is more resilient to outages and Denial-of-Service attacks.

There are several research projects that target to define a new architecture for the future Internet. The most important ones are FIND (Future Internet Design) and GENI (Global Environment for Network Innovations) in the USA, FIRE (Future Internet Research and Experimentation) and 4WARD (The FP7 4WARD Project) in Europe, finally AKARI (Harai, 2009) and Future Internet Forum (Future Internet Forum) in Asia. The importance of locator-identifier separating techniques is clear, as all these projects have adopted some parts of this idea. Due to its wide-scale potential, Host Identity Protocol is a very promising candidate for application in these research activities, despite the fact that the details of the proposed architectures are diverging in means if the routing architecture, mapping system, and namespace design. The common goal is to define general architectures and frameworks for future new generation networks. These projects do not provide solutions to only a subset of problems but they address all requirements identified for those networks and the ID/Loc splitting paradigm seems to be the platform for reaching such goals.

CONCLUSION

In this chapter, we presented the basics of Host Identity Protocol focusing on how it can be utilized to support advanced mobility scenarios and manage complex mobility management problems. The main part of the chapter was devoted to introduce HIP's built-in mobility management capabilities followed by its numerous extensions designed to implement mechanisms like micromobility, network mobility, per-application mobility, simultaneous end-host mobility, dual-stack handover support, location privacy, and application of HIP in 3G and beyond architectures. Our goal was to show that HIP is one of the most promising instances of the ID/Loc separation concept, on which several advanced mobility services and applications can be built, while also keeping the generic security and unique architectural properties of the base protocol. We also tried to prove that HIP and the related research work are quite far from being ready or closed. A lot of effort was put into this research area, but a lot more design and implementation experience will be needed to ensure the widespread usage of Host Identity Protocol and the integration of the mobility management schemes based on it.

ACKNOWLEDGMENT

This work was made in the frame of Mobile Innovation Centre's 'MEVICO.HU' project, supported by the National Office for Research and Technology (EUREKA_Hu_08-1-2009-0043) under the co-operation of the Celtic Call7 Project MEVICO.

REFERENCES

Aydin, Z. G., Chaouchi, H., & Zaim, A. H. (2009). *eHIP: Early update for host identity protocol.* 6th International Conference on Mobile Technology, Application & Systems. Nice, France.

Bokor, L., Nováczki, S., & Imre, S. (2007). *A complete HIP based framework for secure micromobility.* 5th @WAS International Conference on Advances in Mobile Computing and Multimedia, MoMM2007, (pp. 111-122). Jakarta, Indonesia.

Bokor, L., Zeke, L. T., Nováczki, S., & Jeney, G. (2009). *Protocol design and analysis of a HIP-based per-application mobility management platform.* 7th ACM International Symposium on Mobility Management and Wireless Access (MobiWAC 2009), (pp. 7-16). Tenerife, Canary Islands, Spain.

Bryan, D., Matthews, P., Shim, E., Willis, D., & Dawkins, S. (2008, July 7). *Concepts and terminology for peer to peer SIP.* IETF Internet Draft. (draft-ietf-p2psip-concepts-02).

Camarillo, G., Mas, I., & Nikander, P. (2008). *A framework to combine the session initiation protocol and the host identity protocol.* Wireless Communications and Networking Conference (WCNC 2008. IEEE). Las Vegas, USA.

Cooper, E., Johnston, A., & Matthews, P. (2007, December). *A distributed transport function in P2PSIP using HIP for multi-hop overlay routing.* IETF Internet Draft. (draft-matthews-p2psip-hip-hop-00).

Dietz, T., Brunner, M., Papadoglou, N., Raptis, V., & Kypris, K. (2005). *Issues of HIP in an operators networks.* (draft-dietz-hip-operator-issues-00).

Farinacci, D., Fuller, V., Meyer, D., & Lewis, D. (2010, April 25). *Locator/ID separation protocol* (LISP). IETF Internet Draft. (draft-ietf-lisp-07).

Fekete, G. (2009). *Policy based flow management with the host identity protocol for multihomed hosts.* First International Conference on Emerging Network Intelligence. Sliema, Malta.

Future Internet Design. (n.d.). Retrieved April 15, 2010, from http://www.nets-find.net/

Future Internet Forum. (n.d.). Retrieved April 15, 2010, from http://fif.kr/

Future Internet Research and Experimentation. (n.d.). Retrieved April 15, 2010, from http://www.future-internet.eu/index.php?id=54

Global Environment for Network Innovations. (n.d.). Retrieved April 15, 2010, from http://www.geni.net/

Gurtov, A., Korzun, D., Lukyanenko, A., & Nikander, P. (2008). Hi3: An efficient and secure networking architecture for mobile hosts. *Computer Communications*, *38*, 2457–2467. doi:10.1016/j.comcom.2008.03.014

Harai, H. (2009). *AKARI architecture design for new generation network*. Summer Topical Meeting, 2009. LEOSST '09., (pp. 155 - 156). Newport Beach, CA.

Hautakorpi, J., Camarillo, G., & Koskela, J. (2007, November). *Utilizing HIP (Host Identity Protocol) for P2PSIP (Peerto- peer Session Initiation Protocol)*. IETF Internet Draft. (draft-hautakorpi-p2psip-with-hip-01).

Heikkinen, S. (2008). Security and accounting enhancements for roaming in IMS. 6th International Conference on Wired/Wireless Internet Communications (WWIC 2008). [Tampere, Finland/ Berlin, Germany: Springer.]. *Lecture Notes in Computer Science*, *5031*, 127–138. doi:10.1007/978-3-540-68807-5_11

Henderson, T. (2004). *Can SIP use HIP*. HIP Workshop, 61st IETF meeting. Retrieved April 15, 2010, from http://hiprg.piuha.net/workshop/

Herborn, S., Haslett, L., Boreli, R., & Seneviratne, A. (2006). *HarMoNy - HIP mobile networks*. 63rd IEEE Vehicular Technology Conference, VTC.

Herborn, S., Huber, A., Boreli, R., & Seneviratne, A. (2007). *A scheme for host identity delegation*. IEEE/Create-Net/ICST International Conference on COMmunication System softWAre and MiddlewaRE (COMSWARE). Bangalore, India.

Hobaya, F., Gay, V., & Robert, E. (2009). *Host identity protocol extension supporting simultaneous end-host mobility*. Fifth International Conference on Wireless and Mobile Communications (ICWMC '09), (pp. 261-266). Cannes/La Bocca, France.

Iapichino, G., & Bonnet, C. (2009). *Host identity protocol and proxy mobile IPv6: A secure global and localized mobility management scheme for multihomed mobile nodes*. IEEE Global Telecommunications Conference (GLOBECOM'09), (pp. 1-6). Honolulu, Hawaii.

Iapichino, G., Bonnet, C., Herrero, O. d., Baudoin, C., & Buret, I. (2009). *Combining mobility and heterogeneous networking for emergency management: A PMIPv6 and HIP-based approach*. 2009 International Conference on Wireless Communications and Mobile Computing: Connecting the World Wirelessly, (pp. 603-607). Leipzig, Germany.

Jokela, P., Melen, J., & Ylitalo, J. (2006, June). *HIP service discovery*. IETF Internet Draft. (draft-jokela-hip-service-discovery-00).

Jokela, P., Moskowitz, R., & Nikander, P. (2008, April). *Using the encapsulating security payload (ESP) transport format with the host identity protocol* (HIP). (RFC 5202).

Kovácsházi, Z., & Vida, R. (2007). *Host identity specific multicast*. International Conference on Networking and Services (ICNS '07). Athens, Greece.

Laganier, J., & Eggert, L. (2008). *Host identity protocol (HIP) rendezvous extension*. (RFC 5204).

Laganier, J., Koponen, T., & Eggert, L. (2008). *Host identity protocol (HIP) registration extension*. (RFC 5203).

Li, Y., Chen, W., Su, L., Jin, D., & Zeng, L. (2009). *Mobility management architecture based on integrated HIP and SIP protocols*. International Conference on Telecommunications, ICT '09, (pp. 243-247). Marrakech, Morocco.

Luo, H., Qin, Y., & Zhang, H. (2009). A DHT-based identifier-to-locator mapping approach for a scalable Internet. *IEEE Transactions on Parallel and Distributed Systems*, *20*, 1790–1802. doi:10.1109/TPDS.2009.30

Maekawa, K., & Okabe, Y. (2009). *An enhanced location privacy framework with mobility using host identity protocol.* Ninth Annual International Symposium on Applications and the Internet (SAINT '09), (pp. 23 - 29). Bellevue (WA).

Matos, A., Santos, J., Girao, J., Liebsch, M., & Aguiar, R. (2006, March). *Host identity protocol location privacy extensions.* IETF Internet Draft. (draft-matoship-privacy-extensions-01).

Matos, A., Santos, J., Sargento, S., Aguiar, R., Girao, J., & Liebsch, M. (2006). *HIP location privacy framework.* First ACM/IEEE International Workshop on Mobility in the Evolving Internet Architecture, (pp. 57–62). New York, NY: ACM Press.

Melen, J., Ylitalo, J., Salmela, P., & Henderson, T. (2009, May 26). *Host identity protocol-based mobile router* (HIPMR). IETF Internet Draft. (draft-melen-hip-mr-02).

Menth, M., Hartmann, M., & Hoefling, M. (2010, May). *Mapping systems for Loc/ID split Internet routing.* (Technical Report No. 472). University of Würzburg.

Menth, M., Hartmann, M., & Klein, D. (2010, April). *Global locator, local locator, and identifier split* (GLI-Split). (Technical Report No. 470). University of Würzburg.

Moskowitz, R., & Nikander, P. (2006, May). *Host identity protocol (HIP) architecture.* (IETF RFC 4423).

Moskowitz, R., Nikander, P., Jokela, P., & Henderson, T. (2008, April). *Host identity protocol.* (RFC 5201).

Muslam, M., Chan, H., & Ventura, N. (2009). *HIP based micro-mobility management optimization.* Fifth International Conference on Wireless and Mobile Communications, ICWMC '09., (pp. 291–295). Cannes, La Bocca.

Nikander, P., & Arkko, J. (2004). Delegation of signaling rights. (LNCS 2845), (pp. 203-214). Security Protocols 2002.

Nikander, P., Henderson, T., Vogt, C., & Arkko, J. (2008). *End-host mobility and multihoming with the host identity protocol.* (RFC 5206).

Nikander, P., & Laganier, J. (2008, April). *Host identity protocol domain name system extension.* (RFC 5205).

Nordmark, E., & Bagnulo, M. (2009). *Shim6: Level 3 multihoming Shim protocol for IPv6.* (RFC 5533).

Nováczki, S., Bokor, L., & Imre, S. (2006). Micromobility support in HIP: Survey and extension of host identity protocol. [Málaga, Spain.]. *MELECON*, *1*, 651–654.

Nováczki, S., Bokor, L., & Imre, S. (2007). A HIP based network mobility protocol. [Hiroshima, Japan.]. *SAINTWONEMO*, *2007*, 48–52.

Nováczki, S., Bokor, L., Jeney, G., & Imre, S. (2008). Design and evaluation of a novel HIP-based network mobility orotocol. *Journal of Networks*, *3*(1), 10–24. doi:10.4304/jnw.3.1.10-24

Pierrel, S., Jokela, P., & Melen, J. (2006, June 19). *Simultaneous multi-access extension to the host identity protocol.* IETF Internet-Draft. (draft-pierrel-hip-sima-00).

Pierrel, S., Jokela, P., Melen, J., & Slavov, K. (2007). *A policy system for simultaneous multiaccess with host identity protocol.* IEEE ACNM2007. Munich, Germany.

Rothenberg, C. E., Wong, W., Verdi, F. L., & Magalhae, M. F. (2008). *SIP over an identifier/ locator splitted next generation Internet architecture*. 10th IEEE International Conference on Advanced Communication Technology (ICACT08). Republic of Korea.

So, J., & Wang, J. (2008). *Micro-HIP: A HIP-based micro-mobility solution*. IEEE International Conference on Communications Workshops (ICC Workshops '08.), (pp. 430-435). Beijing, China.

So, J. Y., & Wang, J. (2006). *HIP based mobility management for UMTS/WLAN integrated networks*. Australian Telecommunication Networks and Applications Conference.

Stoica, I., Adkins, D., Zhuang, S., Shenker, S., & Surana, S. (2002). Internet indirection infrastructure. [Pittsburgh, PA, USA.]. *ACM SIGCOMM, 02*, 73–88. doi:10.1145/964725.633033

Teraoka, F., Ishiyama, M., & Kunishi, M. (2003, June 24). *LIN6: A solution to mobility and multihoming in IPv6*. IETF Internet Draft. (draft-teraoka-ipng-lin6-02).

The FP7 4WARD Project. (n.d.). Retrieved April 15, 2010, from http://www.4ward-project.eu/

Varjonen, S., Komu, M., & Gurtov, A. (2009). *Secure and efficient IPv4/IPv6 handovers using host-based identifier-locator split*. 17th International Conference on Software, Telecommunications and Computer Networks (SoftCOM 2009), (pp. 111 - 115). Split-Hvar-Korcula.

Varjonen, S., Komu, M., & Gurtov, A. (2010). Secure and efficient IPv4/v6 handovers using host-based identifier-location split. *Journal of Communications Software and Systems, 6*(1).

Wagner, S. (2006, December 1). *Implementation and evaluation of the interaction between host identity protocol and session initiation protocol*. Masterarbeit im Studiengang Angewandte Informatik.

Yang, S., Luo, H., Qin, Y., & Zhang, H. (2009). Design and evaluation of DNS as location manager for HIP. *Journal Wireless Personal Communications, 48*(4), 605–619. doi:10.1007/s11277-008-9550-x

Yang, S., Qin, Y., Luo, H., & Zhang, H. (2008). *P-HIP: Paging extensions for host identity protocol. IEEE Communications Workshops, ICC Workshops '08*. Beijing: Chaina.

Yang, S., Qin, Y., & Yang, D. (2007). Dynamic hierarchical location management scheme for host identity protocol. [*Mobile Ad-Hoc and Sensor Networks Proceedings*.]. *Lecture Notes in Computer Science*, 4864.

Yang, S., Zhou, H., Qin, Y., & Zhang, H. (2009). SHIP: Cross-layer mobility management scheme based on session initiation protocol and host identity protocol. *Journal of Telecommunication Systems, 42*(1-2).

Y.ipv6split, I.-T. (2009, March 5). *Framework of ID/LOC separation in IPv6-based NGN*. ITU-T draft recommendation.

Y.ipsplit, I.-T. (2009, Feb. 6). *General requirements for ID/locator separation in NGN*. ITU-T draft recommendation.

Ylitalo, J. (2005). *Re-thinking security in network mobility*. NDSS Wireless and Security Workshop. San Diego, CA, USA.

Ylitalo, J., & Melé, J. P. S., & Petander, H. (2008). An experimental evaluation of a HIP based network mobility scheme. WWIC 2008, *LNCS, 5031/2008*, (pp. 139–151).

Ylitalo, J., Melén, J., Nikander, P., & Torvinen, V. (2004). Re-thinking security in IP based micro-mobility. *Lecture Notes in Computer Science, 3225/2004*, (pp. 318-329).

Ylitalo, J., & Nikander, P. (2006). BLIND: A complete identity protection framework for end-points. *Lecture Notes in Computer Science*, 3957.

ADDITIONAL READING

Gurtov, A., Host Identity Protocol (HIP): Towards the Secure Mobile Internet, ISBN 978-0-470-99790-1, Wiley and Sons, June 2008.

KEY TERMS AND DEFINITIONS

Dual-Stack Mobility: Dual-stack mobility refers to the special scenario in which a host roams between different addressing realms. A host moving from an IPv4 domain to a network using IPv6 must be able to handle this situation if it wants to keep its ongoing communication sessions. Dual stack mobility proposals are playing a key role in IPv6 deployment, as seamless mobility can be provided across IP address families.

Host Identity Protocol: The Host Identity Protocol is currently the most widespread proposal that implements separation of both main IP address functions (i.e., locator and identifier). It defines a new cryptographic namespace for identifiers and a new protocol layer that handles this namespace and integrates it into the traditional TCP/IP stack. HIP provides built-in security services and makes mobility management simple and fast.

ID/Loc Separation: In the traditional widespread version of the Internet Protocol (IPv4 and IPv6) network entities are assigned to a specially designed number, the IP address. This address has two roles. On one hand, it points to the location of the communicating entity inside the network topology. On the other hand, it is used to identify the entity itself. Thus, the identity and location information is strongly bound to each other, which naturally makes certain scenarios, like mobility management and security services, quite complex to handle. The idea of ID/Loc separation aims to break this tight coupling between locators and identifiers making the current Internet architecture more flexible and scalable.

Micromobility: If wireless networking domains are aggregated and a special protocol is responsible for the local mobility management of this group of domains in order to offer fast and seamless handover control over a limited geographical area, than we speak about micromobility, the aggregated group of domains is called micromobility domain, and the special control protocol is called micromobility protocol.

Network Mobility: A special mobility scenario. It arises when a router–connecting a network to the Internet dynamically– changes its point of attachment to the fixed infrastructure, thereby causing the accessibility of the entire network to be changed in relation to the fixed Internet topology.

Per-Application Mobility: If a mobile user runs several applications (e.g. file downloading, voice communication, video streaming, e-mail) on a device with multiple interfaces and simultaneously available access networks, and the actual connection (i.e., out- and inbound interface) of each application is handled independently according to QoS requirements and environmental parameters, than we are talking about per-application mobility.

Signaling Delegation: A method used by a network entity (the delegator) to securely offer to another one (usually a middlebox, called the delegate) the ability to send signaling messages on behalf of it.

APPENDIX

Table of abbreviations

3G	3rd Generation
AR	Access Router
BEET	Bound End-to-End-Tunnel
BEX	Base Exchange
CN	Core Network
Co-A	Co-Agent
DBSMS	DNS-Based Mapping System
DH-HIP	Dynamic Hierarchical HIP
DMB	Direct Map-Base
DNS	Domain Name System
DoS	Denial-of-Service
EID	Endpoint ID
eHIP	Early Update for HIP
EPS	Evolved Packet System
ESP	Encapsulating Security Payload
FIND	Future Internet Design
FIRE	Future Internet Research and Experimentation
FIRMS	Future InteRnet Mapping System
GENI	Global Environment for Network Innovations
GLI	Global Locator, Local Locator, and Identifier
GRVS	Gateway Rendezvous Server
HarMoNy	HIP and Network Mobility
HI	Host Identifier
Hi3	Host Identity Indirection Infrastructure
HIP	Host Identity Protocol
HIPMR	HIP-based Mobile Router
HISM	HIT Specific Multicasting
HIT	Host Identity Tag
ID	Identifier
I-HSIP	Integrated HIP-SIP
IMS	IP Multimedia Subsystem
IP	Internet Protocol
IPsec	Internet Protocol security
IPv4	Internet Protocol version 4
IPv6	Internet Protocol version 6
ITU	International Telecommunication Union

ITU-T	International Telecommunication Union - Telecommunications
LFN	Local Fixed Node
LINA	Location Independent Network Architecture
LIN6	Location Independent Networking for IPv6
LISP	Locator/ID Separation Protocol
Loc	Locator
LRVS	Local Rendezvous Server
LTE	Long Term Evolution
MIP	Mobile IP
MitM	Man-in-the-Middle
MN	Mobile Node
MNN	Mobile Network Node
MR	Mobile Router
MRFO	Map-Request Forwarding Overlay
mRVS	Mobile Rendezvous Server
NAT	Network Address Translation
NEMO	Network Mobility
P2PSIP	peer-to-peer SIP
PA	Paging Area
P-HIP	Paging Extensions for HIP
PMIP	Proxy Mobile IP
QoS	Quality of Service
RFC	Request for Comments
RLOC	Routing Locator
RTP	Real-time Transport Protocol
RTT	Round Trip Time
RVS	Rendezvous Server
SA	Security Association
SHIM6	Site Multihoming by IPv6 Intermediation
SHIP	SIP and HIP
SIP	Session Initiation Protocol
SPI	Security Parameter Index
TCP	Transmission Control Protocol
THI	Temporary Host Identifiers
TLMS	Two Level Mapping System
UDP	User Datagram Protocol
UMTS	Universal Mobile Telecommunications System
USA	United States of America
WLAN	Wireless Local Area Network

Chapter 13
Overview of IP Multimedia Subsystem Protocols and Communication Services

Sándor Szabó
Budapest University of Technology and Economics, Hungary

László Gyöngyösi
Budapest University of Technology and Economics, Hungary

Károly Lendvai
Budapest University of Technology and Economics, Hungary

Sándor Imre
Budapest University of Technology and Economics, Hungary

ABSTRACT

The IP multimedia subsystem (IMS) provides advanced facilities and standardized service architecture for Next Generation Networks (NGN). IMS enables both cellular and fixed network operators to provide a wide and attractive service portfolio to the customers. In this chapter, we provide an overview of the IMS architecture and of the most popular IMS services. We discuss the basic components and mechanisms defined for IMS, the related IMS services, and we give a survey on the future of the evolving telecommunication networks. We examine the most popular IMS services such as Presence Service (PS), Instant Messaging (IM), and the background of these IMS services. The IMS system enables fixed-mobile convergence, and it bridges the gap between the fixed and mobile communications. To understand and use the new concepts and IMS services, it is important to know what the new IMS services and concepts bring to us. Our goal is to give a brief overview in order to help understanding today's and tomorrow's relevant communication technologies and protocols.

INTRODUCTION

Today's new communications services must be able to be a part of complex multimedia applications and advanced IP-based services must also provide circuit-switched equivalent performance over cellular access. The convergence of fixed and mobile services merges real-time voice, video,

DOI: 10.4018/978-1-60960-732-6.ch013

and messaging applications together with data services onto a single broadband infrastructure, based on IP technology.

IMS supports the principle of access independence and IMS services can be provided over a variety of IP connectivity access networks such as cellular, XDSL, WLAN, and cable systems (3GPP Overview). In IMS services, a wide range of media types can be used to support many Quality of Services (*QoS*) parameters required by the services (3GPP TS 23.107). The IMS services must be capable of supporting multiple multimedia applications within a single session, which requires advanced service control mechanisms and functions. IMS communications services can be discussed in terms of the network model, media combinations and media formats supported by the IMS service (Poikselka, 2006).

3GPP Release 99 made available basic packet services for use in the IMS architecture, while Release 5; defined the architecture over the GPRS service layer and SIP signaling method (RFC 3261). The basic architecture of IMS had been specified in 3GPP Release 5, these specifications were developed in 2001–02. IMS Release 5 made it possible to use VoIP over cellular access technology and the impacts of Release 5 were greater than ever thought.

The fundamentally new IMS specifications have been integrated into practical applications during the past few years; however, these IMS specifications did not define the well-known and widely applied and popular VoIP services (Bang, 2007). The next step was 3GPP Release 6, which defined many advanced functions over 3GPP Release 5. The function of Release 6 integrated the non-GPRS based access technologies and SIP-based Terminal Equipment (RFC 3261; SIPForum, 2010). However, later, 3GPP Release 7 defined the required function to give the IMS operator full control over the procedures of the Quality of Services negotiation procedure.

BACKGROUND

The convergence of fixed and mobile networks has made it possible to use new technologies for end-users and for operators and vendors. The evolution of fixed and mobile technologies results in cost-effective advanced access technologies (Harrowell, 2006). The IP Multimedia Subsystem (*IMS*) architecture is a universal communication platform. Its architecture offers many voice, video, and multimedia services as well as end-user applications.

The IMS architecture provides multimedia services based on Internet technologies and SIP protocol (Burger, 2006; SIP, 2008). SIP is the main signaling protocol for the IMS architecture and, using SIP, any IMS terminal can communicate with any SIP-based Internet terminal (3GPP Overview). The 3GPP Release specifications define the IMS architecture with mobility and differentiated QoS services (Hwang, 2008).

Many services are based on IMS, such as:

- Multimedia Messaging Service (MMS),
- Push-to-Talk,
- Instant Messaging,
- Presence Service,
- Location Services.

The successful collaboration of Open Mobile Alliance (OMA) (OMA Overview), European Telecommunications Standards Institute (ETSI), and Technical Committee within ETSI, which works on standards for Next Generation Networks (TISPAN), has made it possible to use these advanced communication techniques.

According to collaborative joint working, the core network becomes the entity that decides suitable QoS for a certain media stream associated with a service request (3GPP TS 23.107). IMS sessions allow differentiated telecommunication services to be used and they allow IMS operators to make IMS services adapted to the market, with customized QoS services and customized specialized functions (3GPP TS 23.107).

IMS ARCHITECTURE

The IMS services are independent of the current network technology and the network interfaces. In addition, the interconnection between the services is standardized. The IMS architecture makes it easier to construct personalized services and cost-effective IMS applications. IMS uses the benefits of NGN convergence and multiple access technologies. IMS users are localized in different domains and use different access technologies; hence, they require different IMS services for their use.

The standardized service control between fixed and mobile access technologies is the IP-based SIP signaling protocol (Ilyas, 2008; SIP, 2008). In Figure 1, we illustrate the IMS domains and we can conclude that the IMS can be used to provide a Fixed Mobile Convergence (*FMC*) mode. In the FMC model, the Call/Session Control Functions (*CSCFs*) deployed in different domains are interconnected, which makes it possible for domains to access services in other domains (3GPP TR 32.200). The IMS architecture makes it possible to take practical advantage of the benefits of FMC

Figure 1. Fixed mobile convergence in IMS

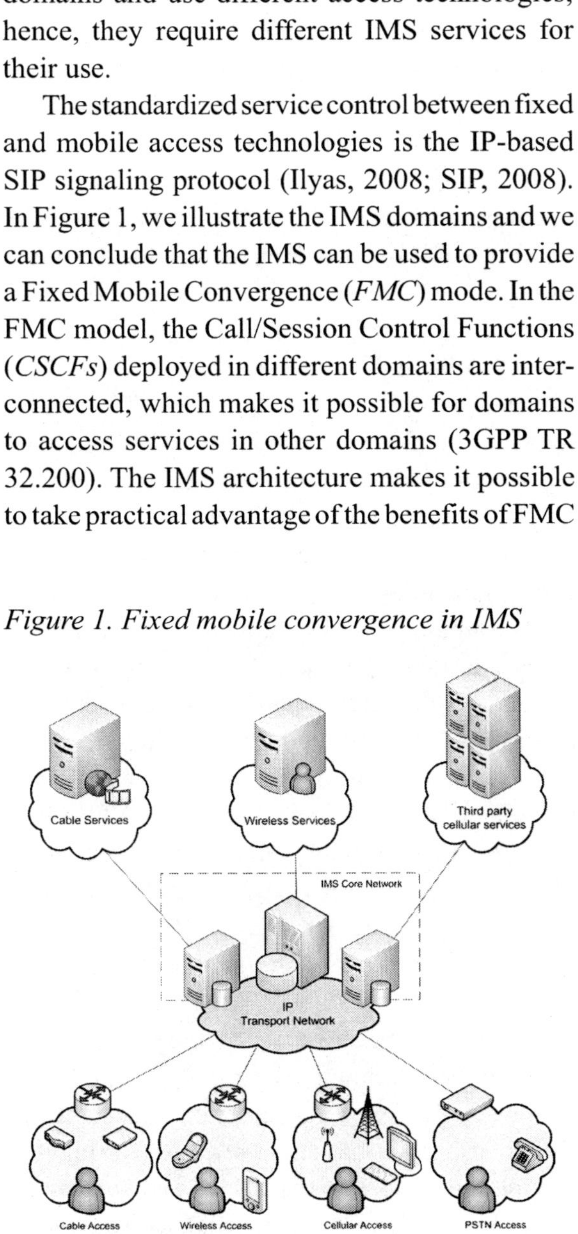

and it can connect fixed and mobile network architectures (3GPP TS 23.002).

3GPP standardizes the logical functions instead of nodes. The IMS functions can be mapped to nodes and the IMS architecture can be described by both methods. The IMS provides many advanced services for users across one user interface; hence, IMS makes it possible to use the advantages of IMS for both users and service providers (Poikselka, 2006).

The 3GPP 2007g standard defines IMS as centralized core network architecture with complex infrastructure, similar to intelligent network architecture. In an intelligent network architecture, the application servers are integrated in order to extend the capabilities of an existing circuit-switched core network architecture (3GPP TS 23.002; Khlifi, 2008). Moreover, 3GPP defines in this specification defines some combination services over IMS and CS networks and it introduces some types of emergency calls.

The IMS is defined on top of current telecom networks and ensures the independence of network and service layers (3GPP TS 23.002). IMS integrates different types of application servers to fulfill the requirements of user applications (Poikselka, 2006). The Application Server (*AS*) provides services to IMS users and IMS supports the communication of these SIP-based and non-SIP-based application servers (Burger, 2006). It uses *Diameter*-based interfaces to communicate with the Serving-CSCF (S-CSCF). The main functions of the *AS* can be summarized as follows:

- provides many value-added IMS services
- provides a *Service* Capability Interaction Manager (*SCIM*) to manage the interactions between users.
- it has a service control function, it can be connected to S-CSCF, and it can interact with the HSS to obtain the IMS profile information for an IMS user.

- supports the most important IMS applications, such as presence service, instant messaging, push-to-talk, and online charging.

In Figure 2, we illustrate the integration of multiple application servers in IMS. The CAMEL provides intelligent networks for the GSM, GPRS, and UMTS network. It also controls the IMS calls and it forms a bridge between the circuit-switched and the IP multimedia networks (Ilyas, 2008).

The standardization of IMS was achieved by several standardization bodies. These standards define the IMS architecture and parts of the IMS architecture. The original 3GPP purpose was to define the standards and technical parameters for 3G communication and 3G mobile devices on GSM core networks and radio access technologies.

Figure 2. Multiple server integration

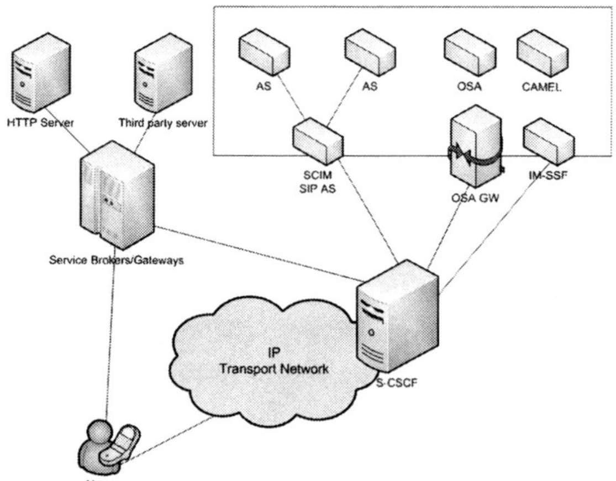

3GPP defined five technical groups, called *Technical Specification Groups* (TSGs). The 3GPP specifications and standards define the whole IMS architecture and these specifications describe completely the services and operations of the IMS architecture (3GPP TR 32.200).

IMS was introduced in 3GPP Release 5 and the work on further releases is currently under way. 3GPP release 5 defined IMS as a core network, with CSCF based on the SIP architecture (RFC 3261) and MGCF for circuit-switched calls. 3GPP Release 6 enhances the IMS core architecture and adds many entities and reference points to the IMS architecture. Release 6 defines High Speed Uplink Packet Access (*HSUPA*) technology, which allows high Mbps data rates, and it increases the maximum number of users in the network. Multiple-output (MIMO) technology was introduced in Release 7, this service having deep relevance to enhancing UMTS network performance (3GPP TS 23.002).

The network architecture for Long Term Evolution (LTE) was proposed in 3GPP Release 8 and later introduced E-UTRAN radio access technology. The LTE architecture has deep relevance to the current IMS service architecture, since LTE and E-UTRAN allow many devices to access IMS centralized services (3GPP TR 32.200).

Originally, the IMS architecture was designed to offer new Internet-based services and multimedia content for mobile devices, using cellular access technologies. In the initial phase of IMS, the main motivation was GPRS/UMTS cellular networking. Later releases of 3GPP extended the technologies to xDSL access, cable access, WLANs, and other types of fixed access network. In Figure 3, we illustrate the evolution of radio access technologies in 3GPP releases.

The technical background of IMS can be analyzed in terms of the phases of evolution of mobile communication techniques and its convergence towards all-IP communication techniques. The story starts with the communication technologies of cellular networks and ends with LTE technologies. The evolution of radio access technologies makes it possible to use higher bit rates and to utilize media-content-based internet communication technologies. The evolution of mobile networks to an all-IP IMS infrastructure can be discussed in similar terms to the evolution of radio access technologies.

The evolution of IMS architecture can be approached through the various releases. The Internet Engineering Task Force (IETF) is an international community (IETF Overview) and, with the estab-

Figure 3. Radio technology evolution

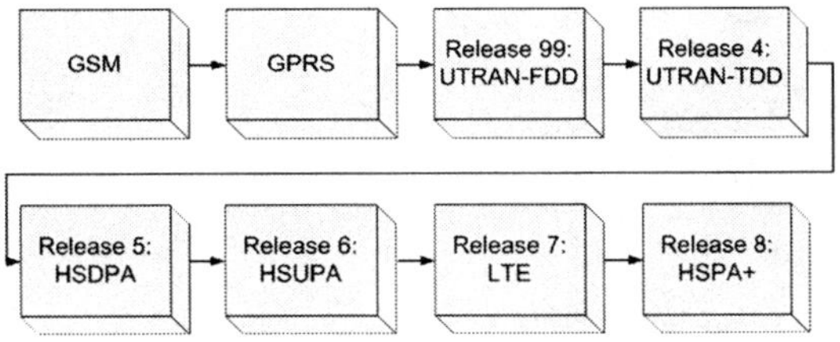

lishment of a set of guidelines, it collaborates with 3GPP. In the IETF, many working groups work on the technical specifications and the results of these groups are published in RFCs. There are many working groups in IETF of relevance to 3GPP and NGN networks, for instance:

- the IPv6 working group,
- Mobile IP, AAA (Authorization, Authentication and Accounting),
- SIP (Session Initiation Protocol),
- and many more (IETF Overview).

3GPP, IETF, ETSI, and ITU play a fundamental role in the development of IMS standards; however, there are many more organizations in the standardization market that have surfaced in recent years.

These standardization organizations define in IMS many standard user profiles and the terms of interoperability of different non-IMS services and IMS technologies (Poikselka, 2006). These new organizations are:

- TISPAN,
- PacketCable,
- Multiservice Forum (MSF),
- IMS Forum,
- and OMA (OMA Overview).

The OMA defines IMS services and functions and keeps a close alignment with 3GPP and 3GPP2.

ETSI TISPAN has developed an architecture for providing multimedia services over fixed accesses (3GPP TR 32.200). This architecture is called the Next Generation Networks (*NGN*) and is based on IMS.

In Figure 4, we illustrate the Core IMS architecture as defined by 3GPP. The most important nodes are:

- the Interconnect Border Control Function (*IBCF*),
- the Interconnect Border Gateway Function (*IBGF*),
- the Access Border Gateway Function (*ABGF*).

The IBGF is controlled by the IBCF (3GPP TR 32.200). The IBGF is on the media plane of the IMS architecture, and this element implements policy control on other networks as well as the Topology Hiding Interwork Gateway (*THIG*) functionality.

Figure 4. Core IMS architecture

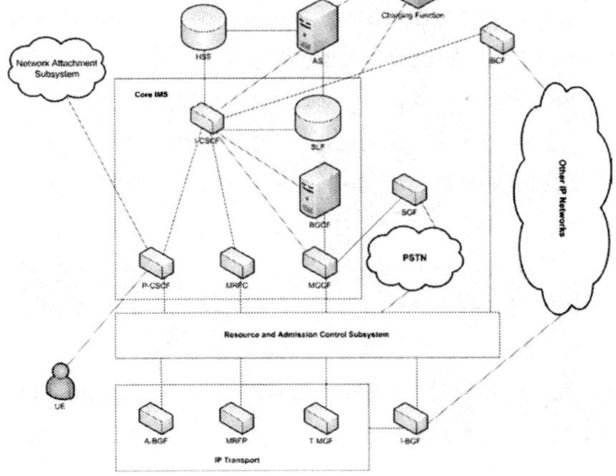

The ABGF is controlled by the *P-CSCF* and it is used as an access gateway element.

The essential network elements of the IMS core network architecture are used to provide real-time IMS services and the IMS uses packet-switched communication for data transport and local mobility management. As illustrated in Figure 4, the IMS can be deployed without a circuit-switched domain, because the core IMS uses a packet-switched domain only. Using NGN, users can enjoy the same set of services regardless of the access they have (3GPP TR 32.200). TISPAN NGN achieves the much-awaited convergence of mobile and fixed services (Mani, 2008).

In NGN, the Resource and Admission Control Sub-system (*RACS*) handles admission control and resource reservation. The Network Attachment Sub-System (*NASS*) handles network-level authentication and IP address provisioning. The network resources and applications of IMS ensure terminal and service mobility and the layered architecture makes it possible to define applications, services, and a transport layer with transport functions. The TISPAN NGN architecture is depicted in Figure 5.

The NGN architecture of TISPAN is based on the IMS architecture of 3GPP Release 6, however the TISPAN architecture is intended to serve operator and network operator demands.

Service-Centric IMS Model

As we have seen in the previous section, the evolution of IMS access technologies leads to an all-IP based network architecture. The convergence in this direction was mainly made possible by the SIP protocol (Garcia-Martin, 2006) and the IMS, and by the fact that the network architecture of IMS can be structured in a horizontally layered architecture (3GPP TR 32.200). In the horizontally structured model, the transport and bearer services are separated from the signaling and session management parts of the network architecture; hence,

Figure 5. NGN architecture of TISPAN

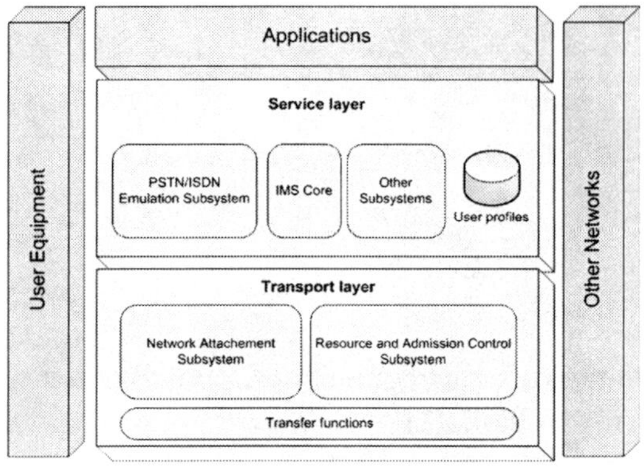

the layers can be discussed as independent and well-separated layers.

Using the horizontally layered view, the IMS separates the service functions and transport technologies. The service capabilities of IMS and the service delivery framework (O'Connell, 2007; TMF, 2009) of IMS can be studied as a service-centric horizontal-layered IMS structure, as illustrated in Figure 6.

The Service integration layer, the Service exposure layer and the Service enabling Layer are the part of IMS Service potential. The Integrator provides the communication between a Third party application server, which is part of the Service Application Layer (Khlifi, 2008). The Service Application Layer consists of another important part, which is called the Application Logic Containers Application Servers.

IMS Functions

Using IMS, developers can focus on the new service-specific logic and on new development. The IMS system will control the session and route the media stream regardless of the access types that are used and regardless of what operators are involved. The IMS architecture is backward compatible with circuit-switched systems and the functional gateway elements. In the functional model of IMS, the Home Subscriber Server (*HSS*) server node is basically a database, which stores

Figure 6. Service-centric representation of IMS

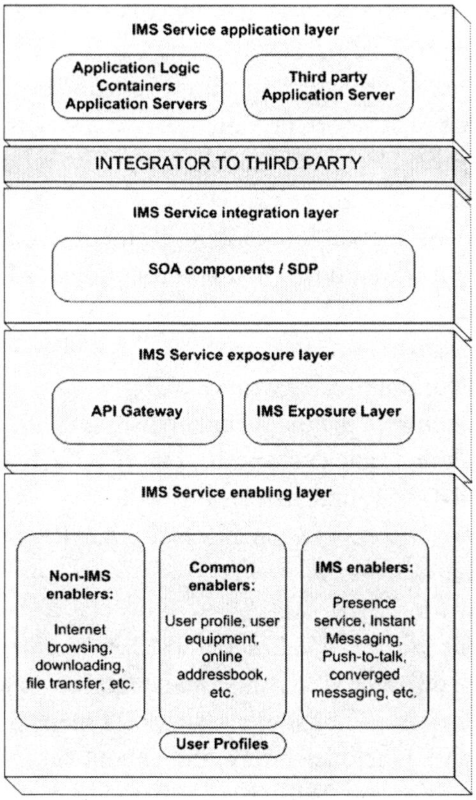

the profile of the IMS users in the given network domain. The Subscription Location Function (*SLF*) is queried by an IMS node to get information about an IMS user. The CSCFs are SIP entities (RFC 3261; SIPForum, 2010); however, they use Diameter-based interfaces.

The central component of the IMS network architecture is the Call Session Control Function (*CSCF*). The functions of the CSCF node are:

- handles the signaling routing information in network communications
- proxies SIP-based signaling traffic.
- provides session control services,
- registration,
- routing and roaming,
- combines different media bearers in an IMS session.
- performs central service-based charging functions,
- controls the secure authentication and the quality of service parameters of IMS services (3GPP TS 23.002).

The CSCF node uses the policy information obtained from the P-CSCF to make policy decisions (3GPP TS 23.002).

The CSCFs are of three types in the IMS architecture, Proxy-CSCF (*P-CSCF*), Serving-CSCF (*S-CSCF*), and Interrogating-CSCF (*I-CSCF*).

The P-CSCF is the first point of contact for the IMS terminal; – basically it is a SIP proxy – located in the visited network or in the home network. The main functions of the P-CSCF are:

- provides subscriber authentication,
- establish security with the IMS terminal,
- inspects the signaling,
- compress and decompress SIP messages,
- authorize media plane resources and QoS over the media plane,
- and generates charging records.

The S-CSCF is a SIP server, performs session control and the central node of the signaling plane. The S-CSCF is located in the home network, and its main functions are:

- communicates with the HSS through Diameter interfaces. The HSS stores the subscribers profile information.
- handles the SIP registrations,
- provides routing,
- enforces the policy of the network operator,
- and decides to which AS the SIP message will be forwarded.

The I-CSCF is located at the edge of an administrative domain, and the remote servers can use it as a forwarding point for SIP packets. The main functions of the S-CSCF are:

- forwards SIP request or response to the S-CSCF,
- queries the HSS to retrieve the address of the S-CSCF,
- performs user registration,
- and it can be used to hide the internal network from the outside world.

The Media Resource Function (*MRF*) enables media servers for next generation services and for media processing. The media severs make it possible to use audio and video content, and conversations between different media types (3GPP TS 23.002).

The MRF can be divided into controller (*MRFC*) and processor (*MRFP*) planes, hence the functions of media control and media processing are well separated. The MRFC can get instructions from an IMS application service and controls the media server using *H.248/Megaco* protocols (3GPP TS 23.002). We note that, in practical applications of IMS, the media server, and media gateway (RFC 3435) can be merged into one component.

The MRFP has many functions connected to MRFC controlled functions and it mixes the incoming media streams and processes, and handles different types of media stream for multiple parties. The full list of MRF functions and MRF-related standards is beyond the scope of this chapter, readers who are not familiar with the question can find details in the reference section (3GPP TS 23.002).

The Breakout Gateway Control Function (*BGCF*) is another important IMS function, it controls calls between IMS and PSTN and it selects the network for PSTN breakout.

The Media Gateway (*IMS-GW*) plays a fundamental role in the communication of the media plane of the Public Switched Telephone Network (*PSTN*) and the Circuit-Switched (*CS*) network. The MGW makes it possible for IMS users to send media content over the classical Real Time Protocol (*RTP*) and it can perform much transcoding between circuit-switched and packet-switched parts of the architecture. The MGW performs many important functions, details can be found in the references.

The Media Gateway Control Function (*MGCF*) has many important functions; it implements the protocol conversion between SIP and other IP-based functions. The *IMS-GW* uses the ITU-T Recommendation H.248 protocol (H.248.1 Recommendation). The main functions of IMS-GW are:

- communicates with MGCF,
- controls the media channels,
- communicates with the S-CSCF and I-CSCF, and also with BGCF,
- converts between SIP and other IP-related protocols.

The profile database of IMS subscribers is managed by the HSS. It also maintains information for session establishment, supports many interfaces, and includes information about subscriber registration and IMS user preferences, etc.

The SLF can be viewed as a simple database inside IMS that maps the addresses of IMS users to HSSs.

3GPP defines several interfaces, which are used by standardized IMS functions. The most important IMS interfaces defined by 3GPP are the *Gm, Mw, Cx, ISC, Sh,* and *Cq* interfaces. We will not describe the properties of these interfaces in detail, we give only a brief introduction to the interfaces supported (3GPP TR 23-992):

- The *Gm* interfaces are used in communications between *UE* and *P-CSCF* in registration and control tasks. The *Gm* interface uses the SIP protocol.
- The *Mw* interface is used between CSCFs.
- The *Cx* interface is implemented between I-CSCF, S-CSCF, and HSS. There are many procedures that require information transfer between these nodes, hence the *Cx* interfaces process many different transfers.

- The ISC interface is integrated between the CSCF and the service IMS platform.
- The *Sh* interface enables communications between the application server and the IMS HSS element.
- Finally, the *Gq* interface makes it possible to use communications between with the *P-CSCF* element.

For further information on these interfaces, refer to the 3GPP standards and related references (3GPP TR 23-992).

In conclusion, the IMS uses several basic functions and interfaces between these functions, as illustrated in Figure 7.

We note that in IMS specifications there are many application servers that can be integrated, for example CAMEL (Customised Applications for Mobile networks Enhanced Logic) (3GPP TS23.278), OSA (3GPP TS 29.198), SIP AS (RFC 3261), as illustrated in Figure 8.

These application server options serve different purposes and support different functions. The Charging Collection (*CC*) element defines two interfaces, called *Rf* for off-line charging and *Eo* for real-time on-line charging inside IMS. These interfaces use the DIAMETER message and are supported by the IMS application service. IMS charging makes it possible to use and to integrate advanced charging techniques, and existing non-IMS-based online technologies.

Figure 7. IMS functions and interfaces (UE: user equipment)

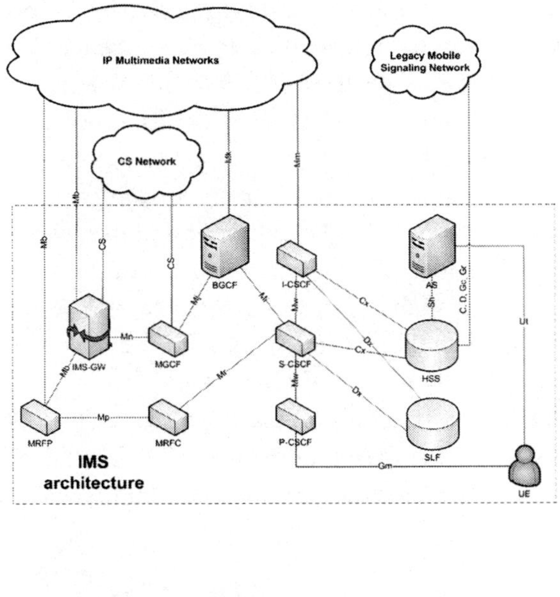

IMS Protocols

IMS services have many advantages over many other internet services, because IMS can provide access to many advanced multimedia services and the attributes of these services can be personalized. The IMS architecture was standardized by the Third Generation Partnership Project in Release 5 (3GPP). IMS-based multimedia communication is based fundamentally on the basic protocols of IMS. The basic protocols of IMS can be classified into session signaling protocols, media control protocols, authentication, and media delivery protocols.

In this section, we discuss the most important IMS protocols, namely:

- SIP (Session Initiation Protocol) protocol,
- SDP (Session Description Protocol) protocol,
- Media transport protocols (RTO and RTCP),

Figure 8. IMS application server preferences

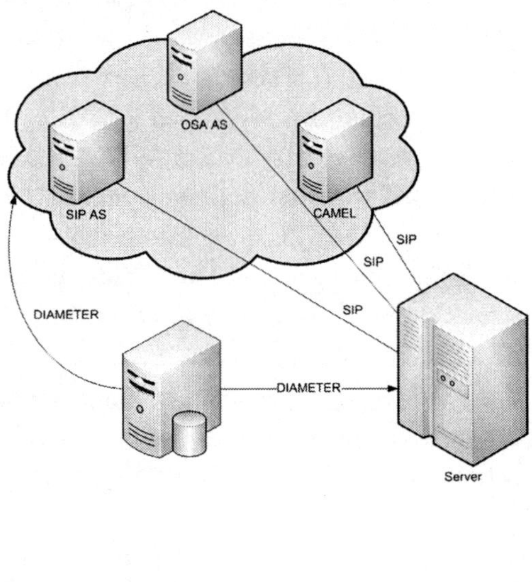

- Policy and charging protocols (DIAMETER),
- Media control protocols (MEGACO).

The SIP protocol was first discussed by IETF and was later accepted as a 3GPP signaling protocol (IETF Overview). The SIP protocol controls IMS sessions, hence it has deep relevance to IMS protocols. It is important to note that SIP is not a standalone protocol; hence, it is implemented as a network component. SIP can be integrated with IMS protocols and can be used independently of current protocols and of controlled IMS session types.

The SDP protocol was originally introduced by the IETF (IETF Overview) and describes the format of multimedia sessions and performance parameters for the session initiation phase. The SDP payload contains some media and transport information, the type of media, the properties of the transport protocol and the format of the media (3GPP TR 23-992). The SDP protocol also handles address, port information, and user information, and it provides resource information for the sessions.

The media control protocols use the MEGACO/H.248 standard to use the different interfaces of media gateway controllers and media gateways, and it originated from the MGCP protocol (RFC 3435). The MEGACO/H.248 protocol enhances the basic MGCP protocol (RFC 3435) and adds much functionality to it. The MEGACO/H.248 standard is based on the fusion of IETF and ITU standards; however, it has also been accepted by 3GPP and TISPAN. The authentication protocol of IMS is the DIAMETER protocol and it was first defined by the IETF in RFC 3588. DIAMETER performs AAA tasks within IMS; hence, it has to be secured from malicious users. The security of DIAMETER is mainly based on the security of IPSEC and TLS.

The services of the Media Delivery Protocol are required for QoS services, hence it involves many transport protocols such as RTP, RTCP, and UDP congestion control and jitter prevention techniques (3GPP TS 23.107). In IMS applications, the RTCP and RTP protocols are integrated together to monitor the quality of different services and to synchronize media transmission (3GPP TR 23-992). The SDP protocol can also be used together with the RTP protocol, however in this case the packets will be sent to well-separated ports.

IMS QoS

The IMS architecture provides the required QoS level for the users; however, the supporting of end-to-end QoS across heterogeneous network parts is a complex task. The 3GPP Release 6 has proposed many new functions and improvements to support the QoS in IMS.

In the model of IMS architecture, the QoS parameters are defined by the PDF. The different QoS parameters for various IMS services are maintained by the PDF and the PEP.

The PDF maps the policies to QoS parameters, and these values are responsible for maintaining the QoS level of the services. The PEP is intended to enforce the policies of the operator and handle the incoming traffic and the QoS usage. If any rules are violated during a call, the PEP alerts the PDF, and it also manages the resource allocation and bandwidth reservation for user traffic.

On the other hand, the *Cq* interface also has a share in providing QoS. It allows dynamic QoS-related service information to be exchanged between the PDF and the P-CSCF, which information is used by the PDF. The PDF uses this information for service-based local policy (*SBLC*) decisions.

The authorization of QoS decisions are also made by the PDF. The authorization of QoS depends on the policies, list of subscribed services, and the availability of resources. The PDF rejects the QoS authorization request if there are not enough resources available for the service or, if the session is not according to policies. The PDF also gives the reason of the rejection. The authorizations are handled by tokens. The tokens are generated by the PDF on request, and include the session ID and the parameters of the PDF, hence if later, the session can be identified using the encapsulated session ID.

The PCRF also provides QoS related functions. It is encompasses the policy control decision and flow-based charging control functions, and provides network control regarding to the QoS. The PCRF uses the service information provided by the P-CSCF, and stores it. The service information is also an important issue to derive the required QoS for the services.

The QoS level is monitored by the RTCP and RTP, and the RTCP implements a feedback on the quality of reception using RTCP reports. In such a RTCP report, the senders report the number of packets they sent and the receivers report the number of packets they received. The RACS also requests and reserves resources from the network, hence the QoS-based real-time services must be able to trigger QoS resource reservation, admission control and policy control capabilities.

IMS SERVICES

IMS makes it possible to re-use parts of the protocol stack that are common to all services, such as the addressing or routing parts. IMS can be seen as a horizontally layered technology platform that enables different (vertical) service realizations. The operation phases of IMS can be discussed in terms of these vertical services and the relations in the vertical and horizontal directions. The IMS infrastructure acts as a service delivery framework and is not limited to a specific set of services or to a specific business model to enable multimedia content and data services (TMF, 2009). The end-to-end SIP interaction between IMS users requires many steps from the IMS architecture, as illustrated in Figure 9. If an IMS user is registered with IMS, then its end-device is discovered by a P-CSCF and IMS authenticates the user. The various operator networks are connected by different interfaces and the end-to-end SIP interaction process can be divided into five phases:

- call initiation,
- service profile-based service routing,
- routing to destination network,
- service profile-based routing for the terminating user,
- delivery to terminating party.

IMS as specified by 3GPP provides the required functionality to both cellular and fixed network technologies (3GPP TR 23-992). IMS was designed to be a general service platform that in theory needs no standardization of services.

In Figure 10, we give an illustrated overview of the standardization of IMS-based services. The standardization communities have incorporated IMS solutions in their specifications making it possible to have true interoperable IP services within managed networks stemming from different standardized technologies.

The functional applications of IMS allow many different applications to be used, integrating many services. The IMS service chaining architecture chains different application servers and their services. The service chaining architecture in-

Figure 9. Operation phases of IMS service architecture

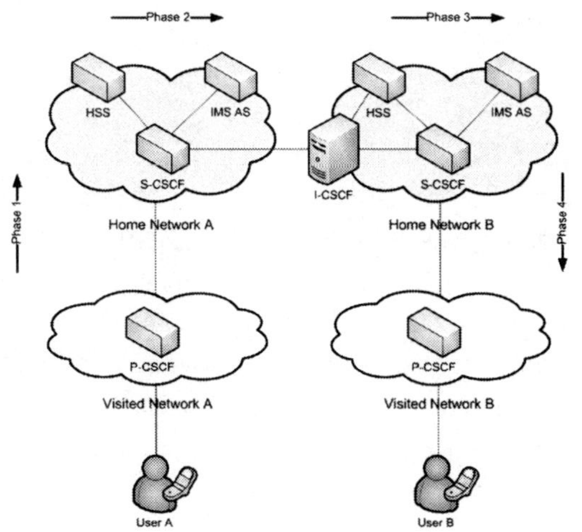

Figure 10. Standardization overview of IMS-based services

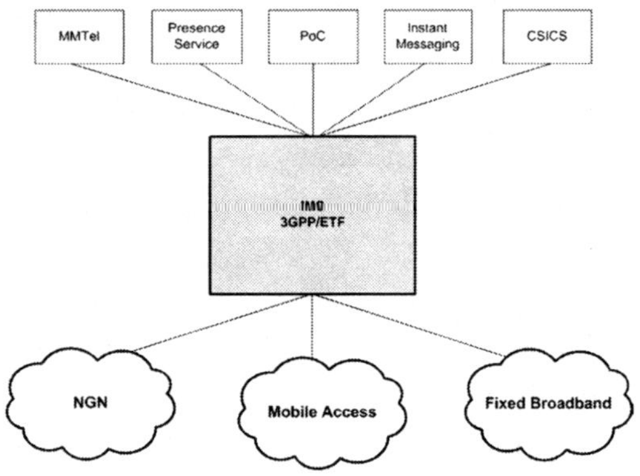

cludes service creation tools and a chaining server that manages the execution of service chaining. The IMS service chaining architecture is illustrated in Figure 11.

The IMS service chaining architecture is based on the key technologies of the IMS service delivery platform. Each service in the chain requires an independent network function; hence, the IMS chaining architecture has to ensure a seamless transition between IMS services.

In the IMS service chaining architecture, the User represents the different Service Interfaces, such as

- Set Top Box,
- Phone,
- PDA,
- PC,
- other service interfaces.

The IMS Services - such as Presence service, Service Control, SMS and E-mail messaging, Conferencing, Location Service, CCXML Server or GLMS - and the User Interfaces are communicate through the Service Delivery Platform.

The Service Delivery Platform consists of three main elements:

- the Service Registry,
- the Content Repository,
- and the Service Controller.

The Service Registry and Content Repository communicate with the Service Controller; however, there is no direct communication between the Service Registry and Content Repository. The communication between the service interfaces and the Service Delivery Platform is realized by the Service Controller element.

Private, Public and Service Identities

IMS provides the convergence of Internet access technologies and the communication techniques of the core network and it also facilities convergence at the level of end-users and services. The service level convergence has deep relevance in the study of the properties of IMS services. The SIP protocols were defined in the IMS standard and were originally designed for IMS to make user-level convergence possible in practical applications.

In IMS, the introduction of IMPU - IMS public user identity made it possible to handle private and public user identities. A typical IMS user can define private and public identities, and multiple private user IDs can use one public user ID within the same IMS identity. This allows to register different multiple user equipment with different private IDs to use the same public user ID.

Figure 11. The service chain of IMS

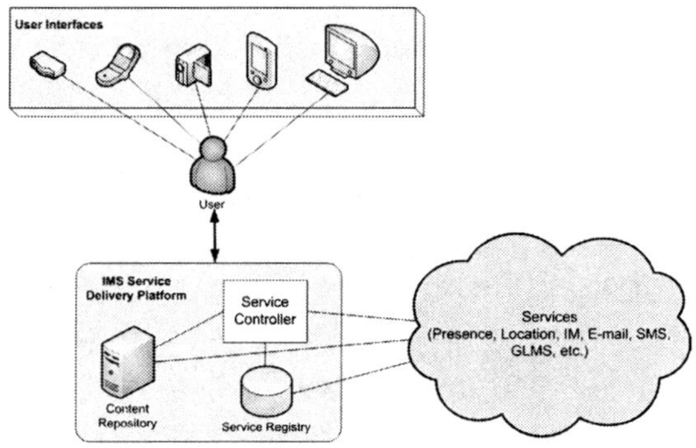

In Figure 12, we illustrate the correspondence between IMS ID and private and public identities. As can be concluded, an IMS ID can support multiple IMS users.

From Release 6, the operators can assign one or more Public User Identities and one or more Private User Identity to the users. The relation between the Private User Identities and the Public User Identities of IMS subscribers has been standardized in 3GPP Release 5, - however in Release 5, an IMS subscriber is allocated only with one Private User Identity [1].

The Public and Private User Identities are handled by the HSS and the S-CSCF. As we have illustrated it, it is possible that some Public User Identities are used in combination with more than one Private User Identities - hence a Public User Identity can be assigned to more than one Private User Identities. As a consequence, a Public User Identity can be used from more than IMS terminals simultaneously - assigned with different Private User Identities [1].

The private user identities and public user identities are handles of the home network operator. The public user identities can be shared across private IMS users within the same IMS subscription. A public user identity can be registered simultaneously from multiple UEs, who have different private user identities and addresses. In Figure 12, public user identities 1 and 2 cannot have the same alias public user identity group, since private users have different service profiles.

Figure 12. Relation of private and public user identities

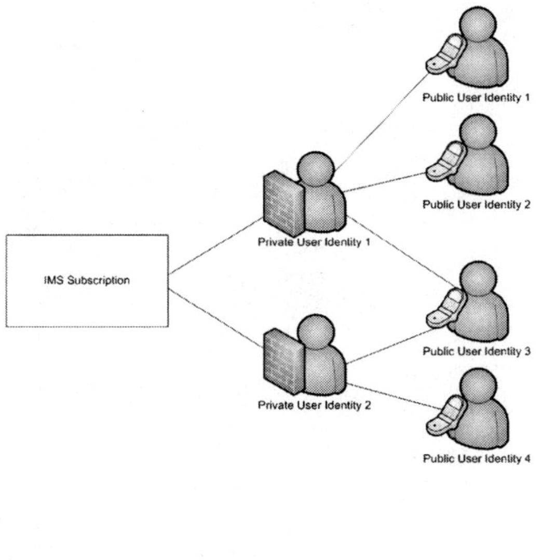

The public service identity is integrated into an SIP URI or Tel URI and identifies a service or service resource. The private service identity is defined by the operator and, unlike the private user identifier, it is not used for registration or AAA processes. The private service identifier makes it possible to associate public service identities with a public service identifier for higher compatibility.

ONLINE AND OFFLINE CHARGING

As IMS sessions may include multiple media components (e.g. audio and video), IMS must provide a method for charging per media component. This would allow the called party to be charged if it adds a new media component to a session. Another requirement is that different IMS networks are able to exchange information on the charging that will be applied to a current session. From an operator or service provider perspective, the ability to charge users is a must for any network.

IMS defines two different paying methods, offline charging for users with permanent subscriptions and online charging for prepaid users.

In Figure 13, we show the IMS offline charging architecture. Information about the session being charged is sent to the Charging Data Function (*CDF*). In the next step, the SIP elements send information over a Diameter-based interface and the CDF generates the Charging Data Records (*CDRs*).

The online charging method uses credit control; hence, if any users run out of credit, they can no longer access the service (3GPP TS 22.115). In Figure 14, we illustrate the IMS online charging architecture. The method uses Application Servers and the MRFC sends information to the Online

Figure 13. IMS offline charging architecture

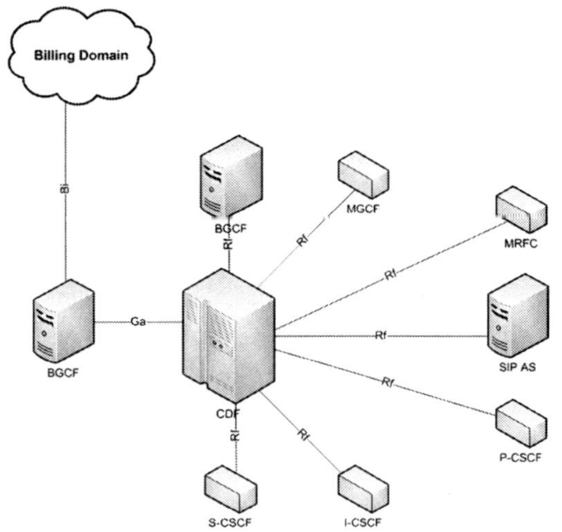

Charging System (*OCS*), using Diameter (3GPP TS 22.115). If a session needs to be terminated, the IMS-GW terminates the session using the ISC interface.

These two architectures are the Service Based Local Policy (*SBLP*) and Flow Based Charging (*FBC*).

The SBLP for the sessions is maintained by the PDF, hence this element responsible for maintaining the operator's policy and user profile-based policy. The policies are rules, which determine how the calls can be handled in IMS.

The FBC model is introduced in Release 6. It makes possible to charge for service data flows, which service data flows are identified by service flow filters. The charging rules makes it possible to filtering the traffic to identify the packets belonging to the IMS, FTP, HTTP, etc, data flow, and the contains methods of how the service data flow could be charged [1].

One part of the IMS evolution in 3GPP release 7 was to merge SBLP and FBC into one Policy Control and Charging (PCC) architecture. This was done to harmonize policy control and charging procedures (3GPP TS 22.115).

PRESENCE SERVICE

IMS makes advanced internet technologies available to everyone from any location or any network connection. In this section, we examine the presence and Instant Messaging services, which have importance in the current IMS market and SIP-based applications. The presence service is a very popular IP-based communication service and the SIP protocol plays a fundamental role in the delivery of this IMS service.

In this section, we give a brief overview of the working mechanism of presence service and of the presence data format. We explain the presence model and its connection with SIP. The IMS presence service makes it possible to send current presence information to users using IM communication technology. The IMS Presence service is based on different presence implementations such as the well known ICQ released in 1996. Many presence applications have been defined since then, but these solutions were incompatible with each other, so a standardized version was required by users (Poikselka, 2006). The IETF and OMA were the first to take a step in this direction. The basis of the presence service is defined in RFC 3856, RFC 3857, and RFC 3858. The presence

Figure 14. IMS online charging architecture

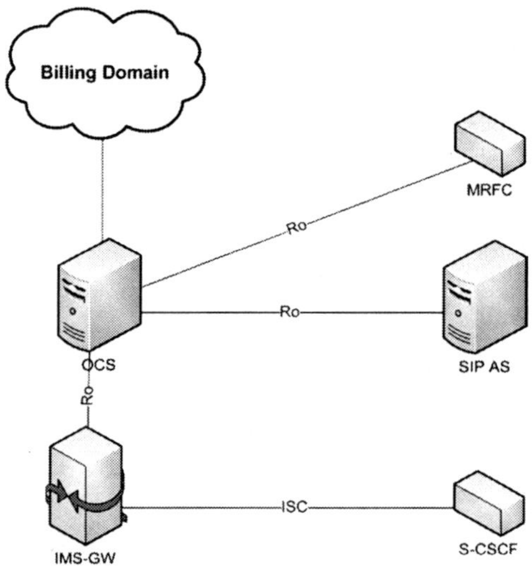

service was extended by 3GPP and OMA from an IM feature to a general IMS feature that can be used across multiple devices and applications.

The Instant Messaging service provides an online communication service, instant communication between the sender and the recipient. In these IM services, both users need to be connected simultaneously and the submission of instant messages can be viewed as opposite to email communication. In IM communication, if the receiver is already connected but currently unable to engage in the communication, the actual presence information can be shared with the originator. The IMS-based presence services add value to the end-user and are today widely used on the Internet. The presence model based on RFC 2778 and RFC 2779 defines three different entities, called presentity, watcher, and presence service entity (Ilyas, 2008). The presence information is received from the presentities and it is distributed among the presence watchers, as illustrated in Figure 15.

If the originator would like to provide its presence information, then its presence information will be distributed among the set of watcher entities. The watcher entities receive the presence information from the presentity entity, which represents the sender. The presentity sends the

Figure 15. Presence service architecture

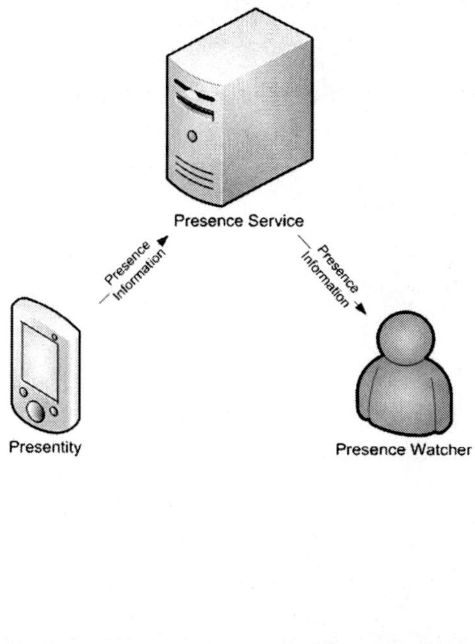

presence information to the presence service entity and then it sends it to the watchers who are interested in the presentity's presence information. The presence traffic itself is in general not time-critical so the presence service should make good use of QoS so as not to disturb e.g. a real-time service like a Multimedia Telephony (*MMTel*) basic voice call (3GPP TS 23.107; Hwang, 2008). The amount of traffic that the presence creates is not only dependent on the presence model (Poikselka, 2006). The number of presence states and type of presence information available may also have an impact the traffic volume.

The presence information can be grouped into a number of categories. Presence information may include person and terminal availability information, communication preferences, terminal capabilities, information about current activity, current location, and currently available services. The standards define two types of presence watchers are

- subscribers
- and fetchers.

The changes in the presentity's presence information are sent to the subscriber watcher, while the fetcher requests only the current presentity presence information (Camarillo, 2006). The fetcher can consist of a special type of fetcher called a *poller*. In the Presence Service model, the watcher requests presence information about one or more presentities or about other watchers. The watchers have three special types:

- subscriber,
- fetcher,
- pooler.

The subscriber requests notification from the presence server if any new presence information is available. The fetcher makes a request for presence information; hence, it requests the presence information of presentity from the presence server.

The poller is a special kind of fetcher, which fetches information a regular basis. The presence watcher model is illustrated in Figure 16.

The presence service allows IMS users to subscribe to presence information and allows it to see other IMS users. The presence users can set the actual presence state, depending on the users online or offline status, the current mood or its location.

Presentity

The presentity combines the words presence and entity (Ilyas, 2008). The Presence Network Agent (*PNA*) or Presence User Agent (*PUA*) makes the presence information of the presentity available to the Presence server. The PNA and PUA are called presence agents.

The PNA can connect to several IMS network elements through several reference points, as illustrated in Figure 17. The network elements are connected to the PNA through with defined reference points. The connected elements can be used by the PNA to collect information about the presentity, if the PUA is not available. If the PUA is not able to send the presence information to the presence server, then a PNA can collect the presentity's presence information. The PNA could use the connected network elements to collect the presence information, - such as the HLR, to get information about the registration status of the presentity, or other elements, such as the MSC, AAA, GMLC, SGSN, PDG, GGSN, S-CSCF. (The abbreviations in the figure can be found in the Abbreviations list).

The PUA can be integrated into the IMS mobile terminal or end user's device and it provides the presence information to the presence server.

Figure 16. Presence watcher presence data model

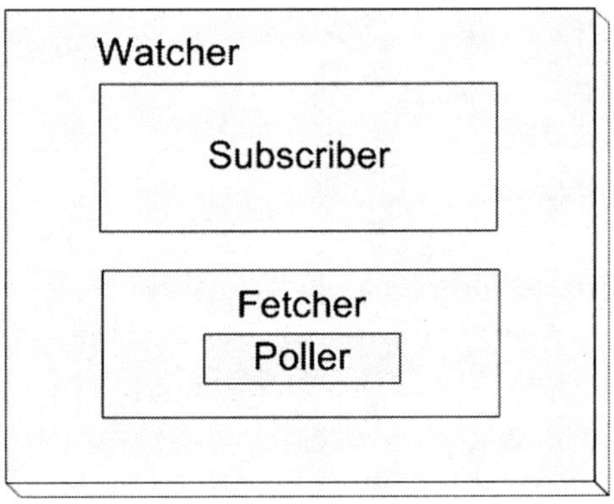

Presence Server

The presence server stores and distributes the subscriber's presence information to the subscribed presence watchers. The presence service architecture defines four functional components:

- presence server,
- presence XML Document Management Server (XDMS),
- Resource List Server (RLS),
- Resource List Server XDMS.

The Presence server collects and distributes the presence information; the XDMS provides presence service specific information in an XML document, such as the list of authorized watchers for the presence information of the presentity's. The RLS manages the presence list subscriptions for a watcher. The XDMS contains the actual list data used by the RLS in an XML document.

The additional functional components of the presence service architecture are:

Figure 17. Relation of presence network agent to presence server

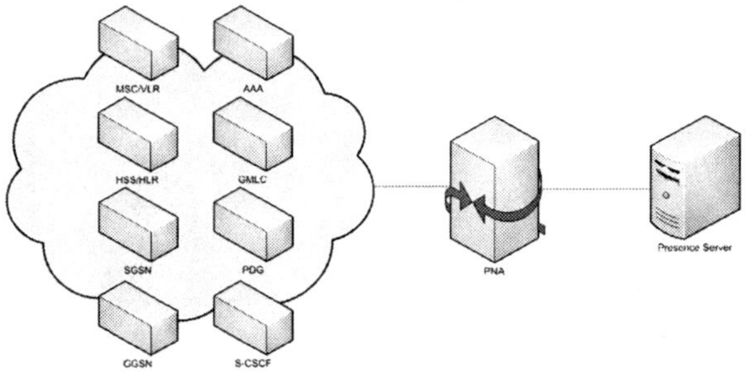

- presence source (PNA or PUA),
- content server,
- watcher entity,
- XDM client.

The content server manages MIME objects; the watcher entity makes a request for presence information or watcher information, while the XDM client manages the XML documents. The presence reference architecture is illustrated in Figure 18.

Watcher

The presence watcher represents the subscribing party who is interested in the presence information of one or more presentities. The presence watcher uses the presence server to subscribe to a presentity's presence information (Poikselka, 2006). If the presentity does not grant permission to the watcher, it is not able to access directly the presentity's presence status information. The communication and authentication steps between presentity and watcher are made through the presence server. The presentity uses the watcher-information package to require new presence information from the presence server (Cuevas, 2006). The watchers can be authorized by presentities, through the presence server presence XDMS component. The XDMS component sends the authorization request to the presence server to apply the authorization steps.

Figure 18. Reference architecture of IMS presence service

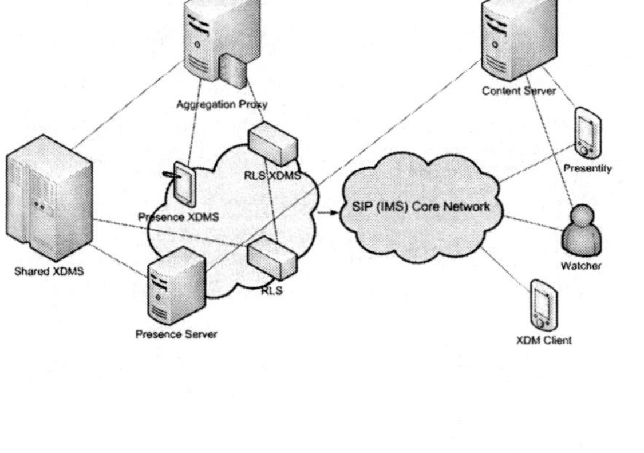

Presence and SIP

The presence service model can be implemented with SIP. The RFC 3856 standard defines a new SIP extension and RFC 3903 defines a mechanism for publication of event state information. The SIP-based presence architecture can be composed of a Presence User Agent (*PUA*), Presence Agent (*PA*), and presence watcher application. The PUA entity allows a presentity to use presence information and it can publish the presence information. The PA can be viewed as a SIP UA, with authorization and authenticating processes (Ilyas, 2008).

The watcher application is a SIP UA module and it can subscribe to the presence information of presentity. The watcher application can also receive notifications about changes in the watched state of the presentity. The presence information is published by the PUA in the SIP PUBLISH message to the Presence Agent and the watcher application is subscribed to presentity's presence information by sending the SIP SUBSCRIBE message provided this be allowed. The SIP NOTIFY message can be used by the PA to notify presence watchers about a change of presentity's state (Camarillo, 2006).

As illustrated in Figure 19, the Presence Server can send SIP NOTIFY messages to the watcher entity and it can receive SIP PUBLISH messages from Presentities. The presence watcher can subscribe to the presence service by sending the SIP SUBSCRIBE message to the presence server. The presence server acts as an IMS application server and, after the user's subscription, and sends SIP NOTIFY message to the watcher about the presence state of the users (Poikselka, 2006). The notification will be sent only about presentities that are in the watcher's presence group.

If the presence state changes, the presence client updates the stored state in the presence server by sending a SIP PUBLISH message with

Figure 19. High-level view of presence service

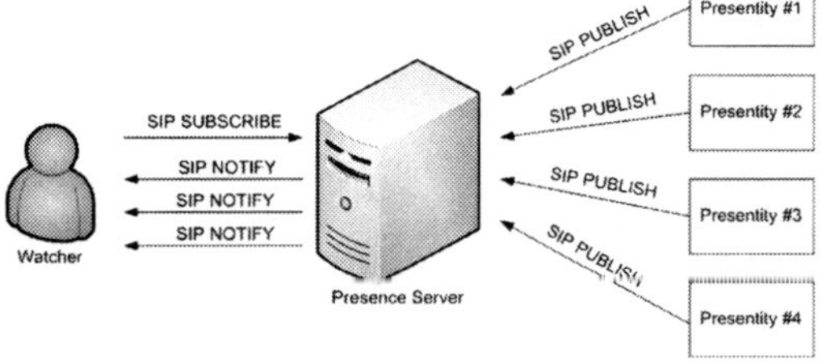

the new presence information. After the SIP PUBLISH message has been sent, the presence watcher is notified about the new presence state (Cuevas, 2006). The watcher will receive a new SIP NOTIFY message, which sends a notification about the new presence information.

The format of presence information is defined in RFC 3863. This reference defines the PIDF in an XML-based presence format for presence information. The presentity is identified by a presence URI and the presence information refers to the presence URI. The presence information consists of one or more presence tuples, such as

- presence identifier
- and presence status.

Optionally, the presence tuple consists of communication address and timestamp, priority, and other readable comment (Ilyas, 2008).

The presentity human readable comment is a free text about the presentity. The address resolution of the presence service is defined in RFC 3859, which defines the URI scheme for presentities and watchers. The resource list format is specified in RFC 4662, which also describes the requesting and conveying mechanism of notification for list of resources. The list of resources is identified by a SIP URI. The XML-based information exchange is managed by the HTTP-based XML Configuration Access Protocol (*XCAP*) protocol. XCAP is defined in RFC 4825 and uses HTTP methods such as GET, PUT, and DELETE. The presence information can be divided into user information, user availability information, mood and location information and reachability information (3GPP TR 23-992).

The popularity of the IMS presence service function is based on the fact that it adds many high value services for end-users and it can be implemented easily on mobile terminals. With the help of the presence service, end-users can communicate their presence information with a predefined granularity and density. The service can be implemented in various network elements; hence, a user can update its current status dynamically from various locations and situations.

INSTANT MESSAGING

The Instant Messaging service of IMS is an online communication-based system. The Instant Messaging service allows users to exchange information with other users in near real-time. The IM service uses short messages and the architecture enables interactive communication between parties. The common profile, consisting of communication formats and protocols for instant messaging, was defined in RFC 3862 and is called the Common Profile for Instant Messaging (*CPIM*). The service enabler supports *"pager mode" and "IM session"* SIP-based IM communication modes (Camarillo, 2006).

In IM pager mode, the complete IM message included in the SIP MESSAGE is requested. The IM session mode splits the signaling from the media and creates signaling among the IM parties. In IM pager mode, the IM messages are independent units and the body of the SIP MESSAGE request contains a MIME object, which contains an object that represents the instant message itself. The Instant Messaging server is an application server and can be used if instant messaging cannot be directly sent between IM users.

The typical functions of an IM server are

- policy enforcement,
- management of chat applications,
- interworking with other IM systems,
- interworking with legacy offline messaging systems (Poikselka, 2006).

The IM server also maintains deferred delivery and can manage the interworking with other services, such as SMS, as illustrated in Figure 20.

The Instant Messaging communication service can be referred to as an OMA SIMPLE IM enabler (OMA Overview), where SIMPLE means SIP for Instant Messaging and Presence Leveraging Extensions. The SIMPLE IM service architecture was defined in the OMA IM document (OMA Overview). The SIMPLE approach is distinguished from the SIP-based IM architecture and OMA specifies and implements extended protocols to SIMPLE (Ilyas, 2008). The OMA IM architecture (OMA Overview) implements different roles in communication, the high-level view of a multi-operator group IM communication is illustrated in Figure 21.

The OMA IM communication service maintains deferred messaging, which is an important service in IM (OMA Overview). It enables communication with an offline IM user or with users who are not registered with the IM service (Chen, 2004).

In Figure 22, we illustrate the relationship between the IM server and deferred messaging service. IM Client *B* is currently not able to use IM services; hence, the IM server defers the messaging function. The IM client's deferred messaging function stores the incoming message sent by IM Client *A*.

The IM media types are defined in 3GPP TS26. The IM service maintains quite a large number of media types; however, the mobile terminals also have to support the current media type. The IM services can provide many media types, such as text, image (FPEG, JFIF), bitmap, vector graphics (GIF, PNG, SVG), speech (AMR coding), audio (AAC+, AMR-WB), video (H.263, H.264), and synthetic audio (MIDI). The Instant Messag-

Figure 20. SMS-based IM message delivery system

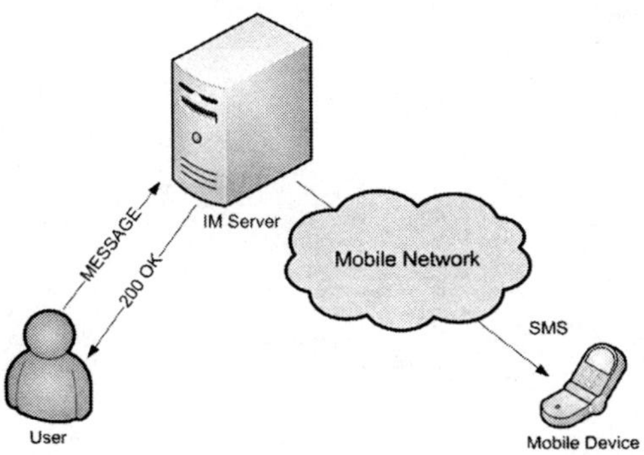

Figure 21. Communication between IM server, controlling IM function, participating IM functions and IM clients

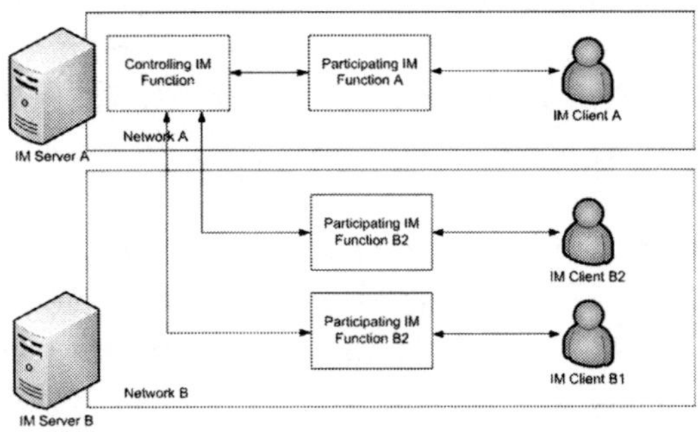

Figure 22. User B is currently unavailable for IM and the IM server activates deferred messaging

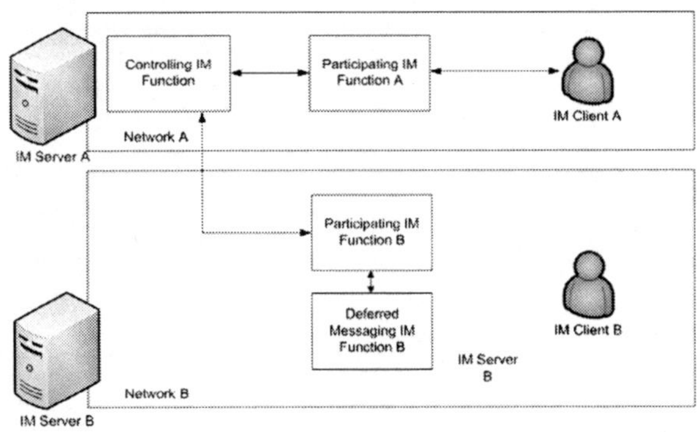

ing functions can be implemented from SIP devices, the service is IP-based and it can be used with mobile phones, digital cell phones, PDAs and desktop computers (Poikselka, 2006). Contrary to ordinary e-mail messaging, with the help of the IM service, clients can determine instantly whether or not other users are currently online and available for communication. IM services can be integrated into enterprising applications, business life, conferencing or meeting schedules, and many more; hence, IM provides an effective and low-cost way of instant communication (Cuevas, 2006).

The success of IMS and IMS-based services originated in Next Generation Networks and multimedia services on the Internet. The convergence between fixed and mobile networks makes it possible for network users to integrate similar services. The Instant Messaging service can be used on most mobile devices, desktop PCs, and IM clients. The message flow of sending IM messages is illustrated in Figure 23. The message flow includes the check for presence status (Chen, 2004).

The PUBLISH method is used when a SIP UA would like to publish its status information. The MESSAGE request transfers instant messages, the body of the SIP MESSAGE being a text message. It could contain an attachment or other media content. The size of the SIP MESSAGE cannot exceed 1300 bytes (Camarillo, 2006).

The interactive nature of the IM services allows them to be used in everyday applications. The interactive nature of an IM service encourages IM users to engage in conversations. Typically, small text messages are exchanged in IM conversations but the OMA Instant Messaging communication service supports content types other than text, such as images, video clips, etc., making it a multimedia IM service enabler (Poikselka, 2006). The OMA Instant Messaging communication service enables a user to send messages to another individual or to a group of users in which everyone can see what everyone else is sending.

IMS on the Market

As we have seen in the section on IMS standards, the IMS architecture evolved from the inside out, based on the requirements of the core service layer. The IMS services are based on the functions of the core service layer and the standards and services of IMS will continue to evolve. The evolution of IMS is based on aspects of the business world and

Figure 23. Message flow for instant messaging

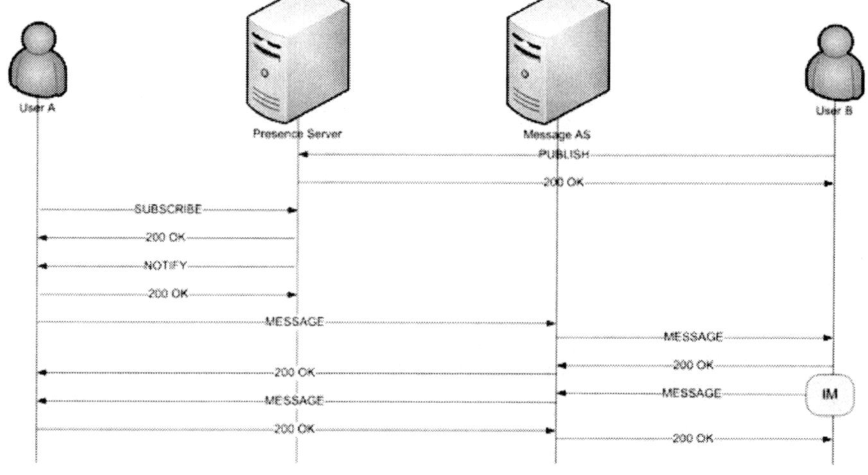

IMS concatenates changes in networks, mobile communications, and media services.

Nowadays, there are many new network implementations that have been designed for IMS users, and these IMS-compatible devices can give access to multiple IMS-based services to users. In the marketplace, IMS comes in many versions of the IMS standards currently in use. These IMS versions seem to be the best solution for many practical applications and these versions work properly with the network devices implemented. IMS and the IMS Multimedia Telephony service are also interesting to operators for several reasons. IMS has deep relevance to the evolution of telecommunications (Poikselka, 2006).

The future of IMS is ensured in the world of telecommunications and it will play a fundamental role in future telecommunication standards (ICT, 2008). IMS was designed to merge cellular and internet technologies. IMS adopts Internet Protocol Version 6 (IPv6) and the core applications of future telecommunications, such as advanced, true peer-to-peer communication technologies and IP-based telephony. The IMS allows new multimedia services and advanced communication techniques to be used and moreover it reduces the cost to end-users. The IMS architecture enhances the services of the classical circuit-switched domain and provides many new services and applications. We note that IMS does not replace all services offered by the circuit-switched or packet-switched architectures, it just combines and uses them (Chen, 2004).

Figure 24. IMS layered service architecture: transport and end-point layer, session control layer and application server layer

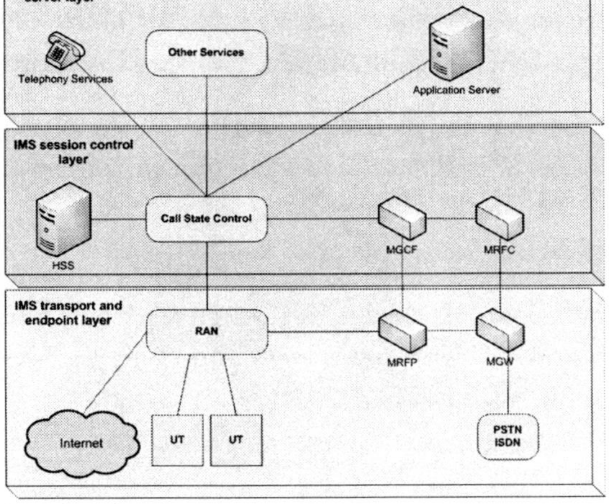

Within IMS, the communication services can be classified into broadband Internet access services, media content services and interpersonal communication services (3GPP TR 23-992). The first group contains mainly broadband fixed and mobile technologies; the second group covers IPTV and other media and content services, while the third group covers the technologies of voice and video calls. IMS does not define a fundamentally new architecture; hence, the services of IMS are not independent of existing communication service architectures.

In Figure 24, we illustrate the three functional layers of IMS. The transport and end-point layer manages the signaling,

- controls the sessions,
- provides bearer services,
- and support for media conversions.

In this layer, UT represents the User Terminal. The second layer is called the session control layer, manages end-points and sessions between them, and combines the transport services. The application service layer allows sessions to interact with other Application servers and can combine multiple sessions (Poikselka, 2006).

IMS works together with existing circuit-switched networks and platforms and extends the layer structure of these network architectures. IMS defines new layers upon the existing layer infrastructure and provides new services and components based on the core circuit-switched and packet-switched network. The layer structure of IMS Applications and their implementation in the service delivery domain is shown in Figure 25.

In conclusion, IMS requires the integration of components of the IP transport networks, elements of the circuit-switched domain and of other enablers for non-IMS applications. IMS Applications integrate the elements of IMS and other enablers.

CONCLUSION

IMS provides ubiquitous access to many services both on mobile and wireless networks and can guarantee quality of service. In this chapter, we have given an overview of the most important IMS communication services existing in standards today, such as Presence Service, Instant Messaging, and Circuit-Switched Combination Service. These IMS services can be offered as standalone IMS communication services; however, they are designed to be common services that have relations to other services. The access technologies of IMS devices cover WiFi, WiMax, xDSL, GSM, UMTS, and cable access; - the convergence of these IMS access technologies is based on the convergence of the IMS service layer. The convergence between fixed and mobile services integrates the core services of fixed and mobile networks. The IMS services were designed to incorporate and fuse the services and techniques of classical cellular systems and advanced methods of the Internet. IMS services provide many supplementary services to classical Internet services and these valuable tools can be expanded and integrated into telecommunication systems. The access technologies and multimedia services of IMS services are based on the classical services of cellular systems and on the fundamental results of the Internet; however, the IMS can provide QoS and many more advanced customized tools, which were not possible in the original cellular systems and Internet services. The IMS in the core network can be viewed as a service delivery system and the well-defined methods of IMS optimize and integrate voice-based services of the core network and packet-switched communications. IMS is able to provide new QoS-based and valuable IP multimedia services and, with the help of IMS, the performance of access services can be optimized. The IMS architecture enables the convergence of mobile and fixed services at different levels, at the transport level, at the core network level, and at the user level.

Figure 25. IMS and service delivery mechanism

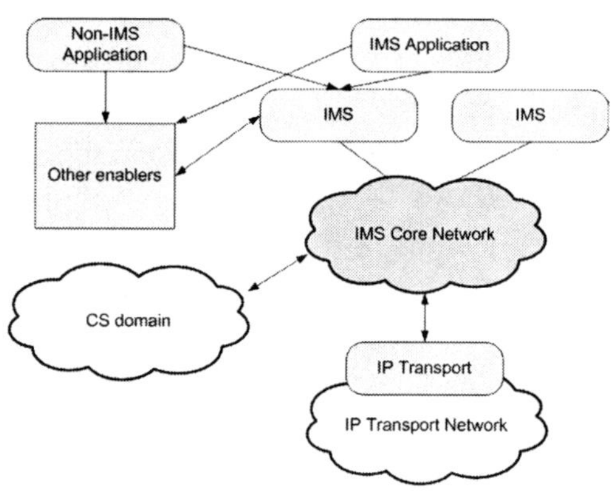

REFERENCES

Andreasen, F., & Foster, B. (2003). *RFC 3435, media gateway control protocol*. MGCP.

Bang, N. D., Jorstad, I., & Jonvik, T. (2007). *Analysis of IMS business requirements*. Retrieved from http://www.icin.biz/files/programmes/Session6A-2.pdf

Burger, E. (2006). *A mechanism for content indirection in session initiation protocol (SIP) messages. RFC 4483*. Internet Engineering Task Force.

Camarillo, G., & García-Martín, M.-A. (2006). *The 3G IP multimedia subsystem (IMS): Merging the Internet and the cellular worlds*. John Wiley and Sons.

Chen, J. C., & Zhang, T. (2004). *IP-based next-generation wireless networks*. John Wiley & Sons. doi:10.1002/0471478253

Cuevas, A., Moreno, J., Vidales, P., & Einsiedler, H. (2006). The IMS service platform: A solution for next-generation network operators to be more than bit pipes. *IEEE Communications Magazine*, *44*(8), 75–81. doi:10.1109/MCOM.2006.1678113

Garcia-Martin, M. (2006). *A session initiation protocol (SIP) event package and data format for various settings in support for the push-to-talk over cellular (PoC) service. RFC 4354*. Internet Engineering Task Force.

3GPP. (n.d.). *Overview*. Retrieved from http://www.3gpp.org

3*GPP. (n.d.)*. TS23.278: Customised applications for mobile network enhanced logic (CAMEL) phase 4, stage 2 IM-CN interworking.

3GPP. (2005). *TR 32.200: Telecommunications management-charging management, charging principles*. September 2005.

3GPP. (n.d.). *TR 23-992: Architecture for an all-IP network*.

3GPP. (n.d.). *TS 22.115: Service aspects, charging and billing*.

3GPP. (n.d.). *TS 23.002: Network architecture*.

3*GPP. (n.d.)*. TS 23.107: Quality of Service (QoS) concept and architecture.

3*GPP. (n.d.)*. TS 29.198: Open service access (OSA), application programming interface (API), part 1–14.

Harrowell, A., & Writer, S. (October 2006). *A pointless multimedia subsystem*. Mobile Communications International.

Hwang, J., Kim, N., Kang, S., & Koh, J. (2008). *A framework for IMS interworking networks with quality of service guarantee*. DC: IEEE Computer Society Washington.

ICT. (2008). *Conference session report on the future of the Internet European Commission Information Society and Media*.

IETF. (n.d.). *Overview*. Retrieved from http://www.ietf.org

Ilyas, M., & Ahson, S. (2008). *IP multimedia subsystem (IMS) handbook*. CRC Press.

ITU-T. (2002). *Recommendation H.248.1: Media gateway control protocol version 2*.

Khlifi, H. (2008). *IMS application servers: Roles, requirements and implementation technologies*. IEEE Internet Computing.

Mani, M., & Crespi, N. (2008). How IMS enables converged services for cable and 3G technologies: A survey. *EURASIP Journal on Wireless Communications and Networking*, *8*(3), 1–14. doi:10.1155/2008/589623

O'Connell, J. (2007). Service delivery within an IMS environment. *IEEE Vehicular Technology Magazine*, *2*(1), 12–19. doi:10.1109/MVT.2007.898099

OMA. (n.d.). *Overview*. Retrieved from http://www.openmobilealliance.org

Poikselka, M., Niemi, A., Khartabil, H., & Mayer, G. (2006). *The IMS: IP multimedia concepts and services*. John Wiley and Sons. doi:10.1002/0470031840

Rosenberg, J., Schulzrinne, H., Camarillo, G., Johnston, A., Peterson, J., Sparks, R., … Schooler, E. (2002). *RFC 3261, SIP: Session initiation protocol*.

Service delivery framework (SDF). (n.d.). *Telemanagement forum*. Retrieved from http://www.tmforum.org/TechnicalPrograms/ServiceDeliveryFramework

SIP. (2008). *Session initiation protocol*. Retrieved from http://www.ietf.org/html.charters/sip-charter.html

SIP. (n.d.). *Forum*. Retrieved from http://www.sipforum.org/.

KEY TERMS AND DEFINITIONS

Fixed Mobile Convergence (FMC): Fixed Mobile Convergence (FMC) is the infrastructure technology and network service to support both the wide-area and local-area access.

Global System for Mobile Communications (GSM): The Global System for Mobile Communications (GSM) is the most popular standard for mobile telephony systems in the world.

High Speed Uplink Packet Access (HSUPA): The High-Speed Uplink Packet Access (HSUPA) is a mobile telephony protocol with high up-link speeds. The 3GPP uses the name Enhanced Uplink (EUL).

Internet Protocol (IP): The Internet Protocol (IP) is the primary protocol that responsible for the communication through the Internet. The IP protocol relays datagrams (packets) across an internetwork and controls the packet routing mechanisms.

IP Multimedia Subsystem (IMS): The IP Multimedia Subsystem (IMS) is an architectural framework for delivering Internet Protocol (IP) multimedia services.

Long Term Evolution (LTE): Long Term Evolution (LTE), is the latest standard in the mobile network technology tree and also a project of the 3rd Generation Partnership Project (3GPP).

Media Gateway (MGW): A Media gateway is a translation device or service that converts digital media streams between disparate telecommunications networks, and enables multimedia communications across Next Generation Networks (NGN) over multiple transport protocols.

Multimedia Telephony (MMTel): The 3GPP/NGN IMS Multimedia Telephony Service (MMTel) is a global standard based on the IP Multimedia Subsystem (IMS), offering converged, fixed and mobile real-time multimedia communication.

Next Generation Network (NGN): The Next Generation Network (NGN) is based on the IP protocol, it describes the key architectural evolutions in telecommunication core and access networks.

Session Initiation Protocol (SIP): The Session Initiation Protocol (SIP) is a signaling protocol, widely used for controlling multimedia communication sessions such as voice and video calls over Internet Protocol.

APPENDIX

Table of abbreviations

3GPP	Third Generation Partnership Project
AAA	Authorization, Authentication and Accounting
ABGF	Access Border Gateway Function
AS	Application Server
BGCF	Breakout Gateway Control Function
CAMEL	Customized Applications for Mobile networks Enhanced Logic
CC	Charging Collection
CDF	Charging Data Function
CDRs	Charging Data Records
CGF	Charging Gateway Function
CPIM	Common Profile for Instant Messaging
CSCF	Call Service Control Function
ETSI	European Telecommunications Standards Institute
FMC	Fixed Mobile Convergence
GGSN	Gateway GPRS Support Node
GMLC	Gateway Mobile Location Center
GSM	Global System for Mobile communications
GUI	Graphical User Interface
GW	Gateway
HSS	Home Subscriber Server
HSUPA	High Speed Uplink Packet Access
IBCF	Interconnect Border Control Function
I-BGF	Interconnect Border Gateway Function
I-CSCF	Interrogating-CSCF
IETF	Internet Engineering Task Force
IM	Instant Messaging
IMS	IP Multimedia Subsystem
IP	Internet Protocol
LAN	Local Area Network
LTE	Long Term Evolution
MGCF	Media Gateway Control Function
MGW	Media Gateway
MGWF	Media Gateway Control Function
MIME	Multipurpose Internet Mail Extensions
MIMO	Multiple-input Multiple-output

MMS	Multimedia Messaging Service
MMTel	Multimedia Telephony
MPLS	Multi Protocol Label Switching
MRF	Media Resource Function
MRFC	Media Resource Function Controller
MRFP	Media Resource Function Processor
MSC	Mobile Switching Center
NASS	Network Attachment Sub-System
NGN	Next Generation Network
OCS	Online Charging System
OMA	Open Mobile Alliance
PCC	Policy Control and Charging
P-CSCF	Proxy-CSCF
PDA	Personal Digital Assistant (Handheld computer)
PDF	Policy Decision Function
PDG	Packet Data Gateway
PEP	Policy Enforcement Point
PLMN	Public Land Mobile Network
PNA	Presence Network Agent
PoC	Push to Talk over Cellular
PS	Presence Service
PSTN	Public Switched Telephony Network
PTT	Push to talk
PUA	Presence User Agent
QoS	Quality of Services
RACS	Resource and Admission Control Sub-system
RLS	Resource List Server
RTP	Real Time Protocol
SBLC	Service-based Local Policy
SCF	Service Switching Function
SCIM	Service Capability Interaction Manager
S-CSCF	Serving-CSCF
SDP	Session Description Protocol
SG	Signaling Gateway
SGSN	Serving GPRS Support Node
SGW	Signaling Gateway
SIP	Session Initiation Protocol
SLF	Subscription Location Function
SSF	Services Switching Function
SSO	Single Sign On
THIG	Topology Hiding Interwork Gateway
TISPAN	Technical Committee within ETSI that works with standards for Next Generation Networks

TSG	Technical Specification Groups
VLR	Visiting Location Register
VoIP	Voice over IP
WCDMA	Wideband Code Division Multiple Access
XCAP	XML Configuration Access Protocol
XDMS	XML Document Management Server
XML	eXtensible Markup Language

Chapter 14
The TFRC Protocol and Its Usage for Wireless Video Transmission

Christos Bouras
Research Academic Computer Technology Institute, Greece & University of Patras, Greece

Vassilis Papapanagiotou
Research Academic Computer Technology Institute, Greece, & University of Patras, Greece

Kostas Stamos
Research Academic Computer Technology Institute, Greece & University of Patras, Greece & Technical Educational Institute of Patras, Greece

Giannis Zaoudis
Research Academic Computer Technology Institute, Greece & University of Patras, Greece

ABSTRACT

The subject of this chapter is to present the TFRC (TCP-Friendly Rate Control) protocol in the area of efficient wireless video transmission and its possible usage in cross-layer power management mechanisms. The basic aspects of TFRC operation are presented, along with the suitability of TFRC usage for video transmission. The chapter examines related work and presents several mechanisms for efficient wireless video transmission using TFRC that have been proposed. These mechanisms utilize cross-layer approaches for adaptation of the power transmission level of the sender and TFRC feedback information regarding the wireless connection status from the receiver for improved transmission statistics, and therefore user experience, without unnecessary power consumption.

INTRODUCTION

Networking complexity has led to the modularization of network architecture in layers. Traditional approaches focus on wired networks and try to separately optimize each network layer such as the physical, the medium access, the routing and the transport layer. This approach reduces the complexity and makes issues more manageable and architectures more flexible and upgradeable,

DOI: 10.4018/978-1-60960-732-6.ch014

but it may lead to suboptimal designs. Under this layered approach, communication occurs between two adjacent layers without taking into consideration the specific characteristics of multimedia applications. Although this layered approach has been the fundamental factor for the growth of the wired networks and the World Wide Web it seems to pose constrains when attempting to adapt protocol's behavior to multimedia applications characteristics and to wireless network conditions. Therefore, a careful cross-layer approach, where selected communication and interaction between layers is allowed, can have performance advantages without negating the successful layer separation that has guided network design so far. A theoretical discussion of the cross-layer problem framework can be found in (Schaar & Shankar, 2005).

Wireless transmission differs in an important way from wired communication, in that the notion of the link is not as fixed and can vary depending on the movement of the communicating nodes, the intermediate interferences and the transmission characteristics of the communicating nodes, most notably their transmission power. While increased power generally correlates with a stronger signal and therefore improved transmission characteristics, in many wireless scenarios reduced power consumption is desired. This trade-off has been explored by various researchers studying TCP (Transmission Control Protocol) modifications (Tsaoussidis & Badr, 2000, Zhang & Tsaoussidis, 2001, Jones et. al, 2001) trying to combine reduced power consumption with increased data throughput. Wireless standards such as IEEE 802.11 specify power saving mechanisms (IEEE 802.11 PSM), although studies have shown that PSM and other similar mechanisms carry a significant performance penalty in terms of throughput (Molta, 2005, Chen & Huang, 2004, Anastasi et. al, 2004, Simunic, 2005).

An important issue for the efficiency of wireless networks is to accurately determine the cause of packet losses. Packet losses in wired networks occur mainly due to congestion in the path between the sender and the receiver, while in wireless networks packet losses occur mainly due to corrupted packets as a result of the low Signal to Noise Ratio (SNR), the multi-path signal fading and the interference from neighboring transmissions. A second difference between wired and wireless networks is the "mobility factor". Mobility in wireless networks introduces a number of additional barriers in multimedia data transmission. Channel fading and handover time are the most important factors that cause packet losses as they introduce additional delays when the mobile user changes its location from one Access Point (AP) to another.

According to its specification, TFRC (Handley et al, 2003) is a congestion control mechanism for unicast flows operating in a best-effort Internet environment. It aims to be reasonably fair when competing for bandwidth with TCP flows, but at the same time achieving a much lower variation of throughput over time compared with TCP, making it thus more suitable for applications such as telephony or streaming media where a relatively smooth sending rate is important. However, TFRC is slower than TCP in responding to the available bandwidth. TFRC congestion control is appropriate for flows that would prefer to minimize abrupt changes in the sending rate, including streaming media applications with small or moderate receiver buffering before playback. TCP-like congestion control, halves the sending rate in response to each congestion event and thus cannot provide a relatively smooth sending rate.

Several researchers have focused on various issues of cross-layer optimization for wireless ad hoc networks, when there is no infrastructure assumed. Also several efforts have been made in order to combine efficiency and TCP fairness. These works are discussed in the relevant sections of this chapter.

RELATED WORK

As noted above, cross-layer optimization for wireless ad hoc networks, when there is no infrastructure assumed is a very active research field. (Radunovic, 2005) proposes a jointly optimal design of the three layers (physical, MAC (Medium Access Control), routing) for wireless ad-hoc networks and studies several existing rate-maximization performance metrics for wireless ad-hoc networks in order to select appropriate performance metrics for the optimization. In (Shan & Zakhor, 2002) the authors propose an application adaptive scheme based on priority based ARQ (Automatic Repeat Request) together with a scheduling algorithm and FEC (Forward Error Correction) coding combined with RLP (Radio Link Protocol) layer granularity. In (Schaar & Shankar, 2005) the need of a cross-layer optimization is examined and an adaptation framework is proposed amongst the application (APP), the Medium Access Control (MAC) and the Physical (PHY) layers. In the same publication a number of different methodologies for cross-layer adaptation are proposed, named "top-down" approach, "bottom-up", "application centric" and "MAC centric".

(Chen et al., 2004) a joint cross-layer design for QoS content delivery is presented. The central concept of the proposed design is that of adaptation. A new QoS-awareness scheduler with a power adaptation scheme is proposed and is applied at both uplink and downlink Medium Access Control (MAC) layer to coordinate the behavior of the lower layers for resource efficiency. The work in (Lin et al., 2006) summarizes the recent developments in optimization based approaches for resource allocation problems in wireless networks using a cross-layer approach. (Li et al., 2008) deals with 802.16 (WiMax) networks. The 802.16 standard provides four kinds of multimedia data services with QoS parameters but does not define any QoS scheduling algorithm. The above mentioned paper presents an adaptive cross-layer scheduling algorithm for the IEEE 802.16 BWA system. The algorithm uses adaptive modulation and coding (AMC) scheme at the physical layer according to the Signal to Noise Ratio (SNR) on wireless fading channels. In addition the cost function is defined for each kind of multimedia connection based on its service status, throughout the deadline in MAC layer. The simulation results provided show that the scheduling algorithm achieved an optimum trade-off between throughput and fairness. In (Warrier, 2007), the gap between existing theoretical cross-layer optimization designs and practical approaches is examined. The trade-off between increased power consumption and improved signal strength has been explored by various researchers studying TCP modifications (Tsaoussidis & Badr, 2000, Zhang & Tsaoussidis, 2001, Jones et. al, 2001) trying to combine reduced power consumption with increased data throughput. Wireless standards such as IEEE 802.11 specify power saving mechanisms, although studies have shown that PSM (Power Saving Mode) and other similar mechanisms carry a significant performance penalty in terms of throughput (Molta, 2005, Chen & Huang, 2004, Anastasi et. al, 2004, Simunic, 2005). Several researchers have dealt with the issue of optimized video transmission using power management techniques (Kim & Kim, 2003, Zamora et al., 2007), but these approaches do not utilize feedback available from protocols such as TFRC.

TFRC can be viewed as a congestion control technique that trades responsiveness to the network conditions for a smoother throughput variation. TCP-Friendly Rate Control offers an alternative to the method in (Floyd & Kempf, 2004). The key differentiator of TFRC, relative to the Additive Increase Multiplicative Decrease (AIMD) method used in TCP and SCTP (Stewart, 2007), is its smooth response to packet loss. TFRC has been implemented as one of the "pluggable" congestion control algorithms for the Datagram Congestion Control Protocol (DCCP, Kohler et al., 2006 and Floyd et. al., 2006) and as a profile for RTP (Lee & Chung, 2006).

Previous work in (Bouras, 2009) introduced a power management mechanism for TFRC, which operates in a MIMD (Multiplicative Increase, Multiplicative Decrease) fashion. In general, when the packet loss increases above a preset threshold, then the power is also increased, else if the packet loss falls below the threshold, the power consumed is decreased. Moreover the power consumed has a lower bound to prevent the base station from halting the transmission and an upper bound to prevent excessive consumption. This mechanism (called MIMD approach) has been extended with a more sophisticated method for quicker selection of the optimal trade-off, called Binary approach.

THE TFRC PROTOCOL

In this section we provide a short summary of the TFRC operation, in order to demonstrate the way that it tries to achieve UDP-levels of efficiency with TCP friendliness. The TFRC (TCP-friendly Rate Control) protocol presents a modern approach to transport layers protocols, which tread protocols as a set of building blocks – independent components, from which transport protocols are assembled. TFRC provides a sending rate within a factor of two of the sending rate a TCP flow would have under the same condition but with relatively more stable throughput which is a desirable characteristic for a streaming service. TFRC is a receiver-based mechanism where the receiver performs some calculation of the congestion control indicators and reports them back to the server. It relies on the underlying transport protocol such as the DCCP (Kohler et al., 2006) to provide means for the exchange of control information between the server and the client.

The algorithm used to calculate the next sending rate depends on whether the sender is still in the initial Slow Start phase or in the Congestion Avoidance phase. In the Slow Start phase, the sender approximately tries to double its sending rate every time a Receiver Report is received in order to reach the maximum throughput the channel can support which can be detected by increasing *RTT* and losses. Once the first loss has been detected, the sender enters the Congestion Avoidance phase. The next sending rate *X* is now determined from the minimum between twice the previous receiving rate and the sending rate as calculated from the TCP throughput equation.

X = min (TCP throughput, 2*receiving rate)

For its congestion control mechanism, TFRC directly uses a throughput equation for the allowed sending rate as a function of the loss event rate and round-trip time. In order to compete fairly with TCP, TFRC uses the TCP throughput equation, which roughly describes TCP's sending rate as a function of the loss event rate, round-trip time, and packet size. Specifically, TFRC's throughput equation is a slightly simplified version of the throughput equation for Reno TCP:

$$X_{TFRC} = \frac{s}{R\sqrt{\frac{2bp}{3}} + t_{RTO}(3\sqrt{\frac{3bp}{8}})p(1+32p^2)}$$

Where:

- X_{TFRC} is the transmit rate in bytes/second.
- *s* is the packet size in bytes.
- *R* is the round trip time in seconds.
- *p* is the loss event rate, between 0 and 1.0, of the number of loss events as a fraction of the number of packets transmitted.
- *t_RTO* is the TCP retransmission timeout value in seconds.
- *b* is the number of packets acknowledged by a single TCP acknowledgement. The value of b is recommended to be set to 1.

Moreover, it would be beneficial to briefly describe some well-known TCP-like congestion control mechanisms like TCP Vegas, TCP Hybla, TCP Tahoe and Reno. In TCP Vegas, timeouts are set and round-trip delays are measured for every packet in the transmit buffer, while in other TCP versions, are based upon only the last transmitted packet in the transmit buffer. TCP Hybla aims to eliminate penalization of TCP connections that incorporate a high-latency terrestrial or satellite radio link, due to their longer round trip times. It stems from an analytical evaluation of the congestion window dynamics, which suggests the necessary modifications to remove the performance dependence on *RTT*. To avoid congestion collapse, TCP uses a multi-faceted congestion control strategy. For each connection, TCP maintains a congestion window, limiting the total number of unacknowledged packets that may be in transit end-to-end. When the congestion window exceeds a certain threshold, the algorithm enters a new state, called congestion avoidance. The congestion avoidance mechanisms of Tahoe and Reno are not the same, and specifically the behavior of Tahoe and Reno differ in how they detect and react to packet loss. In Tahoe, triple duplicate *ACKs* are treated the same as a timeout, while in Reno, if three duplicate *ACKs* are received, Reno will halve the congestion window.

TFRC defines a loss event as one or more lost or marked packets from a window of data, where a marked packet refers to a congestion indication from Explicit Congestion Notification (Ramakrishnan, 2001). TFRC congestion control mechanism works as follows:

- The receiver measures the loss event rate and feeds this information back to the sender.
- The sender also uses these feedback messages to measure the round-trip time (*RTT*).
- The loss event rate and *RTT* are then fed into TFRC's throughput equation, giving the acceptable transmit rate.
- The sender then adjusts its transmit rate to match the calculated rate.

The dynamics of TFRC are sensitive to how the measurements are performed and applied. Specific mechanisms are used to perform and apply these measurements. Other mechanisms are possible, but it is important to understand how the interactions between mechanisms affect the dynamics of TFRC.

For the purposes of the cross-layer mechanisms detailed later in the chapter, it is very important to understand the mechanism and structure of the feedback packets that the TFRC protocol specifies.

The receiver periodically sends feedback messages to the sender. Feedback packets should normally be sent at least once per *RTT*, unless the sender is sending at a rate of less than one packet per *RTT*, in which case a feedback packet should be send for every data packet received. A feedback packet should also be sent whenever a new loss event is detected without waiting for the end of an *RTT*, and whenever an out-of-order data packet is received that removes a loss event from the history. If the sender is transmitting at a high rate (many packets per *RTT*) there may be some advantages to sending periodic feedback messages more than once per *RTT* as this allows faster response to changing *RTT* measurements, and more resilience to feedback packet loss. However, there is little gain from sending a large number of feedback messages per *RTT*.

Each feedback packet sent by the data receiver contains the following information:

- The timestamp of the last data packet received. We denote this by *t_recvdata*. If the last packet received at the receiver has sequence number i, then *t_recvdata* = *ts_i*. This timestamp is used by the sender to estimate the round-trip time, and is only needed if the sender does not save timestamps of transmitted data packets.
- The amount of time elapsed between the receipt of the last data packet at the receiver, and the generation of this feedback report. We denote this by *t_delay*.

- The rate at which the receiver estimates that data was received since the last feedback report was sent. We denote this by *X_recv*.
- The receiver's current estimate of the loss event rate, *p*.

The sender's behavior specified by TFRC when a feedback packet is received is as follows:

The sender knows its current sending rate, *X*, and maintains an estimate of the current round trip time, *R*, and an estimate of the timeout interval, *t_RTO*.

When a feedback packet is received by the sender at time *t_now*, the following actions should be performed:

1. Calculate a new round trip sample.

```
R_sample = (t_now - t_recvdata)
- t_delay.
```

2. Update the round trip time estimate:

```
if no feedback has been received
before
   R = R_sample;
else
   R = q*R + (1-q)*R_sample;
```

TFRC is not sensitive to the precise value for the filter constant *q*, but a default value of 0.9 is recommended.

3. Update the timeout interval:

```
_RTO = 4*R.
```

4. Update the sending rate as follows:

```
if (p > 0)
   Calculate X_calc using the TCP
   throughput equation.
   X = max(min(X_calc, 2*X_recv),
   s/t_mbi);
else
   if (t_now - tld >= R)
      X = max(min(2*X, 2*X_recv), s/R);
      tld = t_now;
```

Note that *if p is equal to zero*, then the sender is in slow-start phase, where it approximately doubles the sending rate each round-trip time until a loss occurs. The *s/R* term gives a minimum sending rate during slow-start of one packet per *RTT*. The parameter *t_mbi* is 64 seconds, and represents the maximum inter-packet backoff interval in the persistent absence of feedback. Thus, when p is greater than zero, the sender sends at least one packet every 64 seconds. The variable tld is an abbreviation for Time Last Doubled.

5. Reset the nofeedback timer to expire after *max(4*R, 2*s/X)* seconds.

In order the sender to receive the feedback analyzed above, the receiver is responsible for the calculation of the Loss Event Rate (*p*).

Obtaining an accurate and stable measurement of the loss event rate is of primary importance for TFRC. Loss rate measurement is performed at the receiver, based on the detection of lost or marked packets from the sequence numbers of arriving packets. We describe this process before describing the rest of the receiver protocol.

TFRC USED FOR VIDEO TRANSMISSION

A very important issue on video transmission is high fluctuations and oscillations which may damage the video transmission, which demands smooth transmission rates. Most video algorithms such as MPEG2 utilize the three major frame types (I-frames, P-frames, B-frames). The video bit rate tends to vary according to the complexity of the frame data, for example an I-frame would be more complex compared to a P-frame as it results in more bits after compression. The same also applies to scene changes and high motion scenes in a video sequence as they tend to incur a higher prediction error which results in a lower compression efficiency. Thus a typical video bit rate will have occasional "pulses". A smoothed transmission rate will reduce these "pulses" and ends up affecting the video quality. To prevent oscillatory behavior in environments with a low degree of statistical multiplexing it is useful to modify sender's transmit rate to provide congestion avoidance behavior by reducing the transmit rate as the queuing delay (and hence *RTT*) increases. To do this the sender maintains an estimate of the long-term *RTT* and modifies its sending rate depending on how the most recent sample of the *RTT* differs from this value. The long-term sample is *R_sqmean*, and is set as follows:

```
if no feedback has been received
before
   R_sqmean = sqrt(R_sample);
else
   R_sqmean = q2*R_sqmean +
   (1-q2)*sqrt(R_sample);
```

Thus *R_sqmean* gives the exponentially weighted moving average of the square root of the *RTT* samples. The constant *q2* should be set similarly to *q*, and a default value of 0.9 is recommended.

The sender obtains the base transmit rate, *X*, from the throughput function. It then calculates a modified instantaneous transmit rate *X_inst*, as follows:

```
X_inst = X * R_sqmean / sqrt(R_sample);
```

When *sqrt(R_sample)* is greater than *R_sqmean* then the queue is typically increasing and so the transmit rate needs to be decreased for stable operation.

This modification is not always strictly required, especially if the degree of statistical multiplexing in the network is high. However, it is recommended that it is done because it does make TFRC behave better in environments with a low level of statistical multiplexing. If it is not done, it is recommend using a very low value of *q*, such that *q* is close to or exactly zero.

Another important issue is the protocol's transmission rate. TFRC computes its maximum transmission rate as the number of packets per second that a TCP application would receive in similar conditions while breaking up its data into 1480-byte chunks. A TFRC application that is using large packets will experience roughly the same transmission rate in bits per second as a TCP application. However, a TFRC application using small packets will experience a lower transmission rate, in bits per second, than a TCP application. The reasoning for this is that bottlenecks can be the bits per second capacity of links, and also the packets per second capacity of routers. In the subsequent sections of this chapter, we present TFRC mechanisms that still remain TCP-friendly, yet their goal is not to contribute too much to network congestion but to achieve a reasonable video quality gain over the conventional method.

POWER MANAGEMENT MECHANISMS USING TFRC REPORTS

Over the last years a number of new protocols have been developed for multimedia applications in the whole OSI layer's scale. The MPEG protocol family includes the encoding and compression of multimedia data. The MPEG-4 protocol with the enhancements of the FGS (Fine Granularity Scalability), AVC (Advanced Video Coding) and SVC (Scalable Video Coding) provides adaptive video coding by taking into account the available bandwidth and is expected to be used by many multimedia applications. Moreover, congestion control and TCP-friendliness pose additional design requirements as highly fluctuating ("shark teeth"-like) transmission rates may be too difficult to be followed by Audio-Video (AV) encoders and decoders. TCP congestion control produces high fluctuations in the transmission rate which are not suitable for the current audio-video codecs which expect predictive and stable bandwidth allocation. Therefore, the development of protocols such as TFRC can be seen as a step to improve multimedia transmission. One way to cope with transient fluctuations of the transmission rate is with the use of buffers at the clients. However, an initial data pre-fetch in a buffer of more than 8 seconds before the player starts playing the stream is not easily accepted by the end user. Moreover, in real time video applications and conversational media large pre-fetch buffers are not acceptable. For multimedia applications smooth and steady transmission rates and low delay are more important attributes than guaranteed and on order delivery of data packets.

The MIMD Mechanism

The first mechanism is henceforth called the MIMD (Multiplicative Increase, Multiplicative Decrease) mechanism (Bouras et al., 2009). The MIMD mechanism uses the TFRC receiver's reports to the sender in order to calculate the packet loss rate percentage. The algorithm considers only a constant number of previous packet losses, so that it is more adaptive to the most recent conditions of the network. In addition, if the packet loss rate increases above a preset threshold, then the power is also increased by a percentage, else if the packet loss falls way below the threshold, the power consumed is decreased for reasons of power efficiency. Moreover the power consumed has a lower bound to prevent the base station from halting the transmission and an upper bound to prevent excessive consumption. This cross-layer mechanism uses information provided by the TFRC protocol which is a transport layer protocol and needs to act upon the physical layer to adjust the transmission power. The parameters involved by each layer include the transmission power at the physical layer, and the packet loss information at the transport layer. The interaction of these parameters is explained in the pseudo code below.

Below the mechanism is expressed in a more compact form using pseudo code (PL stands for Packet Losses (as a percentage) and TP for Transmission Power, while A>1 and 0<B<1):

```
while (true) {
  retrieve last TFRC report
  set PL = Average of last N reports
  if (PL > Threshold_1) and (TP < Upper_
  Bound) then
    set TP = A * TP
  else if (PL < Threshold_2) and (TP >
  Lower_Bound) then
    set TP = B * TP
}
```

After extensive experimentation with the values *A,B* and the thresholds, (Bouras et al., 2009) conclude that the values *A=1.05, B=0.05, Threshold_1=0.1, Threshold_2=0.075* lead to both good PSNR values and limited energy consumption. The values for *Upper_Bound* and *Lower_Bound* are discussed in the Experiments section.

The target of the proposed mechanism is to minimize or eliminate packet losses, since even a small packet loss rate can result to important reduction of multimedia quality in the end user and result to a bad end user experience. Improvements in the above two areas will lead to improved media parameters such as PSNR and MOS, which better represent the end user experience. At the same time, it has to make sure that power consumption will be bounded and will only increase when this results to noticeably improved video quality.

The Binary Mechanism

The second mechanism that is proposed is henceforth called the Binary mechanism. The Binary mechanism uses the TFRC receiver's reports to the sender in order to calculate the packet loss rate percentage. The algorithm considers only a constant number of previous packet losses, so that it is more adaptive to the most recent conditions of the network. This cross-layer mechanism uses information provided by the TFRC protocol which is a transport layer protocol and needs to act upon the physical layer to adjust the transmission power. The parameters involved by each layer include the transmission power at the physical layer, and the packet loss information at the transport layer.

The finite automaton presented in Figure 1 is the mechanism used by the sender of the video via TFRC. Every time the sender receives a TFRC report from the receiver changes its state according to the state it is in and the new data. The mechanism after receiving the first report, if packet loss is not satisfactory, defines a region in which it will try to approximate the optimum power. The optimum power is the one that produces a desired value of packet loss. After defining the region, the sender will increase its power to the maximum possible in that region and send the next TFRC packet with that power (state A). When the sender receives the next report, it tests whether there has been as significant improvement. If there has been an improvement and packet loss is below a predetermined threshold goes to state C or else repeats the actions of state A. In state C, the mechanism sets the power to the middle of the defined region and the sender goes to state D. In state D the algorithm tests whether the packet loss constraints are still satisfied and if this is the case it repeats state C. If this is not the case the algorithm goes to state E where it goes back to the previous known acceptable power value. The mechanism stays at state E while the packet loss value is acceptable, and if not it goes back to state A. Below there is a summary of the states of the automaton:

```
INIT: initializations
A: Expand "power region" and apply region-
maximum power, then goes to state B
B: Improvement and constraint testing. If
qualified, goes to state C, else it goes to
state A
C: Lowers consumption to the middle of the
defined power region and goes to state D
D:If all the constraints are satisfied, goes
to state C, else goes to state E
E: Backtracks to the last known acceptable
power value and stays there while packet
loss is acceptable, else it goes to state A.
```

Testbed Setup

In order to examine the efficiency of the above described mechanisms, the Network Simulator 2 (ns-2.30) was used as a basic tool for simulating multimedia data transmission over wireless networks. The purpose of the experiments was to judge whether TFRC reports could actually

Figure 1. Finite automaton for the proposed mechanism for the sender

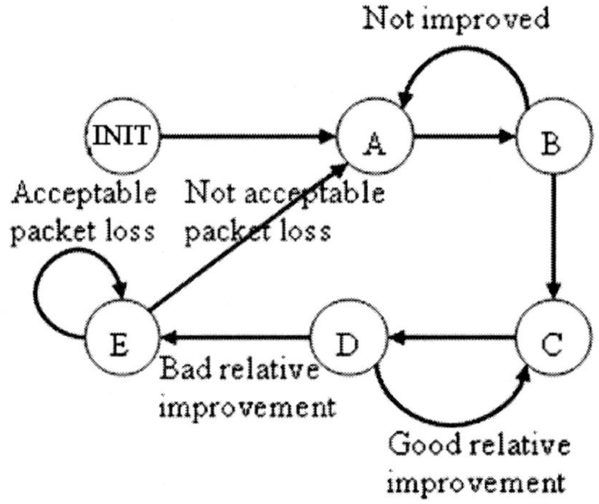

lead to better power management techniques in wireless networks.

In order to simulate MPEG-4 video transmission using ns-2, another software package is needed, namely Evalvid-RA (Lie & Klaue, 2008). Evalvid-RA supports rate-adaptive multimedia transfer based on trace file generation of an MPEG video file. A typical trace file provides information for frame number, frame type, size, fragmentation into segments and timing for each video frame. The multimedia transfer is simulated by using the generated trace file and not the actual binary multimedia content. The simulator keeps its own trace files holding information on timing and throughput of packets at each node during simulation. Combining this information and the original video file, Evalvid-RA can rebuild the video file as it would have been received on a real network. Additionally, by using the Evalvid-RA toolset the total noise introduced can be measured (in dB PSNR) as well as Mean Opinion Score (MOS) can be calculated. An example implementation is illustrated in (Haukass, 2007).

Several modifications of the network simulators were needed in order to build a working instance of the proposed mechanism. Firstly, a module that implements the logic of the proposed mechanism was added in the simulator. Then, the module that implements the TFRC protocol was changed so that it provides information about packet losses to our mechanism. The mechanism calculates the power needed to improve PSNR and then this information is passed to the modified wireless physical layer module that is able to increase or decrease power according to the mechanism.

The next sections of the chapter present some experiments that use a simple network topology, consisting of a transmitting and a receiving node. The transmitting node is located behind a wired connection to a router, while the receiving node is connected through the wireless interface of the intermediate router. The akiyo sample video found in Xiph.org was used for video streaming for the purposes of these experiments.

Firstly, the video file was pre-processed and many video files were produced of different quality and resolution using the ffmpeg tool and shell scripts included in the Evalvid-RA toolset. Then, trace files were generated for all these files and by using these trace files the simulation took place. Ns-2 scripts were created to simulate video transmission over a wireless network over TFRC. After simulating the transfer of the video in several different resolutions, ns-2 trace files were obtained which then were used to reconstruct the videos as it would have been sent over a real network. In this phase, several measurements and calculations can be done involving network and video metrics such as PSNR, MOS, jitter, throughput and delay. By using this procedure and another simulation script and algorithm we can make extensive comparisons and reach conclusions about the efficiency of each algorithm.

Performance Evaluation Experiments

In our ns-2 experiments, we transfer H.264 video over TFRC over wireless links, and in particular over a single hop in a wireless ad hoc network. In order to model various instances of network degradation, we have performed experiments where both nodes are stationary, or where the transmitting node remains stationary, while the receiving node moves with steady speed away from the sender. We then compare the achieved throughput in terms of PSNR, packet losses and power consumption. Objective PSNR measurements can be approximately matched to subjective MOS (Mean Opinion Score) according to the standardized Table 1. The MOS scores reported below are derived from the automatic PSNR to MOS mapping according to Table 1.

In the MIMD mechanism the *Lower_Bound* ranged from *0.02* to *0.04* and the *Upper_Bound* from *0.06* to *0.1*. In Experiments 1 and 2 we ran a set of experiments with different *Lower_Bound* and *Upper_Bound* each time in the above range and increasing by *0.01* in each experiment. The results presented in Table 1 are from the average of these experiments.

In order to model various instances of network degradation, we have performed a series of experiments with various scenarios, with both stationary and mobile nodes:

- Scenario 1: Two nodes, both stationary
- Scenario 2: Two nodes, one stationary, one moving away
- Scenario 3: Two nodes, one stationary, one moving closer and then moving away
- Scenario 4: Two nodes, one stationary, one moving closer
- Scenario 5: Two nodes, one stationary, one moving closer and then moving away and then moving closer again
- Scenario 6: Two nodes, one stationary, one moving away and then stops moving
- Scenario 7: Two nodes, one stationary, one moving closer and then stops moving
- Scenario 8: Two nodes, one stationary, one moving randomly

We repeat each scenario three times, one without any power management, one with MIMD power management algorithm and one with Binary power management algorithm. We then compare the achieved throughput in terms of PSNR, packet losses and power consumption. Objective PSNR measurements can be approximately matched to subjective MOS (Mean Opinion Score) according to the standardized Table 1. The MOS scores reported below are derived from the automatic PSNR to MOS mapping according to Table 1.

Table 1. PSNR to MOS mapping

PSNR [dB]	MOS
>37	Excellent (5)
31-37	Good (4)
25-31	Fair (3)
20-25	Poor (2)
<20	Bad (1)

The MIMD method's performance varied according to the values of the thresholds chosen, while the Binary Method is insignificantly susceptible to thresholds' change. The Binary Method's performance however, depends on the initial desired power that one wants to use.

Experiments

We ran the 7 scenarios described above and took the ratio average PSNR over average power per experiment. The purpose is to maximize this ratio as the larger its value the better the performance. Indeed a large value means larger average PSNR or lower average power or both. The Binary method clearly outperforms the MIMD method and the version without mechanism.

We also present a detailed graph for each scenario, and provide trend lines in order to illuminate the behavior of each mechanism under different conditions. It is worthwhile to note that in many cases as shown in the following figures the Binary method achieves Excellent Mean Opinion Score (see Table 2) whereas the other methods achieve at most Good Mean Opinion Score.

In the first scenario, both nodes are stationary, so power requirements do not vary. Nevertheless, power management mechanisms offer a better ratio of PSNR to transmission power. The proposed mechanism proves especially capable in taking advantage of the available transmission power. For a given amount of transmission power, the proposed mechanism significantly outperforms both the MIMD method and the original transmission approach, in terms of achieved video quality (as measured by the PSNR metric). For example, as shown in the graph, for an average available transmission power of about 0.03 all mechanisms achieve a PSNR value of about 34, which is considered good. When however, average available transmission power is doubled, the simple approach yields almost no benefit, the MIMD mechanism achieves a slightly higher PSNR value of 36, while the binary mechanism excels with a PSNR value of over 41 which qualifies as excellent.

The same observations apply also when one of the nodes is moving away. This time, the MIMD mechanism also displays a noticeable performance advantage over the simple approach. The Binary Method converges faster and closer to an optimum value of power needed to decrease packet, therefore achieves better PSNR values for the same average power. This scenario is one of the most beneficial for the proposed MIMD and Binary mechanisms, because the movement is monotonous and they can easily find an excellent tradeoff between energy consumption and video quality.

Because of the increased proximity of the nodes in Scenarios 3 and 4, the simple transmission approach is able to achieve better performance, without however being able to match either the MIMD or the Binary power management approach because of their adjustment of power according to the packet loss. It is interesting to note that in this scenario, the Binary mechanism clearly outperforms the rest implementations even when average transmission power is low, since the variability in the movement of one of the nodes better suits a quickly adapting algorithm.

When a node is moving closer it is natural to achieve a better PSNR value in all methods. By also using rapid adjustment of power even better results occur. Also, the Binary method achieves Excellent Mean Opinion Score for power over 0.04. In this scenario, the performance gain of the MIMD mechanism over the simple one is reduced because as the moving node increases its proximity, less transmission power is required, and therefore a simple implementation can cope.

In Scenarios with more complicated movement patterns, the basic conclusion seems to be the same: the proposed Binary approach demonstrates a significant performance lead, which often results in the received video quality to be excellent. The MIMD method provides intermediate benefits, while the original transmission approach without active power management lacks behind both in terms of video quality and consumed power.

In the cases where the nodes stop after moving, the MIMD and Binary methods adjust themselves to be as power saving as possible without making a reduction to the quality of video image transmitted. In fact, for the same PSNR values the MIMD and Binary methods consume less energy than when using no mechanism and when the Binary

Table 2. Scenario results

	Normal	**MIMD**	**Binary**
Scenario	**PSNR/Power**	**PSNR/Power**	**PSNR/Power**
1:	669,2	813,1	790,1
2	666,4	769,4	782,5
3:	662,2	759,8	798,8
4:	676,2	798,9	814,8
5:	671,8	800,3	789,7
6:	666,4	769,4	782,5
7:	669,2	813,1	790,1
8:	919,3	902,3	968,4
Average	700,9	803,3	814,6
stddev	88,66	45,06	63,02

Figure 2. Scenario 1: Two nodes, both stationary

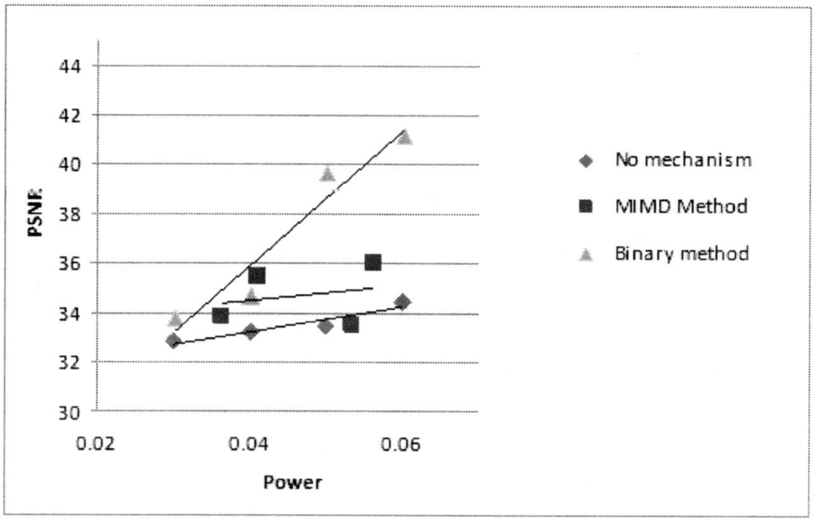

method uses power 0.05 and over achieves excellent results.

When one node moves randomly the results show that all mechanisms tend to display similar behavior, for power values up to 0.04. Above this value the Binary method again gains a significant advantage and achieves excellent results. The fact that in this scenario the performance gains are not as pronounced as in previous scenarios can be attributed to the fact that the adaptive methods need some time to adjust (they adjust every time they receive a TFRC report). Random motions tend to quickly change the assumptions upon which adaptive behavior is based. Therefore, the adaptive methods (MIMD, Binary) tend to perform best in situations where there are movement patterns and changes in movement direction occur slower than the round-trip time of a TFRC report.

Figure 3. Scenario 2: Two nodes, one stationary, one moving away

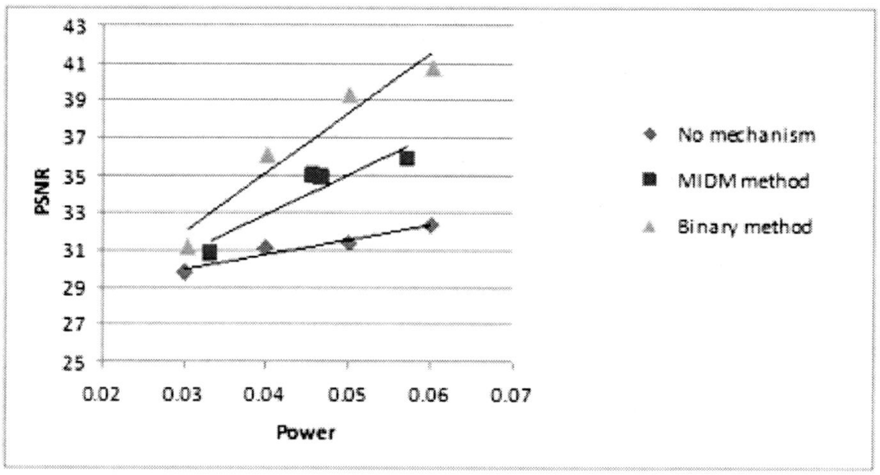

Figure 4. Scenario 8: Two nodes, one stationary, one moving randomly

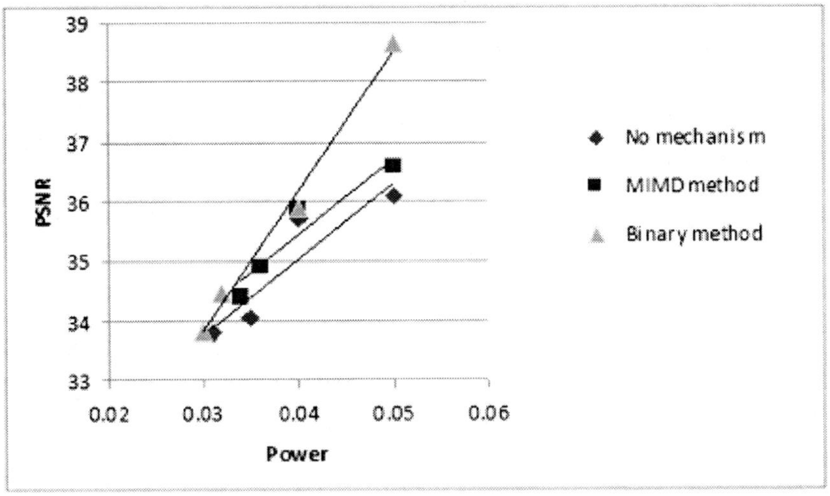

CONCLUSION

In this chapter we have seen how cross-layer optimization techniques can be used to improve quality and reduce power consumption. More specifically, it has been examined how TFRC can be used for video transmission in wireless networks. The basic aspects of TFRC operation have been presented along with the suitability of TFRC usage for video transmission. Additionally, related work has been presented and two cross-layer mechanisms for wireless transmission via TFRC have been proposed that may be utilized for efficient power management, since the trade-off of power consumption and transmission quality is an important factor for wireless devices.

We have seen that by inserting a simple cross-layer mechanism for power management in wireless TFRC transmission, we can significantly improve both the objective quality of the transmitted video, and make a more optimal usage of available power. The complexity cost of the mechanism is quite small, and slightly larger fluctuations in PSNR measurements seem to be the only remaining trade-off. An advanced power management cross-layer mechanism for power management in wireless TFRC transmission seems to significantly improve both the objective quality of the transmitted video, and makes more optimal usage of available power. A simpler MIMD power management approach also has performance benefits, albeit significantly smaller. The complexity cost of the binary mechanism is relatively small, as the implementation in the ns2 simulator has shown. Ongoing research in this area deals with also taking into account the PSNR metric along with packet loss and using the capabilities of H.264 in order to change video quality dynamically so that there can be adaptation of the transmission rate according to the available bandwidth.

REFERENCES

Anastasi, G., Conti, M., Gregori, E., & Passarella, A. (2004, June). A performance study of power-saving polices for Wi-Fi hotspots. *Computer Networks: The International Journal of Computer and Telecommunications Networking*, *45*(3), 295–318.

Bouras, C., Gkamas, A., Kapoulas, V., Papapanagiotou, V., Stamos, K., & Zaoudis, G. (2009, September). *Video transmission over TFRC using cross-layer power management.* 17th International Conference on Software Telecommunications and Computer Nerworks, Hvar, Croatia.

Chen, H., & Huang, C.-W. (2004). *Power management modeling and optimal policy for IEEE 802.11 WLAN systems.* IEEE Vehicular Technology Conference 2004.

Chen, J., Lv, T., & Zheng, H. (2004, May). *Cross-layer design for QoS wireless communications.* 2004 International Symposium Circuits and Systems, ISCAS '04, 23-26 May 2004 Volume 2, (pp. 217-20).

Evalvid-RA. (n.d.). Retrieved on December 23, 2009, from http://www.item.ntnu.no/~arnelie/Evalvid-RA.htm

FFmpeg project. (n.d.). Retrieved on December 23, 2009, from http://www.ffmpeg.org/

Floyd, S., & Kempf, J. (2004, March). *IAB concerns regarding congestion control for voice traffic in the Internet.* (IETF RFC 3714).

Floyd, S., Kohler, E., & Padhye J. (2006, March). *Profile for datagram congestion control protocol (DCCP) congestion control ID 3: TCP-friendly rate control* (TFRC). (RFC 4342).

Handley, M., Floyd, S., Pahdye, J., & Widmer, J. (2003, January). *TCP friendly rate control (TFRC): Protocol specification.* (IETF RFC 3448).

Haukass, T. (2007). *Rate adaptive video streaming over wireless networks explained & explored.* MSc thesis, Norwegian University of Science and Technology, Department of Telematics.

IEEE 802.11. (2009). *PSM standard. Power management for wireless networks, section 11.11.2.*

ISI. (2009). *ns2 network simulator.* Retrieved on December 23, 2009, from http://www.isi.edu/nsnam/ns/

Jones, C. E., Sivalingam, K. M., Agrawal, P., & Chen, J. C. (2001, August). A survey of energy efficient network protocols for wireless networks. *Wireless Networks, 7*(4), 343–358. doi:10.1023/A:1016627727877

Kim, I.-M., & Kim, H.-M. (2003, January). An optimum power management scheme for wireless video service in CDMA systems. *IEEE Transactions on Wireless Communications, 2*(1).

Kohler, E., Handley, M., & Floyd, S. (2006, March). *Datagram congestion control protocol* (DCCP). (IETF RFC 4340).

Lee, S., & Chung, K. (November, 2006). TCP-friendly rate control scheme based on RTP. *Information Networking. Advances in Data Communications and Wireless Networks.*

Li, X., Wu, X., Li, W., & Wang, X. (2008). *An adaptive cross-layer scheduling algorithm for multimedia networks.* 4th International Conference on Intelligent Information Hiding and Multimedia Signal Processing, IIH-MSP 2008, art. no. 4604006, (pp. 52-55).

Lie, A., & Klaue, J. (2008, June). Evalvid-RA: Trace driven simulation of rate adaptive MPEG-4 VBR video. [Berlin/Heidelberg, Germany: Springer.]. *Multimedia Systems, 14*(1), 33–50. doi:10.1007/s00530-007-0110-0

Lin, X., Shroff, N. B., & Srikant, R. (2006, August). A tutorial on cross-layer optimization in wireless networks. *Selected Areas in Communications . IEEE Journal, 24*(8), 1452–1463.

Molta, D. (2005, December). Wi-Fi and the need for more power. *Network Computing.*

Radunovic, B. (2005, July). A cross-layer design of wireless ad-hoc networks. Ph.D. Thesis.

Ramakrishnan, K., Floyd, S., & Black D. (2001, September). *The addition of explicit congestion notification (ECN) to IP.* (IETF RFC 3168).

Shan, Y., & Zakhor, A. (2002). *Cross layer techniques for adaptive video streaming over wireless networks.* IEEE International Conference on Multimedia and Expo, volume 1, (pp 277-280).

Simunic, T. (2005). *Power saving techniques for wireless LANs.* Conference on Design, Automation and Test in Europe - volume 3 (pp. 96-97).

Stewart, R. (2007, September). *Stream control transmission protocol.* (IETF RFC 4960).

Tsaoussidis, V., & Badr, H. (2000 November). *TCP-probing: Towards an error control schema with energy and throughput performance gains.* 8th IEEE Conference on Network Protocols, Japan.

van der Schaar, M., & Sai Shankar, N. D. (2005, August). Cross-layer wireless multimedia transmission: Challenges, principles and new paradigms. *IEEE Wireless Communications.*

Warrier, A., Le, L., & Rhee, I. (September, 2007). *Cross-layer optimization made practical.* 4th International Conference on Broadband Communications, Networks, Systems, BroadNets, art. no. 4550507, (pp. 733-742).

Xiph.org. (n.d.). *Test sequences.* Retrieved on December 23, 2009, from http://media.xiph.org/video/derf/

Zamora, N. H., Kao, J.-C., & Marculescu, R. (2007, April). *Distributed power-management techniques for wireless network video systems.* Design, Automation & Test in Europe Conference & Exhibition, (pp. 1-7).

Zhang, C., & Tsaoussidis, V. (2001). *TCP real: Improving real-time capabilities of TCP over heterogeneous networks.* 11th IEEE/ACM NOSSDAV, NY.

ADDITIONAL READING

Fu Z., Meng X., & Lu S. (2004, December). "A Transport Protocol For Supporting Multimedia Streaming in Mobile Ad Hoc Networks", IEEE JSAC.

Klues, K. (2006). *"Power Management in Wireless Networks", Report, for Advanced Topics in Networking: Wireless and Mobile Networking by R.* Jain, Wash.: Univ. in St. Louis.

Li M., Lee C.-S., Agu E., Claypool M., & Kinicki R (2004, September). "Performance Enhancement of TFRC in Wireless Ad Hoc Networks", Distributed Multimedia Systems (DMS).

Lochin, E., Dairaine, L., & Jourjon, G. (2006). "gTFRC: a QoS-aware Congestion Control Algorithm, International Conference on Networking (ICN)

Rulnick, J. M., & Bambos, N. (1997). Mobile power management for wireless communication networks . *Wireless Networks*, 3. doi:10.1023/A:1019168125436

Tsao S.-C., Lai Y.-C., & Lin Y.-D. (2007, November). "Taxonomy and Evaluation of TCP-Friendly Congestion-Control Schemes on Fairness, Aggressiveness, and Responsiveness", IEEE Network.

Vojnovic, M., & Le Boudec, J.-Y. (2005). "On the Long-Run Behavior of Equation-Based Rate Control", ACM SIGCOMM 2002, also . *IEEE/ACM Transactions on Networking*, 2005.

Welzl M. (2005, June). "User-Centric Evaluation of TCP-friendly Congestion Control for Real-Time Video Transmission", Elektrotechnik und Informationstechnik.

Yan J., May M., Katrinis K., & Plattner B. (2005). "A New TCP-Friendly Rate Control Algorithm for Scalable Video Streams", NETWORKING 2005.

Yan, J., Zheng, X., Liu, J., & Li, J. (2008). *Wireless Loss Detection for TCP Friendly Rate Control Algorithm in Wireless Networks*. Information Networking. Towards Ubiquitous Networking and Services.

KEY TERMS AND DEFINITIONS

B-Frame: A video frame used in a video compression algorithm that requires both previous and forward frames for its decoding (used in MPEG video standards).

Cross-Layer: The method of a network systems design that takes into consideration multiple aspects of system operation across various layers in order to improve functional efficiency.

H.264/MPEG-4: A modern standard for video compression that aims at flexibility and providing good video quality at lower bit rates than previous compression methods.

I-Frame: A video frame used in a video compression algorithm that does not require other frames for its decoding (used in MPEG video standards).

P-Frame: A video frame used in a video compression algorithm that requires previous frames for its decoding (used in MPEG video standards).

TCP-Friendliness: The characteristic of a flow that does not create network congestion so that competing TCP flows are not squeezed out of bottleneck links.

TCP-Friendly Rate Control (TFRC): A: congestion control protocol for TCP-friendly flows that achieve better throughput than TCP.

Wireless: Transmission of information without the usage of wires.

APPENDIX

Table of abbreviations

AIMD	Additive Increase Multiplicative Decrease
AMC	Adaptive Modulation and Coding
AP	Access Point
ARQ	Automatic Repeat Request
AV	Audio-Video
AVC	Advanced Video Coding
DCCP	Datagram Congestion Control Protocol
FEC	Forward Error Correction
FGS	Fine Granularity Scalability
MAC	Medium Access Control
MIMD	Multiplicative Increase, Multiplicative Decrease
MOS	Mean Opinion Score
PL	Packet Loss
PSM	Power Saving Mode
PSNR	Peak Signal-to-Noise Ratio
RLP	Radio Link Protocol
RTT	round-trip time
SNR	Signal-to-Noise Ratio
SVC	Scalable Video Coding
TCP	Transmission Control Protocol
TFRC	TCP-Friendly Rate Control
TP	Transmission Power

Chapter 15
Cross-Layer Protocols for Multimedia Communications over Wireless Networks

Jaydip Sen
Tata Consultancy Services, India

ABSTRACT

In the last few years, the Internet throughput, usage, and reliability have increased almost exponentially. The introduction of broadband wireless mobile ad hoc networks (MANETs) and cellular networks, together with increased computational power, have opened the door for a new breed of applications to be created; namely, real-time multimedia applications. Delivering real-time multimedia traffic over a complex network like the Internet is a particularly challenging task since these applications have strict quality-of-service (QoS) requirements on bandwidth, delay, and delay jitter. Traditional Internet protocol (IP)-based best effort service is not able to meet these stringent requirements. The time-varying nature of wireless channels and resource constrained wireless devices make the problem even more difficult. To improve perceived media quality by end users over wireless Internet, QoS supports can be addressed in different layers, including application layer, transport layer, and link layer. Cross layer design is a well-known approach to achieve this adaptation. In cross-layer design, the challenges from the physical wireless medium and the QoS-demands from the applications are taken into account so that the rate, power, and coding at the physical (PHY) layer can be adapted to meet the requirements of the applications given the current channel and network conditions. A number of propositions for cross-layer designs exist in the literature. In this chapter, an extensive review has been made on these cross-layer architectures that combine the application-layer, transport layer, and the link layer controls. Particularly, issues like channel estimation techniques, adaptive controls at the application and link layers for energy efficiency, priority based scheduling, transmission rate control at the transport layer, and adaptive automatic repeat request (ARQ) are discussed in detail.

DOI: 10.4018/978-1-60960-732-6.ch015

INTRODUCTION

As the wireless networks evolved from circuit-switched voice traffic based 2G networks to an all-IP based packet-switched network catering to a mix of high speed real-time traffic such as voice, multimedia teleconferencing, online gaming etc., and data-traffic such as WWW browsing, messaging, file transfers etc., there has been a dramatic change in the quality-of-service (QoS) requirements in terms of transmission accuracy, delay, jitter, throughput and so on. In order to achieve a successful and profitable commercial market for future wireless technology, network service designers and providers need to pay much attention to efficient utilization of radio resources due to fast growth of the wireless subscriber population, increasing demand for new mobile multimedia services and consequent diverse and more stringent QoS requirements. Traffic on wireless networks is becoming increasingly complex with a mix of real-time traffic such as voice, multimedia teleconferencing, gaming, and data-traffic such as WWW browsing, messaging and file transfers etc. All these applications require widely varying QoS guarantees for different types of traffic. Of late, various mechanisms have been proposed in the literature to support these QoS requirements. However, providing a robust QoS support for multimedia applications over wireless networks is a very challenging task due the following reasons (Jiang et al., 2005).

- Different applications have different QoS requirements. Real-time media such as video and audio is delay-sensitive but capable of tolerating a certain degree of errors. Non-real time media such as web data is less delay-sensitive but requires reliable transmission.
- Wireless channels have high packet loss rate and bit error rate (BER) due to fading and multi-path effects. Resulting packet loss and bit errors can have an adverse effect on the multimedia applications.
- Wireless channels have bandwidth limitation and fluctuations of the available bandwidth, packet loss rate, delay and jitter.
- Traditional transport layer protocols perform poorly in wireless networks since they assume congestion to be the primary cause for packet losses and unusual delay in the network. These protocols reduce the transmission rate whenever they observer packet loss. In wireless networks, the packet may be dropped due to channel errors, thereby resulting in unnecessary reduction in end-to-end throughput.
- The mobile devices are power constrained. Maintaining good media quality and minimizing average power consumption (for processing and communication) are two conflicting requirements.
- Receivers in multimedia delivery systems are quite different in terms of latency requirements, visual quality requirements, processing capabilities, power limitations, and bandwidth constraints. Moreover, multimedia may traverse different types of networks, e.g., wire-line networks, cellular networks, and wireless local area networks (WLAN). Each of these networks has different characteristics such as reliability, delay, jitter, bandwidth, and medium access control (MAC) mechanisms.

In view of the above constraints, a strict modularity and protocol layer independence of the traditional transmission control protocol (TCP) / Internet protocol (IP) or OSI stack will lead to a sub-optimal performance of applications over IP-based wireless networks. For optimization, we require protocol architectures that require modification of the reference layered stack by allowing direct communication between protocols at non-adjacent layers or sharing state variables across different layers to achieve better performance. The goal of a cross layer design is to actively exploit this possible dependence between protocol layers to achieve performance gains. Although the cross layer design is an evolving area of research, considerable amount of work has already been done on this area. The objective of this chapter is to introduce the concept of cross-layer design and discuss the various existing cross-layer protocols for QoS-aware multimedia applications over resource constrained wireless networks.

The chapter is organized as follows. Section 2 describes various QoS parameters such as delay, latency, jitter, packet drop rate etc. those are relevant in multimedia communication. Section 3 discusses various issues in cross-layer design, depicts some generic cross-layer frameworks, and also identifies the relevant protocol layers in which cross-layer design principles may be applied for QoS support in multimedia applications. Section 4 presents various types of adaptations required at different layers of the protocol stack for cross-layer design and optimization. Section 5 describes some important link layer adaptation mechanisms in cross-layer design. Section 6 discusses the role of the transport layer in cross-layer architecture and also presents some transport layer-initiated cross layer protocols. Section 7 describes some of the application layer-specific issues in a cross-layer environment. Section 8 discusses the future trends in research in cross-layer protocol design and the associated challenges. Finally, Section 9 concludes the chapter.

DIFFERENT QoS CLASSES IN MULTIMEDIA APPLICATIONS

One major challenge in multimedia services over wireless networks is QoS provisioning with efficient resource utilization. Heterogeneous multimedia applications in future IP-based wireless networks require a more complex QoS model and more sophisticated management of scarce radio resources. QoS can be classified according to its implementation in the networks, based on a hierarchy of four different levels: bit-level, packet-level, call-level, and application-level (Jiang et al., 2005). Transmission accuracy, transmission rate (i.e., throughput), timeliness (i.e., delay and jitter), fairness, and user perceived quality are the main considerations in this classification:

- Bit-level QoS: to ensure some degree of transmission accuracy, a maximum BER for each user is required. Any transmission with BER greater than the maximum permissible limit is not acceptable for applications, which have a stringent QoS requirement. Data applications are more sensitive to bit errors than video applications.
- Packet-level QoS: for delay-sensitive applications like voice over IP (VoIP) and videoconferencing, each packet should be transmitted within a delay bound. On the other hand, data applications like Internet download can tolerate delay to a certain degree. Throughput is a more pertinent QoS criterion for data applications. Each traffic type can also have a packet loss rate (PLR) requirement.
- Call-level QoS: due to insufficient capacity at a particular instant of time in a wireless system, there is always a chance that a new call may be blocked or a handover is dropped. From the user's point of view, the issue of handover call dropping is more serious than blocking of a new call because the user might be in the middle of an important transaction when the handover takes places. It is necessary to devise an effective call admission control to ensure that handover calls are not disturbed; the new calls, which may arrive during the handover process, may be blocked instead.
- Application-level QoS: the application layer-perceived QoS parameters like the peak signal to noise ratio (PSNR) for video application and the end-to-end throughput

for data application provided by the responsive TCP, more suitably represent the service quality seen by the end user, than bit and packet level QoS.

Another big challenge is to develop an accurate mapping mechanism for application layer QoS parameters to the lower layer (e.g., the physical layer) parameters so that the requirements specified at the application layer are suitably converted to the corresponding requirements in the lower layers before being passed over the carrier. Kumwilaisak et al have proposed one such mapping architecture (Kumwilaisak, et al., 2003). In addition to QoS parameter mapping, an effective link layer packet-scheduling scheme with appropriate power allocation is required to support bit- and packet- level QoS requirements of the applications running on mobile devices. Specifically, the power levels of the mobile devices should be managed in such a way that each mobile device achieves the required bit energy to interference-plus-noise density ratio, and the transmission from/to all the mobile devices are controlled to meet the delay, jitter, throughput, and the PLR requirements.

CROSS-LAYER FRAMEWORKS FOR MULTIMEDIA TRANSMISSION

To handle the challenges mentioned in Section 2, many studies have been performed and a number of cross-layer protocols have been proposed in the literature for multimedia transmission over wireless networks. Most of these protocols involve message communications across various layers, e.g., application, transport and link layers.

Considering the limitation of bandwidth in wireless systems, the most important target in the link layer is to increase link utilization. It is known that real-time transport protocol (RTP) / user datagram protocol (UDP) / IP and TCP/IP have the problem of large header overhead on bandwidth-constrained links. Header compression has been found to be efficient for using those protocols. Unfortunately, many header compression schemes (Casner et al., 1999) do not work well on noisy links, especially the one with high BER and long round-trip time (RTT). Internet Engineering Task Force (IETF) had a working group (WG), called robust header compression (ROHC) to address the header compression issue (Pelletier et al., 2008).

To handle the severe bandwidth and delay fluctuations in wireless Internet, available network condition estimation and congestion control are some of the key issues that need to be addressed in the transport layer. Throughput calculation, packet-pair, and packet-train bandwidth probing are several popular techniques for bandwidth measurement (Lai et al., 1999). Controlling parameters such as packet error rate, delay, and delay jitter are also important. Different congestion and rate control schemes must be implemented so that multimedia such as video and audio can adapt to the estimated network information in a smooth way (Yang et al., 2001).

In the application layer perspective, many studies have been performed to improve media delivery quality. Error protection, power saving, and proxy management are some of the well-known approaches in this regard. To overcome the packet loss and residual bit error in wireless Internet, error control techniques such as forward error correction (FEC) and automatic repeat request (ARQ) are necessary to maintain high-quality media delivery. Unequal error control (Zhang et al., 1999) can be adopted for providing varying degrees of importance to different parts of the media content. To make a tradeoff between power consumption and quality of the delivered media, power control and joint source and channel coding (JSCC) are two effective approaches. Power control is conducted from the group point of view by controlling transmission power and spreading gain for a group of users to reduce interference (Sampath et al., 1995). JSCC is, on the other hand, is conducted from the individual user's point of view to effectively combat the errors occurred during transmission by allocating bits between source and channel (Qian et al., 1999). The heterogeneous networks and different requirements of receivers ask for an efficient proxy-caching mechanism to satisfy different characteristics of receivers. Traditional proxy servers were designed to serve web request for non-continuous media, such as textual and image objects. With the increasing advent of video and audio streaming applications, continuous-media caching has been studied in (Sen et al., 1999). However, the varying wireless Internet condition and different media characteristics impose challenges on how to efficiently cache both continuous and non-continuous media.

In following sub-sections, some of the existing cross layer designs, architectures, and algorithms for multimedia transmission over wireless networks are presented. The salient features of these schemes are discussed, and their specific contributions and areas of applications are highlighted.

A CROSS-LAYER ARCHITECTURE FOR MULTIMEDIA QoS

Zhang, Yang, and Zhu have presented a general architecture that is based on the Universal Mobile Telecommunications System (UMTS) for multimedia delivery over the wireless Internet (Zhu et al., 2005). Figure 1 depicts the architecture, where a multimedia server, a base station (BS) or a gateway with media proxy, and several heterogeneous mobile clients are deployed. Various control mechanisms at the application-layer, the transport-layer, and the link-layer control are taken into account and suitably deployed into this generic architecture, to achieve the desired end-to-end quality of the multimedia services.

In Figure 1, the application is transmitted via TCP or UDP in the Internet segment depending on the characteristics of the traffic. The IP packets arriving in the downlink (BS to the mobile client) in the UMTS network are transported to the radio network controller (RNC). Appropriate header compression techniques are applied to the packets in the packet data convergence protocol (PDCP) layer of the UMTS stack. The compression technique used in the PDCP layer varies depending on the implementation. The PDCP layer compresses each packet, attaches a header, and forwards it further. It uses the services provided by a lower layer called the radio link control (RLC) layer. The RLC layer is employed to support reliable upper layer protocols such as the TCP. It uses sophisticated retransmission schemes to perform partial error recovery at the link layer thereby hiding the transmission errors from the upper layers and reducing the chances of degradation in the performance of the upper layer protocols. The RLC protocol data units (PDU) of a particular IP connection are served by the MAC layer. If deterministic transmission time intervals

Figure 1. A generic architecture for multimedia delivery over wireless Internet

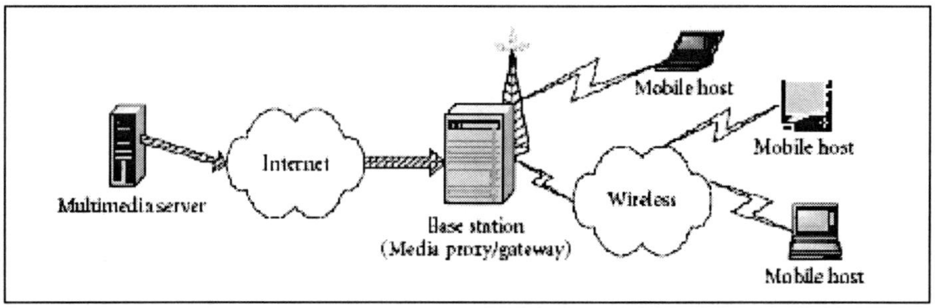

(TTIs) are used, the MAC layer entities request the corresponding RLC layer entities for a certain number of RLC PDUs, which are then transferred through the radio interface in MAC frames. The TTI refers to the length of an independently decodable transmission on the radio link. It is related to the size of the data blocks being passed from the higher network layers to the radio link layer. In order to be able to adapt quickly to the changing conditions in the radio link, shorter TTIs are preferable. However, in order to exploit the advantages from the effect of interleaving and to increase the efficiency of error-correction and compression techniques, the system must have longer TTIs. The determination of an appropriate TTI value is, therefore, an optimization problem.

Figure 2 depicts the cross-layer architecture for the generic framework depicted in Figure 1. The following functionalities of the cross-layer architecture are identified for providing QoS support to multimedia applications.

- Estimating the dynamic wireless Internet conditions: to track the varying wireless Internet conditions, network estimation mechanisms in different layers on the server, the BS, and the mobile hosts have to work together.
- Adapting to the network condition: the cross-layer architecture should adaptively adjust the amount of wireless Internet resources such as bandwidth, time slot etc., according to the varying network conditions. This function is carried out by the congestion control module in the multimedia server and the BS.
- Network-aware media adaptation: in response to the changing network conditions, the media encoding mechanisms and different parts of media should be adaptively adjusted or customized in order to maximize the system efficiency and minimize the end-to-end delay.
- Power efficiency and robustness to errors: the application- and the link-layer error control schemes may be used together for increasing the robustness to errors. The overall power consumption in the mobile hosts should also be minimized.
- Efficient network utilization: to improve the network utilization, especially in wireless channels, header compression should be performed both the BS and the mobile hosts.
- Multi-services support: For supporting multiple types of traffic, each having different types of QoS requirement, employ-

Figure 2. A cross-layer architecture for multimedia delivery over wireless Internet

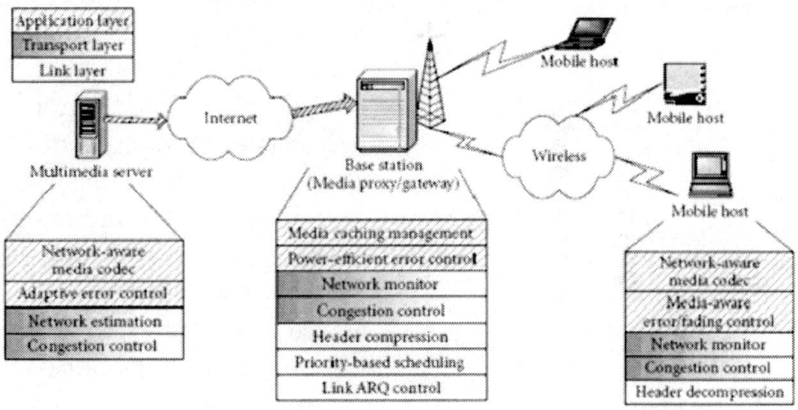

ing a priority-based scheduling is an efficient approach.

- Network and client heterogeneity: heterogeneity in different networks and client devices should be supported by QoS-adaptive proxy caching.

A CROSS-LAYER RESOURCE ALLOCATION IN 3G NETWORKS

Jiang et al. have proposed a cross-layer design approach for real-time video transmission over time-varying 3G CDMA wireless networks, where the link layer resource allocation benefits from information in both the application and physical (PHY) layers (Jiang et al., 2005). Figure 3 depicts the schematic diagram of the inter-layer message communication. The authors have identified three possible cross- layer information flows: (i) from the PHY to the link layer, (ii) from the link to the transport layer and vice versa, and (iii) from the link to the application layer and vice versa. Three modules of the cross-layer framework have been proposed: (i) a channel-aware scheduling, (ii) TCP over CDMA wireless links, and (iii) a joint video source/channel coding and power allocation. In the following, these modules are briefly described.

In channel-aware scheduling, the time-varying characteristics of a wireless channel are exploited by using a multiuser diversity framework to improve system performance. The principle of multi-user diversity is that for a cellular system with multiple mobile stations (MSs) having independent time-varying channels, it is very likely that there exists an MS with instantaneous received signal power close to its peak value. Overall resource utilization can be maximized by providing service at any time only to the MS with the highest instantaneous channel quality. The authors argue that with the capability to support simultaneous transmissions in a CDMA system, multi-user diversity can be employed more effectively and flexibly than traditional channel-aware scheduling schemes for a TDMA system. An MS does not need to wait until it has the best channel quality among all the MSs. It is allowed to transmit as long as its channel quality is good enough. However, for real-time traffic such as voice or video which have delay constraints, if an MS is in a bad channel state for a relatively long period, its packets will be discarded when multiuser diversity is employed, as the MS has to wait until its channel state improves. Hence, it is a challenging task to apply multiuser diversity to real-time traffic. The authors have solved this problem by incorporating packet delay in scheduling decision in the proposed resource allocation framework.

Figure 3. A generic cross-layer design approach

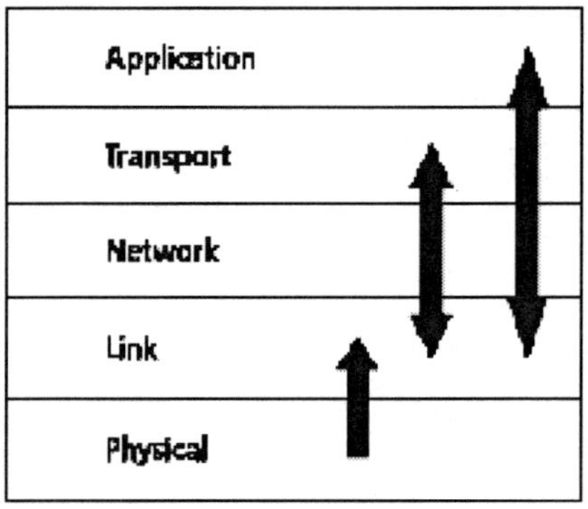

The second module is an adaptive TCP protocol. The traditional TCP in wired networks adjusts its sending rate based on the estimated network congestion status to achieve congestion control or avoidance. In a wireless environment, TCP performance can be degraded severely as it interprets losses due to unreliable wireless transmissions as signs of network congestion and invokes unnecessary congestion control (Jiang et al., 2005). To improve TCP performance over wireless links, several solutions have been proposed to alleviate the effects of non-congestion-related packet losses (Xylomenos et al., 2001). The authors have argued that when a TCP connection is transmitted over CDMA cellular networks, in addition to the issues like congestion control, link errors etc some additional considerations are to be made. First, CDMA capacity is interference-limited. TCP transmission from an MS generates interference to other MSs. It is desired to achieve acceptable TCP performance (e.g., a target throughput) and at the same time introduce minimum interference to other MSs (i.e. to require minimum lower-layer resources). Second, power allocation and control in CDMA can lead to a controllable BER, which affects TCP performance. Keeping in mind these issues, the authors have proposed an adaptive TCP that dynamically adjusts the sending rate of TCP segments (which will be fed back into the link layer transmission queue) according to network congestion status (e.g., packet loss and round-trip delay). A link layer design parameter ultimately determines the packet loss rate and transmission delay over the wireless link and therefore affects the TCP performance. With a proper choice of this link layer design parameter, it will be possible to achieve the target TCP throughput.

The third module is responsible for carrying out JSCC and efficiently allocating power to different applications. It has been shown that for video services over a CDMA channel with limited capacity, an effective way is to pass source significance information (SSI) from the source coder in the application layer to the channel coder in the PHY layer (Jiang et al., 2005). Therefore, more powerful FEC code involving more overhead can be used to protect more important information, while no or weaker FEC may be applied to less important information. This approach of JSCC is a cross-layer approach, and is known as unequal error protection (UEP). The authors have also argued that in case of a shortfall in the system capacity, deployment of UEP schemes results in a more graceful quality degradation producing smaller distortion or higher PSNR than equal error protection (EEP). It has been shown that based on channel capacity, the optimal transmission rate and power allocation for packets of each priority can be found to minimize the average distortion of the received video by means of an optimization formulation over CDMA channels.

A CROSS-LAYER SCHEDULING ALGORITHM

Liu et al. have proposed a scheduling algorithm at the MAC layer for multiple connections with diverse QoS requirements, where each connection employs adaptive modulation and coding (AMC) scheme at the PHY layer over wireless fading channels (Liu et al., 2006). A priority function (PRF) is defined for each connection admitted in the system. This priority function is updated dynamically depending on the wireless channel quality, QoS satisfaction, and services across all layers. The connection with highest priority is scheduled each time. The priority of each connection is updated dynamically based on its channel and service status. At the MAC layer, each connection belongs to a single service class and is associated with a set of QoS parameters that quantify its characteristics. Following IEEE 802.16 standard, four QoS classes are used: (i) unsolicited grant (UGS) services, (ii) real-time polling services (RTPS), (iii) non-real-time polling services (nRTPS), and (iv) best effort (BE) services. The UGS supports constant bit rate (CBR) and fixed throughput connections, and provides guarantees on latency, and jitter. The RTPS provides guarantees on throughput and latency but when compared with UGS it allows for more tolerance on latency. NRTPS can give guarantees only on throughput, and is suitable for data applications, such as file transfer protocol (FTP). The BE service cannot provide any guarantee on delay or throughput, and is used for hypertext transmission protocol (HTTP) and email applications. The cross-layer scheduler has the following features:

- The scheduler utilizes the available bandwidth efficiently so that at no allocation interval, it assigns a time slot to a connection that has a bad channel quality. In other words, it efficiently exploits multi-user diversity.

- Delay bound is provided for applications that are based on RTPS.
- Throughput is guaranteed for NRTPS connections if sufficient bandwidth is available for those connections.
- Implementation complexity of the scheduler is low because it simply updates the priority of each connection per frame and allocates maximum time slots to those connections that have the highest priority.
- The scheduler is flexible as it does not depend on any traffic or channel model.
- The design of the scheduler is scalable. When the available bandwidth decreases due to addition of new connections, the performances of connections with low-priority service classes are degraded before the admission of the high priority classes.

Figure 4 shows the topology of a wireless network, where multiple subscriber stations (SS) are connected to the BS or relay station (RS) over wireless channels. Multiple connections (sessions, flows) can be supported by each SS. All connections communicate with the BS using time division multiplexing (TDM) or time division multiple access (TDMA). The wireless link of each connection from the BS to each SS is depicted in Figure 5. A buffer is implemented at the BS for each connection that operates in first-in-first-out (FIFO) mode. The AMC controller follows the buffer at the BS (i.e., the transmitter) and the AMC selector is implemented at the SS (i.e., the receiver). At the PHY layer, multiple transmission modes are available to each user, with each mode representing a pair of specific modulation format and an FEC code. Based on the channel estimates obtained at the receiver, the AMC selector determines the modulation-coding pair (mode or burst profile), whose index is sent back to the transmitter through a feedback channel for the AMC controller to update the transmission mode. Coherent demodulation and soft-decision Viterbi decoding are employed at the receiver. The decoded bit streams are mapped to packets, which are pushed upwards to the MAC.

A CROSS-LAYER OPTIMIZER IN BROADBAND NETWORKS

Triantafyllopoulou et al. have proposed a cross-layer optimization mechanism for multimedia traffic over IEEE 802.16 standard-based broadband wireless networks (Triantafyllopoulou et al., 2007). The scheme utilizes information provided by the PHY and MAC layers, such as signal quality, packet loss rate and the mean delay, in order to

Figure 4. A wireless network topology

Figure 5. The wireless links from the base station (BS) to the subscriber station (SS)

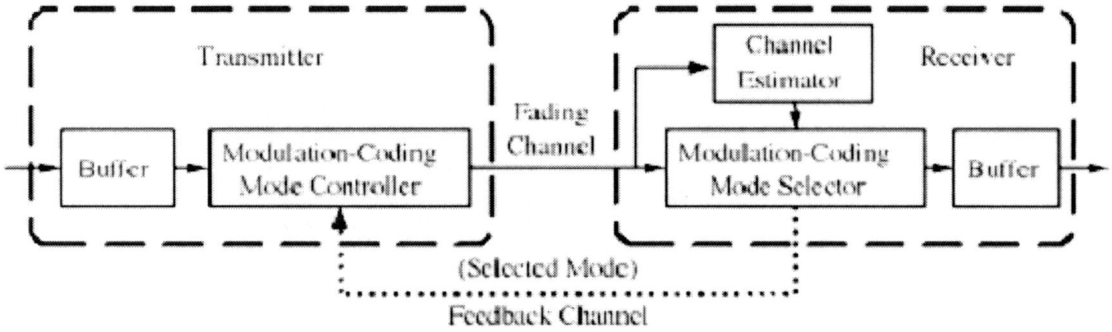

control parameters at the PHY and the application layers and improve the performance of the system. Essentially, the adaptive modulation capability of the PHY layer and the multi-rate data-encoding feature of multimedia applications are combined to achieve an improved end user QoS.

The cross-layer optimizer is split into two parts- one residing at the BS part and the other at the SS. The part residing at the BS accepts an abstraction of layer-specific information regarding the channel conditions and the QoS parameters of active connections provided by the BS PHY and MAC layers. Based on this information, a specific decision algorithm determines the most suitable modulation and/or traffic rate of each SS, separately for each direction (the uplink and the downlink). Finally, the optimizer at the BS informs the corresponding layers of the required modifications. This is depicted in Figure 6 (a). If the decision of the optimizer at the BS involves traffic rate changes, it communicates with the optimizer at the SS through the SS MAC layer. The SS MAC then instructs the SS application layer accordingly. This is depicted in Figure 6 (b). The optimizer at the SS may either accept the suggestions provided by the optimizer at the BS, or may refine them if it has more accurate knowledge of the status of active connections. In the proposed architecture, the optimizer at the SS is designed as a passive module that can only instruct the application layer at the SS based on the suggestion provided by the optimizer at the BS.

The BS decision algorithm relies on the values of two major QoS parameters, i.e., the packet loss rate and the mean delay. The packet loss rate is the sum of (i) the packet error rate (i.e. the percentage of packets that are lost due to channel errors, and (ii) the packet timeout rate (i.e., the percentage of packets that are lost due to expiration). To compute these rates, the optimizer at the BS has to maintain up-to-date information on channel conditions in directions, as well as traffic and QoS status of active connections.

The packet error rate is estimated based on the channel conditions. The channel conditions on the uplink are known from the PHY layer of the BS. The channel conditions on the downlink may be assumed similar to the uplink conditions or may be obtained either by the received channel measurement report response (REP-RSP) message, or through the channel quality information channel (CQICH). Packet timeout rate and mean delay for all active connections in both directions are provided by the BS MAC layer

If at some point an SS faces unacceptable packet loss rates, the optimizer at the BS takes the following actions depending on the nature of the loss:

Figure 6. The cross-layer optimizer at the BS (a) and the SS (b)

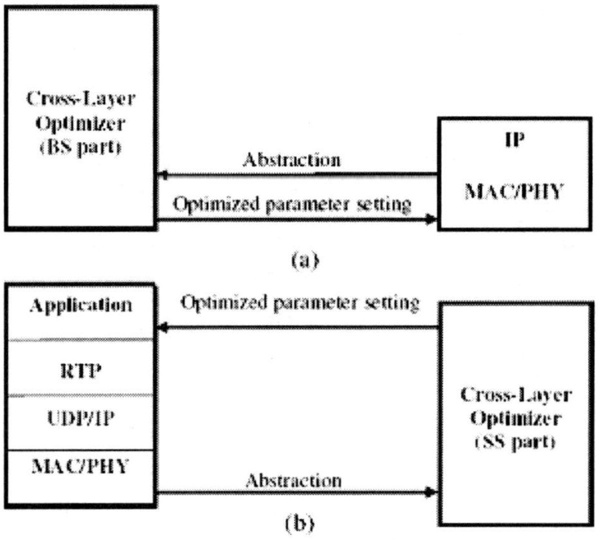

- In case most of the losses are due to poor channel conditions (packet errors), the cross-layer optimizer as the BS instructs the MAC layer for a degradation of the modulation, in order to achieve higher channel error resilience and increase robustness against interference. Thus, the BS optimizer selects the highest modulation that will restore the loss rate to acceptable values and instructs the MAC layer accordingly.
- In case of most of the losses are the result of packet timeouts (unacceptable delays), the action to be performed depends on the contribution of these timeouts to the overall packet loss. If the loss rate is caused almost exclusively by packet timeouts, the optimizer at the BS concludes that the channel is very slow and unable to satisfy the transmission speed requirements. In this case, the optimizer at the BS instructs a modulation upgrade in order to increase the transmission speed and reduce the losses caused by timeouts. On the other hand, when a significant percentage of packet losses are caused by errors due to the poor channel conditions and a modulation upgrade is not possible, the optimizer at the BS instructs the optimizer at the SS for a traffic rate reduction in order to moderate timeouts.

To perform efficiently under all conditions, the cross-layer optimizer has to take proper actions also when the conditions are improved. Thus, when the loss rate decreases significantly, the optimizer at the BS may decide to either switch to a higher modulation and increase the available bandwidth, or instruct the optimizer at the SS to increase the traffic rate and improve the QoS. The specific action depends on the mean delay experienced by the active connections of the SS. If the mean delay is relatively low, compared to the delay bound, the optimizer at the BS instructs for a traffic rate increase to improve the service provided to the user. On the other hand, if the mean delay is close to the delay bound, the optimizer at the BS instructs for a modulation upgrade to increase transmission speed and reduce delays.

ADAPTATIONS AT DIFFERENT LAYERS OF THE PROTOCOL STACK

Various adaptations are necessary at different layers of the standard protocol stack for providing a robust QoS support to multimedia applications over wireless networks. In Section 1, it has already been seen that wireless channels pose a number of challenges in designing such adaptive schemes. Considering the limitation of bandwidth in wireless systems, the most important goal at the link layer is to increase the link utilization. It is known that RTP/UDP/IP and TCP/IP have the problem of large header overhead on bandwidth-limited links. Header compression has been proven efficient for using those protocols. To handle the severe bandwidth and delay fluctuation in wireless Internet, available network condition estimation and congestion control are key issues needed to be addressed in the transport layer. Error protection, power saving, and proxy management are some of the important issues to be handled in the application layers. These layer-specific issues are described in details in the following Sections.

LINK LAYER ADAPTATION MECHANISMS

There are several currently existing approaches for link layer adaptation under varying wireless channel conditions. Four important mechanisms used for this purpose are (i) application adaptive ARQ, (ii) priority-based scheduling, (iii) header compression, and (iv) channel-aware scheduling. In the following, these mechanisms are explained briefly.

(i) Application adaptive ARQ: to overcome packet loss, ARQ is used for packet retransmissions. ARQ uses acknowledgments (ACKs) and timeouts to achieve reliable data transmission. The receiver sends an ACK to the transmitter to indicate that it has correctly received a data frame or packet. The sender waits for a pre-defined period (timeout) for the ACK to arrive. If the ACK arrives then the sender sends the next packet. Otherwise, it resends the previous packet until it receives an ACK or it exceeds a pre-defined number of retransmissions. ARQ can be implemented at any of the layers: application, transport, or link. ARQs implemented at the link layer are more efficient than those implemented at the application or transport layers because – (i) they have a shorter control loop and hence can recover lost data more quickly, (ii) they operate on frames that are much smaller than the IP datagrams and (iii) they might be able to use local knowledge that is not available to end hosts, to optimize delivery performance for the current link conditions. This information can include information about the state of the link and channel, e.g., knowledge of the current available transmission rate, the prevailing error environment, or available transmit power in wireless links (Fairhurst et al., 2002). However, optimal performance cannot be achieved using link-level ARQ as it may result in an undesirably large amount of data retransmission among different layers. This will consequently degrade the performance of the transport layer protocol. A more efficient way of using the link layer ARQ is to make it application QoS-aware on a per packet basis (Jiang et al., 2005). The link layer ARQ can then adjust its behavior accordingly. The effects of the adaptive ARQ are implicitly passed on to the application through packet drops and delay.

(ii) Priority-based scheduling: in priority-based schedulers, packets are grouped into several classes with different priority according to their QoS requirements. In other words, the MAC layer is made aware of the application layer QoS. While the packets belonging to higher priority classes are scheduled to be transmitted first, those in the same class are served in a FIFO manner. Based upon the priority scheduling mechanism, each QoS class gets a guaranteed statistical QoS (Zhu et al., 2005). Liao et al. have proposed a priority packet-scheduling algorithm by relaxing the packet service order (Liao et al., 2003).

Kumwilaisak et al. have proposed a priority-based scheduling policy and have analytically computed the rate constraints for different video sub-streams with different QoS requirements (Kumwilaisak et al., 2003).

(iii) Header compression: The IETF has set up a ROHC working group (WG) to address the header compression issues. The goal of the ROHC is to develop header compression schemes that perform well over links with high error rates and long link RTT. In the ROHC framework, relevant information from past packets is maintained in a context. The context information is used to compress (and decompress) subsequent packets. The compressor and decompressor update their contexts upon certain events. It is known that, impairment events may lead to inconsistencies between the contexts of the compressor and decompressor, which in turn may cause incorrect decompression. Thus, ROHC scheme needs some mechanisms for avoiding context inconsistencies and also mechanisms for making the contexts consistent when they are not. Due to the limited packet loss robustness of the existing real-time traffic compression scheme, CRTP, and the demands of the cellular industry for an efficient way of transporting VOIP over wireless, ROHC has designed an ROHC scheme for IP/UDP/RTP headers (Pelletier et al., 2008), which are generous in size, especially compared to the payloads often carried by packets with such headers. ROHC-RTP has become a very efficient, robust, and capable compression scheme that is able to compress the header down to a total size of one octet only. In addition, transparency is guaranteed to an extremely great extent even when residual bit errors are present in compressed headers delivered to the decompressor. TCP-aware robust compression (TAROC) scheme has been proposed that can significantly improve the compression efficiency in unidirectional link by using congestion window tracking mechanisms and window-based least significant bit (LSB) encoding technique (Liao et al., 2001).

(iv) Channel-aware scheduling: in a multiple access wireless network, the radio channel is normally characterized by time-varying fading. As discussed in Section 3, to exploit the time-varying characteristic of the wireless channel, a kind of channel-state dependent scheduling, called multiuser diversity, can be exploited to improve system performance. For a wireless system with multiple MSs having independent time-varying fading channels, it may be assumed that the channels are either ON i.e. one packet can be transmitted successfully to the MS during the time-slot, or OFF i.e. the channels are unsuitable for transmission. The scheduler at the BS MAC layer gets the channel state information from its PHY layer, and based on that information the scheduler transmits to the MS whose channel is in the ON state. In case more than one user channel is in ON state, the scheduler selects one user channel randomly. No data is sent by the BS when all the channels are OFF state. For a three-user case, all the channels will be in OFF state only for 1/8 of the time on average. Thus, total data rate achieved by the scheduler is (1-1/8) = 7/8 packets per slot. Hence average data rate per user is (7/8)/3 = 7/24 packets/slot. For round robin scheduling with 3 users, each user will get 1/3 slot time. Since the user channels are equally likely to be ON or OFF in each timeslot, each user will get a data rate of (1/3)/2 = 1/6 packets/slot which is almost half that of the channel-aware multi user diversity scheduler. In this manner, the overall resource utilization can be improved by using a channel-aware scheduling mechanism (Jiang et al., 2005) (Shakkottai et al., 2003).

Since different QoS metrics are used in different layers of the protocol stack, some researchers have proposed to move the physical channel models upwards to the link layer and suggested models to convert PHY layer QoS parameters into application-specific QoS metrics (Wu et al., 2003). Wu and Negi have proposed effective capacity (EC) theory for modeling a wireless channel by means of two functions (Wu et al., 2003). These functions are: (i) the probability of non-empty buffer, and (ii) the QoS exponent of a connection that characterizes the queuing behavior in the link layer. The EC model has been effectively used to estimate QoS parameters like delay bound, available bandwidth etc. of various multimedia applications (Kumwilaisak et al., 2003).

Zori et al. have shown through analysis and simulation that a first-order Markov process is a good approximation model for data transmission over fading channels (Zori et al., 1995). Following this model, Zhang, Zhu, and Zhang have addressed the issues of resource allocation for scalable video transmission over 3G wireless networks (Zhang et al., 2004). In their proposed resource allocation model, the authors have first presented a method of estimation of time-varying wireless channels through measurements of throughput and error rate. A distortion-minimized bit allocation scheme with UEP and delay-constrained ARQ is also described that dynamically adapts to the estimated time-varying network conditions. The simulation results show that the proposed scheme can significantly improve the reconstructed video quality even when the network conditions are very much degraded.

In the following subsections, two important link-layer adaptation-based cross-layer design frameworks are described.

EFFECTIVE CAPACITY AND MODELING OF WIRELESS CHANNELS

Wu et al. have developed a link-layer channel model called effective capacity (EC) for modeling wireless channel that can easily translate into connection-level QoS measures such as data rate, delay and delay-violation probability (Wu et al., 2003).

The authors have argued that a major problem in designing QoS provisioning mechanisms is the high complexity in characterizing the relation between the control parameters of QoS provisioning mechanisms, and the calculated QoS measures, based on the existing PHY layer channel models. This is because the PHY layer channel models (e.g. Rayleigh fading model with a specified Doppler spectrum) do not explicitly characterize a wireless channel in terms of the link-level QoS metrics specified by the users, such as data rate, delay and delay-violation probability. Estimating the PHY layer channel model parameters and then extracting the link-level QoS metric from them is a very challenging task. To counter this challenge, the authors have proposed to move up the channel model in the protocol stack from the PHY layer to the link layer. This new model in the link-layer is known as the EC model because it captures a generalized link-level capacity notion of the fading channel. The authors have presented the EC channel model under the setting of a single hop, constant-bit-rate arrivals, fluid-traffic, and wireless channels with negligible propagation delay (Wu et al., 2003). In a later work, the authors have utilized the EC theory to derive QoS measures for more general situations, such as, networks with multiple wireless links, variable-bit-rate sources, packetized traffic, and wireless channels with non-negligible propagation delay (Wu et al., 2004). For better understanding of the EC theory, some of the fundamental concepts are discussed in the rest of this subsection.

Consider a single-hop system, where the user is allotted a single time varying channel. Assume that the user source has a fixed rate r_s and a specified delay bound D_{max}, and requires that the delay-bound violation probability is not greater than a certain value ε, that is,

$$P_r\{D(\infty) > D_{max}\} \leq \varepsilon \qquad (1)$$

$D(\infty)$ is the steady-state delay experienced by a flow, and $P_r\{D(\infty) > D_{max}\}$ is the probability of $D(\infty)$ exceeding a delay bound D_{max}. The user is specified by the statistical QoS triple $\{r_s, D_{max}, \varepsilon\}$.

Even for this simple case, it is not immediately obvious as to which QoS triples are feasible, for the given channel, since a rather complex queueing system (with an arbitrary channel capacity process) will need to be analyzed. The concept of EC allows us to obtain a simple and efficient test, to check the feasibility of QoS triple for a single time-varying channel. Let r(t) be the instantaneous channel capacity at time t. Assume that the asymptotic log-moment generation function of r(t) as in Equation (2) exists for all $u \geq 0$.

$$\Lambda(-u) = \lim_{t\to\infty} \frac{1}{t} \log E[e^{-u\int_0^t r(\tau)d\tau}] \quad (2)$$

Then, the EC function $\alpha(u)$ of r(t) is defined as in Equation (3):

$$\alpha(u) = \frac{\Lambda(-u)}{u}, \forall u > 0. \quad (3)$$

Expressed in a different way, $\alpha(u)$ may also be written as in Equation (4):

$$\alpha(u) = -\lim_{t\to\infty} \frac{1}{ut} \log E[e^{-u\int_0^t r(\tau)d\tau}], \forall u > 0 \quad (4)$$

Figure 7 depicts a queue of infinite buffer size supplied by a data source of constant data rate μ. It has been shown by the authors that if $\alpha(u)$ indeed exists (e.g., for ergodic, stationary, Markovian r(t)), then the probability of $D(\infty)$ exceeding a delay bound D_{max} satisfies following approximate Equation (5) below:

$$P_r\{D(\infty) > D_{max}\} \approx e^{-\theta(\mu)D_{max}} \quad (5)$$

The function $\theta(\mu)$ of source rate μ depends only on the channel capacity process r(t). $\theta(\mu)$ can be considered as a channel model that models the channel at the link layer (in contrast to the physical models specified by Markov process, or Doppler spectra). The approximate Equation (5) is accurate for large D_{max}.

In terms of the EC function defined in Equation (4), the QoS exponent function $\theta(\mu)$ can be written as:

$$\theta(\mu) = \mu\alpha^{-1}(\mu) \quad (6)$$

In Equation (6), $\alpha^{-1}(.)$ is the inverse function of $\alpha(u)$. Once $\theta(\mu)$ has been measured for a given channel, it can be used to check the feasibility of QoS triples. Specifically, a QoS triple $\{r_s, D_{max}, \varepsilon\}$ is feasible if $\theta(r_s) \geq \rho$, where $\rho = -\log \varepsilon / D_{max}$. Thus, we can use the EC model of $\alpha(u)$ (or equivalently, the function $\theta(\mu)$ via Equation (6)) to relate the channel capacity process r(t) to statistical QoS. Since EC method predicts an exponential dependence between ε and D_{max}, one can consider the QoS pair $\{r_s, \rho\}$ to be equivalent to the QoS triple $\{r_s, D_{max}, \varepsilon\}$, with the understanding that $\rho = -\log \varepsilon / D_{max}$.

A CROSS-LAYER QOS MAPPING ARCHITECTURE AND PROTOCOL

Kumwilaisak et al. have proposed cross-layer architecture for video transmission over wireless networks (Kumwilaisak et al., 2003). As shown in Figure 8, the system has several building blocks: (i) QoS interaction between video coding and transmission modules, (ii) QoS mapping mechanism, (iii) video quality adaptation, and (iv) source rate constraint derivation. The authors have argued that to coordinate effective adaptation, cross-layer interaction, and QoS mapping mechanism are essential. However, the design of a good cross-layer QoS mapping and adaptation mechanism is a particularly challenging task, because at the priority transmission layer, QoS is expressed in terms of probability of buffer overflow, and the

Figure 7. A queuing system model

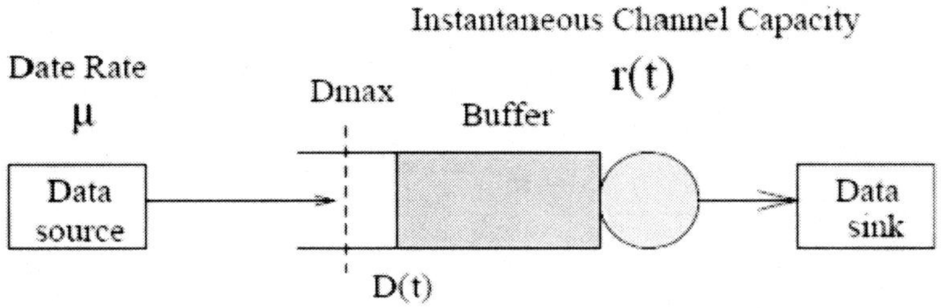

probability of delay violation at the link layer. On the other hand, at the video application layer, QoS is measured objectively by the mean squared error (MSE) and the PSNR.

The authors have identified some critical components in QoS adaptation and mapping:

- An adaptation model that shows how QoS parameters of both priority transmission systems and the video applications should be adjusted based on time-varying wireless channel.
- A coordination mechanism between the priority transmission system and the video applications, which provides interaction between the two layers.
- A resource allocation within the priority transmission system, which provides soft QoS guarantee based on time-varying wireless channel.

To address these issues, the authors have presented a QoS mapping architecture that performs the following functions: (i) derives of the rate constraints of a priority transmission system, (ii) optimally maps video classes to statistical QoS guarantees of a priority transmission system, (iii) incorporates a QoS interaction procedure between video applications and the priority transmission system to provide the best tradeoff between the video application quality and the transmission capability under time-varying wireless channel.

The authors have modeled the wireless channel at the link layer since the link layer modeling more amenable for analysis and simulations of the QoS provisioning system (Wu et al., 2003). The fading, time-varying, and non-stationary characteristics of the wireless channel is modeled by a discrete-time Markov model, where each state represents the available transmission rate under current channel conditions. This channel modeling process is performed by the adaptive channel-modeling module in Figure 8. The adaptive channel-modeling module periodically measures and updates the transition probability matrix of the Markov model to keep track of the current channel characteristics based on the algorithm proposed in (Kumwilaisak et al., 2002). In the link-layer transmission control module, a class-based buffering and scheduling mechanism is employed to achieve differentiated services. Based on the class-based buffering and strict priority scheduling algorithm each QoS priority class have statistical QoS guarantees in terms of probability of packet loss and packet delay. The QoS-mapping and adaptation module is designed to optimally match the video application layer QoS and the underlying link-layer QoS. At the video application layer, each video packet is characterized based on its loss and delay properties, which contribute to the end-to-end video quality and service. The video packets are classified and optimally mapped to classes of link transmission module under the rate constraint. The interaction between the video application layer QoS and the

Figure 8. The schematic architecture of the cross-layer design

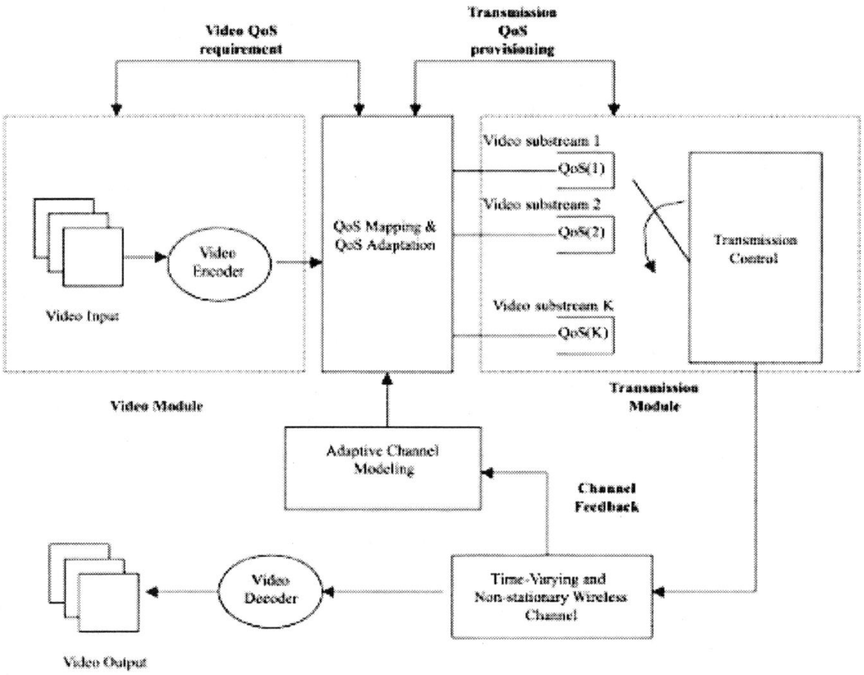

link layer QoS so that adaptation can be achieved based on the wireless channel condition. Simulation results demonstrate that the scheme can provide consistent video service and enhanced end-to-end video quality over time-varying and non-stationary wireless channels.

TRANSPORT LAYER ADAPTATION MECHANISMS

The wireless medium is very dynamic in nature due to the mobility of the devices and the interference and the fading of the wireless signals. The fast changing, small-scale channel variations result in burst error at the receiver. Moreover, large-scale channel variations may also occur where the average channel state condition depends on the location of the user and the interference level of the signals. The dynamic conditions of the channel cause bit errors, frame errors, and packet losses in the wireless networks.

In order to deliver multimedia over wireless networks, it is necessary to estimate the conditions of the underlying network so that QoS requirements of the applications can be satisfied. Congestion may occur within a network when the routers are overloaded with traffic that causes building up of queues and eventual overflows. This causes higher delay and more packet losses in the networks. The network conditions may be assessed by congestion estimation based on -- packet loss (Jiang et al., 2005) (Shakkottai et al., 2003) and currently available bandwidth (Zhu et al., 2005).

As discussed in Section 3, TCP attributes all packet losses in a network to congestion. This is mainly because of the fact that TCP was originally designed for wired networks, which have reliable PHY layers. Packet losses in these networks occur mainly due to network congestion. This characteristic of TCP is unsuitable for wireless networks since losses due to inherent channel errors are also treated as a signal of network congestion. As a result, TCP at the source node to reduce its transmission rate by shrinking its congestion window size, even when there is no congestion in the network resulting in an unnecessary decrease in throughput. In principle, packet loss due to channel errors should result in retransmissions not rate reduction. In order to improve the TCP performance in wireless scenario, it is necessary to differentiate the congestion-related packet losses from non-congestion packet losses. Two well-known protocols to achieve this objective are: (i) Snoop TCP and (ii) TCP with explicit congestion notification (ECN). These protocols are briefly described below.

Snoop TCP

Snoop TCP provides a reliable TCP-aware link layer (Balakrishnan et al., 1997). The mechanism is described in a scenario where data transfer occurs between a fixed host (FH) and a mobile host (MH) with a BS in between them. A snoop agent is created at the BS, which buffers data at its link layer for retransmissions instead of going back to TCP end points at the FH and the MH. Snoop maintains a state for each TCP connection traversing through the BS thus tracking TCP data and the acknowledgements. The protocol also caches unacknowledged TCP packets and uses the loss indications conveyed by duplicate acknowledgments and local timers to transparently retransmit lost data. It hides duplicate acknowledgments indicating wireless losses from the TCP sender, thereby preventing redundant TCP recovery. Snoop exploits the information present in TCP packets to avoid link layer control overhead, and preserves end-to-end TCP semantics. However, it cannot work on encrypted datagrams, and hence, not suitable in virtual private networks (VPNs).

TCP with ECN

ECN is an end-to-end mechanism to notify the sender whenever congestion occurs in a network (Floyd, 1994). TCP with ECN is a protocol that overcomes the inherent insensitivity of the TCP congestion control mechanisms to delay or loss of individual packets. It focuses mainly on minimizing the impact of packet losses from the perspective of throughput in a network. It is tailor-made to improve the QoS such as reducing the delay and packet loss in sensitive multimedia applications e.g., video-conferencing, VOIP etc over wireless networks. In a standard IP packet header, an ECN field is included (Ramakrishnan et al., 2001). Whenever a router detects a persistent congestion in the network, it sets the ECN field and the packet is said to be marked. The marked packet eventually reaches the destination, which in turn informs the source about the congestion by setting the ECN echo flag in the TCP header. The source adapts its transmission rate accordingly using the usual TCP congestion control mechanisms of slow start, fast retransmit and fast recovery. The ECN capability thus overrides any signal of packet losses as imminent congestion indication. However, for TCN with ECN protocol to work, ECN scheme should be enabled at all the intermediate routers on the path from the source to the destination.

In the following subsections, some of the currently existing transport layer adaptation-based cross-layer techniques are discussed, clearly highlighting the fundamental principles of each of the mechanism and its application areas.

AN IMAGE TRANSPORT PROTOCOL FOR THE INTERNET

Raman et al. have proposed an efficient transport layer protocol, called the image transport protocol (ITP) for transmission of images over loss-prone, congested or wireless networks (Raman et al., 2002). The authors have argued that while TCP provides a generic, reliable, and in-order byte-stream abstraction, it is overly restrictive for transporting image data. In order to validate their argument, the authors have analyzed the progression of image quality at the receiver with time and have shown that in-order delivery abstraction provided by a TCP-based approach prevents the receiver application from processing and rendering portions of an image when they actually arrive. As a result, the image is rendered in bursts, interspersed with long idle times rather than in a smooth manner. In the proposed protocol, the application data unit (ADU) boundaries are exposed to the transport module. This enables the transport module to perform out-of-order delivery of packets. As the transport layer is aware of the application framing boundaries, the mechanism utilizes the concept of application level framing (ALF), which uses a one-to-one mapping from an ADU to a network packet or protocol data unit (PDU) (Clark et al., 1990). ITP deviates from the TCP's notion of reliable delivery. Instead, it incorporates selective reliability, where the receiver is in control of deciding on what is to be transmitted from the sender at any instant. ITP runs over UDP, incorporates receiver-driven selective reliability, and uses a congestion manager (CM) to adapt to the network congestion. It also enables a variety of new receiver post-processing algorithms such as error concealment that further improves the interactivity and responsiveness of the reconstructed images. The authors have presented the performance of a user-level implementation of ITP across a range of network conditions that demonstrate that the rate of increase in PSNR with time is significantly higher for ITP compared to TCP-like in-order delivery of images.

AN ADAPTIVE TCP-FRIENDLY STREAMING PROTOCOL

Yang et al. have proposed an end-to-end TCP-friendly multimedia streaming protocol for wireless Internet (Yang et al., 2004). The protocol, known as the wireless multimedia streaming TCP-friendly protocol (WMSTFP), can effectively differentiate erroneous packet losses from congestive losses and filter out the abnormal round-trip time values caused by the highly varying wireless environment. Utilizing the properties of WMSTFP, the authors have proposed a novel loss pattern differentiated bit allocation scheme that applies unequal loss protection for scalable video streaming over the wireless Internet. In order to minimize the expected end-to-end distortion in the video, the authors have also presented a rate-distortion-based bit allocation scheme that takes into account the status of the wired and wireless networks. The global optimal solution for the bit allocation scheme is obtained by a local search algorithm that takes into account the characteristics of the progressive fine granularity scalable video (PFGS). Figure 9 depicts the detailed diagram of the end-to-end scalable video streaming mechanism. The key components in this architecture consist of WMSTFP congestion control, WMSTFP network monitor, unequal loss protection (ULP) channel encoder, and loss differentiated rate distortion-based bit allocation. WMSTFP congestion control and WMSTFP network monitor provide network adaptation at end hosts, which mainly deal with probing and estimating the dynamic network conditions using the TCP-friendly protocol. The WMSTFP congestion control module adjusts the sending rate on the sender side based on the feedback information, and the WMSTFP network monitor module on the receiver side analyzes the erroneous loss rate and congestive loss rate caused in a connection comprising both wired and wireless links and estimates the end-to-end available network bandwidth. The control data consisting of the estimated network bandwidth and other related network status parameters such as congestive packet loss rate, erroneous packet loss rate, and smoothed packet transmission time are fed back to the sender. Network-adaptive ULP channel encoder module protects different layers of PFGS video against congestive packet losses and erroneous losses according to their importance and network status using Reed Solomon (RS) codes (Wu et al., 2001). Loss differential rate distortion-based bit allocation module performs media adaptation control so that the total sending rate is adapted to the estimated network conditions. Based on the feedback information from the receiver, the bit allocation module in the sender side distributes the total sending rate between video bit rate and error protection rate according to the available bandwidth and different packet loss conditions in wired and wireless connections.

The main contributions of WMSTFP are:

- WMSTFP can accurately distinguish between the packet losses caused by the errors in wireless channels using the information acquired at the link-layer. By jointly using the status information at the link-layer and the sequence number of incoming packets, WMSSTFP can effectively differentiate the different types of packet losses in wireless Internet.
- The authors have observed that packets have different loss patterns for different types of losses. They have used two Gilbert models to describe the burstiness of these two types of packet losses respectively. Consequently, the authors have developed a robust technique for estimating the packet loss ratio and the packet error ratio.
- The wireless channel introduces large delay variations and the packet RTT fluctuates sharply. The authors have proposed a method to measure the average RTT during a period of time. As a result, the rate adjustment performs more smoothly, while achieving a very good throughput.

The network simulator ns-2 has been used to study the performance of WMSTFP and the network adaptive bit allocation algorithm for PFGS video streaming. The authors have also analytically evaluated the performance of WMSTFP and compared it with that of TCP-friendly rate control (TFRC) protocol. The results clearly show that WMSTFP has lower packet loss for the same frame error rate (FER) when compared to TFRC.

A CROSS-LAYER ADAPTIVE PROTOCOL IN IP NETWORKS

Ahmed et al. have proposed a media content analysis technique and a network control mechanism for adaptive video streaming over IP networks (Ahmed et al., 2005). The authors have leveraged the characteristics of MPEG-4 and Internet Protocol (IP) differentiated service frameworks, to propose an innovative cross-layer content delivery architecture that is capable of receiving

Figure 9. The system architecture for scalable video streaming over wireless Internet

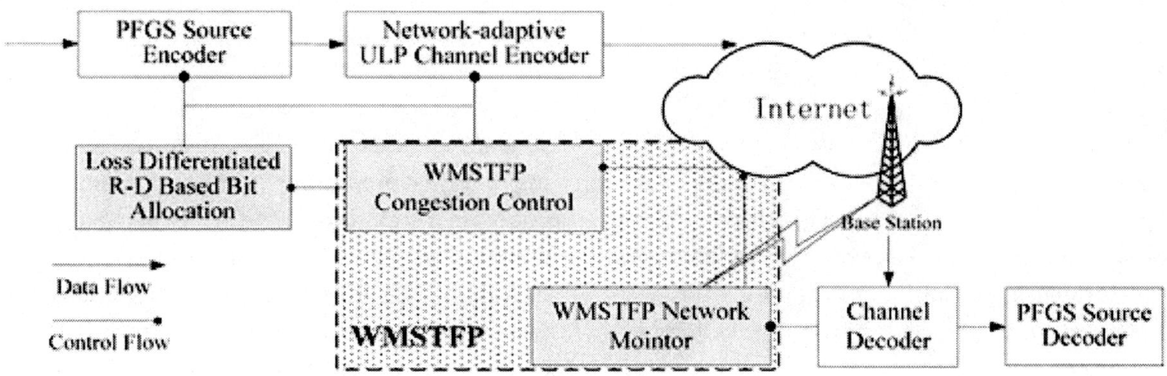

information from the network and adaptively tune the transport layer parameters, bit rates, and QoS mechanisms according to the underlying network conditions. The proposed service-aware IP transport architecture integrates a cognitive layer that consists of three components: (i) a content-based video classification model for automatic translation from video application level QoS (e.g., MPEG-4 object descriptor and/or MPEG-7 meta data framework) to network system level QoS (e.g. IP DiffServ per-hop-behaviors (PHBs)), (ii) a robust and adaptive application level framing (ALF) protocol with video stream multiplexing and unequal forward error protection, (iii) a fine grained TCP-friendly video rate adaptation algorithm.

The cognitive layer is an extension to the MPEG-4 system architecture that makes use of a neural network classification model to dynamically and accurately group audiovisual objects of a scene with the same QoS requirements to create elementary video streams that are subsequently mapped to IP DiffServ PHBs. These MPEG-4 audiovisual objects (AVOs) are classified based on application-level QoS descriptors and MPEG-7 content-descriptive metadata. Thus, MPEG-4 AVOs requiring same QoS from the network are automatically classified and multiplexed within one of the IP DiffServ PHB. Object data packets within the same class are then transmitted over the selected transport layer with the corresponding bearer capability and relative priority score (RPS). The transmitted MPEG-4 streams take also benefit from the cognitive layer by applying an UEP according to the priority score of each object. The amount of recovered data is related to the priority score of the AVOs in the MPEG-4 scene. For faire share of bandwidth and higher user perceived quality, the content-based rate adaptation mechanism for MPEG-4 video streams uses a TFRC protocol. The video rate adaptation is performed by adding and dropping MPEG-4 AVOs according to their subjective relevancy to the service, and the instantaneous network congestion estimations.

The protocol has been evaluated on the network simulator ns-2. The obtained experimental results shows that by introducing cross-layer interactions and injecting content-level semantic and service-level requirements within the transport and traffic control protocols lead to intelligent and efficient support of multimedia services over complex network architectures resulting in a clear gain in terms of audio video quality of a streaming application.

A TCP-COMPATIBLE RATE CONTROL FOR VIDEO TRANSMISSION

Vieron et al. have described a rate control algorithm that takes into account the behavior of TCP's congestion avoidance mechanism and the delay constraints of real-time streams (Vieron et al., 2004). The authors have argued that TFRC model does not take into account the characteristics of multimedia flows: it assumes that the packet size is constant whereas loss resilient video transport often leads to packets of varying size. In case of video flows, TFRC (Floyd et al., 2000) may estimate inaccurately the loss rate, leading to unfair share of bandwidth with conformant TCP flows. Moreover, the predicted bandwidth values are often directly fed into the encoder as a rate constraint translated into a bit budget per frame. The authors have shown that this approach can suffer from severe timeouts effects induced by the real-time constraint of the source.

To address the above issues, the proposed scheme extends the TFRC protocol by designing a TCP-compatible rate control mechanism coupling a source-adaptive TCP-compatible rate control protocol with a source rate control model encompassing timing and buffering models of the source in order to minimize the expected distortion at the receiver. The proposed protocol makes use of RTP and RTP control protocol (RTCP) and takes into account the characteristics of the multimedia flows, like variable packet size, delay etc. Based on the estimated current channel state, the states of the encoder and the decoder buffers as well as the delay constraints of the real-time video source are translated into encoder rate constraints. Both channel and buffer states are periodically updated taking into account the varying RTT over the network. The rate control proposed has been experimented with using H.263+ compatible loss resilient encoder. The source rate control has been further improved by a frame skipping strategy that better trades the frame rate against PSNR even with highly varying rate constraints.

The authors have extensively evaluated the performance the global rate control model and the loss resilient video compression algorithm on various Internet links. The results clearly demonstrate the benefits of the source-adaptive TCP-compatible rate control protocol and the global source rate control model. The coupling of the two mechanisms results in a significant decrease in timeouts phenomenon for a compatible bandwidth utilization, and hence the expected distortion of the decoded signal is also minimized.

APPLICATION LAYER ADAPTATION MECHANISMS

Due to real-time nature, multimedia services typically require QoS guarantees like large bandwidth, stringent delay bound and relatively error-free video/audio/speech quality. Multimedia services over the wireless channels become very challenging due to the dynamic uncertain nature of the channel resulting in variable available bandwidths and random packet losses. The main objectives of the application layer QoS control for multimedia communication over wireless networks are –(i) to avoid bursty losses and excessive delay (caused by network congestion) that have a devastating effect on multimedia presentation quality, and (ii) to maximize multimedia quality even when packet loss occurs in a wireless communication network. A number of approaches for adaptation in the application layer currently exist in the literature in. The in the following subsections some of the well-known mechanisms are discussed in detail. In particular, the joint design of source rate control and QoS-aware congestion control mechanism proposed by Zhu et al (Zhu et al., 2007a)(Zhu et al., 2007b) and the joint design of source coding and link layer FEC/ retransmission proposed by Jiang et al. (Jiang et al., 2005) and Zhu et al. (Zhu et al., 2005) are elaborately discussed. In addition, some other propositions are also described.

JOINT SOURCE RATE CONTROL AND CONGESTION CONTROL

Congestion control for streaming media at the transport layer and source rate control at the application layer are employed to overcome the problems of multimedia communication over the wireless channels. In traditional layered design approach, source rate control and congestion control are designed independently and in isolation with each other. This imposes a limitation on the overall system performance e.g., end-to-end delay constraint and smooth playback quality. Congestion control for streaming multimedia usually needs to smooth its sending rate to help the application achieve smooth playback quality. However, this is not always possible as the source-coding block at application layer can abruptly change the coding complexity and the sending rate based on its QoS requirements unless explicitly notified otherwise by the transport layer. Moreover, source rate control alone cannot guarantee the end-to-end delay constraint due to minimum bandwidth requirement and quality smoothness requirement in the absence of congestion control mechanism at the transport layer.

Zhu et al have proposed a joint source rate control and a cross-layer QoS-aware congestion control mechanism to achieve an improved overall system performance (Zhu et al., 2007a). The authors have argued that if the sending rate is allowed to temporarily violate the TCP-friendliness nature of the transport layer, the quality of the multimedia content is significantly improved. However, the long-term TCP–friendly sending rate is preserved by implementing the rate compensation algorithm (Zhu et al., 2007b). There are two main contributions of the proposition. First, a QoS-aware congestion control mechanism, called TCP-friendly rate control with compensation (TFRCC) has been designed that supports improved multimedia transmission over wireless network than TFRC protocol. Secondly, over the TFRCC protocol, at the application layer, a virtual network buffer management mechanism proposed in (Xie et al., 2004) is used to translate the QoS requirements of the application into the desired source and sending rates.

A middleware component is introduced between the application layer and the transport layer wherein the joint decision of the source rate and the sending rate is done. To make the protocol work effectively in wireless environment, the authors have utilized the analytical rate control (ARC) protocol (Akan et al., 2004). The ARC protocol is intended to achieve high throughput and multimedia support for real-time traffic flows while preserving fairness to the TCP sources, which share the same wired link resources. The sender performs rate control using the ARC protocol to avoid any unnecessary rate reduction due to wireless link errors, thus enabling the system to work optimally in a wireless environment.

The architecture of the system is depicted in Figure 10. At the transport layer, TFRCC is used as the congestion control mechanism. As shown in Figure 11, at the application layer, a virtual network management mechanism (VB) is used to derive the constraint of the source rate and the sending rate according to the QoS requirements of the application. There is a middleware component located between the application layer and the transport layer. At the receiver, the middleware collects information from the application (e.g., the amount of received video data) than feed it back to the sender together with the feedback of TFRCC. At the sender, the joint decision of the source rate and the sending rate is done within the middleware by considering the constraints of the source rate and sending rate provided by VB, and the TCP-friendliness constraint provided by TFRCC.

JOINT SOURCE CODING AND LINK LAYER FEC/RETRANSMISSION

In order to adapt to the varying network conditions like loss, delay, variable bandwidth etc., the media codecs are designed using scalable coding techniques. Scalability in video can be achieved by layered coding technique as in MPEG-4. The adaptation of audio codec, which also has a layered structure, can be achieved in a way similar to that of scalable video codec (Pan, 1995). Speech codecs also allow dynamic rate adaptation, controlled by an in-band signaling procedure (Zhu et al., 2005).

The layered coding technology divides the video into several layers. The incremental reception of the layers increases the media fidelity. Video codecs encode a video sequence into one base layer and multiple enhancement layers based on any of the following three classes of layered coding techniques – temporal, spatial and signal to noise ratio (SNR) scalability (Li, 2001). Different layers in a scalable coder have different importance in video transmission and reception. The correct decoding of the enhancement layers depend on the errorless receipt of the base layer. Therefore, from the video reception point of view, the base layer is more important than the enhancement layers.

FEC and link layer retransmission are the most widely used error correction mechanisms in the link layer. FEC is a channel coding technique used for protecting the source data by adding redundant

Figure 10. The system architecture for source rate control and congestion control

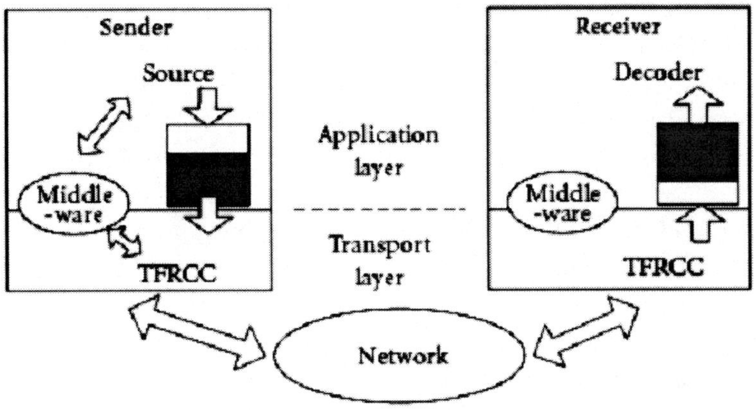

bits during transmission. Therefore, FEC is not bandwidth efficient but very effective in applications which have strict delay requirements such as voice communications. In these applications, retransmission of packets may induce unacceptably high latencies. On the other hand, applications where delay requirements are much relaxed, link layer retransmission is a more suitable technique as it is more bandwidth efficient than FEC.

The packet losses in wireless networks due to traffic congestion and wireless transmission errors invariably have different patterns of loss. Such different loss patterns are reflected as different perceived QoS at the application layer (Jiang et al., 2000). Yang et al. have proposed a loss differentiated rate-distortion based bit allocation protocol that takes into account the different loss patterns due to network congestion and wireless transmission errors, and minimizes the end-to-end video distortions (Yang et al., 2004). The authors have proposed JSCC schemes to achieve the optimal end-to-end quality by adjusting the source and channel coding parameters simultaneously. As discussed in Section 3, a simple JSCC scheme using UEP has been proposed by Jiang et al. (Jiang et al., 2005). UEP can be implemented with Bose Chaudhuri Hocquenghem (BCH) codes, Reed Solomon (RS) codes, and rate compatible punctured convolutional (RCPC) codes with different coding rates for packets with different priorities. A hybrid UEP scheme taking ARQ-based retransmission on the same SSI can also be implemented in which the base layer data may be scheduled for maximum number of retransmis-

Figure 11. The virtual network buffer model

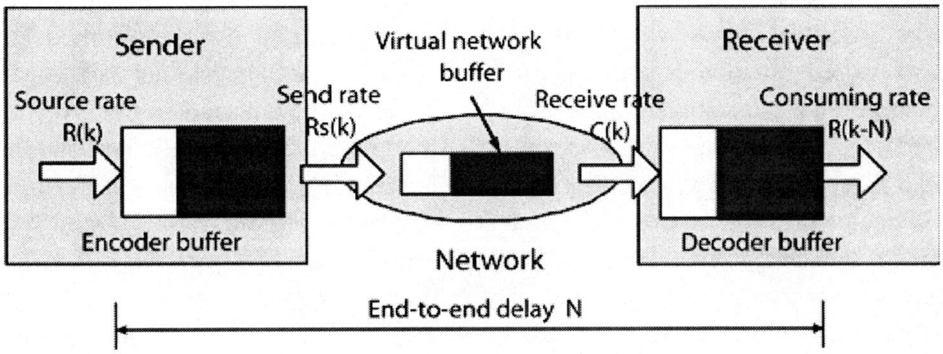

sions with the provision for a minimum number or no retransmissions at all for the enhancement layers. A delay-bound in such a hybrid scenario can be achieved by limiting the number of retransmissions (Zhu et al., 2005).

OTHER ADAPTATION MECHANISMS AT THE APPLICATION LAYER

In subsections 7.1 and 7.2, two important adaptation mechanisms at the application layer: joint design of the source rate control and QoS-aware congestion control and the joint design of source coding and link layer FEC and retransmission techniques have been discussed respectively. In this subsection, three other existing adaptation mechanisms at the application layer are discussed briefly. The first scheme is based on a robust error handling mechanism that efficiently takes care of packet errors in a streaming application over a wireless network. The second scheme is an adaptive video streaming technique that dynamically adapts the sending rate and drops less priority video frames when the available bandwidth is limited. The third scheme is concerned with a cross-layer video streaming mechanism in which multiple user access one server simultaneously over the wireless links.

Superiori et al. have proposed a robust error handling technique for video streaming over mobile networks (Superiori et al., 2007). If larger size packets are used for video streaming, loss of a packet typically affects a rather large area of a picture. On the other hand, use of smaller packets involves a very large overhead. In order to avoid large overhead caused by smaller packets, the authors have proposed a scheme that utilizes the residual redundancy of the encoded video stream. At the decoder side, there is a syntax analyzer that enables exact localization of errors within a packet. In addition, the scheme involves an entropy code resynchronization mechanism that is based on the out-of-band-signalized length indicators. The authors have used the concept of slice in a picture. A slice consists of an integer number of macroblocks belonging to the same picture. A significant portion of correctly received part of a slice may be lost if a whole packet is discarded due to packet errors in transmission. The fraction of code preceding the occurrence of errors can be exploited to reconstruct error-free macroblocks. The decoding process for a damaged slice is segmented into three steps. Staring from the beginning of the slice up to the error occurrence, the macroblocks are correctly decoded. From the error occurrence up to the error detection, the macroblocks are wrongly decoded. From the error detection up to the end of the slice, the macroblocks are concealed. The method does not require any modifications at the encoder side and does not add any overhead in terms of required bandwidth. Experimental results have shown that the protocol provides substantial improvement in PSNR for the same rate compared to the standard packet size reduction techniques.

Burza et al. have described a robust streaming protocol for delivery of combined MPEG audio/video content over in-home wireless networks, where the amount of data transmitted by the sender is dynamically adapted to the available bandwidth by selectively dropping data (Burza et al., 2007). In this way, the perceived quality of the audio/video stream is dynamically adjusted according to the quality of the network link. The transmitted bit rate is constantly adapted to the available network bandwidth by using a packet scheduling technique called I-frame delay (IFD) that performs priority-based frame dropping when the available bandwidth is limited. The basic idea of IFD is that the scheduler will drop video frames when the transmission buffer is full and overflow is imminent due to insufficient bandwidth. The less important frames (B-frames) are dropped in favor of more important frames (I- and P-frames). The transmission of I-frames is delayed when conditions are bad. However, these frames are never dropped even if they are out-of-date with respect to the display time because they can still be used to decode the subsequent inter-predicted frames. Essentially, the IFD scheme has two phases: (i) during the parsing and aggregating the stream into network packets, the stream is analyzed and the packets are tagged with a priority number reflecting the frame type: I, P or B, and (ii) during transmission, the packets are dropped by the IFD scheduler when the available bandwidth is insufficient. The proposed solution has been implemented using the real time transport protocol (RTP) and TCP at the transport layer.

Choi et al. have proposed a cross-layer optimization approach for wireless multi-user video streaming that jointly considers the application layer and the PHY/MAC layer of the protocol stack (Choi et al., 2004). The optimizer maximizes the end-to-end QoS of the video streaming service jointly for all users while efficiently using the wireless resources. The authors have considered a video-streaming server located at the BS and multiple streaming clients. The clients are assumed to be sharing the same air interface and network resources but they request different video content. The service optimization at the BS is achieved by means of the architecture shown in Figure 12. Necessary state information is first collected from the application layers and the radio link layer through the process of parameter abstraction. The process of parameter abstraction results in the transformation of layer specific parameters into parameters that are comprehensible for the cross-layer optimizer. The optimization is carried out with respect to a particular objective function. From a given set of possible cross-layer parameter tuples, the tuple optimizing the objective function is selected. After the decision on a particular cross-layer parameter tuple is made, the optimizer distributes the decision information back to the corresponding layers. The simulation results have demonstrated that even for a small number of users and a fewer degrees of freedom in the optimization, significant improvements in the quality of video streaming can be obtained.

FUTURE RESEARCH DIRECTIONS

Technological advances have brought wireless networking a step forwards towards the goal of service provision on an "anytime, anywhere" basis, while ensuring instantaneous and secure communications. However, such innovation is constrained by the restrictions included in the TCP/IP protocol of the original Internet, which does not include, for example, mobility support, security, and active networking. For this reason, technological advancements were achieved at the cost of increased network complexity and limited performance (Barakat et al., 2000). The fundamental reason for performance inefficiency is the difficulty in configuring and managing network- a task traditionally performed by network operators and technicians (Clark et al., 2003). Recently, self-awareness, self-management, and self-healing characteristics have been proposed in order to optimize network operation, reconfiguration, and management, as well as to improve data transfer performance by bringing intelligence into the network, thereby creating a new paradigm known as cognitive networking, which is expected to become a key part of the fourth generation (4G) wireless networks (Syputa, 2006).

According to Thomas, "A cognitive network has a cognitive process that can perceive current network conditions, and then plan, decide, and act on those conditions. The networks can learn from these adaptations and use them to make future decisions, all while taking into account end-to-end goals" (Thomas, 2007). The term cognitive network is related to the ability of a network to be aware of its operational status and adjust its operational parameters to fulfill specific tasks, such as detecting changes in the environment and user requirements. Cognitions requires support from network elements (routers, switches, base stations etc.), which should host active tasks to perform measurements to reconfigure the network. These characteristics are related to the paradigm of active networks (Tennenhouse et al., 1997),

Figure 12. A cross-layer optimization architecture

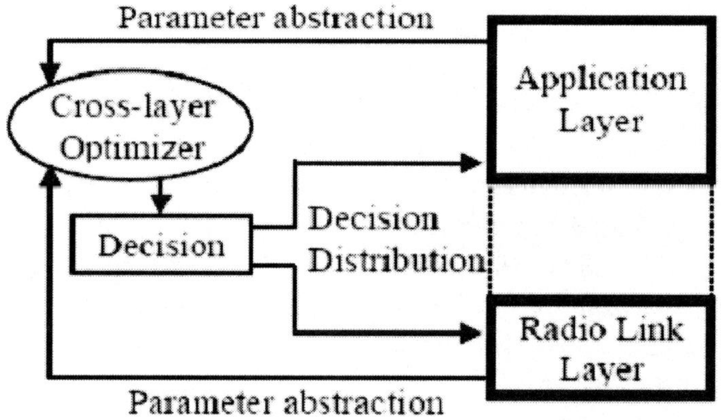

which differ from cognitive networks service in that they do not include cognitive process that considers adaptation and learning techniques.

In recent years, there has been a tremendous driving force for cognitive networks. From the technological perspective, cognitive networking is envisioned as a logical evolution towards the definition of a unified QoS-aware environment, encompassing multiple technologies already available in the wireless network domain (Kliazovich et al., 2009). The diversity of network configurations, involved technologies, and objectives dictated by the requirements of user applications is the main motivation behind cognitive networking. From the business perspective, cognitive networks are envisioned as the way to increase profits for wireless service providers through cost reduction and development of new revenue streams obtained by the offer of heterogeneous wireless access solutions. The benefits enabled by cognitive networking include the possibility to rely on common hardware and software platforms while supporting the evolution of radio technologies, development of new services, minimization of infrastructure upgrades, accelerated innovation, and maximization of return-on-investment through the reuse of already available network equipment (Clark et al., 2003).

The flexibility of the cognitive networks presents an opportunity for researchers to reexamine how network protocol layers operate with respect to providing QoS-aware transmission among wireless nodes. This opportunity is enhanced by the continued development of spectrally responsive devices- ones that can detect and respond to changes in the radio frequency environment. Present wireless network protocols define reliability and other performance-related tasks narrowly within layers. For example, the frame size employed on 802.11 can substantially influence the throughput, delay, and jitter experienced by an application, but there is no simple way to adapt this parameter. Furthermore, while the data link layer of 802.11 provides error detection capabilities across link, it does not specify additional features, such as forward error correction schemes, nor does it provide a means for throttling retransmissions at the transport layer. In fact, currently, the data link layer and the transport layer function counterproductively with respect to reliability of transmission. As has been observed in the previous sections of the chapter, considerable amount of research has been done in the area of cross-layer protocol design for wireless networks. A considerable amount of effort has been spent also on the research in cognitive networks. However, most of the work in the area of cross-layer protocol design focuses on enhancing throughput, QoS and energy consumption (Goldsmith et al., 2002, Barrett et al., 2002, Jiang et al., 2003). These protocols tend to focus mostly on two layers of the protocol stack with the goal of enhancing a specific performance measure. As such, they do not consider multi-factor variation nor do they consider effects of this variation on real-time applications. For addressing this challenge of multi-factor cross layer design to further improve the performance of wireless systems, an integrated approach combining cross-layer optimization with cognitive systems is emerging as a new and exciting research direction (Weingart et al., 2007).

Since an ideal cognitive network should maintain a network-wide scope with the cognitive process operating based on end-to-end goals, the existing cross-layer signaling proposals are not suitable for these networks. As has been mentioned earlier, the existing cross-layer protocols employ signaling between different layers within the protocol stack of a single node. For cognitive networks, an encapsulation of signaling information into packet headers or ICMP messages can be an efficient approach. Another cross-network cross-layering mechanism is the explicit congestion notification (ECN) (Ramakrishnan et al., 2001). It realizes in-band signaling approach by marking in-transit TCP data packet with congestion notification bit. However, due to the limitation of signaling propagation to the packet paths, this notification needs to propagate to the receiver first, which echoes it back in the TCP ACK packet outgoing to the sender node. This unnecessary signaling loop can be avoided with explicit ICMP packets signaling. However, it requires traffic generation capability fro network routers and it consumes bandwidth.

An example of the adaptation of central cross-layer architecture to a cross-network, cross-layer signaling is presented in (Kim, 2001). The proposed mechanism uses a network service, which collects parameter values related to the wireless channel located at the link, as well as at the physical layer and the provisioning of this information to adaptive applications.

A unique combination of local- and network-wide cross-layer signaling approaches called cross-talk is proposed in (Winter et al., 2006). The proposed architecture consists of two cross-layer optimization planes, where one is responsible for the organization of cross-layer information exchange between protocol layers of the local protocol stack and their coordination. The other plane is responsible for network-wide coordination, considered the aggregation of cross-layer information provided by the local plane. It serves as an interface for cross-layer signaling over the network. Most of the signaling is performed in-band, using the packet headers, making it accessible not only at the end host bust at the network routers as well. Cross-layer information received from the network is aggregated and then can be considered for the optimization of local protocol stack operation based on global network conditions.

Main problems associated with the deployment of cross-layer signaling over the network include security issues, problems with non-conformant routers, and processing efficiency (Sarolahti et al., 2007). Security considerations require the design of proper protective mechanisms, avoiding protocol attacks attempted by malicious nodes, which furnish incorrect cross-layer information in order to trigger specific behavior. The second problem addresses misbehavior of network routers. In most of the cases, IP packets with unknown options are dropped in the network or by the receiver protocol stack. Finally, the problem with processing efficiency is related to the additional costs of the routers hardware for cross-layer information processing. While it is not an issue for the low-speed links, it becomes relevant for high speed ones where most of the routers decrement only the TTL field to maintain a high packet processing speed.

CONCLUSION

Cross-layer adaptations are essential for guaranteeing QoS supports in real-time multimedia traffic over wireless networks. This chapter has presented some of the currently existing cross-layer adaptation protocols at the application, the transport and the link layers for multimedia transmission over wireless networks. More specifically, network-aware adaptive media source coding, dynamic estimation of the varying channel, adaptive and energy-efficient application and link-level error control, efficient congestion control, adaptive ARQ, and priority-based scheduling are discussed in detail. However, the designing cross-layer architecture is an extremely challenging task since it involves numerous issues like network characteristics, QoS requirements of applications, adaptability of the protocol being used etc. Providing QoS support in multicast media streaming is one area, which poses a particularly serious challenge (Zhang et al., 2004). Device mobility brings along another dimension of complexity that calls for an efficient handling of the problem related to handover while satisfying the application QoS. In mobile ad hoc networks (MANETs), changes in the topology of the network and the interference due to simultaneous communications of the nodes make design of cross-layer protocol architecture particularly difficult. Multi-path media streaming and QoS-aware MAC design are two cross-layer design approaches proposed in the literature for providing QoS support in MANETs (Mao et al., 2003)(Kumar et al., 2006).

The chapter has also discussed the emerging issues related to evolution towards self-aware, autonomous, and adaptive networks for resolving inefficiencies in network configuration and management. Various issues and challenges for designing cross-layer, cross-network protocols, and these emerging networks are also presented.

However, a good cross layer design should take a cautious and careful approach as some adverse impact on the system performance may occur in certain situations due to cross layer interactions (Kawadia et al., 2005). Unbridled and extensive cross layer interactions can lead to a complex spaghetti design and thwart further innovations. Moreover, such design will lack standardization and compatibility and portability features. A careful impact analysis of the design of any cross layer protocol stack is always necessary before its deployment.

REFERENCES

Ahmed, T., Mehaoua, A., Boutaba, R., & Iraqi, Y. (2005). Adaptive packet video streaming over IP networks: A cross layer approach. *IEEE Journal on Selected Areas in Communications*, *23*(2), 385–401. doi:10.1109/JSAC.2004.839425

Akan, O. B., & Akyildiz, I. F. (2004). ARC: The analytical rate control scheme for real-time traffic in wireless networks. *IEEE/ACM Transactions on Networking*, *12*(4), 634–644. doi:10.1109/TNET.2004.833155

Balakrishnan, H., Padmanabhan, V. N., Seshan, S., & Katz, R. H. (1997). A comparison of mechanisms for improving TCP performance over wireless links. *IEEE/ACM Transactions on Networking*, *5*(6), 756–769. doi:10.1109/90.650137

Barakat, C., Altman, E., & Dabbous, W. (2000). On TCP performance in a heterogeneous network: A survey. *IEEE Communications Magazine*, *38*(1), 40–46. doi:10.1109/35.815451

Barrett, C., Marathe, A., Marathe, M. V., & Drozda, M. (2002). Characterizing the interaction between routing and MAC protocols in ad-hoc networks. In *Proceedings of the 3rd ACM International Symposium on Mobile Ad Hoc Networking and Computing* (MobiHoc'02), (pp. 92-103). Lausanne, Switzerland.

Burza, M., Kang, J., & Van der Stock, P. (2007). Adaptive streaming of MPEG-based audio/video content over wireless networks. [Academy Publisher.]. *Journal of Multimedia*, *2*(2), 17–27. doi:10.4304/jmm.2.2.17-27

Casner, S., & Jacobson, V. (1999). *Compressing IP/UDP/RTP headers for low-speed serial links*. (IETF RFC 2508).

Choi, L.-U., Kellerer, W., & Steinbach, E. (2004). Cross-layer optimization for wireless multi-user video streaming. In *Proceedings of the International Conference on Image Processing* (ICIP'04), *3*(3), 2047-2050.

Clark, D. D., Partridge, C., Ramming, J. C., & Wroclawski, J. T. (2003). A knowledge plane for the Internet. In *Proceedings of the 2003 Conference on Applications, Architectures, and Protocols for Computer Communication*, (pp. 3-10). Karlsruhe, Germany.

Clark, D. D., & Tennenhouse, D. L. (1990). Architectural considerations for a new generation of protocols. *ACM SIGCOMM Computer Communication Review*, *20*(4), 200–208. doi:10.1145/99517.99553

Fairhurst, G., & Wood, L. (2002). *Advice to link designers on link automatic repeat request* (ARQ). (IETF RFC 3366).

Floyd, S. (1994). TCP and explicit congestion notification. *ACM SIGCOMM Computer Communication Review*, *4*(5), 8–23. doi:10.1145/205511.205512

Floyd, S., Handley, M., Padhye, J., & Widmer, J. (2000). Equation-based congestion control for unicast applications. In *Proceedings of ACM SIGCOMM'00*, (pp. 43-56).

Goldsmith, A., & Wicker, S. (2002). Design challenges for energy-constrained ad hoc wireless networks. *IEEE Wireless Communications Magazine*, *9*(4), 8–27. doi:10.1109/MWC.2002.1028874

Jiang, H., Zhuang, W., & Shen, X. (2005). Cross-layer design for resource allocation in 3G wireless networks and beyond. *IEEE Communications Magazine*, *43*(12), 120–126. doi:10.1109/MCOM.2005.1561929

Jiang, R., Gupta, V., & Ravishankar, C. V. (2003). Interactions between TCP and the IEEE 802.11 MAC protocol. In [DISCEX]. *Proceedings of DARPA Information Survivability Conference and Exposition*, *1*, 273–282. doi:10.1109/DISCEX.2003.1194891

Jiang, W., & Schulzrinne, H. (2000). Modeling of packet loss and delay and their effect on real-time multimedia service quality. In *Proceedings of the 10th International Workshop on Network and Operating System Support for Digital Audio and Video* (NOSSDAV'00), Chapel Hill, NC, USA.

Kawadia, V., & Kumar, P. R. (2005). A cautionary perspective on cross-layer design. *IEEE Wireless Communications Magazine*, *12*(1), 3–11. doi:10.1109/MWC.2005.1404568

Kim, B.-J. (2001). A network service providing wireless channel information for adaptive mobile applications. In *Proceedings of the IEEE International Conference on Communications* (ICC), (pp. 1345-1351). Helsinki, Finland.

Kliazovich, D., Granelli, F., & Da Fonseca, N. L. S. (2009). Architecture and cross-layer design for cognitive networks . In *Handbook on sensor networks*. River Edge: World Scientific Publishing Co. Inc.

Kumar, S., Raghavan, V. S., & Deng, J. (2006). Medium access control protocols for ad hoc wireless networks: A survey. *Elsevier Ad-Hoc Network Journal*, *4*(3), 326–358. doi:10.1016/j.adhoc.2004.10.001

Kumwilaisak, W., & Jay Kuo, C.-C. (2002). Adaptive variable length Markov chain for non-stationary fading channel. In *Proceedings of IEEE GLOBECOM'02, vol. 3*, (pp. 2046-2050).

Kumwilaisak, W., Thomas Hou, Y., Zhang, Q., Zhu, W., Jay Kuo, C.-C., & Zhang, Y.-Q. (2003). A cross-layer quality-of-service mapping architecture for video delivery in wireless networks. *IEEE Journal on Selected Areas in Communications*, *21*(10), 1685–1698. doi:10.1109/JSAC.2003.816445

Lai, K., & Baker, M. (1999). Measuring bandwidth. In *Proceedings of INFOCOM'99, vol 1*, (pp. 235-245). New York, USA.

Li, W. (2001). Overview of fine granularity scalability in MPEG-4 video standard. *IEEE Transactions on Circuits and Systems for Video Technology*, *11*(3), 301–317. doi:10.1109/76.911157

Liao, H., Zhang, Q., Zhu, W., & Zhang, Y.-Q. (2001). *TCP-aware robust header compression* (TAROC). (IETF Internet Draft, ROHC WG).

Liao, H., & Zhu, W. (2003). Enable fair queueing with decoupled bandwidth-delay guarantees by relaxing the packet service order. In *Proceedings of IEEE International Conference on Communications (ICC'03), vol. 1*, (pp. 147-151). Anchorage, AK, USA.

Liu, Q., Wang, X., & Giannakis, G. B. (2006). A cross-layer scheduling algorithm with QoS support in wireless networks. *IEEE Transactions on Vehicular Technology*, *55*(3), 839–847. doi:10.1109/TVT.2006.873832

Mao, S., Lin, S., Panwar, S. S., Wang, Y., & Celebi, E. (2003). Video transport over ad hoc networks: Multistream coding with multipath transport. *IEEE Journal on Selected Areas in Communications*, *21*(10), 1721–1737. doi:10.1109/JSAC.2003.815965

Pan, D. (1995). A tutorial on MPEG/audio compression. *IEEE Transactions on Multimedia*, *2*(2), 60–74. doi:10.1109/93.388209

Pelletier, G., & Sandlund, K. (2008). *Robust header compression, version 2: Profiles for RTP, UDP, IP, ESP and UDP-lite*. (IETF RFC 5225).

Qian, L., Jones, D. L., Ramachandran, K., & Appadwedula, S. (1999). A general joint source-channel matching method for wireless video transmission. In *Proceedings of IEEE Data Compression Conference* (DCC'99), (pp. 414-423). Snowbird, Utah, USA.

Ramakrishnan, K., Floyd, S., & Black, D. (2001). *The addition of explicit congestion notification (ECN) to IP*. (IETF RFC 3168).

Raman, S., Balakrishnan, H., & Srinivasan, M. (2002). An image transport protocol for the Internet. *IEEE/ACM Transactions on Networking*, *10*(3), 297–307. doi:10.1109/TNET.2002.1012362

Sampath, A., Kumar, P. S., & Holtzman, J. M. (1995). Power control and resource management for a multimedia CDMA wireless system. In *Proceedings of the 6th IEEE International Symposium on Personal, Indoor and Mobile Radio Communications* (PIMRC'95), vol. 1, (pp. 21-25). Toronto, Ontario, Canada.

Sarolahti, P., & Floyd, S. (2007). *Cross-layer indications for transport protocols*. (Internet draft: draft-sarolahti-tsvwg-crosslayer-00.txt).

Sen, S., Rexford, J., & Towsley, D. (1999). Proxy prefix caching for multimedia streams. In *Proceedings of the 18th IEEE INFOCOM'99, vol. 3*, (pp. 1310-1319). New York, NY, USA.

Shakkottai, S., Rappaport, T. S., & Karlsson, P. C. (2003). Cross-layer design for wireless networks. *IEEE Communications Magazine*, *41*(10), 74–80. doi:10.1109/MCOM.2003.1235598

Superiori, L., Nemethova, O., & Rupp, M. (2007). Robust error handling for video streaming over mobile networks. In *Proceedings of the 18th Annual IEEE International Symposium on Personal, Indoor and Mobile Radio Communications* (PIMRC'07), (pp. 1-5).

Syputa, R. (2006). *Clash of the Titans – WiMAX and 4G: The battle for convergence is joined.* Maravedis Market Research and Analysis Report.

Tennenhouse, D. L., Smith, J. M., Sincoskie, W. D., Wetherall, D. J., & Minden, G. J. (1997). A survey of active network research. *IEEE Communications Magazine*, *35*(1), 80–86. doi:10.1109/35.568214

Thomas, R. (2007). *Cognitive networks*. PhD Dissertation. Department of Computer Engineering, Virginia Polytechnic Institute and State University, Blacksburg, Virginia, USA.

Triantafyllopoulou, D.-K., Passas, N., & Kaloxylos, A. (2007). A cross-layer optimization mechanism for multimedia traffic over IEEE 802.16 networks. *International Journal of Network Management*, *17*(5), 347–361. doi:10.1002/nem.668

Vieron, J., & Guillemot, C. (2004). Real-time constrained TCP-compatible rate control for video over the Internet. *IEEE Transactions on Multimedia*, *6*(4), 634–646. doi:10.1109/TMM.2004.830805

Weingart, T., Sicker, D. C., & Grunwald, D. (2007). Identifying opportunities for exploiting cross-layer interactions in adaptive wireless systems. [Hindawi Publishing Corporation.]. *Journal of Advances in Multimedia*, *1*, 2.

Winter, R., Schiller, J. H., Nikaein, N., & Bonnet, C. (2006). Crosstalk: Cross-layer decision support based on global knowledge. *IEEE Communications Magazine*, *44*(1), 93–99. doi:10.1109/MCOM.2006.1580938

Wu, D., & Negi, R. (2003). Effective capacity: A wireless link model for support of quality of service. *IEEE Transactions on Wireless Communications*, *2*(4), 630–643.

Wu, D., & Negi, R. (2004). Effective capacity-based quality of service measures for wireless networks. In *Proceedings of the First International Conference on Broadband Networks*, (pp. 527-536).

Wu, F., Li, S., & Zhang, Y.-Q. (2001). A framework for efficient progressive fine granularity scalable video coding. *IEEE Transactions on Circuits and Systems for Video Technology*, *11*(1), 332–344.

Xie, B., & Zeng, W. (2004). Rate distortion optimized dynamic bitstream switching for scalable video streaming. In *Proceedings of IEEE International Conference on Multimedia and Expo* (ICME'04), (pp. 1327-1330). Taipei, Taiwan.

Xylomenos, G., Polyzos, G. C., Mahonen, P., & Saaranen, M. (2001). TCP performance issues over wireless links. *IEEE Communications Magazine*, *39*(4), 52–58. doi:10.1109/35.917504

Yang, F., Zhang, Q., Zhu, W., & Zhang, Y.-Q. (2001). An efficient transport scheme for multimedia transport over wireless Internet. In *Proceedings of IEEE International Conference on 3G Wireless and Beyond* (3Gwireless ,01), San Francisco, CA, USA.

Yang, F., Zhang, Q., Zhu, W., & Zhang, Y.-Q. (2004). End-to-end TCP-friendly streaming protocol and bit allocation for scalable video over wireless Internet. *IEEE Journal on Selected Areas in Communications*, *22*(4), 777–790. doi:10.1109/JSAC.2004.826008

Zhang, Q., Zhu, W., & Zhang, Y.-Q. (2004). Channel-adaptive resource allocation for scalable video transmission over 3G wireless networks. *IEEE Transactions on Circuits and Systems for Video Technology*, *14*(8), 1049–1063. doi:10.1109/TCSVT.2004.831966

Zhang, T., & Xu, Y. (1999). Unequal packet loss protection for layered video transmission. *IEEE Transactions on Broadcasting*, *45*(2), 243–252. doi:10.1109/11.796266

Zhu, P., Zeng, W., & Li, C. (2005). Cross-layer QoS support for multimedia delivery over wireless Internet. [Hindawi Publishing Corporation.]. *EURASIP Journal on Applied Signal Processing*, *2*, 207–219.

Zhu, P., Zeng, W., & Li, C. (2007a). Cross-layer design of source rate control and congestion control for wireless video streaming. *Advances in Multimedia, 1*, 3-15, Article ID 6850, Hindawi Publishing Corporation.

Zhu, P., Zeng, W., & Li, C. (2007b). Joint design of source rate control and QoS-aware congestion control for video streaming over the Internet. *IEEE Transactions on Multimedia*, *9*(2), 366–376. doi:10.1109/TMM.2006.886284

Zori, M., Rao, R. R., & Milstein, L. B. (1995). On the accuracy of a first-order Markov model for data transmission on fading channels. In *Proceedings of the 4th IEEE International Conference on Universal Personal Communications* (ICUPC'95), (pp. 211-215). Tokyo, Japan.

KEY TERMS AND DEFINITIONS

Cross-Layer Protocols: The protocols that require message communications across various layers of the stack, e.g., application, transport, link etc., are called cross-layer protocols.

Effective Capacity: Effective capacity is a link-layer channel model for modelling wireless channel that can easily translate into connection-level QoS measures such as data rate, delay and delay-violation probability.

End-to-End Congestion Notification (ECN): ECN is an end-to-end mechanism to notify the sender whenever congestion occurs in a network. TCP with ECN is a protocol that overcomes the inherent insensitivity of the TCP congestion control mechanisms to delay or loss of individual packets.

Joint Source and Channel Coding (JSCC): To make a tradeoff between power consumption and quality of the delivered media, joint source and channel coding is an effective approach. It is conducted from the individual user's point of view to effectively combat the errors occurred during transmission by allocating bits between source and channel.

Multi-User Diversity: In a multiple access wireless network, the radio channel is normally characterized by time-varying fading. To exploit the time-varying characteristic of the wireless channel, a kind of channel-state dependent scheduling, called multi-user diversity is exploited to improve system performance.

Priority-Based Scheduling: In priority-based scheduling, packets are grouped into several classes with different priority according to their quality of service (QoS) requirements. In other words, the medium access control (MAC) layer is made aware of the application layer QoS.

Robust Header Compression (ROHC): The goal of ROHCis to develop header compression schemes that perform well over links with high error rates and long link round trip time (RTT). In ROHC framework, relevant information from past packets is maintained in a context. The context information is used to compress (and decompress) subsequent packets.

Snoop TCP: Snoop TCP is a reliable TCP-aware transport layer protocol where data transfer occurs between a fixed host (FH) and a mobile host (MH) with a base station (BS) between them. A snoop agent is created at the BS which buffers data at its link layer for retransmission instead of gong back to TCP end point at the FH and the MH.

Transmission Time Interval (TTI): TTI refers to the length of an independently decidable transmission of the radio link. It is related to the size of the data blocks being passed from the higher network layers to the radio link layer. In order to be able to adapt quickly to the changing conditions in the radio link, shorter TTIs are preferable.

Unsolicited Grant Service (UGS): UGS is a class of applications that supports constant bit rate (CBR) and fixed throughput connections, and provides guarantees on latency, and jitter.

APPENDIX

Table of abbreviations

2G	2nd Generation
3G	3rd Generation
4G	4th Generation
ACK	Acknowledgment
ADU	Application Data Unit
ALF	Application Level Framing
AMC	Adaptive Modulation and Coding
ARC	Analytical Rate Control
ARQ	Automatic Repeat reQuest
AVO	Audiovisual Object
BCH	Bose Chaudhuri Hocquenghem
BE	Best Effort
BER	Bit Error Rate
BS	Base Station
CBR	Constant Bit Rate
CDMA	Code Division Multiple Access
CM	Congestion Manager
CQICH	Channel Quality Information Channel
CRTP	Compressed RTP
EC	Effective Capacity
ECN	Explicit Congestion Notification
EEP	Equal Error Protection
FEC	Forward Error Correction
FER	Frame Error Rate
FH	Fixed Host
FIFO	First-In-First-Out
FTP	File Transfer Protocol
HTTP	Hypertext Transmission Protocol
ICMP	Internet Control Message Protocol
IEEE	Institute of Electrical and Electronics Engineers
IETF	Internet Engineering Task Force
IFD	I-Frame Delay
IP	Internet Protocol
ITP	Image Transport Protocol
JSCC	Joint Source and Channel Coding
LSB	Least Significant Bit

MAC	Medium Access Control
MANET	Mobile Ad-Hoc Network
MPEG	Moving Picture Experts Group
MS	Mobile Station
MSE	Mean Squared Error
nRTPS	Non Real-Time Polling Services
OSI	Open Systems Interconnection
PDCP	Packet Data Convergence Protocol
PDU	Protocol Data Unit
PFGS	Progressive Fine Granularity Scalable
PHB	Per-Hop Behaviors
PHY	Physical
PLR	Packet Loss Rate
PRF	Priority Function
PSNR	Peak Signal-to-Noise Ratio
QoS	Quality of Service
RCPC	Rate Compatible Punctured Convolution Code
REP-RSP	Report Response
RLC	Radio Link Control
RNC	Radio Network Controller
ROHC	Robust Header Compression
RPS	Relative Priority Score
RS	Reed Solomon
RS	Relay Station
RTCP	RTP Control Protocol
RTP	Real-time Transport Protocol
RTPS	Real-Time Polling Services
RTT	Round Trip Time
SNR	Signal-to-Noise Ratio
SS	Subscriber Station
SSI	Source Significance Information
TAROC	TCP-Aware Robust Compression
TCP	Transmission Control Protocol
TDM	Time Division Multiplexing
TDMA	Time Division Multiple Access
TFRC	TCP-Friendly Rate Control
TFRCC	TCP-Friendly Rate Control with Compensation
TTI	Transmission Time Interval
TTL	Time To Live
UDP	User Datagram Protocol
UEP	Unequal Error Protection
UGS	Unsolicited Grant Services

ULP	Unequal Loss Protection
UMTS	Universal Mobile Telecommunications System
VoIP	Voice over Internet Protocol
VPN	Virtual Private Network
WG	Working Group
WLAN	Wireless Local Area Network
WMSTFP	Wireless Multimedia Streaming TCP-Friendly Protocol
WWW	World Wide Web

Chapter 16
Session Management and Transport Protocols for Multimedia Services over IP Networks

László Lois
Budapest University of Technology and Economics, Hungary

Ákos Sebestyén
Budapest University of Technology and Economics, Hungary

ABSTRACT

In the last couple of years, there has been a rapid increase in the number of users consuming media content through multimedia services over IP networks. This chapter aims at introducing the protocols used in and the application questions related to such services. First, a brief overview is given on an IPTV system, a typical and popular multimedia system. This introductory section is followed by the description of application layer transport protocol (RTP - Real-time Transport Protocol) and session management protocols (RTCP - Real-time Control Protocol and RTSP - Real-time Streaming Protocol) of IPTV (and other similar services). Apart from the protocols themselves, implementation and application issues related to an IPTV environment are also discussed. The discussions presented might be a useful starting point for the analysis of other multimedia services.

DOI: 10.4018/978-1-60960-732-6.ch016

INTRODUCTION

The increase in the number of broadband Internet connections worldwide implies a change in the consumption habits of the average Internet user. While a decade ago, static content requiring relatively low bandwidth constituted most of the Internet traffic, nowadays more and more people access dynamic web sites that offer not only textual information but also multimedia content. Moreover, the Internet and similar IP based closed networks constitute the transmission media for several multimedia applications like multi-party online games, Internet telephony, and other streaming services.

This chapter aims at introducing the corresponding protocols that fulfill session management and transport tasks in the application layer. Please note that the term "application layer" is used in the Internet Protocol Suite sense as described in (Braden, 1989) and does not refer to the appropriate ISO OSI layer. Rather, it includes any protocols that belong to the ISO OSI's application, presentation, and session layers.

As the chapter focuses on the application layer transport and session management protocols themselves, the applications that are implemented by them are more or less unimportant, at least as far as the operation of the protocols are concerned. From a functional perspective, the protocols work almost the same way regardless of the data they carry. However, to come to grips with as many aspects as possible, the Internet Protocol Television was chosen as a typical service to demonstrate streaming services and to help the reader to relate the discussion to a real life application. The choice was not entirely arbitrary: as typical IPTV services involve the simultaneous and real time transmission of video, audio, and other content, they call for more elaborate solutions than other less complex applications like Internet telephony.

We believe that the discussions given here can be easily adapted to other application areas as well. Again, it must be emphasized that it is not the applications, we are focusing on, but the overlaying application layer protocols with session management and transport tasks, therefore certain aspects (e.g. digital rights management, access control, quality of service, etc.) are out of scope of the recent study. IPTV is a pay-TV service that uses an IP infrastructure which is a private and closed system maintained by a particular service provider. As the network is closed, the service provider has freedom to prioritize packets carrying television streams over other packets, and can invest in implementing proprietary solutions for other challenges. (One such challenge is the reduction of time required for television channel change by the user.) IPTV applications are more critical from the quality point of view and require the usage of protocols that are capable of simultaneously and reliably serve a vast number of end users. Two typical applications in an IPTV environment are the broadcast-like, simultaneous (multicast) transmission of television programs to many users and the unicast transmission of selected content in unicast mode to individual subscribers (so-called Video-on-Demand or VOD service).

In a streaming environment not only error free transmission but also timing fidelity is a concern. Although a minor delay is tolerable at the start of playback, once started, playback must be uninterrupted and seamless. Therefore, when choosing or developing a transport protocol both fidelity and delay questions must be considered.

Protocols that meet the strict timing and fidelity requirements include the Stream Control Transport Protocol (SCTP), the Datagram Congestion Control Protocol (DCCP) in the transport layer, and the Real-time Transport Protocol running on top of User Datagram Protocol (UDP) in the application layer. (Again, application layer is meant in the sense as defined in the Internet Protocol Suite.) Since commercial IPTV SCTP and DCCP implementations are not available yet, this chapter covers the details of RTP, and focuses on its IPTV related aspects. Apart from RTP itself, the chapter also introduces the corresponding control protocols, namely RTCP and RTSP. While the former provides out of band statistics and quality metrics for an established RTP session, which can be used for controlling e.g. the bit rate of the transmitted media and for synchronizing the audio and video streams, the later is used for session and playback state management and media description. Finally, the chapter introduces the concept of group management that allows for forming multicast groups from the users consuming the same content, thus reducing the burden on the network backbone.

BACKGROUND

With the appearance of digital television in the early 1990's, the previously separated branches of telecommunications and broadcasting experienced a rapid convergence. Whereas, in television context, telecommunication systems had been used for delivering digitized, high data rate television signals from the content provider to the television transmitters or head-ends, and broadcasting was used for reaching the end users, in the last decade telecommunication companies started to directly supply their subscribers with television content. This was partly ensured by the appearance of sophisticated access technologies like Digital Subscriber Line (xDSL) technologies, and partly by the implementation of very efficient coding methods like MPEG-2 and h.264 AVC. Together with the encoding standards, packetization specifications have also appeared that prescribed how an encoded stream could be transmitted through different media, IP being one of them.

IPTV and other streaming services build upon a vast number of protocols each having their prescribed role in the transmission chain. What these protocols have in common is that they are laid over an IP layer. However, while some of them uses more general transmission protocols (like UDP) and extends them with the required services in the application layer, others provide the timing and quality of service related functionalities in the transmission layer.

Two protocols that belong to this latter group are DCCP and SCTP. DCCP and STCP enhance the basic transport services of Transmission Control Protocol (TCP) and UDP with new mechanisms (such as congestion control and multi-homing support) and form the basis of more recent studies. The fundamentals of DCCP are laid down in (Kohler et al., 2006), and SCTP is described in (Stewart, 2007). Although they have not made their way into commercial applications yet, their behavior in streaming environments are under investigation. For SCTP related papers good examples are (Kim et al., 2007) and (Xu et al., 2009) that evaluate SCTP in IPTV and multimedia streaming applications. Ongoing standardization efforts of SCTP include specifying modular extensions, adding socket API extensions and providing enhanced security features for this protocol.

DCCP provides congestion control mechanisms and is mainly used in open and mobile networks, in which, due to the best effort nature of the medium, it is necessary to ensure stability (Park & Jeong, 2009) (Park et al, 2010). Just like SCTP, DDCP is being extended and is under heavy standardization.

Nevertheless, in this chapter we are going to revisit IPTV from a protocol perspective emphasizing only those application layer protocols that form the basis of commercial products and have passed through extensive testing. Hence, we omit any discussions on SCTP and DCCP, and concentrate on RTP and its sister protocols. There are numerous books that follow the same path. Some of them are meant to serve as a guideline for the average people to understand what is going on in an IPTV environment (Simpson & Greenfield, 2009; O'Driscoll, 2008). As a result, these publications simply do not wish and cannot go into the fine details. They only paint a general picture of this booming industrial sector. Other books follow a different approach and have a different target audience: the industrial experts

(Minoli, 2009). They usually focus on a narrow area within Internet protocol television and web television, and cover that particular segment as deeply as possible by assuming a thorough understanding of the IP basics from the reader. We try to position our chapter somewhere in between, though due to size constraints only some of the IPTV related protocols are discussed.

OVERVIEW OF IPTV RELATED PROTOCOLS AND SERVICES

IPTV (like other digital television transmission solutions) not only provides linear content for the end users, but also offers enhanced interactive services. Services include, among others, time-shifted TV that empowers users to pause live feeds, electronic program guide, video-on-demand capabilities, and digital video recorder application to create personal video libraries. Major IPTV product suppliers include Technicolor (formerly Thomson), Alcatel-Lucent and UTStarcom with end-to-end solution, and Microsoft mainly with its Mediaroom middleware.

Based on five industry analyst predictions, an overall IPTV household penetration of almost 60 million is forecast by the end of 2011.

A simplified IPTV system is depicted in Figure 1. The system consists of a core network component, an aggregation network, and an access network part. The access network typically utilizes xDSL technology to provide a link for the end user. At the edge of access and the core network digital subscriber line aggregation modules (DSLAMs) aggregate the requests coming from and separate the responses flowing to the end users. The DSLAM is connected to the aggregation network through multicast capable edge routers. The core network with the video servers and the aggregation network are interconnected by distribution routers. Content coming from content providers is inserted into the network in national (super) and regional or local head ends.

Figure 1 also shows the different protocols and the relationship between them. In the following sections, the protocols highlighted in the protocol stack will be discussed.

Figure 1. A typical IPTV system and its protocols

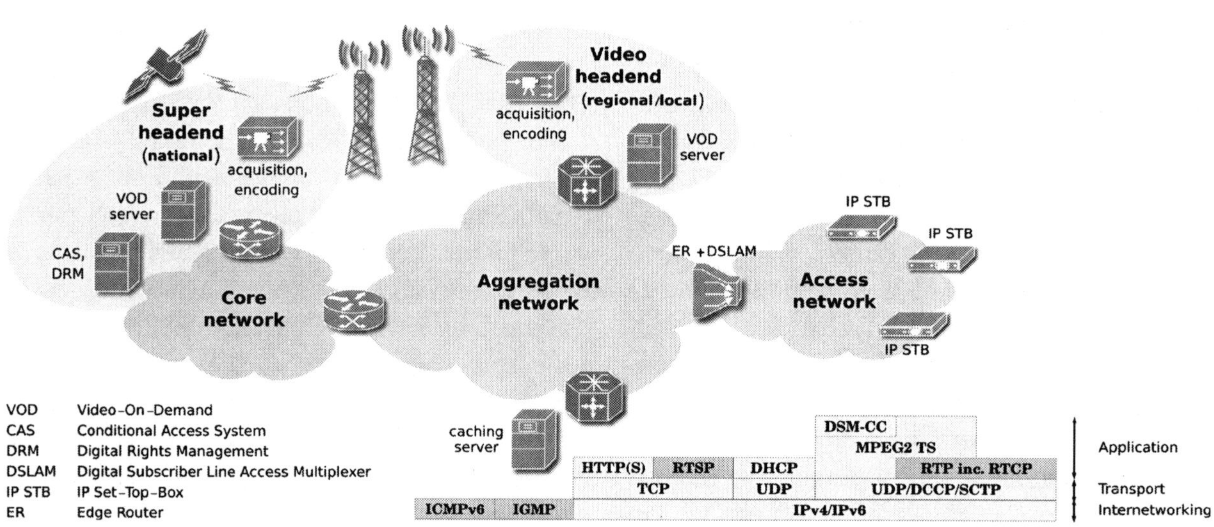

Application Layer Protocols for IPTV Services

In order to understand the mechanism behind the IPTV related application layer transport protocols, it is important to analyze the structure of audio and video content. Audio and video are divided into smaller data units. The data units of a common audiovisual content are aligned so that when they are decoded and perceived, they produce a natural sensation. Alignment is achieved by using a common clock reference on the transmitter side to tag particular segments with a so-called timestamp. So long as this reference clock can be regenerated after transmission, the timestamps can be used to restore the synchronization between the audio and video tracks. (It is obvious that this approach can be used not just for audio and video content, but also for other types of streams, e.g. online game data. Thus, the operation of the described protocols is more or less independent of the content they carry and the application they implement.)

Although the encoded media stream consists of data units each with the corresponding timestamp, these packets cannot be mapped directly to transport layer packets without loss of the alignment; hence, an intermediary layer must be constructed.

The application layer acts as an interface between the media encoder and the transport layer, and implements specific services to ensure real-time media transmission by providing signaling, packetization and transport mechanisms. The application layer protocol must meet the following requirements:

- It must be able to perform TV like delivery of content via a transport layer protocol in different scenarios including, but not limited to peer-to-peer transmission for Video-on-Demand services and multicast transmission for providing live content.

Figure 2. Mapping of encoded segments to transport layer packets

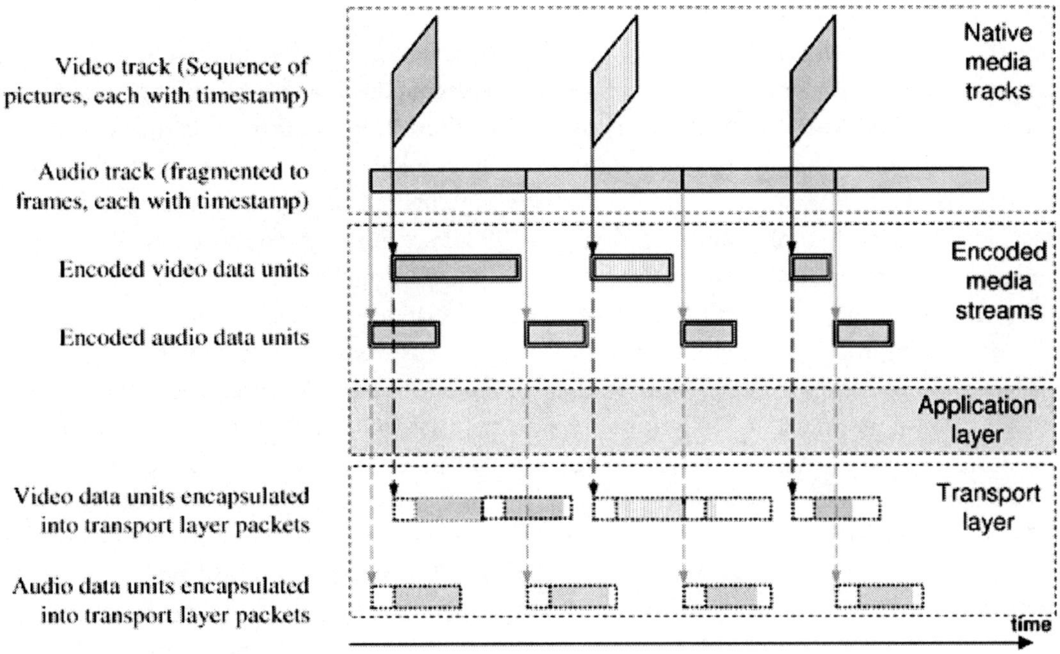

- It must work regardless of what transport layer protocols are used.
- In case of error prone transport layer protocols, it must support error detection solutions.
- It must carry information on media framing and timing to achieve proper clock regeneration and synchronization at the receiver side.
- It must provide a framework to signal the received stream quality and the link capacity to control the encoder and the transmission at the sender side.
- It must be able to signal a change in encoding format.

Most of the IP based applications use TCP as a transport layer protocol since it provides reliable error-free delivery, but does it at the expense of latency. In comparison, UDP, the other widely used protocol, though unreliable by itself, provides fast and balanced delivery of packets.

Some of the functionalities within an IPTV system requires reliable transmission of information, but lacks the strict timing and latency constraints. Such functionalities are session control and media format description mechanisms, which are thus implemented through TCP. The transmission of the media content, on the other hand, must be achieved with as low a latency as possible, therefore audio and video information is usually delivered by UDP.

Though UDP meets the delay and speed requirements, and is inherently able to detect conditions like sender change, incorrect packet size, excessive jitter or improper bit rate, still cannot provide for synchronization by itself. Although with MPEG-2 transport streams UDP is suitable to provide basic media transmission, it lacks some of the functionalities that the application layer RTP protocol can provide.

Real-Time Transport Protocol

RTP (Schulzrinne et al., 2003) performs real-time data transmission by first encapsulating real-time media streams, then delivering them from the sender to the receiver. This delivery mechanism is enhanced by the services offered by RTCP. RTCP is a sub-protocol of RTP that transmits signaling information between the sender and the receiver in both directions. RTCP provides minimal services to RTP in the sense that it only implements functionalities indispensable for the seamless operation of RTP. Among its functions is monitoring of the received RTP stream quality and providing inter-RTP synchronization. For enhanced tasks like session management, a third protocol, RTSP had been created.

Since RTCP carries only supplemental signaling information, whereas RTP carries real-time media having a predetermined rate somewhere around 3 Mbps for a standard definition h.264 encoded video, the frequency of RTCP packets as compared to the frequency of the RTP packets is relatively low.

The RTP Session

In general, a media stream contains one or more video, audio, and other auxiliary data tracks. Each of these tracks can be transmitted separately in unique RTP sessions, or they can be multiplexed together into a common RTP session. Thus, an RTP session consists of an RTP connection for transmitting the useful data (track or multiplexed tracks) and an RTCP connection for controlling the transmission. Each RTP session has a specific, 32-bit unique identifier and a well-defined port pair assigned to it. It is recommended that RTP data packets use an even, the corresponding RTCP packets the next odd port number. In case of hierarchically encoded streams, which comprise of a base layer and several enhancement layers, from which the client can freely choose to access only the base layer for basic quality or the base layer and several enhancement layers for enhanced quality according to its quality requirements (or available bandwidth), the RTP sessions formed from each layer should use a set of contiguous port numbers. In this case, the RTCP port numbers should be signaled by other means.

RTP Data Packet Format

The RTP data packet format is depicted in Figure 3. The most important fields are described below. For a complete reference, see (Schulzrinne et al., 2003).

- P: The flag specifies that the payload contains padding octets at the end. Padding is added and this bit is set if the transport protocol supports only fixed length payload or an encryption algorithm requires fixed length data.
- CC (CSRC Count): The four-bit field specifies the number of CSRC items in the CSRC list.
- SSRC: A unique identifier assigned to the source that produces the RTP stream. In most cases, the source is the encoder. If re-multiplexing, re-encoding or translation is used, the identifier identifies the RTP translator or mixer. In this case, the *CSRC* list will carry the *SSRC* of the module that produced the original RTP stream.
- RTP timestamp: This field specifies the sampling instant of the first octet in the RTP data packet and is used for resynchronization on the receiver side. The timestamp is derived from a reference clock. The granularity of the reference clock depends on the transmitted media and on the desired synchronization resolution. The timestamp is indispensable for the proper decoding and playout of the audio or video content.
- Seq. number (Sequence number): A counter, whose value is incremented by one at the generation of every new RTP packet. It is used for detecting packet loss, bad packet delivery order or packet duplication. Reordering of packets prior to decoding in the receiver buffer is based on the sequence number.
- PT: A 7-bit field used for identifying the format of the encapsulated data stream. For different media formats strict encapsulation rules apply, which are defined in separate standards. The media format can also be signaled using RTSP DESCRIBE response messages (see Section 3.2.1.1).

Figure 3. RTP data packet format

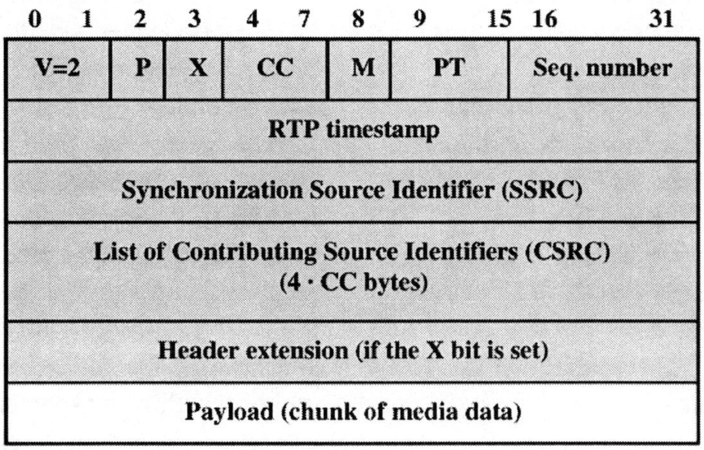

Should this be case, the interpretation of the *PT* and the following *M* field would be application dependent.

- M: A marker bit which is originally used for signaling events such as frame boundaries. Its actual value is application dependent.
- Payload: The *Payload* carries a chunk of the encoded media stream and constitutes the useful portion of the RTP packet. The *Payload* is preceded by an RTP header with the fields defined above. The RTP payload and the RTP header together are encapsulated in the payload of the corresponding transport packet. What the *Payload* portion actually carries depends on the encoding format used for the media. In general, the available payload size (the payload size of the transport packet minus the size of the RTP header) differs from the size of the data units the encoder produces. Depending on the relation between the payload size and the data unit size, three packetization and encapsulation methods exist:
- An application data unit longer than the payload size spans across several RTP packets. The timestamp of the corresponding RTP packets equals to the timestamp of the data unit. Long data units are usually split so that they use up the available payload size. The last RTP packet is usually either shorter or is filled with padding octets, but certainly cannot carry fragments of consecutive data units.
- An RTP packet might carry more than one consecutive smaller data units. In this case, the timestamp in the RTP header equals to the timestamp of the first data unit.
- There can be a one to one mapping between the data units and the RTP packets. Should this be the case, the timestamp of the data unit and the timestamp in the RTP packet are identical.

RTCP Packet Format

Beside the unidirectional RTP data transport, each RTP session contains a bidirectional RTCP control session. The RTCP control session is implemented through 5 different control packet types. Each of these packet types consists of a 32-bit header followed by one or more blocks whose content depends on the type itself. The structure of the 32-bit header in the RTCP packet is quite similar to the first 32-bit portion of the RTP header.

Figure 4 depicts the different predefined (i.e. not application defined) RTCP control packets. The structure of the packets is detailed below. For a more thorough description, please consult Section 6.1 of (Schulzrinne et al., 2003).

For fulfilling different tasks, 5 types of RTCP packets are defined. The type of the packet is identified by the *PT* field in the message:

- PT=200: Sender report (SR) RTCP packet with sender info and several report blocks
- PT=201: Receiver report (RR) RTCP packet with several report blocks
- PT=202: Source description (SDES) RTCP packet with several SDES items
- PT=203: Goodbye RTCP packet with several SSRC or CSRC identifiers
- PT=204: Application defined RTCP packet

The RTCP SR messages are sent by the senders, the RTCP RR messages are sent by the receivers. Both packet types contain a list of 24-bit report blocks. Beyond that, the SR RTCP packets have a 20-bit *Sender info* field.

The *Sender info* field carries the 64-bit NTP (wallclock) time when this report was sent, along with the RTP timestamp corresponding to the NTP time. Based on these pieces of information the receiver can determine the relationship between the NTP time and the RTP timestamps, and can synchronize the different tracks constituting a common audiovisual content. Besides, the *Sender info* field contains two 32-bit counters that specify

Figure 4. RTCP control packet formats

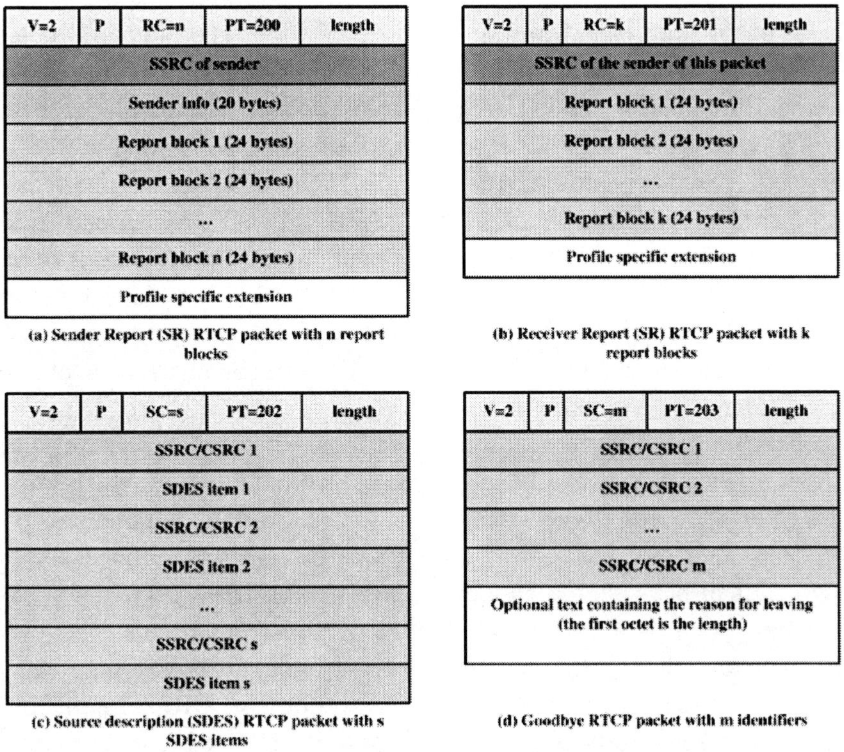

(a) Sender Report (SR) RTCP packet with n report blocks

(b) Receiver Report (SR) RTCP packet with k report blocks

(c) Source description (SDES) RTCP packet with s SDES items

(d) Goodbye RTCP packet with m identifiers

the total number of packets and octets the sender transmitted in the RTP data packets since the last report.

Each *Report block* contains the SSRC of a sender, and provides feedback on the quality of the received RTP data stream coming from that particular sender. Feedback information includes information like packet loss and inter-arrival jitter. The *Report block* can optionally contain the last SR timestamp (LSR) and the delay since the last SR (DLST) for computing the round-trip delay between the sender and the receiver. The round-trip delay is a good indicator of the link quality; hence, its value can be used for e.g. controlling the encoded bitrate of the content.

The SDES and Goodbye RTCP packets are used for manipulating session state information. All participants of an RTP session regularly send the RTCP SDES packet indicating that they are still in the session. The SDES packet carries the mandatory *Canonical End-Point Identifier*, the status of the participant, the used application, phone number, e-mail address, etc. Once a participant wants to terminate its RTP session, it might send a Goodbye RTCP packet or simply quit the session without sending any more messages. In the later case, its session will expire after a predefined timeout period due to lack of activity.

Enhanced Session Management for IPTV Services

While basic session management is implemented by RTCP, more complex situations call for more sophisticated solutions. Complex session management with playout control mechanisms (restart, stop, rewind, forward) is achieved by the Real-time Streaming Protocol.

Real-Time Streaming Protocol

RTSP was developed by the Multiparty Multimedia Session Control Working Group (MMUSIC WG) of the Internet Engineering Task Force (IETF) and was published as an IETF draft in 1998 (Schulzrinne et al., 1998). Its most current version at the time of writing is version 2 published in 2010 (Schulzrinne et al., 2010), which contains several enhancements and modifications. Except where clearly stated, the following description is valid for both RTSP version 1.0 and 2.0.

RTSP can be considered as a negotiator between the clients and the servers. This negotiating role consists of the following phases:

- Presentation description
- Session establishment between the clients and the servers
- Controlling and manipulating the playout of the media as requested by the client
- Session maintenance and termination

RTSP is a stateful protocol in which the client and the server communicate through handshaking by exchanging requests and corresponding replies in a predetermined order.

Each of these requests and responses carries textual information in UTF encoded ISO 10646 character set. The textual nature of the protocol makes it easier to extend the parameters by adding self-describing optional elements. Although the processing of textual information can be cumbersome and resource intensive, this is not a concern due to the low frequency of the messages.

Both request and response messages consist of a start line, optional header fields, an empty line, and an optional message body. The request start line provides the most relevant information on the request: it defines what method to execute on which stream (defined by a Unified Resource Identifier, or URI), and which RTSP version is used for the communication. The response start line provides information about the attempt to process and serve the request in the form of a status code and a corresponding brief textual description. The fields of the RTSP request and response messages are the following:

Method: The method specifies what action needs to be performed on the URI that follows it. The most important methods are DESCRIBE, SETUP, PLAY, PAUSE, and TEARDOWN. These methods are further described in Section 3.2.1.1, 3.2.1.2, 3.2.1.3, and 3.2.1.4. For a complete list of methods and their functions, please refer to Table 7 of (Schulzrinne et al., 2010).

- Request URI: The absolute URI of the media stream, on which the method must be performed. For methods that do not require a URI to be defined, the field carries only an asterisk. RTSP 2.0 understands both IPv4 and IPv6 URIs. Please note that in order to request a stream; the client must know its URI beforehand. This knowledge is gathered e.g. by downloading a web page in case of webTV applications or via proprietary protocols or session announcements in an IPTV environment.
- RTSP version: The field contains the version of RTSP used. The most current version is 2.0. For this, the RTSP version field carries the value of RTSP/2.0.
- Optional request header fields and Optional response header fields: Optional header fields are used for further specifying the parameters of the request or response. In IPTV applications, the most important optional request field is the *<Session>*, whose value is used for identifying which session a particular request belongs to. It is set only after a SETUP request (see Section 3.2.1.2) was successfully processed.
- Message body: The optional message body carries relevant information that cannot be included in the message header. An example of such information is the Session Description Protocol fields (see Section

3.2.1.1). The presence of the <*Message body*> must be signaled in the Content-length header field.

- Status code: The <*Status code*> is a 3-digit code corresponding to the outcome of the attempt to process and execute the requested method. The status codes can be grouped as follows:
  - 1xx: The client understood the request but needs more time to process it.
  - 2xx: The request was received, interpreted, and executed without any error.
  - 3xx: Redirection is required to process and serve the request.
  - 4xx: This group of codes indicates client error due to bad syntax or unfulfilled conditions (e.g. lack of authorization).
  - 5xx: These codes correspond to unsuccessful completion of an apparently correct request.

Whenever possible, RTSP adopts existing HTTP/1.1 status codes. Moreover, the pool of existing status codes can freely be extended. It is not required for the client to understand each and every RTSP status code, but it must be able to classify the response into the five groups above and act accordingly.

- Reason phrase: The <*Reason phrase*> is a brief textual representation of the <*Status code*> received. See Table 4 (Schulzrinne et al., 2010) for a complete list of status codes and corresponding reason phrases.

Request and Response Messages for Presentation Description

The purpose of RTSP is to provide access for different multimedia streams specified by URIs, and do it without prescribing what media types are supported. The streamed content can consist of stored files as well as live presentations requiring real-time processing; can consist of compressed or uncompressed audio and video data. Moreover, the streams can be transmitted in a unicast or a multicast fashion.

For proper transmission and playback of a vast number of video and audio formats, the receiver must be notified about the parameters of the content and the protocols used for delivery. RTSP uses external mechanisms to gather and transmit information about a requested content.

The exchange of stream parameters is done through an RTSP DESCRIBE request and a corresponding response message. The DESCRIBE request carries the URI of the stream that is to be transmitted to the client, and has the following format:

```
DESCRIBE rtsp://152.66.249.21:5554/Test
RTSP/2.0
CSeq: 33
Accept: application/sdp
User-Agent: VLC media player (LIVE555
Streaming Media v2010.01.07)
```

where *CSeq* is the sequence number of an RTSP request-response pair. The value of the *CSeq* is increased by any new requests issued by the client during the session. The *Accept* line of the DESCRIBE request specifies what kind of descriptions the client understands. The most widely used description type is the Session Description Protocol (SDP).

For the DESCRIBE request the server responds with an RTSP OK message:

```
RTSP/2.0 200 OK
Content-type: application/sdp
Server: VLC Server
Content-Length: <n>
CSeq: 33
Cache-Control: no-cache
<sdp record>
```

where the <*sdp record*> carries the actual description of the content.

The description of media streams includes the addresses and ports the media is sent from, as well as information like the type of media, the transport protocol used, the format of the media, the session name and purpose, and the time when the session is active.

Just like RTSP messages, SDP uses textual fields in ISO 10646 character set. The description of the properties are given in the form of *<type>*=*<value>* pairs. <type> is always a one-byte long ASCII character and is case sensitive, whereas the corresponding *<value>* might include any characters from the ISO 10646 character set.

The session description consists of session level information that apply to all media streams in the session, and auxiliary media level information that are related to a particular media stream. A session description typical to IPTV VOD service has the following format. For a complete list of available session and media-level fields, please refer to Section 5 of (Handley et al., 2006).

```
v=0
o=- 3042574000 3 IN IP4 152.66.249.21
c=IN IP4 0.0.0.0
t=0 0
a=range:npt=0-596.461
m=audio 0 RTP/AVP 96
a=rtpmap:96 mpeg4-generic/48000
a=fmtp:96 streamtype=5; profile-level-
id=15; mode=AAC-hbr;
          config=11b0;
SizeLength=13;IndexLength=3;
IndexDeltaLength=3; Profile=1;
a=control:rtsp://152.66.249.21:5554/Test/
trackID=0
m=video 0 RTP/AVP 97
a=rtpmap:97 H264/90000
a=fmtp:97 packetization-mode=1;profile-
level-id=4d401e;
          sprop-parameter-sets=J01AHqkYGw
e83gDUBAQG2wrXvfAQ,KN4JyA==;
a=control:rtsp://152.66.249.21:5554/Test/
trackID=1
```

- v: Describes the version of the SDP.
- o: It gives the originator of the session in *<username>* *<sess-id>* <sess-*version*> *<nettype>* *<addrtype>* *<unicast-address>* format. *<username>* is the login name of the user on the originating host or, if this functionality is not supported, it is "-". *<sess-id>* is a number generated by the process providing the description so that the tuple formed by the *<username>* *<sess-id>* <sess-*version*> *<nettype>* *<addrtype>*, and *<unicast-address>* values uniquely identifies a session. <sess-*version*> is again generated by the host application, and its value must be increased whenever a change in the description occurs. This functionality, however, is hardly ever used in an IPTV environment. *<nettype>* and *<addrtype>* identify the type of transport medium and the addressing scheme used within: IN is for internet traffic. The *<addrtype>* can be either IP4 for IPv4 addresses or IP6 for IPv6 addresses. *<unicast-address>* is the unicast address of the host that created the session. *<unicast-address>* can contain the fully qualified domain name, or, for IPv4 addresses, it can carry the dotted decimal representation of the host IP address, for IPv6 addresses, it can carry the compressed textual representation of the IPv6 host address.
- c: The field carries information on the connection and comprises of the *<nettype>* *<addrtype>* and *<connection-address>* sub-fields. The description must contain one session level *c* field, or one media level *c* field per media. If both the session and the media level *c* fields are present, then the media level *c* field takes precedence. The *<nettype>* and the *<addrtype>* sub-fields are similar to the sub-fields used in the *o* directive.

The *<connection-address>* can be an IPv4 or IPv6 unicast or multicast address. For multicast addresses, a time-to-live (TTL) parameter must be appended to the IPv4 multicast address by a slash. Furthermore, the SDP has a separate notation for multicast layered or hierarchical content. Each of the layers has separate multicast address associated to them. If these addresses are contiguous, then they can be signaled as *<base multicast address>*[/*<ttl>*]/*<number of addresses>*.

- m: This field provides the media description and has a format of m=*<media>* *<port>* *<proto>* *<fmt>* [*<fmt>*…] The *<media>* sub-field defines the media type. For IPTV and webTV applications, the type can be audio or video. *<port>* is the destination port and its interpretation depends on the value of the *<proto>* sub-field. If RTP is used as a transport protocol, then *<port>* defines the base port of the transmission and usually has an even value. The corresponding RTCP can be derived by adding one to the base port. Should this not be the case, the RTCP port must be signaled with the a=rtcp:*<rtcp port>* attribute (see below). If the base port number for RTP is odd, then the usage of the a=rtcp:*<rtcp port>* is mandatory.

The *<proto>* sub-field specifies the protocol of the transmission. In an IPTV or webTV environment, the *<proto>* field can have a value of udp that corresponds to an unspecified protocol running over UDP, or it can have a value of RTP/AVP (RTP Audio Video Profile) or RTP/SAVP (RTP Secure Audio Video Profile).

The *<fmt>* and any subsequent sub-fields provide the format description of the stream. If the *<proto>* sub-field has a value of RTP/AVP or RTP/SAVP, then the *<fmt>* sub-fields contain RTP payload type numbers. If more than one *<fmt>* sub-field is present, the client can freely choose from the available formats. Note, however, that the default format of the session is the first *<fmt>* value. The RTP payload type assignment can be static or dynamic. For static assignment (e.g. m=audio 49232 RTP/AVP 0), the value of the *<fmt>* field directly translates to a particular media type (PCM encoded single channel audio sampled at 8 kHz in this case). In case of dynamic assignment, an a=rtpmap:*<payload type>* *<encoding name>*/*<clock rate>*[/*<encoding parameters>*] attribute field is required, where the *<payload number>* corresponds to the *<fmt>* value in the *m* field. For H.264 encoded video, the dynamic assignment has the following form:

```
m=video <port number> RTP/AVP 97
a=rtpmap:97 H264/9000.
```

Additional attribute fields further describe the encoding format and provide more information about the stream.

If more than one RTP payload number is present, each payload number might have one dynamically assigned value specified by an attribute field.

If the *<proto>* sub-field has a value of udp, then the *<fmt>* value refers to a media type described under the video or audio top-level media types. For UDP carrying MPEG2 transport stream, the *m* field is as follows:

```
m=video <port number> udp 33
```

- a: This field is used to provide attributes and extend the functionality of the core SDP specification (Handley et al., 2006). The attributes are either session-level and refer to the entire session or media-level and refer to a particular media component within the session. In the latter case, the *a* field is preceded by a media field. The field might have a format of a=*<flag>*, in which case simply the presence of the attribute implies that the condition is valid for either the session or the stream. Otherwise, the attribute might take the format of a=*<attribute>*:*<value>*. For a complete list of the available SDP attributes, see Section 6 in (Handley et al., 2006).

- t: The *t* line consists of a *<start time>* and an *<end time>* sub-field which together specify the start and end time for a session, respectively. The description might include multiple *t* lines, in which case each line corresponds to a particular period when the stream is available. For live TV content, both the *<start time>* and *<end time>* sub-fields have a value of 0, indicating that the session is permanent.

Session Establishment

Once the description is available to the client, it can establish a session for delivery of the source streams that suit its purpose the best. Which streams are chosen might depend on the codecs supported by the client, the available bandwidth or resources. IPTV and webTV content consists of video and audio streams, which need to be set up individually by SETUP request and response messages. A SETUP message has the following format:

```
SETUP rtsp://152.66.249.21:5554/Test/
trackID=0 RTSP/2.0
CSeq: 34
Transport: RTP/AVP;unicast;client_
port=1170-1171
User-Agent: VLC media player (LIVE555
Streaming Media v2010.01.07)
```

In the SETUP request message, the client specifies which stream it requests and what transport protocol and server port the server can use for the delivery. In the above example, the SETUP request refers to the first media stream described by the SDP record. (In multicast transmission, no client port is specified.)

In a corresponding SETUP response message the server acknowledges the client parameters and generates a random *Session* identifier that uniquely identifies any forthcoming messages between the server and the client. The format of the SETUP response message is the following:

```
RTSP/2.0 200 OK
Transport: RTP/AVP/UDP;client_port=1170-
1171
Server: VLC Server
Content-Length: 0
Cseq: 34
Cache-Control: no-cache
Session: 19169
```

Any corresponding SETUP requests and other RTSP messages belonging to the session must include the *Session* parameter generated at the first SETUP request.

The established session exists until it is torn down by the client through a TEARDOWN request or until the session times out due to inactivity. The client can keep a session alive by periodically issuing RTSP requests (e.g. GET_PARAMETER messages not detailed here) that contain the *Session* identifier in their *<Message header>*. Moreover, if the transport protocol used is RTP, then RTCP Receiver Reports belonging to any of the media streams can also signal the liveness of the session.

Controlling Playout

Once a session has been established, the client can issue a PLAY request to the server to start sending the content through the protocol and ports specified in the SETUP requests and responses. The PLAY request message has the following format:

```
PLAY rtsp://152.66.249.21:5554/Test
RTSP/2.0
CSeq: 36
Session: 19169
Range: npt=0.000-
User-Agent: VLC media player (LIVE555
Streaming Media v2010.01.07)
```

The PLAY request must contain the *Session* identifier as well as a range that determines the starting and, optionally, the ending position within the stream. The server must position the playtime to the starting position and start playing out the stream until another PLAY or PAUSE request is received, or the end of range or end of stream is reached. In a multicast environment, no further playout control mechanism other than the first PLAY request is available. In such environments, the specification of range is somewhat different, too.

In IPTV and webTV applications, only a so-called aggregate control mechanism is used, which means that the streams comprising the content cannot be controlled individually. The URI used for aggregate control also reflects this behavior. (See the message header of the PLAY request for an example.) A PLAY request is acknowledged by the following PLAY OK message:

```
RTSP/2.0 200 OK
Server: VLC Server
Content-Length: 0
CSeq: 36
Cache-Control: no-cache
Session: 19169;timeout=5
```

A *timeout* parameter can be used to specify the inactivity period after which the server closes the session.

Session Termination

Sessions are usually terminated by the client through a TEARDOWN request message in the following format:

```
TEARDOWN rtsp://152.66.249.21:5554/Test
RTSP/2.0
CSeq: 37
Session: 19169
User-Agent: VLC media player (LIVE555
Streaming Media v2010.01.07)
```

Upon receiving a TEARDOWN request, the server stops transmitting the content and after a grace period that allows the TEARDOWN response message to reach the client, it frees all associated resources. Just like playout of content, session teardown is controlled through aggregate control mechanisms in an IPTV environment. TEARDOWN response messages have the following format:

```
RTSP/2.0 200 OK
CSeq: 37
Server: VLC Server
```

Although less frequently and certainly not in IPTV environment, sessions can be torn down by server-initiated TEARDOWN requests. The reason for such requests might be server maintenance or session timeout due to inactivity. Server initiated TEARDOWN is only supported in RTSP 2.0.

Multicast Transmission in an IPTV Environment

In an IPTV environment, a vast number of clients are connected to the server through intermediate routers. Since the number of programs offered by an IPTV service provider is limited as compared to the number of users, each program is usually consumed by many clients. In such a scenario, multicast transmission proves to be more beneficial (Figure 5).

As opposed to unicast transmission where an individual link is set up between the server and each client, in a multicast environment users consuming the same content form a so-called multicast group. The multicast group can be served with one single stream thus relieving the burden on the network backbone.

Multicast transmission involves several different mechanisms that are provided by different protocols:

- Dynamic host registration (group management): The network must be notified about the mapping of the different hosts to certain multicast groups. Group management is done by the Internet Group Management Protocol (IGMP) for IPv4 networks and by the Multicast Listener Discovery (MLD) protocol embedded in the Internet Control Message Protocol (ICMP) for IPv6 networks.

Figure 5. Comparison of multicast and unicast transmission

- Multicast payload forwarding: Payload forwarding is done by the transport protocols described in the previous sections.
- Multicast routing: For routing the appropriate traffic to the appropriate clients, distribution trees must be constructed that define a unique forwarding path between the server and the different subnets formed by the clients of a multicast group.

As the dynamic host registration is essential for session management, and multicast routing is just a mean to transmit information within the network, only the group management will be considered. For details on multicast routing, please consult (Minoli, 2008).

Group Management in an IPv4 Environment: IGMP

IPv4 based clients can report their group memberships to neighboring multicast aware routers by the Internet Group Management Protocol (IGMP). Moreover, IGMP messages are used by these routers to query which multicast group a particular client is subscribed to. The IGMP protocol is defined by (Cain et al., 2002). Its most current version at the time of writing is version 3.

IGMP messages are encapsulated in IPv4 datagrams with a version number of 2. The following types of messages are considered for group management:

- membership query messages (type number 0x11),
- membership report messages (type number 0x22) sent by the clients and routers to report their membership status.

Membership Query Messages

Membership query messages are sent by the routers to query multicast membership status of their neighbors. The membership query messages have three types:

- With the General Query message, the router queries all of its neighbors about any multicast addresses they are entitled to receive. Both the *Number of sources* and the *Group address* fields defined below contain zero. The source address in the IP header of such IGMP messages is set to the all-system multicast address (224.0.0.1).
- With a Group-Specific Query, the router queries all of its neighbors about the reception state of one particular multicast address. In this case, the *Number of sources* field contains zero, the *Group address* field contains the multicast address in question. The source address in the IP header of such IGMP messages is set to the particular multicast address
- With a Group-and-Source-Specific Query, the router queries whether any of its neigh-

bors want to receive a particular multicast address coming from the source addresses in the message. The source address in the IP header of such IGMP messages is set to the particular multicast address

The format of the membership query messages is depicted in Figure 6(a). The most important fields are described below. For a complete reference, please see Section 4.1 of (Cain et al., 2002).

- Max resp code: Describes the timeout in 1/10 seconds before the client must answer with a membership report message. For values greater or equal to 128, the timeout value is interpreted as $8/10 \cdot (16 + \textit{Max resp code} \ \& \ 0x07) \cdot 2^{0x70 \ \& \textit{Max resp code}}$ seconds.
- Checksum: Checksum is the 16-bit one's complement of the one's complement sum of the entire IGMP message. For computing the checksum, the *Checksum* field is set to zero.
- QRV (Querier's Robustness Variable): Determines how many times the router resends the messages. If the value is other than 0, the receiver must set its own robustness variable to the sent value. The robustness variable determines how many packet losses a connection can tolerate.
- QQIC (Querier's Query Interval Code): The *QQIC* code determines the maximum interval between resending of messages. Resending interval for a particular message is a random value chosen between 0 and this maximum. *QQIC* and *QRV* together determine the level of protection of the IGMP message sequence against packet losses.
- N (Number of sources): The value specifies how many source addresses are present in the message.
- Source Address [i]: The *Source Address [i]* fields form a vector containing *N* unicast IP addresses.

Membership Report Messages

Membership report messages are either sent in response to membership query messages from

Figure 6. (a) Membership query and (b) membership response message format

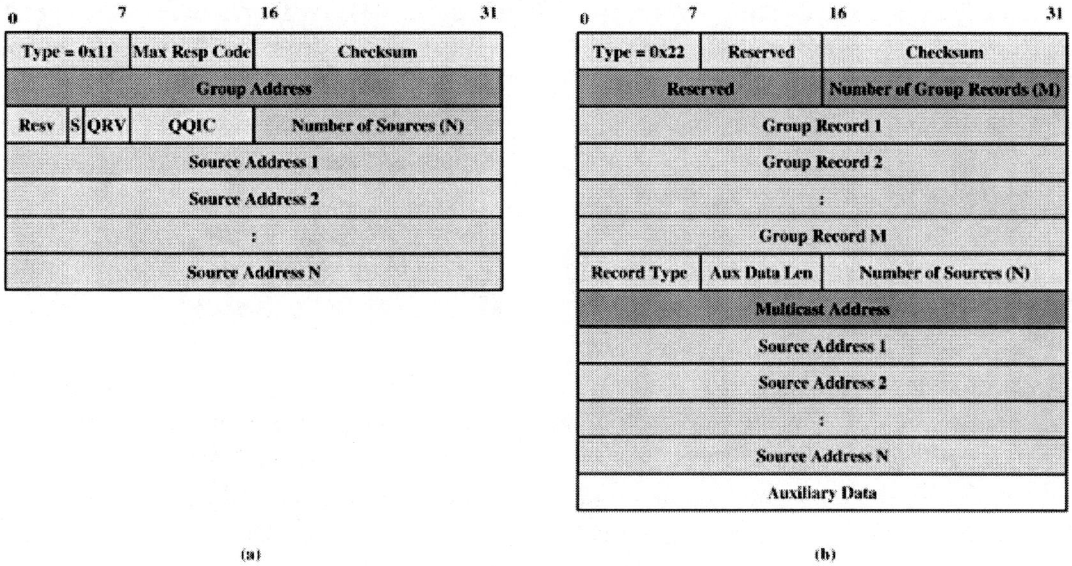

routers, or sent by the clients if their reception state concerning a multicast address changes (e.g. because of a television channel switch by the user). The IP header source address should be a valid address in the subnet from where the report message is sent.

The format of the membership report messages is depicted in Figure 6(b). The most important fields are described below. For a complete reference, please see Section 4.2 of (Cain et al., 2002).

- M (Number of group records): The value specifies how many group records the message contains.
- Group record: *Group record* is a block of data that carries information about the sender's relation to a particular multicast address.
- Record type: The record type indicates the type of the report message. The following record types are allowed:
    - Current-State Record is included in a membership report message as a response to a query. It describes the reception state of the sender with respect to the multicast address in the query. The Current-State record can specify that the client is enrolled or not enrolled to a multicast address, and in both cases, if the source addresses of the multicast packets should be forwarded or not by the router.
    - A Filter-Mode-Change Record is sent by a client if a change in its state with respect to a multicast address occurs, i.e. a client wants to receive packets previously not received, or it wants to terminate reception.
    - A Source-List-Change Record indicates that a change in the list of sources the client wants to hear about occurs. A change might mean an inclusion of new source addresses to the list and the exclusion of old addresses from the list.
- N (Number of sources): The value specifies how many source addresses are present in the message.
- Source Address [i]: The *Source Address [i]* fields form a vector containing *N* unicast IP addresses.
- Multicast address: The field specifies which multicast address the message is related to.

Once a router learnt about the reception states of all of its neighbors, it can decide, which multicast address needs to be forwarded to which client. This information is handed over to whatever routing protocol is used in the particular router to ensure that the required multicast packets are forwarded to the clients. All that the routers usually need to know is that at least one client on the subnet the router is responsible for wants to receive a particular multicast address.

Group Management in an IPv6 Environment: MLD Sent Over ICMPv6

MLD is the translation of IGMPv3 protocol for IPv6 environment. MLD is defined by (Vida & Costa, 2004). Its functionality and operation is more or less identical to IGMPv3 except that every address in the message is in IPv6 form. MLD request and response messages are transmitted in the payload part of ICMPv6 packets.

Message Sequence Charts for Unicast VOD and Multicast IPTV Services

Figure 7 gives an overview of how a Video-on-Demand service is set up and torn down through a handshaking of different RTSP request and response messages. While the session is alive and is in the PLAY state, RTP and corresponding RTCP packets are exchanged.

Figure 8 depicts the message sequence for a multicast IPTV session. Here no RTSP handshaking takes place. The client collects all information necessary to decode the stream from the RTP packets. IGMPv3 is used to join and leave the multicast group that delivers the particular television channel the user is interested in. The IGMP messages are sent two times just to ensure proper delivery.

FUTURE RESEARCH DIRECTION

The advances in transmission and signal processing technologies resulted in a viable form of television that can be easily distributed in IP environment. IPTV environment, though still lacks some of the benefits of cable distribution, is more easily expandable and with lower cost to the service provider. In the not so distant future, IPTV services will find their way to mobile devices. Furthermore, Internet Media Subsystem functionalities will appear in home appliances,

Figure 7. Message sequence chart for a VOD service and (b) a multicast IPTV service

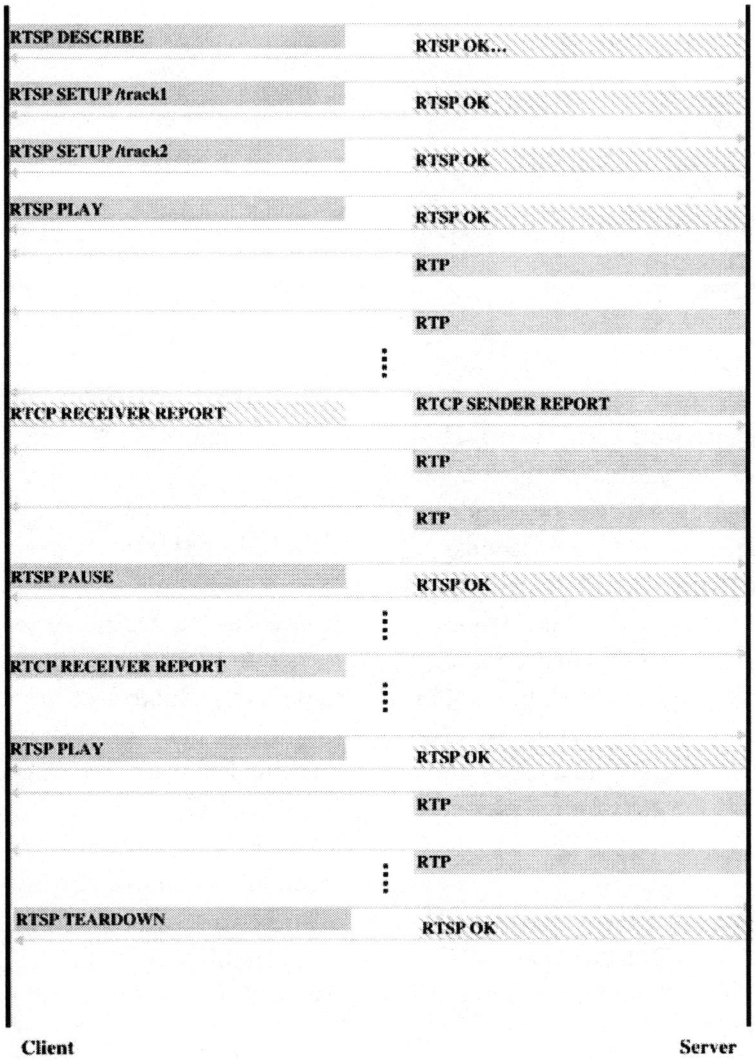

Figure 8. Message sequence chart for a multicast IPTV service

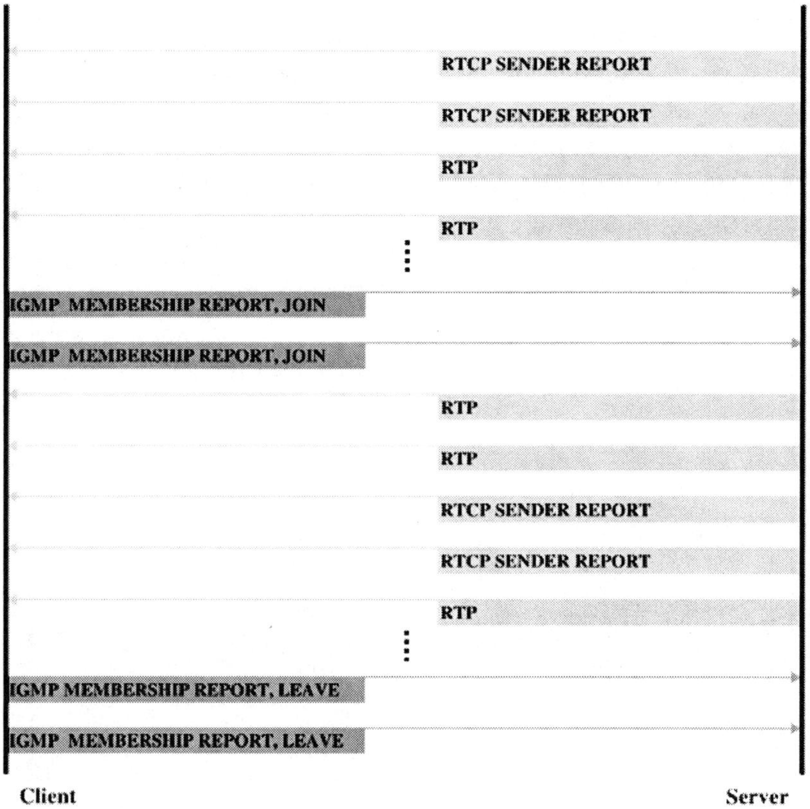

which will ensure seamless handover of IPTV content between home network components. This very subsystem will bring all communication channels (text messages, MMS, Internet content, emails) together and provide comprehensive content delivery anytime, anywhere. For this, DCCP and SCTP solutions are being investigated and are thought of as an extension to present day systems with new capabilities and integrated design. Current open source operating systems feature SCTP and DCCP implementations, thus they may form a basis for commercial products supporting congestion control, quality of service, and multihoming.

These trends were even recognized by the European Digital Video Broadcasting (DVB) project and standardization efforts are being made to converge to the ever-growing DVB world and IP technologies.

CONCLUSION

In this chapter we introduced several protocols related to IPTV applications, and described how these protocols are used in a multicast IPTV environment or in a VOD like service. The protocols presented here are adequate to set up a functional IPTV network. However, in a real life system with a lot of users situated at geographically distant locations, an interaction of these protocols and other, possibly proprietary protocols are needed to provide sufficient quality of service and to enhance the system with interactive services.

REFERENCES

Braden, R. (Ed.). (1989). *Requirements for Internet hosts – communication layers*. IETF. RFC 1122.

Cain, B., Deering, S., Kouvelas, I., Fenner, B., & Thyagarajan, A. (2002). *Internet group management protocol, version 3*. IETF. RFC 3376.

Handley, M., Jacobson, V., & Perkins, C. (2006). *SDP: Session description protocol*. IETF. RFC 4566.

Kim, S. T., Koh, S. J., & Kim, Y. J. (2007). *Performance of SCTP for IPTV applications*. 9th International Conference on Advanced Communication Technology, vol. 3. (pp. 2176-2180). Gangwon-Do, South Korea: IEEE.

Kohler, E., Handley, M., & Floyd, S. (2006). *Datagram congestion control protocol* (DCCP). IETF. RFC 4340.

Minoli, D. (2008). *IP multicast with applications to IPTV and DVB-H*. Hoboken, NJ: John Wiley & Sons. doi:10.1002/9780470260876

O'Driscoll, G. (2008). *Next generation IPTV services and technologies*. Hoboken, NJ: John Wiley & Sons.

Park, S., Haw, R., & Hong, C. S. (2010). *System architecture for IPTV seamless services in mobility*. International Conference on Consumer Electronics (IEEE-ICCE 2010) (pp. 59-60). Las Vegas, NV: IEEE.

Park, S., & Jeong, S.-H. (2009). Mobile IPTV – approaches, challenges, standards and QoS support. *IEEE Internet Computing Magazine*, *3*(13), 23–31. doi:10.1109/MIC.2009.65

Schulzrinne, H., Casner, S., Frederick, F., & Jacobson, V. (2003). *RTP: A transport protocol for real-time applications*. IETF. RFC 3550.

Schulzrinne, H., Rao, A., & Lanphier, R. (1998). *Real-time streaming protocol* (RTSP). IETF. RFC 2326.

Schulzrinne, H., Rao, A., Westerlund, M., Lanphier, R., & Stiemerling, M. (2010). *Real-time streaming protocol 2.0* (RTSP). IETF. (draft-ietf-mmusic-rfc2326bis-23).

Simpson, W., & Greenfield, H. (2009). *IPTV and Internet video: Expanding the reach of television broadcasting* (2nd ed.). Burlington, MA: Focal Press.

Stewart, R. (Ed.). (2007). *Stream control transmission protocol*. IETF. RFC 4960.

Vida, R. & Costa, L. (2004). *Multicast listener discovery version 2 (MLDv2) for IPv6*. IETF. RFC 3810.

Xu, C., Fallon, E., Qiao, Y., Muntean, G.-M., Li, X., & Hanley, A. (2009). *Analysis of real-time multimedia transmission over PR-SCTP with failover detection delay and reliability level differential*. International Conference on Communication Software and Networks (IEEE-ICCSN'09). (pp. 768-772). Macau: IEEE.

ADDITIONAL READING

Cai, L., Shen, X., & Mark, J. W. (2009). *Multimedia Services in Wireless Internet: Modeling and Analysis*. Hoboken, NJ: John Wiley & Sons. doi:10.1002/9780470747759

Perkins, C. (2003). *RTP: Audio and Video for the Internet*. Reading, MA: Addison-Wesley Professional.

Rüngeler, I. (2010). *SCTP - A Next Generation Transport Protocol: Evaluating, Improving and Extending the Protocol for Broader Deployment*. Saarbrücken, Germany: Südwestdeutscher Verlag für Hochschulschriften.

Simpson, W. (2008). Video Over IP, Second Edition: IPTV, Internet Video, H.264, P2P, Web TV, and Streaming: A Complete Guide to Understanding the Technology (Focal Press Media Technology Professional Series). Burlington, MA: Elsevier Inc.

Wuthnow, M., Shih, J., & Stafford, M. (2009). *IMS: a new model for blending applications*. Boca Raton, FL: Auerbach Publications.

KEY TERMS AND DEFINITIONS

Aggregate Control: In RTSP terminology, aggregate control means the common management of different (e.g. audio and video) tracks constituting the audiovisual content.

Layered Encoding: Layered encoding or hierarchical encoding is the act of producing a basic quality material along with several enhancement layers. Enhancement layers cannot be decoded by themselves; they can only be used to enhance the quality of the base layer and any corresponding lower layers. The advantage of layered encoding is that it provides for scaling to available capacity. Clients with lower link quality can access only the base layer, whereas clients with higher capacity link may access higher layers as well.

Multicast Group: A group of receivers simultaneously receiving content sent to the same multicast address.

Payload: Useful data content within a packet.

RTSP Session: A succession of client requests and server responses, which result in an established connection between the parties with resources allocated at both sides. An RTSP session is started by the successful completion of a SETUP request, destroyed by a TEARDOWN request or timeout, and identified by a unique identifier.

Session: A set of senders and receivers together with the data stream being transmitted between them.

Streaming Media: Audio or video content that is simultaneously transmitted and presented to the user. The term refers to the delivery method rather that the media through which the delivery occurs.

Unicast Transmission: The act of transmitting content to a unique receiver, identified by a network address that is unique within the network domain.

APPENDIX

Table of abbreviations

CC	CSRC Count
CSRC	Contributing Source Identifier
DCCP	Datagram Congestion Control Protocol
DLST	delay since the last SR
DSLAM	Digital Subscriber Line Aggregation Modules
DVB	Digital Video Broadcasting
ICMP	Internet Control Message Protocol
IETF	Internet Engineering Task Force
IGMP	Internet Group Management Protocol
LSR	Last SR timestamp
MLD	Multicast Listener Discovery
MMUSIC WG	Multiparty Multimedia Session Control Working Group
QQIC	Querier's Query Interval Code
QRV	Querier's Robustness Variable
RR	Receiver Report
RTCP	Real-time Control Protocol
RTP	Real-time Transport Protocol
RTP/AVP	RTP Audio Video Profile
RTP/SAVP	RTP Secure Audio Video Profile
RTSP	Real-time Streaming Protocol
SCTP	Stream Control Transport Protocol
SDES	Source Description
SDP	Session Description Protocol
SR	Sender Report
SSRC	Synchronization Source Identifier
TCP	Transmission Control Protocol
TTL	time-to-live
UDP	User Datagram Protocol
URI	Unified Resource Identifier
VOD	Video-on-Demand

Chapter 17
A Solution for Evaluating the QoS of Voice over IP:
Measurements, Analysis and Modeling

Homero Toral-Cruz
Center of Research and Advanced Studies, Mexico

Deni Torres-Román
Center of Research and Advanced Studies, Mexico

Leopoldo Estrada-Vargas
Center of Research and Advanced Studies, Mexico

ABSTRACT

In this chapter, we present a solution for evaluating the Quality of Services (QoS) of Voice over Internet Protocol (VoIP). First, we present an introduction to the main concepts and mathematical background relating to QoS and Internet Protocol (IP) traffic nature, which subsequently are used in the measurements, analysis, and modeling of VoIP traffic. Secondly, we analyze network measurements and the result of the simulation in order to characterize the VoIP traffic nature. As results of this analysis, it is shown that VoIP jitter can be modeled by alpha-stable distributions and self-similar processes, with either Short or Long Range Dependence (i.e., SRD or LRD). Thirdly, we investigate the packet loss effects on the VoIP jitter, and present a methodology for simulating packet loss on VoIP jitter. Finally, we found an empirical relationship between the Hurst parameter (H) and the Packet Loss Rate (PLR); this relationship is based on voice traffic measurements and can be modeled by means of a power-law function with three fitted parameters.

DOI: 10.4018/978-1-60960-732-6.ch017

INTRODUCTION

VoIP is one of the most attractive and important service nowadays in communication networks and it demands strict *QoS* levels and real-time voice packet delivery. The *QoS* level of VoIP applications depends on many parameters, such as bandwidth, One Way Delay (*OWD*), jitter, *PLR*, codec type, voice data length, and de-jitter buffer size. In particular, *OWD*, jitter, and *PLR* have an important impact. Therefore, the *QoS* can be seen as a function of these parameters, i.e. *Qos=f(OWD, Jitter, PLR*0.

In order to study the VoIP traffic nature and implement adequate traffic models, it is necessary, to monitor the main *QoS* parameters by means of active and passive measurements.

One of the models, which have been widely applied in classical teletraffic modeling, is the Poisson model. However, the IP networks traffic has different characteristics and the Poisson approximation will be acceptable only under particular conditions. Empirical studies showed that IP traffic exhibits self-similar (Lelan et al, 1994; Park & Willinger, 2000; Sheluhin et al, 2007) and has heavy-tail (Toral et al, 2009; Rizo et al, 2010) nature.

Additionally it has been shown that packet loss is bursty in nature and exhibits a finite temporal dependency (Yajnik et al, 1999), i.e., the probability that the current packet is lost depends on whether the past packets have been received or lost. Therefore, if packet n is lost then normally there is a higher probability that packet $n + 1$ will also be lost. Consequently, there is a strong correlation between consecutive packet losses, resulting in a bursty packet loss behavior. The most generalized model to capture this temporal dependency is a finite Markov chain (ITU-T Recommendation G.1050, 2005). The objective of packet loss modeling is to characterize its probabilistic behavior, because it is relevant for the design and performance analysis of VoIP applications.

Our studies presented in this chapter have revealed that VoIP jitter presents heavy tail and self-similar nature with either *SRD* or *LRD*. Through an extensive analysis, it is shown that VoIP jitter can be modeled by alpha-stable distribution and self-similar processes (Toral et al, 2009). These facts have some implications: (1) the *LRD* nature of the IP traffic has a negative impact on network performance, and (2) the heavy-tail nature has impact on the design of de-jitter buffer size.

Finally, we investigate the packet loss effects on the VoIP jitter, and present a methodology for simulating packet loss on VoIP jitter data traces. In order to generate the packet loss patterns, the *2*-state Markov model is used. We found an empirical relationship between H and *PLR*; this relationship is based on voice traffic measurements and can be modeled by means of a power-law function with three fitted parameters (Toral et al, 2009). Simulations result shows the effectiveness of our models in terms of the Mean Squared Error (*MSE*).

The proposed models can be used as input to the design and evaluation of the network, and the implementation of *QoS* mechanisms on VoIP applications.

BASIC CONCEPTS AND MATHEMATICAL BACKGROUND

This section introduces the basic concepts and mathematical background relating to *QoS*, and IP traffic nature. Initially, the *QoS* parameters and their relationships are described and, finally, a brief overview of self-similar processes and heavy tail distributions are presented.

QoS Parameters and their Relationships

The voice quality of VoIP applications depends on many parameters, such as bandwidth, *OWD*, jitter, packet loss rate, codec, voice data length, and de-jitter buffer size. In particular, *OWD*, jitter, and packet loss have an important impact on voice quality.

One-Way Delay

The delay experienced by a packet across a path consists of several components: propagation, processing, transmission, and queuing delays (Park, 2005). The Internet metric called *one-way delay* (ITU-T Recommendation G.114, 2003) is the time needed for a packet to traverse the network from a source to a destination host. It is described analytically by Equation (1):

$$D^{K}\left(L\right)_{OWD} = \delta + \sigma + \sum_{h=1}^{s}\left(\frac{L}{C_{h}} + X_{h}^{K}\left(t\right)\right) \quad (1)$$

where $D^{K}(L)_{OWD}$ is the *OWD* of a packet *K* of size *L*, δ represents the propagation delay, σ the processing delay, *s* the number of hops, L/C_h the transmission delay, and $X_{h}^{K}\left(t\right)$ the queuing delay of a packet *K* of size *L* at hop *h* (*h*=1,…,*s*) with capacity C_h. The *OWD* variation between two successive packets, *K* and *K*-1 is called *OWD jitter* and is given by the Equation (2):

$$J^{K}\left(L\right) = D^{K}\left(L\right)_{OWD} - D^{K-1}\left(L\right)_{OWD} \quad (2)$$

Jitter

When voice packets are transmitted from source to destination over IP networks, packets may experience variable delay, called *delay jitter*. The packet Inter-arrival Time (*IAT*) on the receiver side is not constant even if the packet Inter-departure Time (*IDT*) on the sender side is constant. As a result, packets arrive at the destination with varying delays (between packets) referred to as jitter. We measure and calculate the difference between arrival times of successive voice packets that arrive on the receiver side, according to RFC 3550 (Schulzrinne et al, 2003).

Let S_K denote the *transmission timestamp* for the packet *K* of size *L*, and R_K the *arrival time* for packet *K* of size *L*. Then for two packets *K* and *K*-1, $J^K(L)$ may be expressed as:

$$\begin{aligned} J^{K}\left(L\right) &= (R_{K} - S_{K}) - (R_{K-1} - S_{K-1}) \\ &= (R_{K} - R_{K-1}) - (S_{K} - S_{K-1}) \end{aligned} \quad (3)$$

$$IDT\left(K, K-1\right) = (S_{K} - S_{K-1}) \quad (4)$$

$$IAT\left(K, K-1\right) = (R_{K} - R_{K-1}) \quad (5)$$

where, *IDT*(*K*,*K*-1) is the Inter-departure Time (in our experiments, *IDT*= {10ms, 20ms, 40ms, and 60ms}) and *IAT*(*K*,*K*-1) is the Inter-arrival Time for the packets *K* and *K*-1. In the current context, *IAT*(*K*,*K*-1) is referred to as jitter. Therefore, the VoIP jitter between two successive packets, i.e., packets *K* and *K*-1, is:

$$IAT\left(K, K-1\right) = J^{K}\left(L\right) + IDT\left(K, K-1\right) \quad (6)$$

Packet Loss

There are two main transport protocols used in IP networks: UDP and TCP. While UDP protocol does not allow any recovery of transmission errors, TCP include an error recovery process. However, the voice transmission over TCP connections is not very realistic. This is due to the requirement for real-time operations in most voice related applications. As a result, the choice is limited to the use of UDP, which involves packet loss problems.

Amongst the different quality elements, packet loss is the main impairment, which makes the VoIP perceptually most different from the public switched telephone network. Packet loss can occur in the network or at the receiver side, for example, due to excessive network delay in case of network congestion.

Owing to the dynamic, time varying behavior of packet networks, packet loss can show a variety of distributions. The packet loss distribution most often studied in speech quality tests is random or Bernoulli-like packet loss. Uniform random loss here means independent loss, implying that the loss of a particular packet is independent of whether or not previous packets were lost. However, uniform random loss does not represent the loss distributions typically encountered in real networks. For example, losses are often related to periods of network congestion. Hence, losses may extend over several packets, showing a dependency between individual loss events. In this work, dependent packet loss is often referred to as bursty. The packet loss is bursty in nature and exhibits temporal dependency (Yajnik et al, 1999). Therefore, if packet *n* is lost then normally there is a higher probability that packet *n*+1 will also be lost. Consequently, there is a strong correlation between consecutive packet losses, resulting in a bursty packet loss behavior. A generalized model to capture temporal dependency is a finite Markov chain (ITU-T Recommendation G.1050, 2005).

In our approach, which we will present in a later Section, the *2*-state model is used to simulate packet loss patterns. In this model, one of the states (*state 1*) represents a packet loss and the other (*state 0*), represents the case where packets are correctly transmitted or received. The transition probabilities in this model, are represented by *p* and *q*. In other words, *p* is the probability of going from state 0 to state 1, and *q* is the probability of going from state 1 to state 0. Different values of *p* and *q* define different packet loss conditions that can occur on the Internet.

The loss distribution related to a *2*-state Markov chain can also be characterized by the number of packets, which are lost or received. For example, the probability that the transition from '1' (lost) to '0' (received) occurs after *n* steps can be expressed as:

$$P\left(n,q\right)=q\left(1-q\right)^{n-1}=q\cdot p^{n-1} \tag{7}$$

According to Equation (7), the number of steps *n* necessary to transit from 'lost' to 'received', that is, the number of consecutively lost packets, is a geometrically distributed random variable. This geometric distribution of consecutive loss events makes the *2*-state Markov chain (and higher order Markov chains) applicable to describing loss events observed in the Internet.

The average number of consecutively lost and received packets can be calculated by *b* and *g*, respectively, as shown in Equation (8).

$$b=\sum_{n=1}^{\infty}n\cdot q\left(1-q\right)^{n-1}=\frac{1}{q},\qquad g=\frac{1}{p} \tag{8}$$

The overall loss probability or *PLR* can be calculated according to Equation (9):

$$PLR=\frac{p}{p+q} \tag{9}$$

Now, if a given data trace is to be modeled using a *2*-state Markov chain, the transition probabilities *p* and *q* can be calculated by Equation (10).

$$p=\frac{q\cdot PLR}{1-PLR}=\frac{\sum_{i=1}^{\max(i)}m_i}{m_0}, \tag{10}$$

where m_i is the number of times *i* packets got lost consecutively, m_0 is the overall number of received packets, $\sum_{i=1}^{\max(i)} m_i$ is the number of consecutive loss events, and $\sum_{i=1}^{\max(i)} i \cdot m_i$ is the overall number of lost packets.

IP Traffic Nature

Many investigations of traffic measurements suggest that the IP traffic is self-similar (Leland et al, 1994; Park & Willinger, 2000; Sheluhin et al, 2007) and has heavy-tail distributions (Toral et al, 2009; Rizo et al, 2010). The fact that IP traffic exhibits self-similarity characteristics means that it is bursty at a wide range of time scales. The main characteristic of a random variable obeying a heavy-tailed distribution is that it exhibits extreme variability. Practically speaking, a heavy-tailed distribution gives rise to very large values with non-negligible probability.

Self-Similar Processes

Traffic processes are said to be self-similar if certain properties are preserved with respect to scaling in space or time. An attractive property of the self-similar processes for modeling a time series of IP traffic is the degree of self-similarity, which is expressed with a single parameter called the *Hurst parameter*. This parameter expresses the speed of decay of the autocorrelation function of the time series.

In order to estimate this parameter there are several methods, for details see (Sheluhin et al, 2007). In this section, a brief overview of self-similar processes is given, based on (Sheluhin et al, 2007; Park & Willinger, 2000; Leland et al, 1994).

The real-valued continuous time series $\{X(t), t \in \Re\}$ is self-similar with the exponent $0<H<1$ if, for any $a>0$ finite-dimensional distributions for $\{X(at), t \in \Re\}$ are identical to finite-dimensional distributions $\{a^H X(t), t \in \Re\}$; i.e., $\{X(at), t \in \Re\} \overset{d}{=} \{a^H X(t), t \in \Re\}$, where $\overset{d}{=}$ denotes equality in distribution.

Let $X_t=(X_t; t\in N)$ denote a discrete time series with mean μ_X, variance σ_X^2, autocorrelation function $r(k)$, and *Auto-covariance Function* (*ACV*) $\gamma(k)$, $k\geq 0$; where X_t can be interpreted as the VoIP jitter.

When considering discrete time series, the definition of self-similarity is given in terms of the aggregated processes, as following:

$$X_k^{(m)} = \left(X_k^{(m)}; k \in N\right) \tag{11}$$

where m represents the aggregation level and $X_k^{(m)}$ is obtained by averaging the original series X_t over non-overlapping blocks of size m, and each term $X_k^{(m)}$ is given by:

$$X_k^{(m)} = \frac{1}{m} \sum_{i=(k-1)m+1}^{km} X_i; \quad k = 1, 2, 3, \ldots \tag{12}$$

Then it is said that X_t is self-similar (*H-ss*) with self-similarity parameter ($0<H<1$) if:

$$X_k^{(m)} \overset{d}{=} m^{H-1} X_t \tag{13}$$

X_t is called *exactly second-order self-similar* with Hurst parameter H if the variance and *ACV* of the aggregated time series are defined by Equations (14) and (15), respectively:

$$\operatorname{var}\left(X_k^{(m)}\right) = \sigma_X^2 \cdot m^{2H-2} \tag{14}$$

$$\gamma_X^m(k) = \frac{\sigma_X^2}{2}\left((k+1)^{2H} - 2k^{2H} + (k-1)^{2H}\right) \quad k\geq 1 \tag{15}$$

The time series X_t is called *asymptotically second-order self-similar* if:

$$\lim_{m\to\infty} \gamma_X^m(k) = \frac{\sigma_X^2}{2}\left((k+1)^{2H} - 2k^{2H} + (k-1)^{2H}\right) \quad (16)$$

Second-order self-similarity (in the exact or asymptotic sense) has been a dominant framework for modeling IP traffic. So far role of second-order self-similarity has been discussed but not much has been mentioned about the role of H and limiting values. The definition of long-range dependence and its interconnection with the correlation factor $r(k)$ will now be discussed.

Let be $r(k) = \gamma(k)/\sigma_X^2$ the autocorrelation function of X_t with self-similarity parameter $0<H<1$, then the asymptotic behavior of $r(k)$, is given by Equation (17):

$$r(k) \sim H(2H-1)k^{2H-2} \; k\to\infty \quad (17)$$

In particular, if $\frac{1}{2} < H < 1$, $r(k)$ asymptotically behaves as $ck^{-\eta}$ for $0<\eta<1$, where $c>0$ is a constant, $\eta=2-2H$, and this also means that the correlations are non-summable: $\sum_{k=-\infty}^{\infty} r(k) = \infty$. That is, the autocorrelation function decays slowly. When $r(k)$ obeys a power-law, the corresponding stationary process X_t is called *long range dependent*. On the other hand, X_t is short range dependent if the sum $\sum_{k=-\infty}^{\infty} r(k) < \infty$ does not diverge.

Generally speaking, time series with long-range dependence have a Hurst parameter $0.5<H<1$, on the other hand, time series with short-range dependence have a Hurst parameter $0<H<0.5$.

Following are some simple facts regarding the value of H and its impact on $\gamma(k)$:

1. $\gamma(k) = \begin{cases} 1, & k=0 \\ 0, & k\neq 0 \end{cases}$ for H=0.5 This is the well-known property of white Gaussian noise.
2. $\gamma(1)<0$ for $0<H<0.5$.
3. $\gamma(1)>0$ for $0.5<H<1$.

Properties 2 and 3 are often termed anti-persistent and persistent correlations, respectively.

Heavy Tail Distributions

Distribution with 'Heavy Tails' (DHT): A random variable X has a 'heavy tail' distribution if:

$$P[X > x] \sim \frac{1}{x^{\alpha}}; x\to\infty;\ 0<\alpha<2 \quad (18)$$

where α is called *tail index*. Note that, heavy tail distribution decays slower than exponential function. It is known that a heavy-tail random variable (RV) has infinite variance. Besides, when $0<\alpha\leq 1$, its mean is infinite (Sheluhin et al, 2007). A very used class of the DHT is the family of α-stable distribution.

Alpha-stable Distribution: A RV X is said to have an alpha-stable distribution if there are parameters $0<\alpha\leq 2$, $\alpha\geq 2$, $-1\leq\beta\leq 1$, and $b \in \Re$, such that its characteristic function has the following form (Karasaridis & Hatzinakos, 2001):

$$\begin{aligned}\Phi(\omega) &= E\left[e^{j\omega X}\right] \\ &= \exp\left\{jb\omega - |a\omega|^{\alpha}\left[1 - j\beta\,\mathrm{sgn}(\omega)\theta(\omega,\alpha)\right]\right\}\end{aligned} \quad (19)$$

where

$$\theta(\omega,\alpha) = \begin{cases} \tan\left(\dfrac{\alpha\pi}{2}\right); & \alpha \neq 1 \\ -\dfrac{2}{\pi}\ln|\omega|; & \alpha = 1 \end{cases} \tag{20}$$

In Equation (19), α is the stability index; β, the skewness parameter; *a*, the scaling parameter, and *b*, the shift parameter.

ANALYSIS AND MODELING OF VOIP TRAFFIC

In this section, we analyzed the jitter and packet loss behavior of VoIP traffic by means of networks performance measurements and simulations results. As result of these analyses, we provide a detailed characterization and accurate modeling of these network performance metrics.

Measurements

In order to analysis the VoIP traffic characteristics and implement adequate traffic models, we monitored the performance metrics by means of active and passive measurements.

For active measurements, we generated VoIP traffic by establishing test calls with the VoIP application "Alliance foreign exchange station" (FXS) (Leyva, 2004).

For passive measurements, we realized the capture of VoIP traffic using Wireshark (Combs et al, 2010) to obtain a set of data traces. The subject of these measurements is to gather traffic patterns of voice packets such as jitter and packet loss for post-processing analysis.

The measurement scenario consists of two different local area networks (with different Internet service provider - ISP) interconnected by the Internet backbone:

- LAN "A" - Local Cable ISP network (Link Speed-3MB).
- LAN "B" - CINVESTAV GDL network (Link Speed-2MB).

The measurement scenario is based on a typical H.323 architecture, composed by two zones interconnected via Internet, as shown in Figure 1. Each zone consists of a single H.323 GK (Willamowius et al, 2009) which acts as the administrator of the zone, and a set of H.323 terminal endpoints (TEs), interconnected via a LAN. The zone "A" is composed by the endpoints A1, A2,

Figure 1. Measurement scenario

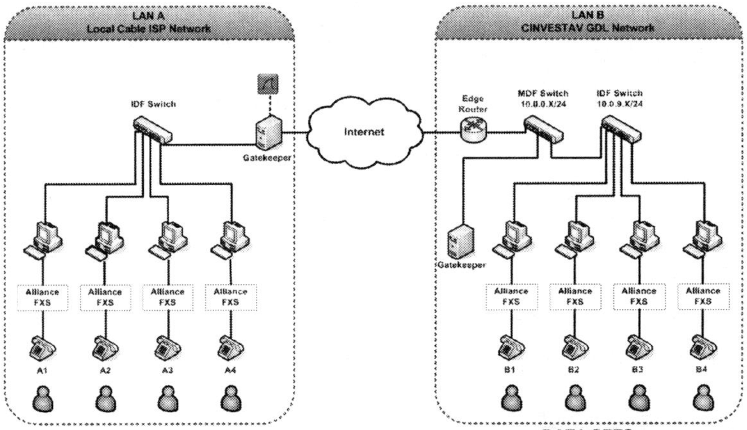

DATA SETS

	A1/B1	A2/B2	A3/B3	A4/B4
SET 1	G.711-10ms	G.711-20ms	G.711-40ms	G.711-60ms
SET 2	G.729-10ms	G.729-20ms	G.729-40ms	G.729-60ms
SET 3	G.711-10ms	G.711-20ms	G.729-10ms	G.729-20ms
SET 4	G.711-40ms	G.711-60ms	G.729-40ms	G.729-60ms

A3, A4, and is administrated by the GK "A". In this zone, the network protocol analyzer Wireshark was installed for collecting the data traces. The zone "B" is composed of the endpoints B1, B2, B3, B4, and the GK "B".

In order to establish the test calls between any endpoints, each endpoint has installed an Alliance FXS peripheral component interconnect (PCI) card and a conventional cord phone. The configuration parameters employed in the test calls are shown in Figure 1.

The measurements corresponding to the data sets described in Figure 1 were collected in the following way:

- Four simultaneous test calls were established between A1/B1, A2/B2, A3/B3, and A4/B4 endpoints.
- The configurations used in the test calls are based on two parameters, codec type (G.711 and G.729) and voice data length (10ms, 20ms, 40ms, and 60ms).
- The measurement periods were 60 minutes for each test call (call duration time).
- For each measurement period (an hour), four jitter and packet loss data traces were obtained.
- The four data sets contain 16.56 millions of RTP packets corresponding to 96 jitter and 96 packet loss data traces, measures in typical working hours.
- Set 1: Friday 09/07/2007 from 10:00 hours until 16:00 hours.
- Set 2: Monday 09/10/2007 from 10:00 hours until 16:00 hours.
- Set 3: Tuesday 09/11/2007 from 10:00 hours until 16:00 hours.
- Set 4: Wednesday 09/12/2007 from 10:00 hours until 16:00 hours.

Our study is based on various data sets that provide valuable information into the behavior of test calls in IP networks. We selected these four data sets, because are representative of packet loss diversity present in all collected data traces.

Packet Loss Effects on the VoIP Jitter

In this work, we used the voice data lengths of 10ms, 20ms, 40ms, and 60ms, and the successive voice packets are transmitted at a constant rate, i.e., 1 packet/10ms, 1 packet/20ms, 1 packet/40ms, and 1 packet/60ms, respectively (see Figure 1). However, when voice packets are transported over IP networks, they may experience delay variations and packet loss.

On the other hand, these impairments are intricately related to each other. These relationships can be used to reveal the overall effects of these impairments and to implement accurate models in order to evaluate the QoS of VoIP applications.

Based on the above concerns, we investigated the packet loss effects on the VoIP jitter, and presented an analytical expression that describes the relationship between these two performance metrics.

Let $IAT(K,K\text{-}1)=J^K(L)+IDT(K,K\text{-}1)$ denote the *jitter* for two packets K and K-1, then, if the packet K-1 is lost,

$$IAT\left(K,K-2\right)=J^{K}\left(L\right)+\left(2\right)IDT \quad (21)$$

Therefore, if n consecutive packets are lost,

$$IAT\left(K,K-n-1\right)=J^{K}\left(L\right)+\left(n+1\right)\left(IDT\right) \quad (22)$$

This behavior is illustrated in Figure 2, where a packet length of 60 ms for voice data is used. In this figure can be observed that: (a) if there is not packet lost, the mean values of the jitter are around of 60 ms, i.e. $IAT(K,K\text{-}1)=J^K(L)+60$; (b) if one packet is lost, the mean values of the jitter are around of 120 ms, i.e. $IAT(K,K\text{-}2)=J^K(L)+120$; (c) if two consecutive packets are lost, the mean

Figure 2. Packet loss effects on VoIP jitter

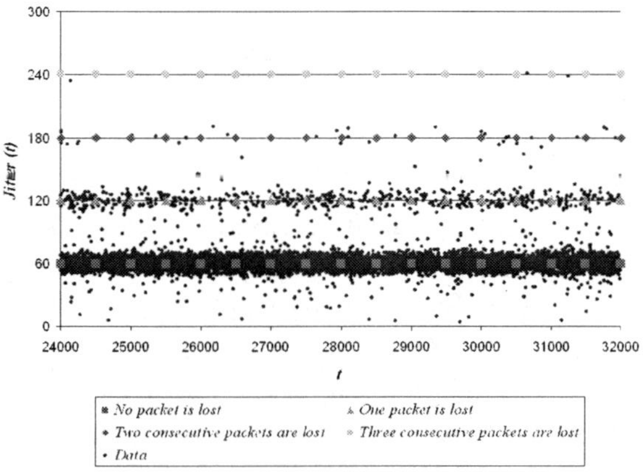

Figure 3. Hurst parameter for VoIP jitter data traces

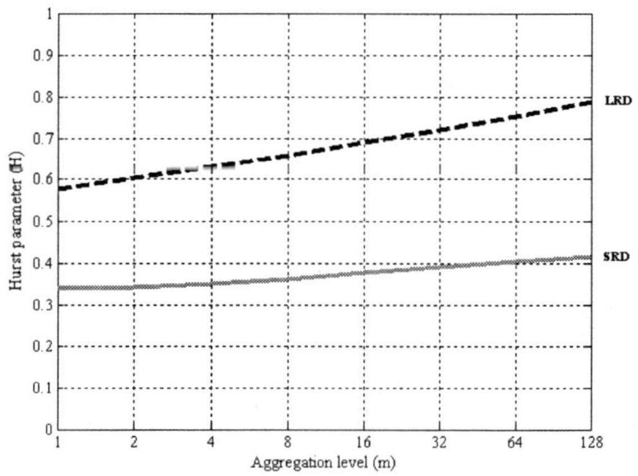

values of the jitter are around of 180 ms, i.e. $IAT(K,K\text{-}3)=J^K(L)+180$; and (d) if three consecutive packets are lost, the mean values of the jitter are around of 240 ms, i.e. $IAT(K,K\text{-}3)=J^K(L)+240$.

Therefore, the Equation (22) describes the packet loss effects on the VoIP jitter. The relationship expressed by this equation can be used for simulating packet loss on VoIP jitter data traces and to observe the impact of this impairment when the packet loss occur for different levels of burstiness.

Jitter Modeling

Jitter Modeling by Self-Similar Processes

In this section, we realized a self-similar analysis on VoIP traffic. Several statistical methods are used for testing and estimation of the degree of self-similarity from the time series. The Hurst parameter was estimated at seven aggregation levels, ($m=\{2,4,8,16,32,62,128\}$) by means of six methods implemented in SelQoS (Ramírez & Torres, 2005) and a wavelet based method (Estrada et al, 2009). They are: Rescaled Adjusted Range (R/S), Absolute Moment (AM), Variance Method (VAR), Modified Allan Variance (MAVAR), Periodogram (PER (10%)), Local Whittle (WHI), and a method based on Haar wavelet (HWAV). Figure 3 shows the Hurst Parameter of a representative set of VoIP jitter data traces to different aggregation levels m.

We observe the phenomenon of scale invariance at different time scales when the H Parameter took values of $0.35<H<0.41$, and $0.58<H<0.8$, these results indicate that the analyzed VoIP jitter data traces have self-similar characteristics with *SRD* and *LRD*, respectively. In this figure, each point of each curve represents the Hurst parameter of aggregation level m for a particular jitter data trace.

In order to complement the above discussion, we study the behavior of the *ACV* of representatives VoIP jitter data traces with short and long-range dependence.

Figure 4 shows the comparison between the *ACV* of a measured data trace with H=0.4223 and the theoretical $\gamma(k)$ function defined by Equation (15). It can be observed that the *ACV* of the measured data trace behaves similarly to the ideal model and decays quickly to zero. This result confirms the presence of short-range dependence or memoryless property in VoIP jitter data traces. In addition, we compared the *ACV* of a measured data trace with H=0.7607 and the theoretical $\gamma(k)$ function, as shown in Figure 5, in this figure it is observed a similar behavior between these *ACV* and a very slow decaying from them. This behavior indicates the presence of a high degree of long-range dependence or long memory property.

These results show that VoIP jitter can be modeled by self-similar processes with short and long-range dependence.

Jitter Modeling by Alpha-Stable Distributions

Interesting results were obtained in the analysis of VoIP jitter distribution. We found that there is evidence of heavy-tail behavior on VoIP jitter distribution.

Through percentile-percentile plot (*P-P Plot*), it is evaluated how well a particular distribution describes the empirical data traces. For each value that appears in the data trace, the *P-P Plot* shows the actual rank of the value versus the expected rank of the value in the model distribution. Perfect agreement with the model distribution yields the 45° line from the origin.

The distribution of VoIP jitter data traces is compared with alpha-stable, Laplace and t-Student distributions. The alpha-stable parameters have been computed by means of Nolan's code (Nolan, 2006). Figure 6 shows the *P-P Plot* between empirical distributions of VoIP jitter data traces and three theoretical distributions; in this figure, we can see that alpha-stable distribution gives the best goodness of fit.

Figure 4. Auto-covariance function for VoIP jitter data traces: SRD

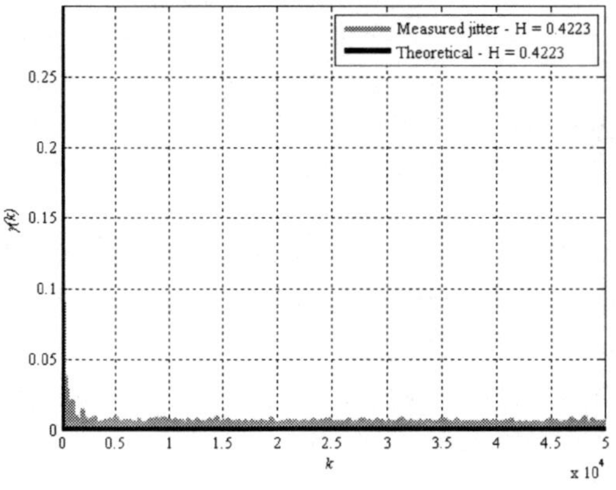

Figure 5. Auto-covariance function for VoIP jitter data traces: LRD

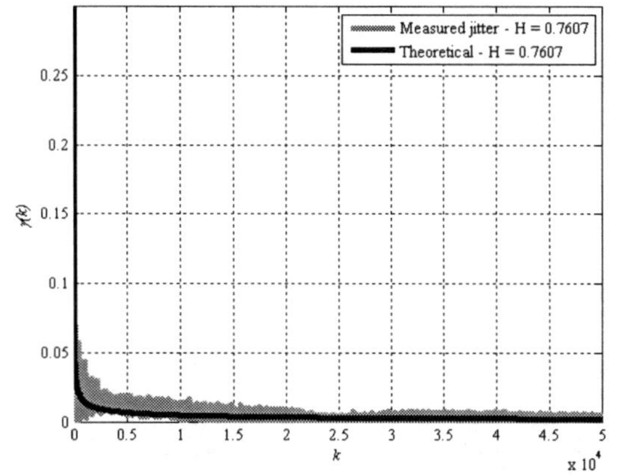

The differences between the empirical and theoretical distributions are measured in terms of *MSE*. The alpha-stable model achieves MSE=2.02·10^{-4}, more than 4 times better than Laplace model (MSE=9.72·10^{-4}) and more than twenty two times better than t-Student model (MSE=4.46·10^{-3}). This analysis shows that alpha-stable model is the most suitable to approximate the VoIP jitter, which means that extremely large values of VoIP jitter occur with non-negligible probability. This is described analytically by Equation (23):

$$P\left[Jitter > x\right] \sim \frac{1}{x^{\alpha}};x\rightarrow\infty;\ 0<\boldsymbol{\alpha}<2 \qquad (23)$$

In order to transfer voice requiring real-time delivery over a packet network, an important design parameter, is the de-jitter buffer size, since it influences the packet loss probability and *OWD*. The de-jitter buffer size is function of the maximum amount of time a packet spends in the

Figure 6. P-P Plot for empirical distribution of VoIP jitter traces compared with alpha-stable, Laplace, and t-Student distributions

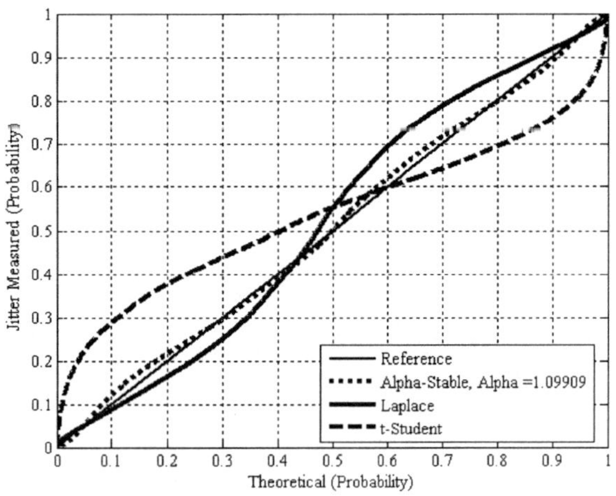

de-jitter buffer before being played out. On the other hand, heavy-tailed behavior on VoIP jitter has implications on the design of de-jitter buffer size:

- If it is too small, as the probability of extremely large values occurrence is non-negligible, then many packets would miss the play out deadline, and thereby increasing the packet loss probability.
- If it is too large, then the *OWD* would increase.

Therefore, there is a trade-off between packet loss and *OWD* when it is designed the de-jitter buffer size and it is important to consider the heavy-tailed behavior of VoIP jitter, as it is expressed in Equation (23).

Packet Loss Modeling

Packet Loss Model Framework

In this work, we used a description of VoIP packet loss based on microscopic and macroscopic packet loss behaviors (Raake, 2006):

- The microscopic packet loss behavior refers to a type of degradation that is clearly perceived as time varying, without leading to time-varying quality, i.e., without considerable level changes in instantaneous quality, during a certain "time window" around particular loss events.
- The macroscopic packet loss behavior refers to time-varying distortions where the quality perceived instantaneously is time varying over longer periods of time. Accordingly, macroscopic refers to a packet loss behavior, which is associated with clearly recognized changes in speech quality.

In our studies, we analyzed packet loss patterns or packet loss data traces with different packet loss levels over particular periods or "time windows". Each "time window" can be represented by a period of microscopic packet loss and all overall periods are represented by a macroscopic packet loss period.

The microscopic packet loss behavior can be modeled by Markov chain with a low number of states. Correspondingly, the macroscopic packet loss behavior can be modeled by Markov chains with a higher number of states. Here, substates represent phases of a given microscopic loss behavior.

Ideally, an *n*-state Markov chain is required in order to capture the macroscopic packet loss behavior. Especially for network planning, a trade off is desirable between very accurate modeling of data traces and a low number of model input parameters, in order to yield a model still usable for network planners with reasonable effort.

Figure 6 shows some packet loss patterns extracted from VoIP test calls. The packet loss pattern is represented as a binary sequence, when a packet is lost takes the value "1" and "0" when a packet is received.

In Figure 7 we can see that packet loss behavior is uniform random, i.e., the packet loss pattern is represented by a microscopic packet loss period. On the other hand, in Figure 8 the packet loss is bursty, i.e., the packet loss pattern is represented by a concatenation of microscopic packet loss periods. In order to simplify this packet loss description, the microscopic packet loss periods can be classified in two sets, one for lower, and one for higher packet loss rates.

In Figure 6, the microscopic packet loss period with lower packet loss rate are delimited by the dotted square, while the microscopic packet loss period with higher packet loss rate are delimited by the continuous square.

Methodology for Simulating Packet Loss

In this subsection, we define a methodology for simulating packet loss on VoIP jitter data traces. In order to simplify this methodology and achieve the trade off between very accurate modeling of data traces and a low number of model input parameters, we consider the following:

- The relationship between packet loss rate and jitter, which is summarized by Equation (22).
- The packet loss description of VoIP based on microscopic and macroscopic packet loss behaviors.

Figure 7. Packet loss patterns from VoIP test calls: Uniform random PLR

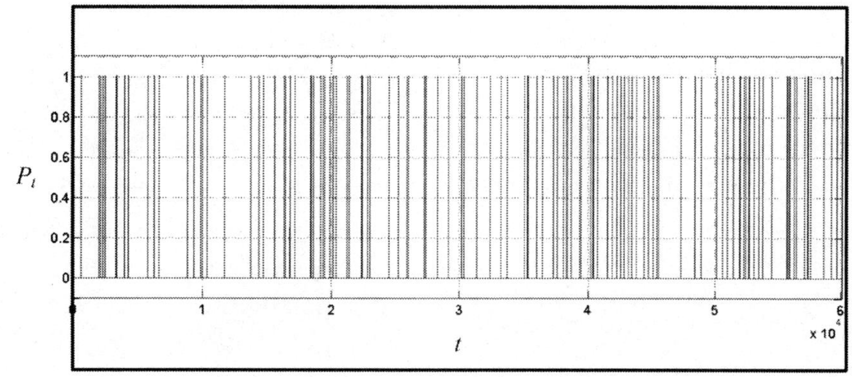

Figure 8. Packet loss patterns from VoIP test calls: Bursty PLR

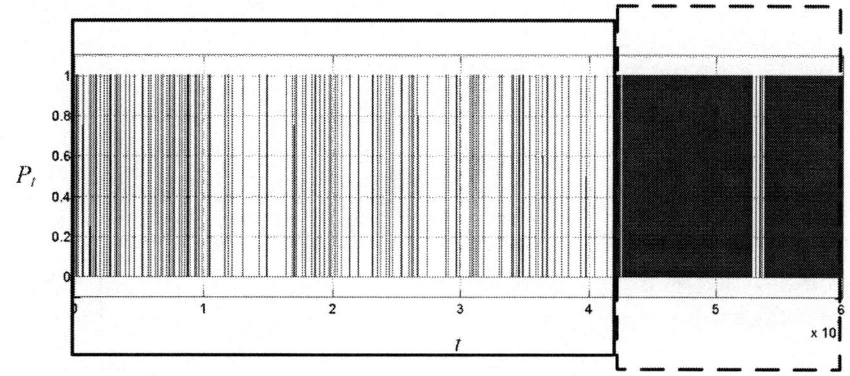

- A set of real jitter and packet loss data traces are used. The jitter data traces exhibit self-similar characteristics and the behavior of packet loss data traces is uniform random, as shown in Figure 7. Therefore, these packet loss patterns are represented for only one microscopic packet loss period with low level of *PLR*.
- In order to incorporate microscopic packet loss periods with high level of *PLR*, as it is shown in Figure 8 we generated packet loss patterns with different levels of *PLR* over a "time window". The packet loss patterns are generated by means of the *2*-state Markov model. So the model input parameters are the "time window" size and the transitions probabilities: p and q.

Let $X=\{X_t;t=1,\ldots,N\}$ be a VoIP jitter data trace with length N, self-similar (H parameter $0<H_0<0.5$), and with a packet loss rate PLR_0.

In order to represent the packet loss pattern, the *2*-state Markov model is used. The packet loss pattern is represented as a binary sequence $P=\{P_t;t=1,\ldots,wN\}$, where $P_t=1$ means a packet loss, $P_t=0$ means a success or received packet, and $w=0.1,0.2,\ldots,1$, is the burstiness level. In this model, different values of p and q define different packet loss patterns.

Since, the packet losses occur in bursts on the Internet, we applied "*T*" different packet loss patterns over a time window W_l^u of size wN (see Equation 24) on the jitter data trace X_t to simulate packet loss with various burstiness levels.

$$W_l^u = \left\{X_l, X_{l+1}, \ldots, X_u : l = 1, 2, \ldots, N - \lfloor wN \rfloor + 1, \right. \\ \left. u = l + \lfloor wN \rfloor - 1,\ l < u \right\} \tag{24}$$

where X_l and $X=u$ are the *l-th* and *u-th* element of time series (X_t) and represent, respectively, the beginning and the ending of the window.

The relationship between jitter and packet loss from Equation (22) is used to apply the packet loss patterns to X_t by means of the algorithm shown in Table 1.

By means of the above algorithm the new time series $\hat{X}^\tau$ are obtained, for $\tau=0,1,2,\ldots,T-1$. For each $\hat{X}^\tau$ the *PLR* and *H* parameter were calculated, and the function $f_w(PLR_\tau,H_\tau)$ was generated.

Simulation Results

In this section, applying the methodology proposed for simulating packet loss, simulation results are presented. The simulations are accomplished over VoIP jitter data traces corresponding to collected data sets. Figure 9 illustrates the relationships between packet loss rate and Hurst parameter.

This figure shows the empirical functions $f_w(PLR_\tau,H_\tau)$ that are obtained by simulations results and the function $f_{REAL}(PLR_\varepsilon,H_\varepsilon)$ that is generated by "E" VoIP jitter data traces. The functions $f_w(PLR_\tau,H_\tau)$, result from applying "*T*" packet loss patterns to representative VoIP jitter data traces X_t over a "time window" of size wN; in these functions, each point represents the correspondent parameters PLR_τ and H_τ of a new time series $\hat{X}^\tau$. On the other hand, in the function $f_{REAL}(PLR_\varepsilon,H_\varepsilon)$, each point represents the correspondent param-

Table 1. Algorithm for simulating packet loss: (a) Generating packet loss pattern, and (b) Applying packet loss pattern

(a)	(b)
FOR n=1 *to* l-1	FOR n=2 *to* N
$P[n]$=0	IF ($P[n]$=1)
END FOR	$X[n]$-$X[n]$+$X[n$-1]
FOR n=1 *to* u	END IF
IF (packet was lost)	END FOR
$P[n]$=1	i=1
ELSE	FOR n=2 *to* N
$P[n]$=0	IF ($P[n]\neq 1$)
END IF	
END FOR	$\hat{X}[i] = X[n-1]$ i=i+1
FOR n=u+1 *to* N	END IF
$P[n]$=0	END FOR
END FOR	

Figure 9. Relationship between PLR and H parameter: $f_w(PLR_\tau,H_\tau)$ vs. $f_{REAL}(PLR_\varepsilon,H_\varepsilon)$

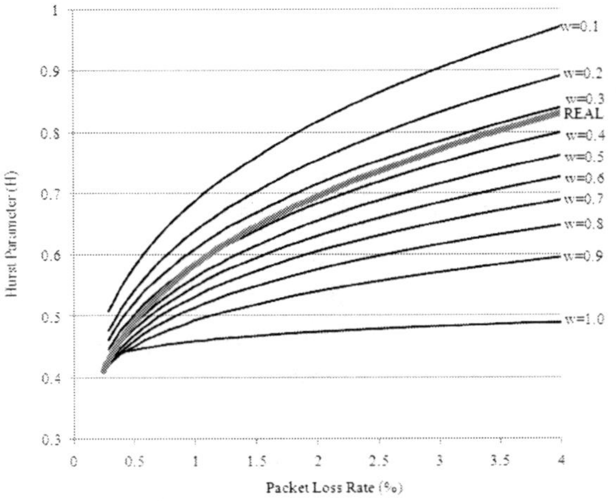

eters PLR_ε and H_ε of a jitter data trace X^ε, where ε=1,2,…,E, and "E" is the total number of jitter data traces collected.

The respective differences between the functions corresponding to simulation results $f_w(PLR_\tau,H_\tau)$ and the function $f_{REAL}(PLR_\varepsilon,H_\varepsilon)$, were quantified in terms of *MSE*.

Table 2 shows the fitted parameters and *MSE* between $f_{REAL}(PLR_\varepsilon,H_\varepsilon)$ and $f_w(PLR_\tau,H_\tau)$.

The results are summarized in Figure 9 and Table 2, where it is shown that a power-law model is suitable for describe the relationships between Hurst parameter and packet loss rate. In addition, it is shown that the function $f_{0.4}(PLR_\tau,H_\tau)$ provides the best match to the function $f_{REAL}(PLR_\varepsilon,H_\varepsilon)$. As result of this analysis we conclude that the presented methodology for simulating packet loss with various burstiness levels, achieved the trade off between a very accurate modeling and a low number of model input parameters (*p*,*q* and *wN*). Based on these results we proposed that the relationships found between *PLR* and *H* parameter can be well modeled by a power-law function with three fitted parameters ($\hat{H}_0$, $\hat{a}$, and $\hat{b}$).

Proposed Model

From our simulations, we found an empirical relationship between the Hurst parameter (*H*) and the Packet Loss Rate (*PLR*); this relationship is

Table 2. Fitted parameters

$f_w(PLR_\tau,H_\tau)$	$\hat{H}_0$	$\hat{a}$	$\hat{b}$	*MSE*
w=0.1	0.042	0.645	0.262	0.0144
w=0.2	0.042	0.596	0.254	0.0032
w=0.3	0.042	0.567	0.245	0.0004
w=0.4	0.042	0.543	0.239	0.0003
w=0.5	0.042	0.521	0.232	0.0019
w=0.6	0.042	0.506	0.216	0.0047
w=0.7	0.042	0.490	0.200	0.0091
w=0.8	0.042	0.472	0.178	0.0155
w=0.9	0.173	0.319	0.201	0.0261
w=1	0.042	0.416	0.049	0.0540
$f_{REAL}(PLR_\varepsilon,H_\varepsilon)$	0.042	0.541	0.271	

based on voice traffic measurements and can be modeled by means of a power-law function with three fitted parameters, as following:

$$H_M = \hat{H}_0 + \hat{a}\left(PLR\right)^{\hat{b}} \quad (25)$$

where H_M is the H parameter of the found model; $\hat{H}_0$, $\hat{a}$, and $\hat{b}$ are the fitted parameters; $\hat{H}_0$ is the H parameter, when PLR=0. The fitted parameters are estimated by linear regression. The strategy to find, $\hat{H}_0$, $\hat{a}$, and $\hat{b}$ is such that it minimizes the mean squared error, i.e. $MSE = \int_r (\hat{H}_0 + \hat{a} \cdot r^{\hat{b}} - H)^2 dr$, where r=PLR and H is the Hurst parameter of the data trace.

FUTURE RESEARCH DIRECTIONS

In this section, we present the future research directions, summarized as following:

1. The methodology presented in this work, for simulating packet loss on VoIP jitter data traces is restricted to incorporate only one microscopic packet loss period with different burst levels. In order to generate this microscopic loss period, we used the *2*-state Markov chain. Therefore, to generalize this methodology, we proposed to incorporate "*n*" microscopic packet loss periods by means of *4*-state Markov chain.
2. In the *4*-state Markov chain a "good" and "bad" state are distinguished and represent phases of lower and higher packet loss respectively. Both for the 'bad' and the 'good' state, an individual *2*-state Markov chain represents the dependency between consecutively lost or received packets. Therefore, it is necessary to find the accurate transitions probability between the *2*-state Markov chains, in order to incorporate representative microscopic loss periods.

CONCLUSION

Several parameters influencing voice quality in IP networks, particularly delays and packet loss have an important impact. Therefore, we analyzed the jitter and packet loss behavior of VoIP traffic by means of network performance measurements and simulations results, in order to provide a detailed characterization and accurate modeling of these main network performance metrics.

As result of these analyses, we proposed that VoIP jitter could be properly modeled by means of alpha-stable distributions and self-similar processes with either *SRD* or *LRD*.

Besides, we have presented a methodology for simulating packet loss on self-similar VoIP jitter. This methodology is based on:

- The relationship between *PLR* and jitter, which is summarized by Equation (22).
- The self-similar behavior of VoIP jitter data traces.
- The packet loss description of VoIP based on microscopic and macroscopic packet loss behavior.
- The microscopic loss behavior may be represented by a simple *2*-state Markov model.

We proposed a new model that allows relating the H parameter with the *PLR* by means of a power-law function with three fitted parameters, summarized by Equation (25). Our results show that the presented methodology for simulating packet loss with various burstiness levels, achieved the trade off between a very accurate modeling and a low number of model input parameters (p,q and wN). Simulation results show the effectiveness of our models in terms of *MSE*.

The proposed characterizations and models between main performance metrics can be used by other researches as input to the design, evaluation of QoS, and implementation of QoS mechanisms of VoIP applications.

REFERENCES

Camarillo, G. (2002). *SIP demystified*. USA: McGraw-Hill Companies, Inc.

Combs, G., et al. (2010). *Wireshark: A network protocol analyzer*. Retrieved from http://www.wireshark.org/

Estrada, L., Torres, D., & Toral, H. (2009). *Variance error for finite-length self-similar time series*. Paper presented at the 7th International Conference on Computing, Communications and Control Technologies (CCCT) - In the Context of The 2nd International Multi-Conference on Engineering and Technological Innovation (IMETI), Orlando, Florida, USA.

ITU-T Recommendation G.114. (2003). *One-way transmission time*. Geneva, Switzerland: International Telecommunications Union.

ITU-T Recommendation G.1050. (2005). *Network model for evaluating multimedia transmission performance over Internet protocol*. Geneva, Switzerland: International Telecommunications Union.

Karasaridis, A., & Hatzinakos, D. (2001). Network heavy traffic modeling using α-stable self-similar processes. *IEEE Transactions on Communications*, *49*(7), 1203–1214. doi:10.1109/26.935161

Leland, W. E., Taqqu, M. S., Willinger, W., & Wilson, D. V. (1994). On the self-similar nature of Ethernet traffic (extended version). *IEEE/ACM Transactions on Networking*, *2*(1), 1–15. doi:10.1109/90.282603

Leyva, L. (2004). *Alliance FXO/FXS/E1 VoIP system*. Retrieved from http://www.cts-design.com

Nolan, J. P. (2006). *Stable MathLink package*. Retrieved from www.robustanalysis.com

Park, K., & Willinger, W. (2000). *Self-similar network traffic and performance evaluation*. New York, NY: John Wiley & Sons, Inc. doi:10.1002/047120644X

Park, K. I. (2005). *QoS in packet networks*. Boston, MA: Springer Science + Business Media, Inc.

Raake, A. (2006). Short- and long-term packet loss behavior: Towards speech quality prediction for arbitrary loss distributions. *IEEE Transactions on Audio, Speech, and Language Processing*, *14*(6), 957–1968. doi:10.1109/TASL.2006.883231

Ramírez, J., & Torres, D. (2005). *A tool for analysis of Internet metrics*. Paper presented at the 2nd International Conference on Electrical and Electronics Engineering (ICEEE) and XI Conference on Electrical Engineering (CIE), Mexico, D.F.

Rizo, L., Torres, D., Muñoz, D., & Vargas, C. (2010). Jitter in IP networks: A cauchy approach. *IEEE Communications Letters*, *14*(2), 190–193. doi:10.1109/LCOMM.2010.02.090702

Schulzrinne, H. (2003). *RTP: A transport protocol for real-time applications (RFC 3550)*. Internet Engineering Task Force.

Sheluhin, O. I., Smolskiy, S. M., & Osin, A. V. (2007). *Self-similar processes in telecommunications*. Chichester, England: John Wiley & Sons Ltd.doi:10.1002/9780470062098

Toral, H., Torres, D., & Estrada, L. (2009). *Simulation and modeling of packet loss on α-Stable VoIP traffic*. Paper presented at the 9th International Conference on Multimedia, Internet & Video Technologies (WSEAS MIV), Budapest, Hungary.

Toral, H., Torres, D., & Estrada, L. (2009). Simulation and modeling of packet loss on VoIP traffic: A power-law model. *WSEAS Transactions on Communications*, *8*(10), 1053–1063.

Toral, H., Torres, D., Hernandez, C., & Estrada, L. (2008). *Self-similarity, packet loss, jitter, and packet size: Empirical relationships for VoIP*. Paper presented at the 18th International Conference on Electronics, Communications, and Computers (IEEE CONIELECOM), Puebla, Mexico.

Willamowius, J., et al. (2009). *OpenH323 gatekeeper: The GNU gatekeeper*. Retrieved from http://www.gnugk.org/

Yajnik, M., Moon, S., Kursoe, J., & Towsley, D. (1999). *Measurement and modelling of the temporal dependence in packet loss*. Paper presented at the 18th International Conference on Computer Communications (IEEE INFOCOM), New York, NY.

ADDITIONAL READING

Chen, X., Wang, C., Xuan, D., Li, Z., Min, Y., & Zhao, W. (2003). Survey on QoS Management of VoIP. Proceedings of the International Conference on Computer Networks and Mobile Computing. (pp. 69-77). Washington, DC: IEEE Computer Society.

Davidson, J., Peters, J., Bhatia, M., Kalidindi, S., & Mukherjee, S. (2007). *Voice over IP Fundamentals*. Indianapolis, IN: Cisco Systems, Inc.

Gao, J., Cao, Y., Tung, W.-W., & Hu, J. (2007). *Multiscale Analysis of Complex Time Series: Integration of Chaos and Random Fractal Theory, and Beyond*. Hoboken, NJ: Wiley-Interscience.

ITU-T Recommendation G.711, (1993). Pulse Code Modulation (PCM) of Voice Frequencies. International Telecommunications Union, Geneva, Switzerland.

ITU-T Recommendation G.108, (1999). Application of the E-model: A planning guide. International Telecommunications Union, Geneva, Switzerland.

ITU-T Recommendation G.729, (2008). Coding of speech at 8 kbit/s using Conjugate-Structure Algebraic-Code-Excited Linear-Prediction (CS-ACELP). International Telecommunications Union, Geneva, Switzerland.

ITU-T Recommendation G.107, (2009). The E-model, a computational model for use in transmission planning. International Telecommunications Union, Geneva, Switzerland.

ITU-T Recommendation P.800, (1996). Methods for subjective determination of transmission quality. International Telecommunications Union, Geneva, Switzerland.

Keagy, S. (2002). *Integrating Voice and Data Networks*. Indianapolis, IN: Cisco Systems, Inc.

Law, D., (2002). Local Area Network protocols (IEEE 802.3). IEEE 802.3 Ethernet Working Group.

Postel, J. (1980). *UDP: User Datagram Protocol (RFC 768)*. Internet Engineering Task Force.

Postel, J. (1981). *IP: Internet Protocol (RFC 791)*. Internet Engineering Task Force.

Singh, R., & Ortega, A. (2002). Modeling of Temporal Dependence in Packet Loss Using Universal Modeling Concepts. Paper presented at the 12th Packet Video Workshop, Pittsburgh, PA.

Spanias, A. S. (1994). Speech Coding: A Tutorial Review. [Tempe: IEEE.]. *Proceedings of the IEEE*, *82*, 1541–1582. doi:10.1109/5.326413

Taqqu, M., & Teverovsky, V. (1998). *On estimating the intensity of long range dependence in finite and infinite variance time series. A Practical guide to heavy Tails* (pp. 177–217). Cambridge, MA: Birkhauser Boston Inc.

Veitch, D., & Abry, P., (1999). A wavelet based joint estimator of the parameters of long-range dependence. IEEE transactions on information theory - special issue on "Multiscale statistical signal analysis and its applications", 45(3), 878-897.

KEY TERMS AND DEFINITIONS

Hurst Parameter: This parameter expresses the speed of decay of the autocorrelation function of the self-similar time series.

Jitter: Jitter is the difference between two delay values. In this work, we called jitter to the difference between arrival times of successive voice packets that arrive on the receiver side.

Packet Loss Rate (PLR): Refers to packet loss probability and can be caused by a number of factors, for example, due to excessive network delay in case of network congestion.

Quality of Service (QoS): Is defined from two points of view: user and network. From the end user's perspective, the *QoS* is the collective effect of service performances, which determine the degree of satisfaction of a user of service. On the other hand, from the network's perspective, the term *QoS* refers to the network's capabilities to provide the *QoS* perceived by the end user.

Voice Over Internet Protocol (VoIP): Is the real-time transmission of voice between two or more terminals, by using IP technologies over packet-switched networks. A basic VoIP system consists of three parts, the sender, the IP networks and the receiver.

APPENDIX

Table of abbreviations

ACV	Auto-covariance Function
AM	Absolute Moment
DHT	Distribution with Heavy Tails
FXS	Foreign Exchange Station
HWAV	Haar wavelet
IAT	Inter-arrival Time
IDT	Inter-departure Time
IP	Internet Protocol
LW	Local Whittle
MAVAR	Modified Allan Variance
MSE	Mean Squared Error
OWD	One Way Delay
PCI	Peripheral Component Interconnect
PER	Periodogram
PLR	Packet Loss Rate
QoS	Quality of Services
R/S	Rescaled Adjusted Range
RV	Random Variable
TE	Terminal Endpoint
VAR	Variance Method
VoIP	Voice over Internet Protocol

Chapter 18
SCTP:
Solution for Transport Layer Mobility and Multihoming

Árpád Huszák
Budapest University of Technology and Economics, Hungary

Sándor Imre
Budapest University of Technology and Economics, Hungary

ABSTRACT

Numerous protocols were introduced in the transport layer, which can be very different depending on the provided services. Beside the traditional TCP and UDP, new transport protocols (SCTP, DCCP) have appeared in recent years to overcome limitations of the conventional protocols. The unique features of SCTP like multihoming and multistreaming make this protocol very attractive for reliable data delivery of streams, even in a mobile environment. It can be also used for applications where monitoring and detection of loss is required. SCTP is the only transport protocol that is able to manage mobility issues and handle handovers in the transport layer. The multihoming feature allows an endpoint of a SCTP association to be mapped to multiple IP addresses, and change the delivery path according to the link conditions. The handover process is hardly influenced by several protocol parameters that can be adjusted by the user. The effects of different protocol settings are investigated in details in this chapter. We have studied the performance of multihomed SCTP hosts through experimental studies in an integrated heterogeneous environment. SCTP will also play a significant role in future LTE–EPS architecture, because it can also be used for core network signaling purposes, not just for user data delivery.

DOI: 10.4018/978-1-60960-732-6.ch018

INTRODUCTION

The transport layer is essential part of the ISO/OSI reference model, as well the TCP/IP protocol stack. Different transport layer protocols were already introduced, which can be very diverse depending on the provided services. The responsibilities of the transport protocols include end-to-end message transfer capabilities independent of the underlying network, along with error control, segmentation, flow control, congestion control, and application addressing (port numbers). In order to select the most appropriate protocol to effectively fulfill the users' requirements, the properties and characteristics of the different transport protocols must be studied. For delay sensitive multimedia applications simple and fast protocols are recommended, while for reliable data transfer the information must be delivered ordered and without any error.

In the next-generation mobile network, the need for mobility management even in the transport layer has been revealed. Mobility is becoming increasingly popular feature for the Internet users; however, the developers of the early network protocols did not take into consideration this possibility. For the next-generation Internet, one of the most essential requirements is to make the roaming possible without loosing the connection between the corresponding hosts. Anywhere, anytime communications is an indispensable need in the future networks. Moreover, with the extensive growth of the Internet and mobile/wireless systems, the user demand for high-speed data access caused the introduction of many different kinds of access technologies. Therefore, future telecommunication architectures could easily appear as an integration of multiple wireless access technologies (e.g. Bluetooth, UMTS, WLAN, WiMAX, etc.). Mobile hosts equipped with multiple network interfaces can be connected to the Internet via different ISPs. Failures in one network cannot easily break ongoing communication sessions if hosts are capable of switching over to another connection. Moreover, if both connections are active at once, but higher packet loss and delay is experienced on one path, multihoming capabilities can be used to hand over current sessions to the connection offering better values of Quality of Service (QoS) parameters.

Mobility and multihoming support can be provided in different layers of the ISO/OSI architecture (Ratola, M., 2004; Eddy, W. M., 2004). Mobile IP (Johnson, D & Perkins, C. & Arkko, J., 2004) with Multiple Care-of Addresses (MCoA) (R. Wakikawa et al., 2007) extension is a layer 3 solution, while several proposals exist also in the transport (Stewart, R., 2007) and even in the application layer (Rosenberg, J et al., 2002). Novel solutions for mobility handling based on HIP (Host Identity Protocol) were also appeared in the recent years (Bokor, L. et al., 2007; Bokor, L., et al, 2009).

In this chapter, we introduce SCTP and its unique multihoming and multistreaming features implemented in the transport layer. The most well known transport protocols are the TCP (Transmission Control Protocol) (Postel, J., 1981) and UDP (User Datagram Protocol) (Postel, J., 1980) standardized in the early years of the Internet age. In the last decade, new protocols were investigated with enhanced features. SCTP is one of these protocols, which uniquely provides multihoming and multistreaming. With multihoming capabilities, SCTP is the only transport layer protocol, which can be utilized in mobile networks, where the hosts are continuously changing their access points to the network.

In this chapter, we focus on SCTP and its multihoming and multistreaming performance over heterogeneous IP networks. We used a native IPv6 UMTS–WLAN environment to analyze the behavior of the protocol from mobility point of view. Building our SCTP testbed based on a native IPv6 architecture resides in the fact that IP is considered as the best solution to integrate heterogeneous wireless access networks, and IPv6 will actually be the main networking protocol of the next generation Internet. In order to study multihoming in an integrated, all-IP heterogeneous mobile environment, we realized a native IPv6 UMTS–WLAN testbed and examined the main performance measures of an SCTP-based Layer 4 mobility/multihoming architecture. Besides the empirical results of the SCTP mobility handling performance, we also made analytical models to estimate the handover process efficiency.

SCTP can be used in the core network for signaling purposes, too. In a GSM, 3G UMTS, or 4G networks, the MSC (Mobile Switching Centre) may connect to other SSPs (Service Switching Point), STPs (Service Transfer Point), or SCPs (Service Control Point) via a traditional SS7 protocol stack or via SIGTRAN (Signaling Transport). SIGTRAN consists of three components: a standard IP layer, an SCTP layer, and user adaptation layers. The Evolved Packet Core (EPC) is the mobility core solution associated with the Evolved Universal Terrestrial Radio Access Network (E-UTRAN), which was formally known as Long Term Evolution (LTE). EPC and E-UTRAN are defined by 3GPP's Release 8 specifications. The combination of E-UTRAN and EPC is called Evolved Packet System (EPS). In the EPS, SCTP is used over the usual IP network layer in order to provide reliable and efficient transport between the e-NodeB and the MME (Mobility Management Entity) core entities. Based on the SCTP's multistreaming capabilities, all dedicated EPS procedures, which include all functions, which apply to a specific communication context, can be supported over a limited number of SCTP streams and with improved transmission efficiency. In this chapter, we introduce the role of SCTP in future LTE–EPS architectures, too.

BACKGROUND

In the early years of the Internet era, two transport protocols were introduced for wired network communication. The first one is the UDP (User Datagram Protocol) (Postel, J., 1980), which uses a simple transmission model without handshaking connection setup, reliability, ordering, or data integrity. The other well-known transport protocol is the TCP (Transmission Control Protocol) (Postel, J., 1981), which a more complex protocol offering reliable, connection oriented and ordered delivery of data. Other important features of TCP are flow control, retransmission of lost data and congestion avoidance. New transport protocols (DCCP, UDPLite, and SCTP) appeared in the recent years to overcome the limitations of the traditional protocols. The DCCP (Datagram Congestion Control Protocol) (Kohler, E. & Handley, M. & Floyd, S., 2006) is a newly defined transport protocol by the IETF that implements bidirectional, unicast connections of congestion controlled, unreliable datagrams. For real-time applications the time constraints are more important than reliability, so media transmissions typically use transport protocols like UDP, where no retransmission occurs, providing minimal packet delay. Presently the reliable TCP, SCTP, and the unreliable DCCP are the only alternative protocols to provide congestion control. DCCP combines the best features of UDP and TCP protocols within media transmission context, supporting congestion control mechanisms. It may be useful to think of DCCP as TCP minus bytestream semantics and reliability, or as UDP plus congestion control, handshakes, and acknowledgements. DCCP, similarly to UDPLite (Larzon et al., 2004), is designed to provide a partial checksum that only covers as much of the user data that the sending application specifies in the DCCP Generic Header. Errors in the rest of the packet are ignored because they are assumed acceptable for the destination application. DCCP connections are congestion controlled, but unlike in TCP, DCCP applications have a choice of congestion control mechanism. DCCP uses Congestion Control Identifiers (CCID) to determine the congestion control mechanism. Currently two identifiers are being defined: CCID2 that implements a TCP-like Congestion Control and CCID3 that implements a TCP-Friendly Rate

Control (TFRC), but DCCP is easily extensible to further forms of unicast congestion control.

Other alternative for end-to-end, real-time, transfer of multimedia data is RTP (Real-time Transport Protocol) (Schulzrinne, H. et al., 2003). The protocol provides facility for jitter compensation and detection of out of sequence arrival in data that are common during transmissions on an IP network. RTP typically use UDP, but may use other transport protocols (most notably, SCTP and DCCP) as well, as the protocol design is transport independent.

Recent applications running on mobile hosts may found TCP too limiting due to disconnections during handovers and parallel connections to the same corresponding node. Some applications need reliable transfer without sequence maintenance (e.g. transmission of control data), while others may process received packets strict sequence order (e.g. file downloading). In both of these cases, the head-of-line blocking causes unnecessary delay if TCP is used. The stream-oriented nature of TCP is often an inconvenience, because applications must add their own record marking to delineate their messages. Developers of SCTP were motivated to fulfill the needs of these applications.

New mobile equipments integrate several access technologies to make connections possible to different type of networks. Simultaneous usage of these interfaces may increase bandwidth availability or to select the most appropriate access technology according to the type of flow or choices of the user. A multihomed endpoint determinates a host with several network interfaces. To manage these interfaces, a novel feature, called multihoming support must be introduced to efficiently control these interfaces for improved connectivity (See Figure 1).

In order to increase the reliability of the end-to-end connection, multihoming is an advantageous technique, because it makes possible to reach the multihomed host on more then one IP address. The traffic from one node to another may be forwarded on physically different paths (different subnetworks and thus different destination IP addresses are used) by configuring the communicating endpoints and the IP network accordingly. In this case, associations may become tolerant against physical network failures. When one of the interfaces/paths becomes unavailable, the other paths are still ready to deliver data. Besides, multi-access provides ubiquitous access to offer an extended coverage area for the mobile hosts. It is also a good opportunity to spread network traffic load among several routes, and distribute traffic load among different connections.

Figure 1. Multihomed hosts

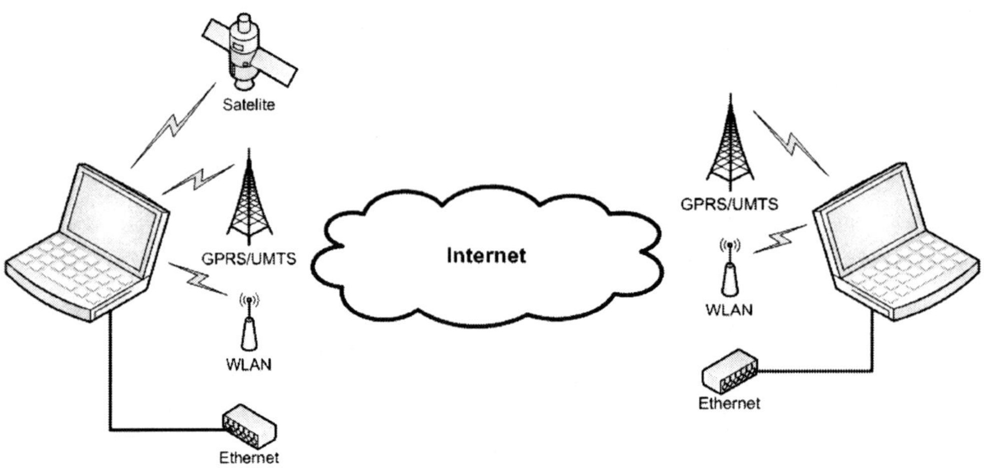

For the efficient interface management of a multihomed host, the corresponding paths must be monitored continuously. When the condition of the actual interface/path is getting worse or fail, a new interface should be assigned for the connection immediately. One of the most important issues is to change the primary path seamlessly; therefore, the endpoint must recognize the link failures as soon as possible and change the active IP address corresponding to a reachable path. However, instantaneous delay increase or temporal channel errors should not cause the change of the primary path. The IP change trigger may depend on several link parameters and protocol settings. It must be defined in the protocol settings, when should a path considered broken and change to another path. The number of retransmission attempts, the packet loss ratio and different time constrains may affect the performance of the interface/path changes.

Alternative Multihoming Solutions

Besides SCTP, other alternative solutions also exist. Based on the IPv6 protocol different approaches have been appeared.

The simplest solution is named *host multihoming*, where the host can have multiple global addresses, one assigned for each of the site's upstream providers with different IP addresses on each interface. While the host knows its addresses, it can choose which source address to use. Multihoming is managed by the application.

The other IPv6-based solution is *site multihoming*, where a host is unaware it is multihomed, allowing the site gateway routers to handle to multiple routing. Site Multihoming by IPv6 Intermediation (SHIMP6) (Nordmark & Bagnulo, 2009) specifies a network layer approach and protocol for providing locator agility below the transport protocols. Multihoming can be provided for IPv6 with failover and load spreading properties, without assuming that a multihomed site will have a provider independent IPv6 address prefix, which is announced in the global IPv6 routing table. The SHIM6 protocol stack uses constant endpoint identities to refer to both itself and to the remote protocol stack. The SHIM6 layer provides a set of associations between endpoint identity pairs and locator sets. The hosts in a site that has multiple provider-allocated IPv6 address prefixes can use the SHIM6 protocol to set up state with peer hosts so that the state can later be used to failover to a different locator pair in case of link errors.

IP addresses can serve only as short term identifiers, because a considerable amount of hosts are portable devices and they change their IP addresses when moved from one network to another. Short-term identifiers disrupt long-term transport layer connections, such as VoIP phone calls, and make locating the peer host more difficult. Therefore, mobility and multihoming are hard to implement securely in the present Internet. Host Identity Protocol uses Host Identifier, to mark uniquely all the hosts which connect to the Internet. The Host Identifier is global unique. The main purpose is to disconnect the close connection between network layer and transport layer so that the function of IP will be concentrated on IP routing. The mark of service in the upper layer will rely on HIP layer.

SCTP

SCTP (Stream Control Transmission Protocol) is a general-purpose transport protocol for the Internet. It was defined by the IETF Signaling Transport (SIGTRAN) working group and described in RFC 4960 in year 2000. SCTP is new protocol operating on top of a connectionless packet network such as IP. Similarly to TCP (Transmission Control Protocol), it also provides connection oriented, reliable and ordered delivery of data between two endpoints.

The developers of SCTP aimed to create a transport protocol that would overcome the limitations of TCP and UDP. SCTP offers the following services to its users:

- acknowledged error-free, non-duplicated transfer of datagrams
- data fragmentation to conform to discovered path MTU size
- multistreaming: sequenced delivery of user messages within multiple streams, with an option for order-of-arrival delivery of individual user messages
- optional bundling of multiple user messages into a single SCTP packet
- multihoming: network-level fault tolerance through supporting of multihoming at either or both ends of an association.

Detection of data to provide reliability, a selective retransmission mechanism is applied. SCTP can also be used for applications where monitoring and detection of loss is required. For such applications, the SCTP failure detection mechanisms will actively monitor the session. These mechanisms are also used to manage the SCTP handover process in case of multihomed hosts.

The used end-to-end window based flow- and congestion control mechanisms in SCTP are similar to the one that is used in TCP. SCTP also uses an Additive Increase, Multiplicative Decrease (AIMD) algorithm, but the congestion control mechanism of SCTP has been modified and adapted for multihoming. SCTP endpoints may be reachable by more than one transport address through different data paths. For each possible path a discrete set of flow and congestion control parameters are maintained. The major feature that confers a distinction on SCTP is that SCTP provides multihoming and multistreaming. The multihoming feature enables to be used for mobility support, without any special router agents or anchor points in the network, while multistreaming supports multiple independent streams within an association. SCTP has several features that are unique, like multistreaming and multihoming.

The comparison of SCTP and other transport layer protocols are summarized in Table 1.

SCTP packets have a simpler basic structure than TCP packets. An SCTP packet is composed of a common header, which occupies the first 12 bytes and chunks. A chunk contains either control information or user data. The SCTP packet format is shown in Figure 2.

Similarly, to other transport protocols, the port numbers identify the association to which this packet belongs. The Verification Tag is used to

Table 1. Comparison of transport protocols

Feature	SCTP	TCP	UDP	UDP Lite	DCCP
Connection oriented	yes	yes	no	no	yes
Reliable	yes	yes	no	no	no
Ordered delivery	yes/no	yes	no	no	no
Checksum	yes	yes	yes	yes	yes
Checksum size (bits)	32	16	16	16	16
Partial checksum	no	no	no	yes	yes
Path MTU	yes	yes	no	no	yes
Congestion control	yes	yes	no	no	yes
Flow control	yes	yes	no	no	yes
Multistreaming	yes	no	no	no	no
Multihoming	yes	no	no	no	no

Figure 2. SCTP header

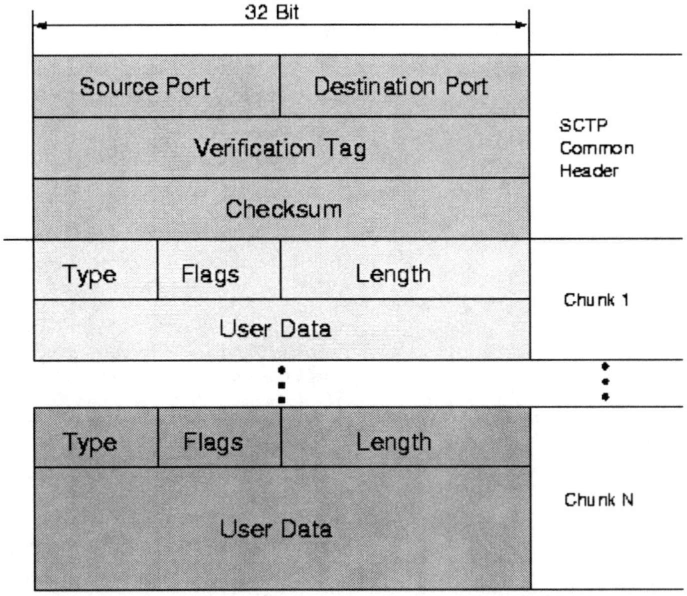

validate the sender of this SCTP packet. The chunk begins with a chunk type field, which is used to distinguish data chunks and different types of control chunks. Multiple chunks can be bundled into one SCTP packet up to the MTU size, except for some special chunk types for initialization and shutdown. The defined chunk types are presented in Table 2.

Security considerations were also important for the SCTP investigators. The protocol was designed with features for improved security, such as 4-way handshake (TCP uses 3-way handshake) to prevent against SYN-flooding attacks.

One of the key aspects was reliability that is also strengthened with the multihoming feature. Multihoming enables an SCTP association to stay open even when some links and interfaces are down. This feature is particularly important for SIGTRAN; because it carries SS7 over TCP/IP network using SCTP, and requires strong resilience during link outages to maintain telecommunication service even when enduring network anomalies.

From mobility point of view, multihoming is the most important feature of the SCTP protocol, while other unique feature is multistreaming. Both of these elements are introduced in details in the following subsections.

Multistreaming in SCTP

SCTP makes it possible to deliver independent streams within one association. Each stream has a stream number that is included inside SCTP packets' chunk headers. Multistreaming eliminates unnecessary head-of-line blocking, as opposed to TCP byte-stream delivery. On the receiving side, SCTP ensures that messages are delivered to the SCTP user in sequence within a given stream. However, while one stream may be blocked waiting for the next sequence user message, delivery from other streams may proceed. User messages sent in independent streams are delivered to the SCTP user as soon as they are received. In other words, a blocked stream does not affect the other streams in an association.

Multiple data and control chunks may be bundled by the sender into a single SCTP packet for transmission, as long as the final size of the

Table 2. SCTP chunk types

Chunk type	Description
Payload Data (DATA)	Used for data transfer.
Initiation (INIT)	Initiates an SCTP association between two endpoints.
Initiation Acknowledgement (INIT ACK)	Acknowledges the receipt of an INIT chunk. The receipt of the INIT ACK chunk establishes an association.
Selective Acknowledgement (SACK)	Acknowledges the receipt of the DATA chunks and also reports gaps in the data.
Cookie Echo (COOKIE ECHO)	Used during the initiation process. The endpoint initiating the association sends the COOKIE ECHO chunk to the peer endpoint.
Cookie Acknowledgement (COOKIE ACK)	Acknowledges the receipt of the COOKIE ECHO chunk. The COOKIE ACK chunk must take precedence over any DATA chunk or SACK chunk sent in the association. The COOKIE ACK chunk can be bundled with DATA chunks or SACK chunks
Heartbeat Request (HEARTBEAT)	Tests the connectivity of a specific destination address in the association.
Heartbeat Acknowledgement (HEARTBEAT ACK)	Acknowledges the receipt of the HEARTBEAT chunk.
Abort Association (ABORT)	Informs the peer endpoint to close the association. The ABORT chunk also informs the receiver of the reason for aborting the association.
Operation Error (ERROR)	Reports error conditions. The ERROR chunk contains parameters that determine the type of error.
Address Configuration Change Chunk (ASCONF)	Request configuration changes.
Address Configuration Acknowledgment Chunk (ASCONF-ACK)	Configuration changes must be acknowledged.
Shutdown Association (SHUTDOWN)	Triggers a graceful shutdown of an association with a peer endpoint.
Shutdown Acknowledgement (SHUTDOWN ACK)	Acknowledges the receipt of the SHUTDOWN chunk at the end of the shutdown process.
Shutdown Complete (SHUTDOWN COMPLETE)	Concludes the shutdown procedure.

packet does not exceed the current path MTU. The receiver will unbundle the packet back into the original chunks. The application may also request a stream to be delivered unordered, which can reduce blocking effects in case of message loss, since the reordering mechanism of one stream is not affected by another stream. SCTP multistreaming is particularly effective in cases, when independent control and data channels are considered in the communication. In TCP, control and data typically share the same connection, which can be problematic, because control packets can be delayed behind data packets. If control and data messages are delivered within independent streams, control data could be dealt with in a timelier manner, resulting better utilization (See Figure 3).

Multihoming in SCTP

Besides multistreaming, the other most important enhancements in SCTP over traditional transport layer protocols is the multihoming capability. A multihomed SCTP endpoint is represented to its peers as a combination of a set of eligible destination transport addresses to which SCTP packets can be sent and a set of eligible source transport addresses from which SCTP packets can be received. A multihomed host can be reached using more than one IP addresses, usually through more than one network interfaces. This feature allows an endpoint of a SCTP association to be mapped to multiple IP addresses. One of the possible IP addresses is selected as Primary Address, while the Primary Path is considered as the network

Figure 3. Multiple streams within one SCTP association

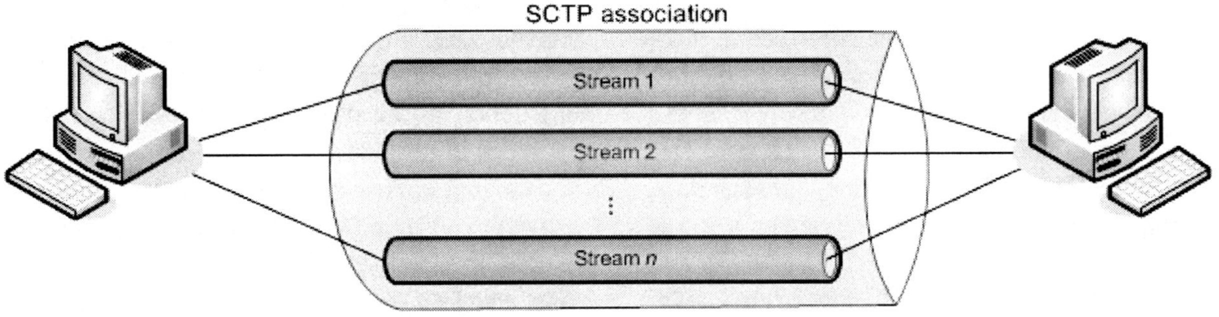

path that leads to the Primary Address. Unless specified otherwise by the SCTP user, an endpoint should always transmit on the Primary Path. The sender may change the Primary Address if the number of failures on the Primary Path exceeds a certain threshold. Retransmissions should be done on different paths as well, so when one link is overloaded, retransmissions do not affect it.

SCTP was developed to take full advantage of such a multihomed host to provide a fast failover and association survivability in the face of such hardware failures. In order to provide fast failover all the available addresses must be bound to the association. While mobile host are able to change the IP address of an interface due to mobility management of the network layer protocol, the SCTP must dynamically add the new IP address to the association. The dynamic addition and subtraction of IP addresses allows an SCTP association to continue to function through host and network reconfigurations. These changes, brought on by provider or user action, may mean that the peer would be better served by using the newly added address; however, this information may only be known by the endpoint that had the reconfiguration occur. If a client is multihomed, it informs the server about all its IP addresses with the INIT chunk's address parameters during the 4-way handshake connection setup. Hence, the client is required to know only one IP address of the server, because the server provides all its IP addresses to the client in the INIT-ACK chunk. After the connection setup a new addresses can be added or removed within an ASCONF chunk. The implementations of Linux 2.6.x versions of SCTP are able to handle both IPv4 and IPv6 addresses. There are numerous protocol parameters (heartbeat interval, retransmission timeout, path maximum retransmission, etc.) that can be adjusted by the user in order to change the SCTP behavior during failovers/handovers. In Linux/BSD operation systems, the *sysctl* is used to modify kernel parameters at runtime. *Sysctl* is an interface for examining and dynamically changing parameters in Linux or BSD.

An SCTP instance monitors all transmission paths to the peer instance of an association. An important feature of the SCTP regarding multihoming is the heartbeat mechanism, which detects failures in idle paths and endpoints. The aim of this mechanism is to detect whether a destination address is active or passive. HEARTBEAT chunks are sent periodically to all idle destinations, and the number of sent HEARTBEAT messages without receipt of a corresponding HEARTBEAT-ACK is maintained. The timing of the HEARTBEAT chunks for destination i is determined by the *HB.Interval* and *RTO* (Retransmission Timeout) protocol parameters according to the following equation:

$$H_i = RTO_i + HB.Interval(1 + \delta) \qquad (1)$$

where δ is a random value between -0.5 and 0.5 and *RTO* is the Retransmission Timeout for path *i*.

An address or path is considered active if acknowledgement is received from its peer within a defined time period (*RTO*). In this way *RTO* is a prediction of the upper limit of *RTT* (Round Trip Time). Separate *RTO* is maintained for each path, which is calculated using the smoothed average of the periodically measured RTT (*SRTT*) and the RTT variation (*RTTVAR*). From these, *RTO* is computed as

$$RTO = SRTT + 4RTTVAR \quad (2)$$

The address with the corresponding path is inactive if the number of consecutive transmission timeouts exceeds the Path Maximum Retransmission (*PMR*) protocol parameter. Every time when timeout occurs, the *PMR* counter is incremented and the value of the *RTO* is doubled for that path. If the new *RTO* is less than *RTO.Min*, it will be set to *RTO.Min*, if it is greater than *RTO.Max*, it will be set to *RTO.Max*. According to the previously presented procedure, the delay of the path failure recovery and the handover process can be calculated as follows:

$$\Delta = \sum_{i=0}^{PMR-1} 2^i \cdot RTO \quad (3)$$

In mobile environment where handovers may frequently occur, the failover time using default SCTP settings is likely to be unacceptable to users. Shortening the failover time can be achieved by setting the relevant protocol parameters (e.g. *RTO.Min, RTO.Max, and PMR*) to smaller values. When the Primary Address becomes unreachable or the link conditions are not acceptable, alternative IP address should be used if the endpoint is multihomed. With the heartbeat mechanism, the SCTP knows which other addresses are active or not and thus, can avoid using another path that has a failure.

Regarding to SCTP, mobility means the ability to change the endpoints and IP addresses while keeping the end-to-end connection. The change of communication link should be done with minimal disruption to the data transmission in progress in order to provide seamless communication in a changing environment. The multiple interface concept can be utilized in dynamic mobile networks, in which e.g., the host terminal has only one interface but the connection is frequently changed. If the possible IP addresses are known, they can be bound even at connection setup. Using the early IP address bindings, the handovers can be handled in such way, but with the dynamical IP binding, the new addresses can be bound during the data transfer.

Configuring SCTP Parameters

SCTP has a large number of adjustable association and path parameters that can be modified using *setsockopt*() linux kernel function.

Previous works examining SCTP mobility and multihoming performance are mainly based on simulations (Jungmaier, A. et al., 2006) or using emulator tools to imitate the behavior of different type of networks (Österdahl, H., 2005). Some of the first previous works dealing with SCTP mobility and multihoming performance evaluation in real life testbeds were (Ravier, T. et al., 2001) and (Jong-Shik Ha et al., 2005), where authors mainly confined their work to study how SCTP deals with packet loss and throughput in quite simple multihoming experiments. Authors of (Fallon, S. et al., 2008) examined SCTP switchover performance in a pure WLAN environment and showed that with the default parameters, SCTP implementations behave in a counterintuitive manner allowing more time for switchover when network conditions degrade. In addition, a pure WLAN testbed was used by authors of (Wakikawa, R. et al., 2006) in order to present a smooth handover scheme for Mobile IPv6 based on SCTP failover mechanism.

In order to increase the multihoming performance of the SCTP the endpoint must recognize the link failures as soon as possible and change the active IP address, but the handover delays or temporal channel errors should not cause the change of the primary path. The IP change trigger may depend on several protocol settings. It must be defined when should a path considered broken. The number of retransmission attempts, the packet loss ratio and different time constrains may affect the performance of the interface/path changes. From the handover point of view, the most important SCTP protocol parameters are *RTO.Min*, *RTO.Max,* and *Path.Max.Retransmission* (*PMR*). These parameters and the continuously calculated *RTO* value determinate the speed of the handover. By setting *RTO.Min* = *RTO.Max*, the *RTO* is kept on a constant value disabling to redouble it every time when timeout occurs, as equation (3) defines it. By adjusting these parameters, the handover process can be speed up. The delay of the path failure recovery and the handover process in this case can be calculated as follows:

$$\Delta = PMR \cdot RTO \qquad (4)$$

To analyze the effect of *RTO* on the handover process we have measured the DATA transmission interrupt in our IPv6-capable UMTS–WLAN heterogeneous testbed. This interrupt is considered as the elapsed time between the last DATA chunk sent on the failed primary (WLAN) interface and the first DATA chunk arrived at the continuously available secondary (UMTS) interface. In order to seamlessly forward real-time data, the handover delay must be decreased, which highly depends on the current *RTO* (See Figure 4).

To analyze the exclusively the *RTO* parameter, all the other parameters were kept constant (*HB. Interval* = 10s, *PMR* = 5) in our examinations. According to our previous considerations, the delay was rising linearly when the *RTO* parameter was incremented. As the results show, the transmission interrupt was lower then 10s when the *RTO* was less then 5s. Using the default SCTP parameters (*RTO.Min* = 1s, *RTO.Max* = 60s) the delay would rise exponentially due to *RTO* re-

Figure 4. DATA transmission interrupt measurements (WLAN→3G UMTS handover)

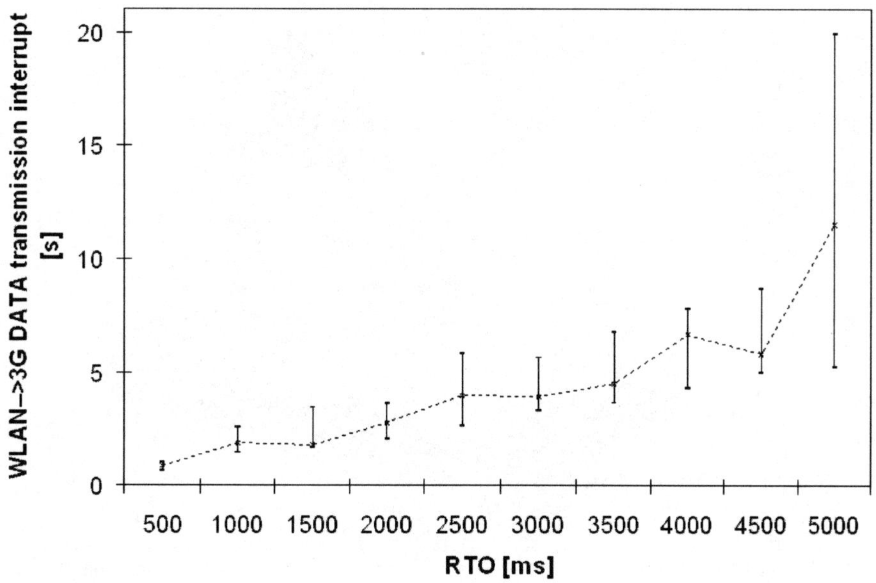

doubling and reach 60s. The average DATA transmission interrupt with default SCTP parameters was 227s, which is not acceptable for the users, especially for real time application users. If the *RTO* is forced to be less then 5s, the transmission delay can be significantly decreased.

The impact of *RTO* parameter is significant on the handover performance of SCTP, but the number of retransmission attempts on a link is also an important parameter. The *PMR* (Path Max. Retransmission) is the other key parameter of the SCTP handover. Parameter *PMR* contains the maximum number of retransmissions before the link shall be considered unreachable. Using the default parameters of the protocol, the link failure recovery time is rising exponentially according to equation (3). In the following figure, the joint impact of *RTO.Max* and *PMR* is introduced, based on our measurements in our native IPv6 UMTS–WLAN testbed (See Figure 5).

SCTP periodically sends HEARTBEAT chunks to idle destinations, or alternate addresses to identify a path failure. The heartbeat mechanism is also responsible for detecting when the primary path has recovered. The HEARTBEAT chunks are sent periodically as defined by the *HB.Interval* and the corresponding equation, see (1). SCTP maintains a counter to store the number of heartbeats that are sent to the inactive destination, without receiving a corresponding HEARTBEAT-ACK chunk. When the counter reaches the specified maximum value, SCTP also declares the destination address as inactive. SCTP notifies the application about the inactive destination address and starts using an alternate address for sending the DATA chunks. To discover that the primary path becomes available (link availability recognition time), it is desirable to keep the interval between HEARTBEATs relatively small. The obtained measurement results confirm this theory as Figure 6 shows.

SCTP continues to send HEARTBEATs to the inactive destination address until it receives a HEARTBEAT-ACK chunk. On receipt of the acknowledgement, SCTP considers the destination address as active again. When changing the com-

Figure 5. Link failure recovery time analysis in function of RTO and PMR (WLAN→3G UMTS handover)

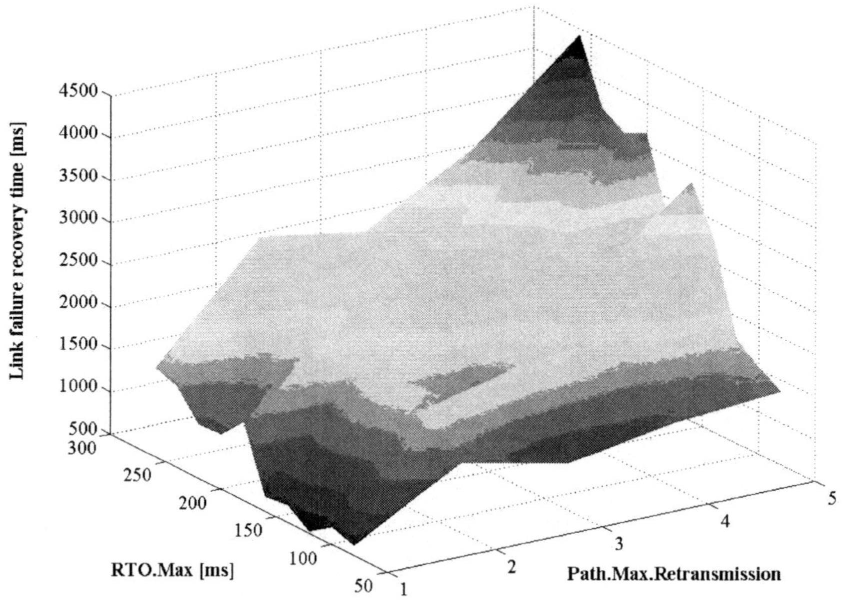

munication path back to the primary path, the first packet on the primary link is always a HEARTBEAT chunk. The DATA chunk can be delivered only if HEARTBEAT-ACK is received. The primary link failure detection is independent from the *HB.Interval* parameter, while the recovery depends on it.

SCTP IN LTE NETWORKS

In mobile networks, the handovers cannot be avoided, but its negative effects can be minimized. SCTP multihoming and multstreaming feature can be effective in reducing of the disadvantages caused by handovers and lossy channels. The mobile terminal can switch between the base stations and between different access technologies (3G to WLAN, 3G to Ethernet, etc.) as well. The SCTP protocol can effectively reduce the handover delay and seamlessly continue the data delivery.

Stream Control Transmission Protocol was defined by the Signal Transport (SIGTRAN) working group of the Internet Engineering Task Force (IETF) and besides user data delivery, SCTP can be also used for signaling purposes. In a GSM, UMTS, or 4G network, the MSC (Mobile Switching Centre) may connect to other SSPs (Service Switching Point), STPs (Service Transfer Point), or SCPs (Service Control Point) via a traditional SS7 protocol stack or via SIGTRAN (Signaling Transport). SIGTRAN, as defined in RFC 2719 and RFC 4166, is a set of protocols that allow circuit-switched telephony messages, such as Media Gateway control and SS7 messages, to be reliably transported over an IP network. SIGTRAN consists of three components: a standard IP layer, an SCTP layer, and user adaptation layers.

The Evolved Packet Core (EPC) is the mobility core solution associated with the Evolved Universal Terrestrial Radio Access Network (E-UTRAN), which was formally known as Long Term Evolution (LTE). EPC and E-UTRAN are

Figure 6. The impact of the HEARTBEAT interval (HB.Interval) on the link availability recognition time (3G UMTS→WLAN handover)

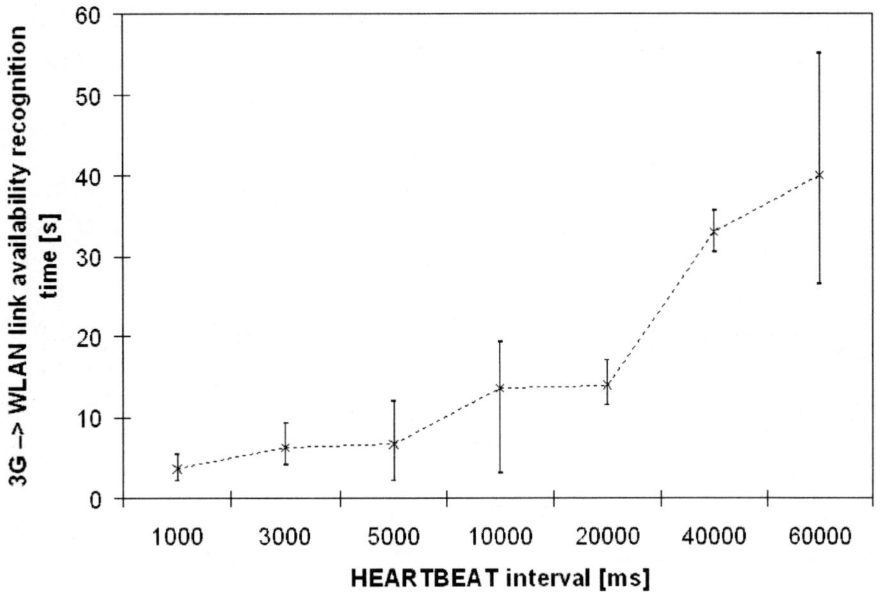

defined by 3GPP's Release 8 specifications, in particular (3GPP TS 23.401, 2007; 3GPP TS 23.402, 2007; 3GPP TS 36.300, 2007). The combination of E-UTRAN and EPC is called Evolved Packet System (EPS) (See Figure 7).

In the EPS, SCTP is used over the usual IP network layer in order to provide reliable and efficient signal messaging between the e-NodeB and the MME (Mobility Management Entity) core entities. Based on the SCTP's multistreaming capabilities, all dedicated EPS procedures, which include all functions applying to a specific communication context, can be supported over a limited number of SCTP streams and with improved transmission efficiency. E-UTRAN introduces a new radio interface technology. A base station that supports this radio interface technology is called Evolved NodeB (e-NodeB). A logical interface (X2) is used to communicate between two e-NodeBs to deliver control information (See Figure 8).

The X2 signaling bearer provides the following functions, based on SCTP services:

- Provision of reliable transfer of X2-AP message over X2 interface.
- Provision of networking and routing function
- Provision of redundancy in the signaling network
- Support for flow control and overload protection

The S1 Application Part (S1-AP) is the control plane signaling protocol between the e-NodeB and the Mobility Management Entity (MME). LTE S1-AP implementation supports the S1-MME interface and utilizes SCTP in the transport layer. SCTP is for the control plane, which guarantees delivery of signaling messages between the MME and e-NodeB.

Unfortunately, SCTP may have negative features as well. With small-cell wireless deployments, the sheer number of SCTP connections that the EPC needs to manage has also increased. While the built-in reliability mechanisms of SCTP are attractive for providing carrier grade networks, they create substantial overhead. This behavior can lead to scalability problems in the EPC's core network.

Figure 7. LTE architecture

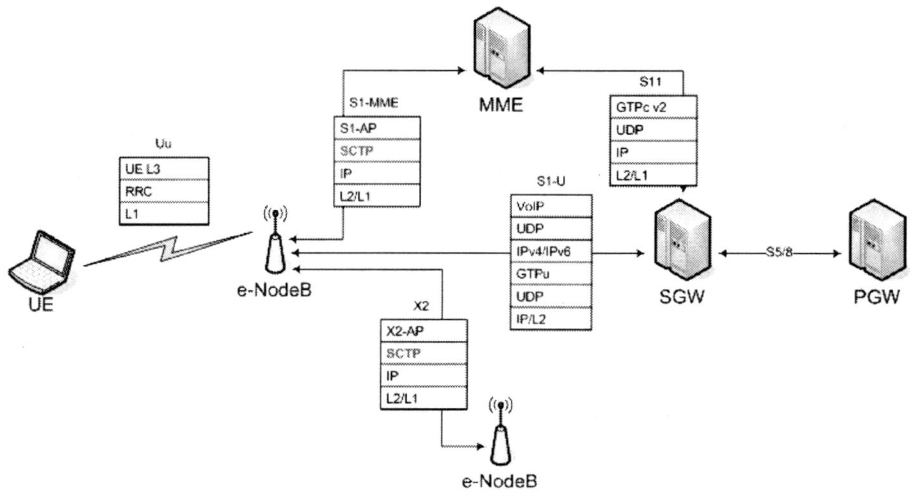

Figure 8. Logical interface (X2) between eNBs

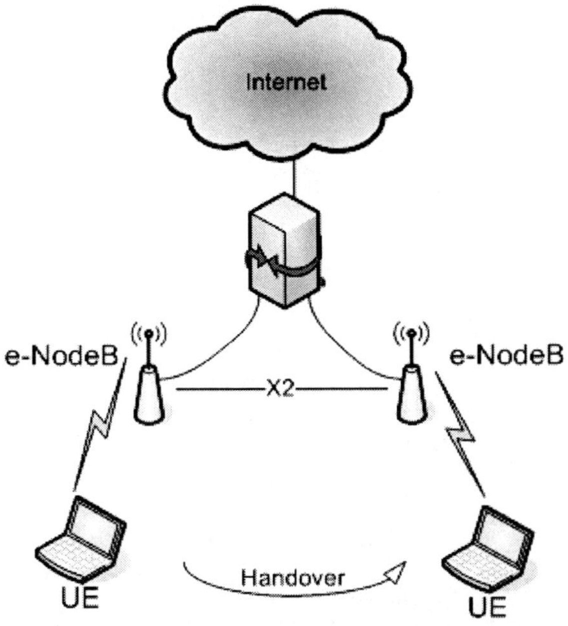

SIP OVER SCTP

SCTP is the transport protocol specified by next generation network architectures and is used also by SIP, Diameter for AAA services, GCP-Gateway Control Protocol (H.248/MEGACO/MGCP), and SIGTRAN User Adaptation (UA) protocol layers. It provides secure and reliable transport, which is a must to fulfill the promise of next generation network services like multimedia messaging.

The use of SIP (Session Initiation Protocol) (Rosenberg, J., 2002) in the IP Multimedia Subsystem (IMS) requires servers that are capable to handle a large number of call requests. The signaling traffic associated to such requests could explode, if an intelligent congestion control were not introduced. SCTP as transport of SIP signaling might be useful in some situations, where usual transport protocols (e.g. TCP and UDP) suffer performance degradation. SIP itself is independent from the transport protocol, and can run over any reliable or unreliable message or stream oriented protocol.

If traffic is generated by a moderate number of sender-receivers sessions, no significant risk of congestion arises. In this case, UDP performs well, but as the number of sessions increases, the SIP retransmission mechanism increases the risk of causing a flood of retransmissions. In order to avoid the overload of the network, a transport protocol with congestion avoidance must be used.

A reliable transport layer as TCP guaranties the successful delivery, so retransmissions are not needed at an application level. However, TCP can cause the Head of Line (HOL) blocking problem when the signaling associated with multiple sessions is sent over a single TCP connection between two servers. The loss of one message stops the immediate delivery to the SIP layer of further messages; therefore, all other messages in the same flow, even if they belong to unrelated sessions, are affected by the loss of a message in a single session.

SCTP can overcome the problems of congestion and HOL. SIP transactions need to be mapped into SCTP streams to avoid HOL blocking. There are three possible ways to use SCTP with SIP:

- mapping of SIP sessions into Streams
- mapping of SIP transactions into Streams
- using Stream 0 and the unordered flag

It is important to note that most of the benefits of SCTP for SIP occur under loss conditions. Therefore, under a zero loss condition, SCTP transport of SIP should perform similarly to SIP/TCP transport.

CONCLUSION

In this chapter, we introduced the unique SCTP features, its utilization possibilities, and experimental analysis of SCTP in wireless environment in terms of handover effectiveness, throughput, and transmission delay. The multihoming behavior of SCTP is an advantageous feature of the protocol; however, its performance highly depends on the parameter settings. In our measurements, we have analyzed numerous settings to justify the analytical correlations of the protocol parameters and the different connection characteristics. Using accurate SCTP parameter setup, the handover delay, and the data transmission interrupt can be significantly decreased. SCTP will play important role in the future LTE networks as the transport layer protocol of signaling messages between the e-NodeBs and the Mobility Management Entity. Besides signaling purposes SCTP can also attractive as the transport protocol of SIP-based applications. Due to the special features of the SCTP protocol, it will become more and more popular in the IP-based networks.

ACKNOWLEDGMENT

This work was supported by the Mobile Innovation Center. The authors would like to express their appreciation to László Bokor and Gábor Jeney for their essential work on this research.

REFERENCES

Bokor, L., Nováczki, S., & Imre, S. (2007). *A complete HIP based framework for secure micromobility*. 5th @WAS International Conference on Advances in Mobile Computing and Multimedia, MoMM2007. (pp. 111-122). Jakarta, Indonesia, 3-5 December 2007. ISBN 978-3-85403-230-4

Bokor, L., Nováczki, S., Zeke, T. L., & Jeney, G. (2009). Design and evaluation of host identity protocol (HIP) simulation framework for INET/OMNeT++. In the *Proceedings of the 12-th ACM International Conference on Modeling, Analysis and Simulation of Wireless and Mobile Systems* (MSWIM 2009), Tenerife, Canary Islands, Spain, 26-30 October 2009.

D-ITG. (2009). *Official website*. Retrieved from http://www.grid.unina.it/software/ITG/

Eddy, W. M. (2004, October). At what layer does mobility belong? *IEEE Communications Magazine*, 155–159. doi:10.1109/MCOM.2004.1341274

Fallon, S., Jacob, P., Qiao, Y., Murphy, L., Fallon, E., & Hanley, A. (2008). *SCTP switchover performance issues in WLAN environments*. 5th IEEE Consumer Communications and Networking Conference (CCNC'08), (pp. 564-568).

3GPP TS 23.401 v8.0.0. (2007). *GPRS enhancements for E-UTRAN access (release 8).*

3GPP TS 23.402 v8.0.0. (2007). *Architecture enhancements for non-3GPP accesses (release 8).*

3GPP TS 36.300 v8.3.0. (2007). *Evolved universal terrestrial radio access (E-UTRA) and evolved universal terrestrial radio access network (E-UTRAN).*

Ha, J. S., Kim, S. T., & Koh, S. J. (2005). *Performance comparison of SCTP and TCP over Linux platform. Lecture Notes in Computer Science, 3645/2005* (pp. 396–404). Berlin/Heidelberg, Germany: Springer.

Johnson, D., Perkins, C., & Arkko, J. (2004). *Mobility zupport in IPv6*. IETF RFC 3775, June 2004.

Jungmaier, A., & Rathgeb, E. P. (2006). On SCTP multi-homing performance. *Telecommunication Systems*, *31*(2-3), 141–161. doi:10.1007/s11235-006-6517-7

Kohler, E., Handley, M., & Floyd, S. (2006). *Datagram congestion control protocol*. Internet Engineering Task Force, RFC 4340, March 2006.

Larzon, L. A., Degermark, M., & Pink, S. (2004). *The lightweight user datagram protocol* (UDP-Lite). RFC 3828, Internet Engineering Task Force, July 2004.

Nordmark, E., & Bagnulo, M. (2009). *Shim6: Level 3 multihoming Shim protocol for IPv6*. RFC 5533, Internet Engineering Task Force, June 2009.

Österdahl, H. (2005). *A comparison of TCP and SCTP performance using the HTTP protocol*.

Postel, J. (1980). *User datagram protocol*. RFC 768, Internet Engineering Task Force, August 1980.

Postel, J. (1981). Transmission control protocol. RFC 792, Internet Engineering Task Force, September 1981.

Ratola, M. (2004). *Which layer for mobility? Comparing mobile IPv6, HIP and SCTP.* Helsinki University of Technology, Telecommunications Software and Multimedia Laboratory 2004.

Ravier, T., Brennan, R., & Curran, T. (2001). *Experimental studies of SCTP multi-homing*. First Joint IEI/IEE Symposium on Telecommunications Systems Research, 2001.

Rosenberg, J., Schulzrinne, H., Camarillo, G., Johnston, A., Peterson, J., Sparks, R., ... Schooler, E. (2002). *SIP: Session initiation protocol*. IETF RFC 3261, June 2002.

Schulzrinne, H., Casner, S., Frederick, R., & Jacobson, V. (2003). *RTP: A transport protocol for real-time applications*. RFC 3550, Internet Engineering Task Force, July 2003.

Stewart, R. (2007). *Stream control transmission protocol*. RFC 4960, Internet Engineering Task Force, September 2007.

Wakikawa, R., Devarapalli, V., Ernst, T., & Kagami, K. (2007). *Multiple care-of addresses registration*. (IETF Draft, draft-ietf-monami6-multiplecoa-09.txt), August 2008.

Wakikawa, R., Nishida, Y., & Murai, J. (2006). *The use of SCTP failover mechanism for efficient network handover on mobile IPv6*. 3rd International Symposium on Wireless Communication Systems (ISWCS'06) (pp. 133-137).

ADDITIONAL READING

Stevens, W. R., Fenner, B., & Rudoff, A. M. (2003). *UNIX Network Programming* (*Vol. 1*). Pearson Education.

Stewart, R., & Xie, Q. (2002). *Stream Control Transmission Protocol (SCTP): A Reference Guide*. Addison-Wesley.

KEY TERMS AND DEFINITIONS

SCTP (Stream Control Transmission Protocol): SCTP is a general-purpose transport protocol for the Internet, defined by the IETF and described in RFC 4960. It provides reliable, connection oriented, congestion controlled packet delivery with multihoming and multistreaming capability.

Transport Layer: The fourth layer of the OSI Reference Model protocol stack is the transport layer. It provides necessary functions to enable communication between software application processes on different hosts. Another key function of the transport layer is to provide connection services for the protocols and applications. It may have reliability and flow-control functions as well.

Multihoming: A multihomed host can be reached using more than one IP addresses, usually through more than one network interfaces. This feature allows an endpoint of an association to be mapped to multiple IP addresses.

Multistreaming: Multistreaming makes possible to deliver independent streams within one association. In case of SCTP, each stream has a stream number that is included inside SCTP packets' chunk headers. Multistreaming eliminates unnecessary head-of-line blocking, as opposed to TCP byte-stream delivery.

LTE (Long Term Evolution): LTE, which often marketed as 4G, is the latest standard in the mobile network technology tree. The LTE specification provides downlink peak rates of at least 100 Mbps, an uplink of at least 50 Mbps.

APPENDIX

Table of abbreviations

AIMD	Additive Increase, Multiplicative Decrease
AP	Application Part
CCID	Congestion Control Identifiers
DCCP	Datagram Congestion Control Protocol
e-NodeB	Evolved NodeB
EPC	Evolved Packet Core
EPS	Evolved Packet System
E-UTRAN	Evolved Universal Terrestrial Radio Access Network
HIP	Host Identity Protocol
HOL	Head of Line
IETF	Internet Engineering Task Force
IMS	IP Multimedia Subsystem
LTE	Long Term Evolution
MCoA	Multiple Care-of Addresses
MME	Mobility Management Entity
MSC	Mobile Switching Centre
PMR	Path Maximum Retransmission
QoS	Quality of Service
RTO	Retransmission Timeout
RTP	Real-time Transport Protocol
RTT	Round Trip Time
RTTVAR	Round Trip Time Variation
SCP	Service Control Point
SCTP	Stream Control Transmission Protocol
SHIMP6	Site Multihoming by IPv6 Intermediation
SIGTRAN	Signaling Transport
SIP	Session Initiation Protocol
SRTT	Smoothed average of Round Trip Time
SSP	Service Switching Point
STP	Service Transfer Point
TCP	Transmission Control Protocol
TFRC	TCP-Friendly Rate Control
UA	User Adaptation
UDP	User Datagram Protocol

Chapter 19
IPv6 Routing in a Special Context:
Serving Efficient Data Aggregation

Zoltán Kanizsai
Budapest University of Technology and Economics, Hungary

Gábor Jeney
Budapest University of Technology and Economics, Hungary

ABSTRACT

This chapter gives an overview of the IPv6 routing techniques focusing on some special IPv6 routing mechanisms, which can ease and optimize IP level data aggregation. Data aggregation should become an important step in information collecting from sensors or other data sources in present-day, IP based networks. In general, a simple feedback message only costs a few bit, or bytes; therefore, using a whole IP packet for every one of them is rather wasteful according to the present resources. Our objective is to introduce a special IPv6 routing method called IPv6 anycasting, which makes possible to identify various services with an IPv6 network address. Basically, by the standard, anycasting is not able to maintain stateful communication sessions, but feedback data aggregation does not require stateful operation, therefore anycasting is an ideal solution to provide feedback routing. The chapter will concentrate on IPv6 advanced routing schemes, and introduce a new IPv6 anycast based system-level data aggregation mechanism.

DOI: 10.4018/978-1-60960-732-6.ch019

INTRODUCTION

IPv6 is the new version of the Internet Protocol (IP) and it is expected to be introduced in the next few years in public networks. IPv6 comes with a huge amount of improvements compared to IPv4; however, it keeps the conceptual basics. For instance, IPv6 has built-in functionality for secure communications, address resolution, and host mobility management, while IPv4 has only a few extensions for these purposes. IPv6 was built on the same fundaments as IPv4: both represent a not reliable, best effort service over a packet switched network. Since IPv6 cannot be a global replacement of IPv4, because they are not compatible, it is expected that IPv6 and IPv4 will live together for a while.

The main task of the IP is to deliver packets from a node to another node on the Internet even if the destination node is on the other side of the world. IP is a common "language" on interconnected packet switched networks.

This chapter is devoted to introduce the main properties and improvements of IPv6. Then we will show the meaning of packet routing with providing different groupings of today's IPv6 routing protocols which will be discussed in details for better understanding. After we got familiar with the routing techniques, a new trend is presented in IP based networks with introducing the problem caused by multimedia and measurement feedbacks. Then we will propose a solution for this problem using a brand new feature of IPv6 called anycasting.

IPV6 BASICS

A few "visionary" in 1987 predicted that the number of networks connected to the Internet may reach one hundred thousand in a few years; however, most people did not care about their opinion. Most Internet users expected that it could happen only a few decades later. Well, it was not the case; the 100,000th network was connected to the Internet in 1996.

Additionally the IPv4 has been developed in the early 1970s for the high-tech industry, universities, and national security institutions. No one expected that over time the Internet will evolve to a commercial product and will provide services to millions of people. In this trend the appearance of an important appliance, called the Personal Computer (PC) had a significant role. In the 1980s and later, the continuous development of PCs and network devices sped up the growth of internetworking. Thus, it became necessary to create a protocol that not only solves the present problem of small address space but it is able to efficiently adapt to the users' present and future expectations.

The Internet Engineering Task Force (IETF) in 1990 began developing a new version of the Internet Protocol. The brief goals were the following:

- to support billions of hosts even if the address space assignment is not optimal
- to reduce the size of the routing tables
- to simplify of the protocol, to make packet processing faster in the routers
- to provide better IP based security (authentication and authorization)
- to devote greater attention to the type of services, especially for real-time streams (Quality of service: QoS)
- to help the usage of multicasting by defining scopes right in the IP layer
- to open the opportunity of roaming for the hosts (mobility support)
- to enable the further development of the protocol
- to enable the coexistence of the old and the new protocol for many years until the conversion process finishes

After much debate and examination of several drafts, IETF finally chose the revised and combined proposal of Steven Deering and Paul

Francis while the protocol's name became Simple Internet Protocol Plus (SIPP). Then the IETF has started a new task force which released the first specification of the IPv6; called Internet Protocol, Version 6 (IPv6) Specification; and released it in RFC1883 (Deering, 1995) in late 1995. The current specification of IPv6 is in RFC2460 (Deering, 1998), compiled by Steven Deering and Robert Hinden, which phased out RFC1883 in 1998.

In the following, we present a short list about the major improvements of IPv6 compared to IPv4.

The Extended Address Space

The primary objective was achieved; the old 32-bit IP addresses were replaced with 128-bit addresses. The new address space contains approximately 3.4×10^{38} individual addresses! This means a huge amount of addresses that probably never will be fully allocated, but the preliminary condition may be realized that address allocation can be ensured for all the hosts even if using an ineffective address allocation method.

Simpler IP Packet Header

The new protocol header has been significantly redesigned. The main point was the header simplification to facilitate the task of the routers, since the smaller but more effective header allows faster packet processing and forwarding in spite of the longer addresses. The new IPv6 header contains only 7 fields while the one of IPv4 contains 13! For special features, IPv6 uses optional header fields.

Hierarchical Addressing Architecture

Classless Inter-Domain Routing (CIDR) (Fuller, 1993) was released for IPv4 creating an opportunity for using classless address allocation and aggregated routing entries. CIDR did not result in a perfect solution. As a method, it works well although it is inefficient and not scalable. In IPv6, thanks to the large address space, it is feasible to create a hierarchical structure of network topology. Learning from the past's mistakes and taking advantage of the situation, address prefixes are allocated to organizations with large number of usable IP addresses and these organizations can reallocate more specific prefixes (inside the original prefix) to their underlying customers, and even if these reallocations leave millions of IP addresses unusable the hierarchical manner of network addressing has its own benefits.

Every level of the hierarchy can aggregate the IP addresses of the lower levels into one single prefix thus the border routers of each level only have to deal with these aggregated prefixes instead of the lower level subnet prefixes.

Support for Routing Protocols

During the protocol design, it was an important aspect to widely support the routing protocols. To achieve scalable routing, IPv6 supports both Interior and Exterior Gateway Protocols (IGPs and EGPs). The main supported routing protocols are: Routing Information Protocol – Next Generation (RIPng), Open Shortest Path First version 3 (OSPFv3), Intermediate System to Intermediate System version 6 (IS-ISv6) and Multiprotocol Border Gateway Protocol version 4 (MP-BGP4). These protocols will be discussed later in this chapter.

Integrated Security Features

IPv6 uses IP Security (IPSec) which is a network layer based security framework supporting many security protocols including a NULL algorithm for those who do not wish security. IPv6 provides unique IP address for every individual node and it supports security. All these mean that IPv6 can afford highly efficient secure end-to-end services.

Autoconfiguration and Plug'n'Play Functions

A completely new feature in IPv6 is host autoconfiguration. After attaching the host to a network architecture, autoconfiguration makes possible to gain IP address without any active interaction with other hosts on the network. The host simply listens for the Router Advertisement (RA) messages, which are sent by the local router periodically. According to the advertised prefix information and the node's MAC address, the node is able to create a globally unique IPv6 address. To check the uniqueness of this address the host uses Duplicate Address Detection (DAD) mechanism (Thomson, 2007). The host also discovers the nodes on the same link using a very important IPv6 specific protocol called Neighbor Discovery (Narten, 2007).

Support for IP Based Host Mobility

Mobility management became an integral part of IPv6. It is known as Mobile IPv6 (Johnson, 2004). This protocol makes possible the roaming between networks while keeping the same IP address (called Home Address, HoA). A special router called Home Agent (HA), which is placed in the Home Network (HN), is used to keep the Mobile Node (MN) available via its original HoA. Every time the MN changes its location, its new temporary IP address (Care-of Address, CoA) is reported to HA. If a packet addressed from a Correspondent Node (CN) to the MN reaches the HN and the HA, it is forwarded to the temporal location of the MN in a secure tunnel. After this initial step, the MN and the originator of the IP packet can reach each other directly. This mechanism is depicted in Figure 1.

Quality of Service

In IPv4, several attempts appeared to ensure the necessary conditions for different types of network services to improve the Quality of Service (QoS). IPv6 packet header contains two special fields to support QoS: the Traffic Class and Flow Label fields.

The Traffic Class field is similar to the one that is used in IPv4. However, the Flow Label field is completely new and it makes possible to identify the various data streams (Voice over IP, video streaming, etc). The flow label field specifies how the intermediate nodes should treat the stream. The labels' contents are set by the sender of the packet.

IPv6 has three types of addresses: unicast, multicast, and anycast. Unicast and multicast addresses had been used in IPv4, as well. However, anycast is a new type. IPv6 unicast addresses are unique identifiers of interfaces and a unicast address must not be assigned to multiple interfaces. The unicast address type realizes the one-to-one communication scheme. The multicast addresses are different from unicast, since a multicast address can be assigned to multiple interfaces at a time. The nodes (interfaces) with the same multicast address are members of a set called multicast group. If a packet is sent to a multicast address, all members of that multicast group receive the packet. Essentially the packet proliferates in the network. It is often termed as one-to-many communication scheme. Like in the case of multicast, anycast addresses can be assigned to multiple interfaces too, though there is a difference. If a packet is sent to an anycast address, the packet will not be multiplied by the network. It will be delivered to only one node (interface) of the anycast group. It is called one-to-nearest or one-to-(one of many) communication scheme. Anycast addresses are usually assigned to various services. It does not mean assignment directly to hosts, but to nodes providing the same service. The follow-

Figure 1. The basic operation of Mobile IPv6

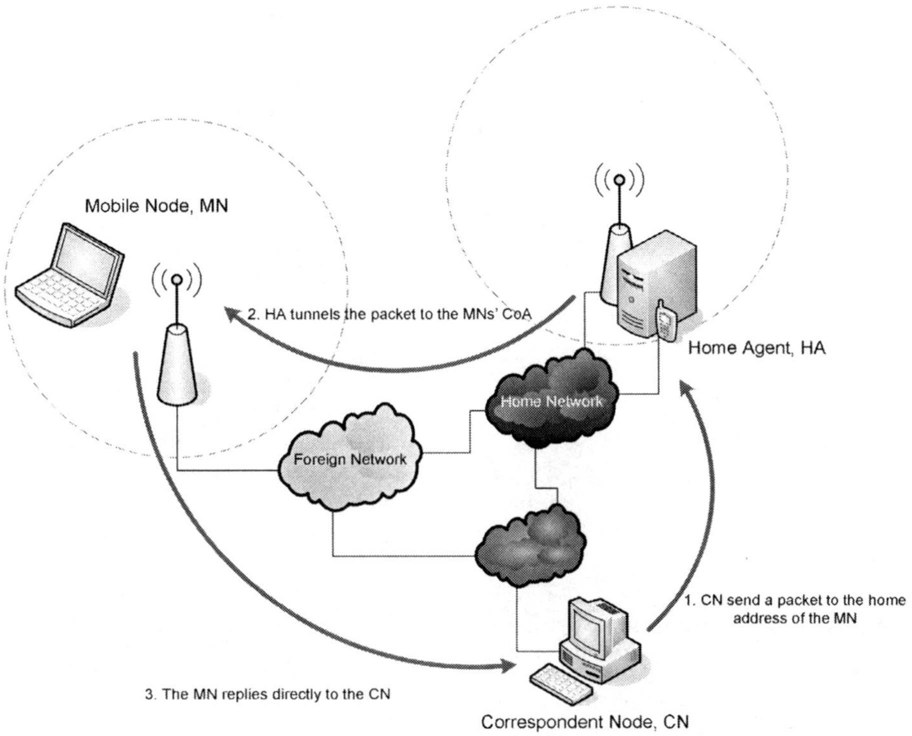

ing capabilities and restrictions are defined for the anycast addressing mechanism in RFC2373 (Hinden, 1998):

- Anycast addresses must be undistinguishable from global unicast addresses
- An anycast address can be assigned to multiple interfaces in a network
- "An anycast address must not be used as the source address of an IPv6 packet.
- An anycast address must not be assigned to an IPv6 host, that is, it may be assigned to an IPv6 router only." (Hinden, 1998)

These last two restrictions were only recommendations. Since that, more experience has been gained and solutions have been agreed. Thus, later they have been modified by *draft-jabley-v6-anycast-clarify-00.txt* as follows:

- "An anycast address MAY be used as the source address of an IPv6 packet.
- An anycast address MAY be assigned to an IPv6 host." (Abley, 2005)

It is an advantage of anycast communication that the user's node does not need to know the network topology and architecture to use a service provided by anycasting. It only has to know the IPv6 (anycast) address of the chosen service and send the service request packet to that (anycast) address. The node does not even have to know that this is an anycast address either, this information is only known by the anycast service providers and the anycast routers. The packets addressed to anycast addresses are routed to the "closest" anycast service provider according to a metric value, which can be calculated from link states, CPU or memory load of the anycast servers, etc. A simplified case of anycast communication is shown in Figure 2.

Figure 2. The principle of anycast communication

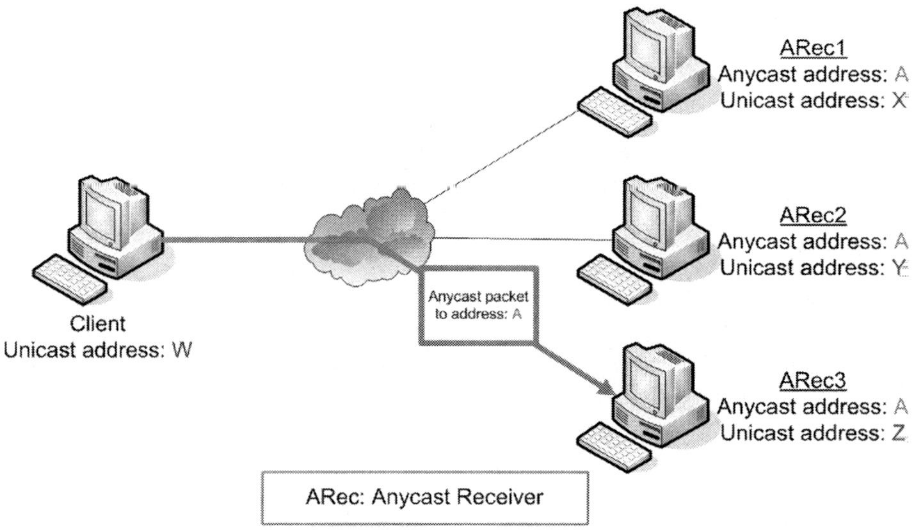

A good overview on applications and services of IPv6 anycasting can be found in (Dudas, 2007), while a novel, anycast-based micromobility management solution is defined in (Bokor, 2005) and (Bokor, 2007).

THE IPV6 NETWORK ARCHITECTURE

As we stated in the previous section IPv6 provides a hierarchical network structure. At the bottom of this hierarchy there are the elementary networks called subnets. The user's host is connected to the Internet through a subnet. If it wants to transmit a packet to another host, it has to determine whether the destination is on-link or off-link. If the recipient is on-link, the packet is sent directly to the destination. If it is off-link, the packet goes through a special network node called router. Then, the router, according to its routing table, routes the packet to the appropriate direction towards the destination host.

Sets of subnets administered and controlled by separate authorities are called Autonomous Systems (AS). Routers of an IPv6 network can be divided into two types, internal and external routers, whether they route packets within an AS or between ASs. Internal routers exchange their routing information using Interior Gateway Protocol (IGP) while Exterior Gateway Protocol (EGP) is used by external routers. The hierarchical architecture of IPv6 networks is shown in Figure 3. IPv6 routing protocols are discussed in the next section in details.

IPV6 ROUTING

Routing in IPv6 is very similar compared to the one in IPv4. What is routing? As we said in the previous subsection if a destination node for a packet is off-link, means cannot be reached on the same link where the originator of the packet is, the packet is sent out from the local network through a special node called router. This node stores a table, which contains pairs of information about other networks: their network addresses and the direction (interface ID) where these networks can be reached. This table is called routing table.

The process when a router decides, according to its routing table, to send out a packet on which

Figure 3. The hierarchical topology of IPv6 networks

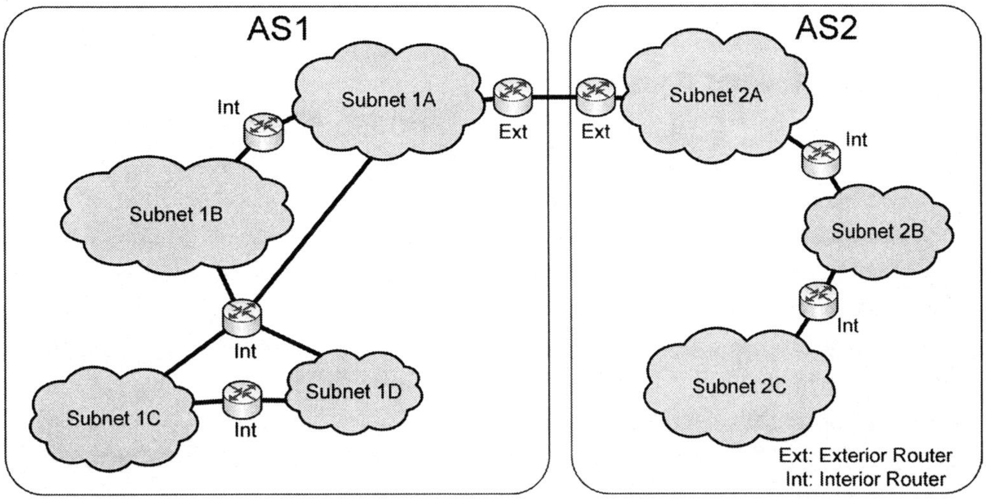

interface towards the destination network is called packet routing or simply routing. The composition of the routing table can be done by manually and in this case, it is called static routing, or by special programs called routing daemons what is called dynamic routing. In this subsection, we will introduce the main properties of static and dynamic routing techniques.

Static Routing

Static routing always requires a network administrator to write the routing tables manually. In this case, the administrator has overall control of the traffic flow but it is also needed if a reconfiguration needed because of some kind of network failure, so static routing cannot adapt to network topology changes without human intervention.

Another disadvantage is that over a particular network size manual editing of routing tables become so complex that it cannot be solved with human resources, especially if the network topology often changes.

However, there are still such network topologies where using static routing is the best option. If a network segment is connected to other networks with only one link, (this is called a stub network) and no other alternative routing paths available, the routing table can be simplified by using only one static entry for all unknown destinations. In every other case, using dynamic routing is preferred.

Dynamic Routing

Dynamic routing has the ability to adapt to changes in the network topology without external intervention. An application called routing daemon continuously exchanges routing information with other routers and maintains the routing table according to the data collected from the other routers and from own measurements.

As dynamic routing is more widespread as static routing, several dynamic routing protocols have been developed in the last few decades, and therefore many types were introduced. Some of the possible groupings of dynamic routing protocols are Internal or External protocols.

As mentioned before, networks are divided into Autonomous Systems (ASs), according to the different controlling authorities of these subnets. The difference of the internal and the external routing protocols is that while internal protocols carry packet routing information only for inside an

AS, then external protocols handle packet routing between the ASs, in the Backbone Area.

Centralized or Distributed Protocols

In today's networks, the most-used routing algorithms are the distributed ones, and that means there is no such designated node in the network, which calculates the routing tables and send them to the routers (centralized case).s In the distributed routing algorithms routers are calculating and using their own routing tables.

All of the routing protocols discussed in this chapter are distributed routing protocols.

Distance Vector or Path Vector or Link State Based Protocols

Distance vector based algorithms are the simplest dynamic routing protocols to be implemented. Routers are not expected to know the whole topology of the network; they exchange their routing information only with the adjacent routers. They store information pairs in their routing table which contains a metric (distance) for a given networks prefix and the interface name where the destination network can be reached with the given metric (direction alias vector). The metric can be hop count, which means the number of routers along the path to the destination network, or a cost value, which is the sum of the cost of the links along the path.

Algorithms using path vectors compare whole paths (list of routers where the packets must go through) in order to select the best one for a destination network prefix. The appropriateness of a path is measured with metric values.

In the case of link state routing protocols the routers have to advertise the status (state) of the links (bandwidth, load, etc.) where they are directly connected to. The scope of advertising is not limited to the adjacent routers, link state information is sent for all routers in the network. Hence, every router comes to know every link in the network including their status information, so a minimum cost spanning tree can be calculated to reach all network segments.

Unicast, Multicast or Anycast Routing Protocols

This alignment means whether a routing protocol is designed for one-to-one, one-to-many or one-to-(one of many) communication scheme. The rest of this subsection is organized according to grouping.

Unicast Routing Protocols

Routing Information Protocol: Next Generation (RIPng)

The Routing Information Protocol was originally designed by Xerox and later it was introduced to the Transmission Control Protocol/Internet Protocol (TCP/IP) protocol stack by the University of California at Berkeley at 1982. The first standard of RIP was RFC1058 (Hendrik, 1988) which was later updated by RFC1388 (Malkin, 1993) and the protocol's name became RIP version 2 (RIPv2). The latest standard of RIPv2 is defined in RFC2453 (Malkin, 1998). Even though it has obvious limitations, RIP became a popular routing protocol in PC networks and many other routing protocols are based on it.

The IPv6 version of RIP is called RIP Next Generation (RIPng) just like IPv6 was called IPng before. RIPng (defined in RFC2080 (Malkin, 1997)) is still the same RIP, it has no remarkable new features; however it has several improvements compared to RIPv2.

RIPng is a distance vector based Interior Gateway Protocol (IGP), which uses hop count metric. A RIPng router sends its whole routing table to its adjacent routers in every 30±15 seconds. The proper amount of the seconds is determined by a random variable and unique for every router in every period. This random update time prevents the network from burst overloading. After receiving the updates, RIPng uses the Bellman-Ford algorithm to calculate the best distance vectors for other networks, which are inserted into the routing table.

RIPng communicates with other routers over User Datagram Protocol (UDP) and uses port number 521 to exchange routing information. Unfortunately, it allows a maximum of 15 hops to reach a foreign network, so prefixes farther than 15 hops are marked as unreachable, and the value of 16 is used for them.

The major disadvantages of RIP based protocols, including RIPng, are the slow convergence time, it does not take into account the bandwidth of the links, and that it can be used only in small networks due to the low hop count limit. Convergence is slow because changes are spreading periodically in a hop-by-hop basis, so it may take several minutes for all routers to become aware of topology changes.

Contrarily the advantages of RIP are that the protocol itself can be implemented easily, it has little demand on resources and does not yield a large overhead at all, therefore in several situations using RIP, or RIPng is still a better solution than a complex dynamic routing protocol.

Enhanced Interior Gateway Routing Protocol (EIGRP)

Enhanced Interior Gateway Routing Protocol (EIGRP) (Albrightson, 1994) is a Cisco Systems proprietary protocol based on Cisco's former Interior Gateway Routing Protocol (IGRP) and used within ASs so it is an IGP. The protocol's main objective was to eliminate the defects of the distance vector based routing protocols (such as RIPng), without creating a new link state based protocol (such as OSPFv3). OSPF and other link state based protocols require a significant CPU power and memory capacity from the routers because of the complexity of the algorithms and the large internal databases. Although EIGRP was developed as a hybrid protocol, which tries to combine the advantages of the two approaches, we will treat it as a distance vector based protocol. Here we must notice that EIGRP is capable of handling IPv4 and IPv6 routing as well and only works on Cisco routers.

The main differences compared to RIPng are the routing update message and the metric. Routing updates are not sent periodically just only when a topology change is detected and instead of RIPng, only the changed routing table entry (RTE) is sent to other routers. The metric became more complex and now it takes into consideration several properties of the connected links. The metric for a path towards a destination network can be calculated with the following formula:

$$\left[\left(K_1 \cdot Bandwidth + \frac{K_2 \cdot Bandwidth}{256 - Load} + K_3 \cdot Delay\right) \cdot \frac{K_5}{K_4 + Reliability}\right] \cdot 256$$

The K constants (K_1 through K_5) are modifiers for different properties and the network administrator can manipulate these values to achieve better routing performance. The functions of these parameters are as follows (their default values are given in brackets):

- **K_1:** Bandwidth modifier (1)
- **K_2:** Load modifier (0)
- **K_3:** Delay modifier (1)
- **K_4:** Reliability modifier (0)
- **K_5:** Reliability modifier (0)

The other variables dynamically represent the properties of a path towards a network:

- **Bandwidth:** the speed of the slowest link in route to the destination (predefined, in kbps)
- **Load:** the load of a link (integer from 1 to 255, where 255 means the link is saturated)
- **Delay:** sum of the delays of every link on the path to the destination (microseconds)

- **Reliability:** represents the probability of a link failure (integer from 1 to 255; 255 means the most reliable)

If we use the formula with the default values of the K modifiers the above formula is effectively reducing to *(Bandwidth+Delay)*256*, since if the K5 variable is set to zero, the term $\frac{K_5}{K_4 + Reliability}$ is interpreted as 1. In EIGRP Bandwidth means *(10,000,000/smallest bandwidth kbps)* and Delay means *(sum of link delays/10)*, thus high-value link metrics are not uncommon.

During the operation of EIGRP, it maintains three different types of tables: the Neighbor table, the Topology table, and the Routing table. The Neighbor table stores information about the adjacent routers, which can be directly accessed through directly connected interfaces. These routers are discovered with an EIGRP Hello packet exchange when the router is connected to a network.

The name of the Topology table can be a little bit misleading since it does not contain information about the whole network topology, rather includes the exchanged routing table entries of adjacent EIGRP routers. The table stores a list of all available destination networks with their respective metrics. EIGRP uses the Diffusing Update Algorithm (DUAL) (Garcia-Luna-Aceves, 1993) to choose a successor (next hop router) and, if there is enough data and multiple possible routes, a feasible successor (backup next hop router) for a destination network. In a converged state, when the routing has stabilized, the entries are marked as Passive, but if something changes in the topology the related routing entries become Active until a new successor is found. DUAL always provides loop-free routes according to a term called *feasibility condition* (Garcia-Luna-Aceves, 1993).

The third EIGRP table type is the Routing Table, which usually stores successors as next hop routers for every destination networks. EIGRP has a special feature, which provides load balancing if multiple paths are present for a destination, even if the costs of these paths are not equal. This is called *un-equal cost load balancing*. If this feature is enabled by the network administrator, the feasible successors as alternative next hop routers are also presented in the Routing table.

Open Shortest Path First version 3 (OSPFv3)

The Open Shortest Path First version 3 (OSPFv3) (Coltun, 2008) is a link state based interior routing protocol, which uses Dijkstra's algorithm to calculate the shortest paths to the destinations networks. The OSPFv3 routers are exchanging information regarding the state of network connections (which is referred as "link state"). Thus, based on all routers' information, a weighted-edge graph is established in the router's memory representing the current state of the network. Then Dijkstra's algorithm is used on this graph, which results a shortest path spanning tree graph, where the root of the tree is the router itself. This tree graph will be the basis of the routing table entries.

To create the weighted-edge graph representation of the network's topology the routers collect Link State Advertisements (LSA) sent by the other routers, which contain information about the state of network connections. If all the LSAs would be advertised in the whole AS, in case of large ASs, this would lead to high overhead and the resources of routers (CPU, memory) would be insufficient to create the Link State Database (LSDB) from the received LSAs.

To avoid this OSPF uses a hierarchical topology. The AS is divided into areas and among these, there is a special one called the Backbone Area (BA), which interconnects the other areas. Areas are identified by a 32-bit identifier (Area ID) which can be a random number but must be unique in the AS.

Non-Backbone Areas

The individual areas should be physically contacted blocks in the topology, and all these areas must have a unique identification number (Area ID) which differs from 0. In addition, each area has a special router, the Area Border Router (ABR). The task of an ABR is to connect the Area to the Backbone Area so a physical or virtual connection to a Backbone Area Router is required. The ABR advertises all the known intra-area routes to the backbone and vice versa, and can be configured (if possible) that way to aggregate the inter-area prefixes into one advertised prefix. This can significantly reduce the number of advertisements to the Backbone Area and saves resources in the routers.

Backbone Area

The Backbone Area is an accentuated area in the AS and its Area ID always equals zero. All ABRs are included in the backbone, which also, as other areas, must be a physically or virtually connected block inside the AS.

External Paths

A router can acquire routing information via protocols other than OSPF as well, such as RIPng, BGP, IS-IS, etc. In terms of OSPF, these are called external routes and can be considered as to be imported into OSPF. The routers, which have at least one interface configured via OSPF, are called Autonomous System Border Routers (ASBR) and can import some extra routes outside from OSPF. The function of these routers is to provide adequate metric to the external paths and then disseminate this information by a special LSA called the AS-external-LSA inside the AS. The ASBRs are also taking place in the Backbone Area.

Stub Areas

Stub areas are known as the areas where AS-external-LSAs are not distributed. Consequently, these areas cannot contain ASBRs, so they do not have direct connection to the external networks. Only one default route is defined as the direction of packets addressed to unknown destinations. The OSPFv3 hierarchy is shown in Figure 4.

The advertisement of different network states is solved by a method called LSA flooding. OSPFv3 defines LSAs with different scopes according to the hierarchical topology and all LSA is advertised in a well-defined range. The different LSAs from the one with the smallest scope to the one with the largest are: *Link-LSA, Router-LSA, Network-LSA, Intra-Area-prefix-LSA, Inter-Area-prefix-LSA, Inter-Area-router-LSA,* and *AS-external-LSA* (Coltun, 2008). As the size of an IPv6 address is quite large (128 bit), OSPFv3 defines only a few necessary LSAs, which may contain IPv6 addresses and prefixes to prevent the network from overhead. These LSAs are the *Link-LSA*, the *Intra-Area-prefix-LSA,* and the *Inter-Area-prefix-LSA*.

LSAs are flooded in the network topology according to the scope values, thus all routers are able to collect information about the state of the network. Then Dijkstra's algorithm is used on the collected data, which defines a network topology graph, to create a minimal cost spanning tree graph to all available network prefixes in the topology. The next hops of all destination networks then written into the routing table. We must notice that the proper definition of link cost (metric) is not defined in the RFCs.

The popularity of OSPFv2 and OSPFv3 is due to their advantages, like fast convergence time, relatively small overhead and that they can handle external paths from other routing protocols. Moreover a router can run OSPFv2 and OSPFv3 simultaneously in a dual-stack environment where IPv4 and IPv6 networking coexists. Unfortunately, OSPF protocols have some disadvantages too as

Figure 4. OSPFv3 hierarchy inside an Autonomous System

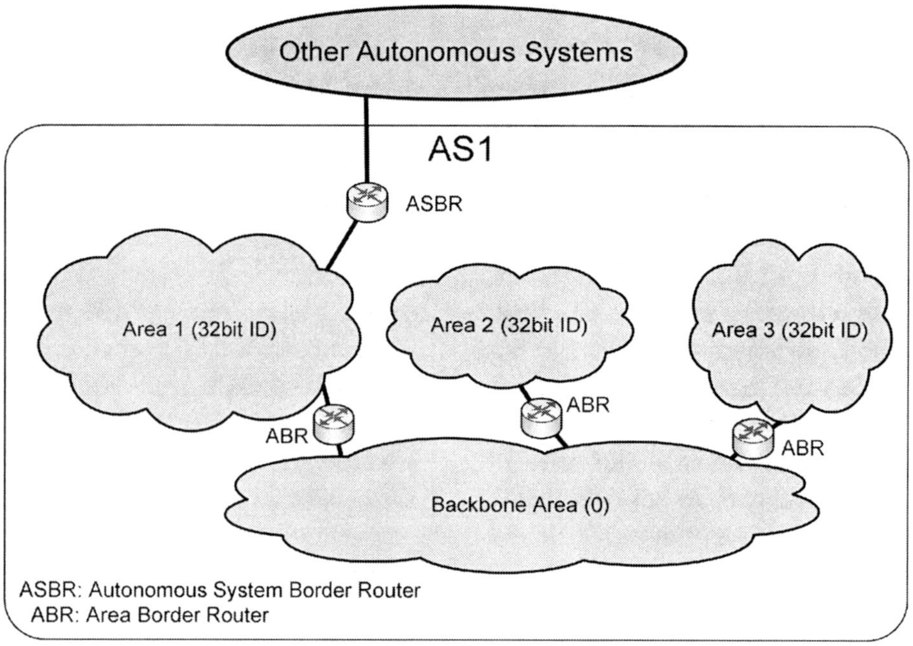

they are really complex protocols and they require well-qualified network administrators to operate a network using OSPF as routing protocol.

Intermediate System: Intermediate System Version 6 (IS-ISv6)

In contrast to previous approaches the IS-IS protocol is not specifically for the TCP/IP protocol stack but developed for the ISO-OSI model in general. Hence the name, the IS, is the abbreviation for Intermediate System which term is used in the OSI network layer concept. Of course, it is also suitable for the dissemination of IPv4 and IPv6 routing information.

The IS-IS approach is quite similar to the OSPF, the main difference is the terminology but for this RFC1195 (Callon, 1990) provides a good basis. The IPv6 protocol support for IS-IS is described in RFC5308 (Hopps, 2008).

Multiprotocol Extension for Border Gateway Protocol Version 4 (MP-BGP4)

In previous subsections, we have seen RIPng and OSPFv3 as an example on how routing protocols operate inside an Autonomous System (AS). In this subsection, a routing protocol is described that was designed to solve the routing information dissemination between Autonomous Systems. So Border Gateway Protocol version 4 (BGP4) (Rekhter, 2006) belongs to the family of Exterior Routing Protocols (ERP). These protocols, as they operate at a higher level of the hierarchy, require a different approach and it is also important to keep in mind that at this level routing is affected by the interests and preferences of the authorities as well.

Each AS uses some kind of preferred Interior Routing Protocol (IRP) for the internal routing. The task of BGP is to disseminate such routing information between ASs and optimize routing in the inter-AS area. Every AS is identified with a unique identification number, which has been allocated by the relevant authority. This makes possible to trace which special prefix can be reached through which AS and to avoid the formation of routing loops.

Three types of ASs should be distinguished from each other in order to a better understanding.

- **Transit AS:** The Transit AS is connected to multiple ASs and forward traffic between them; consequently, the incoming routing information is disseminated to the other ASs as well. Larger Internet Service Providers (ISPs) operate such ASs.
- **Stub AS:** A stub AS has only one connection to other network segments and all traffic goes through this link. In general, the smaller service providers, companies, and university campuses have this type of ASs. Often they do not have a unique identification number either, simply treated as a part of the parent AS. Stub ASs usually connected to parent ASs with static routing (e.g. default route to the parent AS) since the overhead of dynamic routing protocols (like BGP4) would simply not worth using them.
- **Multihomed Nontransit AS:** A multihomed nontransit AS, as the name suggests, also has more connections towards multiple ASs. However, it does not transmit traffic and routing information between them. The more connections are only for the distribution of incoming and outgoing loads of the AS.

The BGP protocol operates over TCP. Two routers first establish a TCP connection and then a BGP connection (they become BGP peers) and after that they can start transferring the routing information inside UPDATE messages. These messages propagate BGP paths, which consist of an IP prefix and other attributes containing route specific information. Such attributes can be the next hop information and the list of ASs, through which a prefix is available (path vector). Although there are many other functions, including the IPv6 support itself, which are solved with using these attributes.

The dissemination of routing information in BGP is based on different policies. The inbound policy determines which paths to be accepted from a given neighbor, while the outbound policy defines which paths to be advertised to a certain direction. It is a general rule that only those paths are advertised which are used by the router itself.

As mentioned before BGP is essentially made for IPv4 and other protocols, including IPv6, are supported with multiprotocol extension which is defined in RFC4760 (Bates, 2007). The brief summary of the characteristics of IPv6 support in BGP are as follows.

In the case of BGP, routers are uniquely identified with the IPv4 address of one of its interfaces. This is a basic role in the BGP standard and even when using the protocol in IPv6 environment routers must have at least one IPv4 address.

The BGP UPDATE message is only suitable to carry IPv4 prefixes and next hop information according to the standard structure of this message type. If we want to disseminate IPv6 routing information it can only be done with a special attribute defined in RFC4760 (Bates, 2007) and RFC2545 (Marques, 1999). This is the MP_REACH_NLRI attribute, which can contain IPv6 next hop information and if it is needed, multiple instance of IPv6 prefixes. Path withdrawal is a similar problem so a similar attribute has been defined as well called MP_UNREACH_NLRI.

For further information about BGP4 and MP-BGP4 please refer to their standards: RFC4271 (Rekhter, 2006), RFC4760 (Bates, 2007) and RFC2545 (Marques, 1999).

Multicast Routing Protocols

The purpose of multicast routing is to forward all messages from a server to multiple clients without the need of sending all messages separately. The main property of multicasting is that a multicast address identifies a multicast group (collection of clients) and messages to the group members should be sent to this group address. A range of addresses have been designated to be used as multicast addresses from the whole IPv6 address space, and this range is the addresses beginning with FF bytes in hexadecimal form. In this subsection, we will shortly introduce the most common multicast protocols, including the listener announcement protocols and the multicast data distribution protocols.

Multicast listener announcement protocols are the Internet Group Management Protocol (IGMP) and the Multicast Listener Discovery (MLD) protocol. IGMP is a traditional protocol, which allows users to sign up to or sign off from a multicast address in the local network (IGMPv2 (Fenner, 1997)). Moreover, it allows the router to query, whether if there is anybody left in the multicast group in a network segment connected to one of the router's interfaces. IGMP was developed to IPv4 and MLD takes over its role in IPv6 environment. MLD has quite similar functionality as IGMPv2; perhaps the main difference is that it no longer appears as a separate protocol as it became a part of the ICMPv6 specification (Conta, 2006).

Before we go any further with the protocols used between multicast routers, we must become familiar with the concept of multicast distribution trees. The leaves of the distribution tree are always the clients signed up for a multicast service, and routers transmitting multicast packets all the time downwards in the distribution tree thus avoiding formation of loops. Depending on the location of the tree's root, we distinguish two types of distribution trees. If the root of the tree is the source itself, (source based distribution tree) then the paths between the source and the clients are optimal, but separate distribution tree for each source is required and this imposes a significant overhead on the routers. The other type of forwarding tree is the shared distribution tree where the root is always a designated router in the network. In this case, first of all, this selected router gets the message from the source and then sends towards the clients. A shared distribution tree is able to forward multicast packets for multiple multicast groups simultaneously while the routers have to deal with only one multicast tree.

It is an important question that how a routers decides the right direction of forwarding. A simple but effective idea is to check the source IP address of the multicast packet and if it arrived through an interface where the network of the source IP can be reached according to the unicast routing table, the packet is forwarded out through every interface that participates in the distribution tree. This method is called Reverse Path Forwarding (RPF) and based on the unicast routing tables but it is independent from the unicast routing protocol (OSPFv3, RIPng, etc.). Thus, multicast routing protocols using this technique are called Protocol Independent Multicast (PIM).

Two of these protocols are used in practice. The PIM-Dense Mode (PIM-DM) (Adams, 2005) uses the method of packet flooding. Starting from the source all routers forward the packets out of their interfaces and this can only be terminated if no one claims the multicast group packages. The advantage of the protocol is the simplicity and that it does not require a common point (Rendezvous Point, RP) among the routers whose failure would compromise the operation. The disadvantage is that it only works with a source based distribution tree and generates a significant overhead due to flooding.

The other solution of the problem is the PIM Sparse-Mode (PIM-SM) protocol (Fenner, 2006). This concept is the opposite of the one used in the case of PIM-DM: packets are only forwarded to those directions where clients explicitly requested the service. Basically, packet distribution starts with using shared distribution tree, but the router in order to optimize the routes can come over to the source based distribution. The main approach of the protocol is using a RP router, where the need for packets of a multicast group towards a subnet can be indicated. Initially the source sends the packages to be distributed to this router first, and it starts the packet forwarding downwards inside the tree.

Anycast Routing Protocols

In most cases anycast routing protocols are basically derived from popular unicast or multicast routing protocols. We must notice that these protocols are still only proposals and none of them became a standard yet. The most commonly mentioned anycast extensions are for RIPng (Malkin, 1997), OSPFv3 (Coltun, 2008) and PIM-SM (Fenner, 2006). In this subsection we will introduce anycast routing principles by way of Anycast RIP (ARIP) (Doi, 2004), Anycast OSPF (AOSPF) (Doi, 2004) and Protocol Independent Anycast – Sparse Mode (PIA-SM) (Matsunaga, 2005). Although there are many proposals to achieve anycast routing in IPv6 networks we chose these three protocols for demonstration because they based on previously discussed routing protocols and they provide a relatively simple solution for anycasting.

The first two (ARIP and AOSPF) are based on unicast routing and we will discuss them in details. The operation of anycast routing protocols are consist of three steps and the difference between ARIP and AOSPF is in Step 2. The three steps are as follows:

- **Step 1:** Collecting anycast member information The anycast receivers (servers, service provider nodes) must be able to notify the nearest anycast router of their joining or leaving status, when they are participating in or leaving an anycast group. The anycast router collects information about anycast receivers and their status.
- **Step 2:** Constructing and updating the routing table, exchanging information with other routers. Using the collected and exchanged information, the anycast router constructs or updates its routing table to achieve proper dynamic routing.
- **Step 3:** Forwarding anycast packets The packet-forwarding algorithm (anycast routing) is executed according to the routing table created in the second step. This works in the same way as in the unicast case. Since the anycast addresses are indistinguishable from unicast addresses, they are kept in the unicast routing tables as host routes (128-bit entries).

In the following, we will discuss Step 2.

The appropriateness of a route to an anycast receiver is indicated by metric. The literature of anycasting defines two kinds of metrics: the receiver metric (cannot be modified by the anycast routers) and the link metric (modified by every anycast routers). If the receiver metric is used in ARIP or AOSPF, the two protocols use a quite similar method for constructing and updating the routing table (Figure 5.). If the anycast routers use only the receiver metric to select the "nearest" anycast receiver, they can use the unicast routing to forward an anycast packet to the appropriate anycast receiver.

The link metric can also be used to create the routing table entries in the anycast routers. In this case, the link metric information is collected in different ways in ARIP and in AOSPF.

Anycast Routing Information Protocol (ARIP)

ARIP (Doi, 2004) is based on RIPng and operates with the link metric as described below. The construction and updating steps of the routing table in ARIP is shown in Figure 6.

- **Step 2.1:** Maintaining membership information by Anycast Routing Discovery (ARD) query/report messages. All anycast routers advertise ARD query messages in a periodical manner to challenge the anycast

Figure 5. IPv6 anycast routing with the receiver metric (Adapted from (Doi, 2004))

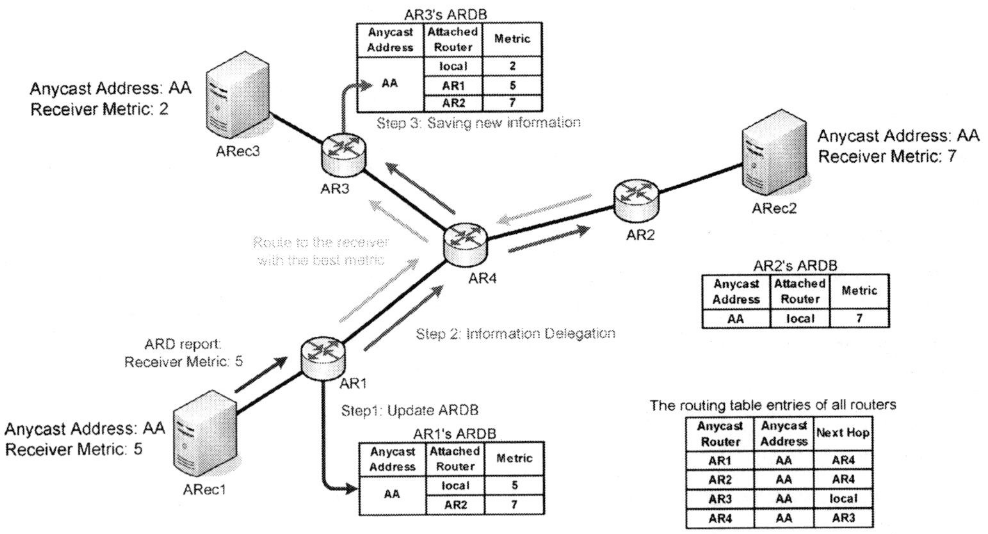

receivers in the subnet. The anycast receivers send back an ARD report message, with their unicast address and the anycast address for the anycast group they are going to participate in. The anycast receivers can send unsolicited ARD report messages if no ARD query was arrived for a predefined interval.

- **Step 2.2:** Send the Anycast Route Information (ARI) message. After collecting the ARD reports from the anycast receivers, the router constructs an ARI message, which includes at least one anycast address-metric pair. Then the router sends the ARI to its adjacent anycast routers. The ARI message is flooded across the anycast

Figure 6. The detailed operation steps ARIP using the link metric. (Adapted from (Doi, 2004))

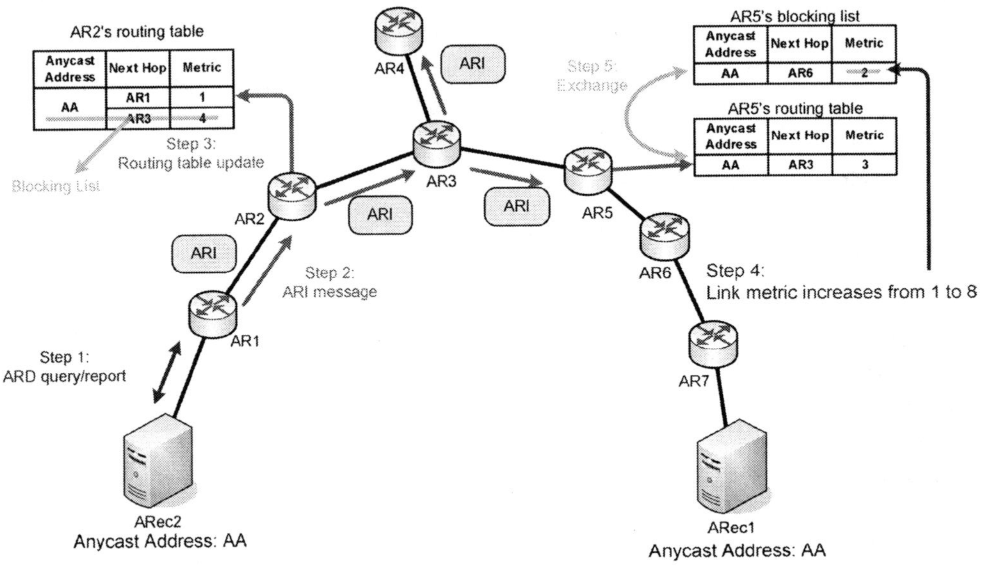

routers in a hop-by-hop basis, and every router who gets the ARI, sends it out on all of its interfaces except where it came from. However, before forwarding an ARI message the router changes the link metric value to the one of the outgoing interface. Note that the link metric from the anycast receiver (alias server or service provider) to the client is more important than the other way.

- **Step 2.3:** Constructing or updating the routing table and/or blocking list. When an ARI message is received by an anycast router, it first checks if the anycast address in the ARI message already exists in the routing table or not. If not, a routing entry is created. Otherwise, the anycast router examines if the new information provides better link metric or not. If the new metric is better, the router updates the information in the routing table, and if this new value belongs to a new anycast receiver (the ARI arrived on a different interface), the old one is moved to the blocking list for backup purposes. If the ARI's metric is not better than the one in the routing table, the router stores this alternative receiver in the blocking list.

Anycast Open Shortest Path First (AOSPF)

AOSPF (Doi, 2004) is based on Open Shortest Path First version 3 (OSPFv3) and using Dijkstra's algorithm to calculate the shortest path between two nodes in the network. The operation of AOSPF is described below and shown in (Figure 7.):

- **Step 2.1:** Constructing the topology graph of anycast routers. Every anycast router creates a Router/Network Link State Advertisement (R/N LSA), which contains metric information about the links, which the router is directly connected to. After creating these R/N LSA messages, the anycast router sends them to all other AOSPF routers. In every router, the collected LSA information is stored in a topological database, and from that database, the router is able to construct the weighted-edge topological graph of the anycast routers, and using Dijkstra's algorithm minimal cost

Figure 7. AOSPF uses AM-LSAs to advertise link metrics. (Adapted from (Doi, 2004))

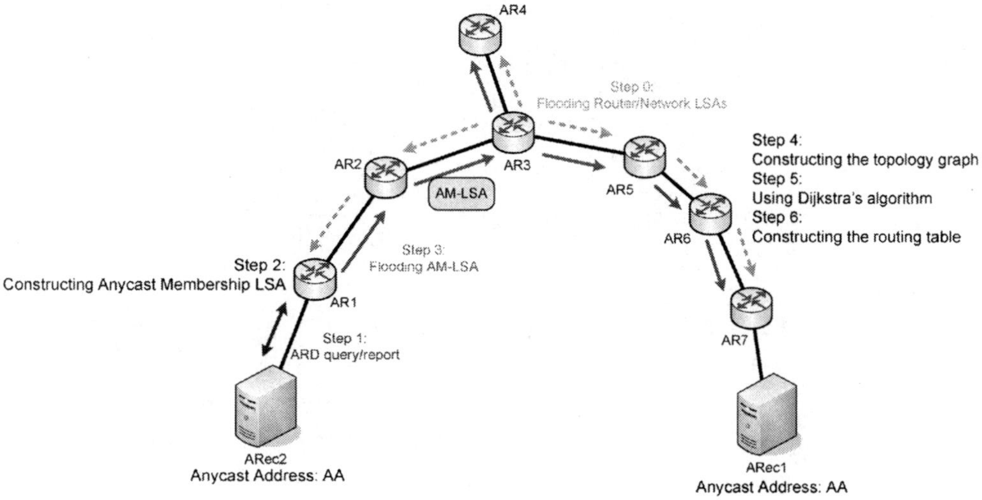

spanning tree can be constructed. This method is the same as used in OSPFv3.

- **Step 2.2:** Maintaining membership information by Anycast Routing Discovery (ARD) query/report messages. This step is the same as used in the first step of ARIP operation.
- **Step 2.3:** Creating and sending the Anycast Membership - Link State Advertisement. After receiving an ARD report, the anycast router constructs an Anycast Membership - LSA (AM-LSA) message which stores information about the anycast address and the anycast receivers in the subnet. Then, the anycast router sends the AM-LSA to other anycast routers by LSA flooding.
- **Step 2.4:** Constructing or updating the routing table and/or blocking list. The AM-LSA message is treated like any other LSAs. The information from the AM-LSAs is collected into the topology database and a new weighted-edge graph representation of the anycast network is calculated but this time including the anycast receivers as well. Then the minimal cost spanning tree is constructed with Dijkstra's algorithm and from this, the entries for the routing table and blocking list (backup entries) are created.

Protocol Independent Anycast: Spare Mode (PIA-SM)

Like the PIM-SM, PIA-SM (Matsunaga, 2005) creates a shared distribution tree for the anycast addresses and this tree has a root node, a designated router, called the Rendezvous Point (RP).

The RP is a router configured to dynamically manage the distribution tree (RPT) for the anycast address. However, unlike PIM-SM, PIA-SM's tree forwards an anycast packet to only one leaf of the RPT. The appropriate leaf anycast receiver is chosen by the metric value (Matsunaga, 2005). After building the RPT, the packets addressed to the RPT's anycast address are delivered into the tree on the closest point (anycast router) to the originator. From the closest point to the RP, the anycast packet is transferred in an encapsulated packet, which is addressed, to the RP's unicast address. After the RP receives the packet, it is

Figure 8. The operation of PIA-SM with encapsulation. (Adapted from (Matsunaga, 2005))

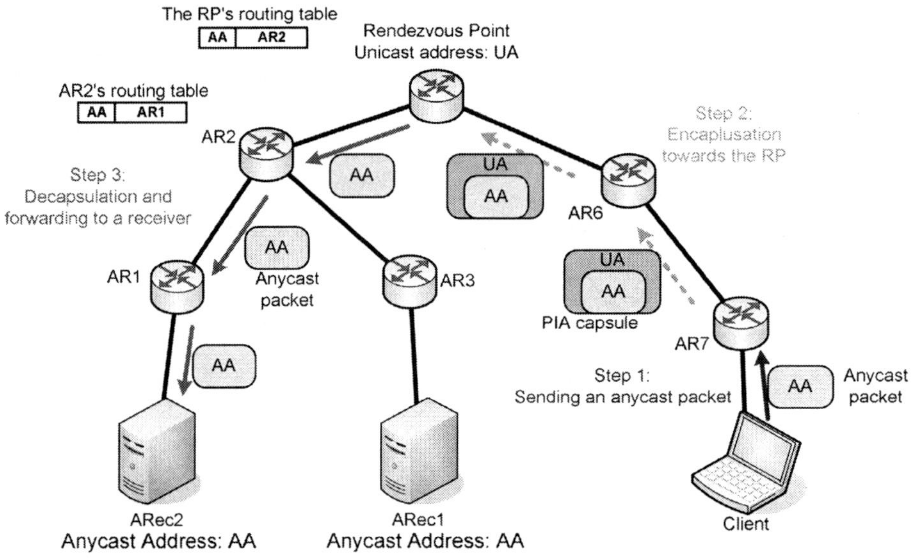

decapsulated and forwarded to the selected anycast receiver on the tree by addressing the packet to the anycast address. Packet forwarding in PIA-SM is depicted in (Figure 8.).

SERVING EFFICIENT DATA AGGREGATION

In present-day networks, mobile phones and other devices using wireless interfaces became more widespread than ever. Mobility turns into part of our life, thus mobile service providers offers more and more applications and services for our mobile devices (e.g. multimedia streaming (IPTV), Voice-over-IP (VoIP), etc.). Another important territory of wireless communication is the sensor networks, where tiny little devices provide measurement information usually in a periodical manner. An example for a sensor network could be a mine where carbon-monoxide sensors are placed in given distances into the tunnels for the safety of the miners. In both cases, the nodes usually have limited resources: battery power, computational capacity, and available wireless bandwidth, therefore the efficiency of communication is always important in wireless and mobile architectures.

According to the previously introduced multimedia services and measurement results, a new type of information flow appeared in the core networks. This flow contains feedback information from the individual devices. Feedback is addressed to the central node, which sends the multimedia flow. The central node processes these feedbacks: it may collect these data for creating statistics or it can change the quality of a given service according to the feedbacks, like in a system where the multimedia stream adapts to the available bandwidth.

As the number of devices and available services grows, the usage of feedback information could generate a significant overhead in the IP core network. To avoid this situation we introduced a method called System-level (or Network-level) feedback aggregation for IPv6 networks which is based on the previously discussed anycast routing.

In general, a data aggregation scheme is responsible to compose, recompose, transform, analyze data and basically consists of a variety of adaptive methods, which can merge, and aggregate information from wide scale of possible data sources and data pieces into a set of well-organized and refined set of information. The routing-centric aggregation mechanisms mainly cover routing problems, e.g., when and where two (or more) information pieces (i.e. datagram) can meet each other in order to be aggregated.

The Intention of Using System-Level Data Aggregation

Mobile devices and wireless sensors are generating feedback information for a central application controller or processing server. In usual network architecture all feedback data, which is generally only a few bytes, should be sent in individual IP packets. Considering that this feedback information generated very often, this would create significant overhead in the core network, since the IP header is much longer than the payload. In the case of IPv6, the fixed part of the IP header is 40 bytes excluding the header of the transport layer protocol! Network overhead causes latency; therefore, the central application controller cannot adapt the multimedia stream parameters correctly in time. Not to mention that latency in the network not only affects this service but all network traffic.

System-level or in another name network level feedback data aggregation tries solve this problem by aggregating feedback data from individual wireless devices into a common IPv6 payload. So the aim of using System-level aggregation is to achieve low latency for feedbacks to the central application controller (and thus for other services), and for that reason to keep a closely real-time monitoring system (Quality-of-Service). The other goals were robustness and minimal overhead on the core network.

Data Aggregation Using Anycast

System-level aggregation requires two new types of entities in the core network to work properly: the Anycast Capable Router (ACR) and the Feedback Aggregation Server (FAS).

When the network-level aggregation is in operation the wireless devices and/or mobile nodes are required to address their IPv6 feedback packets to the anycast address of the Feedback Aggregation Servers, called the Feedback Anycast Address (FAA).

These packets are delivered to the „closest" aggregation server (or right to the central application controller, since this central unit also must be a member of the anycast group) using an anycast routing protocol (AOSPF, ARIP, etc.) which is implemented in the intermediate ACRs. It is not necessary for all routers to be anycast capable, because anycast routing creates an overlay network over the physical network.

The Feedback Aggregation Servers (FAS) are joined into the same anycast group, have the same anycast address (while they have their unicast addresses too) and provide the System-level aggregation which collects the feedback information from the arrived IPv6 packets and sends the collected data in one IPv6 packet to the central application controller by unicast addressing. This method is very similar to multicast, but in the reverse order, while multicast is a one-to-many protocol, this should be a many-to-one.

In accordance to the literature (Jelinek 2008), the aggregation ratio in this level is determined by the MTU on the aggregator server's uplink. On average, an aggregation ratio between 2:1 and 10:1 can be achieved. A data aggregation scenario using anycast is shown in Figure 9.

FUTURE RESEARCH DIRECTIONS

The method introduced in the previous subsection tries to decrease the number of IPv6 packets with small feedback payloads while it still wants to guarantee the real-time monitoring, therefore the impact of timing is very important. The timers in the aggregation servers have to be defined to generate not too much latency while it keeps the feedback messages in its queues, but also to keep

Figure 9. Anycast based feedback data aggregation in IPv6 networks

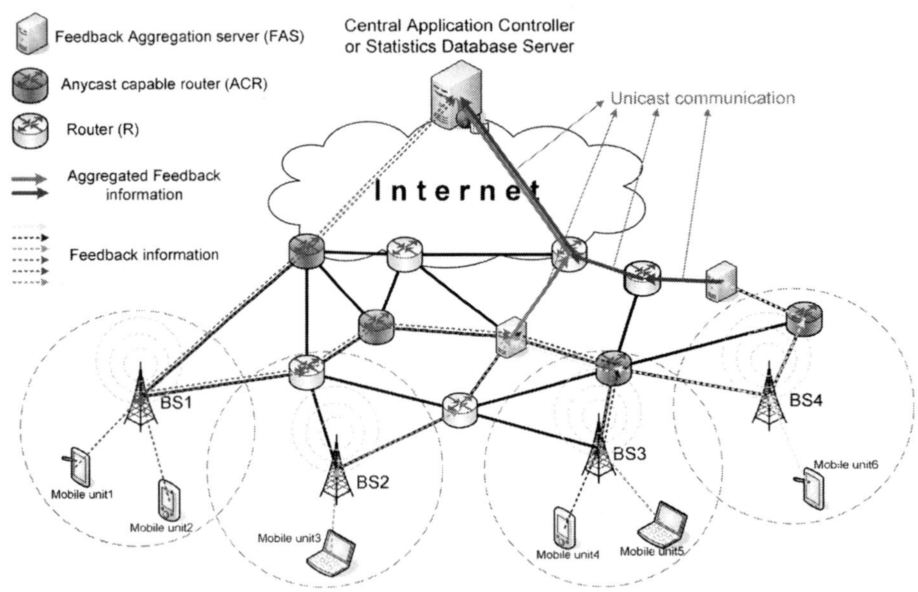

the aggregation in an appropriate level. Therefore, we have to find the proper mechanism to rate the timers, according to the properties of the service and the condition of the IPv6 network.

Simultaneously we will provide mathematical analysis to prove that using network-level feedback aggregation prevents the IP core network from overheads caused by feedback information, thus more service subscribers can be served using the same bandwidth.

CONCLUSION

In the beginning of this chapter, we presented the basics of the Internet Protocol version 6 and we introduced the significant new features according to version 4. Then we presented the hierarchical nature of the IPv6 network topology and later discussed the role of packet exchange alias IP packet routing with defining groupings by different points of view.

We introduced the most popular unicast, multicast and anycast IPv6 routing protocols and then our proposal how to aggregate multimedia and measurement feedbacks with anycast routing protocols in today's and tomorrow's IP based packet switched networks.

REFERENCES

Abley, J. (2005). *Anycast addressing in IPv6*. IETF Internet draft. Retrieved April 29, 2010, from http://tools.ietf.org/html/draft-jabley-v6-anycast-clarify-00

Adams, A., Nicholas, J., & Siadak, W. (2005). *Protocol independent multicast - dense mode (PIM-DM): Protocol specification* (revised). RFC3973. Retrieved April 29, 2010, from http://tools.ietf.org/html/rfc3973

Albrightson, R., Garcia-Luna-Aceves, J. J., & Boyle, J. (1994). EIGRP - a fast routing protocol based on distance vectors. In *Proceedings of Networld/Interop 94*. Berlin, Germany. Retrieved April 29, 2010, from http://www.cse.ucsc.edu/research/ccrg/publications/interop94.pdf

Bates, T., Chandra, R., Katz, D., & Rekhter, Y. (2007). *Multiprotocol extensions for BGP-4*. RFC4760. Retrieved April 29, 2010, from http://tools.ietf.org/html/rfc4760

Bokor, L., Dudas, I., & Imre, S. (2005). Anycast-based micromobility. In *Proceedings of SoftCOM 2005* (pp. 125–130.), Split, Marina-Frapa, Croatia, (2005.09.15–17).

Bokor, L., Dudas, I., Szabo, S., & Imre, S. (2005) Anycast-based micromobility: A new solution for micromobility management in IPv6. In *Proceedings of MoMM2005* (pp. 68–75.). Malaysia, Kuala Lumpur, (2005.09.17–21). ISBN: 3-85403-195-5

Bokor, L., Simon, V., Dudás, I., & Imre, S. (2007). Anycast subnet optimization for efficient IPv6 mobility management. In *Proceedings of IEEE GIIS 2007* (pp. 187–190). Marrakesh, Morocco, (2007.07.02–06). DOI: 10.1109/GIIS.2007.4404188, ISBN 978-1-4244-1376-8

Callon, R. (1990). *Use of OSI IS-IS for routing in TCP/IP and dual environments*. RFC1195. Retrieved April 29, 2010, from http://tools.ietf.org/html/rfc1195

Coltun, R., Ferguson, D., Moy, J., & Lindem, A. (2008). *OSPF for IPv6*. RFC5340. Retrieved April 29, 2010, from http://tools.ietf.org/html/rfc5340

Conta, A., Deering, S., & Gupta, M. (2006). *Internet control message protocol (ICMPv6) for the Internet protocol version 6 (IPv6) specification*. RFC4443. Retrieved April 29, 2010, from http://tools.ietf.org/html/rfc4443

Deering, S., & Hinden, R. (1995). *Internet protocol, version 6 (IPv6) specification*. RFC1883. Retrieved April 29, 2010, from http://tools.ietf.org/html/rfc1883

Deering, S., & Hinden, R. (1998). *Internet protocol, version 6 (IPv6) specification*. RFC2460. Retrieved April 29, 2010, from http://tools.ietf.org/html/rfc2460

Doi, S., & Miyahara, H. (2004). *Design, implementation and evaluation of routing protocols for IPv6 anycast communication*. Master's thesis, Department of Information Networking, Graduate School of Information Science and Technology Osaka University, Osaka, Japan.

Dudas, I., Bokor, L., & Imre, S. (2007). Survey and extension of applications and services in IPv6 anycasting . In Taniar, D. (Ed.), *Encyclopedia of mobile computing & commerce*. New York, NY: Idea Group Inc.

Fenner, R., Handley, M., Holbrook, H., & Kouvelas, I. (2006). *Protocol independent multicast - sparse mode (PIM-SM): Protocol specification* (revised). RFC4601. Retrieved April 29, 2010, from http://tools.ietf.org/html/rfc4601

Fenner, W. (1997). *Internet group management protocol, version 2*. RFC2236. Retrieved April 29, 2010, from http://tools.ietf.org/html/rfc2236

Fuller, V., Li, T., Yu, J., & Varadhan, K. (1993). *Classless inter-domain routing (CIDR): An address assignment and aggregation strategy*. RFC1519. Retrieved April 29, 2010, from http://tools.ietf.org/html/rfc1519

Garcia-Luna-Aceves, J. J. (1993). Loop-free routing using diffusing computations. *IEEE/ACM Transactions On Networking, 1*(1), 130–141. Retrieved April 29, 2010, from http://www.cse.ucsc.edu/research/ccrg/publications/jj.dual.ton93.pdf

Hedrick, C. (1988). *Routing information protocol*. RFC1058. Retrieved April 29, 2010, from http://tools.ietf.org/html/rfc1058

Hinden, R., & Deering, S. (1998). *IP version 6 addressing architecture*. RFC2373. Retrieved April 29, 2010, from http://tools.ietf.org/html/rfc2373

Hopps, C. (2008). *Routing IPv6 with IS-IS*. RFC5308. Retrieved April 29, 2010, from http://tools.ietf.org/html/rfc5308

Johnson, D., Perkins, C., & Arkko, J. (2004). *Mobility support in IPv6*. RFC 3775. Retrieved April 29, 2010, from http://tools.ietf.org/html/rfc3775

Malkin, G. (1993). *RIP version 2, carrying additional information*. RFC1388. Retrieved April 29, 2010, from http://tools.ietf.org/html/rfc1388

Malkin, G. (1998). *RIP version 2*. RFC2453. Retrieved April 29, 2010, from http://tools.ietf.org/html/rfc2453

Malkin, G., & Minnear, R. (1997). *RIPng for IPv6*. RFC2080. Retrieved April 29, 2010, from http://tools.ietf.org/html/rfc2080

Marques, P., & Dupont, F. (1999). *Use of BGP-4 multiprotocol extensions for IPv6 inter-domain routing*. RFC2545. Retrieved April 29, 2010, from http://tools.ietf.org/html/rfc2545

Matsunaga, S., Ata, S., Kitamura, H., & Murata, M. (2005). Design and implementation of IPv6 anycast routing protocol: PIA-SM. *Proceedings of the 19th International Conference on Advanced Information Networking and Applications* (AINA'05) (pp. 839–844). Taipei, Taiwan. ISBN 0-7695-2249-1

Narten, T., Nordmark, E., Simpson, W., & Soliman, H. (2007). *Neighbor discovery for IP version 6* (IPv6). RFC4861. Retrieved April 29, 2010, from http://tools.ietf.org/html/rfc4861

Rekhter, Y., Li, T., & Hares, S. (2006). *A border gateway protocol 4* (BGP-4). RFC4271. Retrieved April 29, 2010, from http://tools.ietf.org/html/rfc4271

Thomson, S., Narten, T., & Jinmei, T. (2007). *IPv6 stateless address autoconfiguration*. RFC4862. Retrieved April 29, 2010, from http://tools.ietf.org/html/rfc4862

ADDITIONAL READING

Doi, S., Ata, S., Kitamura, H., & Murata, M. (2003). Protocol design for anycast communication in IPv6 network. In Proceedings of 2003 IEEE Pacific Rim Conference on Communications, Computers and Signal Processing (PACRIM'03) (pp. 470–473.)

Doi, S., Ata, S., Kitamura, H., & Murata, M. (2004). IPv6 anycast for simple and effective communications. *IEEE Communications Magazine*, *42*, 163–171. doi:10.1109/MCOM.2004.1299362

Doi, S., Ata, S., Kitamura, H., & Murata, M. (2005). Design, Implementation and Evaluation of Routing Protocols for IPv6 Anycast Communication. Proceedings of the 19th International Conference on Advanced Information Networking and Applications (AINA'05) (pp. 833-838.) ISBN 0-7695-2249-1, Taipei, Taiwan

Jelinek, M., Komosny, D., & Burget, R. (2008). Hierarchical feedback aggregation in IPTV. Proceedings of the 12th WSEAS international conference on Computers (pp. 852-856.) ISBN: 1790-5109, Heraklion, Greece

Matsunaga, S. (2004). PIA-SM: Design and implementation of core-based routing protocol for ipv6 anycast. Bachelor's thesis, School of Engineering Science, Osaka University, Japan

KEY TERMS AND DEFINITIONS

Anycast: IPv6 supports three communication forms by the standard; anycast is one of these and realizes a one-to-(one-of-many) communication scheme.

Feedback Aggregation: As feedback information is usually only a few bytes long, using one IP packet for one feedback fragment is rather wasteful. Feedback aggregation means a method when multiple feedback fragments are sent in a common IP packet.

IPv6: IPv6 is the latest version of the Internet Protocol, which provides many improvements according to its predecessor, IPv4. Some of these new features are: host mobility management, supporting secure communications, host autoconfiguration, etc.

Metric: Metric is a value, which defines the appropriateness of a path or a data receiver in routing protocols. This appropriateness can be defined many ways: the metric can be simply hop count, but it can be represented by a cost either, which is a result of a complex formula.

Multicast: IPv6 supports three communication forms by the standard; multicast is one of these and realizes a one-of-many communication scheme.
Routing protocols: These protocols are responsible for delivering a packet from one node to another through other special nodes called routers. These protocols are running continuously on the routers and according to the state of the network, they decide to send out a packet on which interface towards the destination host.

Unicast: IPv6 supports three communication forms by the standard; unicast is one of these and realizes the basic, one-to-one communication scheme.

APPENDIX

Table of abbreviations

ABR	Area Border Router
ACR	Anycast Capable Router
AM-LSA	Anycast Membership - Link State Advertisement
AOSPF	Anycast Open Shortest Path First
ARD	Anycast Routing Discovery
ARI	Anycast Route Information
ARIP	Anycast Routing Information Protocol
AS	Autonomous Systems
ASBR	Autonomous System Border Routers
BA	Backbone Area
CIDR	Classless Inter-Domain Routing
CN	Correspondent Node
CoA	Care-of Address
DAD	Duplicate Address Detection
DUAL	Diffusing Update Algorithm
EGP	Exterior Gateway Protocol
EIGRP	Enhanced Interior Gateway Routing Protocol
ERP	Exterior Routing Protocol
FAA	Feedback Anycast Address
FAS	Feedback Aggregation Server
HA	Home Agent
HN	Home Network
HoA	Home Address
IETF	Internet Engineering Task Force
IGMP	Internet Group Management Protocol
IGP	Interior Gateway Protocol
IGRP	Interior Gateway Routing Protocol
IP	Internet Protocol
IPsec	IP Security
IRP	Interior Routing Protocol
IS-ISv6	Intermediate System to Intermediate System version 6
ISP	Internet Service Provider
LSA	Link State Advertisement
LSDB	Link State Database
MLD	Multicast Listener Discovery
MN	Mobile Node

MP-BGP4	Multiprotocol Border Gateway Protocol version 4
OSPFv3	Open Shortest Path First version 3
PC	Personal Computer
PIA-SM	Protocol Independent Anycast – Sparse Mode
PIM	Protocol Independent Multicast
PIM-DM	PIM-Dense Mode
PIM-SM	PIM Sparse-Mode
QoS	Quality of Service
R/N LSA	Router/Network Link State Advertisement
RA	Router Advertisement
RIP	Routing Information Protocol
RIPng	Routing Information Protocol – Next Generation
RP	Rendezvous Point
RPF	Reverse Path Forwarding
RTE	Routing Table Entry
SIPP	Simple Internet Protocol Plus
TCP/IP	Transmission Control Protocol/Internet Protocol
UDP	User Datagram Protocol

Chapter 20
Multiprotocol Label Switching Virtual Private Networks:
Problems, Protocols, Possibilities

Jan Schankin
Christchurch Polytechnic Institute of Technology, New Zealand

Eduardo Correia
Christchurch Polytechnic Institute of Technology, New Zealand

ABSTRACT

Multiprotocol Label Switching (MPLS) was originally designed with the intention of improving the speed with which routers could forward packets in Internet Protocol (IP) networks. Due to significant improvements in packet forwarding, this is no longer an advantage, but the technology has found large-scale industry-wide acceptance because of its greatly widened scope and application. Multiple extensions and enhancements to the protocol make it capable of solving an array of current service provider and customer network requirements for a converged network in an IP dominated world. The chapter considers the use of MPLS for the provisioning of a virtual private network over a shared physical infrastructure, and discusses the logic and functionality of key protocols associated with MPLS. It then goes on to explore the problems, protocols, and possibilities of these technologies in current environments.

INTRODUCTION

Multiprotocol Label Switching (MPLS) is by no means new. The concept was introduced in 1996 in "cut through routing" (Newman, Edwards et al, 1996), which involved 'switching' IP over ATM networks by applying and using a label to forward packets. Toshiba and IBM also introduced technologies that in some way used labels to forward data (Davie, 2000). Cisco, an organization with a focus on Ethernet and Internet Protocol (IP), not ATM, then took the concept and adapted it for Ethernet. Initially a proprietary protocol, it was later rebranded as 'Label Switching' under the banner

DOI: 10.4018/978-1-60960-732-6.ch020

of the Internet Engineering Task Force (IETF). A few years later, it was once again rebranded as MPLS, with support for Layer 2 (L2) and Layer 3 (L3) virtual private networks (VPNs) that are capable of transporting a wide range of protocols over IP. In fact, MPLS populates lookup tables and distributes labels in a variety of ways, and is not so much a protocol as it is a set of protocols that make IP truly universal, both for local and service provider networks. As such, it is considered by many people to be the rightful successor to Frame Relay and Asynchronous Transfer Mode (ATM) (Sosinsky, 2009).

It is not cost-effective for providers, whether they are telecommunications companies (telcos) or Internet Service Providers (ISPs), to offer each service on its own separate network (Xu, 2010). These providers have found MPLS VPNs a viable solution to the growing demand for highly available, highly secure, and highly cost effective services. This chapter describes the purpose, problems, and possibilities of MPLS and associated enhancements and extensions used to create provider provisioned virtual private networks (PPVPNs), and their support for various connection scenarios, traffic requirements, and applications. While MPLS is associated with large networks and providers, it also affects the very nature of wide area network (WAN) links and therefore impacts customers, who, for example, cannot easily propagate an Open Shortest Path First (OSPF) default route across an MPLS WAN.

The chapter first covers the basics of MPLS and the ways, in which several other protocols and extensions interact to create L3 VPNs across a shared physical infrastructure. The standard for this particular type of PPVPN, referred to as "BGP/MPLS IP VPN", is outlined in RFC 4364 (Rosen & Rekhter, 2006). The chapter then goes on to explore the problems, protocols and possibilities of a BGP/MPLS IP VPN or, as it is referred to, an MPLS L3 VPN, starting with an overview of various customer-to-service provider connection scenarios, including internet access and multihomed environments. It concludes with some reflections on what MPLS has become, where it will take the industry and considers the implications for the role and future of IP.

BACKGROUND

Service provider networks face the problem of connecting multiple customers at various locations across a shared physical infrastructure. Legacy networks followed a strict connection-oriented approach in which data is sent across a pre-established physical or logical path. This deterministic approach followed the method used by the original analogue public switched telephone network (PSTN) and led to the development of other circuit switched technologies such as Integrated Services Digital Network (ISDN) and in its original form Synchronous Optical Networking (SONET) or Synchronous Digital Hierarchy (SDH). Some packet switched technologies follow the connection-oriented approach by establishing virtual circuits between two or multiple points across a shared network, including the old but reliable X.25 and unreliable Frame Relay, as well as ATM and MPLS (Perros, 2005).

These technologies function at either Layer 1 (L1) or L2 but on a fundamental level, they actually serve the same purpose as the now ubiquitous IP protocol, which functions at L3. It seems strange, especially for a new generation of network engineers for whom networks are automatically IP and ATM or Frame Relay legacy protocols, to view IP as a protocol that simply connects multiple Ethernet-based networks. That, though, is the actual fundamental purpose of IP: connecting local area networks through a hierarchical addressing scheme. The great flexibility, ease of implementation and maintenance plus overall cost effectiveness have led to a near total dominance of IP in local area network sector with Ethernet as the protocol of choice for the Data Link layer.

The same cannot be said about the service provider sector because the great flexibility of IP does give rise to certain problems. On its own, IP is a connection-less protocol, with every device along the path expected to make its own forwarding decision based on a routing table lookup. This results in various packets from the same stream of information taking different paths and indeed different times to reach their destination, with the result that packets arrive in the incorrect sequence. IP did not achieve industry-wide acceptance because it made it difficult for service providers to make optimal, cost-effective use of the network in a connection-less environment. In fact, during the "IP vs. ATM in all-out-war" (Dawson, 1998), many experts expected ATM to replace, even in local area networks, because of its quality of service (QoS) capabilities and predominance in the WAN market. While this did not eventuate, in truth ATM and Frame Relay offered benefits IP could not rival. For example, both ATM and Frame Relay are connection-oriented and easily support robust implementations of QoS technologies (Kosonen, 1999), so that the time, order and cost of packets in a stream of information can be predetermined. This makes suitable for converged video, audio and data networks.

Why then, the question has to be asked, has MPLS become so successful? Frame Relay and ATM were designed with reference to the OSI model and not TCP/IP as such. This is particularly true for ATM, where TCP/IP support adds considerable overhead to ATM cells (Cavanaugh, 1994). As a complete solution that is not reliant on IP, ATM is not ideal for connecting IP-based customer networks. Another disadvantage is that in the case of overlay ATM and Frame Relay VPNs, customers need to maintain their own inter-site links, as the service provider merely creates a transparent connection between two or more points. It is the customer's responsibility to establish and maintain adjacencies between sites. This increases complexity, administrative costs, and the risk of less than optimal forwarding decisions, since the inner service provider network is transparent to the customer. Customers often favor the lower economic cost of peer-to-peer VPNs, where the customer edge establishes an adjacency with the provider edge and the provider's core network takes an active part in the adjacencies and optimal forwarding of customers' inter-site traffic. This approach, though, does not clearly separate each customer and makes it complex and difficult to support multiple customers using the same subnet.

The solution is a technology that mixes overlay and peer-to-peer VPNs, with full support for IP and its extensions. It combines the robust predictability of a connection-oriented protocol with the flexibility of connection-less IP, and doing this not by adding functions at a higher layer, therefore degrading performance, but by exploiting L2 technologies. The solution is viable because it reduces resources required to forward packets on the Internet rather than focus on sheer speed, now that forty Gigabit interfaces are common on modern routers. Enter MPLS, a solution that addresses all these issues, though not without introducing some challenges and difficulties of its own, as the chapter later shows.

PART 1: AN OVERVIEW OF THE BGP/MPLS IP VPN

The MPLS VPN is a solution for most converged customer network requirements and service provider capabilities. In an MPLS VPN, the Customer Edge (CE) routers form a peer relationship with the Provider Edge (PE) router, through the use of a CE to PE dynamic routing protocol, like Routing Information Protocol (RIP), Enhanced Interior Gateway Routing Protocol (EIGRP) or Open Shortest Path First (OSPF). This simplifies setup and maintenance, because the service provider is responsible for customer inter-site adjacencies. In addition, the CE to PE peer relationship allows optimal routing across the service provider core network, as the virtual circuits between customer sites or in the case of MPLS, Label Switched Paths (LSPs), are based on the service provider Internal Gateway Protocol (IGP)

Although an L3 MPLS VPN is clearly specified as a peer-to-peer VPN, since there is only a peer relationship between the CE and PE Router and not between the two CE routers, it is still able to deliver the benefits of an overlay VPN. Virtual instances of routing tables ensure that customers cannot access or view other customer networks, including customer routing information, and in so doing enable multiple customers to use the same subnet and private addresses without interfering with other customer networks. It is the use of labels within the provider core router that makes this possible.

On the one hand, an MPLS VPN is primarily based on IP, a connectionless protocol with no deterministic capabilities. On the other, because the provider core is using MPLS, a connection-oriented protocol to forward packets along LSPs in reference to the IP routing table, the connections across the service provider are connection-oriented without losing the flexibility and cost efficiency of IP. Moreover, a legacy data link technology like ATM can be used with MPLS, by using the Virtual Path Identifier and Virtual Circuit Identifier of ATM as the MPLS label values (Alwayn, 2001). This enables service providers to integrate MPLS with existing ATM networks and gradually replace the older infrastructure. (An ATM based MPLS implementation is referred to as Cell mode MPLS as opposed to Frame mode MPLS for frame based L2 protocols, like Ethernet.)

The advantage in speed of reading and forwarding label information as opposed to IP address information is negligible; the real benefit of MPLS over traditional IP networks is that it reduces resource usage significantly. In order to forward customer routing information, an MPLS VPN employs an extended community of the Border Gateway Protocol (BGP). The PE takes the routing information from the CE and redistributes it into BGP, which then transports it to the egress PE Router that in turn then redistributes it back to the PE to CE routing protocol. Not all the Provider routers in transit need to run the BGP routing protocol, with the result that routing tables remain small and efficient. In other words, although IGP is required to build the label forwarding information table and BGP to create the tunneling mechanism through the service provider core network, the use of labels to forward packets within an MPLS core network conserves resources.

The MPLS protocol

In essence, MPLS is a simple technology and not especially new: to use labels instead of IP addresses to connect networks at L2. When a Label Switched Router (LSR) receives a packet with a label of 20, it will look up that label in its Label Forwarding Information Base (LFIB), which is little more than a switching table for MPLS, but one that links up to the IP routing table. Once an LSR finds a local entry for labels coming in with a value of y in the LFIB it will send the packet out of an interface x with an outgoing label of x, where x is defined in the LFIB. These Labels are only significant locally, so on any segment along the path, the value of the labels is significant, but the same labels could be used at a later segment along the path. The label itself consists of an additional header that is inserted between the data link header and the IP address information, and it is for this reason that MPLS is defined as a Layer 2.5 OSI stack protocol. The MPLS header is completely independent of the L3 header but to indicate a following label, the L2 protocol uses a type value of 0x8847 for a unicast IP packet and 0x8848 for a multicast IP packet with at least one following label (See Figure 1).

The first twenty-bit field carries the actual value of the label and is used to determine the next hop and exit interface along the path according to the LFIB. The next field, Traffic Class (TC) was once reserved for experimental purposes and

Figure 1. MPLS Header

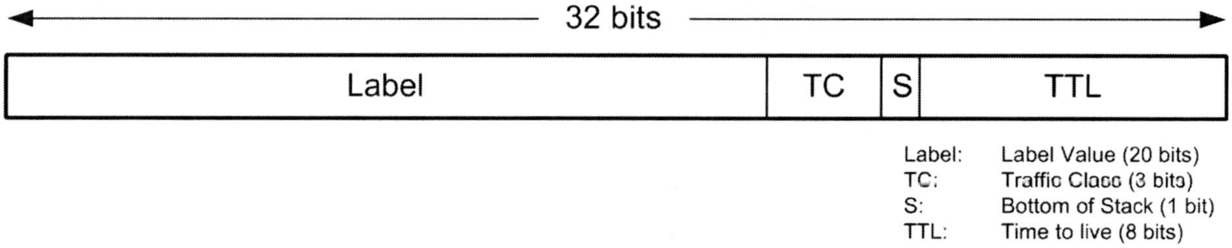

therefore called EXP but with RFC 5462 (Andersson & Asati, 2009) the field now refers to its actual use, QoS and can therefore store Precedence or Differentiated Services Code Point (DSCP) values. The S field indicates the bottom of a stack, where a value of 1 indicates the last label in a stack. This is an important piece of information, as only the topmost label is used to switch packets in an MPLS network and multiple labels can be used on top of one other in a Label Stack. (Later this chapter shows how MPLS is used to create VPN tunnels where the VPN is identified by one label (the inner label S bit set to 1) and the actual path throughout the MPLS switching domain by another (the outer label).) The last field is Time to Live, designed, as in any other protocol, to avoid endless loops.

A label distribution protocol populates the LFIB dynamically and establishes agreement among LSRs as to the meaning of labels and the action to perform. In fact, a number of protocols can be used to build the label-switching table: Tag Distribution Protocol (TDP), Label Distribution Protocol (LDP), as well as BGP and Resource Reservation Protocol - Traffic Engineering (RSVP-TE). These last two protocols, BGP and RSVP-TE, are only used under certain circumstances, for example traffic engineering in an MPLS network through the MPLS Traffic Engineering extension of the RSVP protocol and BGP can be used for Inter-Autonomous System MPLS VPN links, an MPLS based VPN link between two service provider autonomous systems. Tag Distribution Protocol (TDP) is a Cisco proprietary protocol and was used in the early stages of MPLS when it was still called tag switching. LDP, standardized under RFC 5036 (Andersson, Minei & Thomas, 2007), is the protocol of choice for exchanging label information between LSRs and populating the LFIB. In fact, Cisco routers these days make LDP available with default settings when MPLS is enabled. RFC 5036 (Andersson, Minei & Thomas, 2007) effectively made RFC 3036 (Andersson, Doolan, Feldman, & Thomas, 2001) obsolete.

Like many other network protocols, LDP uses the notion of neighbors that exchange information. It establishes neighbors by sending "hello" discovery messages every five seconds for directly connected peers on the well-known (UDP) port 646 to the well-known all router multicast address 224.0.0.2. For non-directly connected neighbors, routers send unicast messages every ten seconds, something known as targeted LDP. For example, targeted LDP or tLDP is used in order to exchange the inner label of an MPLS-based VPN. In both cases the discovery message includes the LDP identifier, a six-byte unique value to the LSR, which comprises of an IP address (loopback 0) plus integers starting with 0 representing the remaining 2 bytes (for example: 1.1.1.1:0).

After establishing hello adjacency through UDP, LSRs then attempt to establish the actual neighbor relationship to exchange information over TCP. To form the neighbor adjacency, LSRs first determine the active/passive or master/slave roles by means of LDP identifiers sent through hello discovery messages. The higher numeric value will be the active peer for a particular segment and it is this active peer that initiates the neighbor adjacency over TCP. The active LSR sends out an initialization message that includes information about timers, LDP version, and label distribution method. The passive LSR matches the LDP identifier with a hello adjacency and either accepts or denies the given parameters. LSRs maintain neighbor adjacency though keepalive messages, which are additional to the hello messages and sent out per neighbor relationship. In short, hellos are per interface keepalives - the actual 'label information' exchange happens after the TCP neighbor relationship has been established and label advertisement messages are exchanged. Moreover, the LDP protocol is tasked with sending notification messages to inform neighbors of changes to the label switching topology or adjacency updates. These mechanisms are essential for the creation of an LDP session between two LSRs.

LDP enables peers to exchange not just label information, but the information relating to the *connection* between labels and the IP routing table, called Forwarding Equivalence Class (FEC). Many young network engineers new to MPLS may be unfamiliar with the term FEC, yet it relates to the main purpose of any router, namely determining the action to perform on an incoming packet. As the name implies, the term Forwarding Equivalence Class describes the action taken on a group of packets with the same destination network or QoS classes. For IP this is usually to send the packet out interface *x* with a destination *y* and use queue *z*. In a generic IP-routed network, the FEC has to be determined at each hop along the path, but with MPLS, the FEC has to be determined only at the Label Edge Router (LER), which is the same as the PE in terms of VPNs. All LSRs along the path simply switch the messages according to their LFIB. This lies at the core of MPLS and explains why it is a little faster than plain IP, even though router manufacturers have developed similar technologies for IP. One example of this is Cisco Express Forwarding, which caches the FEC of the first packet in a stream in the Forwarding Information Base (FIB), so that instead of determining the FEC of a packet according to the IP routing table, an incoming packet is matched to a cached value in the FIB (Doyle, 2008).

In MPLS, the FEC combines the intelligence of a L3 routing protocol with the forwarding speed and simplicity of the data link layer. The smarts of an MPLS network is therefore in the Edge, with the core left to the simple task of forwarding/switching packets as fast as possible. The LER takes much of the load, as it has to determine the FEC of an incoming packet. Once it has determined the label, interface and queue for an outgoing message, the LER has to push (or impose) a label between the L2 and L3 headers of the packet. The LSRs in the MPLS core network then have to perform a swap operation, which involves removing the incoming label and replacing it with the appropriate outgoing label, according to the LFIB. The egress LSR then looks up the label value in the LFIB and determines that it is the destination of the LSP. The LER pulls (or disposes) the label by removing it, only to perform another lookup, this time in the IP routing table. Since the first lookup is unnecessary, most MPLS implementations use Implicit Null Labels to inform the second to last LSR of an LSP to pop one label of a packet or label stack. This reduces the number of lookups an LER has to perform and is called Penultimate Hop Popping (PHP). The last LSR along an LSP sends out the Implicit Null label which has a value of 3 and falls into the reserved range of values (0-15) that cannot be used for generic LSR switching operations (Ghein, 2007).

In order to examine how MPLS uses LDP to create a label switched network, we set up a basic routing infrastructure, which uses LFIB to switch instead of route packets (figure 2). In this example of Frame Mode MPLS, Intermediate System-to-Intermediate System (IS-IS) is the Internal Gateway Protocol (IGP) that converges the Ethernet-based local area networks. Once the network is fully converged, an LSP or

Figure 2. An Example of MPLS

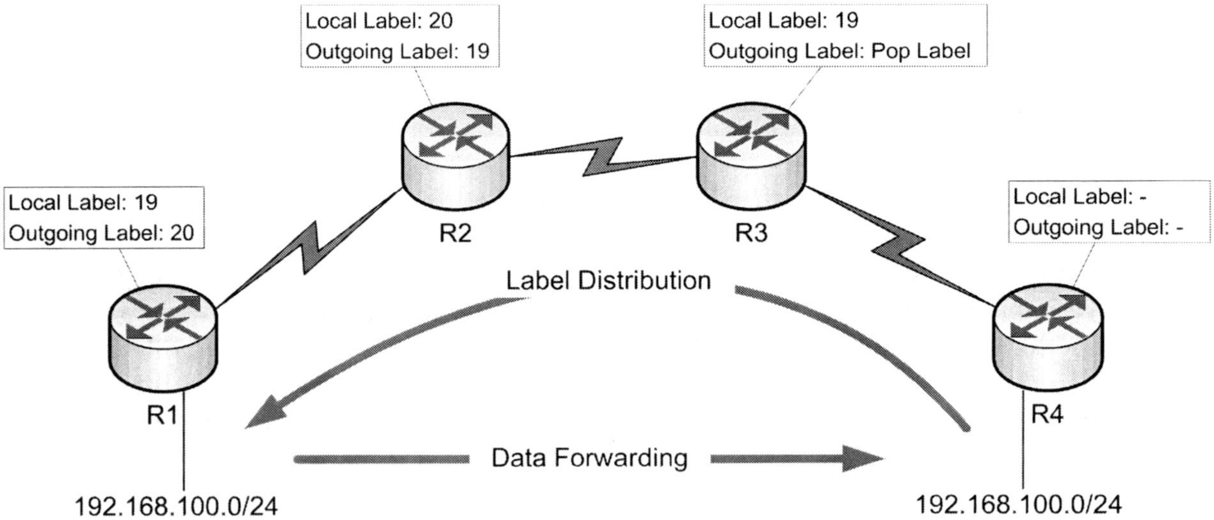

virtual circuit has to be created for the source network 192.168.100.0/24tothedestinationnetwork192.168.200.0/24, and as LSPs are unidirectional, another LSP is required for the return path. If LDP is used to distribute and synchronize label information and resulting FECs, then an LSR will first announce their LDP-aware presence on each segment, by sending out LDP Hello messages. Once discovery messages have been exchanged, the active peer, the LSR with the higher LDP ID, will send out the initialization method, in which the method of label distribution specified in RFC 3037 (Thomas & Gray, 2001) will be determined for every LDP session. For example, LSR R1 and LSR R2 could use the Downstream on-Demand label distribution method, in which case R1, the upstream router, would request a label binding for a particular FEC (source 192.168.100.0/24destination192.168.200.0/24) from the downstream LSR R2. In the next segment, R2 and R3 could use the downstream unsolicited method, where the downstream LSR R3 advertises the label bindings for a particular FEC to the upstream LSR R2. Both label distribution methods may be used in one network, but only one method can be used for any single LDP adjacency. Most implementations default to the downstream unsolicited method, with the result that FECs are advertised and created from the downstream to the upstream LSR. Once the method of label distribution is determined, the label bindings for the equivalent IP FECs can be exchanged.

Since LER R4 is directly connected to the non-MPLS network 192.168.200.0/24, the network is not included in the LFIB of R4 and an implicit Null Label is advertised to R3. The remaining LSRs use this information to populate their LFIBs by announcing their local label value for a particular destination network to their neighbors. Once the LSR R1 receives the LDP advertisements, the LSP from R1 to R4 for the destination network of 192.168.200.0/24 is created. When R1 receives a packet for the destination network, it needs to determine the FEC. Since it has information about the network in the LFIB, it will push a label onto the message and send it out to R2. Because an LSP is a virtual circuit, R2 will not determine the FEC for a labeled message arriving with a label of 20, but use the information stored in the LFIB to swap the topmost label with 19 and send it to R3. R3 then pulls the label and switches the message directly to R4. Depending on the number of labels in the Label stack, R4 will receive either a labeled packet and perform a lookup in the LFIB, or a generic IP packet and route it according to its IP routing table.

Virtual Routing Instances

MPLS provides a switching network with the intelligence and flexibility of IP, while keeping the forwarding (data plane) and the intelligence (control plane) separate. A BGP/MPLS IP VPN or L3 MPLS VPN connects multiple networks at the edge of the service provider through an 'intelligence-independent' MPLS core. This is achieved through the use of virtual networks for each VPN customer at the edge of the MPLS core. BGP connects the virtual networks logically (at the control plane of a router), whereas MPLS connects edge devices (at the data plane of a router).

The separated virtual network consists of a Virtual Routing and Forwarding (VRF) table or more specifically when used for MPLS VPNs, the VPN Routing and forwarding table (VRF). A VRF is a separate instance of the default-forwarding table and multiple VRFs can exist on any PE device without interfering with other VRFs. Since the core of an MPLS network primarily makes use of labels, it is only Provider Edge devices that require virtualized routing tables, and each VRF instance is connected through the MPLS core. Before populating the VRF table of a particular customer, the interface connecting the PE and CE devices needs to be added to the VRF. This is important, since the interface will no longer be part of the default routing table and any packet coming into the interface or, as it is also called, VRF attachment circuit, will be forwarded according to the VPN routing and forwarding table. Once the interface has been added to the VRF, the virtualized routing table can be treated like the default routing table and populated manually through static routes or dynamically through RIP, EIGRP, OSPF, or another routing protocol.

Connecting VRFS

A PE router can be configured with multiple VRFs for multiple customers, where every virtual forwarding instance is independent of other virtual forwarding instances. While the routing table is independent, the logical connection between PE devices and therefore the connection between different VRF instances belonging to one customer is not. The Multiprotocol Extension of BGP (MP-BGP) or more specifically MP-iBGP for internal BGP sessions distributes the VRF routes of one site to VRFs of other sites. In short, the purpose of MP-iBGP for MPLS VPNs is to define a format to differentiate between individual customer routes, changing them into a globally unique value and then to distribute the routes to MP-iBGP neighbors.

To allow multiple customers to use the same RFC 1918 (Rekhter, Moskowitz, Karrenberg, de Groot & Lear, 1996) private address ranges for their VPN routes, an 8-byte Route Distinguisher (RD) has to be added in front of the four-byte IPv4 destination network address, turning it into a 96 bit VPN-IPv4 address (figure 3). The global uniqueness for the VPN-IPv4 address family or VPNv4 is based on the RD, which is of the conventional format Autonomous System Number (ASN):nn, or IP-address:nn, where n is an arbitrary number. For MPLS networks with a large number of PE devices, it is best practice to use the ASN format, for example, the RD of the AS 100 could be 100:1 for the first VPN. However, as the RD for one PE can be different to another belonging to the same VPN, the format of IP (loopback of PE):nn is a better choice in smaller networks. For example, using 200.0.0.1:1 as an RD for a particular VRF on a PE with a Loopback address of 200.0.0.1 enables troubleshooting to be easier, because it makes the originator of a route in the MP-BGP routing domain fairly obvious.

It is important to understand that the RD is not responsible for the separation of routes belonging to different VPNs. On the contrary, MP-BGP peered and enabled iBGP neighbors exchange

Figure 3. MP-iBGP update components

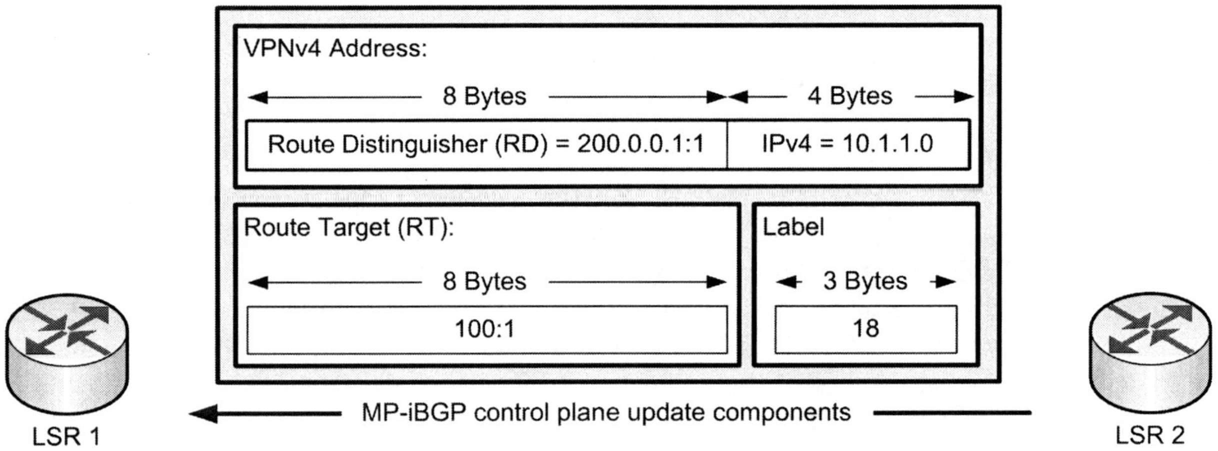

customers' IPv4 routes as distinguished VPNv4 routes through the same 'channel'. The component of MP-BGP that enables the PE router to distinguish between different VRFs and therefore VPNs is the Route Target (RT). When a PE router receives IPv4 unicast routes from the CE via a routing protocol, the PE device needs to have a separate routing protocol process for each VRF. The updates from the VRF are then redistributed into a VRF specific IPv4 address family instance of BGP, and from there advertised to other MP-BGP neighbors until it arrives at the egress PE redistributed from BGP back into the PE-CE routing protocol. This process can and must be controlled with the RT, which is an extended community of BGP with the same format as the RD, and must be applied in both directions. The first direction is export, which assigns an RT of for example 100:1 for the AS 100, to routes being redistributed from the customer VRF into BGP. The second controls which routes are being imported from BGP back into the VRF. As figure 3 shows, because this is a control plane packet, the label size is three bytes and not four bytes (as in figure 1), as it is used for labeled packets.

The first step towards an MPLS VPN is a converged MPLS core network. For our example the IGP is IS-IS but other routing protocols can be used. Enabling LDP for the MPLS core network, in this case LER PE1, will impose or push an MPLS label of 20 based on the IGP to reach the LER PE2, which will be swapped by the LSR P1 with a label of 19 and disposed or popped by the LSR P2 to send it to PE2. Note that when the downstream unsolicited LDP distribution method is in use these values are advertised in reverse order, so PE2 via P2 and P1 to PE1 for the FEC PE1 to PE2 and because LSPs are unidirectional, another LSP needs to be created on PE2 with an FEC for PE2 to PE1.

Once the MPLS network is fully converged, VRFs can be created on PE1 and PE2. In figure 4, only one VRF called A is created with an RD for PE1 of 165.84.0.1:1 and for PE2 of 165.84.0.2:1. The RT is 45138:1 and this will be exported and imported on both PE routers. The interfaces connecting the PE and CE routers have to be added to the VRF next and a VRF OSPF instance created for Customer A. The routes from VRF A on PE1 are then distributed to the VRF A on PE2 by creating the actual tunnel. R1 and R4 will be peered in an iBGP session from and to the loopbacks, in order to create the MP-iBGP VPNv4 address family neighbor relationship between the same. Finally, routes from the VRF A OSPF instance need to be

Figure 4. BGP/MPLS IP VPN

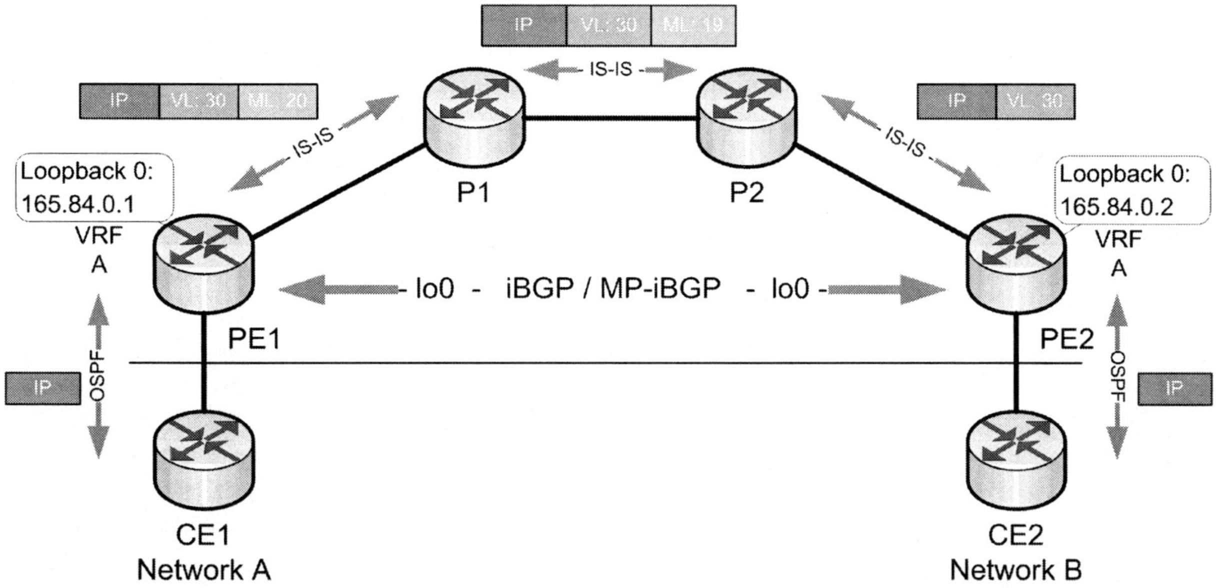

redistributed into the IPv4 address family VRF A BGP instance and vice versa.

For an IPv4 packet from Network A to Network B, the CE1 router will have the network B in its routing table through OSPF from PE1 pointing to PE as the next hop address. PE1 will perform a lookup in the VRF A forwarding table and because PE2 has announced itself as the next hop for Network B, PE1 will determine that the packet should be sent to PE2 via the iBGP session. PE1 therefore pushes a label onto the packet with a VPN value of 30, which was also announced by PE2, the iBGP next hop, representing a route membership of the VRF/VPN A during the exchange of BGP prefixes. PE1 then has to determine the FEC to reach PE2, pushes another label onto the label stack with an "MPLS" value of 20, and sends it out the interface connected to P1. The LSR P1 then looks up the label value 20 in the LFIB, swaps the outer label with a label value of 19, and sends it out to the LSR P2, which pops the outer label of the label stack because of PHP. Notice that P1 and P2 only swap the labels, they do not have any information about the VRFs, the iBGP, and MP-iBGP processes. All the intelligence of the MPLS VPN is located at the Provider Edge and the core 'simply' forwards labeled packets. Once PE2 receives the MPLS message with a label value of 30 it will associate it with the VPN A and forward it according to the virtual routing table A out to CE2.

PART 2: PROBLEMS, PROTOCOLS, POSSIBILITIES

Connection Scenarios

A major problem for MPLS VPNs is the provisioning of an Internet connection for individual VPN customers. Since an MPLS VPN is intended to be a tunnel between two sites, a connection from the tunnel to the outside world can only be implemented through workarounds (Guichard, Pepelnjak, Apcar, 2003). The following three options exist:

- **Option 1:** leaking. A static route can be created for the customer VRFs that point to the global routing table. This, though, makes managing the network edge difficult, especially in terms of security. Moreover, the return path can be hard to implement, as a static route pointing to a VRF cannot be configured in most VRF implementations. Only a route pointing out the interface where the CE is connected can be configured, which can be complex to achieve with static routes for multiple networks.

- **Option 2:** a shared VRF. An additional 'Internet' virtual network can be configured, so that a default route can be exported to customer VRFs with the appropriate route targets. Again, the return path is difficult to implement, as the 'Internet' VRF needs to be aware of every possible source for outgoing traffic coming from customer VRFs.
- **Option 3:** NAT. Here a NAT-PE is used, with a dedicated NAT PE router or per VRF NAT, and NAT performed between customer VRFs and the global routing table, at either a central dedicated PE router or per VRF. Again, this is difficult to implement and manage, especially as NAT needs to be frequently reconfigured by the customer.

Two solutions are sometimes advocated. The first solution entails the use of separate links for the external (Internet) and internal VPN networks. The service provider uses either individual links and Provider devices or sub interfaces to create a global network edge at one of the Customer Edge routers. The major benefit of this first solution is that the customer is in control of the connection to the external world. This may, though, result in less than optimal routing and unnecessary traffic load on the VPN links, as traffic destined for external networks is sent across the MPLS VPN network to the site with the internet connection. (One workaround is the use of multiple internet connections for many if not all sites, increasing complexity, manageability, and cost.) The second and better solution entails firewall provisioning. Here a software or hardware router, configured as an enterprise firewall, connects internally to the particular VPN, creating a PE like device within the MPLS Core.

When a customer uses two ISPs, the VPN has to span multiple autonomous systems. One way of achieving this is to create a back-to-back VRF where two service providers are connected through one physical or logical link per VRF, both treat the opposite site as a CE device, and no labels are exchanged. This does not scale well for obvious reasons. One possibility is the use of Multihop VPNv4 MP-eBGP, where the two PE routers that connect customer sites form a VPNv4 MP-eBGP neighbor relationship in order to distribute route information for customer VRFs. This is achieved by an exchange of labels via an IPv4 eBGP session between the service providers PE routers. This, though, requires service providers to share internal routing information represented by labels, something that has serious resource and security implications. One viable solution is to peer the two service provider routers via a VPNv4 MP-eBGP session, allowing the exchange of labels and creating end-to-end LSPs. It is simple to configure, since most requirements are already in place, but because there has to be a VPNv4 peering with all internal routers, scalability can be an issue for larger implementations.

OPTIMIZATION

The Traffic Engineering extension of the Resource Reservation Protocol (RSVP-TE) has made MPLS more flexible and reliable (Xu, 2010). RSVP-TE, outlined in RFC 3209 and updated for inter AS support by RFC 5151, creates LSPs based on parameters such as available bandwidth and explicit hops (Awduche, Berger, Gan, Li, Srinivasan & Swallow, 2001). While basic implementation is straightforward, enabling the protocol on each MPLS interface and setting the bandwidth and explicit hops, makes it difficult to deploy in larger networks. It makes it possible to utilize available bandwidth more efficiently and enable unequal cost load balancing and ameliorates some limitations of IGP. As RSVP requires an overview of the available bandwidth in the topology, which is signaled to other LSRs along the LSP, the routing protocol of choice for RSVP-TE is either OSPF or IS-IS. It is important to remember that in RSVP-TE bandwidth reservations are only made for the control plane, which means that the actual data plane does not necessarily respect these reservations. A fully functional and manageable RSVP-TE implementation therefore requires the additional configuration of a differentiated services (DiffServ) QoS model.

One reason for the success of MPLS is its compatibility with standard QoS models, something that is an integral part of any modern network. The actual QoS functionality remains the same in an MPLS environment, with the exception of the amount of possible classes. Since MPLS uses the TE field in the MPLS header with a length of three bits, 'only' eight classes can be configured, which matches the amount of eight classes for the IP precedence classes but not the sixty four possible classes of Differentiated Services Code Point (DSCP) format. Multiple QoS methods for MPLS, though, do exist, depending on if and how customer classes are used within the service provider network. For Uniform Mode the IP precedence value (Prec./DSCP) is copied from the IP packet coming from the customer into the TE field of the MPLS header. On the egress PE router the Prec./DSCP value of the IP packet is ignored and the label value is copied to the outgoing packet and sent to the customer edge. In Pipe mode the ingress Prec./DSCP value of the IP packet from the customer is not changed and not copied into the TE field of the MPLS header and more importantly the underlying IP Prec./DSCP values remain on the egress PE router. The service provider and customer QoS implementation are therefore independent of one another. In Short Pipe mode the Prec./DSCP value from the customer is copied into the MPLS header, which might change during transit through the MPLS network, but the Prec./DSCP value of the label at the egress PE router is not copied onto the underlying IP packet as it is in Uniform Mode. Instead, the Prec./DSCP value of the IP packet is maintained during transit and can be used at the egress customer edge.

CONCLUSION

While the deployment of IPv6 has been slow, MPLS VPNs can connect IPv6 customer sites. IPv6-based MPLS core networks, though, do not work, since LDP and other label distribution protocols are not available for IPv6. One approach is therefore to retain the IPv4 core, and either avoid or tunnel through the MPLS core. 6PE implementations, where IPv6 LSPs are created using IPv4-based protocols, are not uncommon (6net, 2008). The mapping between IPv4 and IPv6 is done through MP-iBGP and an IPv6 address family has to be created so that the IPv4-mapped IPv6 address is the next hop. The result is essentially an IPv6 tunnel created between two PE routers, which retain the underlying setup and logic of IPv4. Another solution is to use a L2 protocol like Ethernet or ATM across the MPLS backbone, with the advantage that CE routers are directly adjacent and IPv6 can easily be deployed.

MPLS is a mature protocol. It is a solution for both service provider networks and larger companies' internal networks (Minei & Lucek, 2008). One advance from the original MPLS protocol is MPLS Transport Profile (MPLS-TP), currently still being standardized, although initial industry deployments have been successful. MPLS-TP is a true carrier class transport network protocol that uses the same data/forwarding plane architecture as MPLS. It adds end-to-end Operations, Administration and Management (OAM) and QoS functionalities while removing some traditional

MPLS options, including PHP, LSP Merging, and Equal Cost Multiple Path (ECMP) (UTStarcom, 2009). MPLS-TP is one of two L2 technologies that compete for the lucrative carrier transport networks segment and will facilitate the move away from circuit switched networks. The other L2 technology is Provider Backbone Bridges-Traffic Engineering (PBB-TE), but it appears that presently MPLS-TP is the more popular choice (Telecom Asia, 2008; Infonetics Research, 2008).

MPLS has made IP a truly universal protocol for customer and carrier networks alike. Its scope and application has increased immensely over the years as it meets nearly every demand applications make on it. While the actual purpose of MPLS is to adapt carrier networks to IP-dominated traffic and customer networks, it does raise the question of whether the increase in complexity at L2 and L3 is justified. The future belongs to MPLS, at least for now, but its growth in the main is due to customer demand. This obliges providers to use a compound of MPLS and LDP with IP and above it BGP and IGP as opposed to a simple L2 protocol like Frame Relay or ATM. This introduces greater complexity and cost in terms of management. In fact, the slow deployment of IPv6 as a solution to the problems inherent in IPv4 raises doubts as to whether the future even lies in IP itself.

Before the rise of the Internet, the focus for communication was the path. It was an end-to-end communication made possible by the kind lady of the telco sitting in front of switchboards. With advent of the packet, the paradigm shifted from path to source and destination. The exponential increase in network traffic and growing demand for new applications, though, has led technologies based on a source and destination to become excessively complex. Attempting to secure packets along the path and ensuring they come from the correct source is problematic, since guaranteeing source and confidentiality of a packet does not in itself make it secure. In the future, providers need to move away from a model of source and destination based communication. This requires a new IP, not IPv6 that is based on the same paradigm as IPv4. Perhaps MPLS can lead the way for a new way of communicating, roughly similar to virtualized networking environments. Here the data itself becomes the focus; much like routing information is exported and imported into a virtual network for an MPLS VPN that is based on a simple L2 technology.

REFERENCES

Alwayn, V. (2001). *Advanced MPLS design and implementation*. Indianapolis, IN: Cisco Press.

Andersson, L., & Asati, R. (2009). *Multiprotocol label switching (MPLS) label stack entry: "EXP" field renamed to traffic class field*. RFC 5462. Retrieved May 6, 2010, from http://www.ietf.org/rfc/rfc5462.txt

Andersson, L., Doolan, P., Feldman, N., Fredette, A., & Thomas, B. (2001). *LDP specification*. Retrieved May 6, 2010, from http://www.ietf.org/rfc/rfc3036.txt

Andersson, L., Minei, I., & Thomas, B. (2007). *LDP specification*. RFC 5036. Retrieved May 6, 2010, from http://www.apps.ietf.org/rfc/rfc5036.html

Awduche, D., Berger, L., Gan, D., Li, T., Srinivasan, V., & Swallow, G. (2001). *RSVP-TE: Extensions to RSVP for LSP tunnels*. RFC 3209. Retrieved May 6, 2010, from http://www.ietf.org/rfc/rfc3209.txt

Cavanaugh, J. D. (1994). *Protocol overhead in IP/ATM networks*. Retrieved May 6, 2010, from http://citeseerx.ist.psu.edu/viewdoc/download?doi=10.1.1.45.2564&rep=rep1&type=pdf

Davie, B. S. (2000). *MPLS: Technology and applications*. San Francisco, CA: Morgan Kaufmann.

Dawson, F. (1998). *It's IP vs. ATM in all-out-war*. Retrieved May 6, 2010 from http://www.ced-magazine.com/its-ip-vs-atm-in-all-out-war.aspx

De Ghein, L. (2007). *MPLS fundamentals: Forwarding labeled packets*. Indianapolis, IN: Cisco Press.

Doyle, J. (2008). *Understanding the role of FECs in MPLS*. Retrieved March 22, 2010, from http://www.networkworld.com/community/node/24093

Farrel, A., Ayyangar, A., & Vasseur, J. P. (2008). *Inter-domain MPLS and GMPLS traffic engineering - resource reservation protocol-traffic engineering (RSVP-TE) extensions*. RFC 5151. Retrieved May 6, 2010, from http://www.ietf.org/rfc/rfc5151.txt

Guichard, J., Pepelnjak, I., & Apcar, J. (2003). *MPLS and VPN architectures* (*Vol. II*). Indianapolis, IN: Cisco Press.

Infonetics Research. (2008). *Service providers lean toward MPLS-TP over PBB-TE in new packet transport survey*. Retrieved May 6, 2010, from http://www.infonetics.com/pr/2008/pots.survey.2008.nr.asp

Kosonen, V. (1999). *QoS and frame relay*. Retrieved May 6, 2010, from http://www.netlab.tkk.fi/~vkosonen/vesan_esitelma_1999.pdf

Minei, I., & Lucek, J. (2008). *MPLS-enabled applications: Emerging developments and new technologies*. Chichester, UK: Wiley and Sons. doi:10.1002/9780470751725

6net. (2008). *IPv6 deployment guide*. San Francisco, CA: Javvin Press.

Newman, P., Edwards, W. L., Hinden, R., Hoffman, E., Ching Liaw, F., Lyon, T., & Minshall, G. (1996). *Transmission of flow labelled IPv4 on ATM data links*. RFC1954. Retrieved May 6, 2010, from http://www.apps.ietf.org/rfc/rfc1954.html

Perros, H. G. (2005). *Connection-oriented networks*. Indianapolis, IN: Wiley. doi:10.1002/0470016361

Rekhter, Y., Moskowitz, B., Karrenberg, D., de Groot, G. J., & Lear, E. (1996). *Address allocation for private Internets*. RFC 1918. Retrieved May 6, 2010, from http://www.ietf.org/rfc/rfc1918.txt

Rosen, E., & Rekhter, Y. (2006). *BGP/MPLS IP virtual private networks* (VPNs). RFC 4364. Retrieved May 6, 2010, from http://www.ietf.org/rfc/rfc4364.txt

Sosinsky, B. (2009). *Networking bible*. Indianapolis, IN: Wiley.

Telecom Asia. (2008). *Transport wars: MPLS-TP vs PBB-TE*. Retrieved May 6, 2010, from http://www.telecomasia.net/print/1187

Thomas, B., & Gray, E. (2001). *LDP applicability*. RFC 3037. Retrieved May 6, 2010, from http://www.rfc-archive.org/getrfc.php?rfc=3037&tag=LDP-Applicability

UTStarcom. (2009). *MPLS-TP based packet transport networks: Delivering new value through next-generation transport networks*. Retrieved May 6, 2010 from http://www.utstar.com/Document_Library/CP-WP-MPLS-TP_PTN-1109.pdf

Xu, Z. (2010). *Designing and implementing IP/MPLS -based Ethernet layer 2 VPN services: An advanced guide for VPLS and VLL*. Indianapolis, IN: Wiley and Son.

KEY TERMS AND DEFINITIONS

Forwarding Equivalent Class (FEC): A group of packets that are forwarded and treated in the same way due to common properties like destination IP address or QoS class.

Label Switched Path (LSP): A label switched path through an MPLS network set up by a signaling protocol like LDP or RSVP-TE.

Label Distribution Protocol: A commonly used signaling protocol that creates label switched paths through an MPLS network.

Multiprotocol Border Gateway Protocol (MP-BGP): Extends the BGP4 protocol so that it can carry routing information for multiple Network Layer protocols (IPv4/6 and VPNv4)

Push (Impose), POP (Dispose) and Swap (Impose and Dispose): An ingress label edge router (LER) pushes a label onto the Frame; the Label Switching Router (LSR) then swaps the label and the egress LER pops or disposes the topmost label. If penultimate hop popping is enabled, the second to last MPLS device along a Label Switched Path (LSP) pops the topmost label.

Label Information Base: (LIB): A control plane entity of MPLS. The information stored in the LIB originates from a label distribution protocol like LDP and maps IP destination prefixes (from the routing table) and labels received from the next hop device. The LIB also stores local labels generated by LDP.

Label Forwarding Information Base: (LFIB): A data plane entity of MPLS. It maps the locally assigned label which was sent out to other MPLS devices and through which incoming packets are being identified with next hop labels along with the outgoing interface.

Virtual Routing and Forwarding: (VRF): Also known as VPN routing and forwarding (Cisco MPLS) is a virtual routing table. Multiple VRFs can co-exist on one physical device with no ensuing conflicts between private IP addresses.

APPENDIX

Table of abbreviations

BGP	Border Gateway Protocol
CE	Customer Edge
DSCP	Differentiated Services Code Point
eBGP	external Border Gateway Protocol
ECMP	Equal Cost Multiple Path
FEC	Forwarding Equivalence Class
FIB	Forwarding Information Base
iBGP	internal Boarder Gateway Protocol
IGP	Internal Gateway Protocol
L2	Layer 2
L3	Layer 3
LAN	Local Area Network
LDP	Label Distribution Protocol
LER	Label Edge Router (MPLS Term)
LFIB	Label Forwarding Information Base
LIB	Label Information Base
LSP	Label switched path
LSR	Label Switch Router (MPLS Term)
MPLS	Multiprotocol Label Switching
OAM	Operations Administration Management
OSPF	Open shortest path first
P	Provider Router (VPN Term)
PAR	Previous Access Router
PBB	Provider Backbone Bridges
PE	Provider Edge Router (VPN Term)
PHP	Penultimate Hop Popping
PPVPN	Provider Provisioned Virtual Private Network
PrRtAdv	Proxy Router Advertisement
RADIUS	Remote Authentication Dial-In User Service
RD	Route Distinguisher
RH	Routing Header
RIP	Routing Information Protocol
RO	Route Optimization
RRH	Reverse Routing Header
RS	Remote Subscription
RSVP	Resource Reservation Protocol
RSVP	Resource Reservation Protocol

RT	Route Target
RtSolPr	Router Solicitation for Proxy
SHO	Soft Handover
TDP	Tag switching protocol
TE	Traffic Engineering
tLDP	targeted LDP
VPNv4	Address family used in BGP to distribute MPLS VPN routes
VRF	Either Virtual Routing and Forwarding or VPN in the case of MPLS
WAN	Wide Area Network
WWW	World Wide Web

Section 4
Protocol Applications and Technologies

Chapter 21
Time Synchronization in Wireless Sensor Networks

Gyula Simon
University of Pannonia, Hungary

Gergely Vakulya
University of Pannonia, Hungary

ABSTRACT

Time synchronization services are often required to support coordinated operation of the nodes in sensor networking applications, potentially containing hundreds or thousands of elements. The synchronization service must provide application-specific performance (in terms of accuracy and overhead), and must be scalable and robust. For the design of suitable algorithms, the error sources of both the applied synchronization models and the physical devices must be understood and taken into account. In this chapter, various time synchronization models will be introduced, and their potential accuracy and complexity will be shown. The error sources of real devices and communication channels will also be analyzed. Based on the models, several synchronization primitives will be reviewed, and complex synchronization algorithms will be introduced and analyzed.

INTRODUCTION

Sensor networking solutions are used in various data collecting, surveillance, or monitoring applications. Potentially hundreds or thousands of sensors perform distributed measurements and send their collected data to base stations for further processing. Such networks are designed for months or years of autonomous operation.

Time synchronization in wireless sensor networks is one of the most important middleware services. Since sensor networks are closely linked to real physical phenomena, a naturally occurring question from user side is "when did it happen?" Various sensor networking services and protocols also require the knowledge of time, e.g. data

DOI: 10.4018/978-1-60960-732-6.ch021

fusion, calibration, in-network processing, and localization, just to name a few.

Typical tasks and operation modes in sensor networks often require synchronization between nodes. This is especially important when low duty cycle operation is required to conserve energy. Typically, in such networks nodes are in low-power (sleep) mode most of the time, and wake up for short periods, to perform measurements and communication tasks. To minimize the necessary awake interval, nodes must accurately synchronize their wakeup times.

Timing information is also important in data collecting applications, where the required precision depends on the application needs, e.g. acoustic localization or beamforming requires sub-millisecond precision, while in an agricultural monitoring application a few seconds of slip is probably acceptable.

Properties of the wireless communication channel, especially the physical and MAC layers do not enable the use of traditional synchronization methods (e.g. NTP) for fine-grain synchronization needs. Power efficiency and resource constraints in sensor nodes are other driving forces towards special synchronization algorithms tailored for sensor networks.

In this chapter, a model-based approach will be used to introduce and analyze time synchronization algorithms for sensor networks. Apart from the high-level models of the algorithms, the error sources present in wireless networks will also be modeled, and based on these models the most influential solutions will be reviewed. These methods contain several reusable building blocks and ideas that other synchronization schemes and methods use. This chapter is not intended to provide a full overview on all of the available methods but rather focuses on the basic, motivating, and compelling ideas present in the reviewed algorithms, with special emphasis on the constraints and needs arising in sensor networking applications.

TIME AND CLOCKS

Clock Models

A clock can be modeled by the following equation:

$$C(t) = \frac{1}{f_{nom}} \int_{t_0}^{t} f(\tau)\,d\tau + C(t_0),$$

Where $C(t)$ is the local time shown by the clock at time instant t, f is the instantaneous frequency of the clock's time base and f_{nom} is the nominal time-base frequency. A clock is called accurate if $C(t)=t$. The difference between the clock value and the real time $C(t)-t$ is called clock offset.

Digital clocks use hardware (usually quartz) oscillators as time bases, and digital counters to approximate the integrator. Ideally, the counter $c(t)$ is incremented by one in every $1/f_{nom}$ second, so the local time of a digital clock is

$$C(t) = \frac{1}{f_{nom}} c(t) + C(t_0),$$

where $c(t_0)=0$.

The frequency of the clock is $\frac{dC(t)}{dt}$, which should ideally be exactly one, while the real time-base frequency is $\frac{dc(t)}{dt}$, which ideally should be equal to f_{nom}. The frequency of real clocks somewhat differs from the nominal frequency. This phenomenon is called drift and defined as

$$\delta(t) = \frac{dC(t)}{dt} - 1,$$

or in terms of the time-base frequency

$$\delta(t) = \frac{f(t)}{f_{nom}} - 1.$$

If the drift value is -1 then the clock is stopped, for values smaller than -1 the clock is going backwards. Thus a reasonable assumption for real clocks is

$$\delta(t) > -1 \text{ for all } t.$$

If the drift is known to be bounded by

$$-\delta_{\max} \leq \delta(t) \leq \delta_{\max}$$

then bounds for measured time differences can be calculated. If two events occur at t_a and t_b, and the corresponding measured local time instants are $C(t_a)$ and $C(t_b)$, respectively, where $t_a < t_b$ then, taking into account the drift of the clock, the following bound can be calculated for the time difference:

$$\frac{C(t_b) - C(t_a)}{1 + \delta_{\max}} \leq t_b - t_a \leq \frac{C(t_b) - C(t_a)}{1 - \delta_{\max}}.$$

Since digital clocks change their values at discrete time instants, an important property called time resolution must be introduced. Time resolution is the smallest time difference the clock can represent, which is obviously $1 / f_{nom}$. Thus, a higher clock frequency provides higher resolution.

Drift in clocks using quartz oscillators results from several sources, the most important ones being manufacturing inaccuracies, temperature changes and ageing. The drift of commonly used, inexpensive quartz oscillators is around 10-100 ppm (parts per million). This means that in case of a 10ppm crystal, the clock will drift 1 second in 100.000 seconds, i.e. almost 28 hours. The time resolution of a clock using a 32 kHz crystal is 30.5 microseconds, while if a 4MHz crystal is used then the resolution is 0.25 microseconds.

Time Synchronization

Depending on the application requirements, two main types of time synchronization methods are used in sensor networks, called external and internal synchronization methods.

The clocks in the network may be synchronized to one or more external reference clocks, thus all the clocks is the network will have the same global time. For example, GPS devices can be used as external references and the clocks of the sensor nodes can be synchronized to show (approximately) the same time. Thus, when an event is detected the sensor can immediately attach the global time stamp. This kind of synchronization mechanism is called external synchronization. In certain applications, synchronization to global time is not important, but the sensor clocks must show the same time. In this case, nodes in the network are synchronized to each other only, and this method is called internal synchronization. In case of internal synchronization, the clock of any node can be used as time base.

Certain networks must constantly be in synchronized state, while other networks are synchronized only on-demand. On-demand synchronization is useful when measurements are infrequent and between measurements, synchronization is not required. For example, in intrusion detection applications intrusions happen very rarely (e.g. one or two events per day), thus it is enough to synchronize the network only after an event of interest is detected, to decrease overhead due to synchronization. For the same reason, sometimes only a part of a network is synchronized (e.g. the area where an event was detected, or an area of interest where measurements must be made).

Extreme approaches do not synchronize nodes in the strict sense, but measure and use the age of messages travelling in the system in order to compute times of events that were earlier detected by the sensors.

In external synchronization schemes, the quality of time synchronization can be measured as the difference between the global time and the local times of sensor nodes, as follows:

$$\Delta_{ext} = \max_{\forall i} \left| C_i(t) - t \right|$$

where $C_i(t)$ is the local time of sensor node i, and t is the injected global time.

In case of internal synchronization, the error is defined using the differences between any two nodes in the network, as follows:

$$\Delta_{int} = \max_{\forall i,j} \left| C_i(t) - C_j(t) \right|.$$

Time synchronization algorithms use implicit, built-in models on the behavior of the clock devices; and thus the clock errors as well. Common clock models are the Known Rate, Constant Rate, Bounded Drift, and Bounded Drift-variation models.

In the Known Rate model, all clocks are assumed to have the same constant frequency but there may be offset between them. The clock model is the following:

$$C_i(t) = \frac{1}{f_{nom}} c + C_i(t_0).$$

The drift in Known Rate model is constant and the same for all nodes:

$$\delta_i(t) = \frac{f}{f_{nom}} - 1.$$

The offset between clocks i and j is constant in time:

$$C_i(t) - C_j(t) = C_i(t_0) - C_j(t_0).$$

Note that in Known Rate model once the clocks are internally synchronized, they remain synchronized forever. In such systems, external synchronization would also be easy, since the drift is constant and thus can be compensated for.

The more realistic Constant Rate model allows different, but constant frequencies for the clocks in the system, as the following clock model shows:

$$C_i(t) = \frac{f_i}{f_{nom}}(t - t_0) + C_i(t_0).$$

The drift of clock i is constant:

$$\delta_i(t) = \frac{f_i}{f_{nom}} - 1,$$

and the offset between two clocks is linearly changing in time:

$$C_i(t) - C_j(t) = C_i(t_0) - C_j(t_0) + \frac{f_i - f_j}{f_{nom}} t..$$

Once two clocks in Constant Rate model are synchronized at time instant t_{sync}, clocks drift apart from each other in time as follows:

$$C_i(t) - C_j(t) = \frac{f_i - f_j}{f_{nom}}(t - t_{sync}),$$

In Bounded Drift model, the drifts are not constant but are bounded:

$$C_i(t) = \frac{1}{f_{nom}} c_i(t) + C_i(t_0), \text{ and}$$
$$-\delta_{\min} \le \delta_i(t) \le \delta_{\max}, \forall t,$$

while the Bounded Drift-variation model includes the realistic assumption that drift change is slow in real cases:

$$-\vartheta_{\max} \le \frac{d\delta_i(t)}{dt} \le \vartheta_{\max}, \forall t.$$

Most synchronization algorithms use simple built-in clock models (Known Rate or Constant Rate). The corresponding algorithms can handle clock offset, or both offset and drift, thus provide solutions with different complexity and accuracy. Real clocks do not have known or constant rate, but rather Bounded Drift and Bound Drift-variation models describe their behavior reasonably well. The discrepancy between the built-in model and the real behavior results in synchronization errors Δ_{ext} or Δ_{int}.

If only clock offset is handled, (i.e. the built-in clock model is Know rate), the drift, always present in real clocks, is not compensated for. Such synchronization schemes set the local time of the synchronized clock to the reference time (global time in case of external synchronization or another node's local time in case of internal synchronization) at the time of synchronization, e.g.:

$$C_i(t_{sync}) = t_{sync}.$$

The clock is accurate at the time of synchronization but with time, the synchronization error increases, depending on the drift. The synchronization error is bounded:

$$\left| C_i(t) - t \right| \le (t - t_{sync})\delta_{\max}.$$

If there is a maximum synchronization error requirement $\Delta_{\max}$ then the synchronization must be performed periodically, with period T_{sync}, so that the error due to drift remains below $\Delta_{\max}$. In worst case, the synchronization error at the end of a synchronization period is $T\delta_{\max}$, so for the synchronization period the following constraint holds:

$$T_{sync} < \frac{\Delta_{\max}}{\delta_{\max}}.$$

In real situations, the synchronization step cannot provide zero error, but a small Δt_{sync} initial error is present:

$$C_i(t_{sync}) - t_{sync} = \Delta t_{sync} < \Delta t_{sync,\max},$$

and thus the bound for the synchronization error has the following form:

$$\left| C_i(t) - t \right| \le \Delta t_{sync,\max} + (t - t_{sync})\delta_{\max}.$$

The error bound is illustrated in Figure 1. If the maximum allowed synchronization error is $\Delta_{\max}$ then the minimum synchronization period is the following:

$$T_{sync} < \frac{\Delta_{\max} - \Delta t_{sync,\max}}{\delta_{\max}}.$$

Example 1

If the requirement is to keep the synchronization error below 1ms ($\Delta_{\max}=1ms$), the clock has a quartz crystal with $\delta_{\max}=100ppm$ drift, and the synchronization mechanism provides $10\mu s$ initial

Figure 1. The bound of synchronization error, using periodic offset compensation

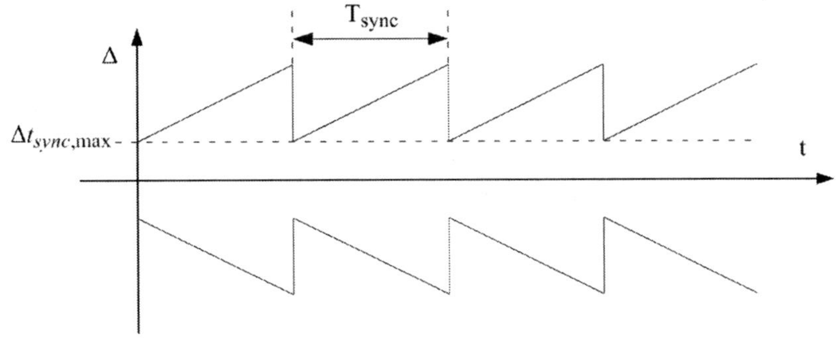

accuracy, then the synchronization period must be no greater than 9.9sec.

If the built-in clock model is Constant Rate, the algorithm compensates for the drift and offset of real clocks. The drift is estimated using multiple synchronization steps. The simplest solution is when the node registers two local timestamps, $C_i(t_1)$ and $C_i(t_2)$, for external triggers emitted by the reference node at t_1 and t_2, respectively. The reference node also sends t_1 and t_2 to node i, which can estimate its drift as follows:

$$\delta_i \cong \frac{C_2(t_1) - C_1(t_1)}{t_2 - t_1} - 1.$$

Node i also compensates offset, e.g. by setting $C_i(t_{sync})=t_{sync}$. After offset compensation, the local time can be modified to compensate for drift, as follows:

$$C_i'(t) = C_i(t) - \delta_i(t - t_{sync})$$

Assuming that both the offset compensation and the drift estimation are correct, and the drift of the node is constant, the above synchronization method provides perfect time estimate $C_i'(t)$ for node i. Although in real clocks drift is never constant, the Bounded Drift-variation model is a good approximation. The drift variation $\vartheta(t)$ is small, thus for short time periods the drift can be considered almost constant. With the variation of the drift taken into account, the bound for the synchronization error can be expressed as follows:

$$\left|C_i(t) - t\right| \leq \frac{\vartheta_{\max}}{2}\left(t - t_{synch}\right)^2.$$

If the offset synchronization is imperfect, but the error is bounded, i.e.

$$C_i\left(t_{sync}\right) - t_{sync} = \Delta t_{sync} < \Delta t_{sync,max},$$

then the drift estimation will be imperfect but bounded, too. The upper bound for drift estimation error is the following:

$$\Delta\delta_i(t) < \frac{2\Delta t_{sync,max}}{t_2 - t_1}.$$

Also, the imperfect offset correction results in a constant offset, with upper bound of $\Delta t_{sync,max}$. Thus, the synchronization error has three terms, representing the drift variation error, the drift error, and the initial offset. From the three terms, the upper bound for the synchronization error can be expressed as follows:

$$\left|C_i(t) - t\right| \le \frac{\vartheta_{\max}}{2}\left(t - t_{synch}\right)^2 + \frac{2\Delta t_{sync,max}}{t_2 - t_1}\left(t - t_{synch}\right) + \Delta t_{sync,max}$$

The above formula clearly shows that periodic resynchronization is necessary again, but with lower rate than in case of the simple offset compensation. If the maximum synchronization error allowed is $\Delta_{\max}$, then in worst-case the synchronization period is the following:

$$T_{sync} < \frac{\sqrt{\left(\frac{2\Delta t_{sync,max}}{t_2 - t_1}\right)^2 + 2\vartheta_{\max}\left(\Delta_{\max} - \Delta t_{sync,max}\right)} - \frac{2\Delta t_{sync,max}}{t_2 - t_1}}{\vartheta_{\max}}.$$

If the synchronization error dominates over drift change, ϑ can be ignored and the constraint for the synchronization period can be expressed by the following simpler inequality:

$$T_{sync} < \frac{\Delta_{\max} - \Delta t_{sync,max}}{2\Delta t_{sync,max}}\left(t_2 - t_1\right)$$

Example 2

The clock of Example1 is now synchronized by drift compensation as well. The drift estimation is done with two measurements, where $t_2 - t_1 = 1sec$. The drift change is bounded by $\vartheta_{\max} = 0.1ppm / sec$. The maximum synchronization period in this case is 44.5sec (ignoring the drift change it would be 49.5sec).

COMMUNICATION MODEL

To design high accuracy synchronization algorithms, a valid implicit clock model, and also a good understanding of the error sources, affecting the synchronization, are needed. The error sources originate from the nature of the inter-node communication, and the software and hardware structure of the nodes. To understand the possible sources, all layers participating in message transmission and reception must be analyzed.

A message is posted at the application layer, and then it goes through the MAC and the physical layers of the sender, then it propagates through the channel, and is received by thc physical and MAC layers of the receiver, and finally it is handled at the application layer again. While crossing several layers, various software, hardware, and physical effects delay the message. Some of these effects are constant, some are random. Some are measurable and thus can be compensated for, some are not; some are negligible, some have considerable effect on the synchronization. The main error sources, based on (Kopetz and Schwabl, 1989) and (Maróti, Kusy, Simon, and Lédeczi, 2004), are shown in Figure 2.

Figure 2. Delays of a wireless communication channel. Deterministic times are denoted by straight lines, non-deterministic times are denoted by wavy lines

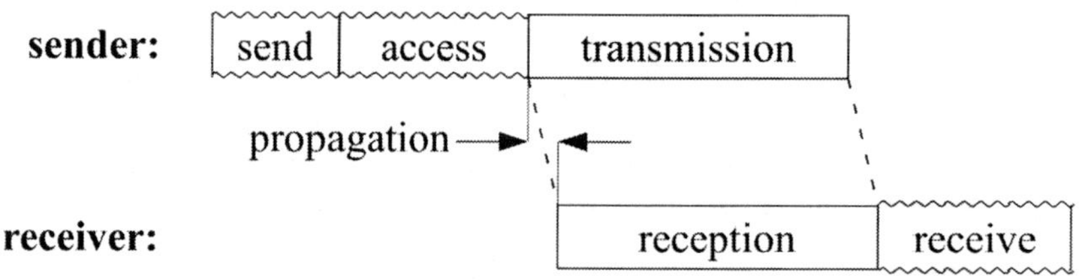

Send Time

On the transmitter side, the message must be assembled and then passed to the MAC layer for transmission. This process usually takes a nondeterministic length of time, depending on the message size, current system load, and events being handled in the system. Send time may be several hundreds of milliseconds.

Access Time

The message is waiting for the start of transmission at the MAC layer. Channel access, however, is highly nondeterministic in most practical MAC layers. The probabilistic nature of the access time can be illustrated by the CSMA/CA protocol, which is often used in sensor networking nodes. In CSMA/CA, after a random initial waiting time, the node listens to the channel for a short time to check for ongoing activity. If the channel is busy then the node waits for a random backoff-time, and then checks the channel again. When the channel is sensed to be idle then transmission begins. The access time depends on the actual channel load and the random values generated in the process, and can vary from milliseconds to seconds.

Transmission Time

This is the time required to transmit the message by the sender radio. Transmission time depends on the message length, speed of the radio, and the message overhead. E.g. using a 256kbps radio the transmission of a 32-byte long message ideally takes one millisecond, but this value may be several times longer, depending on the channel coding utilized. Transmission time is deterministic.

Propagation Time

This is the length of time it takes for the message to reach the receiver after being transmitted. It is deterministic and depends on the distance between the two nodes. E.g., propagation time is approximately one microsecond for a distance of 300 meters.

Reception Time

This is the time required to receive the message by the receiver radio. It is equal to the transmission time.

Receive Time

The time it takes for the MAC layer to process the incoming message and to notify the receiver application. This time depends on the system load, and thus it is nondeterministic. Receive time may be a few hundreds of milliseconds.

To overcome the possible error sources, several, often inter-layer synchronization primitives have been proposed.

SYNCHRONIZATION PRIMITIVES

Synchronization primitives provide means for two nodes i and j to establish the relationship between their clocks C_i and C_j, respectively, ideally providing $C_i(t)=C_j(t)$, for some $t=t_{sync}$. These primitives compensate for offset only, but two or more offset compensation steps can be used to estimate drift as well, as has been shown in Section Time and Clocks.

In practice, compensation is usually done by using software clocks. In this approach, the hardware clock is not adjusted at all but the local time is calculated based on the measured offset and drift. In this chapter, we simply use the general notion of clock adjustment, which can mean either hardware clock correction or software clock calculation.

Figure 3. Unidirectional time synchronization of two nodes

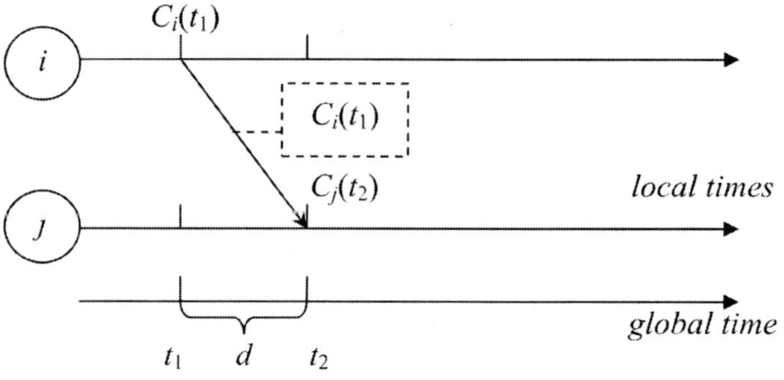

Unidirectional Synchronization

Node i sends a message containing a local time-stamp $C_i(t_1)$ at time t_1, which is received by node j at local time $C_j(t_2)$, as shown in Figure 3. If the message delay $d=t_2-t_1$ is known a priory then node j can synchronize its clock to node i, as follows:

$$C_j(t_2) \leftarrow C_i(t_1)+d.$$

If the delay is not known then the best node j can do is the following:

$$C_j(t_2) \leftarrow C_i(t_1).$$

In this case the maximum synchronization error is

$$\Delta t_{sync,max}=d.$$

If upper and lower bounds on d are known, i.e. $d_{min}<d<d_{max}$ then the following formula can be used:

$$C_j(t_2) \leftarrow C_i(t_1) + \frac{d_{min}+d_{max}}{2},$$

Where the synchronization error in worst case will be as follows:

$$\Delta t_{sync,max} = \frac{d_{min}+d_{max}}{2}.$$

Clearly, this method can effectively be used if the delay is known, at least with satisfying accuracy. If the delay is unknown then a synchronization error will be present, its magnitude depending on the actual delay. Naturally, with the unidirectional synchronization primitive all neighbors of broadcasting node i can synchronize themselves to node i at the same time.

Round-Trip Synchronization

The bidirectional or round-trip synchronization is illustrated in Figure 4. Node i sends a request message to node j at time instant t_1, which is received by node j at time t_2. Node j responds with a message at a later time t_3. The response message contains local time $C_j(t_2)$ and the time difference between receiving the request and sending the reply, i.e. $\delta=C_j(t_3)-C_j(t_2)$. Node i receives the message at local time $C_i(t_4)$ and calculates the roundtrip delay $D=t_4-t_1=d_1+\delta+d_2 \cong C_i(t_4)-C_i(t_1)$. Assuming that the delays d_1 and d_2 are approximately the same, node j can estimate the message delays and synchronize its clock to node i as follows:

$$C_i(t_4) \leftarrow C_j(t_2)+\delta+d_2 = C_j(t_2)+\frac{D+\delta}{2}$$

Figure 4. Roundtrip time synchronization of two nodes

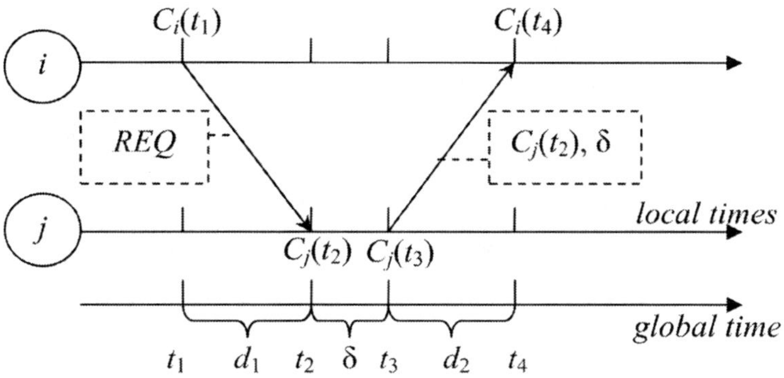

Round-trip synchronization has the implicit assumption that message delays in both directions are the same. This is unfortunately not true in general, thus if the delays are different then the synchronization error will be

$$\Delta t_{sync} = \frac{d_1 - d_2}{2}.$$

In the worst case the synchronization error is:

$$\Delta t_{sync,\max} = \frac{D - \delta}{2}.$$

If a priori bounds on the message delay exist, i.e. $d_{\min} \le d_1, d_2 \le d_{\max}$ then the worst-case synchronization error is:

$$\Delta t_{sync,\max} = \min\left(d_{\max}, D - \delta\right) - \frac{D - \delta}{2}.$$

The fact that inequalities $d_{\min} \le \frac{D - \delta}{2} \le d_{\max}$ trivially must hold, offers a simple way to keep the maximum synchronization error at a low value. Several synchronization steps may be performed until the measured round-trip delay D is sufficiently close to $2d_{\min} + \delta$. If $D - \left(2d_{\min} + \delta\right) < \varepsilon$ then the maximum synchronization error is:

$$\Delta t_{sync,\max} = \frac{\varepsilon}{2}$$

Note that the round-trip synchronization primitive can synchronize more than two nodes with anonymous synchronization. If node k is in the neighborhood of both node i and node j then node k can eavesdrop on the messages sent by node i and node j. Assuming that the request message is received approximately at the same time by nodes j and k, and the reply is received approximately at the same time by nodes i and k, then node k has local timestamps $C_k(t_2)$ and $C_k(t_4)$, for which the following equality holds:

$$D' = \delta + d_2 \cong C_k\left(t_4\right) - C_k\left(t_2\right)$$

Thus, node k can synchronize itself to node j, without sending any messages, as follows:

$$C_k\left(t_4\right) \leftarrow C_j\left(t_2\right) + D'$$

An alternative solution for roundtrip synchronization is shown in Figure 5, where node i does not need to store state information. As shown in

Figure 5. Alternative solution for roundtrip time synchronization

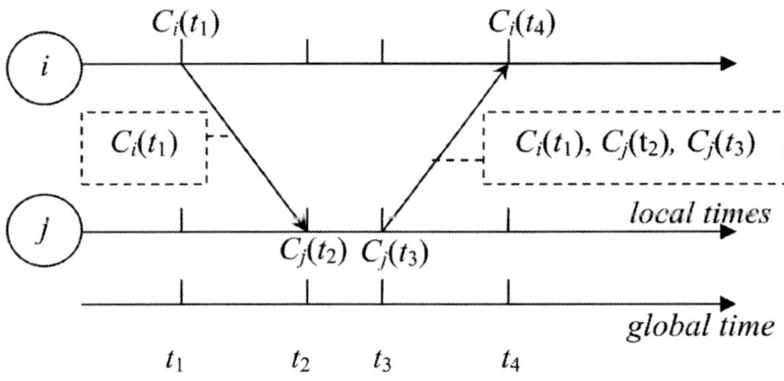

Figure 5, the messages are slightly longer here. In this case, the following equation holds:

$$C_j(t_2) - C_i(t_4) = C_i(t_1) + d_1 + \Delta - \left(C_j(t_3) + d_2 - \Delta\right) \approx C_i(t_1) - C_j(t_3) + 2\Delta,$$

Where $\Delta = C_j(t) - C_i(t)$ is the offset between the two clocks. The offset can be calculated as follows:

$$\Delta = \frac{C_j(t_2) - C_i(t_4) - C_i(t_1) + C_j(t_3)}{2},$$

thus, the synchronization is done by

$$C_i(t_4) \leftarrow C_i(t_4) + \Delta.$$

Reference Broadcasting

In the reference broadcasting primitive an additional beacon node *i* is necessary to synchronize nodes *j* and *k*, as shown in Figure 6. The beacon node broadcasts a synchronization (SYNC) message at time t_1, which is received approximately at the same time t_2 by both nodes. Node *j* eventually sends a message to node *k* with timestamp

Figure 6. Synchronization of nodes j and k with reference broadcast from node i

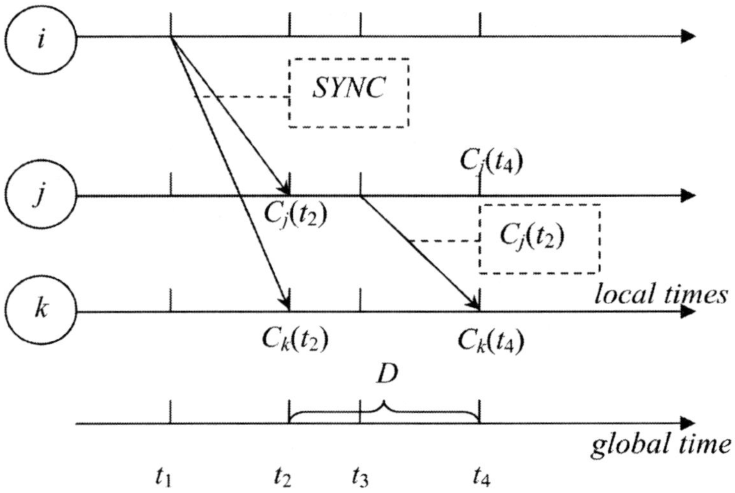

$C_j(t_2)$, which is received at time t_4. Node k then calculates time difference $D=t_4-t_2=C_k(t_4)-C_k(t_2)$ and synchronizes itself to node j as follows:

$$C_k\left(t_4\right) \leftarrow C_j\left(t_2\right) + D$$

Note that it is possible that node j cannot communicate directly with node k, but assuming symmetrical links, it is able to send messages to the beacon. Thus in an alternative solution node j sends the timestamp message back to the beacon, which forwards it to node k. The algorithm is trivially extendable to synchronize all neighbors of the beacon to node j. In this case, all nodes receive the SYNC message and also the forwarded timestamp $C_j(t_2)$ from the beacon node.

The accuracy of the reference broadcasting primitive, implemented on Mica2 motes, was tested (Elson, Girod, and Estrin, 2002). In their test, they measured the inter-receiver phase offset of two nodes, receiving the same beacon messages. Due to interrupt handling delays, nodes do not perform time-stamping at the same time, causing a synchronization error. In the tests, the time differences followed a normal distribution with zero mean and 11.1 microseconds of standard deviation. This property is very advantageous, since by repeated synchronization steps, the error cannot only be decreased statistically but the level of the expected error can also be estimated.

The main advantage of reference broadcasting is that it completely eliminates all the uncertainties on the transmitter side, and also cancels all the deterministic delays on the receiver side as well. With reference broadcasting it is easier to achieve higher accuracy than with unidirectional or roundtrip synchronization methods. The price is that for every synchronization step an additional beacon is required, thus network-wide synchronization is more difficult to organize.

TIME-STAMPING

Synchronization primitives assume that the time instant of message reception and message transmission can be measured, according to the local clock. This measurement activity, which highly affects the accuracy of synchronization, is called time-stamping.

Since message transmission and reception takes a longer time, a particular piece (practically a byte or bit) of the message is selected as a reference point. Theoretically, the reference point can be any part of the message, e.g. first byte, last byte, end of preamble, etc. It will be shown that for practical reasons it is more advantageous to choose the reference point at the beginning of the message.

Time-stamping records the local time of either the transmission or the reception of the reference point. To avoid confusion about what actual transmission and reception means, we define the ideal transmission time as the physical transmission of the electromagnetic signal at the antenna of the transmitter, and similarly the ideal reception time is the arrival of the electromagnetic signal to the antenna of the receiver. Clearly, the difference between ideal transmission and reception times is the propagation time.

Real devices, depending on the hardware and software support, provide different means to approximate ideal time-stamps. Naturally, it is not possible to perform measurements on the antenna, thus delays between the occurrence and the detection of the time-stamping event are present. If these delays are constant, then they can be compensated for, otherwise the time-stamps will have errors.

Time-stamping can be performed at different layers; the simplest solution being application layer time-stamping, which is available on all platforms. Here time-stamping of the transmitted messages is performed when the high-level transmission command (e.g. send, broadcast) is issued, as shown in Figure 7 (a). Note that in most

Figure 7. Pseudo-code of time-stamping at the application layer. (a) time-stamping of the transmission command issue, (b) time-stamping of the completed message transmission, (c) time-stamping of the completed message reception

```
...
SendMessage();
TimeStamp = Now();
...
```

or

```
...
TimeStamp = Now();
SendMessage();
...
```

(a)

```
...
SendMessage();
...

MessageSent() {

  ...
  TimeStamp = Now();
  ...

}
```

(b)

```
MessageReceived() {

  ...
  TimeStamp = Now();
  ...

}
```

(c)

systems message sending passes the message to the MAC layer thus there is no significant difference between the two solutions Figure 7 (a) shows; in either case nondeterministic send and access times will cause error in time-stamping (see also Figure 2). In the solution, shown in Figure 7 (b), time-stamping is performed when the message transmission has been completed; eliminating uncertainties of medium access, but random software delays are still present. Time-stamping of the completed message transmission provides more accurate results than the simple command-issue time-stamping, but it requires some support from the lower layers, i.e. notification when the message is actually sent. Similarly, message reception can be time-stamped at the application layer after the notification from the MAC-layer arrives, as shown in Figure 7 (c). This solution involves the nondeterministic receive time, as shown in Figure 2, thus it is not suitable for high-accuracy synchronization.

If low-layer support is available, most uncertainties of send, access, and receive times can be eliminated. MAC-layer time-stamping utilizes the radio chip's low-level signals to perform time-stamping. For example at the transmitter side, the radio chip raises an interrupt after the transmission of each data byte. One of these interrupts can be used as a reference point, to read the local clock and produce a time-stamp, as shown in Figure 8. Small delay variations may be present due to interrupt handling jitters, but these errors can either be ignored or reduced. MAC-layer time-stamping at the receiver side is quite similar: e.g. if the receiver raises an interrupt after each received byte, the interrupt associated with the reference point can be used to generate the time-stamp.

If the transmitter allows the modification of data after its transmission is started, the actual time-stamp of the transmission can be placed in the transmitted message itself, as illustrated in Figure 8. If the transmitter does not allow modification of data under transmission then another message needs to be transmitted with the time-stamp.

Due to the uncertainties of send, access, and receive times, the accuracy of application-level time-stamping is limited, and the synchronization error can easily be as high as a few hundreds of milliseconds. Using carefully implemented MAC-

Figure 8. Time-stamping of the transmitted message at the MAC- layer

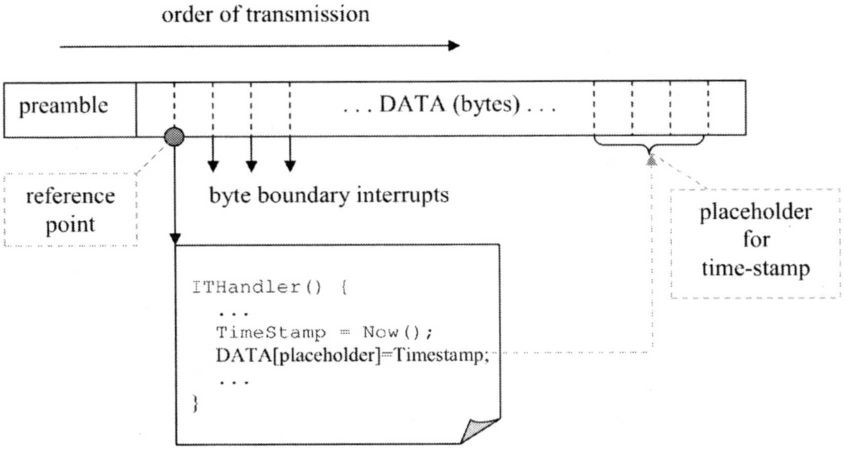

layer time-stamping, the synchronization error can be reduced to the range of microseconds, even with inexpensive hardware solutions.

SYNCHRONIZATION ALGORITHMS

In this section, the most influential and successful time-synchronization algorithms will be reviewed, and their performance will be analyzed. The algorithms and their main features are the following:

- Reference Broadcast Synchronization (RBS)

Using RBS two nodes can synchronize their clocks using the service of a third broadcast node. The broadcast node emits a message, which is received by the recipients (neighbors of the broadcast node) at the same time, if propagation delay is ignored. Using additional communication the recipients can exchange their reception times and can agree on a common time. Extensions of the algorithm can be used to synchronize multihop networks.

- Tree-based pair-wise synchronization protocols

Timing-Sync Protocol for Sensor Networks (TPSN) and Lightweight Time Synchronization (LTS) protocol synchronize the whole network using spanning trees and then pair-wise synchronization between parents and children, starting from the root node. Time stamping in the MAC layer increases accuracy of pair-wise synchronization steps.

- Flooding Time Synchronization Protocol (FTSP)

FTSP is a massively distributed algorithm, which synchronizes the whole network to a leader node at the same time rather than performing pair-wise synchronization steps. The algorithm uses several tricks to provide high accuracy, robustness, and still low overhead: time stamping in the MAC layer, linear regression from multiple sources, and dynamic leader election.

- Elapsed Time on Arrival (ETA)

This post facto synchronization method does not synchronize the whole network, only those nodes that transmit the message to the sink from the source node that generated the message. The algorithm uses MAC layer time stamping support to accurately measure the time that a message spends in the network. At the base station the transmission time of each received message can be calculated, thus the times of the triggering events can be determined. This scheme requires no additional messages and its overhead is very low.

Reference Broadcast Synchronization Protocol

The RBS protocol, proposed by (Elson, Girod, and Estrin, 2002), synchronizes two or more receivers by sending reference broadcasts by beacon nodes. Reference broadcasting completely eliminates sender-side uncertainties (send and access times) and possibly the receiver-side uncertainties (receive time), too, by applying MAC-layer time-stamping. Similarly to the scheme shown in Figure 6, a beacon node broadcasts beacon messages at time instants $t_k\{t_1,t_2,...\}$. The receiving nodes record the reception times, according to their local clock, and then exchange their time-stamps. Node j, knowing its own time-stamps $C_j(t_k)$, and the corresponding time-stamps $C_i(t_k)$ of node i, calculates a linear regression (by excluding outliers) and establishes the relationship between clocks C_j and C_i, by estimating the drift and offset between them. This relationship can be expressed as a function

$$C_i(t) = TCONV_{i,j}\left(C_j(t)\right).$$

The simple pair-wise synchronization scheme can be extended to multihop networks. Using the service of multiple beacons in the network, nodes can perform pair-wise synchronization steps. The synchronization graph of the network can be defined as follows: the vertices of the graph are the nodes. An edge exists between two vertices if the corresponding nodes can be synchronized, i.e. they are in the neighborhood of a common beacon. A synchronization graph of a network with four beacons is shown in Figure 9. Multihop synchronization, called Time Routing, can be performed along the synchronization graph. When an event is detected at node i, the local time-stamp $C_i(t)$ is generated. When the message is forwarded to node j, which is a neighbor of node i in the synchronization graph, node j converts the time-stamp to its own local clock by applying $TCONV_{ij}$ to the received time-stamp.

The accuracy of RBS depends on the accuracy of the receiver-side time-stamping. In the tests presented in (Elson, Girod, and Estrin, 2002),

Figure 9. (a) Network topology and (b) the corresponding synchronization graph

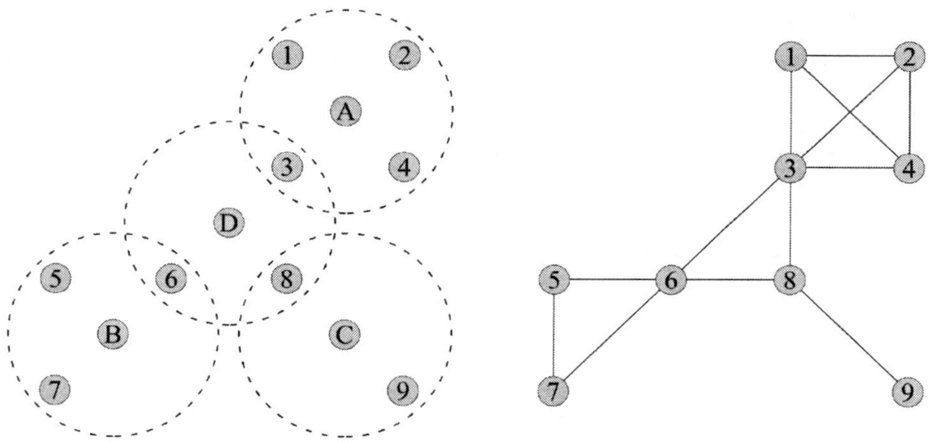

the algorithm was implemented on various platforms and the measured accuracy was in the range of only a few microseconds per hop, with MAC-layer time-stamping support.

An obvious drawback of the algorithm is that the message exchange and coordination between nodes requires extra communication. Another disadvantage is that the automatic beacon selection and the control of the synchronization graph are not defined.

Tree-Based Pair-Wise Synchronization Protocols

Timing-Sync Protocol for Sensor Networks (TPSN), proposed in (Ganeriwal, Kumar, and Srivastava, 2003), uses pair-wise synchronization along a spanning tree, and provides global synchronization to the root node. TPSN consists of two phases: In the level discovery phase, a spanning tree is constructed, where the root node is the time reference; while in the synchronization phase all nodes are synchronized pair-wise along the edges of the tree, using the round-trip synchronization primitive shown in Figure 5.

The level discovery phase runs when the sensors are deployed. In this process there is a level number assigned to every node, the root node having level 0. At first, the root broadcasts a level discovery packet, which contains its unique ID and level L. The nodes receiving this packet assign level $L+1$ to themselves (in this case 1), and they also broadcast a level discovery packet with their IDs and level numbers; this process continues until all nodes in the network will have a level assigned. Nodes react only to the first level discovery packet they receive, the further ones are ignored. This way a spanning tree is constructed, where the root is the reference node with level 0 and all the child nodes have one bigger level than they parent.

After the spanning tree is constructed the synchronization is done using the pair-wise round-trip synchronization primitive shown in Figure 5. The packet transmitted by node *i* is called synchronization pulse and in addition to $C_i(t_1)$ it also contains the level number of *i*. The packet sent by node *j* is called acknowledgement and, in addition to time-stamps $C_i(t_1)$, $C_j(t_2)$, $C_j(t_3)$, it also contains the level of the sender. The protocol is illustrated in Figure 10. The synchronization process is initiated by the root node r, which broadcasts a time sync packet. After a random delay, level 1 neighbors l_{11} and l_{12} synchronize themselves with the root node. Level 2 nodes overhear the synchronization pulse messages and after a random delay they also

Figure 10. The TPSN protocol

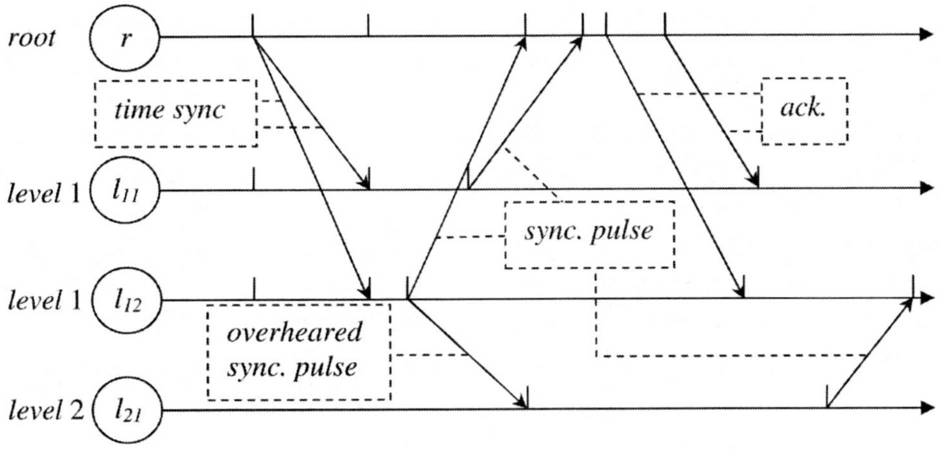

synchronize themselves with level 1 nodes (l_{21} to l_{12} in Figure 10). This process continues until all nodes are synchronized. Despite of the random delays, collisions may occur, thus nodes not hearing the acknowledgement within a timeout period, will try to synchronize again by transmitting a new synchronization pulse.

In a real application the topology of the sensor network can often change, e.g., motes can die randomly, new ones can be deployed, and communication errors are also possible. If a node joins after the level discovery phase, or is unable to receive any level discovery packets due to collisions or other communication errors, then after a timeout, it broadcasts a level request packet. Each neighbor replies with its own ID and level number, and the requesting node will assign to itself a level number one bigger than the smallest one it received.

If a node with level n leaves the network, there will possibly be some nodes with level $n+1$, which cannot communicate with any node with level n. These nodes will never receive the acknowledgement for their synchronization pulse messages. After a fixed number of tries, these nodes will try to assign a new level to themselves, starting a level discovery process. Their child nodes also have to do level discovery in a similar way.

Another tree-based solution is the Lightweight Time Synchronization protocol, proposed in (van Greunen and Rabaey, 2003). The main driving force behind the design of LTS was that in most sensor networking applications the low energy consumption is much more important than the accuracy of the synchronization, so this algorithm is optimized to the former, still providing the required synchronization accuracy.

The algorithm also uses the modified round-trip synchronization primitive shown in Figure 5 for pair-wise synchronization along the edges of a spanning tree, to synchronize the whole network or certain nodes to one or more global reference nodes.

There are two variations of the LTS protocol. In the Centralized Multi-Hop LTS, first a spanning tree is formed, and then the synchronization is controlled by the root element, which plays the role of the reference node. The root regularly performs pair-wise synchronization steps with its neighbors and then the neighbors also synchronize themselves with their child nodes. This procedure continues towards the leaves of the spanning tree.

The other variation is the Distributed Multi-Hop LTS. In this approach, the synchronization is initiated by the nodes. Each node can calculate its synchronization period, based on the required accuracy, the drift of the local clock and the distance from the nearest reference, and the synchronization requests are generated on demand. Synchronization is done using the pair-wise method along the routing path between the requesting node and the reference node, and all intermediate nodes are also synchronized. Assuming equal accuracy requirements and the same drift for each node, nodes furthest away from the references will have the shortest synchronization period, thus the resynchronization is always initiated by the leaves of the spanning tree. The main advantage of this LTS variation is that the spanning tree is constructed implicitly, therefore the algorithm can adapt to changes in network topology.

In tree-based synchronization protocols the maximum synchronization error grows with the distance from the reference node as a linear function of the number of hops, thus the largest error in the system is proportional to the depth of the spanning tree. The smallest depth spanning tree, i.e. the smallest maximum synchronization error can be achieved by breadth-first-search.

If the round-trip error is Gaussian with standard deviation σ, and the synchronization error is independent in each pair-wise step then, in a multihop network, the standard deviation of the time synchronization error of a node n hops away from the root is $\sqrt{n}\sigma$. Thus, for example, the synchronization error is smaller than $3.3\sqrt{n}\sigma$ with 99.9%

confidence. The accuracy of the pair-wise synchronization greatly depends on the accuracy of the MAC-layer time-stamping. Depending on the implementation, σ is reported to be in the range of 1-20 microseconds, using inexpensive hardware.

Flooding Time Synchronization Protocol

The algorithm was proposed in (Maróti, Kusy, Simon, and Lédeczi, 2004), and since then it has been successfully used in various sensor networking applications, and it is the de facto time synchronization method of the popular TinyOS sensor networking operating system. The FTSP can provide high accuracy internal synchronization services for multihop networks with low overhead. The algorithm is robust against node and link failures.

FTSP uses an improved version of the MAC-layer time-stamping scheme shown in Figure 8. Since the interrupt handling time has jitter, the accuracy of the time-stamps in FTSP is increased by averaging results from measuring transmission times of multiple bytes. Interrupts corresponding to the transmission (on the transmitter side) or reception (on the receiver side) of k consecutive bytes are used to produce time measurements t_1, t_2, …, t_k, quite similarly on both the transmitter and the receiver. The measurements are normalized by the T_{byte} transmission/reception time of one byte as follows:

$$t_i' = t_i - (i-1)T_{byte}.$$

From the normalized byte transmission times (after possible filtering), the timestamp for the message is calculated by averaging:

$$t_{sync} = average(t_i').$$

According to the test performed by the authors of FTSP on Mica2 motes with clock frequency of 7.37 MHz, the improved MAC-layer time-stamping method can produce an average time-stamping error of 1.4 microseconds, while the simple method using only transmission/reception time of a single byte (illustrated in Figure 8), produces errors in the range of a few tens of microseconds.

The transmitter embeds its local timestamp $t_{trans}(t)$ in the transmitted message and the receiver produces its local timestamp $t_{rec}(t)$ and thus can establish the relationship between the two local clocks, producing a synchronization point $\left(t_{trans}(t), t_{rec}(t)\right)$. This information itself could be used to eliminate the offset between the two clocks, but instead, FTSP uses multiple synchronization points to eliminate both offset and drift. Based on a set of synchronization points (eight points were used in the tests), called regression table, the receiver calculates a linear regression to approximate the transmission time-stamps with the reception time-stamps. If the linear regression has the form of

$$t_{trans}(t) = at_{rec}(t) + b$$

then a+1 approximates the drift δ_{rec} with respect to the clock of the transmitter, and b approximates the offset between the transmitter and the receiver.

Note that this scheme allows the utilization of synchronization points from multiple sources: a receiver may receive synchronization messages from multiple transmitter nodes and can construct its regression table from multiple sources. Assuming that the sources are synchronized, the receiver node synchronizes itself to the common clock of the set of transmitters. Using this idea, the multihop synchronization of networks requires no additional organization (e.g. construction of synchronization trees). The network synchronization is started from a root (reference) node, which periodically broadcasts synchronization messages. When one of its neighbor nodes has enough synchronization points in its regression table it synchronizes itself. Synchronized nodes also periodically broadcast synchronization messages, similarly to the root. Thus, the whole network will eventually be synchronized, similarly to a flood, originating from the root node.

The synchronization message contains a round number. The root increases the round number in each broadcasted synchronization message. Other nodes record the highest received round number and include it in their broadcasted synchronization messages. In the regression tables each synchronization point must have different round number; this restriction ensures that the linear regression is calculated for points distributed over a longer period of time, thus the accuracy of offset and drift calculation will be higher.

The root node is the reference in the network, thus it is a single point of failure. To provide fault tolerant operation, FTSP provides a robust root election service. The election is based on the unique node identifiers (node IDs): the root node is the node with the smallest node ID. Each node locally stores the ID of the root, according to its best knowledge, in variable myRootID, and also includes myRootID in each broadcasted synchronization message. If a node has no information about who the root is (e.g. at network startup) then the node declares itself root and behaves accordingly (broadcasts synchronization messages with increasing round number and its own node ID as root ID.) When a node receives a message with root ID smaller than the receiver's myRootID and a new (higher) round number then the node updates its myRootID. If the received root ID is larger than the node's own ID then the node drops the message. When the root dies or the network breaks into multiple segments then there will be nodes who do not receive new synchronization messages for a period longer than RootTimeout. If this happens then the node assumes that the root died and immediately declares itself the new root and, by transmitting synchronization messages with new root ID and increasing round numbers, it initiates a new election process. When multiple roots exist in the system then those with higher ID give up their root status when they receive a synchronization message with smaller root ID. This simple mechanism ensures that (1) after a period of time longer than RootTimeout there will be new root(s) in the network, and (2) after an additional transient period there will be only one root in the network and all nodes will have the same myRootID.

The FTSP protocol was tested (Maróti, Kusy, Simon, and Lédeczi, 2004), in an 11-hop wide network containing 60 nodes. With resynchronization time of 30 seconds, the network provided 0.5 microseconds/hop average accuracy. The network got into synchronized state in approximately 12 minutes after cold startup. The network was tested under harsh conditions, including switching off and then on half of the nodes and switching off the root node. The maximum error during the tests was below 2.3 microseconds/hop.

Elapsed Time on Arrival

The Elapsed Time on Arrival (ETA) synchronization method was proposed by (Kusy et al, 2006). This synchronization protocol does not synchronize the whole network, only those nodes that are required to deliver the message. The synchronization process is illustrated in Figure 11. When node 1 detects an event at t_E, it records the local event timestamp $C_1(t_E)$. Node 1 then transmits the message to node 2, at time t_{TR1}. Using MAC-layer time-stamping, similarly to Figure 8, node 1 generates timestamp $C_1\left(t_{TR1}\right)$, but instead of $C_1\left(t_{TR1}\right)$, it places the elapsed time $\delta_1 = C_1\left(t_{TR1}\right) - C_1\left(t_E\right)$ into the transmitted message. The subsequent node 2 receives the message and, also using MAC-layer time-stamping, gener-

Figure 11. Elapsed Time on Arrival synchronization

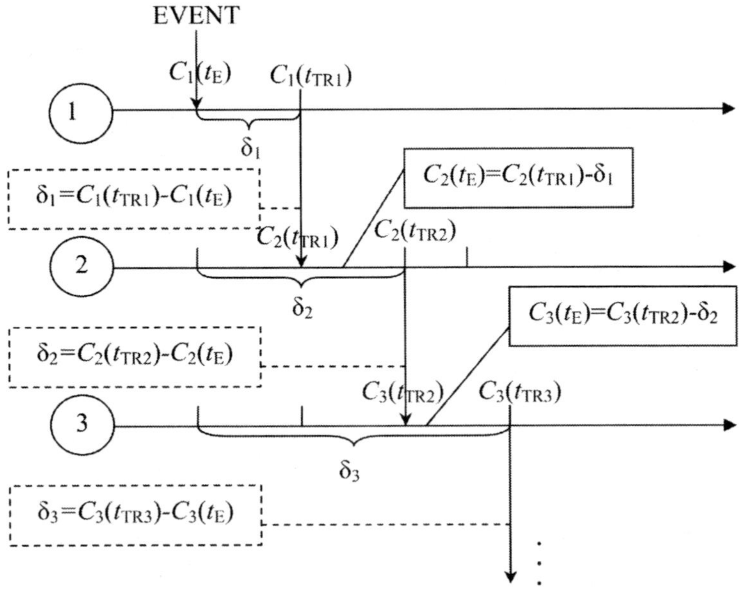

ates $C_2\left(t_{TR1}\right)$. Node 2 calculates the time of event in its local time as $C_2\left(t_E\right) = C_2\left(t_{TR1}\right) - \delta_1$. Similarly, as the message is transmitted from one node to another, the message always includes the elapsed time between the event occurrence and the last message transmission. Each receiving node j translates the elapsed time δ_{j-1} to the time of event $C_j(t_E)$ in its local time, and translates it back to elapsed time δ_j when the message is retransmitted. Thus, the destination node also can calculate the time of event, according to its own local clock.

The ETA algorithm does not require any additional message to be sent in the network for synchronization purposes, the only overhead it requires is the time stamp placed in the message. Kusy et al tested the accuracy of the algorithm in a network where the distance between sources and the sink varied between two and six hops. In this setup, the average synchronization error was less than 8 microseconds.

The ETA can be used for fast network-wide synchronization as well. For this purpose, the time of event t_E is the global time of the reference clock itself. This global time is also included in the message and is never changed. The message is broadcast to all nodes in the network. Every recipient node j can calculate $C_j(t_E)$, and having t_E it can synchronize itself to the reference clock. Using multiple synchronization messages the drift can also be estimated and compensated for. In their experiment, Kusy et al measured 2.7 microseconds/hop average error in an 11-hop network, with only 30 seconds of initial synchronization time after startup.

FUTURE RESEARCH DIRECTIONS

Successful synchronization solutions agree on that the precondition of good accuracy is the precise time stamping, possibly carried out in the MAC-layer. Since different implementations on the same platforms show notable differences in accuracy, possible improvements are expected in this field. New hardware platforms allow the creation of new time-stamping solutions, too.

The proposed solutions differ on how the local (mainly pair-wise) synchronization can be extended to provide network-wide synchronization. Solutions utilizing heavy and potentially central control (e.g. maintenance of a spanning tree) do not scale well and these solutions tolerate topology changes poorly. Completely distributed solutions, utilizing only local information, are more advantageous from these aspects. New, truly distributed solutions may provide more robust and scalable solutions.

An important and yet mainly unanswered question is the security of the synchronization protocols. Clearly, various attacks can be utilized against sensor networking applications, one of them being the assault against the synchronization mechanism. Studies on this topic have been published but there is much work left to do in order to create robust and attack-resilient synchronization mechanisms.

CONCLUSION

Sensor networking applications often require time synchronization services to aid coordinated action or to provide common time base for observed events. The performance (accuracy, overhead) of the time synchronization highly affects the performance of the whole application, thus the right choice of the algorithm is essential.

In this chapter, we introduced various time synchronization models and showed the potential accuracy of the synchronization mechanisms that use these models. We illustrated that simple implicit models provide modest accuracy and potentially simpler algorithms; while the utilization of more precise models results in better accuracy, but also potentially higher complexity, both in computation and communication. We also modeled and reviewed the potential error sources, which are present when the synchronization is carried out using real physical devices and real communication channels.

We introduced typical synchronization primitives, which can be used as building blocks for synchronization algorithms. These primitives utilize various clock models and eliminate or reduce various error sources, providing solutions with different accuracy and complexity. The primitives offer tradeoffs that the designer can use to create systems, to fulfill various requirements.

Finally, synchronization algorithms, which, based on the basic synchronization primitives, provide network-wide services, were discussed, and analyzed. Different approaches were presented with different complexity and potential accuracy, fulfilling the needs of various requirements of different application fields. The ideas and solutions present in these algorithms can be used to develop specially tailored algorithms in the presence of challenging constraints, potentially arising in future sensor networking applications.

REFERENCES

Elson, J., Girod, L., & Estrin, D. (2002). Fine-grained network time synchronization using reference broadcasts. *SIGOPS Operating Systems Review, 36(*SI), 147-163.

Ganeriwal, S., Kumar, R., & Srivastava, M. B. (2003). Timing-sync protocol for sensor networks. In *Proceedings of the 1st international Conference on Embedded Networked Sensor Systems SenSys '03* (pp. 138-149).

Kopetz, H., & Schwabl, W. (1989). *Global time in distributed real-time systems*. (Technical Report 15/89), Technische Universitat Wien.

Kusy, B., Dutta, P., Levis, P., Maroti, M., Ledeczi, A., & Culler, D. (2006). Elapsed time on arrival: A simple and versatile primitive for canonical time synchronization services. *International Journal of Ad Hoc and Ubiquitous Computing, 1*(4), 239–251.

Maróti, M., Kusy, B., Simon, G., & Lédeczi, Á. (2004). The flooding time synchronization protocol. In Proceedings of the 2nd international Conference on Embedded Networked Sensor Systems, SenSys '04. (pp. 39-49).

Van Greunen, J., & Rabaey, J. (2003). Lightweight time synchronization for sensor networks. In *Proceedings of the 2nd ACM international Conference on Wireless Sensor Networks and Applications* (pp. 11-19).

ADDITIONAL READING

Elson, J., & Römer, K. (2003). Wireless sensor networks: a new regime for time synchronization. *SIGCOMM Comput. Commun. Rev.*, *33*(1), 149–154. doi:10.1145/774763.774787

Elson, J. E. (2003). Time Synchronization in Wireless Sensor Networks. PhD dissertation in Computer Science, University of California, Los Angeles, California.

Ganeriwal, S., Čapkun, S., Han, C., & Srivastava, M. B. (2005). Secure time synchronization service for sensor networks. In Proceedings of the 4th ACM Workshop on Wireless Security (pp. 97-106)

Mills, D. L. (1991). Internet time synchronization: The network time protocol. *IEEE Transactions on Communications*, *39*(10), 1482–1493. doi:10.1109/26.103043

Mills, D. L. (1991). Improved algorithms for synchronizing computer network clocks. *IEEE/ACM Transactions on Networking*, *3*(3), 245–254. doi:10.1109/90.392384

Precise Networked Clock Synchronization Working Group. (2008). *1588-2008 IEEE Standard for a Precision Clock Synchronization Protocol for Networked Measurement and Control Systems*. New york, NY: IEEE.

Römer, K., Blum, P., & Meier, L. (2005). Time Synchronization and Calibration in Wireless Sensor Networks . In Stojmenovič, I. (Ed.), *Handbook of Sensor Networks: Algorithms and Architectures* (pp. 199–237). Hoboken, NJ: John Wiley & Sons.

Schmid, U., & Schossmaier, K. (2003). Interval-based clock synchronization with optimal precision. *Information and Computation*, *186*(1), 36–77. doi:10.1016/S0890-5401(03)00103-2

Su, W., & Akyildiz, I. F. (2005). Time-diffusion synchronization protocol for wireless sensor networks. *IEEE/ACM Transactions on Networking*, *13*(2), 384–397. doi:10.1109/TNET.2004.842228

Sun, K., Ning, P., & Wang, C. (2006). TinySeRSync: secure and resilient time synchronization in wireless sensor networks. In Proceedings of the 13th ACM Conference on Computer and Communications Security (pp. 264-277)

Sundararaman, B., Buy, U., & Kshemkalyani, A. D. (2005). Clock synchronization for wireless sensor networks: a survey. *Ad Hoc Networks*, (3): 281–323.

KEY TERMS AND DEFINITIONS

Clock Drift: Clock drift is the relative difference between the frequencies of two clocks.

Clock Offset: Clock offset is the difference between the real time and the time shown by a clock. Offset between two clocks is the difference between the times shown by the clocks.

Event: An event of interest that can either be an application-specific external event (e.g. detection of an acoustic pattern), or an internal event. (Particularly interesting ones for time synchronization purposes are transmission/reception times of a reference point in the message.)

Medium Access Layer (MAC): The MAC layer is a network protocol (sub)layer, which provides addressing and also channel access mechanisms to allow the utilization of a common medium by multiple communicating devices.

Sensor Network: A sensor network is a group of spatially distributed cooperating sensor nodes to solve distributed sensing and monitoring problems.

Sensor Node: A sensor node is a small autonomous device with computing, sensing, and wireless communicating capabilities. In most cases, a sensor node is built around a microcontroller and a low power radio, and use embedded operating systems.

Synchronization Primitives: Synchronization primitives are the building blocks of synchronization algorithms. They usually provide means for synchronization of a pair of nodes.

Time Reference: Time reference is an external (e.g. GPS) or internal source (e.g. a particular sensor node), to which the other nodes are synchronized.

Time Stamping: Time stamping is the method of measuring and storing the time of an event.

Time Synchronization: Time synchronization is an act resulting in a common sense of time on independent clocks of the system.

APPENDIX

Table of abbreviations

ETA	Elapsed Time on Arrival
FTSP	Flooding Time Synchronization Protocol
GPS	Global Positioning System
LTS	Lightweight Time Synchronization
MAC	Medium Access
ppm	parts per million
RBS	Reference Broadcast Synchronization
SYNC	synchronization message
TPSN	Timing-Sync Protocol for Sensor Networks

Chapter 22
Application-Driven Routing in Wireless Sensor Networks

Gyula Simon
University of Pannonia, Hungary

ABSTRACT

Sensor networks are built from tiny, resource limited nodes, which are able to communicate with each other, and thus, provide distributed services. The quality of communication, and especially of routing, is one of the major enablers of good performance, energy efficiency, and longevity in resource-deprived sensor networks. Since the advent of wireless sensor networks, literally hundreds of routing algorithms, tailored especially for this domain, have been proposed in the literature. This chapter highlights the main ideas and illustrates how these solutions help to reach various design goals. After a general overview, the taxonomy of routing algorithms in sensor networks will be presented, and then the mainstream algorithms, with the greatest impact on the field, will be introduced and analyzed. Through typical routing algorithms, we will show how and what kind of quality of services can be provided for various application needs.

INTRODUCTION

Sensor networks, constituting a special realm of wireless ad-hoc networks, offer a diverse set of exciting new ubiquitous application domains. The network is composed of tiny autonomous sensor nodes, which are possibly embedded or otherwise distributed in our environment in large quantities, and can perform distributed sensing, with spatial resolution and coverage that was impossible with ordinary measurement systems before. The nodes in the network communicate with each other and form a distributed, collaborative sensing and computing domain.

The application of sensor networks is motivated by both civil and military needs, including environment and habitat monitoring, intelligent agriculture, industrial monitoring, healthcare

DOI: 10.4018/978-1-60960-732-6.ch022

applications, traffic surveillance in the civilian domain; and various military applications, e.g. battlefield surveillance, intruder detection and tracking, and shooter localization, just to name a few. Sensor nodes typically contain a low-power microcontroller, radio transceiver, and energy resources (typically batteries, but energy harvesting is also possible, e.g. from solar energy or motion). Sensing capabilities include temperature, humidity, pollution, motion, sound, but any application-specific sensor can also be applied. The hardware and software architecture of the nodes is designed to support long, autonomous operation. A comprehensive overview on sensor networks can be found in (Yick, Mukherjee, Ghosal, 2008).

The communication between nodes is typically performed by radio in one of the free ISM bands, using multi-hop, ad-hoc networking. Sensor networks typically contain special devices (base stations) to provide gateways to external devices or networks. The majority of applications send collected data to the base stations, requiring unicast (also called convergecast) services. User commands or system maintenance messages are typically sent from base stations to all nodes or a set of nodes, using multicast routing services. Peer-to-peer routing is relatively rare, but occasionally present in sensor networking applications. No matter what types of services are used in the network, routing is one of the major middleware services that define the performance of the overall network. Due to the special architecture and requirements present in sensor networks, traditional solutions, developed for wired or general wireless networking domains, are not applicable here. Thus literally hundreds of routing algorithms, specially tailored for sensor networks, can be found in the literature, and the number is still counting. An excellent review on various protocols can be found in (Akkaya and Younis, 2005). This chapter highlights the main and basic ideas present in various routing algorithms, and illustrates how these solutions help to reach various design goals.

First, the special challenges, resulting from the hardware/software constraints and the application domain, are discussed, and then the taxonomy of the typical routing algorithms in sensor networks is briefly presented. Finally, the mainstream algorithms with the greatest impact on the field will be introduced and analyzed. Through typical routing algorithms, we will show how different quality of services can be provided for various application needs.

CHALLENGES IN SENSOR NETWORKS

Routing algorithms are affected by the special needs, constraints, and requirements of sensor networking applications. Some of the requirements are derived from physical constraints; others are driven by the special needs of the applications.

- **Resource-limited nodes.** Large sensor networks necessitate the usage of inexpensive devices. This results in limited processing power (8-bit microcontrollers are common), small storage resources (both for program and data), and limited communication capabilities. Apart from expenses, the other limiting factor is energy. Most devices are driven from batteries, from which the system should operate for months or years, without user intervention. Thus low energy consumption is necessary. Since nodes in wireless sensor networks serve as both sensors and routing devices, the routing protocol has great impact on the nodes' energy consumption. Another serious limitation is the wireless communication channel's limited bandwidth and its traditional problems (fading, large error rate).
- **Ad-hoc deployment, unmanaged networks.** In many applications deployment of the nodes is not manageable, thus self-organizing network solutions are required.

In addition, the operation of the network must be maintained without human services for possibly the whole lifetime of the system. Some of the sensor nodes inevitably fail, thus the network (and thus the network protocols) must be fault tolerant.

- **Large number of nodes.** Senor networks may contain several thousands of resource-limited nodes. Traditional global addressing schemes with large routing tables cannot be applied. Furthermore, the protocols must be scalable.
- **Data-centric networks.** In many applications data is important and not the nodes themselves. (E.g. if only data from areas with high noise are of interest then only those nodes need to reply, which detect noise level above the specified limit.)
- **Application-driven data flow.** Data flow in sensor networks depends on the application. In most cases, many sources (sensors) and only a few sinks (base stations) are present. Command and control result in reverse chain: information must be delivered from base stations to specific or all sensors. Data reporting is also application-specific: data collection may be time-driven, event-driven, or query-driven. Requirements also change from one application to another, typical requirements being robustness, longevity, low latency, or their combination. In addition, data properties (e.g. redundancy) depend on the application, and may be exploited by the routing algorithm (e.g. aggregation). In-network processing is also possible in several cases.
- **Location-awareness.** A sensor node may or may not have information on its own (absolute or relative) location. If such information is available (e.g. using acoustic-, radio-, or GPS-based self-localization methods) the routing algorithm may use it to improve its performance. In many cases, however, location information is very limited.
- **Mobility.** Some sensor networks are inherently mobile, or partially mobile: sensor nodes, and/or sink nodes may change their locations. A typical example for massively mobile networks is when sensors are attached to animals. In these cases, dedicated algorithms are used for data gathering and routing to (possibly mobile) base stations.

ROUTING ALGORITHMS IN SENSOR NETWORKS

Some requirements (ad-hocness, robustness, fault tolerance) are general in most sensor networks but the specific application-dependent requirements may differ from one application to another. The most common application-driven requirements are the following:

- **Guaranteed delivery.** The application requires safe data delivery. The routing protocol must provide means to ensure small (if any) data loss.
- **Real-time delivery.** The most important requirement is guaranteed delivery times. All measurements or detected events must reach a base station within a given time interval. In practice, mostly soft-real time guarantees are used.
- **Maximum network lifetime.** The emphasis is made on network longevity, even if other performance metrics are sacrificed.
- **Data reporting model.** The data reporting model strongly depends on the application. Traditional data collecting applications are usually time-driven; surveillance applications are event-driven, while data-centric applications are usually query-driven.
- **Support Mobility.** Certain applications require routing services when sensor nodes and/or base stations are mobile. Sometimes

the mobility of some nodes (typically the base stations) can be controlled by the user to improve performance.

- **Combined requirements.** Applications often need to fulfill several (often contradicting) requirements. In these cases, tailored combined requirements must be used. A real-world example is when the majority of measurement (detected events) must be collected at the base station as soon as possible. Late arrivals are also valuable within a certain time frame.

TAXONOMY AND THE ANALYZED ALGORITHMS

Routing algorithms are classified using various factors. Based on the network structure, the algorithms can be either flat or hierarchical. Flat (or single-tier) algorithms have nodes with equal capabilities and responsibilities. The majority of the routing protocols are flat, serving well in typical applications with reasonable network sizes. The main problem with flat routing algorithms is that they do not scale well with the increase of network size. Hierarchical routing protocols provide scalable solutions by creating clusters with manageable size, and providing a higher hierarchical layer for inter-cluster communication. Each cluster has a special node, called cluster head, which is responsible for the inter-cluster communication, by providing a gateway between the upper layer and the nodes in the cluster. Cluster heads also can organize the internal operation of the cluster and provide aggregation or fusion services as well. Cluster heads are either dedicated devices with enhanced capabilities (e.g. more energy supply, long-distance radio), or the role of the cluster head is dynamically changed in the network to provide equal load on the sensor nodes.

Location-based routing services maintain and use geographical information while forwarding packets. In these protocols messages are intended to be forwarded to locations (as opposed to specific nodes), thus addresses are geographical regions instead of node identifiers. During routing, the location of nodes is used to find efficient routes towards the destination region. These protocols naturally match the philosophy of many typical sensor-networking applications, where at the application level the goal is to gather information from an area. However, these protocols can only be applied if the nodes are location-aware (by means of on-board GPS or other localization services).

According to the protocol operation, there are negotiation based, muti-path based, query based or QoS based solutions. Negation-based protocols spread short description information (meta-data) on the available data, before actually sending the data itself. This approach is especially useful when the data has large size and thus the distribution of it requires considerable amount of resources (e.g. bandwidth, energy). Based on the available meta-data the actual data is forwarded only to where and when it is necessary. Multi-path based solutions maintain multiple possible paths between the source and destination, providing robust operation in case of node or link failures, and also load balancing becomes possible. Query-based routing protocols establish routes when a query at a sink node is generated. The query is used to find a route to the source, and then the actual data is routed back on the reverse path. QoS routing protocols provide guaranteed services, in terms of e.g. message delay, delivery speed, or delivery ratio. These protocols are useful e.g. in mission-critical applications or real-time systems.

In the following sections some notable and typical routing algorithms will be reviewed to illustrate the various types of protocols, and also show how (even special) application needs can be fulfilled. The discussed algorithms are the following:

- **Flooding and gossiping protocols** provide simple means to disseminate information in the network without any a piori knowledge. They may be suboptimal in many senses but due to their simplicity and robustness, they are useful building blocks of more complex protocols.

- **Sensor Protocols for Information via Negotiation.** This typical negotiation-based protocol family is able to effectively disseminate information in the network, based on information advertisement and on demand broadcast.
- **Directed Diffusion** is a classical scalable and robust query-based multi-path algorithm. A typical application area is multi-rate data collection.
- **Gradient-Based Routing** is a single-path, query-based protocol, capable of load balancing and energy-aware adaptation.
- Geographical and Energy-Aware Routing Protocol can deliver messages to a destination region, based on location information of nodes, and the delivery cost values learned and adapted during message delivery.
- **Low Energy Adaptive Clustering Hierarchy.** A hierarchical protocol working with dynamically elected cluster heads and dynamic cluster formation, used for proactive periodic data collection.
- **Threshold-sensitive Energy Efficient Sensor Network Protocol** and its adaptive versions are also cluster-based and are suitable for event-based, energy-aware applications.
- **SPEED** is a QoS-based routing protocol, providing guaranteed delivery speed and load balancing. May be used in soft real-time applications.

Flooding and Gossiping

Flooding is probably the simplest of all existing and utilized routing approaches. Despite (and mostly because) of its simplicity flooding primitives can be found in many sophisticated routing protocols as building blocks. Flooding can be used for dissemination of information in route discovery or topology update, e.g. in AODV routing (Perkins, Belding-Royer, & Das, 2003) and the increasingly popular Zigbee protocol stack.

The principle of flood routing is the following:

- **Step 1.** The source node broadcasts the message
- **Step 2.** Every node in the network, which receives a message for the first time, retransmits it, otherwise drops duplicated messages.

An important practical question is how a node can determine whether a received message have been heard before or not. In practice messages are identified by short message identifiers so that receiving nodes can determine whether the received message is a new one or not, without the need of analyzing the content of the message, or storing previous messages. Other trivial, although somewhat more complicated solutions for message identification are simple hash functions.

In principle, all nodes in a connected network could receive the flooding message, but in practice, message collisions may cause severe degradation in delivery ratio: nodes transmitting messages at the same time may cause collisions and thus certain nodes may not receive the message at all. Note that typical collision avoidance schemes used in sensor networks (e.g. CSMA) cannot eliminate all collisions, due to hidden terminal effect. A related problem is the excessive amount of generated messages, often called broadcast storm, which can possibly cause congestion and interference with existing traffic in the network, and also unnecessarily consumes energy. These problems, originating from the highly redundant network traffic, are especially emphasized when network size and density increases.

Despite the obvious drawbacks, flooding has its advantages as well. The main beauties of the algorithm are its simplicity, robustness and its conformity to changing network topology. The algorithm requires very few resources thus it can be implemented on virtually any platform. The implicit redundancy offers resilience when network links change their quality in time. No information on topology is required to be known by the algorithm, apart from the sound assumption that the network is connected, thus flooding can be utilized in even highly mobile networks.

To alleviate problems stemming from broadcast storms, several modifications to basic flooding were proposed. A simple solution to reduce the number of collisions is the application of random delays before each retransmission. With careful tuning of the distribution of the delays the majority of the collisions can be avoided, but the problem of the excess number of messages is still present.

A family of flooding algorithms, maintaining the simplicity of the basic algorithm, reduces the number of generated messages in probabilistic ways. The simplest probabilistic solution is the following:

- **Step 1.** The source node broadcasts the message with probability one.
- **Step 2.** Every node in the network, which receives a message for the first time, retransmits it with probability p, otherwise silently drops the message.

The rationale behind the probabilistic solution is that in dense environment there is no need for every sensor to rebroadcast the message; it will eventually get to the majority of the nodes anyway. The number of messages is obviously proportional to p, and the delivery ratio also highly depends on the value of the parameter: p = 1 obviously provides the original flooding algorithm, while a small probability parameter may result in a premature flood death with low delivery ratio. The choice of p depends on the actual topology and can be tuned experimentally to keep the delivery ratio at a high level, while reducing the number of messages as low as possible. Several empirical modifications were proposed by (Haas, Halpern, and Li, 2006), to improve the delivery ratio and allow low message number. One of their algorithms forces nodes near the source to forward messages with higher probability, to avoid premature death. Another algorithm uses higher retransmission probabilities for nodes with few neighbors. In a third algorithm a node, which decides not to retransmit a message, monitors the transmissions of its neighbors, and if does not hear the retransmitted message within some time period, retransmits it. An adaptive solution, proposed by (Vakulya and Simon, 2009), is based on percolation theory and eliminates the need of experimentation with magic constants in the algorithms, and also provides effective solution even if the node density varies in the network. The algorithm uses the easily computable node degree (i.e. the number of neighbors) to tune the probability parameter of each node as follows:

$$p_n = \begin{cases} \dfrac{K}{K_n} & K_n > K \\ 1 & \text{otherwise} \end{cases}$$

Where p_n and K_n are the retransmission probability and the degree of node n, respectively, and K is a design parameter with typical values around 6-8.

Other solutions, utilizing location information of one-hop or two-hop neighborhood, optimize retransmissions and efficiently reduce the number of sent messages. A notable example is the algorithm of (Liu, Jia, Wan, Liu, and Yao, 2007), which uses 1-hop neighborhood location information and is able to provide full coverage with minimal number of messages. These solutions, however, tend to be much more complex than the original, or the simple probabilistic flooding algorithms, and the required location information is rarely available in typical inexpensive networks.

An alternative approach, often called gossiping, handles the problems of broadcast storm differently: a gossiping node does not utilize broadcast but rather randomly selects a neighbor and sends the message to it, and the recipient behaves similarly. This way collisions resulting from rebroadcasts at the same time can be eliminated, but the algorithm has significant delay, compared to flooding.

Various flooding algorithms have widespread use in sensor networking, either as standalone protocols, or as building blocks in more sophisticated routing schemes. Flooding can effectively be used for information dissemination in the whole network. Typically commands, destined to all or a large portion of nodes, are flooded through the network. Flooding is also used to find a destination node with unknown location. Other routing protocols use flooding exactly for this purpose: the sequence of nodes retransmitting a message from a source to the destination node is recorded during flooding, and thus a route between the two nodes is formed for later usage.

Flooding algorithms are simple and require very few resources. No or very little information on topology is required by the algorithms. They are robust and can adapt to changing network topology and varying communication channels. The main drawback of flooding algorithms is the possibly large amount of unnecessarily broadcasted message, causing large energy consumption and heavy network traffic, which may result in performance degradation as well. Variants of the algorithm address this problem, but flooding-based algorithms are still lavish and waste resources. The algorithms are successfully used when occasionally a fast, robust, and best effort method is required for information dissemination or network discovery, with no special emphasis on energy efficiency.

Sensor Protocols for Information via Negotiation (SPIN)

The SPIN family, proposed by (Kulik, Heinzelman, and Balakrishnan, 2002) contains negotiation-based protocols to efficiently disseminate information among sensors in energy-constrained wireless sensor networks. The typical application scenario behind SPIN is the following: nodes in the network detect events, which are potentially of interest of all other nodes. Furthermore, nearby nodes potentially have the same data. Information on detected events is disseminated in the whole network and thus any node can be queried by the user to immediately receive the requested information. The protocols eliminate the transmission of redundant data and also support energy-aware behavior, allowing sensors to efficiently distribute data, given a limited energy supply.

The basic ideas of SPIN algorithms are the negotiation protocol and adaptation, based on available resources. Collected data is described by short meta-data, which is advertised throughout the network. The rationale is that disseminating large amount of data is expensive, but disseminating short meta-data is much cheaper. Based on the meta-data, other nodes may decide to request the data itself. Actual data is transmitted only if and to where another node really needs it. Due to the negotiation, phase large amount of data is never sent unnecessarily or redundantly in the network, thus data dissemination is efficient. Each node also monitors its own energy status and may reduce its participation in the protocol: using a conservative approach, a node will only participate in a stage of the protocol if its energy status allows completion of all the other stages of the protocol.

In SPIN, three types of messages are used:

- **advertise new data (ADV).** When a node has new data to share, it transmits a short ADV message containing the meta-data.
- **request data (REQ).** When a node wants to receive the actual data, it sends a short REQ message.
- **data message (DATA).** DATA messages contain actual sensor data with a meta-data header. DATA messages may be long and numerous, depending on the size of the actual data.

Figure 1. The operation of the SPIN protocol. (a) Node A advertises data (b) Node B sends request (c) Node A broadcasts data (d) Node B advertises data

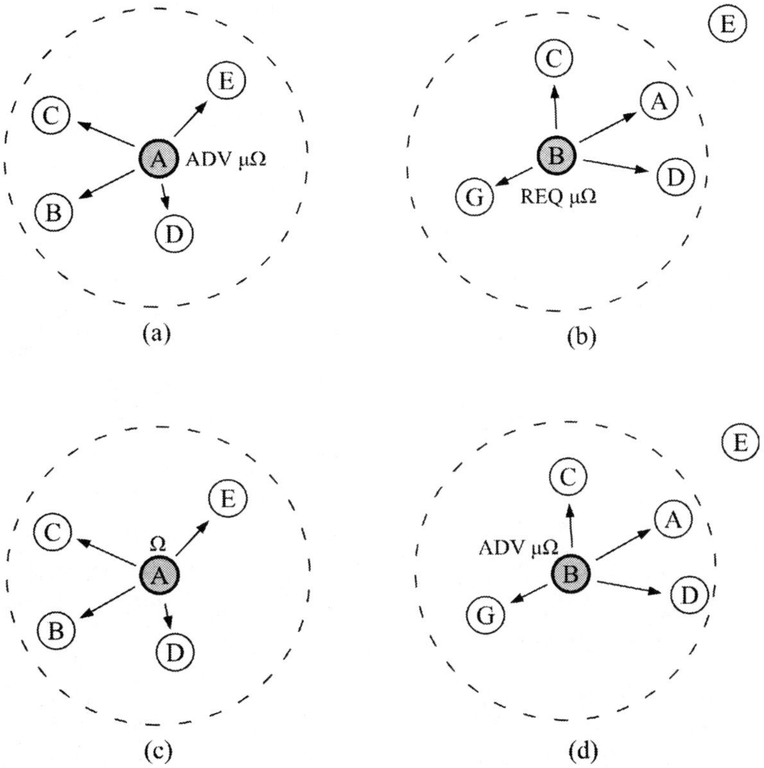

The operation of the protocol is illustrated in Figure 1. Node A starts advertising a new data Ω to its neighbors by sending ADV messages, including meta-data μΩ. Suppose all neighbors are interested in Ω. The neighbors delay their answer randomly, so one of them, assume B, will be the first to send a REQ(μΩ) to A. Some neighbors of A (nodes C and D), overhearing REQ(μΩ), suppress their requests and listen to A. Node A broadcasts Ω, which is received by all neighbors, thus node E also suppresses its request. When Ω is received, nodes B, C, D, and E broadcast ADV(μΩ) to their neighbors.

Meta-data is application specific and its format is not defined by SPIN. It may describe position, time, and type of the detected event (e.g. movement at position x,y @ t). The application layer can decide, whether the data is interesting or not (e.g. if a data from a movement at position x+dx,y+dy @ t is already available from elsewhere, then the new data probably describes the same event and is of no interest.)

The SPIN protocol was proven more efficient than flooding in different aspects. In a test performed by (Kulik et al. 2002), SPIN was found at least as fast as flooding, while requiring only 25% of the energy. Both the dissemination rate and the energy usage of SPIN were close to the ideal optimum. A drawback of SPIN is that its data advertising mechanism cannot guarantee the dissemination of data. When nodes between a data source node A and a potential consumer node B are not interested in the offered data then B will not receive the data offered by A. Still, SPIN is a successful data-centric protocol for many sensor-networking applications.

Directed Diffusion

Directed Diffusion, proposed by (Intanagonwiwat, Govindan, Estrin, Heidemann, and Silva, 2003), is a typical data-centric and application-aware routing scheme, which is able to provide scalable and robust routing services for data collecting applications in an energy efficient manner.

According to the application model, several potential data sources (sensors) and some sinks are present in the network. Sensors can generate named data using attribute-value pairs, while sinks may express their interest for such named data. The fundamentals of the data collection process are illustrated in Figure 2. An interest from sink node A is disseminated through the network hop-by-hop, for example using flooding. While the interest is propagated, each node stores the interest and, at the same time, sets up and stores gradients toward nodes where it received the interest from. A sensor, observing an event matching a stored interest, generates a data message and sends it to those neighbor nodes, which have gradients with the same interest, forming a multi-path route towards the sink. Based on the received messages, the sink may send reinforcement and negative reinforcement messages to the network, thus it can reduce the communication routes to an empirically good (reliable and fast) path. Note that the reinforcement mechanism allows the establishment of a primary (e.g. high data rate) route and alternative routes (e.g. with low data rates) to provide more robust and fail-safe routing framework.

A gradient in Directed Diffusion has a direction and a value. The semantics of the value are not defined, giving the designer freedom to attach application-specific meaning to it. Possible examples may be simple binary values showing which neighbors to use for data propagation towards the sink, or reporting rate, as proposed by (Intanagonwiwat et al., 2003).

Each node in the network maintains an interest cache. Items in the interest cache contain the type, attributes of the interest, and also contain up to one field for each neighbor, describing the gradient and its duration. The interest cache contains only one item for every interest, thus similar interests (with same type and overlapping attributes) are

Figure 2. The operation of Directed Diffusion. (a) Sink node propagates interest in the network, while (b) gradients are set. (c) Event source delivers data to sink along gradients. (d) Sink sends reinforcement messages and (e) the reinforced path is used to deliver data

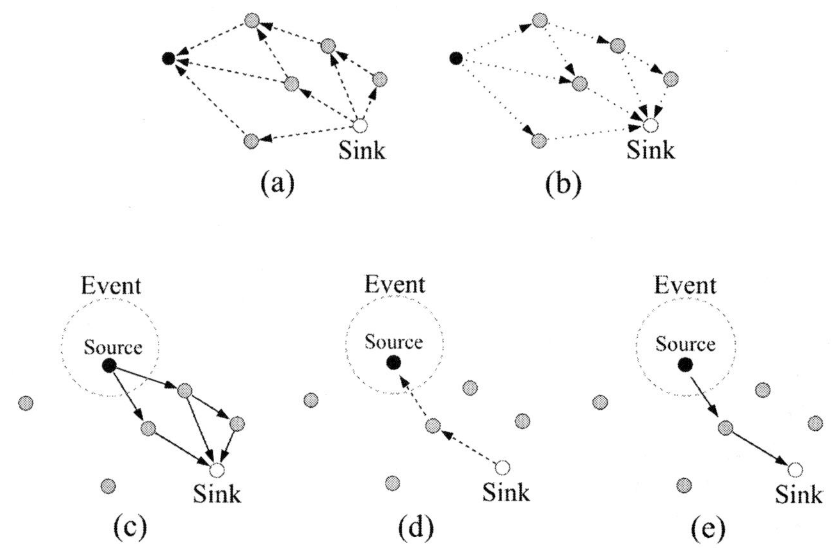

aggregated. The interest cache is updated when a node receives an interest: a new item may be created, or a new node may be added to an existing interest, or simply the duration of an interest may be updated. Expired gradients and interest entries are eventually removed from the cache.

Every node also maintains a message cache, in which each received data message has an entry. The main role of data cache is loop prevention, but it can be used for application-specific purposes as well (e.g. data interpolation).

When a node receives a message, it tries to find a corresponding interest in its event cache. If no such interest is found, the message is dropped. If a match exists, the node also checks its data cache. If the received data message has a matching data cache entry, the message is dropped in order to prevent message loops; otherwise, the message is stored in the data cache and is forwarded according to the gradients stored in the interest cache.

Directed Diffusion naturally handles multiple sources and multiple sinks. The reinforcement mechanism can be used to locally repair failed or degraded paths and always provide robust and high quality connection.

Despite its significant maintenance overhead (interest propagation, maintenance of multiple paths, reinforcement messages, and the interest matching process at the sensors), Directed Diffusion can achieve significant energy savings, compared to flooding or even omniscient multicast. In the experiments of (Intanagonwiwat, Govindan, Estrin, Heidemann, and Silva, 2003), the energy consumption of Directed Diffusion was 50-70% less than that of flooding.

In terms of message delay, Directed Diffusion also outperformed flooding by a factor of 6-10, depending on network size. Good performance is provided by caching, aggregation, and the on demand nature of the algorithm. Due to its flexibility, performance, and robustness Directed Diffusion is a popular candidate in applications where query-driven data access is required. However, for applications that require continuous data delivery (e.g. monitoring and data collection) Directed Diffusion is not the adequate choice.

Gradient-Based Routing

Gradient-Based Routing, proposed by (Schurgers and Srivastava, 2001), can be considered a variant of Directed Diffusion, with practical new concepts to aid traffic balancing in the network. As its name suggests, Gradient-Based Routing uses the concept of gradients to construct data paths from data sources towards sinks. Similarly, to Directed Diffusion, the sinks propagate their interests in the network, but here the nodes calculate their minimum hop distance to the sink, called height. Unlike in Directed Diffusion, the semantics of gradient are explicitly defined: the gradient between two nodes is calculated as the difference between the heights of the nodes. While data is propagated from data source to the sink, messages always are forwarded towards the smallest (or one of the smallest) gradient. This results in a single path between a node and a sink with minimum number of hops between them. During message propagation, nodes also use aggregation, to reduce energy consumption.

The height of a node is determined by the node itself and in normal circumstances; it reflects the node's hop-distance to the sink. The height is then broadcasted to its neighbors to allow the calculation of gradients. This scheme gives freedom to a node to alter its own height and tell a higher number to its neighbors in order to discourage them from forwarding messages through it. An artificially elevated height will push traffic to other direction (if possible), allowing energy-aware load balancing, and traffic balancing.

(Schurgers and Srivastava, 2001) proposed three techniques to spread network traffic. In the Stochastic scheme a node can select the next hop randomly, if there are two or more neighbors with the same lowest gradient. This simple solution spreads the network traffic while maintains the same shortest path for each data flow. In the

Energy-based scheme, a node can increase its height when it detects that its energy reserve has dropped below a certain threshold. This increased height is broadcasted to all the neighbors of the node, thus traffic in its direction is discouraged. Note that the increased height may affect other nodes' height as well and the change is propagated through the network. In Stream-based scheme, a node can divert away new streams from itself if it is already part of the path of another stream. In this case, a node tells an elevated height value to all its neighbors, except to the one from where the existing stream originates. This scheme leaves existing streams unaffected, while new streams possibly find a new path.

The experimental results presented in (Schurgers and Srivastava, 2001) show that the proposed spreading techniques can decrease the energy consumption of the network by as much as 25%, compared to the simple scheme used Directed Diffusion. Unfortunately, the delivery ratio of the algorithm occasionally decreases, depending on the network topology. The Gradient-Based Routing, due to its good performance and simple implementation is a popular candidate for query-based applications.

Geographical Routing

When the nodes in the network are aware of their location, by means of on-board GPS or other localization techniques, the routing algorithm can effectively utilize this knowledge.

In geographical routing messages are always routed to a location instead of a specific node. The basic forwarding algorithm is quite simple: a node forwards the message to one of its neighbors, which is closer to the destination. If this can be done at every node then the message is routed to the destination in finite steps. However, when the network contains holes the basic approach obviously fails. In fact, circumvention of holes is the main challenge of geographical routing algorithms. Various solutions have been proposed to solve the hole-problem; one of the most successful and inspiring is Geographical and Energy-Aware Routing (GEAR), proposed by (Yu, Govindan, and Estrin, 2001).

GEAR addresses the common routing primitive present in location aware networks: a message is forwarded to all nodes in a target region. The address (region) can be specified, for example, as a rectangle. The routing consists of two phases: in phase 1, messages are forwarded towards the target region R, and in phase 2, messages are disseminated inside region R. The process is illustrated in Figure 3.

Phase 1

A node selects one of its neighbors to forward the message to, based on a geographical and energy aware selection process. If node A has at least one neighbor closer to the destination region then A forwards the message to the neighbor, which is closer to the destination and has the smallest learned cost (node B in Figure 3). When all neighbors of node C are further away then C is at the perimeter of a hole. In this case, C selects a neighbor node with the minimal learned cost (node D in Figure 3) and forwards the message to it. Whenever a node N selects a neighbor N_i to forward a message to, it updates its learned cost $C_{learned}(N, R)$ as follows:

$$C_{learned}(N,R) = C_{learned}(N_i, R) + C(N, N_i),$$

where $C_{learned}(N_i, R)$ is the learned cost of node N_i to region R, and $C(N, N_i)$ is, for example, a combination of the cost of sending a message from N to N_i, the energy levels of the two nodes, and the distance between them. (In the most obvious case $C(N, N_i)$ is simply the distance between the nodes.) The node also broadcasts its new learned cost value to its neighbors.

Figure 3. Operation of Geographical and Energy-Aware Routing. Message is forwarded towards target region R and finally disseminated in R

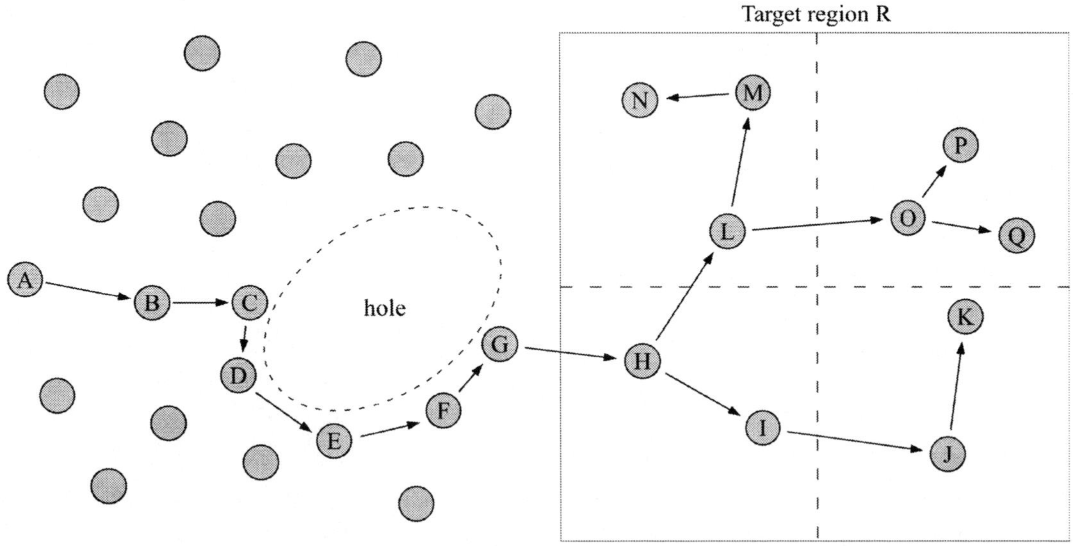

The learned cost reflects the total cost of forwarding a message to the destination through a given node. It contains distance and energy components as well. Initially the learned cost value of node N is initialized to the estimated cost value $C_{estimated}(N, R)$, as follows:

$$C_{estimated}(N,R) = \alpha \frac{d(N,R)}{\max\limits_{\text{for all neighbors } j} d(N_j,R)} + (1-\alpha)\frac{C(N)}{\max\limits_{\text{for all neighbors } j} C(N_j)}$$

where $d(N, R)$ is the distance of node N from the centroid of the target region R, $C(N)$ is the node's energy depletion level, and $0 \leq \alpha \leq 1$ is a weighting parameter.

When the first message reaches the destination, the last relay node learns its true cost value. With each delivered message the correct learned value is propagated one hop backward, thus a node, n hops away from the destination, must forward n messages to the same destination to have its learned cost converged.

Phase 2

When the message reaches a node in the destination region, the message can be propagated in the region using either flooding or Recursive Geographic Forwarding. The latter approach splits the target region into smaller sub-regions, and sends a copy of the message to each sub-region, using GEAR, as illustrated in Figure 3. The recursive splitting and forwarding is continued until there is only one node in the target region.

The performance of GEAR was investigated by (Yu, Govindan, and Estrin, 2001). According to their results, GEAR provides longer network lifetime than non-energy-aware routing algorithms. The difference between flooding and GEAR clearly shows the efficiency of GEAR: the same simulated network was able to deliver as much as 100 times more messages using GEAR than with flooding.

Hierarchical Routing

(Heinzelman, Chandrakasan, and Balakrishnan, 2000) proposes a hierarchical, cluster-based protocol for sensor networks, designed for periodic data collection, called Low Energy Adaptive Clustering Hierarchy (LEACH). Although LEACH has several assumptions, which question its usability in real applications in its original form, the protocol contains ideas that boosted up the design of successive routing protocols. LEACH has the following, not too realistic assumptions:

Each node can increase its transmission power so that it can talk to the base station, or can decrease it to communicate with its small neighborhood only.

The nodes can select from multiple CDMA codes.

During the operation of LEACH, rounds are performed after each other, where each round begins with a short setup, followed by a long steady-state phase. During setup phase nodes are organized into clusters, each cluster having a dedicated cluster-head, while during steady-state phase data collection is performed by the nodes, data is forwarded to the cluster-heads, and cluster-heads forward data to the base station after aggregation. Since nodes consume much more energy when they are cluster-heads, this role is rotated in each round, to evenly distribute power consumption in the network.

Setup Phase

In the setup phase, clusters are created: first cluster heads are selected, and then the rest of the nodes join to one of the cluster-heads, forming clusters for the next steady-state phase. The process is illustrated in Figure 4. Cluster heads are selected randomly, so that the expected number of cluster-heads is a predetermined percentage P of the total number of nodes. Each node decides whether it will

Figure 4. Operation of the LEACH protocol. (a) Nodes elect themselves as cluster-heads and broadcast their decision. (b) Follower nodes join a cluster and notify their cluster-head. (c) Cluster heads broadcast the TDMA schedule and the CDMA channel of the cluster to the members. (d) Communication during the steady state phase of LEACH. Nodes inside a cluster use TDMA schedule, while clusters use different CDMA codes, to reduce inter-cluster interference

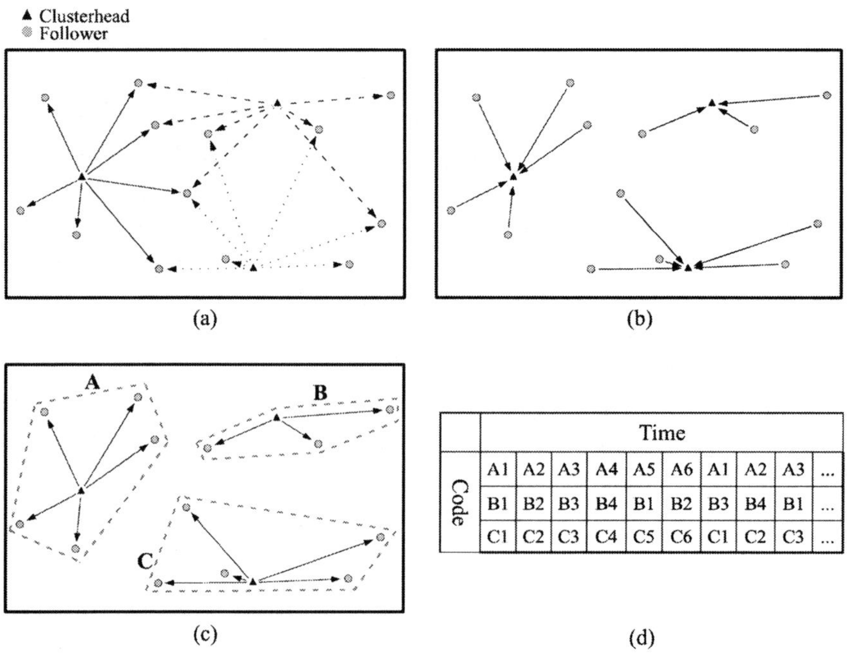

	Time									
Code	A1	A2	A3	A4	A5	A6	A1	A2	A3	...
	B1	B2	B3	B4	B1	B2	B3	B4	B1	...
	C1	C2	C3	C4	C5	C6	C1	C2	C3	...

be a cluster-head or a follower, in the probabilistic decision process taking into account the required percentage P, and also assuring that each node becomes head exactly once in every 1/P rounds. Nodes that will be cluster-heads advertise their decision. A follower node, based on the signal strength of the received advertisements, joins the strongest (and possibly closest) cluster-head. Each follower notifies the cluster-head that it will be a member of the cluster. The cluster head, knowing the members, creates a TDMA schedule, and selects a CDMA channel for the cluster. The cluster-head broadcasts their schedule and the channel identifier to the members.

During setup phase, all nodes use a common CDMA channel and CSMA MAC protocol for channel access, while during the steady-state phase each cluster will use the selected CDMA channel and the cluster's associated TDMA schedule.

Steady-State Phase

Nodes in a cluster perform their measurements and transmit the results to the cluster-head in their allocated time slice, according to the TDMA schedule, as shown in Figure 4 (d). Data transmission is done using low-power mode, since the cluster head is nearby. The nodes must be awake during their time slice; otherwise, they can switch off their radio or the whole node, to spare energy. Cluster-heads aggregate received measurements in an application-specific way and then forward the measurements to the base station, using high-energy transmission. Cluster-heads are awake all the time and use considerably more energy than followers use. After the end of the steady-state phase, a new election process begins and other cluster-heads will be elected.

LEACH is basically a proactive protocol, but several applications fit better to the reactive model, where nodes send information only when an important event is observed. Manjeshwar and Agrawal proposed algorithms Threshold-sensitive Energy Efficient sensor Network protocol (TEEN) and Adaptive Periodic Threshold-sensitive Energy Efficient sensor Network protocol (APTEEN), in 2001 (Manjeshwar, A., & Agrawal, D.P., 2001) and in 2002 (Manjeshwar, A., & Agrawal, D.P., 2002), respectively, to adjust the cluster-oriented hierarchy of LEACH to reactive applications.

The cluster formation is done using a centralized variant of LEACH, where clusters are formed by the base station, based on available energy and location information. When clusters are formed, nodes can broadcast measurements to the cluster head in their associated time-slice (APTEEN) or instantaneously after the detection (TEEN). An important improvement is that nodes transmit measurements only if a significant event is detected. To define significant events, cluster heads broadcast parameters in each setup phase. Parameters include hard and soft thresholds. Hard threshold defines when a measured parameter is to be reported (e.g. report when temperature is above 25 degrees Celsius). Soft threshold is used to quantize data and thus reduce transmission rate: a measured value is sent only if the current value differs from the previous report by an amount equal to or greater than the soft threshold (e.g. retransmit temperature value only if the change is at least 5 degrees Celsius). APTEEN also utilizes a proactive feature: the minimum time interval between two reports can be defined (e.g. report at least in every 10 minutes, regardless of the measured value). Using the parameters, the user can control the trade-off between energy efficiency and data accuracy.

LEACH was found very effective in increasing network lifetime, as shown in the simulations done by (Heinzelman, Chandrakasan, and Balakrishnan, 2000). In the experiments, the network lifetime was orders of magnitude higher than that of direct (one-hop) and minimum transmission energy communication protocols. APTEEN and TEEN outperforms even LEACH, in terms of energy efficiency and network lifetime. However, there are serious issues concerning the practical applicability of LEACH-based protocols: the radios must be able to reach the base station in one hop (when the node is cluster head) therefore the algorithm cannot be used in large-scale applications with ordinary inexpensive radio chips. In addition, the combined TDMA/CDMA modes require complex hardware and software support.

SPEED: Soft Real-Time Routing

In (He, Stankovic, Lu, and Abdelzaher, 2003) the SPEED algorithm is proposed, which is able provide a soft real-time communication service with a guaranteed delivery speed across the sensor network. Due to the guaranteed delivery speed, the end-to-end delay is proportional to the distance between the source and destination nodes. SPEED is a location-aware protocol, thus each node is assumed to know its own location.

The algorithm provides three communication services: real-time unicast (e.g. messages from sensors are routed to the base station), real-time area multicast (e.g. commands from base station are routed to sensors in a given area), and real-time area-anycast (e.g. query from the base station to any node in a given area).

In addition to the provided real-time services, SPEED satisfies the following design objectives as well:

- The algorithm is virtually stateless, i.e. maintains only immediate neighborhood information and thus its memory requirements are proportional to the number of neighbors.
- The behavior of the algorithm is local in general, i.e. all node actions affect the neighborhood only and not the whole network (e.g., there is no flooding for route discovery). Locality and statelessness ensures that SPEED scales well with network size.
- The algorithm works on all existing best effort MAC layers; it does not require any real-time MAC layer support.
- SPEED can balance traffic load, is able to detect and avoid congested regions as well as holes in the network.

Each node in SPEED maintains a neighbor table, in which the identifier and the position of the neighbor are stored, along with the SendToDelay quantity. $SendToDelay_i^j$ is the estimated delay between initiating a packet send from node i, and its actual reception time at neighbor node j. This single hop delay is in fact used as a metric to approximate the load of node j. The delay is measured by the sender node as follows: The packet, when entered to the network send-queue, and the corresponding received acknowledgement (ACK) are timestamped, and thus the round-trip time is calculated as the difference between the two timestamps. Moreover, the receiver measures the duration of the acknowledgement processing, and sends it back in the ACK message, thus the sender can adjusts the round-trip time to have a better delay estimate. SendToDelay is then calculated from many measured message delays, using exponential averaging. From the single hop delay, the relay speed is calculated as follows:

$$relaySpeed_i^j(Dest) = \frac{d(i, Dest) - d(j, Dest)}{SendToDelay_i^j},$$

where *Dest* is the destination node, and d(.,.) is the distance between two nodes. The required message delivery speed in the network is denoted by $speed_{req}$.

The heart of SPEED is the Stateless Non-deterministic Geographic Forwarding (SNGF) scheme. Using SNGF, a node forwards a packet, if possible, to a neighbor, which is closer to the destination than it is. The set of such eligible neighbors of node *i* is $FS_i(Dest)$, the Forwarding Candidate Set. If in FS there is at least one node with $relaySpeed \geq speed_{req}$ then the message is forwarded to one of these nodes. When multiple candidates exist, SNGF chooses one of them probabilistically, the nodes with higher relay speed having higher probability. This selection scheme provides automatic traffic balancing, since packets are routed with higher probability to nodes that have smaller delay.

If FS_i (*Dest*) is empty, indicating a network hole, the packet is dropped and a backpressure message is broadcasted, with content $(i, Dest, \infty)$, indicating that node i cannot forward packets towards node Dest. All neighbors k of node i, for which $i \in FS_k(Dest)$, modify their neighbor table by $SendToDelay_k^i = \infty$ and stop sending messages to node i.

If in FS_i (*Dest*) every node has $relaySpeed < speed_{req}$ then there is a congestion in the network. In this case, two mechanisms are triggered to reduce traffic in the congested area. First, a backpressure message is issued with content (*i*, *Dest*, *AvgSendToDelay*), where

$$AvgSendToDelay = \frac{\sum_{j \in FS_i(Dest)} SendToDelay_i^j}{|FS_i(Dest)|},$$

and $|FS_i(Dest)|$ is the size of set $FS_i(Dest)$.

In response to the backpressure message, nodes k for which $i \in FS_k(Dest)$, modify their neighbor table by $SendToDelay_k^i = AvgSendToDelay$. Second, the number of injected messages is decreased by dropping some messages going towards the congested area, if necessary. Message drops are controlled probabilistically, by the relay ratio. The probability of a message drop is

$$P_{drop} = 1 - RelayRatio$$

The relay ratio is computed from the miss ratios, as follows:

$$\mathrm{Re}\,layRatio_i = \begin{cases} 1 - K\dfrac{\sum_{j \in FS_i(Dest)} MissRatio_i^j}{|FS_i(Dest)|} & \text{if } \forall MissRatio_i^j > 0 \\ 1 & \text{if } \exists MissRatio_i^j = 0 \end{cases}$$

The $MissRatio_i^j$, measured in the MAC layer, is the ratio of messages unsuccessfully delivered from node i to node j, and K is a proportional gain.

SNGF effectively controls packet relays towards the target area, e.g. defined by its center point and radius. The further processing of the message inside the target area is called last mile processing, and it is basically a restricted flooding. SPEED provides anycast, multicast, and unicast services, where the type of service is indicated by the packet-type field. When a node inside the destination area receives an anycast packet, it will deliver the packet to the application and will not forward it. If a multicast packet is received inside the destination area then the first node receiving it will set a TTL (Time To Live) field and rebroadcasts it as a last mile packet. Nodes inside the destination area rebroadcast last mile packets and deliver them to the application layer, while the TTL allows. Nodes outside the target region omit last mile packets. If a packet is unicast, it is processed similarly to multicast messages, but only the target node will deliver the packet to the application layer.

SPEED is a successful protocol tailored for sensor nodes that are constrained in computing power, bandwidth, and memory. The algorithm provides real-time quality of service: it is able to maintain a desired delivery speed across the network. The algorithm was also found very effective in terms of end-to-end delay and miss ratio, in the tests of (He, Stankovic, Lu, and Abdelzaher, 2003).

FUTURE RESEARCH DIRECTIONS

Due to the resource-limited nature of sensor networks, energy efficiency is a key question for real applications. Traditional routing protocols do not take into account the limited energy supply of nodes, but in sensor networks, it becomes an important issue. Energy-aware protocols try to reduce the number of messages (e.g. by aggregation), the amount of time the radios are switched on, and are also able to adapt the routing paths according to energy levels of the nodes, to prolong network lifetime. Although promising results are available, routing algorithms, suitable for low and extreme low duty cycle operation with low latency, require further research.

As sensor networking applications become more and more widespread, the question of security must be answered, especially in safety-critical environments. An obvious point of attack is message delivery, thus secure routing protocols must be applied where safety and security of the application and the data is of concern. There are numerous ways of attacking communication in sensor networks (e.g. sniffing, denial of service, black hole attack, jamming, just to name a few), thus comprehensive defense is required. Reliable, secure, and attack resilient routing protocols still have several open questions to answer.

CONCLUSION

Routing algorithms are one of the most important middleware services in sensor networks; the performance of the application is greatly affected by the quality of the utilized routing protocol. Applications may require the use of either time-driven, event-driven, or query-driven data reporting models. Certain applications may need guaranteed delivery or real-time performance, others may demand mobility support. The resource-constrained sensor nodes, the ad-hoc deployment, the changing topology (due to node or communication failures), the possibly very large number of nodes makes the design and even the application of routing protocols challenging, especially for autonomous, unmanaged, and possibly mobile networks.

For sensor networks, hundreds of specially tailored algorithms have been proposed. In this Chapter we introduced some of the most typical ones to illustrate the possible ways routing can be performed in the resource-constrained nodes, fulfilling various requirements of the application domain. The selected protocols included flat and hierarchical algorithms, location-aware, negotiation based, muti-path based, query based and QoS based solutions. Although the presented set cannot represent all the available solutions, the selected protocols illustrate the main concepts and ideas.

REFERENCES

Akkaya, K., & Younis, M. (2005). A survey on routing protocols for wireless sensor networks. *Elsevier Ad Hoc Network Journal*, *3*(3), 325–349. doi:10.1016/j.adhoc.2003.09.010

Haas, Z. J., Halpern, J. Y., & Li, L. (2006). Gossip-based ad hoc routing. *IEEE/ACM Transactions on Networking*, *14*(3), 479–491. doi:10.1109/TNET.2006.876186

He, T., Stankovic, J. A., Lu, C., & Abdelzaher, T. (2003). SPEED: A stateless protocol for real-time communication in sensor networks. In *Proceedings of the 23rd International Conference on Distributed Computing Systems ICDCS* (pp. 46-55).

Heinzelman, W. R., Chandrakasan, A., & Balakrishnan, H. (2000). Energy-efficient communication protocol for wireless micro sensor networks. In *Proc. 33rd Hawaii International Conference on System Sciences, vol. 8* (pp. 3005-3014).

Intanagonwiwat, C., Govindan, R., Estrin, D., Heidemann, J., & Silva, F. (2003). Directed diffusion for wireless sensor networking. *IEEE/ACM Transactions on Networking*, *11*(1), 2–16. doi:10.1109/TNET.2002.808417

Kulik, J., Heinzelman, W., & Balakrishnan, H. (2002). Negotiation-based protocols for disseminating information in wireless sensor networks. *Wireless Networks*, *8*(2/3), 169–185. doi:10.1023/A:1013715909417

Liu, H., Jia, X., Wan, P., Liu, X., & Yao, F. (2007). A distributed and efficient flooding scheme using 1-Hop information in mobile ad hoc networks. *IEEE Transactions on Parallel and Distributed Systems*, *18*(5), 658–671. doi:10.1109/TPDS.2007.1023

Manjeshwar, A., & Agrawal, D. P. (2001). TEEN: A routing protocol for enhanced efficiency in wireless sensor networks. In *Proceedings of the 15th International Parallel & Distributed Processing Symposium* (pp. 2009–2015).

Manjeshwar, A., & Agrawal, D. P. (2002) APTEEN: A hybrid protocol for efficient routing and comprehensive information retrieval in wireless sensor networks. In *Proceedings of the 2nd International Workshop on Parallel and Distributed Computing Issues in Wireless Networks and Mobile computing* (pp. 195–202).

Perkins, C., Belding-Royer, E., & Das, S. (2003). *Ad hoc on-demand distance vector (Aodv) routing*. RFC3561. RFC Editor. Retrieved February 3, 2010, from http://www.rfc-archive.org/getrfc.php?rfc=3561

Schurgers, C., & Srivastava, M. B. (2001). Energy efficient routing in wireless sensor networks. In *Proc. IEEE Military Comm. Conference, vol. 1* (pp. 357-361).

Vakulya, G., & Simon, G. (2009). Percolation driven flooding for energy efficient routing in dense sensor networks. *Journal of Telecommunications and Information Technology*, *3*(2), 5–12.

Yick, J., Mukherjee, B., & Ghosal, D. (2008). Wireless sensor network survey. *Computer Networks*, *52*(12), 2292–2330. doi:10.1016/j.comnet.2008.04.002

Yu, Y., Govindan, R., & Estrin, D. (2001). *Geographical and energy aware routing: A recursive data dissemination protocol for wireless sensor networks*. (UCLA Computer Science Department Technical Report UCLA/CSD-TR-01-0023), May 2001.

ADDITIONAL READING

Akyildiz, I. F., Su, W., Sankarasubramaniam, Y., & Cayirci, E. (2002). A survey on sensor networks. *IEEE Communications Magazine*, *40*(8), 102–114. doi:10.1109/MCOM.2002.1024422

Al-Karaki, J. N., & Kamal, A. E. (2004). Routing Techniques in Wireless Sensor Networks: A Survey. *Wireless Communications*, *11*(6), 6–28. doi:10.1109/MWC.2004.1368893

Chipara, O., et al. (2006). Real-time Power-Aware Routing in Sensor Networks. In Proceedings of the14th IEEE International Workshop on Quality of Service. (pp. 83-92).

Culler, D., Estrin, D., & Srivastava, M. (2004). Guest Editors' Introduction: Overview of Sensor Networks. *Computer*, *37*(8), 41–49. doi:10.1109/MC.2004.93

Gu, Y., & He, T. (2007). Data forwarding in extremely low duty-cycle sensor networks with unreliable communication links. In Proceedings of the 5th international Conference on Embedded Networked Sensor Systems, (pp. 321-334)

Hui, J. W., & Culler, D. E. (2008). IP is dead, long live IP for wireless sensor networks. In Proceedings of the 6th ACM Conference on Embedded Network Sensor Systems (pp. 15-28)

Levis, P., Patel, N., Culler, D., & Shenker, S. (2004). Trickle: a self-regulating algorithm for code propagation and maintenance in wireless sensor networks. In Proceedings of the 1st Conference on Symposium on Networked Systems Design and Implementation: Vol. 1. (pp. 1-13).

Maróti, M. (2004). Directed flood-routing framework for wireless sensor networks. In Proceedings of the 5th ACM/IFIP/USENIX international Conference on Middleware: Vol. 78. (pp. 99-114). Springer-Verlag.

Parno, B., Luk, M., Gaustad, E., & Perrig, A. (2006). Secure sensor network routing: a clean-slate approach. In Proceedings of the 2006 ACM CoNEXT Conference (pp. 1-13)

Tilak, S., Abu-Ghazaleh, N. B., & Heinzelman, W. (2002). A taxonomy of wireless micro-sensor network models. *ACM SIGMOBILE Mobile Computing and Communications Review*, *6*(2), 28–36. doi:10.1145/565702.565708

Yang, D., Li, X., Sawhney, R., & Wang, X. (2009). Geographic and energy-aware routing in Wireless Sensor Networks. *International Journal of Ad Hoc and Ubiquitous Computing*, *4*(2), 61–70. doi:10.1504/IJAHUC.2009.023897

KEY TERMS AND DEFINITIONS

Ad-Hoc Network: An Ad-hoc network is a network created in a spontaneous manner, without predefined network structure and topology.

Delivery Ratio: The delivery ration is the ratio of successfully delivered messages.

Flooding: Flooding is the simple dissemination protocol to deliver messages to all nodes in the network.

Multi-Hop: Multi-hop means the message is delivered from the source to the destination through multiple relay nodes.

Node: A node is an element in the network. In sensor networks nodes have sensing (and optionally actuating), computing, and communication capabilities.

Routing: Routing is the delivery of a message from a source to a destination node, through a possible multi-hop path.

Sensor Network: A sensor network is an ad-hoc network constructed from sensor nodes, in order to perform distributed sensing.

Sink/Destination: The sink/destination is the target node where the message is to be delivered.

Source: The source is the node where the message is originated from.

APPENDIX

Table of Abbreviations

APTEEN	Adaptive Periodic Threshold-sensitive Energy Efficient sensor Network protocol
GEAR	Geographical and Energy-Aware Routing
LEACH	Low Energy Adaptive Clustering Hierarchy
SNGF	Stateless Non-deterministic Geographic Forwarding
SPIN	Sensor Protocols for Information via Negotiation
TEEN	Threshold-sensitive Energy Efficient sensor Network protocol
TTL	Time To Live

Chapter 23
Radio Frequency Identification

Róbert Schulcz
Budapest University of Technology and Economics, Hungary

Gábor Varga
Budapest University of Technology and Economics, Hungary

ABSTRACT

Radio frequency identification (RFID) is a group of technologies used to identify or track electronic devices (RFID tags or transponders), typically applied to or incorporated into products, animals, or even persons, using radio waves or electromagnetic fields. RFID systems have various applications, with the most prevalent one being enterprise supply chain management to improve the efficiency of inventory tracking and management, as RFID transponders complement bar codes (allowing the reading from several meters, even beyond the line of sight), and may one day even take their place. Other uses include payment services (for example public transportation, toll roads, season parking tickets, ski tickets, proximity smart cards), asset and inventory tracking, patient identification in hospitals, animal identification, library systems, anti-theft technologies, or even race timing. The main reasons for blocking its even more widespread utilization are privacy and security related issues, originating from the wireless nature of the technology. Technological challenges and development directions include robust collision resolution (caused by simultaneously transmitting transponders or readers), the mitigation of the privacy concerns, and the creation of more intelligent chips (by adding cryptographic algorithms, new sensors, more memory, and processing power to the transponders) while retaining miniaturization and cost-efficiency efforts.

In this chapter, we will start by briefly summarizing the history of radio frequency identification systems. After that, we will introduce the components of such systems and classify them based on programmability, data capacity, frequency, and reading distance, as well as power supplement and reply transfer methods. We will describe the various coupling types used in RFID systems, present the common coding schemes and modulations, and give an overview of the standardization efforts. This chapter will focus on collision detection and resolution algorithms and conclude by practical suggestions on RFID system selection for different tasks.

DOI: 10.4018/978-1-60960-732-6.ch023

HISTORY OF RFID

The first radio frequency based identification technology (currently known as RFID) was developed during World War II. In 1939, Great-Britain, it was observed by chance that when a pilot makes a certain swinging maneuver with his aircraft, the shape of the reflected waves change, thus friendly airplanes became distinguishable from enemy ones on the radar screen. This can be considered the very first passive RFID system.

In 1948, Harry Stockman predicted the advent of applications based on reflected radio waves in his work titled "Communications by Means of Reflected Power", but first he attracted attention to a large number of unresolved issues and the necessity and inevitability of extensive research in the field. As it turned out, some 30 years of research and development was needed until the effective deployment of the RFID technology.

The 1960s was the prelude to the technological explosion of RFID in the 1970s. The papers of R. F. Harrington regarding electromagnetic fields ("Field measurements using active scatterers" and "Theory of loaded scatterers", 1963 and 1964) grounded the future popularity of RFID. In the 1970s serious development was conducted in both the US and Europe.

By the 1980s, the research and development phase was supplanted by the implementation of new results and their application in products. In the 1990s, the application of RFID technology broadened: electronic toll collection on toll roads, immobilizers, (ski) tickets, entry systems, and access control systems started to harness the possibilities provided by RFID. Products based on RFID solutions began to spread in other regions of the world too.

The first microwave Schottky diodes integrated on CMOS circuits made simple single-IC microwave RFID transponders possible. This meant greater reading distance and faster data transmission rates. UHF RFID got a boost in 1999 when the Uniform Code Council, EAN International, and industrial partners founded the Auto-ID Center at the Massachusetts Institute of Technology.

PARTS OF AN RFID SYSTEM

A wide spectrum of demands can determine the set-up of RFID systems. A configuration consisting of a single transponder and a reader, as used in electronic article surveillance, may be regarded as the simplest RFID system. A more complex system can contain thousands of transponders, a number of networked readers, controlling computers and back-end databases.

Radio based identification needs at least two devices: one to identify and another that identifies the first. The identifying device initiates some kind of data connection with the other, during which data is transmitted in one or both directions. Communication is transmitted over radio frequencies, so both devices need to have radio interfaces. Therefore, a basic RFID system consists of at least the following two components:

- A transponder that is attached to the object to identify
- And a reader that is able to read and/or write the transponder

The above system, depending on the application, may be augmented with a controlling computer, which might be harmonizing the work of multiple readers and linking the readers to back-end databases. The devices between the reader and the end application are referred to as middleware. The back-end databases store information about the transponders in the range of the readers. Information can be queried, but also updated during data transmissions. (Want, 2006)

Reader

The reader's goal is to establish a radio connection with the transponder, identify it, and to establish, maintain and close the data connection, during which the reader and the transponder exchange information. According to this the reader always contains an RF module (transmitter and receiver), a control unit and connector interfaces to interoperate with other systems. Common connector types include RS-232, RS-485, Wi-Fi, USB, etc. Most readers have internal antennas, but the more expensive ones have ports to attach external antennas too. Readers might also be referred to as interrogators, emphasizing their initiative role in the communication.

Transponder

The RFID transponder holds the information assigned to the product or tagged item. It consists of a microchip and an antenna, and it is usually mounted on some rigid or flexible carrier object. The information can be a unique identifier, a manufacturer ID, product code, or any other data. The transponder might also be called an RFID tag or label (Figure 1).

CLASSIFICATION

RFID devices are produced by a great number of different manufacturers, and a wide range of devices is in existence. To clearly see the differences among them, we categorize them by their most important features (Finkenzeller, 2003).

Programmability

RFID systems can be classified by the way data is written into the transponder. To most basic chips are programmed during the production process, usually with a simple immutable serial number. The contents of writable transponders can be modified by the reader at any time; data can be written to three different kinds of memory chips. The most prevalent ones are the EEPROM modules used in inductively coupled RFID systems. Their disadvantages are their limited write count (usually between 100,000 and 1,000,000 cycles) and the high-energy requirement of writing. In contrast to this ferroelectric RAMs (FRAMs) have lower energy requirements (approximately one hundredth part) and their write time is one thousandth part of that of EEPROMs, however they are not widely used due to yet existing difficulties in their manufacturing process. Static RAM (SRAM) modules used in microwave systems are also popular for their fast writing cycles, although they need a continuous external power source.

Data Capacity

RFID systems may also be classified by the information and data management functions of the

Figure 1. Parts of an RFID system

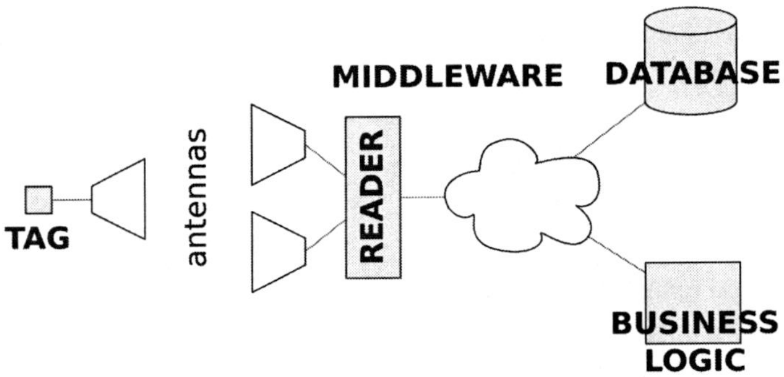

transponder, and its memory capacity. The data storage capabilities of RFID transponders range from a few bytes to a few kilobytes, although 1-bit transponders also exist for use in special applications. 1 bit is just enough for the reader to know if there is a transponder in the zone or not. Since these transponders do not even require a microchip, their production is exceptionally simple and cheap. These are preferred in anti-theft applications (Electronic Article Surveillance). If someone tries to leave the store (more precisely, the sensor gate at the exit) the reader detects the transponder in the zone and signals an alarm. Usually during the payment process, the transponder is removed or deactivated.

Low-end read-only transponders equipped with a microchip may only hold a few bytes of preprogrammed information, in the form of a serial number. As such, a transponder detects the RF field of the reader; it begins to broadcast its serial number. The reader is not able to address one selected transponder. In these systems, generally, only one transponder is allowed in the reader's zone, the presence of two or more transponders would cause a collision and the reader would be incapable of detecting the transponders. Despite these deficiencies, read-only transponders are very popular, since in many applications reading the serial number is sufficient. Simple functionality results in low power consumption and small chip surface, meaning low production costs. Low-end systems are able to operate at any frequencies due to their low power requirements. Because of these features, low-end transponders are well suited for bar code replacement applications, areas of process control, container tracking (ISO 18000) and animal identification (ISO 11785).

Mid-range systems constitute the middle class with their greater sized, writable memories and collision resolution algorithms. The size of available memory varies from a few bytes to even 100 kilobytes; the media can be EEPROM (in case of passive transponders) or SRAM (active transponders). These transponders are able to interpret simple commands coming from the reader like selective reading or writing from the data memory. The majority of systems offer the use of some kind of collision resolution algorithm, so the transponders residing in the zone are able to avoid interference, and the reader becomes capable of selective addressing. The support of cryptographic methods is also widespread, and the operating frequency can still be of any domain.

The essential requirement of high-end systems is microprocessor controlled operation and having a smart card operating system. The use of microprocessors enables the application of more complex cryptographic and authentication algorithms. At the top of this class are the dual interface smart cards having cryptographic coprocessors. The resources liberated by the coprocessor and shortened computing times make high security applications possible.

Frequency and Distance

The operating frequency is defined as the carrier frequency of the signals transmitted by the reader. The transmission frequency of the transponder is usually not important, most of the time it is equal to that of the reader (load and backscatter coupling). The various frequencies are classified into four domains: LF (30 to 300 kHz), HF (3 to 30 MHz), UHF (0.3 to 3 GHz) and microwave bands (greater than 3 GHz). We can also differentiate close-coupled (less than 1 centimeter), vicinity coupled (less than 1 meter), and remote-coupled, long-range systems (more than 1 meter). Vicinity coupled transponders with a range of up to 10 centimeters (such as entry cards) are sometimes referred to as proximity systems, however they still use the electromagnetic near-fields of the readers for communication (proximity cards are defined in ISO 14443, vicinity cards in ISO 15693). Passive transponders may have a maximum reading range of 15 meters. (Aroor, 2007)

The most important parameters of RFID systems are the operating frequency of the reader, its range, and the type of coupling used. Reading distance is dependent on multiple factors, but the most important element is the power supply mechanism of the transponder and the way it sends data to the reader. For this purpose, different types of coupling are used utilizing electric, magnetic or electromagnetic fields. In general, the higher the frequency used the longer the reading distance and the more susceptible the communication is for being disturbed by metallic or liquid materials.

In case of close coupled systems the transponder has to be inserted into a sensor slot or placed directly onto the reader. The operating frequency can be chosen freely from DC to 30 MHz, the coupling may be electric or magnetic. The proximity of the transponder to read allows higher energy levels to be used, so even a more lavish processor could be acceptable. The scope of application of these systems includes mainly security-focused services like electronic door entry systems. At present, contactless smart cards use this principle, although the market share of the technology is decreasing.

Systems with ranges under 1 meter use inductive coupling in general, hence they are also called inductive radio systems. Capacitive coupled systems are also in existence. Of all the RFID systems sold on the market to date, 90% is inductively coupled; their operation is defined by a number of international standards.

Long-range RFID systems use electromagnetic waves with no exception. They transmit in the UHF or microwave bands. Their physical operation is based on the backscatter principle. Systems with ranges below 3 meters generally employ passive transponders, while over 15 meters active transponders are more common. Generally, the external power source of active transponders drives only the microprocessor and the memory modules; the energy required for transmission is still supplied by the field of the reader. However, the category of active transponders also includes devices that use battery assisted or battery powered transmission. The reading range of such active transponders can be as high as 100 meters. As beacons broadcast data periodically, their operation is also independent of readers.

Low frequency systems use inductive coupling, so these have the lowest reading range. Although their production costs are increased by the copper coils used, they are still the most prevalent type of RFID systems due to the matureness of LF technology. Low frequency waves are the least absorbed in liquids and metals.

Among high frequency systems, inductive and capacitive coupling is general; they provide low to medium reading ranges. They present a cost efficient solution for smart card applications, but they are also used in tracking and in libraries. HF waves also propagate quite well through metallic materials and liquids.

While LF and HF frequency ranges can be used unlicensed universally, there exists no global standard for UHF systems. Because of different regional regulations, they operate at 868 MHz in Europe and at 915 MHz in the US. They are usually realized with capacitive coupling, so ultra-high frequency systems have great reading distances and fast data transmission rates. They provide an optimal solution even in the case of reading multiple transponders. UHF transponders have to be line of sight in order to read them without difficulties. Because of the development of IC technology, they are cheaper than LF and HF transponders.

Microwave systems have similar properties to their UHF counterparts, but they provide even higher data transmission rates. They operate at microwave frequencies of 2.45 and 5.8 GHz. Their performance is the most sensitive to the vicinity of metals and liquids, they are the most difficult to read in such environments.

Power Supplement Method

Another important aspect is the question of energy. Passive RFID transponders do not have their own power source, so they draw all the required energy from the reader's electric or magnetic field. They are cheaper than active transponders with a typical price between 20 and 40 cents per unit, but they can only be read from a much shorter distance. Passive RFID tags usually operate at low to ultra-high frequencies.

In contrast to passive transponders, active ones have built-in power sources. Active tags may be further categorized as transponders and beacons. Active transponders only awaken when they receive a signal from the reader. These are used for toll collection or setting up checkpoints: when a vehicle equipped with an active transponder approaches a sensor gate, upon the reception of the signal of the reader integrated into the gate it responds with its unique identifier. By only broadcasting their data when in range of a reader, active transponders conserve battery power.

Beacons are generally used in real time location services, where the position of valuable items must be continuously tracked. In such systems, beacons transmit their unique identifiers periodically at definable intervals. This timespan can range from seconds to days depending on the importance and required precision of the location information. The signal of the beacon is received by at least three readers; its position is calculated by some kind of triangulation technique. Real time location systems are usually deployed outdoors, but they can also be utilized in large factory halls.

Reply Transfer Method and Reply Frequency

Transponders can transmit in three different ways:

- Reflection or backscatter (the frequency of the reflected wave is equal to the readers frequency)
- Load modulation (the transponder influences the RF or EM field of the reader)
- Using subharmonics

The reply frequency can be the same as the one used during reading, or can be different.

Modes of Operation

In contrast to 1-bit transponders, which use simple physical processes like oscillating or harmonic excitement (with diodes or non-linear hysteresis curves of metals), most transponders utilize microchips for communication. These chips have a few kilobytes of memory capacity. To read or write data into the memory, data exchange is necessary between the reader and the transponder. Two modes of operation are possible: full duplex (FDX) and half duplex (HDX). Half-duplex operation can be further classified as continuous transmission or sequential operation.

In half duplex systems, the direction of data transmission is alternated between reader to transponder and transponder to reader paths. At frequencies below 30 MHz, this method is often used in conjunction with both regular and sub-carrier load modulation, which allows very simple circuits to be used. Load modulation directly affects the electromagnetic field generated by the reader.

In full duplex systems data transmission from the transponder to the reader happens simultaneously to the transmission from the reader to the transponder. This method employs such processes that the transponder to reader transmission is carried by subharmonics of the reader's transmitted signal or it uses completely different frequencies.

Both methods (FDX and continuous HDX) have in common that the energy radiation from the reader to the transponder is continuous, so it is independent of the direction of the data streams. In contrast to this, HDX sequential systems restrict energy transmission to short periods of time for reader to transponder communication; transponder to reader transmission occurs during the gaps or pauses between them. For this reason, sequential operation is categorized as being a kind of half-duplex solution.

COUPLING TYPES

Coupling methods of RFID systems define the way the reader and the transponder interact with each other. The various coupling methods may be classified by the range of coupling (close, vicinity and long-range coupling), but also by the physical properties of coupling. In this manner inductive, capacitive, magnetic and backscatter coupling methods are differentiated. Capacitive and magnetic coupling is usually used in close coupled systems, inductive coupling in vicinity or proximity systems, while backscatter coupling is generally employed in long-range systems.

Besides the range of operation, the chosen coupling method is also in tight connection with the frequency used by the transponder. Inductively coupled systems typically perform best in the LF and HF domain; the ideal operating frequency for backscatter coupling is higher: UHF and microwave bands are better suited. Magnetic coupling is the most effective at 110 MHz, while capacitively coupled systems yield an adequately fast clock rate for communication at about 10 MHz.

Backscatter Coupling

Long-range RFID systems operating in the UHF or microwave bands require smaller and more efficient antenna systems than low frequency communications, due to the shorter wavelengths at higher frequencies.

Only a small fraction of the power radiated by the antenna of the reader reaches the antenna of the transponder (due to path loss and the lack of directed beams). The high frequency voltage induced by the power reaching the transponder is suitable to bring its circuits to life after rectification. A part of the incoming power is reflected back to the reader. The reflection properties of the antenna can be influenced by altering the connected load. In accordance to this, a toggleable resistor is connected to the antenna in a series circuit, during transponder to reader communication a digital data stream can be transmitted by switching it on and off. So the amplitude of the reflected power can be modulated (lower and higher amplitudes, respectively), this is called modulated backscatter. The reflected energy is further attenuated according to the laws of free space path loss, so only a smaller fraction of it is returned back to the reader. The reader separates the modulated signal from the original electromagnetic waves, and processes it as useful data from the transponder (Figure 2).

As the communication happens on the same frequency for both reader to transponder and transponder to reader directions, only half-duplex operation is viable. The reader and transponder

Figure 2. Backscatter coupling

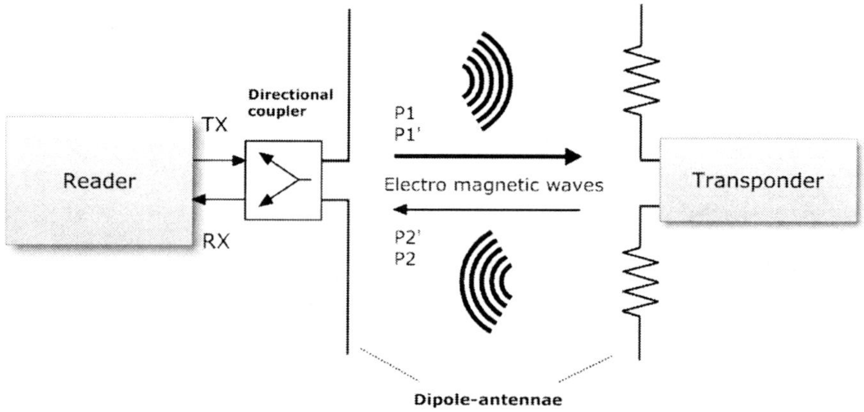

transmit in turns, while the reader continuously supplies the transponder with power.

Inductive Coupling

Inductively coupled transponders are almost always passive, that is they do not contain internal power sources, all the power required for communication is received from the reader. The two main components of the transponder are the chip responsible for communication and the antenna coil. To supply enough power to the transponder, the antenna of the reader generates a very strong, high frequency EM field.

When the transponder enters the EM field produced by the reader, voltage is inducted in its coil (Figure 3). The inducted voltage serves as a power source for the transponder's chip. To create a strong EM field the reader's antenna constitutes a resonant circuit with a capacitor connected in a parallel circuit, the resonant frequency of which matches the data transmission frequency of the reader. The transponder's antenna coil and the paralleled capacity are tuned to the frequency of the reader, thus the voltage inducted in the coils is maximized.

The two antenna coils can be interpreted as a transformer, where the transformer coupling is very weak. The efficiency of power transfer is the function of the operating frequency, the number of turns in the coils, the area enclosed by the transponder's antenna coil and the angle of the two coils. By raising the frequency the needed inductance, and thus the number of rounds can be decreased.

Load Modulation

Inductively coupled systems are only possible when the distance between the two antennas is in the near-field range, meaning it is shorter than the wavelength multiplied by 0.16. A transponder placed in the near-field of the reader's antenna gains its energy from the EM field generated by the reader's antenna. The answer of the transponder can be seen as changes in the load impedance of the reader's antenna by switching a load resistor on and off; the result is the amplitude modulation of voltage in the reader's antenna coil.

This modulation type does not use sub-carriers, so decoding the information only requires demodulation of the voltage at the receiver antenna by a diode. After rectification and applying a low-pass RC filter, the signal reaches the IC of the reader, where it gets converted to TTL level by a Butterworth low-pass filter, an amplifier and a Schmitt trigger, and can be processed by a microprocessor.

Sub-Carrier Load Modulation

Due to the weak coupling between the reader and the transponder, the change of voltage caused by the transponder at the receiver antenna is multiple orders of magnitude lower than the voltage on the

Figure 3. Inductive coupling

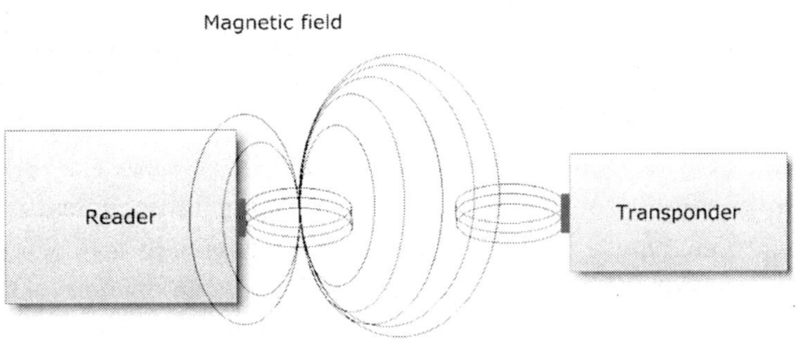

antenna coil, so detection can only be done using very complex circuits. For this reason, sub-carriers generated by the amplitude modulation are used in detection.

When load modulating a sub-carrier, two sidebands arise. Modulation sidebands can be easily detected even in the presence of stronger signals by band-filtering one of the sub-carrier frequencies. After adequate amplification, the modulated sub-carrier can be demodulated and the carried data can be decoded. Because of the high bandwidth requirement, it can only be used in the ISM bands.

Subharmonic Method

The subharmonics of sinusoid signals are also sine waves, the frequencies of which are integer fractions of the original. In communication methods based on subharmonics, usually the half of the original frequency is used, which is produced by digital division of the reader's frequency. The signal on the output of the digital divisor is ready to be modulated with the transponder's data stream. The modulated signal is then fed back to the antenna of the transponder by an output driver. The operating frequency of such systems is typically 128 kHz, so the transponder communicates with the reader at 64 kHz.

Both sub-carrier and subharmonic methods use FDX mode during the communication. It is possible because the reader and the transponder transmit at different frequencies.

Magnetic Coupling

Magnetic coupling is very similar to inductive coupling, as the antennas of the reader and the transponder form an electric transformer in both cases. The difference is that in magnetic coupling the reader's antenna coil is a ferrite core equipped with circular or U-shaped windings.

The transponder has to be in a 1 cm range of the reader in the air gap of the transformer. Since the distance is very short and the coupling is exceptionally strong, for the sake of simplicity direct ASK modulation may be used. This coupling type can supply power even to the most lavishing chips.

Capacitive Coupling

The other type of close coupling is capacitive (or electric) coupling. It is primarily used in smart cards (covered by ISO 10536). In order to form capacitive coupling, we do not need antennas but a pair of electrodes. Both the reader and the transponder have a conductive plate, which form a capacitor together. Naturally, the two plates are required to be parallel.

The chip of the transponder is located between its two plates, one of which forms a capacitor with the reader's plate, the other with the Earth. The potential difference across its two plates serves as the power source for the transponder chip.

In capacitively coupled systems, besides the usual load modulation, capacitive modulation can also be used. In this case, the transponder switches a capacitor instead of a resistor, thus alternating the resonance frequency of the transponder between two values. This modulates the voltage at the reader's antenna both in amplitude and in phase.

COUPLING PARAMETERS

The most important common parameter of the coupling methods is the coupling coefficient. It was introduced to numerically express the coupling strength independent of the dimensions of the conductive loops. The value k is defined as the quotient of the mutual inductance and the square root of the product of the self-inductances. Its domain ranges from 0 to 1, with 0 being totally uncoupled (in case of large distances or shielding) and 1 being fully coupled, with the same magnetic flux belonging to both antenna coils.

There is an inverse relationship between the transponder's resonant frequency and its bandwidth. The quality factor is defined as $Q = f_0 / b$. When implementing data communication methods one of our goals is the maximization of this value.

In case of load modulation, the change is reflected in the current of the transponder's coil. The modulation sidebands of the modulated current are attenuated as a function of the bandwidth of the transponder's resonant circuit. The bandwidth is the frequency band below 3 dB amplitude loss from the resonant frequency. If the Q factor is too high, the modulation sidebands of the current are attenuated due to the low bandwidth, so the range of the transponder is shortened.

The development of inductively coupled RFID systems is always a compromise between the desired range and the transmission speed. Systems that need short data exchange times (meaning fast transmission and high bandwidth) are often bounded to a few centimeters, while systems that tolerate longer transmission times (with low data speed and low bandwidth requirements) might even have multiple orders of magnitude longer ranges. Reading range is of course also determined by the emitted power, but this factor is limited by national and international standards and regulations. Some regulations also govern the levels of modulation products that are not elemental parts of the communication, like harmonic frequencies.

ANTENNAS

The antennas of the most often used UHF systems are usually either linearly or circularly polarized. In the case of linearly polarized antennas, the electromagnetic waves have only one energy field. Linearly polarized antennas have a narrower beam, but thanks to this, they also have longer reading range than circularly polarized antennas. Due to the narrower beam, readability is highly dependent on direction and orientation.

Linearly polarized antennas are sensitive to the orientation of transponders: they are only able to read the ones with orientation matching their polarization. Because of this fact, these antennas are primarily used in applications where the transponders have a fixed orientation.

On the other hand, with circularly polarized antennas, electromagnetic waves have two energy fields, which are equal in amplitude, but their phase difference is 90 degrees. Therefore, when one field is at its minimum, the other is at its maximum, this in turn makes readability independent of the orientation of the transponders.

CODING SCHEMES AND MODULATION METHODS

Communication in RFID systems can be divided into three major blocks: the reader, the transponder and the medium or channel between them. To implement data transmission, the data to be transferred needs to be encoded, and the coded information has to be modulated. On the receiver end, the opposite happens: after demodulating the signal, it has to be decoded to extract the information that was sent.

The coding of data must not be confused with modulation. The encoder assigns a signal representation to the message according to the properties of the transmission medium. An appropriate coding scheme makes a certain degree of error detection and recovery possible. Coding is done in baseband, which is then modulated to the chosen carrier frequency. Modulation is the process when the baseband signal generated by the coding algorithm is transposed to a certain carrier frequency.

Coding Schemes of the Data Transmission

Coding determines the way in which the transponder and the reader interpret the digital data represented by the changes of the analog carrier signal. So coding is an agreement between the transmitter and the receiver according to the used keying methods. Coding schemes used in RFID systems are the following (Stallings, 2007):

- Manchester coding
- Non-return-to-zero (NRZ) coding
- Unipolar return-to-zero (RZ) coding
- Differential bi-phase (DBP) coding
- Miller coding
- Modified Miller coding
- Differential coding
- Pulse-pause coding (PPC)

Modulation Methods

The transmission medium carries the modulated information from one end to the other. However due to the channel's attenuation and other parameters causing transmission errors (interference, external noise sources) the maximum possible distance of communication is finite. In RFID systems, the stored data can be transmitted by appropriately influencing magnetic or electromagnetic fields.

Demodulation is the counterpart of modulation; it mixes the received signal back to baseband. The demodulated signal yields the originally sent data after the process of decoding. The function of the decoder is to correct or mark possible transmission errors.

Naturally communication can be directed both from the reader to the transponder and from the transponder to the reader.

In classic radio technologies, the analog modulation methods (amplitude, frequency, and phase modulation) were built on altering the appropriate properties of the EM waves. All the other modulation methods can be deduced from these basic procedures. In RFID systems data is transmitted by digital modulation. In digital systems, there is no pre-existing signal to transfer with its shape preserved, but the codeword symbols are modulated onto the channel as elementary signals.

The three utilized modulation methods are:

- ASK: Amplitude Shift Keying
- FSK: Frequency Shift Keying
- PSK: Phase Shift Keying

All of these modulation methods use a modulation product that is symmetrical to the carrier: sidebands. We can differentiate between lower and upper sidebands. Most modulation procedures used in RFID systems are antipodal, meaning that the symbol set consists of two opposing signals.

Digital modulation systems are best described by their bit error rate, which is expressed as a function of the signal to noise ratio (SNR). In practice the quotient of the power carried by a symbol and the noise in a bandwidth unit are also generally used to characterize systems. The benefit of this indicator is that if the spectral density function of the noise in the frequency domain of interest can be regarded as constant (which is true in the case of Gaussian noise), then we don't need to know the bandwidth of the noise source. However if we know the noise bandwidth then the SNR can be expressed too.

In case of transponders operating at frequencies below 135 kHz, load modulation is usually controlled by a baseband coded signal series. Modulated signals can be restored by examination of the voltage amplitude at the reader's antenna. Digital demodulation incorporates a decision of the most probable sent symbol based on the signal at the channel's output. We might even say that digital demodulation is in fact only a decision making process. In RFID systems, operating at higher frequencies load modulation is controlled by a modulated sub-carrier signal. All the sidebands arising during modulation carry the same data; the reader filters out an arbitrarily selected one and demodulates it, recovering the baseband signals.

STANDARDIZATION

Auto-ID and other notable companies, in collaboration with the Uniform Code Council, developed the Electronic Product Code (EPC) as the future standard for product identification. The goal of the establishment of EPC was to replace the current bar code standards with a future-proof identification and numbering system. The existing structure of the well-known Global Trade Item Number (GTIN) was reused, so in theory it would be usable instead of all the regular bar codes issued so far. In essence, EPC expands the current numbering system to a broader range, and extends it with a serial number for unique item identification. EPC as an identification number is a reference to data stored in a standardized joint network of data banks, similar to the familiar "license plate" system of operation of bar codes. To sum up, EPC is a complex system capable of completely fulfilling the functions of known identification systems faster and more efficiently, augmented with extra functionalities. The ability of EPC to identify any single unit makes it an appropriate technology for tracking and shipping too, in contrast to GTIN.

EPC is a number constituted of an addressing header and three data fields. The header contains the version number of EPC, referring to the different types or lengths of EPC. The second part identifies the owner or handler of the code range, in most cases the manufacturer of the product, as indicated on the label. The third part denotes the stock-keeping unit, or product type, and the last one is a unique serial number identifying a distinct product item. The most prevalent EPC scheme is the 96 bits long GID-96, in which the header is 8 bits long and there are 28 bits for the EPC manager, 24 bits for the object class and 36 bits for the serial number.

In the beginning, RFID was only a patented technology without standardization. Today numerous standards cover the different applications, ranging from logistics to livestock and product identification. The two most prominent developers of supply chain standards are the International Organization for Standardization (ISO) and EPCglobal Inc., a subsidiary of the GS1 non-profit organization handling Universal Product Codes and Uniform Code Council standards. Most national and industry standards are based on the standards of ISO and EPCglobal. EPCglobal's Generation 2 UHF standard of was handed over to ISO to be part of the 18000 series of standards, which became the most widely used RFID technology standard.

Related ISO standards cover four main areas: technology, data content, conformance and efficiency, and application standards. However, they tend to focus on interfaces, not the structure of data, so they are more general, supporting different systems in different environments, independent of the type of data. Standards describe physical properties, RF power and signal interfaces, initialization, collision management and transmission protocols. However not only communication between transponders and readers is standardized, but also the operation of contactless cards, RFID data management functions, data structures and coding schemes, logical memory functions, radio interfaces, testing procedures, agriculture electronics, and other technologies regarding containers, logistics and supply chain applications are documented in ISO and EPCglobal standards.

COLLISION DETECTION AND RESOLUTION

Because of the popularity of RFID applications in all areas of the supply chain, it became necessary to be able to read large amounts of goods (transponders) in very short times. It is not uncommon in warehouse and logistics applications to read hundreds of transponders per second. These tasks require the integration of modern collision detection and collision resolution algorithms into RFID systems. Some of them are new ideas, while others may be known from earlier applications such as Ethernet or ALOHAnet. However, the chance of collision is not restricted to reading multiple transponders, the same problem may arise in the presence of more than one reader.

Figure 4. Collision of readers

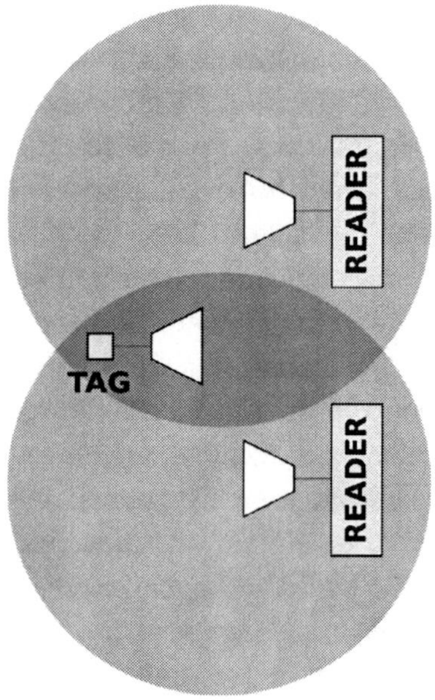

Collision of Readers

A collision occurs in RFID systems when two or more readers cover a common space with a transponder in it (Figure 4). The most conspicuous negative effect of collisions is the decrease in reading speed. Therefore, the acceleration of the reading speed of RFID systems may be done either by increasing the bandwidth or by eliminating collisions between readers (Klair, 2008).

Multiple solutions exist to resolve this problem. The reader may register itself in a central access point and send transmission requests along with its neighbors' data. Then the AP examines the states of the neighbors, and responds accordingly. Another solution is centralized graph coloring. All endpoints send their information about their neighbors to the AP, which calculates the positions of free time slots with a graph-coloring algorithm.

The following three methods are usually used as a final solution strategy:

- NAK-based resolution
- Beacon-based solution
- Usage of adjustable range readers

Negative Acknowledgment Based Collision Resolution

The negative acknowledgment (NAK) based collision resolution assumes that each reader has its own control channel. The range of this control channel is long enough to reach every other reader in its vicinity that may cause interference in the data channel.

In accordance with this, the reader first broadcasts a ready to send (RTS) message to all other readers. If any of them is busy, it responds with a NAK message. If the reader receives at least one NAK, it restricts access to the data channel.

Beacon Based Collision Resolution

The conditions are the same as for NAK based solutions: an own long distance control channel. While reading, the reader periodically transmits beacon signals on its control channel. If another reader wishes to read, first it checks the time elapsed since the last beacon. If it is lower than a certain value (the period time of the beacon) the reading is probably still in progress, so it waits to see if there is another beacon during that time.

Power Adjustment of Readers

The adaptive adjustment of the reader's range is an obvious solution to the collision problem. This method doubles the covered area every few reading cycles, or in case of collisions optimizes reading by decreasing its size. Thus, space division multiple access is implemented by altering the power levels.

Figure 5. Collision of transponders

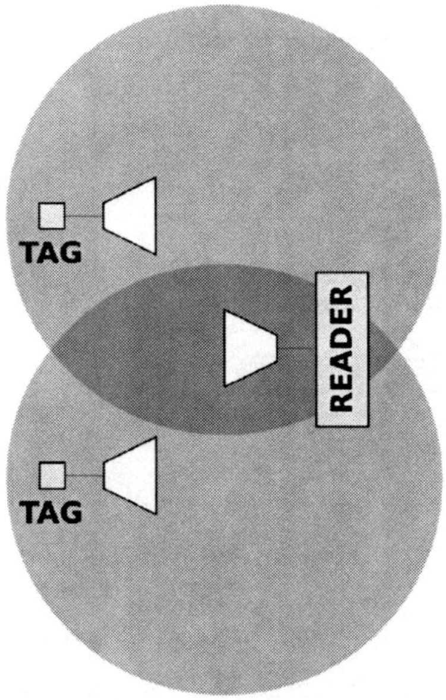

Collision of Transponders

Multiple transponders may be present in the range of a reader. When reading, the reader broadcasts a request message to the tags. If only one tag is in the covered space, the reader receives only one response (Figure 5). However if there are multiple transponders in the system, all of them send their responses back to the reader. This causes a collision in the radio channel, so the responses will be impossible to understand. To avoid these circumstances, numerous collision avoidance algorithms were developed.

In most cases collision resolution can be reduced to some kind of multiple access problem, so pre-existing algorithms can be reused. The effect of coding must also be considered in collision detection tasks.

Multiple Access Methods

Since the transponders share a common medium when sending their responses, the solution of the problem can be reduced to a multiple access exercise. Four types of multiple access methods are distinguished in RFID systems too (Lee, 2005):

- Space division multiple access (SDMA)
- Frequency division multiple access (FDMA)
- Time division multiple access (TDMA)
- And code division multiple access (CDMA), also known as spread spectrum multiple access (SSMA)

In RFID systems, a collision resolution method is such an algorithm or protocol that makes multiple accesses without interference possible. Some vendors do not publicize their proprietary collision resolution algorithms, while others use standardized solutions. In general, custom solutions are also derived from these standards with the addition of application-specific extensions.

ALOHA, Contention Based Random Multiple Access

ALOHA is a transponder-driven stochastic TDMA method. Generally, this method can only be used in conjunction with read-only transponders because of their small packet sizes (typically identifiers). This data is sent in a cyclic sequence with long pauses between transmissions, while the iteration intervals of individual transponders are very different. This means that the data transmission needs of transponders arise at different moments in time with a very high probability, so the chance of collision or congestion remains low. The load is of course proportional to the number of simultaneously active transponders. The maximum efficiency is at 18.4% transmission time: smaller load would leave the channel idle; more packets would dramatically increase the number of collisions so more than 80% of the channel capacity seems to be unused.

The simple implementation of ALOHA makes it useful as a collision resolution algorithm. This method is used in digital broadcast networks like

Figure 6. Slotted ALOHA

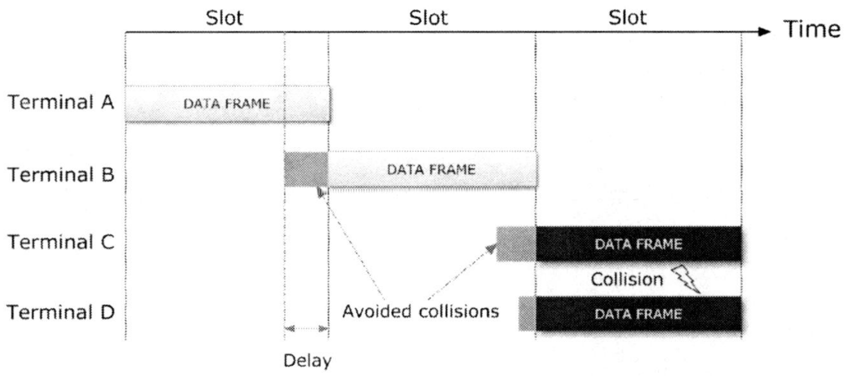

packet radio, which is a worldwide popular communication form among amateur radio operators.

Slotted ALOHA was developed to increase the relatively low throughput of ALOHA. The transponders using this algorithm may only initiate communication at allotted times (sync points or time slots). Synchronization is the task of the reader, which makes it a stochastic reader-driven TDMA collision resolution algorithm (Figure 6). Using this algorithm and equal-sized packets, collisions may only appear in half the periods as ALOHA, making 36.8% transmission time efficiency possible.

Dynamic Frame Slotted ALOHA

Slotted ALOHA systems have a maximum capacity where the number of transponders reaches the number of available time slots, because when too many transponders are placed in the interrogation zone, the throughput rate soon drops near to zero due to collisions, and in the worst case we won't be able to detect any identifiers after any number of attempts. This effect can be avoided by defining a sufficient number of time slots; however, the efficiency of communication will still be decreasing. The reason behind this is that the system has to constantly listen for all available transponders in all time slots even if there is only one transponder in the zone.

Dynamic frame slotted ALOHA tries to overcome this by dynamically changing the number of time slots. One possibility for this is the broadcasting of the number of time slots as an argument of the REQUEST command. If the large number of transponders is the bottleneck in communications, each REQUEST message increases the number of available time slots until a unique transponder becomes clearly detectable. After this, the reader broadcasts a BREAK message, which makes all transponders but the selected one enter a blocked state.

Binary Search Algorithm

The binary search algorithm requires precise bit positioning in the reader in case of an incidental collision. This can be achieved by the appropriate coding scheme (NRZ or Manchester). The algorithm consists of predefined interactions (commands and replies) that allow the reader and the transponders to make contact and select one single transponder for data transmission. The transponders still have unique identifiers, represented as bit patterns. In each iteration step, all transponders whose identifiers are lower or equal to the identifier in the REQUEST command respond with their identifiers. As the first step of the algorithm, the reader broadcasts a REQUEST command with the all-one pattern, which is the highest possible serial number value; all transponders in the zone will reply because all their identifiers must be lower or equal to this.

By being in precise bit sync, the algorithm is able to mark individual bit positions with collisions, as different bit values are superimposed. A collision means that there are at least two transponders in the interrogation zone. The reader sets the first colliding bit to "0" and sends another REQUEST message. This iteration process continues until only one transponder remains, which then can be selected for uninterrupted communication. After finishing data transmission, the reader silences this transponder by issuing an UNSELECT command for it. The number of remaining transponders decreases as more and more transponders are switched to the UNSELECTED state. When only one transponder is in the zone, the number of iterations is of course one; when there are multiple transponders, this number can quickly rise quite high (Figure 7).

Dynamic Binary Search

The main negative property of the previous algorithm is its wasteful operation, since both the query parameters and the serial numbers are fully transmitted in every step. Such a system transfers a huge amount of unnecessary data.

Instead of transmitting the whole serial numbers, only their already fixed prefixes are included in the REQUEST commands. All transponders whose serial numbers begins with these sequences respond with the remaining bits in their unique identifiers. This way the number of total transmitted bits is reduced, the more transponders exist with the longer serial numbers, the more this gain becomes.

Slotted Binary Tree

The algorithm works as follows: if a collision happens in a time slot, all transponders that did not participate, wait until the collision is resolved. The colliding transponders are randomly distributed into two groups; the first group retransmits in the following time slot, the other waits until the successful execution of this first step. If there is no collision in the following time slot (empty channel or successful transmission), the members of the second group take turn in the subsequent time slot, otherwise the whole process is restarted.

Figure 7. Binary search algorithm

	send		receive	responders	selected (1)	(2)	(3)
REQUEST	1111	→	XX1X	(1,2,3)	×	×	×
REQUEST	0111	→	0X10	(2,3)	×	×	×
REQUEST	0010	→	0010	(3)	×	×	×
(3) DATA OK							✓
UNSELECT	0010				×	×	
REQUEST	1111	→	XX1X	(1,2)	×	×	
REQUEST	0111	→	0110	(2)	×	×	
(2) DATA OK						✓	✓
UNSELECT	0111				×		
REQUEST	1111	→	1011	(1)	×		
(1) DATA OK					✓	✓	✓
UNSELECT	1011						
REQUEST	1111	→	Ø				
SELECT	1111	(all)			×	×	×

transponder ID	
(1)	1011
(2)	0110
(3)	0010

Bit-by-Bit Binary Tree Algorithm

At its start, the bit-by-bit algorithm requests for the first bit of the transponders' identifiers. After the transponders have sent their "0" or "1" bits, the reader checks to see if there was a collision before requesting the next bit. If a collision occurred, the colliding transponders are separated into two groups. One of these groups is selected, and all other transponders are silenced. The next bit position of the selected group is queried, and the algorithm is iterated until the whole binary tree constructed of the serial numbers is traversed, thus all transponders are identified.

Secure Collision Resolution

The previously introduced simple binary tree algorithms have an inherent security flaw due to the asymmetric nature of the uplink and downlink channels' security. The communication directed from the reader to the transponders can be heard by everyone. The reader broadcasts messages with high signal strength to ensure they reach all transponders. With an appropriate device, a remote eavesdropper is able to monitor this direction of data communication from even a hundred meters at certain frequencies. The main security issue is the signal strength of the downlink channel. Depending on the frequency used, remote attackers can listen to traffic from hundreds of meters, ascertaining the contents of the transponders. This is possible because the private serial numbers of the transponders are broadcast in clear. However, this data is sensitive, because it can be used to identify unique items as being present in the zone. The following two algorithms were developed to patch this security hole (Weis, 2003).

Blinded Tree Walking

The blinded tree-walking algorithm is a modified version of the general binary tree algorithm that eliminates the need to broadcast insecure identifiers on the downlink channel without decreasing throughput speed.

Initially it takes a group of transponders with a common prefix (manufacturer ID, common product code, etc.). To distinguish these transponders the reader requests the subsequent bit. If there is no collision, all transponders have sent the same bit. Remote attackers can only hear the reader's stronger signals; they cannot listen to the transponders' answers. This way the transponders and the reader share a common secret, the prefix of the identifiers so far. In case of collisions, the reader has to decide which group of transponders it wants to address in the following messages. But instead of broadcasting the chosen bit in clear, it XORs it with the previously established common secret, unknown to the attacker. After this, the algorithm goes just as the common binary tree protocol did, by requesting the subsequent bit in every step and traversing the binary tree.

Of course, this method does not provide protection against eavesdropping inside the range of the return channel, but increases security against remote listeners at the price of minimal extra complexity. Its efficiency is the same as regular tree walking, since a single transponder can still be identified by sending all the bits of its serial number.

Random Tree-Walking

This protocol deceives the remote eavesdropper with random meta-identifiers. Each transponder to be scanned chooses a random identifier. The reader traverses the binary tree made up of these randomly selected identifiers instead of the real serial numbers. After a transponder has been identified and selected, its real serial number gets transmitted on backward channel that is secure against remote attackers. Therefore, the randomly generated identifier can be found out, but the real one remains protected. The great disadvantage of this method is that transponders need to be equipped with random number generators, which raises production costs.

Rivest's idea (Weis & Rivest, 2004) simplifies the previously discussed blinded tree walking algorithm by assigning random pseudo-identifiers to the transponders at tree traversal, thus omitting the need for key based channel encryption. The full-length collision of the randomly chosen identifiers can be avoided by appending the real serial numbers, while secrecy is retained by not requiring the protocol to run until the last bit (uniqueness is presumed after a few iterations without collisions, still colliding transponders will be recognized and the algorithm continued from where it left off). However, due to the extra necessary communication the effective speed of the algorithm is lower than in regular tree walking. Another negative feature of the algorithm is that the reader must sustain the energy field until the successful end of the protocol run, because its interruption would mean the erasure of the pseudo-identifiers from the transponders. The third disadvantage is the transponders' need to store the bit positions where they were temporarily silenced.

SYSTEM SELECTION

The popularity of RFID systems has grown remarkably after the downturn of recent years. Maybe the best examples for this are the smart card based entry systems used in public transportation. The use of millions of RFID tickets would have been unimaginable even a few years ago.

The proper technical background has to be chosen for each application, be it entry systems, animal tracking, or access control. Although in some areas the requirements are overlapping, usually every service has a distinct set of demands. To make things even more complicated, except for a few special cases like tracking livestock or close-coupled smart cards, there exists no universal standard for the operation of RFID systems. Choosing the appropriate system is therefore not an easy task, the following section should aid in the decision process by describing the main parameters of RFID systems.

Frequency

The inductively coupled systems usually have an operating frequency between 100 kHz and 30 MHz, while their electromagnetically coupled and microwave counterparts operate in the 2.45 to 5.8 GHz range. The attenuation of water and non-conductive materials measured at 100 kHz is one hundredth part of that at 1 GHz. HF systems, on the other hand, have better penetration characteristics; this feature also has to be taken into account.

The range of microwave equipments is far greater than inductive systems, typically 2 to 15 meters. However, transponders may need to have an external power source, since the power received from the reader is usually not enough for the operation of the transponder. In the planning process the electromagnetic disturbance and interference of the surrounding equipments must be considered too, for example powerful electric motors may have adverse effects on the communication links. The inductive transponders have low tolerance for these electromagnetic disturbances; microwave systems are more suited for these environments.

Range

One of the most important features of a system is its range of operation, which is measured as the maximal physical distance between the reader and the transponder where data communication is still possible.

Its magnitude is determined by the following factors:

- Positioning and relative placement of the transponder
- Minimal distance between transponders during normal operation

- Velocity of moving transponders inside the zone

First, let us consider contactless payment applications like public transportation entry systems. The speed of positioning is extremely low, since the transponder is held to the reader by hand. The minimal distance between transponders is the minimal distance of two passengers, not less than 5 to 10 centimeters. A range greater than this would cause unintended multiple readings, since more tickets would be concurrently present in the zone, which would impede correct personal authentication. In these cases, the directed beams of microwave systems are often an advantage over the omnidirectional fields of inductive systems.

The maximal speed of transponders relative to readers together with the maximal read/write distances gives the amount of time spent in the readers' zone. The range of vehicle identification systems is calculated from the maximum speed of vehicles and the minimum time spent in the zone sufficient for reliable data transmission.

Memory Capacity

The size (and price) of a chip is mainly determined by the capacity of the memory installed. Price sensitive applications and those that have low informational demands utilize solely readable transponders. Such systems may be complemented with a controlling computer and a back-end database, so that the low memory capacity would not cause problems even at more complex tasks.

If the data needs to be rewritten, transponders equipped with EEPROM or RAM memory chips may be the solution. EEPROM modules mainly used in inductive systems can vary in capacity from 16 bytes to 8 kilobytes. SRAM memories used in microwave systems, supplemented with an external power source, can hold 256 bytes to 64 kilobytes of data.

Electromagnetic Interference

Electromagnetic interference is a common problem in poorly designed RFID systems. The weak signals of transponders might be partly or totally suppressed by radio waves from other sources in the same frequency band. The same can happen to readers too. In case of inappropriate channel separation and direction, these can also interfere with each other. The main issue with electromagnetic interference sources is that they are unpredictable, and disturbance is difficult to foretell. Adjacent production lines, older WLAN terminals, and GSM phones can also cause interference. Protection against the adverse effects of electromagnetic interference can be implemented by electromagnetically shielding the sources of interference.

Security Considerations

From the security point of view, RFID systems can be divided into two groups:

- Industrial or closed, proprietary applications
- Public applications related to money or goods transfer

Deployment of RFID systems in high security applications such as payment systems or patient identification is getting more and more common. These applications necessitated the integration of security extensions to existing systems to prevent unauthorized access.

Modern authentication systems operate by verifying the knowledge of a secret value (typically key-based methods). The communication algorithm has to prevent the attacker from retrieving this secret. Today's high security RFID systems must have this capability in order to avoid the following attacks:

- Unauthorized access to data medium, duplication or modification of stored data
- Placement of alien data medium by circumvention of authentication algorithms
- Interception and play-back of radio communications in order to impersonate a valid data medium ("replay and fraud")

When choosing an RFID system, we have to consider the security needs of the application. For an application that does not require a high level of security (for example industrial automation), integrating these functions would only make deployment needlessly expensive. On the other hand, for security sensitive applications the existence of these functions is of utmost importance, even at the expense of higher production costs.

CONCLUSION

In this chapter, we presented the history of radio frequency identification, and we enumerated the components of RFID systems. We explored different aspects of classification, reviewed coupling types, coding schemes, modulation techniques, as well as collision detection and resolution algorithms. Finally we concluded with some practical advice to consider when planning or deploying RFID systems.

Although the underlying principles had been discovered for decades, the possibilities of radio frequency identification were only realized in the 1980s. The following decade brought wider adoption and many new uses for RFID, with the end of the century demanding for standardization of both hardware and software, as well as product type registries and business processes related to logistics and asset management. RFID technology has not yet prevailed over traditional bar codes in the entire supply chain, but it complements them with valuable extra enterprise level functions; and it has also found innovative uses in other fields of life, ranging from door entry systems to contactless payment methods, or from electronic passports to indoor positioning and navigation services.

The future may hold an even more widespread adoption of RFID technology, assuming the current privacy and security concerns are met. The chips in the transponders are getting more and more intelligent, with added storage capacity and processing power, embedded cryptographic algorithms and mounted sensors, while the middleware systems and databases behind the readers are getting more and more complex and networked. In the short term, it can mean more integrated and automatized business processes, and in the long run this may even promise a world of annotated objects linked to the global information infrastructure, or a vision of augmented reality.

REFERENCES

Aroor, S. R., & Deavours, D. D. (2007). Evaluation of the state of passive UHF RFID: An experimental approach. *Systems Journal*, *1*(2), 168–176. doi:10.1109/JSYST.2007.909179

Chawla, V., & Dong Sam, H. (2007). An overview of passive RFID. *Communications Magazine*, *45*(9), 11–17. doi:10.1109/MCOM.2007.4342873

Finkenzeller, K. (2003). *RFID handbook*. Munich, Germany: John Wiley & Sons. doi:10.1002/0470868023

Heidrich, J., Brenk, D., Essel, J., Schwarzer, S., Seemann, K., Fischer, G., & Weigel, R. (2010). The roots, rules, and rise of RFID. *Microwave Magazine*, *11*(3), 78–86. doi:10.1109/MMM.2010.936075

ISO/IEC FDIS 10536-1. (2000). *Identification cards – contactless integrated circuit(s) cards – close-coupled cards – part 1: Physical characteristics*. Geneva, Switzerland: International Electrotechnical Commission, International Organization for Standardization.

ISO/IEC FDIS 10536-2. (1995). *Identification cards – contactless integrated circuit(s) cards – part 2: Dimensions and location of coupling areas*. Geneva, Switzerland: International Electrotechnical Commission, International Organization for Standardization.

ISO/IEC FDIS 11785. (1996). *Radio frequency identification of animals – technical concept*. Geneva, Switzerland: International Electrotechnical Commission, International Organization for Standardization.

ISO/IEC FDIS 14443-1. (2008). *Identification cards – contactless integrated circuit(s) cards – proximity cards – part 1: Physical characteristics*. Geneva, Switzerland: International Electrotechnical Commission, International Organization for Standardization.

ISO/IEC FDIS 14443-2. (2001). *Identification cards – contactless integrated circuit(s) cards – proximity cards – part 2: Radio frequency power and signal*. Geneva, Switzerland: International Electrotechnical Commission, International Organization for Standardization.

ISO/IEC FDIS 14443-3. (2001). *Identification cards – contactless integrated circuit(s) cards – proximity cards – part 3: Initialization and anticollision*. Geneva, Switzerland: International Electrotechnical Commission, International Organization for Standardization.

ISO/IEC FDIS 14443-4. (2008). *Identification cards – contactless integrated circuit(s) cards – proximity cards – part 4: Transmission protocol*. Geneva, Switzerland: International Electrotechnical Commission, International Organization for Standardization.

ISO/IEC FDIS 15693-1. (2000). *Identification cards – contactless integrated circuit(s) cards – vicinity cards – part 1: Physical characteristics*. Geneva, Switzerland: International Electrotechnical Commission, International Organization for Standardization.

ISO/IEC FDIS 15693-2. (2006). *Identification cards – contactless integrated circuit cards – vicinity cards – part 2: Air interface and initialization*. Geneva, Switzerland: International Electrotechnical Commission, International Organization for Standardization.

ISO/IEC FDIS 15693-3. (2009). *Identification cards – contactless integrated circuit cards – vicinity cards – part 3: Anticollision and transmission protocol*. Geneva, Switzerland: International Electrotechnical Commission, International Organization for Standardization.

ISO/IEC FDIS 18000-1. (2008). *Information Technology – radio frequency identification for item management – part 1: Reference architecture and definition of parameters to be standardized*. Geneva, Switzerland: International Electrotechnical Commission, International Organization for Standardization.

ISO/IEC FDIS 18000-2. (2009). *Information Technology – radio frequency identification for item management – part 2: Parameters for air interface communications below 135 kHz*. Geneva, Switzerland: International Electrotechnical Commission, International Organization for Standardization.

ISO/IEC FDIS 18000-3. (2008). *Information Technology – radio frequency identification for item management – part 3: Parameters for air interface communications at 13.56 MHz*. Geneva, Switzerland: International Electrotechnical Commission, International Organization for Standardization.

ISO/IEC FDIS 18000-4. (2008). *Information Technology – radio frequency identification for item management – part 4: Reference parameters for air interface communications at 2.45 GHz.* Geneva, Switzerland: International Electrotechnical Commission, International Organization for Standardization.

ISO/IEC FDIS 18000-6. (2004). *Information Technology – radio frequency identification for item management – part 6: Parameters for air interface communications at 860 MHz to 960 MHz.* Geneva, Switzerland: International Electrotechnical Commission, International Organization for Standardization.

ISO/IEC FDIS 18000-7. (2009). *Information technology – radio frequency identification for item management – part 7: Parameters for active air interface communications at 433 MHz.* Geneva, Switzerland: International Electrotechnical Commission, International Organization for Standardization.

Klair, K., Chin, K., & Raad, R. (2008). *A survey and tutorial of RFID anti-collision protocols. Communications Surveys & Tutorials*. IEEE.

Lee, S. W. (2005). *A multiple access algorithm for passive RFID tags*. Unpublished thesis, School of Electrical and Electronic Engineering, College of Engineering, Yonsei University, Seoul, South Korea.

Phillips, T., Karygiannis, T., & Kuhn, R. (2005). Security standards for the RFID market. *Security & Privacy*, *3*(6), 85–89. doi:10.1109/MSP.2005.157

Preradovic, S., Karmakar, N., & Balbin, I. (2008). RFID transponders. *Microwave Magazine*, *9*(5), 90–103. doi:10.1109/MMM.2008.927637

Ranky, P. G. (2007). Engineering management-focused radio frequency identification (RFID) model solutions. *Engineering Management Review*, *35*(2), 20–30.

Rieback, M. R., Crispo, B., & Tanenbaum, A. S. (2006). The evolution of RFID security. *Pervasive Computing*, *5*(1), 62–69. doi:10.1109/MPRV.2006.17

Stallings, W. (2007). Signal encoding techniques . In *Data and computer communications* (8th ed., pp. 138–179). Upper Saddle River, NJ: Prentice Hall.

Tesoriero, R., Gallud, J., Lozano, M., & Penichet, V. (2008). Using active and passive RFID technology to support indoor location-aware systems. *IEEE Transactions on Consumer Electronics*, *54*(2), 578–583. doi:10.1109/TCE.2008.4560133

Want, R. (2006). An introduction to RFID technology. *Pervasive Computing*, *5*(1), 25–33. doi:10.1109/MPRV.2006.2

Weis, S. A., Sarma, S. E., Rivest, R. L., & Engels, D. W. (2004). Security and privacy aspects of low-cost radio frequency identification systems . In *Security in pervasive computing* (pp. 50–59). Heidelberg, Germany: Springer. doi:10.1007/978-3-540-39881-3_18

KEY TERMS AND DEFINITIONS

Collision: Since wireless communication uses a common medium (for example an electromagnetic frequency channel) for numerous devices, and they are generally independent of each other, there is no way to predetermine when they (or the other transponders) wish to use the medium for communication. When two or more transponders are transmitting at the same time, the resultant signal becomes noise uninterpretable by the reader, as it is the sum of all the now indistinguishable unique signals. This effect is referred to as a collision.

Coupling: Coupling methods of RFID systems define the way the reader and the transponder interact with each other. The various coupling methods may be classified by the range of coupling (close, vicinity and long range coupling), but also by the physical phenomena used for data communication (inductive, capacitive, magnetic and backscatter coupling methods). The chosen coupling method is also in tight connection with the frequency used by the transponder.

Reader: The reader's goal is to establish a radio connection with the transponder, identify it, and to establish, maintain and close the data connection, during which the reader and the transponder exchange information. According to this the reader always contains an RF module (transmitter and receiver), a control unit and connector interfaces to inter-operate with other systems. Most readers have internal antennas, but the more expensive ones have ports to attach external antennas too. Readers might also be referred to as interrogators, emphasizing their initiative role in the communication.

RFID: Radio frequency identification is the common name for a group of closely related technologies, which were the topic of this chapter. RFID systems consist of readers and transponders, which communicate using electromagnetic phenomena. Their primary use is to identify or track electronic devices (RFID tags or transponders), typically applied to or incorporated into products, animals, or even persons, using radio waves or electromagnetic fields. RFID systems can be classified based on programmability, data capacity, frequency and reading distance, as well as power supplement and reply transfer methods; other key properties include the coupling type, modulation and the coding scheme. Many aspects of these systems are standardized by international standardization bodies.

Transponder: The RFID transponder holds the information assigned to the product or tagged item. It consists of a microchip and an antenna, and it's usually mounted on some rigid or flexible carrier object. The information can be a unique identifier, a manufacturer ID, product code or any other data; or the microchip may even be able to execute code on its own. The transponder might also be called an RFID tag or label.

APPENDIX

Table of abbreviations

AP	Access Point
ASK	Amplitude Shift Keying
CDMA	Code Division Multiple Access
CMOS	Complementary Metal-Oxide Semiconductor
DBP	Differential Bi-Phase
DC	Direct Current
EAN	European Article Number
EEPROM	Electrically Erasable Programmable Read-Only Memory
EM	Electromagnetic
EPC	Electronic Product Code
FDMA	Frequency Division Multiple Access
FDX	Full Duplex
FRAM	Ferroelectric Random Access Memory
FSK	Frequency Shift Keying
GID	Global Identification
GSM	Global System for Mobile Communications
GTIN	Global Trade Item Number
HDX	Half Duplex
HF	High Frequency
IC	Integrated Circuit
ID	Identification
ISM	Industrial, Scientific, Medical
ISO	International Organization for Standardization
LF	Low Frequency
NAK	Negative Acknowledgment
NRZ	Non-Return-To-Zero
PPC	Pulse-Pause Coding
PSK	Phase Shift Keying
RAM	Random Access Memory
RF	Radio Frequency
RFID	Radio Frequency Identification
RS	Recommended Standard
RTS	Ready To Send
RZ	Return-To-Zero
SDMA	Space Division Multiple Access
SNR	Signal-to-Noise Ratio
SRAM	Static Random Access Memory

continued on following page

SSMA	Spread Spectrum Multiple Access
TDMA	Time Division Multiple Access
TTL	Transistor-Transistor Logic
UHF	Ultra High Frequency
US	United States
USB	Universal Serial Bus
Wi-Fi	Wireless Fidelity
WLAN	Wireless Local Area Network
XOR	eXclusive OR

Compilation of References

Abley, J. (2005). *Anycast addressing in IPv6*. IETF Internet draft. Retrieved April 29, 2010, from http://tools.ietf.org/html/draft-jabley-v6-anycast-clarify-00

Aboba, B., et al. (2004, June). *RFC 3748 extensible authentication protocol*. Retrieved from http://www.ietf.org/rfc/rfc3758.txt

Abramson, N. (1985). Development of the ALOHANET. *IEEE Transactions on Information Theory*, *31*, 119–123. doi:10.1109/TIT.1985.1057021

Adams, A., Nicholas, J., & Siadak, W. (2005). *Protocol independent multicast - dense mode (PIM-DM): Protocol specification* (revised). RFC3973. Retrieved April 29, 2010, from http://tools.ietf.org/html/rfc3973

Ahmed, T., Mehaoua, A., Boutaba, R., & Iraqi, Y. (2005). Adaptive packet video streaming over IP networks: A cross layer approach. *IEEE Journal on Selected Areas in Communications*, *23*(2), 385–401. doi:10.1109/JSAC.2004.839425

Akan, O. B., & Akyildiz, I. F. (2004). ARC: The analytical rate control scheme for real-time traffic in wireless networks. *IEEE/ACM Transactions on Networking*, *12*(4), 634–644. doi:10.1109/TNET.2004.833155

Akkaya, K., & Younis, M. (2005). A survey on routing protocols for wireless sensor networks. *Elsevier Ad Hoc Network Journal*, *3*(3), 325–349. doi:10.1016/j.adhoc.2003.09.010

Al Ja'afreh, R., Mellor, J., & Awan, I. (2009). *A comparison between the tunneling process and mapping schemes for IPv4/IPv6 transition*. International Conference on Advanced Information Networking and Applications Workshops, 2009. WAINA '09 (pp. 601-606). Bradford.

Albrightson, R., Garcia-Luna-Aceves, J. J., & Boyle, J. (1994). EIGRP - a fast routing protocol based on distance vectors. In *Proceedings of Networld/Interop 94*. Berlin, Germany. Retrieved April 29, 2010, from http://www.cse.ucsc.edu/research/ccrg/publications/interop94.pdf

Alcatel-Lucent., et al. (2010, February 11). *Beyond FMC interworking – UDC to FMC*. 3GPP/BBF FMC Workshop, February 18-19, San Francisco, USA.

AlJa'afreh, R., Mellor, J., Kamala, M., & Kasasbeh, B. (2008). *Bi-directional mapping system as a new IPv4/IPv6 translation mechanism*. Tenth International Conference on Computer Modeling and Simulation, 2008. UKSIM 2008 (pp. 40-45). Cambridge, UK.

Alwayn, V. (2001). *Advanced MPLS design and implementation*. Indianapolis, IN: Cisco Press.

An, H., Luo, W., Li, X., Zhang, X., & Yan, B. (2009). *A new IPv6 tunneling protocol: Escort*. International Symposium on Computer Network and Multimedia Technology, 2009. CNMT 2009 (pp. 1-6). Wuhan.

Anastasi, G., Conti, M., Gregori, E., & Passarella, A. (2004, June). A performance study of power-saving polices for Wi-Fi hotspots. *Computer Networks: The International Journal of Computer and Telecommunications Networking*, *45*(3), 295–318.

Andersson, L., & Asati, R. (2009). *Multiprotocol label switching (MPLS) label stack entry: "EXP" field renamed to traffic class field*. RFC 5462. Retrieved May 6, 2010, from http://www.ietf.org/rfc/rfc5462.txt

Andersson, L., Doolan, P., Feldman, N., Fredette, A., & Thomas, B. (2001). *LDP specification*. Retrieved May 6, 2010, from http://www.ietf.org/rfc/rfc3036.txt

Andreasen, F., & Foster, B. (2003). *RFC 3435, media gateway control protocol*. MGCP.

Anemone, I. S. T. (n.d.). Retrieved from http://www.ist-anemone.eu

Anttalainen, T. (2003). *Introduction to telecommunications network engineering*. Artech House.

Aoul, Y. H., Negru, D., Nafaa, A., & Mehaoua, A. (2003). M3G: A mobile multicast multimedia gateway for seamless IPv4/IPv6 transition. *Lecture Notes in Computer Science*, 2839.

Aoun, C., & Davies, E. (2007, July). *Reasons to move the network address translator - protocol translator (NAT-PT) to historic status*. (IETF RFC 4966).

Aroor, S. R., & Deavours, D. D. (2007). Evaluation of the state of passive UHF RFID: An experimental approach. *Systems Journal*, *1*(2), 168–176. doi:10.1109/JSYST.2007.909179

Awduche, D., Berger, L., Gan, D., Li, T., Srinivasan, V., & Swallow, G. (2001). *RSVP-TE: Extensions to RSVP for LSP tunnels*. RFC 3209. Retrieved May 6, 2010, from http://www.ietf.org/rfc/rfc3209.txt

Aydin, Z. G., Chaouchi, H., & Zaim, A. H. (2009). *eHIP: Early update for host identity protocol*. 6th International Conference on Mobile Technology, Application & Systems. Nice, France.

Balakrishnan, H., Padmanabhan, V. N., Seshan, S., & Katz, R. H. (1997). A comparison of mechanisms for improving TCP performance over wireless links. *IEEE/ACM Transactions on Networking*, *5*(6), 756–769. doi:10.1109/90.650137

Bang, N. D., Jorstad, I., & Jonvik, T. (2007). *Analysis of IMS business requirements*. Retrieved from http://www.icin.biz/files/programmes/Session6A-2.pdf

Barakat, C., Altman, E., & Dabbous, W. (2000). On TCP performance in a heterogeneous network: A survey. *IEEE Communications Magazine*, *38*(1), 40–46. doi:10.1109/35.815451

Barrett, C., Marathe, A., Marathe, M. V., & Drozda, M. (2002). Characterizing the interaction between routing and MAC protocols in ad-hoc networks. In *Proceedings of the 3rd ACM International Symposium on Mobile Ad Hoc Networking and Computing* (MobiHoc'02), (pp. 92-103). Lausanne, Switzerland.

Bates, T., Chandra, R., Katz, D., & Rekhter, Y. (2007). *Multiprotocol extensions for BGP-4*. RFC4760. Retrieved April 29, 2010, from http://tools.ietf.org/html/rfc4760

Baumgarten, B., & Giessler, A. (1994). *OSI conformance testing methodology and TTCN*. Elsevier Science Ltd.

Bellman, R. E. (1957). *Dynamic programming*. Princeton, NJ: Princeton University Press.

Bi, J., Leng, X., & Wu, J. (2007). Scalable IPv4/IPv6 transition: A peer-to-peer based approach. *Lecture Notes in Computer Science*, 406–416. doi:10.1007/978-3-540-71661-7_39

Bokor, L., Dudas, I., Szabo, S., & Imre, S. (2005) Anycast-based micromobility: A new solution for micromobility management in IPv6. In *Proceedings of MoMM 2005* (pp. 68–75.). Malaysia, Kuala Lumpur, (2005.09.17–21). ISBN: 3-85403-195-5

Bokor, L., Nováczki, S., & Imre, S. (2007). *A complete HIP based framework for secure micromobility*. 5th @WAS International Conference on Advances in Mobile Computing and Multimedia, MoMM2007. (pp. 111-122). Jakarta, Indonesia, 3-5 December 2007. ISBN 978-3-85403-230-4

Bokor, L., Nováczki, S., Zeke, T. L., & Jeney, G. (2009). Design and evaluation of host identity protocol (HIP) simulation framework for INET/OMNeT++. In the *Proceedings of the 12-th ACM International Conference on Modeling, Analysis and Simulation of Wireless and Mobile Systems* (MSWIM 2009), Tenerife, Canary Islands, Spain, 26-30 October 2009.

Bokor, L., Simon, V., Dudás, I., & Imre, S. (2007). Anycast subnet optimization for efficient IPv6 mobility management. In *Proceedings of IEEE GIIS 2007* (pp. 187–190). Marrakesh, Morocco, (2007.07.02–06). DOI: 10.1109/GIIS.2007.4404188, ISBN 978-1-4244-1376-8

Bokor, L., Zeke, L. T., Nováczki, S., & Jeney, G. (2009). *Protocol design and analysis of a HIP-based per-application mobility management platform.* 7th ACM International Symposium on Mobility Management and Wireless Access (MobiWAC 2009), (pp. 7-16). Tenerife, Canary Islands, Spain.

Bound, J., Toutain, L., & Richier, J. (2005, October). *Dual stack IPv6 dominant transition mechanism* (DSTM). IETF Internet-Draft. (draft-bound-dstm-exp-04).

Bouras, C., Gkamas, A., Kapoulas, V., Papapanagiotou, V., Stamos, K., & Zaoudis, G. (2009, September). *Video transmission over TFRC using cross-layer power management.* 17th International Conference on Software Telecommunications and Computer Nerworks, Hvar, Croatia.

Braden, R. (Ed.). (1989). *Requirements for Internet hosts – communication layers.* IETF. RFC 1122.

Braden, R., et al. (1997). *Resource reservation protocol (rsvp) – version 1 functional specification.* (RFC 2205/2750).

Bryan, D., Matthews, P., Shim, E., Willis, D., & Dawkins, S. (2008, July 7). *Concepts and terminology for peer to peer SIP.* IETF Internet Draft. (draft-ietf-p2psip-concepts-02).

Buchanan, W. (1999). *Mastering networks. Houndmills, Basingstoke, Hampshire & London.* UK: MacMillan Press Ltd.

Budkowski, S., Cavalli, A., & Najm, E. (1998). *Formal description techniques and protocol specification, testing and verification.* Boston, MA/ Dordrecht, Germany/ London, UK: Kluwer Academic Publishers.

Burger, E. (2006). *A mechanism for content indirection in session initiation protocol (SIP) messages. RFC 4483.* Internet Engineering Task Force.

Burza, M., Kang, J., & Van der Stock, P. (2007). Adaptive streaming of MPEG-based audio/video content over wireless networks. [Academy Publisher.]. *Journal of Multimedia, 2*(2), 17–27. doi:10.4304/jmm.2.2.17-27

Cain, B., Deering, S., Kouvelas, I., Fenner, B., & Thyagarajan, A. (2002). *Internet group management protocol, version 3.* IETF. RFC 3376.

Cain, B., et al. (2002, October). *RFC 3376 Internet group management protocol.* Retrieved from http://www.ietf.org/rfc/rfc3376.txt

Calhoun, P., et al. (2003, September). *RFC 3588 diameter base protocol.* Retrieved from http://www.ietf.org/rfc/rfc3588.txt

Calhoun, P., et al. (2009, March). *RFC 5415 control and authentication of wireless access points (CAPWAP) protocol specification.* Retrieved from http://www.ietf.org/rfc/rfc5415.txt

Callon, R. (1990). *Use of OSI IS-IS for routing in TCP/IP and dual environments.* RFC1195. Retrieved April 29, 2010, from http://tools.ietf.org/html/rfc1195

Camarillo, G., & García-Martín, M.-A. (2006). *The 3G IP multimedia subsystem (IMS): Merging the Internet and the cellular worlds.* John Wiley and Sons.

Camarillo, G. (2002). *SIP demystified.* USA: McGraw-Hill Companies, Inc.

Camarillo, G., Mas, I., & Nikander, P. (2008). *A framework to combine the session initiation protocol and the host identity protocol.* Wireless Communications and Networking Conference (WCNC 2008. IEEE). Las Vegas, USA.

Carpenter, B., & Jung, C. (1999, March). *Transmission of IPv6 over IPv4 domains without explicit tunnels.* (IETF RFC 2529).

Carpenter, B., & Moore, K. (2001, February). *Connection of IPv6 domains via IPv4 clouds.* (IETF RFC 3056).

Casner, S., & Jacobson, V. (1999). *Compressing IP/UDP/RTP headers for low-speed serial links.* (IETF RFC 2508).

Cavanaugh, J. D. (1994). *Protocol overhead in IP/ATM networks.* Retrieved May 6, 2010, from http://citeseerx.ist.psu.edu/viewdoc/download?doi=10.1.1.45.2564&rep=rep1&type=pdf

Chawla, V., & Dong Sam, H. (2007). An overview of passive RFID. *Communications Magazine, 45*(9), 11–17. doi:10.1109/MCOM.2007.4342873

Chen, W.-E., Lin, Y.-B., & Pang, A.-C. (2005). An IPv4-IPv6 translation mechanism for SIP overlay network in UMTS all-IP environment. *IEEE Journal on Selected Areas in Communications*, *23*(11), 2152–2160. doi:10.1109/JSAC.2005.856836

Chen, W.-E., Su, C.-Y., & Lin, Y.-B. (2005). NCTU SLT: A socket-layer translator for IPv4-IPv6 translation. *IEEE Communications Letters*, *9*(10), 865–867. doi:10.1109/LCOMM.2005.10001

Chen, J. C., & Zhang, T. (2004). *IP-based next-generation wireless networks*. John Wiley & Sons. doi:10.1002/0471478253

Chen, H., & Huang, C.-W. (2004). *Power management modeling and optimal policy for IEEE 802.11 WLAN systems*. IEEE Vehicular Technology Conference 2004.

Chen, J., Lv, T., & Zheng, H. (2004, May). *Cross-layer design for QoS wireless communications*. 2004 International Symposium Circuits and Systems, ISCAS '04, 23-26 May 2004 Volume 2, (pp. 217-20).

Cheng-Wei, L., & Sun, Y. (2008). HiMIP-NEMO: Combining cross-layer network mobility management and resource allocation for fast QoS-handovers. *IEEE Proceedings,* (2282-2286).

Chiba, M., et al. (2003, July). *RFC 3576 dynamic authorization extensions to remote authentication dial in user service* (RADIUS). Retrieved from http://www.ietf.org/rfc/rfc3576.txt

Choi, L.-U., Kellerer, W., & Steinbach, E. (2004). Cross-layer optimization for wireless multi-user video streaming. In *Proceedings of the International Conference on Image Processing* (ICIP'04), *3*(3), 2047-2050.

Chown, T. (2006, June). *Use of VLANs for IPv4-IPv6 coexistence in enterprise networks*. (IETF RFC 4554).

Clark, D. D., & Tennenhouse, D. L. (1990). Architectural considerations for a new generation of protocols. *ACM SIGCOMM Computer Communication Review*, *20*(4), 200–208. doi:10.1145/99517.99553

Clark, D. D., Partridge, C., Ramming, J. C., & Wroclawski, J. T. (2003). A knowledge plane for the Internet. In *Proceedings of the 2003 Conference on Applications, Architectures, and Protocols for Computer Communication*, (pp. 3-10). Karlsruhe, Germany.

Coltun, R., Ferguson, D., Moy, J., & Lindem, A. (2008). *OSPF for IPv6*. RFC5340. Retrieved April 29, 2010, from http://tools.ietf.org/html/rfc5340

Combs, G., et al. (2010). *Wireshark: A network protocol analyzer*. Retrieved from http://www.wireshark.org/

Conta, A., & Deering, S. (1998, December). *Generic packet tunneling in IPv6 specification*. (IETF RFC 2473).

Conta, A., Deering, S., & Gupta, M. (2006). *Internet control message protocol (ICMPv6) for the Internet protocol version 6 (IPv6) specification*. RFC4443. Retrieved April 29, 2010, from http://tools.ietf.org/html/rfc4443

Cooper, E., Johnston, A., & Matthews, P. (2007, December). *A distributed transport function in P2PSIP using HIP for multi-hop overlay routing*. IETF Internet Draft. (draft-matthews-p2psip-hip-hop-00).

Csopaki, G., Dibuz, S., & Tarnay, K. (1999). *Testing of communicating systems: Methods and applications*. Dordrecht, The Netherlands: Kluwer Academic Publishers.

Cuevas, A., Moreno, J., Vidales, P., & Einsiedler, H. (2006). The IMS service platform: A solution for next-generation network operators to be more than bit pipes. *IEEE Communications Magazine*, *44*(8), 75–81. doi:10.1109/MCOM.2006.1678113

Davie, B. S. (2000). *MPLS: Technology and applications*. San Francisco, CA: Morgan Kaufmann.

Davies, D. W., Barber, D. L. A., Price, W. L., & Solomonides, C. M. (1982). *Computer networks and their protocols*. John Wiley and Sons Ltd.

Dawson, F. (1998). *It's IP vs. ATM in all-out-war*. Retrieved May 6, 2010 from http://www.cedmagazine.com/its-ip-vs-atm-in-all-out-war.aspx

De Ghein, L. (2007). *MPLS fundamentals: Forwarding labeled packets*. Indianapolis, IN: Cisco Press.

Deering, S. (1991). *ICMP router discovery messages*. (RFC 1256).

Deering, S., & Hinden, R. (1998, December). *Internet protocol, version 6 (IPv6) specification*. (IETF RFC 2460).

Deering, S., & Hinden, R. (1995). *Internet protocol, version 6 (IPv6) specification*. RFC1883. Retrieved April 29, 2010, from http://tools.ietf.org/html/rfc1883

Deering, S., & Hinden, R. (1998). *Internet protocol, version 6 (IPv6) specification*. RFC2460. Retrieved April 29, 2010, from http://tools.ietf.org/html/rfc2460

Devarapalli, V., et al. (2005). *Network mobility (NEMO) basic support protocol*. (IETF, RFC 3963).

Dietz, T., Brunner, M., Papadoglou, N., Raptis, V., & Kypris, K. (2005). *Issues of HIP in an operators networks*. (draft-dietz-hip-operator-issues-00).

Dijkstra, E. W. (1959). A note on two problems in connexion with graphs. *Numerical Mathematic*, *1*, 269–271. doi:10.1007/BF01386390

Diot, C., Dabbous, W., & Crowcroft, J. (1997). Multipoint communication: A survey of protocols, functions, and mechanisms. *IEEE Journal on Selected Areas in Communications*, *15*(3), 277–290. doi:10.1109/49.564128

D-ITG. (2009). *Official website*. Retrieved from http://www.grid.unina.it/software/ITG/

Doi, S., & Miyahara, H. (2004). *Design, implementation and evaluation of routing protocols for IPv6 anycast communication*. Master's thesis, Department of Information Networking, Graduate School of Information Science and Technology Osaka University, Osaka, Japan.

Doyle, J. (2008). *Understanding the role of FECs in MPLS*. Retrieved March 22, 2010, from http://www.networkworld.com/community/node/24093

Droms, R. (1997, March). *RFC 2131 dynamic host configuration protocol*. Retrieved from http://www.ietf.org/rfc/rfc2131.txt

Dubuisson, O. (2000). *ASN.1 communication between heterogeneous systems*. Morgen Kaufmann.

Dudas, I., Bokor, L., & Imre, S. (2007). Survey and extension of applications and services in IPv6 anycasting . In Taniar, D. (Ed.), *Encyclopedia of mobile computing & commerce*. New York, NY: Idea Group Inc.

Durand, A., Fasano, P., Guardini, I., & Lento, D. (2001, January). *IPv6 tunnel broker*. (IETF RFC 3053).

Eddy, W. M. (2004, October). At what layer does mobility belong? *IEEE Communications Magazine*, 155–159. doi:10.1109/MCOM.2004.1341274

Ellsberger, J., Hogrefe, D., & Sarma, A. (1997). *SDL: Formal object-oriented language for communicating systems*. Englewood Cliffs, NJ: Prentice Hall.

Elson, J., Girod, L., & Estrin, D. (2002). Fine-grained network time synchronization using reference broadcasts. *SIGOPS Operating Systems Review, 36(*SI), 147-163.

Estrada, L., Torres, D., & Toral, H. (2009). *Variance error for finite-length self-similar time series*. Paper presented at the 7th International Conference on Computing, Communications and Control Technologies (CCCT) - In the Context of The 2nd International Multi-Conference on Engineering and Technological Innovation (IMETI), Orlando, Florida, USA.

Estrin, D., et al. (1998, June). *RFC 2362 protocol independent multicast-sparse mode (PIM-SM): Protocol specification*. Retrieved from http://www.ietf.org/rfc/rfc2362.txt

ETSI. (1998). *11). GSM 01.02. General description of a GSM public land mobile network*. France: Sophia Antipolis.

ETSI. (1999). *04). TS 100.524 (GSM 03.04). Signalling requirements relating to routing of calls*. France: Sophia Antipolis.

ETSI. (2006, February). *Technical report 180 000 telecommunications and Internet converged services and protocols for advanced networking (TISPAN)- NGN terminology*. Retrieved from http://www.tech-invite.com/Ti-etsi-standards-TISPAN.html#tr-180-000

Ettikan, K., & Abdullah, R. (2002). *An analysis of anycast architecture*. In Communications and Computer Networks Conference 2002. Cambridge, MA: ACTA Press.

Fairhurst, G., & Wood, L. (2002). *Advice to link designers on link automatic repeat request* (ARQ). (IETF RFC 3366).

Fallon, S., Jacob, P., Qiao, Y., Murphy, L., Fallon, E., & Hanley, A. (2008). *SCTP switchover performance issues in WLAN environments*. 5th IEEE Consumer Communications and Networking Conference (CCNC'08), (pp. 564-568).

Farinacci, D., Fuller, V., Meyer, D., & Lewis, D. (2010, April 25). *Locator/ID separation protocol* (LISP). IETF Internet Draft. (draft-ietf-lisp-07).

Farrel, A., Ayyangar, A., & Vasseur, J. P. (2008). *Inter-domain MPLS and GMPLS traffic engineering - resource reservation protocol-traffic engineering (RSVP-TE) extensions*. RFC 5151. Retrieved May 6, 2010, from http://www.ietf.org/rfc/rfc5151.txt

Fekete, G. (2009). *Policy based flow management with the host identity protocol for multihomed hosts*. First International Conference on Emerging Network Intelligence. Sliema, Malta.

Fenner, R., Handley, M., Holbrook, H., & Kouvelas, I. (2006). *Protocol independent multicast - sparse mode (PIM-SM): Protocol specification* (revised). RFC4601. Retrieved April 29, 2010, from http://tools.ietf.org/html/rfc4601

Fenner, W. (1997). *Internet group management protocol, version 2*. RFC2236. Retrieved April 29, 2010, from http://tools.ietf.org/html/rfc2236

Finkenzeller, K. (2003). *RFID handbook*. Munich, Germany: John Wiley & Sons. doi:10.1002/0470868023

Floyd, S. (1994). TCP and explicit congestion notification. *ACM SIGCOMM Computer Communication Review*, *4*(5), 8–23. doi:10.1145/205511.205512

Floyd, S., & Kempf, J. (2004, March). *IAB concerns regarding congestion control for voice traffic in the Internet*. (IETF RFC 3714).

Floyd, S., Handley, M., Padhye, J., & Widmer, J. (2000). Equation-based congestion control for unicast applications. In *Proceedings of ACM SIGCOMM'00*, (pp. 43-56).

Floyd, S., Kohler, E., & Padhye J. (2006, March). *Profile for datagram congestion control protocol (DCCP) congestion control ID 3: TCP-friendly rate control* (TFRC). (RFC 4342).

Ford, L. R. Jr, & Fulkerson, D. R. (1962). *Flows in networks*. Princeton, NJ: Princeton University Press.

Fuller, V., Li, T., Yu, J., & Varadhan, K. (1993). *Classless inter-domain routing (CIDR): An address assignment and aggregation strategy*. RFC1519. Retrieved April 29, 2010, from http://tools.ietf.org/html/rfc1519

Ganeriwal, S., Kumar, R., & Srivastava, M. B. (2003). Timing-sync protocol for sensor networks. In *Proceedings of the 1st international Conference on Embedded Networked Sensor Systems SenSys '03* (pp. 138-149).

Garcia-Luna-Aceves, J. J. (1993). Loop-free routing using diffusing computations. *IEEE/ACM Transactions On Networking*, *1*(1), 130–141. Retrieved April 29, 2010, from http://www.cse.ucsc.edu/research/ccrg/publications/jj.dual.ton93.pdf

Garcia-Martin, M. (2006). *A session initiation protocol (SIP) event package and data format for various settings in support for the push-to-talk over cellular (PoC) service. RFC 4354*. Internet Engineering Task Force.

Gilligan, R., & Nordmark, E. (2000, August). *Transition mechanisms for IPv6 hosts and routers*. (IETF RFC 2893).

Goldsmith, A., & Wicker, S. (2002). Design challenges for energy-constrained ad hoc wireless networks. *IEEE Wireless Communications Magazine*, *9*(4), 8–27. doi:10.1109/MWC.2002.1028874

Govil, J. (2007). *IPv6: Mobility management and roaming between IPv6 and IPv4*. International Conference on Convergence Information Technology, 2007 (pp. 1553-1558). Gyeongju.

Guichard, J., Pepelnjak, I., & Apcar, J. (2003). *MPLS and VPN architectures* (*Vol. II*). Indianapolis, IN: Cisco Press.

Gundavelli, S., et al. (2008, August). *Proxy mobile IPv6*. Retrieved from http://www.ietf.org/rfc/rfc5213.txt

Gurtov, A., Korzun, D., Lukyanenko, A., & Nikander, P. (2008). Hi3: An efficient and secure networking architecture for mobile hosts. *Computer Communications*, *38*, 2457–2467. doi:10.1016/j.comcom.2008.03.014

Ha, J. S., Kim, S. T., & Koh, S. J. (2005). *Performance comparison of SCTP and TCP over Linux platform. Lecture Notes in Computer Science, 3645/2005* (pp. 396–404). Berlin/Heidelberg, Germany: Springer.

Haas, Z. J., Halpern, J. Y., & Li, L. (2006). Gossip-based ad hoc routing. *IEEE/ACM Transactions on Networking*, *14*(3), 479–491. doi:10.1109/TNET.2006.876186

Hagino, J., & Yamamoto, K. (2001, June). *An IPv6-to-IPv4 transport relay translator*. (IETF RFC 3142).

Handley, M., Floyd, S., Pahdye, J., & Widmer, J. (2003, January). *TCP friendly rate control (TFRC): Protocol specification.* (IETF RFC 3448).

Handley, M., Jacobson, V., & Perkins, C. (2006). *SDP: Session description protocol.* IETF. RFC 4566.

Harai, H. (2009). *AKARI architecture design for new generation network.* Summer Topical Meeting, 2009. LEOSST '09., (pp. 155 - 156). Newport Beach, CA.

Harrowell, A., & Writer, S. (October 2006). *A pointless multimedia subsystem.* Mobile Communications International.

Haukass, T. (2007). *Rate adaptive video streaming over wireless networks explained & explored.* MSc thesis, Norwegian University of Science and Technology, Department of Telematics.

Hautakorpi, J., Camarillo, G., & Koskela, J. (2007, November). *Utilizing HIP (Host Identity Protocol) for P2PSIP (Peerto- peer Session Initiation Protocol).* IETF Internet Draft. (draft-hautakorpi-p2psip-with-hip-01).

He, T., Stankovic, J. A., Lu, C., & Abdelzaher, T. (2003). SPEED: A stateless protocol for real-time communication in sensor networks. In *Proceedings of the 23rd International Conference on Distributed Computing Systems ICDCS* (pp. 46-55).

Hedrick, C. (1988). *Routing information protocol.* RFC1058. Retrieved April 29, 2010, from http://tools.ietf.org/html/rfc1058

Hei, Y., & Yamazaki, K. (2004). *Traffic analysis and worldwide operation of open 6to4 relays for IPv6 deployment.* International Symposium on Applications and the Internet, (pp. 265-268).

Heidrich, J., Brenk, D., Essel, J., Schwarzer, S., Seemann, K., Fischer, G., & Weigel, R. (2010). The roots, rules, and rise of RFID. *Microwave Magazine, 11*(3), 78–86. doi:10.1109/MMM.2010.936075

Heikkinen, S. (2008). Security and accounting enhancements for roaming in IMS. 6th International Conference on Wired/Wireless Internet Communications (WWIC 2008). [Tampere, Finland/ Berlin, Germany: Springer.]. *Lecture Notes in Computer Science, 5031*, 127–138. doi:10.1007/978-3-540-68807-5_11

Heine, G. (2002). *GPRS signaling & protocol analysis.* Karlsruhe, Germay: INACON.

Heinzelman, W. R., Chandrakasan, A., & Balakrishnan, H. (2000). Energy-efficient communication protocol for wireless micro sensor networks. In *Proc. 33rd Hawaii International Conference on System Sciences, vol. 8* (pp. 3005-3014).

Held, A., & Ziegert, T. (1999). Service mobility - a new feature of mobile computing. *Integrated Computer-Aided Engineering, 6*(2), 131–142.

Helsinki University of Technology. (2006). *Mobile IPv6 for Linux website.*

Henderson, T. (2004). *Can SIP use HIP.* HIP Workshop, 61st IETF meeting. Retrieved April 15, 2010, from http://hiprg.piuha.net/workshop/

Herborn, S., Haslett, L., Boreli, R., & Seneviratne, A. (2006). *HarMoNy - HIP mobile networks.* 63rd IEEE Vehicular Technology Conference, VTC.

Herborn, S., Huber, A., Boreli, R., & Seneviratne, A. (2007). *A scheme for host identity delegation.* IEEE/Create-Net/ICST International Conference on COMmunication System softWAre and MiddlewaRE (COMSWARE). Bangalore, India.

Hinden, R., & Deering, S. (1998). *IP version 6 addressing architecture.* RFC2373. Retrieved April 29, 2010, from http://tools.ietf.org/html/rfc2373

Hirorai, R., & Yoshifuji, H. (2006). *Problems on IPv4-IPv6 network transition.* International Symposium on Applications and the Internet Workshops, 2006. SAINT Workshops 2006 (pp. 37-42). Phoenix, AZ.

Hoare, C. A. R. (1985). *Communicating sequential processes.* Prentice Hall.

Hobaya, F., Gay, V., & Robert, E. (2009). *Host identity protocol extension supporting simultaneous end-host mobility.* Fifth International Conference on Wireless and Mobile Communications (ICWMC '09), (pp. 261-266). Cannes/La Bocca, France.

Hoffman, P., & Schaad, J. (2010). *New ASN.1 modules for PKIX.* IETF Internet-Draft.

Hogrefe, D., Koch, B., & Neukirchen, H. (2001). *Some implementation of MSC, SDL and TTCN time extensions for computer-aided test generation*. In SDL 2001: Meeting UML (pp. 168-181). Heidelberg, Germany: Springer.

Holzmann, G. J., & Pehrson, B. (2003). *The early history of data networks*. Hoboken, NJ: John Wiley & Sons Inc.

Holzmann, G. J. (1991). *Design and validation of computer protocols*. Murray Hill, NJ: Prentice Hall.

Hopps, C. (2008). *Routing IPv6 with IS-IS*. RFC5308. Retrieved April 29, 2010, from http://tools.ietf.org/html/rfc5308

Huang, S.-M., Wu, Q., & Lin, Y.-B. (2006). Enhancing Teredo IPv6 tunneling to traverse the symmetric NAT. *IEEE Communications Letters*, *10*(5), 408–410. doi:10.1109/LCOMM.2006.1633339

Huang, L.-W., Pan, J.-Y., Yu, K.-S., & Chiao, C. (2005). *Hierarchical mobility management in IP4/IP6 co-existed networks*. International Conference on Wireless Networks, Communications and Mobile Computing, 2005, 2, (pp. 1023-1028).

Huitema, C. (2001, June). *An anycast prefix for 6to4 relay routers*. (IETF RFC 3068).

Huitema, C. (2006, February). *Teredo: Tunneling IPv6 over UDP through network address translations* (NATs). (IETF RFC 4380).

Hwang, J., Kim, N., Kang, S., & Koh, J. (2008). *A framework for IMS interworking networks with quality of service guarantee*. DC: IEEE Computer Society Washington.

Iapichino, G., & Bonnet, C. (2009). *Host identity protocol and proxy mobile IPv6: A secure global and localized mobility management scheme for multihomed mobile nodes*. IEEE Global Telecommunications Conference (GLOBECOM'09), (pp. 1-6). Honolulu, Hawaii.

Iapichino, G., Bonnet, C., Herrero, O. d., Baudoin, C., & Buret, I. (2009). *Combining mobility and heterogeneous networking for emergency management: A PMIPv6 and HIP-based approach*. 2009 International Conference on Wireless Communications and Mobile Computing: Connecting the World Wirelessly, (pp. 603-607). Leipzig, Germany.

IEEE. (2004). *IEEE 802.1D local and metropolitan area networks: Media access control (MAC) bridges*. Retrieved from http://standards.ieee.org/getieee802/download/802.1D-2004.pdf

IEEE. (2004). *IEEE 802.1X local and metropolitan area networks: Port-based network access control*. Retrieved from http://standards.ieee.org/getieee802/download/802.1X-2004.pdf

IEEE. (2005). *IEEE 802.1Q local and metropolitan area networks: Virtual bridged local area networks*. Retrieved from http://standards.ieee.org/getieee802/download/802.1Q-2005.pdf

IEEE. (2007). *IEEE 802-2001 local and metropolitan area networks: Overview and architecture*. Retrieved from http://standards.ieee.org/getieee802/download/802-2001.pdf

IEEE. (2008). *IEEE 802.3-2008 local and metropolitan area networks: Carrier sense multiple access with collision detection (CSMA/CD) access method and physical layer specifications*. Retrieved from http://standards.ieee.org/getieee802/download/802.3-2008_section1.pdf

Ilyas, M., & Ahson, S. (2008). *IP multimedia subsystem (IMS) handbook*. CRC Press.

Infonetics Research. (2008). *Service providers lean toward MPLS-TP over PBB-TE in new packet transport survey*. Retrieved May 6, 2010, from http://www.infonetics.com/pr/2008/pots.survey.2008.nr.asp

Intanagonwiwat, C., Govindan, R., Estrin, D., Heidemann, J., & Silva, F. (2003). Directed diffusion for wireless sensor networking. *IEEE/ACM Transactions on Networking*, *11*(1), 2–16. doi:10.1109/TNET.2002.808417

International Telecommunication Union. (1994/07). ITU-T recommendation X.200: ***I**nformation technology - open systems interconnection - basic reference model: The basic model.

International Telecommunication Union. (1997). *Information Technology – open systems interconnection – common management information protocol: Specification*. Retrieved April 27, 2010, from http://www.itu.int/rec/T-REC-X.711-199710-I/en

International Telecommunication Union. (2000). *Series M: TMN and network maintenance: International transmission systems, telephone circuits, telegraphy, facsimile, and leased circuits*. Retrieved April 26, 2010, from http://www.itu.int/rec/t-rec-m.3400-200002-I/en

ISI. (2009). *ns2 network simulator*. Retrieved on December 23, 2009, from http://www.isi.edu/nsnam/ns/

ISO/IEC FDIS 10536-1. (2000). *Identification cards – contactless integrated circuit(s) cards – close-coupled cards – part 1: Physical characteristics*. Geneva, Switzerland: International Electrotechnical Commission, International Organization for Standardization.

ISO/IEC FDIS 10536-2. (1995). *Identification cards – contactless integrated circuit(s) cards – part 2: Dimensions and location of coupling areas*. Geneva, Switzerland: International Electrotechnical Commission, International Organization for Standardization.

ISO/IEC FDIS 11785. (1996). *Radio frequency identification of animals – technical concept*. Geneva, Switzerland: International Electrotechnical Commission, International Organization for Standardization.

ISO/IEC FDIS 14443-1. (2008). *Identification cards – contactless integrated circuit(s) cards – proximity cards – part 1: Physical characteristics*. Geneva, Switzerland: International Electrotechnical Commission, International Organization for Standardization.

ISO/IEC FDIS 14443-2. (2001). *Identification cards – contactless integrated circuit(s) cards – proximity cards – part 2: Radio frequency power and signal*. Geneva, Switzerland: International Electrotechnical Commission, International Organization for Standardization.

ISO/IEC FDIS 14443-3. (2001). *Identification cards – contactless integrated circuit(s) cards – proximity cards – part 3: Initialization and anticollision*. Geneva, Switzerland: International Electrotechnical Commission, International Organization for Standardization.

ISO/IEC FDIS 14443-4. (2008). *Identification cards – contactless integrated circuit(s) cards – proximity cards – part 4: Transmission protocol*. Geneva, Switzerland: International Electrotechnical Commission, International Organization for Standardization.

ISO/IEC FDIS 15693-1. (2000). *Identification cards – contactless integrated circuit(s) cards – vicinity cards – part 1: Physical characteristics*. Geneva, Switzerland: International Electrotechnical Commission, International Organization for Standardization.

ISO/IEC FDIS 15693-2. (2006). *Identification cards – contactless integrated circuit cards – vicinity cards – part 2: Air interface and initialization*. Geneva, Switzerland: International Electrotechnical Commission, International Organization for Standardization.

ISO/IEC FDIS 15693-3. (2009). *Identification cards – contactless integrated circuit cards – vicinity cards – part 3: Anticollision and transmission protocol*. Geneva, Switzerland: International Electrotechnical Commission, International Organization for Standardization.

ISO/IEC FDIS 18000-1. (2008). *Information Technology – radio frequency identification for item management – part 1: Reference architecture and definition of parameters to be standardized*. Geneva, Switzerland: International Electrotechnical Commission, International Organization for Standardization.

ISO/IEC FDIS 18000-2. (2009). *Information Technology – radio frequency identification for item management – part 2: Parameters for air interface communications below 135 kHz*. Geneva, Switzerland: International Electrotechnical Commission, International Organization for Standardization.

ISO/IEC FDIS 18000-3. (2008). *Information Technology – radio frequency identification for item management – part 3: Parameters for air interface communications at 13.56 MHz*. Geneva, Switzerland: International Electrotechnical Commission, International Organization for Standardization.

ISO/IEC FDIS 18000-4. (2008). *Information Technology – radio frequency identification for item management – part 4: Reference parameters for air interface communications at 2.45 GHz*. Geneva, Switzerland: International Electrotechnical Commission, International Organization for Standardization.

ISO/IEC FDIS 18000-6. (2004). *Information Technology – radio frequency identification for item management – part 6: Parameters for air interface communications at 860 MHz to 960 MHz*. Geneva, Switzerland: International Electrotechnical Commission, International Organization for Standardization.

ISO/IEC FDIS 18000-7. (2009). *Information technology – radio frequency identification for item management – part 7: Parameters for active air interface communications at 433 MHz*. Geneva, Switzerland: International Electrotechnical Commission, International Organization for Standardization.

ITU-T Recommendation G.114. (2003). *One-way transmission time*. Geneva, Switzerland: International Telecommunications Union.

ITU-T Recommendation G.1050. (2005). *Network model for evaluating multimedia transmission performance over Internet protocol*. Geneva, Switzerland: International Telecommunications Union.

ITU-T Y.2001. (2004). *General overview of NGN*. Retrieved from http://www.itu.int/rec/T-REC-Y.2001/en

ITU-T. (1988). *Recommendation Q.702. Signalling data link*. Geneva, Switzerland.

ITU-T. (2001). *Recommendations Q.711-Q.714. Specifications of signalling system no. 7 – signalling connection control part*. Geneva, Switzerland.

ITU-T. (2002). *Recommendation H.248.1: Media gateway control protocol version 2*.

ITU-T. (2004). *Recommendation Y.2001 general overview of NGN*. Retrieved from http://www.itu.int/rec/T-REC-Y.2001/en

ITU-T. (2004). *Recommendation Y.2011 general principles and general reference model for next generation networks*. Retrieved from http://www.itu.int/rec/T-REC-Y.2011/en

Jankiewicz, E., Green, D., & Fiuczynski, M. E. (2006). *IPv6 translation for IPv4 embedded systems*. 2006 IEEE Sarnoff Symposium, (pp. 1-4). Princeton, NJ.

Janneteau, C. et al. (2004). *IPv6 multicast for mobile networks with MLDProxy*. (IETF Internet Draft, draft-janneteau-nemo-multicast-mldproxy-00.txt).

Jaskó, S. (2005). A new mathematical formalism for the TTCN 3 core language. *Periodica Polytechnica Series . Electrical Engineering*, *49*(3), 1–12.

Jayanthi, J., & Rabara, S. (2010). *IPv6 addressing architecture in IPv4 network*. Second International Conference on Communication Software and Networks, 2010. ICCSN '10 (pp. 461-465). Singapore.

Jeong, S., Shin, M.-K., & Kim, H.-J. (2007). *A handover scheme supporting IPv4-IPv6 traversal over network-based mobility management domains*. The 9th International Conference on Advanced Communication Technology, (pp. 1033-1036). Gangwon-Do.

Jiang, H., Zhuang, W., & Shen, X. (2005). Cross-layer design for resource allocation in 3G wireless networks and beyond. *IEEE Communications Magazine*, *43*(12), 120–126. doi:10.1109/MCOM.2005.1561929

Jiang, R., Gupta, V., & Ravishankar, C. V. (2003). Interactions between TCP and the IEEE 802.11 MAC protocol. In [DISCEX]. *Proceedings of DARPA Information Survivability Conference and Exposition*, *1*, 273–282. doi:10.1109/DISCEX.2003.1194891

Jiang, W., & Schulzrinne, H. (2000). Modeling of packet loss and delay and their effect on real-time multimedia service quality. In *Proceedings of the 10th International Workshop on Network and Operating System Support for Digital Audio and Video* (NOSSDAV'00), Chapel Hill, NC, USA.

Johnson, D., Perkins, C., & Arkko, J. (2004). *Mobility support in IPv6*. RFC 3775. Retrieved April 29, 2010, from http://tools.ietf.org/html/rfc3775

Jokela, P., Melen, J., & Ylitalo, J. (2006, June). *HIP service discovery*. IETF Internet Draft. (draft-jokela-hip-service-discovery-00).

Jokela, P., Moskowitz, R., & Nikander, P. (2008, April). *Using the encapsulating security payload (ESP) transport format with the host identity protocol* (HIP). (RFC 5202).

Jones, C. E., Sivalingam, K. M., Agrawal, P., & Chen, J. C. (2001, August). A survey of energy efficient network protocols for wireless networks. *Wireless Networks*, *7*(4), 343–358. doi:10.1023/A:1016627727877

Jungmaier, A., & Rathgeb, E. P. (2006). On SCTP multihoming performance. *Telecommunication Systems*, *31*(2-3), 141–161. doi:10.1007/s11235-006-6517-7

Juniper Networks. (2009). *VLAN design for IPTV/multiplay networks.* Retrieved from www.juniper.net/us/en/local/pdf/whitepapers/2000186-en.pdf

Kaffle, V. P., Kamioka, E., & Yamada, S. (2006). CoMoRoHo: Cooperative mobile router based handover scheme for long-vehicular multihomed networks. *IEICE Transactions on Communications, 89*(10).

Kaliski, B. S., Jr. (1993). *A layman's guide to a subset of ASN.1, BER, and DER.* (NIST/OSI Implementors' Workshop document SEC-SIG-91-17).

Kamel, G., Mihailovic, & Aghvami, A. H. (2008). Case analysis of a cost-optimal QoS aggregation policy for network mobility. *IEEE Communications Letters*, *12*(2), 130–132. doi:10.1109/LCOMM.2008.071662

Kamel, G., Mihailovic, A., Pangalos, P., & Aghvami, A. H. (2007). *Cost-optimal QoS aggregation for network mobility* (pp. 5006–5010). IEEE Globecom.

Karasaridis, A., & Hatzinakos, D. (2001). Network heavy traffic modeling using α-stable self-similar processes. *IEEE Transactions on Communications*, *49*(7), 1203–1214. doi:10.1109/26.935161

Kawadia, V., & Kumar, P. R. (2005). A cautionary perspective on cross-layer design. *IEEE Wireless Communications Magazine*, *12*(1), 3–11. doi:10.1109/MWC.2005.1404568

Kempf, J., Khalil, M., & Pentlan, B. (2005). *IPv6 fast router advertisement.* (draft-mkhalil-ipv6-fastra-05.txt), January 2005.

Kent, S., & Atkinson, R. (1998). *Security architecture for the Internet protocol.* (RFC 2401).

Kent, S., & Seo, K. (2005, December). *RFC 4301 security architecture for the Internet protocol.* Retrieved from http://www.ietf.org/rfc/rfc4301

Khlifi, H. (2008). *IMS application servers: Roles, requirements and implementation technologies.* IEEE Internet Computing.

Kim, I.-M., & Kim, H.-M. (2003, January). An optimum power management scheme for wireless video service in CDMA systems. *IEEE Transactions on Wireless Communications*, *2*(1).

Kim, B.-J. (2001). A network service providing wireless channel information for adaptive mobile applications. In *Proceedings of the IEEE International Conference on Communications* (ICC), (pp. 1345-1351). Helsinki, Finland.

Kim, S. T., Koh, S. J., & Kim, Y. J. (2007). *Performance of SCTP for IPTV applications.* 9th International Conference on Advanced Communication Technology, vol. 3. (pp. 2176-2180). Gangwon-Do, South Korea: IEEE.

Kitamura, H. (2001, April). *A SOCKS-based IPv6/IPv4 gateway mechanism.* (IETF RFC 3089).

Klair, K., Chin, K., & Raad, R. (2008). *A survey and tutorial of RFID anti-collision protocols. Communications Surveys & Tutorials.* IEEE.

Kliazovich, D., Granelli, F., & Da Fonseca, N. L. S. (2009). Architecture and cross-layer design for cognitive networks . In *Handbook on sensor networks.* River Edge: World Scientific Publishing Co. Inc.

Kohler, E., Handley, M., & Floyd, S. (2006, March). *Datagram congestion control protocol* (DCCP). (IETF RFC 4340).

König, H. (1990). *Kommunikationsprotokolle.* Berlin, Germany: Akademie-Verlag.

König, H. (2003). *Protocol Engineering Prinzip, Beschreibung und Entwicklung von Kommunikationsprotokollen.* Wiesbaden, Germany: B.G. Teubner Verlag.

Koodli, R. (2005). *Fast handovers for mobile IPv6.* (IETF RFC 4068).

Kopetz, H., & Schwabl, W. (1989). *Global time in distributed real-time systems.* (Technical Report 15/89), Technische Universitat Wien.

Kosonen, V. (1999). *QoS and frame relay.* Retrieved May 6, 2010, from http://www.netlab.tkk.fi/~vkosonen/vesan_esitelma_1999.pdf

Kovácsházi, Z., & Vida, R. (2007). *Host identity specific multicast*. International Conference on Networking and Services (ICNS '07). Athens, Greece.

Kulik, J., Heinzelman, W., & Balakrishnan, H. (2002). Negotiation-based protocols for disseminating information in wireless sensor networks. *Wireless Networks*, *8*(2/3), 169–185. doi:10.1023/A:1013715909417

Kumar, S., Raghavan, V. S., & Deng, J. (2006). Medium access control protocols for ad hoc wireless networks: A survey. *Elsevier Ad-Hoc Network Journal*, *4*(3), 326–358. doi:10.1016/j.adhoc.2004.10.001

Kumwilaisak, W., Thomas Hou, Y., Zhang, Q., Zhu, W., Jay Kuo, C.-C., & Zhang, Y.-Q. (2003). A cross-layer quality-of-service mapping architecture for video delivery in wireless networks. *IEEE Journal on Selected Areas in Communications*, *21*(10), 1685–1698. doi:10.1109/JSAC.2003.816445

Kumwilaisak, W., & Jay Kuo, C.-C. (2002). Adaptive variable length Markov chain for non-stationary fading channel. In *Proceedings of IEEE GLOBECOM'02, vol. 3*, (pp. 2046-2050).

Kusy, B., Dutta, P., Levis, P., Maroti, M., Ledeczi, A., & Culler, D. (2006). Elapsed time on arrival: A simple and versatile primitive for canonical time synchronization services. *International Journal of Ad Hoc and Ubiquitous Computing*, *1*(4), 239–251.

Kwon, T. T., et al. (2002, October). *Mobility management for VoIP service: Mobile IP vs. SIP. IEEE wireless communications.*

Laganier, J., & Eggert, L. (2008). *Host identity protocol (HIP) rendezvous extension*. (RFC 5204).

Laganier, J., Koponen, T., & Eggert, L. (2008). *Host identity protocol (HIP) registration extension*. (RFC 5203).

Lai, K., & Baker, M. (1999). Measuring bandwidth. In *Proceedings of INFOCOM'99, vol 1*, (pp. 235-245). New York, USA.

Lan, K., et al. (2005). *MOBNET: The design and implementation of a network mobility testbed for NEMO protocol*. 14th IEEE Workshop on Local and Metropolitan Area Networks.

Larmouth, J. (1999). *ASN.1 complete*. Morgen Kaufmann.

Larzon, L. A., Degermark, M., & Pink, S. (2004). *The lightweight user datagram protocol* (UDP-Lite). RFC 3828, Internet Engineering Task Force, July 2004.

Lee, I. (2009). *Handbook of research on telecommunications planning and management for business*. Hershey, PA & New York, NY: IGI Global.

Lee, S. W. (2005). *A multiple access algorithm for passive RFID tags*. Unpublished thesis, School of Electrical and Electronic Engineering, College of Engineering, Yonsei University, Seoul, South Korea.

Lee, S., & Chung, K. (November, 2006). TCP-friendly rate control scheme based on RTP. *Information Networking. Advances in Data Communications and Wireless Networks.*

Lee, S., Shin, M.-K., Kim, Y.-J., Nordmark, E., & Durand, A. (2002, October). *Dual stack hosts using "bump-in-the-API"* (BIA). (IETF RFC 3338).

Lee, S.-D., Shin, M.-K., & Kim, H.-J. (2006). *The implementation of ISATAP router*. 8th International Conference on Advanced Communication Technology (ICACT'06), (p. 1163). Phoenix Park, Republic of Korea.

Leech, M., Ganis, M., Lee, Y., Kuris, R., Koblas, D., & Jones, L. (1996, March). *SOCKS protocol version 5*. (IETF RFC 1928).

Leland, W. E., Taqqu, M. S., Willinger, W., & Wilson, D. V. (1994). On the self-similar nature of Ethernet traffic (extended version). *IEEE/ACM Transactions on Networking*, *2*(1), 1–15. doi:10.1109/90.282603

Leyva, L. (2004). *Alliance FXO/FXS/E1 VoIP system*. Retrieved from http://www.cts-design.com

Li, W. (2001). Overview of fine granularity scalability in MPEG-4 video standard. *IEEE Transactions on Circuits and Systems for Video Technology*, *11*(3), 301–317. doi:10.1109/76.911157

Li, X., Wu, X., Li, W., & Wang, X. (2008). *An adaptive cross-layer scheduling algorithm for multimedia networks*. 4th International Conference on Intelligent Information Hiding and Multimedia Signal Processing, IIH-MSP 2008, art. no. 4604006, (pp. 52-55).

Li, Y., Chen, W., Su, L., Jin, D., & Zeng, L. (2009). *Mobility management architecture based on integrated HIP and SIP protocols*. International Conference on Telecommunications, ICT '09, (pp. 243-247). Marrakech, Morocco.

Liao, H., & Zhu, W. (2003). Enable fair queueing with decoupled bandwidth-delay guarantees by relaxing the packet service order. In *Proceedings of IEEE International Conference on Communications (ICC'03), vol. 1*, (pp. 147-151). Anchorage, AK, USA.

Liao, H., Zhang, Q., Zhu, W., & Zhang, Y.-Q. (2001). *TCP-aware robust header compression* (TAROC). (IETF Internet Draft, ROHC WG).

Lie, A., & Klaue, J. (2008, June). Evalvid-RA: Trace driven simulation of rate adaptive MPEG-4 VBR video. [Berlin/Heidelberg, Germany: Springer.]. *Multimedia Systems*, *14*(1), 33–50. doi:10.1007/s00530-007-0110-0

Lin, X., Shroff, N. B., & Srikant, R. (2006, August). A tutorial on cross-layer optimization in wireless networks. *Selected Areas in Communications . IEEE Journal*, *24*(8), 1452–1463.

Linn, R. J., & Umit Uyar, M. (1994). *Conformance testing methodologies and architecture for OSI protocols* (pp. 93–152). IEEE Computer Society Press.

Liu, T., Guan, X., Zheng, Q., & Qu, Y. (2009). A new worm exploiting IPv6 and IPv4-IPv6 dual-stack networks: Experiment, modeling, simulation, and defense. *IEEE Network*, *23*(5), 22–29. doi:10.1109/MNET.2009.5274918

Liu, Q., Wang, X., & Giannakis, G. B. (2006). A cross-layer scheduling algorithm with QoS support in wireless networks. *IEEE Transactions on Vehicular Technology*, *55*(3), 839–847. doi:10.1109/TVT.2006.873832

Liu, H., Jia, X., Wan, P., Liu, X., & Yao, F. (2007). A distributed and efficient flooding scheme using 1-Hop information in mobile ad hoc networks. *IEEE Transactions on Parallel and Distributed Systems*, *18*(5), 658–671. doi:10.1109/TPDS.2007.1023

Liu, M., Wu, X., Jin, M., & Li, D. (2005, July). *Tunneling IPv6 with private IPv4 addresses through NAT devices*. IETF Internet-Draft. (draft-liumin-v6ops-silkroad-03).

Liu, X., Cao, X., & Li, H. (2008). *An IPv4-IPv6 translation mechanism for IMS network*. 4th International Conference on Wireless Communications, Networking and Mobile Computing, 2008. WiCOM '08 (pp. 1-4). Dalian.

Ludwig, R., & Meyer, M. (2003). *The Eifel detection on algorithm for TCP*. (IETF RFC3522).

Luo, H., Qin, Y., & Zhang, H. (2009). A DHT-based identifier-to-locator mapping approach for a scalable Internet. *IEEE Transactions on Parallel and Distributed Systems*, *20*, 1790–1802. doi:10.1109/TPDS.2009.30

Maekawa, K., & Okabe, Y. (2009). *An enhanced location privacy framework with mobility using host identity protocol*. Ninth Annual International Symposium on Applications and the Internet (SAINT '09), (pp. 23 - 29). Bellevue (WA).

Malik, M. A., Kanhere, S. S., Hassan, M., & Benatallah, B. (2004). On-board RSVP: An extension of RSVP to support real-time services in on-board IP networks. *Lecture Notes in Computer Science*, *3326*, 264–275. doi:10.1007/978-3-540-30536-1_31

Malkin, G. (1993). *RIP version 2, carrying additional information*. RFC1388. Retrieved April 29, 2010, from http://tools.ietf.org/html/rfc1388

Malkin, G. (1998). *RIP version 2*. RFC2453. Retrieved April 29, 2010, from http://tools.ietf.org/html/rfc2453

Malkin, G., & Minnear, R. (1997). *RIPng for IPv6*. RFC2080. Retrieved April 29, 2010, from http://tools.ietf.org/html/rfc2080

Mani, M., & Crespi, N. (2008). How IMS enables converged services for cable and 3G technologies: A survey. *EURASIP Journal on Wireless Communications and Networking*, *8*(3), 1–14. doi:10.1155/2008/589623

Manjeshwar, A., & Agrawal, D. P. (2001). TEEN: A routing protocol for enhanced efficiency in wireless sensor networks. In *Proceedings of the 15th International Parallel & Distributed Processing Symposium* (pp. 2009–2015).

Manjeshwar, A., & Agrawal, D. P. (2002) APTEEN: A hybrid protocol for efficient routing and comprehensive information retrieval in wireless sensor networks. In *Proceedings of the 2nd International Workshop on Parallel and Distributed Computing Issues in Wireless Networks and Mobile computing* (pp. 195–202).

Mao, S., Lin, S., Panwar, S. S., Wang, Y., & Celebi, E. (2003). Video transport over ad hoc networks: Multistream coding with multipath transport. *IEEE Journal on Selected Areas in Communications*, *21*(10), 1721–1737. doi:10.1109/JSAC.2003.815965

Maróti, M., Kusy, B., Simon, G., & Lédeczi, Á. (2004). The flooding time synchronization protocol. In Proceedings of the 2nd international Conference on Embedded Networked Sensor Systems, SenSys '04. (pp. 39-49).

Marques, P., & Dupont, F. (1999). *Use of BGP-4 multiprotocol extensions for IPv6 inter-domain routing*. RFC2545. Retrieved April 29, 2010, from http://tools.ietf.org/html/rfc2545

Matos, A., Santos, J., Sargento, S., Aguiar, R., Girao, J., & Liebsch, M. (2006). *HIP location privacy framework*. First ACM/IEEE International Workshop on Mobility in the Evolving Internet Architecture, (pp. 57–62). New York, NY: ACM Press.

Matsunaga, S., Ata, S., Kitamura, H., & Murata, M. (2005). Design and implementation of IPv6 anycast routing protocol: PIA-SM. *Proceedings of the 19th International Conference on Advanced Information Networking and Applications* (AINA'05) (pp. 839–844). Taipei, Taiwan. ISBN 0-7695-2249-1

McCarthy, B., Edwards, C., & Dunmore, M. (2009). *Using NEMO to support the global reachability of MANET nodes*. IEEE INFOCOM'09.

Melander, B., et al. (2007). *FMC support in fixed access architecture, technical report*. Retrieved from http://www.ist-muse.org/Deliverables/TF1/MUSE_DTF1.8_v1.0.pdf

Melen, J., Ylitalo, J., Salmela, P., & Henderson, T. (2009, May 26). *Host identity protocol-based mobile router* (HIPMR). IETF Internet Draft. (draft-melen-hip-mr-02).

Melsen, T., & Blake, S. (2006, June). *RFC 4562 MAC-forced forwarding: A method for subscriber separation on an Ethernet access network*. Retrieved from http://www.ietf.org/rfc/rfc4562.txt

Menth, M., Hartmann, M., & Hoefling, M. (2010, May). *Mapping systems for Loc/ID split Internet routing*. (Technical Report No. 472). University of Würzburg.

Menth, M., Hartmann, M., & Klein, D. (2010, April). *Global locator, local locator, and identifier split* (GLI-Split). (Technical Report No. 470). University of Würzburg.

Metro Ethernet Forum. (2006). *Metro Ethernet services – a technical overview*. Retrieved from http://metroethernetforum.org/PDF_Documents/metro-ethernet-services.pdf

Mickelsson, H., et al. (2006 December 11-14). *Nomadism/FMC use cases and AAA impact*. Geneva, Switzerland: BroadBand Europe.

Minei, I., & Lucek, J. (2008). *MPLS-enabled applications: Emerging developments and new technologies*. Chichester, UK: Wiley and Sons. doi:10.1002/9780470751725

Minoli, D. (2008). *IP multicast with applications to IPTV and DVB-H*. Hoboken, NJ: John Wiley & Sons. doi:10.1002/9780470260876

Molta, D. (2005, December). Wi-Fi and the need for more power. *Network Computing*.

Moore, N. (2005). *Optimistic duplicate address detection for IPv6*. (draft-ietfipv6-optimistic-dad-05.txt), February 2005.

Moskowitz, R., & Nikander, P. (2006, May). *Host identity protocol (HIP) architecture*. (IETF RFC 4423).

Moskowitz, R., Nikander, P., Jokela, P., & Henderson, T. (2008, April). *Host identity protocol*. (RFC 5201).

Motorola. (2008). *LTE inter-technology mobility – enabling mobility between LTE and other access technologies*. Retrieved from http://www.motorola.com/staticfiles/Business/Solutions/Industry%20Solutions/Service%20Providers/Network%20Operators/LTE/_Document/Static%20Files/InterTech%20Mobility%20White%20Paper.pdf

Muslam, M., Chan, H., & Ventura, N. (2009). *HIP based micro-mobility management optimization*. Fifth International Conference on Wireless and Mobile Communications, ICWMC '09., (pp. 291–295). Cannes, La Bocca.

Na, J., et al. (2004). *Route optimization scheme based on path control header*. (IETF Internet Draft, draft-na-nemo-path-control-header-00.txt).

Naoe, H., Wetterwald,M. & Bonnet, C. (2007). *IPv6 soft handover applied to network mobility on heterogeneous access networks*. IEEE PIMRC, September 2007.

Narten, T., Nordmark, E., & Thomson, W. (1998). *Neighbor discovery for IP version 6* (IPv6). (IETF RFC 2461).

Narten, T., Nordmark, E., Simpson, W., & Soliman, H. (2007). *Neighbor discovery for IP version 6* (IPv6). RFC4861. Retrieved April 29, 2010, from http://tools.ietf.org/html/rfc4861

Nautilus6. (n.d.). Retrieved from http://www.nautilus6.org

Newman, P., Edwards, W. L., Hinden, R., Hoffman, E., Ching Liaw, F., Lyon, T., & Minshall, G. (1996). *Transmission of flow labelled IPv4 on ATM data links*. RFC1954. Retrieved May 6, 2010, from http://www.apps.ietf.org/rfc/rfc1954.html

Ng, C., et al. (2007). *Network mobility route optimization solution space analysis*. (RFC 4889). Ng, C., Thubert, P., Ohnishi, H., & Paik, E. (2005). *Taxonomy of route optimization models in the NEMO context*. (IETF Internet Draft, draft-thubert-nemo-ro-taxonomy-04.txt).

Nikander, P., & Arkko, J. (2004). Delegation of signaling rights. (LNCS 2845), (pp. 203-214). Security Protocols 2002.

Nikander, P., & Laganier, J. (2008, April). *Host identity protocol domain name system extension*. (RFC 5205).

Nikander, P., Henderson, T., Vogt, C., & Arkko, J. (2008). *End-host mobility and multihoming with the host identity protocol*. (RFC 5206).

Nolan, J. P. (2006). *Stable MathLink package*. Retrieved from www.robustanalysis.com

Nordmark, E. (2000, February). *Stateless IP/ICMP translation algorithm* (SIIT). (IETF RFC 2765).

Nordmark, E., & Bagnulo, M. (2009). *Shim6: Level 3 multihoming Shim protocol for IPv6*. (RFC 5533).

Nordmark, E., & Gilligan, R. (2005, October). *Basic transition mechanisms for IPv6 hosts and routers*. (IETF RFC 4213).

Nováczki, S., Bokor, L., & Imre, S. (2006). Micromobility support in HIP: Survey and extension of host identity protocol. [Málaga, Spain.]. *MELECON*, *1*, 651–654.

Nováczki, S., Bokor, L., & Imre, S. (2007). A HIP based network mobility protocol. [Hiroshima, Japan.]. *SAINT-WONEMO*, *2007*, 48–52.

Nováczki, S., Bokor, L., Jeney, G., & Imre, S. (2008). Design and evaluation of a novel HIP-based network mobility orotocol. *Journal of Networks*, *3*(1), 10–24. doi:10.4304/jnw.3.1.10-24

O'Connell, J. (2007). Service delivery within an IMS environment. *IEEE Vehicular Technology Magazine*, *2*(1), 12–19. doi:10.1109/MVT.2007.898099

O'Driscoll, G. (2008). *Next generation IPTV services and technologies*. Hoboken, NJ: John Wiley & Sons.

Ohnishi, H., Sakitani, K., & Takagi, Y. (2003). *HMIP based route optimization method in a mobile network*. (IETF Internet Draft, draft-ohnishi-nemo-ro-hmip-00.txt).

OMA. (n.d.). *Overview*. Retrieved from http://www.openmobilealliance.org

Österdahl, H. (2005). *A comparison of TCP and SCTP performance using the HTTP protocol*.

Palet, J., Olvera, C., & Fernandez, D. (2003, October). *Forwarding protocol 41 in NAT boxes*. IETF Internet Draft. (draft-palet-v6ops-proto41-nat-03).

Pan, D. (1995). A tutorial on MPEG/audio compression. *IEEE Transactions on Multimedia*, *2*(2), 60–74. doi:10.1109/93.388209

Park, S., & Jeong, S.-H. (2009). Mobile IPTV – approaches, challenges, standards and QoS support. *IEEE Internet Computing Magazine*, *3*(13), 23–31. doi:10.1109/MIC.2009.65

Park, K., & Willinger, W. (2000). *Self-similar network traffic and performance evaluation*. New York, NY: John Wiley & Sons, Inc.doi:10.1002/047120644X

Park, K. I. (2005). *QoS in packet networks*. Boston, MA: Springer Science + Business Media, Inc.

Park, S., Haw, R., & Hong, C. S. (2010). *System architecture for IPTV seamless services in mobility*. International Conference on Consumer Electronics (IEEE-ICCE 2010) (pp. 59-60). Las Vegas, NV: IEEE.

Pelletier, G., & Sandlund, K. (2008). *Robust header compression, version 2: Profiles for RTP, UDP, IP, ESP and UDP-lite*. (IETF RFC 5225).

Peng, W., Zhou, Y., Wang, C., & Yang, Y. (2009). *Research on IPSec-based NAT-PT transition mechanism*. IEEE International Conference on Network Infrastructure and Digital Content, 2009. IC-NIDC 2009 (pp. 222-226). Beijing.

Perkins, C. (1997). Mobile IP. *IEEE Communications Magazine*, *35*(5), 84–99. doi:10.1109/35.592101

Perkins, C. (Ed.). (2002). *IP mobility support for IPv4*. (IETF RFC 3344).

Perkins, C., Belding-Royer, E., & Das, S. (2003). *Ad hoc on-demand distance vector (Aodv) routing*. RFC3561. RFC Editor. Retrieved February 3, 2010, from http://www.rfc-archive.org/getrfc.php?rfc=3561

Perros, H. G. (2005). *Connection-oriented networks*. Indianapolis, IN: Wiley. doi:10.1002/0470016361

Petander, H., Perera, E., & Seneviratne, A. (2005). *Multiple interface handoffs: A practical method for access technology independent make-before-break handoffs*. NICTA Technical Report.

Petrescu, A., Olivereau, A., Janneteau, C., & Lach, H-Y. (2004). *Threats for basic network mobility support* (NEMO threats). (IETF Internet Draft,draft-petrescunemo-threats-01.txt), January 2004.

Phillips, T., Karygiannis, T., & Kuhn, R. (2005). Security standards for the RFID market. *Security & Privacy*, *3*(6), 85–89. doi:10.1109/MSP.2005.157

Pierrel, S., Jokela, P., & Melen, J. (2006, June 19). *Simultaneous multi-access extension to the host identity protocol*. IETF Internet-Draft. (draft-pierrel-hip-sima-00).

Pierrel, S., Jokela, P., Melen, J., & Slavov, K. (2007). *A policy system for simultaneous multiaccess with host identity protocol*. IEEE ACNM2007. Munich, Germany.

Poikselka, M., Niemi, A., Khartabil, H., & Mayer, G. (2006). *The IMS: IP multimedia concepts and services*. John Wiley and Sons. doi:10.1002/0470031840

Postel, J. (1980). *User datagram protocol*. RFC 768, Internet Engineering Task Force, August 1980.

Postel, J. (1981). Transmission control protocol. RFC 792, Internet Engineering Task Force, September 1981.

Preradovic, S., Karmakar, N., & Balbin, I. (2008). RFID transponders. *Microwave Magazine*, *9*(5), 90–103. doi:10.1109/MMM.2008.927637

Qian, L., Jones, D. L., Ramachandran, K., & Appadwedula, S. (1999). A general joint source-channel matching method for wireless video transmission. In *Proceedings of IEEE Data Compression Conference* (DCC'99), (pp. 414-423). Snowbird, Utah, USA.

Raake, A. (2006). Short- and long-term packet loss behavior: Towards speech quality prediction for arbitrary loss distributions. *IEEE Transactions on Audio, Speech, and Language Processing*, *14*(6), 957–1968. doi:10.1109/TASL.2006.883231

Radunovic, B. (2005, July). A cross-layer design of wireless ad-hoc networks. Ph.D. Thesis.

Ramakrishnan, K., Floyd, S., & Black D. (2001, September). *The addition of explicit congestion notification (ECN) to IP*. (IETF RFC 3168).

Ramakrishnan, K., Floyd, S., & Black, D. (2001). *The addition of explicit congestion notification (ECN) to IP*. (IETF RFC 3168).

Raman, S., Balakrishnan, H., & Srinivasan, M. (2002). An image transport protocol for the Internet. *IEEE/ACM Transactions on Networking*, *10*(3), 297–307. doi:10.1109/TNET.2002.1012362

Ramani, I., & Savage, S. (2005). *Syncscan: Practical fast handoff for 802.11 infrastructure networks*. IEEE INFOCOM'05.

Ramírez, J., & Torres, D. (2005). *A tool for analysis of Internet metrics*. Paper presented at the 2nd International Conference on Electrical and Electronics Engineering (ICEEE) and XI Conference on Electrical Engineering (CIE), Mexico, D.F.

Ranky, P. G. (2007). Engineering management-focused radio frequency identification (RFID) model solutions. *Engineering Management Review*, *35*(2), 20–30.

Ratola, M. (2004). *Which layer for mobility? Comparing mobile IPv6, HIP and SCTP.* Helsinki University of Technology, Telecommunications Software and Multimedia Laboratory 2004.

Ravier, T., Brennan, R., & Curran, T. (2001). *Experimental studies of SCTP multi-homing.* First Joint IEI/IEE Symposium on Telecommunications Systems Research, 2001.

Rekhter, Y., Li, T., & Hares, S. (2006). *A border gateway protocol 4* (BGP-4). RFC4271. Retrieved April 29, 2010, from http://tools.ietf.org/html/rfc4271

Rekhter, Y., Moskowitz, B., Karrenberg, D., de Groot, G. J., & Lear, E. (1996). *Address allocation for private Internets.* RFC 1918. Retrieved May 6, 2010, from http://www.ietf.org/rfc/rfc1918.txt

Rieback, M. R., Crispo, B., & Tanenbaum, A. S. (2006). The evolution of RFID security. *Pervasive Computing, 5*(1), 62–69. doi:10.1109/MPRV.2006.17

Rigney, C., et al. (1997, April). *RFC 2138 remote authentication dial in user service.* Retrieved from http://www.ietf.org/rfc/rfc2138.txt

Rigney, C., Willens, S., Rubens, A., & Simpson, W. (2000). *Remote authentication dial in user service* (RADIUS). (IETF RFC 2865), June 2000.

Rizo, L., Torres, D., Muñoz, D., & Vargas, C. (2010). Jitter in IP networks: A cauchy approach. *IEEE Communications Letters, 14*(2), 190–193. doi:10.1109/LCOMM.2010.02.090702

Rosen, E., & Rekhter, Y. (2006). *BGP/MPLS IP virtual private networks* (VPNs). RFC 4364. Retrieved May 6, 2010, from http://www.ietf.org/rfc/rfc4364.txt

Rosen, E., Viswanathan, A., & Callon, R. (2001). *RFC 3031 MPLS architecture.* Retrieved from http://www.ietf.org/rfc/rfc3031.txt

Rosenberg, J., et al. (2002, June). *RFC 3261 SIP: Session initiation protocol.* Retrieved from http://www.ietf.org/rfc/rfc3261.txt

Rothenberg, C. E., Wong, W., Verdi, F. L., & Magalhae, M. F. (2008). *SIP over an identifier/locator splitted next generation Internet architecture.* 10th IEEE International Conference on Advanced Communication Technology (ICACT08). Republic of Korea.

Sabir, M., Fahiem, M., & Mian, M. (2009). *An overview of IPv4 to IPv6 transition and security issues.* WRI International Conference on Communications and Mobile Computing, 2009. CMC '09, 3, (pp. 636-639). Yunnan.

Sampath, A., Kumar, P. S., & Holtzman, J. M. (1995). Power control and resource management for a multimedia CDMA wireless system. In *Proceedings of the 6th IEEE International Symposium on Personal, Indoor and Mobile Radio Communications* (PIMRC'95), vol. 1, (pp. 21-25). Toronto, Ontario, Canada.

Sanguankotchakorn, T., & Jaiton, P. (2008). *Effect of triangular routing in mixed IPv4/IPv6 networks.* Seventh International Conference on Networking, 2008. ICN 2008 (pp. 357-362). Cancun.

Sarolahti, P., & Floyd, S. (2007). *Cross-layer indications for transport protocols.* (Internet draft: draft-sarolahti-tsvwg-crosslayer-00.txt).

Schaad, J., & Turner, S. (2010). *Additional new ASN.1 modules.* IETF Internet-Draft.

Schiller, J. (2000). *Mobile communications.* Addison Wesley.

Schulzrinne, H. (2003). *RTP: A transport protocol for real-time applications (RFC 3550).* Internet Engineering Task Force.

Schulzrinne, H., Casner, S., Frederick, F., & Jacobson, V. (2003). *RTP: A transport protocol for real-time applications.* IETF. RFC 3550.

Schulzrinne, H., Casner, S., Frederick, R., & Jacobson, V. (2003). *RTP: A transport protocol for real-time applications.* RFC 3550, Internet Engineering Task Force, July 2003.

Schulzrinne, H., et al. (1996, January). *RFC 1889 RTP: A transport protocol for real-time applications.* Retrieved from http://www.ietf.org/rfc/rfc1889.txt

Schulzrinne, H., Rao, A., & Lanphier, R. (1998). *Real-time streaming protocol* (RTSP). IETF. RFC 2326.

Schulzrinne, H., Rao, A., Westerlund, M., Lanphier, R., & Stiemerling, M. (2010). *Real-time streaming protocol 2.0* (RTSP). IETF. (draft-ietf-mmusic-rfc2326bis-23).

Schurgers, C., & Srivastava, M. B. (2001). Energy efficient routing in wireless sensor networks. In *Proc. IEEE Military Comm. Conference, vol. 1* (pp. 357-361).

Sen, S., Rexford, J., & Towsley, D. (1999). Proxy prefix caching for multimedia streams. In *Proceedings of the 18th IEEE INFOCOM'99, vol. 3*, (pp. 1310-1319). New York, NY, USA.

Service delivery framework (SDF). (n.d.). *Telemanagement forum*. Retrieved from http://www.tmforum.org/TechnicalPrograms/ServiceDeliveryFramework

Shakkottai, S., Rappaport, T. S., & Karlsson, P. C. (2003). Cross-layer design for wireless networks. *IEEE Communications Magazine, 41*(10), 74–80. doi:10.1109/MCOM.2003.1235598

Shan, Y., & Zakhor, A. (2002). *Cross layer techniques for adaptive video streaming over wireless networks*. IEEE International Conference on Multimedia and Expo, volume 1, (pp 277-280).

Sheluhin, O. I., Smolskiy, S. M., & Osin, A. V. (2007). *Self-similar processes in telecommunications*. Chichester, England: John Wiley & Sons Ltd. doi:10.1002/9780470062098

Shi, W., Huang, C., Wang, Q., Chen, Y., Huang, Y., & Cheng, Y. (2007). *A novel IPv4/IPv6 translation mechanism based on NAT-PT*. The 9th International Conference on Advanced Communication Technology, (pp. 1037-1041). Gangwon-Do.

Simpson, W., & Greenfield, H. (2009). *IPTV and Internet video: Expanding the reach of television broadcasting* (2nd ed.). Burlington, MA: Focal Press.

Simpson, W. (1994, July). *RFC 1661 the point-to-point protocol*. Retrieved from http://www.ietf.org/rfc/rfc1661.txt

Simunic, T. (2005). *Power saving techniques for wireless LANs*. Conference on Design, Automation and Test in Europe - volume 3 (pp. 96-97).

SIP. (2008). *Session initiation protocol*. Retrieved from http://www.ietf.org/html.charters/sip-charter.html

So, J. Y., & Wang, J. (2006). *HIP based mobility management for UMTS/WLAN integrated networks*. Australian Telecommunication Networks and Applications Conference.

So, J., & Wang, J. (2008). *Micro-HIP: A HIP-based micro-mobility solution*. IEEE International Conference on Communications Workshops (ICC Workshops '08.), (pp. 430-435). Beijing, China.

Soliman, H. (2007, November). *Mobile IPv6 support for dual stack hosts and routers* (DSMIPv6). (draft-ietf-mip6-nemo-v4traversal-06).

Soliman, H., Catelluccia, C., Malki, K. E., & Bellier, L. (2005). *Hierarchical mobile IPv6 mobility management* (HMIPv6). (IETF RFC 4140).

Sosinsky, B. (2009). *Networking bible*. Indianapolis, IN: Wiley.

Stallings, W. (1998). *SNMP, SNMPv2, SNMPv3 and RMON 1 and 2*. Indianapolis, IN: Addison-Wesley Professional.

Stallings, W. (2002). *Wireless communications and networks*. NJ: Prentice Hall.

Stallings, W. (2006). *Cryptography and network security: Principles and practice*. NJ: Prentice Hall.

Stallings, W. (2007). Signal encoding techniques. In *Data and computer communications* (8th ed., pp. 138–179). Upper Saddle River, NJ: Prentice Hall.

Stewart, R. (2007). *Stream control transmission protocol*. RFC 4960, Internet Engineering Task Force, September 2007.

Stewart, R. (2007, September). *Stream control transmission protocol*. (IETF RFC 4960).

Stewart, R. (Ed.). (2007). *Stream control transmission protocol*. IETF. RFC 4960.

Stoica, I., Adkins, D., Zhuang, S., Shenker, S., & Surana, S. (2002). Internet indirection infrastructure. [Pittsburgh, PA, USA.]. *ACM SIGCOMM, 02*, 73–88. doi:10.1145/964725.633033

Superiori, L., Nemethova, O., & Rupp, M. (2007). Robust error handling for video streaming over mobile networks. In *Proceedings of the 18th Annual IEEE International Symposium on Personal, Indoor and Mobile Radio Communications* (PIMRC'07), (pp. 1-5).

Syputa, R. (2006). *Clash of the Titans – WiMAX and 4G: The battle for convergence is joined.* Maravedis Market Research and Analysis Report.

Taib, A., & Budiarto, R. (2007). *Security mechanisms for the IPv4 to IPv6 transition.* 5th Student Conference on Research and Development, 2007. SCOReD 2007 (pp. 1-6). Selangor, Malaysia.

Tais, M., & Labiod, H. (2005). *Resource reservation for NEMO networks.* International Conference Wireless Networks, Communications and Mobile Computing (pp. 232–237).

Tanenbaum, A. S. (2003). *Computer networks.* Upper Sadle River, NJ: Prentice Hall.

Tarnay, K. (1991). *Protocol specification and testing.* New York, NY & London, UK: Plenum Press.

TelecomAsia. (2008). *Transport wars: MPLS-TP vs PBB-TE.* Retrieved May 6, 2010, from http://www.telecomasia.net/print/1187

Templin, F., Gleeson, T., Talwar, M., & Thaler, D. (2005, October). *Intra-site automatic tunnel addressing protocol* (ISATAP). (IETF RFC 4214).

Tennenhouse, D. L., Smith, J. M., Sincoskie, W. D., Wetherall, D. J., & Minden, G. J. (1997). A survey of active network research. *IEEE Communications Magazine, 35*(1), 80–86. doi:10.1109/35.568214

Teraoka, F., Ishiyama, M., & Kunishi, M. (2003, June 24). *LIN6: A solution to mobility and multi-homing in IPv6.* IETF Internet Draft. (draft-teraoka-ipng-lin6-02).

Tesoriero, R., Gallud, J., Lozano, M., & Penichet, V. (2008). Using active and passive RFID technology to support indoor location-aware systems. *IEEE Transactions on Consumer Electronics, 54*(2), 578–583. doi:10.1109/TCE.2008.4560133

Thomas, B., & Gray, E. (2001). *LDP applicability.* RFC 3037. Retrieved May 6, 2010, from http://www.rfc-archive.org/getrfc.php?rfc=3037&tag=LDP-Applicability

Thomas, R. (2007). *Cognitive networks.* PhD Dissertation. Department of Computer Engineering, Virginia Polytechnic Institute and State University, Blacksburg, Virginia, USA.

Thomson, S., Narten, T., & Jinmei, T. (2007). *IPv6 stateless address autoconfiguration.* RFC4862. Retrieved April 29, 2010, from http://tools.ietf.org/html/rfc4862

Thubert, P., & Molteni, M. (2004). *IPv6 reverse routing header and its application to mobile networks.* (IETF Internet Draft, draft-thubert-nemo-reverse-routing-header-05.txt).

Toral, H., Torres, D., & Estrada, L. (2009). Simulation and modeling of packet loss on VoIP traffic: A power-law model. *WSEAS Transactions on Communications, 8*(10), 1053–1063.

Toral, H., Torres, D., & Estrada, L. (2009). *Simulation and modeling of packet loss on α-Stable VoIP traffic.* Paper presented at the 9th International Conference on Multimedia, Internet & Video Technologies (WSEAS MIV), Budapest, Hungary.

Toral, H., Torres, D., Hernandez, C., & Estrada, L. (2008). *Self-similarity, packet loss, jitter, and packet size: Empirical relationships for VoIP.* Paper presented at the 18th International Conference on Electronics, Communications, and Computers (IEEE CONIELECOM), Puebla, Mexico.

Triantafyllopoulou, D.-K., Passas, N., & Kaloxylos, A. (2007). A cross-layer optimization mechanism for multimedia traffic over IEEE 802.16 networks. *International Journal of Network Management, 17*(5), 347–361. doi:10.1002/nem.668

Tsaoussidis, V., & Badr, H. (2000 November). *TCP-probing: Towards an error control schema with energy and throughput performance gains.* 8th IEEE Conference on Network Protocols, Japan.

Tsirtsis, G., & Srisuresh, P. (2000, February). *Network address translation - protocol translation* (NAT-PT). (IETF RFC 2766).

Tsuchiya, K., Higuchi, H., & Atarashi, Y. (2000, February). *Dual stack hosts using the "bump-in-the-stack" technique* (BIS). (IETF RFC 2767).

Tsukada, M., Mehani, O., & Ernst, T. (2008). *Simultaneous usage of NEMO and MANET for vehicular communication.* 1st Workshop on Experimental Evaluation and Deployment Experiences on Vehicular networks.

Turner, J. S. (1986). New directions in communications (or which way to the information age). *IEEE Communications Magazine*, *24*, 8–15. doi:10.1109/MCOM.1986.1092946

UTStarcom. (2009). *MPLS-TP based packet transport networks: Delivering new value through next-generation transport networks*. Retrieved May 6, 2010 from http://www.utstar.com/Document_Library/CP-WP-MPLS-TP_PTN-1109.pdf

Vakulya, G., & Simon, G. (2009). Percolation driven flooding for energy efficient routing in dense sensor networks. *Journal of Telecommunications and Information Technology*, *3*(2), 5–12.

van der Schaar, M., & Sai Shankar, N. D. (2005, August). Cross-layer wireless multimedia transmission: Challenges, principles and new paradigms. *IEEE Wireless Communications*.

Van Greunen, J., & Rabaey, J. (2003). Lightweight time synchronization for sensor networks. In *Proceedings of the 2nd ACM international Conference on Wireless Sensor Networks and Applications* (pp. 11-19).

Varjonen, S., Komu, M., & Gurtov, A. (2009). *Secure and efficient IPv4/IPv6 handovers using host-based identifier-locator split*. 17th International Conference on Software, Telecommunications and Computer Networks (SoftCOM 2009), (pp. 111 - 115). Split-Hvar-Korcula.

Vida, R., & Costa, L. (Eds.). (2004). *Multicast listener discovery version 2 (MLDv2) for IPv6*. (IETF RFC 3810).

Vieron, J., & Guillemot, C. (2004). Real-time constrained TCP-compatible rate control for video over the Internet. *IEEE Transactions on Multimedia*, *6*(4), 634–646. doi:10.1109/TMM.2004.830805

Wagner, S. (2006, December 1). *Implementation and evaluation of the interaction between host identity protocol and session initiation protocol*. Masterarbeit im Studiengang Angewandte Informatik.

Wakikawa, R., Devarapalli, V., Ernst, T., & Kagami, K. (2007). *Multiple care-of addresses registration*. (IETF Draft, draft-ietf-monami6-multiplecoa-09.txt), August 2008.

Wakikawa, R., Nishida, Y., & Murai, J. (2006). *The use of SCTP failover mechanism for efficient network handover on mobile IPv6*. 3rd International Symposium on Wireless Communication Systems (ISWCS'06) (pp. 133-137).

Wallace, C., & Gardiner, C. (2010). *ASN.1 translation*. IETF Internet-Draft.

Want, R. (2006). An introduction to RFID technology. *Pervasive Computing*, *5*(1), 25–33. doi:10.1109/MPRV.2006.2

Warrier, A., Le, L., & Rhee, I. (September, 2007). *Cross-layer optimization made practical*. 4th International Conference on Broadband Communications, Networks, Systems, BroadNets, art. no. 4550507, (pp. 733-742).

Weingart, T., Sicker, D. C., & Grunwald, D. (2007). Identifying opportunities for exploiting cross-layer interactions in adaptive wireless systems. [Hindawi Publishing Corporation.]. *Journal of Advances in Multimedia*, *1*, 2.

Weis, S. A., Sarma, S. E., Rivest, R. L., & Engels, D. W. (2004). Security and privacy aspects of low-cost radio frequency identification systems. In *Security in pervasive computing* (pp. 50–59). Heidelberg, Germany: Springer. doi:10.1007/978-3-540-39881-3_18

Weiser, M. (1991, September). The computer for the twenty first century. *Scientific American*, 94–104. doi:10.1038/scientificamerican0991-94

Weske, M. (2007). *Business process management*. Heidelberg, Germany: Springer.

White, J. (1989). ASN.1 and ROS: The impact of X.400 on OSI. *IEEE Journal on Selected Areas in Communications*, *7*(7). doi:10.1109/49.44554

Willamowius, J., et al. (2009). *OpenH323 gatekeeper: The GNU gatekeeper*. Retrieved from http://www.gnugk.org/

Willcock, C., Deiss, T., Tobias, S., Keil, S., Engler, F., & Schulz, S. (2005). *An introduction to TTCN-3*. Wiley. doi:10.1002/0470017317

Winter, R., Schiller, J. H., Nikaein, N., & Bonnet, C. (2006). Crosstalk: Cross-layer decision support based on global knowledge. *IEEE Communications Magazine*, *44*(1), 93–99. doi:10.1109/MCOM.2006.1580938

Wu, J., Cui, Y., Li, X., & Metz, C. (2006). The transition to IPv6 part 1 - 4over6 for the China Education and Research Network. *IEEE Internet Computing, 10*(3), 80–85. doi:10.1109/MIC.2006.67

Wu, D., & Negi, R. (2003). Effective capacity: A wireless link model for support of quality of service. *IEEE Transactions on Wireless Communications, 2*(4), 630–643.

Wu, F., Li, S., & Zhang, Y.-Q. (2001). A framework for efficient progressive fine granularity scalable video coding. *IEEE Transactions on Circuits and Systems for Video Technology, 11*(1), 332–344.

Wu, D., & Negi, R. (2004). Effective capacity-based quality of service measures for wireless networks. In *Proceedings of the First International Conference on Broadband Networks*, (pp. 527-536).

Wu, J., Cui, Y., Li, X., Xu, M., & Metz, C. (2010, March). *4over6 transit solution using IP encapsulation and MP-BGP extensions*. (IETF RFC 5747).

Wu, X., Chan, M. C., & Ananda, A. L. (2007). *TCP handoff: A practical TCP enhancement for heterogeneous mobile environments*. IEEE ICC 2007.

Xianfeng, Y., & Zheng, C. (2006). *IPv4-IPv6 multicast transition technologies*. Retrieved April 15, 2010, from http://wwwen.zte.com.cn/endata/magazine/ztecommunications/2006year/no3/articles/200610/t20061010_162407.html

Xiaodong, Z., Yan, M., & Yumei, Z. (2009). *Research on the next-generation Internet transition technology*. Second International Symposium on Computational Intelligence and Design (ISCID '09), (pp. 380-382). Changsha, China.

Xie, J., & Narayanan, U. (2009). Performance analysis of mobility support in IPv4/IPv6 mixed wireless networks. *IEEE Transactions on Vehicular Technology, 59*(2).

Xie, B., & Zeng, W. (2004). Rate distortion optimized dynamic bitstream switching for scalable video streaming. In *Proceedings of IEEE International Conference on Multimedia and Expo* (ICME'04), (pp. 1327-1330). Taipei, Taiwan.

Xu, Z. (2010). *Designing and implementing IP/MPLS -based Ethernet layer 2 VPN services: An advanced guide for VPLS and VLL*. Indianapolis, IN: Wiley and Son.

Xu, C., Fallon, E., Qiao, Y., Muntean, G.-M., Li, X., & Hanley, A. (2009). *Analysis of real-time multimedia transmission over PR-SCTP with failover detection delay and reliability level differential*. International Conference on Communication Software and Networks (IEEE-ICCSN'09). (pp. 768-772). Macau: IEEE.

Xylomenos, G., Polyzos, G. C., Mahonen, P., & Saaranen, M. (2001). TCP performance issues over wireless links. *IEEE Communications Magazine, 39*(4), 52–58. doi:10.1109/35.917504

Y.ipsplit, I.-T. (2009, Feb. 6). *General requirements for ID/locator separation in NGN*. ITU-T draft recommendation.

Y.ipv6split, I.-T. (2009, March 5). *Framework of ID/LOC separation in IPv6-based NGN*. ITU-T draft recommendation.

Yajnik, M., Moon, S., Kursoe, J., & Towsley, D. (1999). *Measurement and modelling of the temporal dependence in packet loss*. Paper presented at the 18th International Conference on Computer Communications (IEEE INFOCOM), New York, NY.

Yang, S., Luo, H., Qin, Y., & Zhang, H. (2009). Design and evaluation of DNS as location manager for HIP. *Journal Wireless Personal Communications, 48*(4), 605–619. doi:10.1007/s11277-008-9550-x

Yang, S., Qin, Y., Luo, H., & Zhang, H. (2008). *P-HIP: Paging extensions for host identity protocol. IEEE Communications Workshops, ICC Workshops '08*. Beijing: Chaina.

Yang, S., Qin, Y., & Yang, D. (2007). Dynamic hierarchical location management scheme for host identity protocol. [*Mobile Ad-Hoc and Sensor Networks Proceedings*.]. *Lecture Notes in Computer Science*, 4864.

Yang, S., Zhou, H., Qin, Y., & Zhang, H. (2009). SHIP: Cross-layer mobility management scheme based on session initiation protocol and host identity protocol. *Journal of Telecommunication Systems, 42*(1-2).

Yang, F., Zhang, Q., Zhu, W., & Zhang, Y.-Q. (2004). End-to-end TCP-friendly streaming protocol and bit allocation for scalable video over wireless Internet. *IEEE Journal on Selected Areas in Communications, 22*(4), 777–790. doi:10.1109/JSAC.2004.826008

Yang, F., Zhang, Q., Zhu, W., & Zhang, Y.-Q. (2001). An efficient transport scheme for multimedia transport over wireless Internet. In *Proceedings of IEEE International Conference on 3G Wireless and Beyond* (3Gwireless ,01), San Francisco, CA, USA.

Yan-ge, C., Shui-mu, T., & Jun-ying, G. (2009). *IPv4/IPv6 intercommunication technology research*. International Conference on Environmental Science and Information Application Technology, 2009. ESIAT 2009 (pp. 626-630). Wuhan.

Yick, J., Mukherjee, B., & Ghosal, D. (2008). Wireless sensor network survey. *Computer Networks*, *52*(12), 2292–2330. doi:10.1016/j.comnet.2008.04.002

Ylitalo, J., & Nikander, P. (2006). BLIND: A complete identity protection framework for end-points. *Lecture Notes in Computer Science*, 3957.

Ylitalo, J. (2005). *Re-thinking security in network mobility*. NDSS Wireless and Security Workshop. San Diego, CA, USA.

Ylitalo, J., & Melé, J. P. S., & Petander, H. (2008). An experimental evaluation of a HIP based network mobility scheme. WWIC 2008, *LNCS, 5031/2008*, (pp. 139–151).

Ylitalo, J., Melén, J., Nikander, P., & Torvinen, V. (2004). Re-thinking security in IP based micro-mobility. *Lecture Notes in Computer Science, 3225/2004*, (pp. 318-329).

Yu, Y., Govindan, R., & Estrin, D. (2001). *Geographical and energy aware routing: A recursive data dissemination protocol for wireless sensor networks*. (UCLA Computer Science Department Technical Report UCLA/CSD-TR-01-0023), May 2001.

Yuan-hang, C., & Lijuan, H. (2009). *Transition from IPv4 to IPv6 based on socket applications*. International Conference on Networking and Digital Society, 2009. ICNDS '09, 2, (pp. 40-43). Guiyang, Guizhou.

Zamora, N. H., Kao, J.-C., & Marculescu, R. (2007, April). *Distributed power-management techniques for wireless network video systems*. Design, Automation & Test in Europe Conference & Exhibition, (pp. 1-7).

Zander, J., Dai, Z. R., Schieferdecker, I., & Din, G. (2005). *From U2TP models to executable tests with TTCN-3 - an approach to model driven testing. Lecture Notes in Computer Science, 3502*. Berlin/Heidelberg, Germany: Springer.

Zhang, Q., Zhu, W., & Zhang, Y.-Q. (2004). Channel-adaptive resource allocation for scalable video transmission over 3G wireless networks. *IEEE Transactions on Circuits and Systems for Video Technology*, *14*(8), 1049–1063. doi:10.1109/TCSVT.2004.831966

Zhang, T., & Xu, Y. (1999). Unequal packet loss protection for layered video transmission. *IEEE Transactions on Broadcasting*, *45*(2), 243–252. doi:10.1109/11.796266

Zhang, C., & Tsaoussidis, V. (2001). *TCP real: Improving real-time capabilities of TCP over heterogeneous networks*. 11th IEEE/ACM NOSSDAV, NY.

Zhu, P., Zeng, W., & Li, C. (2005). Cross-layer QoS support for multimedia delivery over wireless Internet. [Hindawi Publishing Corporation.]. *EURASIP Journal on Applied Signal Processing*, *2*, 207–219.

Zhu, P., Zeng, W., & Li, C. (2007b). Joint design of source rate control and QoS-aware congestion control for video streaming over the Internet. *IEEE Transactions on Multimedia*, *9*(2), 366–376. doi:10.1109/TMM.2006.886284

Zhu, P., Zeng, W., & Li, C. (2007a). Cross-layer design of source rate control and congestion control for wireless video streaming. *Advances in Multimedia*, *1*, 3-15, Article ID 6850, Hindawi Publishing Corporation.

Zhu, Y., Chen, M., Zhang, H., & Li, X. (2008). *Stateless mapping and multiplexing of IPv4 addresses in migration to IPv6 Internet*. IEEE Global Telecommunications Conference, 2008. IEEE GLOBECOM 2008 (pp. 1-5). New Orleans, LO.

Zhuang, Z., Wu, J., & Xia, Q. (2006). *Study & implementation of proxy-based IPv4/IPv6 transition mechanism*. Asia-Pacific Conference on Communications (APCC '06), (pp. 1-5). Busan, South Korea.

Zori, M., Rao, R. R., & Milstein, L. B. (1995). On the accuracy of a first-order Markov model for data transmission on fading channels. In *Proceedings of the 4th IEEE International Conference on Universal Personal Communications* (ICUPC'95), (pp. 211-215). Tokyo, Japan.

About the Contributors

Katalin Tarnay is Professor Emeritus at the University of Pannonia (Hungary). She received her M.Sc. degree at the Technical University of Budapest in 1956, Ph.D. degree at the Hungarian Academy of Sciences in 1974. She developed software tools and methods for protocol engineering first in the Central Research Institute for Physics, and later in Nokia Hungary. At the same time, she was appointed full Professor at the University of Pannonia and honorary Professor at the Budapest University of Technology. She was author of the book "Protocol Specification and Testing" (Plenum Press, New York, 1991.)

Gusztáv Adamis is a master teacher at the Budapest University of Technology and Economics (Hungary), Department of Telecommunications and Media Informatics. He received his M.Sc. degree from the Technical University of Budapest in 1989, and a specialized engineer degree in the field of telecommunications in 1991. He developed various software tools in the fields of mobile telecommunications and telecommunications protocols. He has more than 20 years of teaching experience in the fields of digital technology, switching theory, telecommunications protocols, and mobile systems. He has developed several course materials and held courses about Signaling System 7, GSM and other mobile protocols at various companies, like Nokia, T-Com, and the Hungarian National Telecommunications Authority.

Tibor Dulai graduated from the University of Pannonia (Hungary) and received his M.Sc. degree in Information Technology. After graduation, he started working on his Ph.D. thesis, which covers the research of new methods in telecommunications protocol design. By the years parallel with his Ph.D. studies, he has worked as Professor assistant at the university. This job gave him the opportunity to find the weaknesses of tutorials and educational books dealing with protocol management. Based on the feedbacks he has received from his students, he has clear ideas how to improve the education materials in this field. At present, his research topic is the effect of vehicle cooperation in the vehicle routing problem.

* * *

László Bokor graduated in 2004 with M.Sc. degree in computer engineering from the Budapest University of Technology and Economics (Hungary), Department of Telecommunications and Media Informatics. In 2006, he received an M.Sc.+ degree in bank informatics from the same university's Faculty of Economic and Social Sciences. He is a Ph.D. candidate at the Budapest University of Technology and Economics, member of the IEEE, member of the Mobile Communications and Computing Laboratory (MC2L) and the Mobile Innovation Center of Hungary, where he participates in researches of wireless protocols and works on advanced mobility management related projects (as FP6-IST PHOENIX and ANEMONE, EUREKA-Celtic BOSS, FP7-ICT OPTIMIX, EURESCOM P1857, EUREKA-Celtic MEVICO). His research interests include IPv6 mobility, mobile computing, next generation networks, mobile broadband networking architectures, network performance analyzing, and heterogeneous future networks. He is author of several referred papers in international scientific journals and conferences.

Christos Bouras obtained his Diploma and Ph.D. from the Computer Science and Engineering Department of Patras University (Greece). He is currently a Professor in the above department. He is also scientific advisor of Research Unit 6 in the Research Academic Computer Technology Institute in Patras (Greece). His research interests include analysis of performance of networking and computer systems, computer networks and protocols, telematics and new services, QoS and pricing for networks and services, e-learning, networked virtual environments, and WWW issues. He has extended professional experience in design and analysis of networks, protocols, telematics, and new services. He has published 300 papers in various well-known refereed conferences and journals. He is co-author of eight books in Greek. He has been PC member and referee in various international journals and conferences.

Eduardo Correia is a senior lecturer at the School of Computing at the Christchurch Polytechnic Institute of Technology (New Zealand). He is a Microsoft Certified IT Professional (Enterprise Administrator) as well as a Microsoft Certified Systems Engineer in Windows NT4, 2000, and 2003 and lectured at Rhodes University and the University of the Western Cape in South Africa. When Dr. Correia emigrated to New Zealand, he worked in the IT industry as a systems engineer and trainer for a number of years before returning to the academic environment. At CPIT, he teaches courses in operating systems and networking, particularly those focusing on Microsoft systems and technologies.

Leopoldo Estrada-Vargas was born in Jalisco, Mexico in 1981. He received his B.Sc. degree in electronic engineering from the Universidad de Guadalajara (Mexico), and M.Sc. degree in electrical engineering from the CINVESTAV, I.P.N. (Mexico), in 2003 and 2007, respectively. He has been research assistant of the Electrical Engineering Department, Telecommunications Section at CINVESTAV, I.P.N. He is a Ph.D. student at the Department of Electrical Engineering and Computer Sciences of CINVESTAV, I.P.N. His research interests include Internet traffic measurements and long memory stochastic processes.

Péter Fazekas is a full time lecturer and researcher at the Budapest University of Technology and Economics (Hungary), Department of Telecommunications, since 2001. He is with the Mobile Communications and Computing Lab since its founding, first as student, later as Ph.D. student and employee. His scientific interests cover radio resource management and medium access questions of wireless networks, as well as the capacity and performance analysis of broadband cellular networks. His recent research activities are targeting energy efficiency of cellular networks as well. He has around 20 journal and conference publications. Over the years, he led the activities of the Lab members within several international and national research projects and industrial cooperation. He held several B.Sc. and M.Sc. courses about cellular mobile systems.

László Gyöngyösi is a Ph.D. student since 2008 at the Budapest University of Technology and Economics (Hungary). He received his M.Sc. degree in computer science with Honors at the Budapest

University of Technology and Economics in 2008. His research interests are quantum computation and communication, quantum cryptography, quantum channels, and quantum information theory. He obtained the Sándor Csibi Ph.D. Researcher Scholarship, and he won Best Paper Awards on international conferences related to quantum computing and quantum information processing, both of them at the University of Harvard (USA).

Róbert Horváth has M.Sc. degree in electrical engineering. He is a member of the Budapest University of Technology and Economics (Hungary), Department of Telecommunications and Media Informatics. He has more than 13 years of experience in telecommunications software design and testing. His main focus is on automated testing of telecommunications protocols. Besides of his research interests, he also took part of several industrial projects including the planning and implementation of an early VoIP solution in Hungary, which consisted of authentication, authorization, accounting, interconnection of different networks, and all the necessary underlying protocols. He also took part in the test generation process for several protocols for Nokia. His publications cover the major aspects of protocol design patterns, testing methods, and formal languages used in connection with telecommunications protocol software.

Árpád Huszák received his M.Sc. degree in 2003 as an electrical engineer at the Budapest University of Technology and Economics (Hungary), Department of Telecommunications and Media Informatics. He received his Ph.D. degree in electrical and computer engineering in 2010. Currently he is working at the Department of Telecommunications as an Assistant Professor; he also joined the Mobile Innovation Center of Hungary. Since 2007, he is a member of the IEEE. Recently, he has been involved in many European projects (FP6-IST, FP7-ICT, and Celtic). His research interests focus on network protocols, mobile computing, and adaptive multimedia communication over wireless networks.

Sándor Imre was born in Budapest in 1969. He received his M.Sc. degree in electrical engineering from the Budapest University of Technology and Economics (Hungary) in 1993. Next, he started his Ph. D. studies at the same university and obtained dr. univ. degree in 1996, Ph.D. degree in 1999, and D.Sc. degree in 2007. Currently he is carrying his activities as Professor and Head of Department of Telecommunications. He is member of the Telecommunications Systems Committee of the Hungarian Academy of Sciences. He participates in the editorial board of two journals: Infocommunications Journal and Hungarian Telecommunications. He was invited to join the Mobile Innovation Center of Hungary as R&D director in 2005. His research interest includes mobile and wireless systems. Especially he has contributions on different wireless access technologies, mobility protocols, security and privacy, and reconfigurable systems.

Szilárd Jaskó was born in 1979. He received his M.Sc. degree in Information Technology engineering in 2002 at the University of Pannonia (Hungary). Since then, he has been a Ph.D. student at the same university. Since 2002, he has taught at the Department of Electrical Engineering and Information Systems of the University of Pannonia. He is interested in protocol modeling and testing, self-adaptive technologies, mobile telecommunications, and formal languages.

Gábor Jeney was born in 1975. He received his M.Sc. degree in electrical engineering in 1998 at the Budapest University of Technology and Economics (Hungary) in the field of mobile and optical communications. He received his Ph.D. degree in electrical engineering in 2005 from the same university in the field of multi-user detection. He also holds an M.Sc.+ engineer-economist degree from the Budapest University of Economic Sciences and Public Administration (Hungary). He is currently a research fellow at the Mobile Innovation Center of Hungary. He is involved in several national and European Commission funded project. He was the author/co-author of several dozens of articles published in conferences and journals. His research interest includes neutral networks, mobile and wide-band communication systems, networking, and transport protocols. Dr. Jeney is a member of HTE (Scientific Association for Info-communications), Hungary since 1999, and he is a member of IEEE since 2000. Recently, he was involved in the Technical Program Committee (TPC) of ICC 2007.

Zoltán Kanizsai is a member of the IEEE. He graduated with M.Sc. degree in computer engineering from the Budapest University of Technology and Economics (Hungary), Department of Telecommunications in 2006. He is a Ph.D. student at the same department and member of the Mobile Communications and Computing Laboratory (Hungary) and the Mobile Innovation Center of Hungary. He assists in IPv6, mobility, and multimedia related works, e.g. in former FP6-IST ANEMONE and now in FP7 ICT-OPTIMIX project. His research topics are: anycast-based IPv6 and MIPv6 optimization, Session Initiation Protocol with IMS over native IPv6 3G UMTS networks, and computer simulations of wired and wireless IPv6 networks.

Gábor Kovács has Ph.D. in electrical engineering. He works for the Department of Telecommunications and Media Informatics in the Budapest University of Technology and Economics (Hungary) as a researcher. Between 2004 and 2007, he participated in the European Union funded research project MUSE, which targeted fixed access technologies in NGN and fixed-mobile convergence, and contributed to standardization forums as TISPAN and DSL-Forum. His current research interests include business process modeling and formal modeling of telecommunications protocols.

Károly Lendvai was born in 1985. He received his M.Sc. degree in electrical engineering from the Budapest University of Technology and Economics (Hungary) in 2008. Next, he started his Ph.D. studies at the same university. Currently he is carrying his teaching and research activities as a Ph.D. student at the Department of Telecommunications of the Budapest University of Technology and Economics. He is member in several research projects at the Mobile Innovation Center of Hungary. His research interests include mobile and wireless systems, mobility modeling, MESH networks, and different security topics.

László Lois was born in 1971. He received his M.Sc. and Ph.D. degree in software engineering from the Budapest University of Technology and Economics (Hungary) in 1995 and 2005, respectively. He is currently with the Department of Telecommunications at the same university, and contributes to research, development, and educational activities. His research interests include source-coding applications, media communications, and IP television.

Dániel Muhi was born in 1979. He received his M.Sc. degree in Information Technology from the University of Pannonia (Hungary) in 2002. Since 2002, he has been associated with research work in the same university as teaching assistant. His research areas are service discovery protocols, e-learning technologies, and process-oriented modeling. He received the Pollák-Virág award from the Scientific

Association for Infocommunications (Hungary). He is a member of the Communication Working Committee of the Veszprém Regional Center of the Hungarian Academy.

Gábor Árpád Németh received his M.Sc. degree from the Budapest University of Technology and Economics in 2007. He is a Ph.D. student at the Department of Telecommunications and Media Informatics of the same university. His current research topic is efficient testing of telecommunications protocols in case of changing specifications.

Szabolcs Nováczki was born in 1982. He received his M.Sc. degree in electrical engineering in 2006 at the Budapest University of Technology and Economics (Hungary) in the field of mobile communications and computer architectures. He is currently a Ph.D. student at the Budapest University of Technology and Economics and a member of the Mobile Communications and Computing Laboratory (Hungary) and the Mobile Innovation Center of Hungary. He is also a student member of IEEE. His research interests include advanced network layer mobility solutions, new networking architectures, network simulation programming, and network performance analysis.

Vassilis Papapanagiotou entered the Computer Engineering and Informatics Department of the University of Patras (Greece) in 2003 and graduated in 2008; then he entered the graduate program of the same university, Department of Computer Science and Technology. He was a member of Research Unit 6 from October 2006 to December 2009. His research interests include cross-layer architectures for wireless networks, network protocols, Layer 2 and Layer 3 Quality of Service, ad-hoc networks, and network programming. He has extensive knowledge of C, C++, Java, PHP, database manipulation, Web technologies, wide area network management, and configuration of Cisco devices. He has worked for the Greek Research Network, developing modules for automatic management of QoS services and managing its networking equipment in the city of Patras. He has also participated in the GN3 Project for advancing the European GEANT3 network. Additionally, he is the co-author of three papers in international refereed conferences, one article in an international journal and two chapters in technical books.

Zoltán Pap received his M.Sc. degree in electrical engineering in 2000 at the Budapest University of Technology and Economics (Hungary). In 2002, he received another M.Sc. degree in business administration at the Corvinus University of Budapest (Hungary). From 2000 to 2006, he worked at the Budapest University of Technology and Economics on various research fields, initially as a Ph.D. student and later as research fellow. His main research topics are telecommunications networks and protocols, next generation networking, peer-to-peer networks, grid computing, advanced software technologies, and testing. He received his Ph.D. degree at the Budapest University of Technology and Economics in 2006. Since 2006, he is working at Ericsson Hungary as systems engineering expert, and in part time at the above university as research fellow.

Jan Schankin is a lecturer in the School of Computing at the Christchurch Polytechnic Institute of Technology (New Zealand). After graduating with a diploma in computer networking with distinction and completing various Cisco and Microsoft certifications, he has worked extensively on switches, routers, and servers at CPIT, and done some consulting work and research. Mr. Schankin's interests focus principally on technical aspects of networking technologies and protocols, and certain areas of server provider networks as well as the practical implementation of various aspects of networking in Cisco, Juniper, and Microsoft systems.

Róbert Schulcz graduated in 2000 with M.Sc. degree in electrical engineering from the Budapest University of Technology and Economics (Hungary), Department of Telecommunications. In 2009, he received the Master of Business Administration (MBA) degree from the same university. Between 2000 and 2003, he participated in the Ph.D. course at the Department of Telecommunications and he is currently a Ph.D. candidate at the Budapest University of Technology and Economics. He is the managing director of the Mobile Innovation Center of Hungary, where he was the testbed director previously, responsible for the setup and operation of the testbed of the Mobile Innovation Center of Hungary. He works as a researcher at the Mobile Communications and Computing Laboratory (Hungary) since 2003. His research interests include mobility, mobile computing, IMS, security of next generation wireless and heterogeneous networks, NFC, and RFID. He is member of the IEEE and the Scientific Association for Infocommunications (Hungary). He is an author of several referred papers in international scientific journals and conferences.

Ákos Sebestyén was born in 1977. He received his M.Sc. degree in electrical engineering at the Budapest University of Technology and Economics (Hungary), Department of Telecommunications in 2002. Now as a member of the staff of the same department, he contributes to the educational, research, and development activities. His research areas include digital broadcasting (DVB-T/S/C/H/T2), media encoding, and IP television.

Jaydip Sen obtained his B.Sc. degree in electrical engineering with honors at the Jadavpur University (India) in 1993, Master of technology (M.Tech.) in computer science with honors at the Indian Statistical Institute (India) in 2001, and Ph.D. in network security at the Indian Institute of Technology (India) in 2007. He has 17 years experience in the field of networking, communication, and security. He has worked in reputed organizations like Oil and Natural Gas Corporation Ltd., India, Oracle India Pvt. Ltd., and Akamai Technology Pvt. Ltd. Currently, he is leading the research and development activities in wireless communication at Tata Consultancy Services, Kolkata (India) for the last three years. He has over 16 years of research and development experience. His research areas include security in wired and wireless networks, intrusion detection systems, secure routing protocols in wireless ad hoc and sensor networks, secure multicast and broadcast communication in next generation broadband wireless networks, trust and reputation based systems, quality of service in multimedia communication in wireless networks and cross layer optimization based resource allocation algorithms in next generation wireless networks, sensor networks, and privacy issues in ubiquitous and pervasive communication. He has more than 75 publications in reputed international journals and referred conference proceedings. He has delivered expert talks and keynote lectures in a various international conferences and symposia. He is a member of ACM and IEEE. He is also an active member of the security group of IEEE 802.16 standard body and has submitted a number of proposals for the evolving 802.16m standard and ETSI. His biography has been listed in Marquis "Who's Who in the World", 2010 edition.

Gyula Simon is an Associate Professor at the University of Pannonia (Hungary), where he is the member of the Department of Computer Science and Systems Technology. He graduated from the

Budapest University of Technology and Economics (Hungary) in 1991 with M.Sc. degree in electrical engineering. In 1998, he earned a Ph.D. with highest honors in electrical engineering at the Budapest University of Technology and Economics. After graduation, he joined the faculty at the same university. From 2002 to 2004, he was a research Assistant Professor at the Institute of Software Integrated Systems of the Vanderbilt University (USA). He joined the faculty of the University of Pannonia in 2005. Gyula Simon's teaching reflects his multiple areas of expertise. He has taught courses on programming, operating systems, software engineering, measurement technology, and sensor networks. His research interest includes measurement systems, sensor networking, signal processing, and information security.

Kostas Stamos obtained his diploma, M.Sc., and Ph.D. from the Computer Engineering and Informatics Department of Patras University (Greece). He has worked for the Networking Technologies Sector of the Research Academic Computer Technology Institute, Research Unit 6 of the Research Academic Computer Technology Institute and the Greek Research and Education Network. He is also teaching at the Computer Engineering and Informatics Department at the University of Patras, and at the Technological Educational Institute of Patras. He has published eight articles in journals and 30 papers in well-known refereed conferences. He is also co-author of two technical books, several encyclopaedia articles, and of a Global Grid Forum (GGF) standard document.

Sándor Szabó was born in 1977. He received his M.Sc. degree in electrical engineering at the Budapest University of Technology and Economics (Hungary) in 2000. Next, he started his Ph. D. studies at the same university. Currently he is carrying his teaching activities as Assistant Professor at the Department of Telecommunications of the Budapest University of Technology and Economics. He leads research projects, and is also a project leader at the Mobile Innovation Center of Hungary. His research interests include mobile and wireless systems, mobility modeling, mobility management, and the integration of heterogeneous wired and wireless networks.

Homero Toral-Cruz was born in Oaxaca, Mexico, in 1980. He received his B.Sc. degree in electronic engineering at the Instituto Tecnologico de la Laguna, Torreon, Coahuila (Mexico) in 2002, and his M.Sc. and Ph.D. degrees at the Electrical Engineering Department, Telecommunications Section at the Centro de Investigacion y Estudios Avanzados del IPN, (Mexico) in 2006 and 2010, respectively. He has been research assistant of the Electrical Engineering Department, Telecommunications Section at the Centro de Investigacion y Estudios Avanzados del IPN. Presently, he is Professor at the University of Quintana Roo (Mexico). His research interests include VoIP systems, quality of service in IP networks and Internet traffic modeling.

Deni Torres-Román was born in Oriente, Cuba in 1950. He received his Ph.D. degree in Telecommunications at the Technical University of Dresden (Germany) in 1986. He is co-author of the book "Data Transmission." He received the Telecommunications Research Prize in 1993 from the AHCIET Association. Since 1996, he is Professor and researcher at the Department of Electrical Engineering and Computer Sciences of the Centro de Investigacion y Estudios Avanzados del IPN (Mexico). His main research involves Internet traffic measurement and modeling, network quality of service, digital communication of medium and high speed, software and hardware design for telecommunications, and modern telephony systems.

Arijit Ukil is currently working in the Innovation Labs, Tata Consultancy Services Ltd., (India) as research scientist. He is primarily engaged with the research activity on ubiquitous computing, security and privacy, and wireless networking. Before joining Tata Consultancy Services Ltd. in 2007, he has worked as scientist C in the Deference Research and Development Organization for four years, where his primary focus area was digital signal processing, embedded systems, and wireless communication for radar applications. He was mainly involved in naval-based radar systems. He has published a number of conference and journal papers of national and international repute. He has published one book chapter titled "Advanced Scheduling Schemes in 4G Systems" in the book "Fourth-Generation Wireless Networks: Applications and Innovations" published by IGI-Global in December, 2009. He has been reviewer of a number of IEEE conferences, like IEEE VTC and IEEE WCNC. He has been invited to deliver keynote lectures in many international and national conferences, like ETCC'08 in NIT Hamirpur, ICCET'09 in Singapur, NCERDM-IT'09, Jaipur, and NCETAC2010-IT, Trichy. He has delivered tutorial in ICWMC'09 on "Ubiquitous Computing". He is enlisted in 2010 Marquis' "Who's Who in the World" as a renowned contributor in the field of computer science and Information Technology. He has done his B.Tech degree in Electronics and Telecommunications in 2002.

Gergely Vakulya received his M.Sc. in Information Technology at the University of Pannonia (Hungary) in 2008. Currently he is pursuing his Ph.D. in Information Science and Technology at the University of Pannonia. His main research interest is sensor networking. He gives courses on operating systems, information security, and sensor networks, and also coordinates projects of undergraduate students.

Gábor Varga graduated in 2010 with M.Sc. degree in technical informatics from the Budapest University of Technology and Economics (Hungary), Department of Telecommunications. He has also been studying economics at the Corvinus University (Hungary). He is currently working as an engineer at the Mobile Innovation Center of Hungary, where his main research interests are positioning and security in wireless networks, and indoor navigation systems. He is co-author of papers published in Hungarian and international scientific journals, and conference papers.

Giannis Zaoudis obtained his diploma at the Computer Engineering and Informatics Department of the Patras University (Greece). He is currently in the graduate program of the Computer Science and Technology department. He has worked for the Research Unit 6 of the Research Academic Computer Technology Institute and the Greek Research and Education Network. He is also participating in the GN3 Project for advancing the European GEANT3 network. His research interests include cross-layer architectures for wireless networks, network protocols and ad-hoc networks. He has extensive knowledge of C, C++, Java, PHP, database manipulation, Web technologies, and Web development. He has yet published one article in a journal and three papers in well-known refereed conferences.

Index

Symbols

A

B

C

D

I

L

M

N

O

P

Q

R

S

T

U

V

W

X

CPSIA information can be obtained at www.ICGtesting.com
263891BV00006BB/2/P

9 781609 60732

$195.00